A Companion to Calculus

A Companion to Calculus

Second Edition

Dennis Ebersole
Northampton Community College

Doris Schattschneider
Moravian College

Alicia Sevilla
Moravian College

Kay Somers
Moravian College

Australia • Brazil • Japan • Korea • Mexico • Singapore • Spain • United Kingdom • United States

A Companion to Calculus, Second Edition

Dennis Ebersole, Doris Schattschneider, Alicia Sevilla, Kay Somers

Publisher: Bob Pirtle

Assistant Editor: Stacy Green

Editorial Assistant: Katherine Cook

Technology Project Manager: Earl Perry

Marketing Manager: Karin Sandberg

Marketing Communications Manager: Bryan Vann

Project Manager, Editorial Production: Cheryll Linthicum

Art Director: Vernon T. Boes

Print Buyer: Barbara Britton

Permissions Editor: Joohee Lee

Production Service: Scratchgravel Publishing Services

Copy Editor: Carol Reitz

Cover Designer: Lisa Berman

Cover Photo: Cameron Davidson

Compositor: Scratchgravel Publishing Services

Library of Congress Control Number: 2005923114

ISBN-13: 978-0-495-01124-8
ISBN-10: 0-495-01124-X

Brooks/Cole
10 Davis Drive
Belmont, CA 94002-3098
USA

Cengage Learning is a leading provider of customized learning solutions with office locations around the globe, including Singapore, the United Kingdom, Australia, Mexico, Brazil, and Japan. Locate your local office at: **www.cengage.com/global**.

Cengage Learning products are represented in Canada by Nelson Education, Ltd.

To learn more about Brooks/Cole, visit **www.cengage.com/brookscole**.

Purchase any of our products at your local college store or at our preferred online store **www.ichapters.com**.

Printed in the United States of America
3 4 5 6 7 8 9 10 12 11 10 09

Preface to the Second Edition

This edition of the *Companion* builds on the strengths of the first edition, providing help and support for students in Calculus I. There are many new examples and exercises throughout the text. Chapter 8, "Review of Trigonometric Functions," and the Appendix have been expanded. All examples give detailed solutions to typical problems encountered in Calculus I, demonstrating how to analyze and set up problems and how to use algebraic techniques in their solutions. Exercises are designed to develop both conceptual understanding and important algebra skills. Some exercises are designed to be carried out with a graphing utility; these are signaled by a calculator icon. Throughout the text, warning boxes emphasize how to avoid common errors.

Please contact us with your comments and suggestions.

Dennis Ebersole, Doris Schattschneider,
Alicia Sevilla, Kay Somers

Acknowledgments

We wish to express our appreciation to colleagues at Moravian College, Northampton Community College, and the Fund for the Improvement of Post-Secondary Education who encouraged us to develop *A Companion to Calculus.* We especially appreciate the cooperation of colleagues at Moravian College who helped with initially class-testing the text and who participated in the comparison testing of students in our Calculus I with Review with those in the one-semester Calculus I. We are also grateful for the cooperation of colleagues and students at several other institutions who class-tested the materials in draft and published versions. We note our special thanks to David Cox and Pamela Crawford for their helpful suggestions for improvements in the first edition. We appreciate the work of Jeff Young and Beth Kuehner, who assisted in preparation of solutions and graphs for the first edition. Finally, we thank the reviewers of the first edition for providing suggestions for the second edition.

Throughout the various stages of development and production of the first and second editions of the *Companion*, we had the constant careful and cheerful assistance of department secretary Kathleen Burkert, who served in the role of production manager—as typist, proofreader, copyeditor, and printing supervisor.

Our thanks to the helpful staff at Brooks/Cole and Scratchgravel Publishing Services for their efficient production of the first and second editions of *A Companion to Calculus.*

A Word to Students

This text has been written with you foremost in mind. Its only purpose is to help you succeed in your first course in calculus. It provides review of the mathematical concepts and techniques that are needed in the study of calculus. As its *Companion* title suggests, it is designed to be used along with a calculus text. Typically, material from a chapter in this text will be covered before a new topic is introduced from the calculus text. Each chapter contains lots of examples—and these examples are the primary method used to review concepts, illustrate techniques, and show applications.

To succeed, you must be an active reader, with pencil and paper in constant use as you carefully read the text and go through the examples. Here are some specific suggestions to make your study worthwhile:

1. Be sure you understand the definitions, terminology, and notations. When a new term is defined, it appears in italics. It is a good idea to prepare your own glossary of terms for easy reference and review—set up a section of your notebook to record terms and their definitions and any special notation for them.

2. Read each example and first try to solve it yourself. When you read the solution, ask yourself: Do I understand all parts of the solution? Ask questions about what you do not understand (ask fellow students, tutors, or the instructor).

3. Exercises at the ends of sections and chapters are based on the ideas in the text and examples. Do not just try to mimic the solution to an example; rather, use the ideas and techniques illustrated in the examples to find your own solution.

4. When you have completed the solution to an exercise, ask the following questions of yourself: Is this result a reasonable answer to the question(s) asked? What have I learned from this problem that I could apply to other problems?

5. Carefully study the illustrations. In calculus, visual understanding is extremely important; you need to know how to read graphs and figures, and you will also be expected to produce them.

6. Have a calculator handy for computations. Whenever possible, use a graphing calculator or a computer with a graphing program to help you investigate graphs of functions or verify the sketches you make by hand. (Most computer-drawn graphs will not show you the fine details of a graph, but they are useful to check the reasonableness of your sketches.)

7. Regular practice with exercises is extremely important. It is often helpful to work with your classmates in small groups to discuss the exercises and examples.

A Word to Instructors

The purpose of this text is to review mathematical concepts and techniques that are prerequisites for understanding and doing calculus. It is not a stand-alone text; rather, it is intended to be used along with a calculus text in a first course in calculus. In addition to providing review of precalculus material, it sets the stage for calculus topics and uses calculus terminology and notation. The material in this *Companion* is intended to be interleaved with material in a calculus text, to provide a review of precalculus concepts and techniques as they are needed for calculus topics.

The *Companion* communicates mathematical information in four distinct modes: words (description), symbols (algebraic and analytic forms), pictures (graphs), and numerical patterns (data). Students should be able to understand information and translate it from one mode to another. Some have difficulty with a particular mode; knowing how to recast the information can be a key to understanding.

Each chapter in the *Companion* is organized around one topic, and the chapters follow a fairly standard order in which topics are introduced in a first calculus course. You should look over the subtopics in each chapter as well as the ordering of the chapters when working out how (and when) you will use the material prior to covering a topic in the calculus text. In most instances, you should be able to rearrange the order of topics, except for those that require a logical order (such as the review of coordinates before the review of functions). An outline of how material from the *Companion* interleaves with the topics in a calculus text follows.

The preparation of the first edition of this text was supported by a grant from the United States Department of Education's Fund for the Improvement of Post-Secondary Education. It is the combined effort of faculty members at Moravian College and Northampton Community College. The *Companion* has been used successfully with several different calculus texts (covering the spectrum from "reform" to "traditional") in a variety of courses at several institutions, including integrated one-year courses such as Calculus I with Review, traditional one-semester courses (in which it was used as a supplement), special courses such as calculus for business or social sciences, and technology-based courses.

You can find information on the rationale for an integrated Calculus I with Review course and results of evaluations of student success rates in these courses at several institutions in the article "College Precalculus Can Be a Barrier to Calculus: Integration of Precalculus with Calculus Can Achieve Success" by Doris Schattschneider, in the MAA Notes volume *Rethinking Precalculus* as well as on the web site www.calculus-with-precalc.org. For additional information, contact Kay Somers, Mathematics and Computer Science Department, Moravian College, 1200 Main Street, Bethlehem, PA 18018-6650; e-mail: SomersK@moravian.edu.

The following parallel columns outline how *Companion* chapters can be integrated with chapters in a calculus text:

Section in *Companion*	Section in Calculus Text
Introduction	
Symbols and Notation	
Modes of Communication	
Cartesian Coordinates	
The Cartesian Coordinate Plane	
Graphs	*Coordinate Geometry and Lines*
Lines and Their Equations	
Parallel and Intersecting Lines	
Distance between Two Points	
The Circle	
Functions	
Function Notation	
Domain and Range of a Function	
Different Ways to Represent Functions	*Functions and Graphs*
The Graph of a Function	
Special Classes of Functions	
Transformations of Graphs	
Companion to Limits	
Combinations of Functions	*Limit of a Function*
Algebraic Simplification of Functions	*Calculating Limits Using Limit Laws*
Inequalities: Linear Inequalities; Absolute Value: Equations and Inequalities	*Precise Definition of Limit*
If–Then Statements	
Companion to Continuous Functions	*Continuity*
Polynomials	
Zeros of a Function: Finding Zeros of a Polynomial; The Approximations of Zeros of Continuous Functions	
More on Domains of Functions: Composite Functions	
The Role of Infinity	
Graphical Interpretation: Horizontal Asymptotes; Vertical Asymptotes	*Limits at Infinity, Horizontal Asymptotes*
Algebraic Manipulations: Finding Asymptotes	*Infinite Limits, Vertical Asymptotes*
Problem-Solving and Rates of Change	
Problem-Solving	
Applications: Average Rates of Change	*Tangents, Velocity, and Other Rates of Change*
Secant and Tangent Lines	
Companion to Rules of Differentiation	
Negative and Rational Exponents	*Derivatives*
Decomposition of Functions	*Differentiation Formulas*; *Chain Rule*
Simplifying Derivatives	
Review of Trigonometric Functions	
Angle Measures	
Definition and Evaluation of the Trigonometric Functions	*Review of Trigonometry*
Properties and Identities for the Trigonometric Functions	
Domain, Range, and Graphs of the Trigonometric Functions	*Derivatives of the Trigonometric Functions*
Combining Functions with the Trigonometric Functions	

Section in *Companion*	Section in Calculus Text
Companion to Implicit Differentiation Implicitly Defined Functions Solving Equations That Contain dy/dx	*Implicit Differentiation*
Companion to Repeated Differentiation Iteration and Patterns in Higher Derivatives Rate of Change of Rate of Change	*Higher Derivatives*
Companion to Related Rates Setting Up Equations for Related Rates Problems Problem-Solving Strategies for Related Rates Problems	*Related Rates*
Linear Approximations and Differentials Tangent Line Approximation The Differential	*The Differential and Tangent Approximation*
Companion to Exponential Functions Rules of Exponents The Natural Exponential Function	*Exponential Functions* *Derivatives of Exponential Functions*
Companion to Inverse Functions One-to-One Functions Properties of a Function and Its Inverse: Domain, Range, Graph Finding the Inverse of a Function: When the Function Is One-to-One When the Function Is Not One-to-One	*Inverse Functions*
Companion to Logarithmic Functions Definition and Properties of Logarithmic Functions Graphs of Logarithmic Functions Solving Equations with Logarithmic and Exponential Functions	*Logarithmic Functions* *Derivatives of Logarithmic Functions* *Exponential Growth and Decay*
Companion to Extreme Values of a Function Extreme Values and Critical Values Setting Up Functions to Solve Extreme Value Problems	*Maximum and Minimum Values* *Applied Maximum and Minimum Problems*
Companion to Curve Sketching Solving Inequalities Graphical Interpretation Putting It All Together	*The First Derivative Test* *Concavity and Points of Inflection* *Curve Sketching*
Companion to Antidifferentiation Antidifferentiation as the Inverse of Differentiation Recognizing Antiderivatives Substitution for Antiderivatives	*Antiderivatives*
Companion to Area and Riemann Sums Exact Areas as Sums of Basic Geometric Shapes Approximations of Areas Riemann Sums and Their Interpretations	*Area*
Companion to the Definite Integral Area under a Curve as a Definite Integral Other Interpretations of the Definite Integral The Fundamental Theorem of Calculus Change of Variable in Definite Integrals	*The Definite Integral* *Properties of the Definite Integral* *The Fundamental Theorem of Calculus* *The Substitution Rule*
Appendix (review sections as needed)	

Contents

CHAPTER 0

Introduction

Calculus is the study of change. How fast is the population of whales decreasing? What speed should the probe maintain to stay in an orbit around Mars? Is the demand for DVDs increasing? At what rate? Calculus allows exploration and analysis of the rates at which varying quantities change.

If you drive 100 miles in 2 hours, your average speed is 50 miles per hour (mph) but your car's speedometer will not stay steady at 50 mph during the whole trip. At each moment of the trip, the speedometer reading gives the speed in miles per hour at that moment. The speedometer illustrates one of the principal concepts in calculus, called *differentiation*, that gives the instantaneous rate of change of a quantity that varies. A second principal concept in calculus is *integration*; integration measures the net accumulation of a quantity that can increase or decrease. The total distance you travel at a variable speed during an interval of time can be found by integration. Integration may be thought of as the process inverse to differentiation, just as subtraction is the process inverse to addition and division is the process inverse to multiplication. Differentiation and integration are each essential techniques used to solve a wide range of problems in a variety of fields. The concept of *limit* is fundamental in calculus because differentiation and integration of functions are defined using particular kinds of limits.

The concept of *function* is fundamental in all of mathematics. When one quantity is described as a function of one or more other quantities, it means that the first quantity depends on the others. For example, the time spent in making the 100-mile car trip is a function of the speed of the car during the trip. (It is a function of other quantities also, such as how many stops are made.) When we want to study how one quantity is affected as another quantity is varied, we express the relationship as a function and then ask how the function changes.

0-A SYMBOLS AND NOTATION

Symbols are used in many fields, including mathematics, for clarity, for exactness, and for brevity. The word *symbol* is derived from the Greek word *symbolon*, which means "a token" or "a sign." We calculate with symbols and we interpret symbols. To be useful, the meaning of each symbol or combination of symbols must be understood, agreed upon universally, and unambiguous. Signs or symbols are used to represent astronomical objects and terms and phases of the moon. Letters or symbols are used in chemistry to designate the chemical elements, and all common units of measurement have standard symbolic abbreviations. Stylized silhouettes on road signs, the familiar biological symbols ♀ and ♂, and religious symbols are all examples of symbolic notation.

In mathematics, we use symbols to represent operations, to signify relations, to abbreviate, to represent quantity, and to be precise. Each symbol is read as a word or phrase. Common symbols of operation, together with their spoken form, include the following:

$+$	"plus" to signify addition
$-$	"minus" to denote subtraction
$\times$, $\cdot$, or $*$	(or omitted symbol) "times" to signify multiplication
$\div$ or /	"divided by" to signify division
$\sqrt{\ }$	"square root"

Symbols that are used to express relationships include:

$=$	"equals"
$\approx$	"approximately equals"
$<$	"less than"
$>$	"greater than"
$\neq$	"not equal to"
$\in$	"belongs to" or "is an element of"
$\parallel$	"is parallel to"

These and other symbols are discussed in more detail in the Appendix.

Functions act on variables that represent numbers; the functions are often given letter names such as f, g, or F. The symbolic notation $f(x)$ indicates that the function f acts on the variable x. Special functions are given standard abbreviated names. For example, the sine function and the natural exponential function are often abbreviated on calculator buttons as **sin** and **exp**, respectively. Functions and their notations are discussed in detail in Chapter 2 and in the Appendix.

Symbols are used to abbreviate. For example, % signifies percent, $ represents dollars, ° signifies degrees, and " represents inches. Letter symbols are used to replace words to express formulas succinctly. For example, the formula "distance equals rate × time" is usually written in abbreviated form as the algebraic equation $d = r \cdot t$. Note that the abbreviations r for rate, t for time, and d for distance appear in this equation, as well as the multiplication operation symbol · and the relation symbol =.

Letters are often used as symbols to represent numbers, or quantity. Arithmetic involves manipulation of numbers according to certain rules of operations; algebra involves manipulation of numbers, as well as of letters that represent numbers, according to the same rules of operations. Unknown quantities are generally represented by the letters x, y, z, s, t at the end of the alphabet or by letters that suggest the unknown quantities. For example, if it is known that the total distance to a particular destination is 55 miles, and if a road sign says our destination is 6 miles away, we can let d denote the (unknown) distance we have traveled so far and express this information as an algebraic equation: $d + 6 = 55$, or $d = 55 - 6 = 49$.

Letters are also used to represent numbers when describing a general rule that is true for a whole class of numbers. For example, to write the rule "Zero added to any real number equals that real number," we could write the following: For all real numbers a, $0 + a = a$. This means that we can substitute any real number for the letter a, and the rule holds true; that is, $0 + 6 = 6$, $0 + \frac{3}{4} = \frac{3}{4}$, $0 + (-3.578) = -3.578$, and so on. In a similar way we write the following: If $a < b$, then $a + c < b + c$, for any real numbers a, b, and c. By using letters to symbolize numbers in this way, we can write general rules more succinctly.

Letters are sometimes used to represent special constants. The Greek letter pi, π, represents the constant ratio of the length of the circumference of any circle to the length of its diameter; the numerical value of π is approximately 3.14159. Symbolically, $\pi \approx 3.14159$. The letter e denotes the natural base for exponential and logarithmic functions; $e \approx 2.71828$. Exponential functions and the base e are discussed further in Chapter 13.

Parentheses ( ), brackets [ ], and braces { } are symbols that appear mainly for clarity in computations and formulas. They can be used to specify the order of operations; expressions enclosed by them are to be combined. For example, in the following expression, 4 + 5 is added and then multiplied by 3:

$$3 \cdot (4 + 5) = 3 \cdot 9 = 27.$$

In more complicated expressions, we use nested parentheses and brackets for clarity and precision; for example,

$$3 + [(4 + 5) \cdot 6] = 3 + [9 \cdot 6] = 3 + 54 = 57.$$

It is especially important to use parentheses when entering expressions in calculators and computers, since parentheses specify the order, or hierarchy, of operations to take place. To evaluate such an expression, you work from the inside out; the expression inside the innermost parentheses is evaluated first.

Parentheses are used for a different purpose in function notation like $f(x)$. Here the notation tells us that the function f acts on the variable x to produce a single output, denoted $f(x)$.

Subscripts on letter variables, like a_1 or a_2, are sometimes used for clarity. For example, when we discuss two lines, the first could be called l_1 and the second l_2. The slope of l_1 would be denoted as m_1 and the slope of l_2 as m_2. Similar quantities (slopes) of the two distinct lines are given similar symbols: The letter m consistently denotes the slope, and the subscript identifies of which line. The Appendix contains further discussion of the use of symbols.

0-B MODES OF COMMUNICATION

The language of mathematics includes not only symbolic formulations but also words, pictures, and numbers. Relationships between quantities are often described in words. For example, we may know that the height of a cylindrical tank is three times its base radius. For a descriptive formulation to be useful, the meaning of each term in the context in which it is used must be understood. To interpret the statement that relates the height and base radius of a cylindrical tank, we need to know what a cylindrical tank is and understand the meaning of height and base radius. When solving problems, we often need to sketch pictures; organize graphs, charts, or tables; and produce formulas, equations, or inequalities using letters and symbols to represent the descriptive sentences. In a particular situation, some forms may depict relationships more clearly than others. It is also crucial to be able to explain in words the expressions or solutions that are given in symbols. The first example below illustrates a relationship expressed in words, clarification of the words in the context used, and organization of the relationships into symbolic notation. The second example shows the interpretation of a picture and formula into words. The third example illustrates how to produce a table of numerical information from a symbolic formula.

EXAMPLE 0.1 Translate the following descriptive sentence into symbolic notation: The ratio of the length of the circumference of any circle to the length of its diameter is the constant π.

Solution We first clarify the terms used—ratio, circumference, and diameter of a circle and the constant π. We assign letters to the terms that appear in the sentence. Let C denote circumference of the circle and let D denote its diameter. We draw a picture (Figure 0.1) that shows these quantities. Then $\frac{C}{D} = \pi$. ■

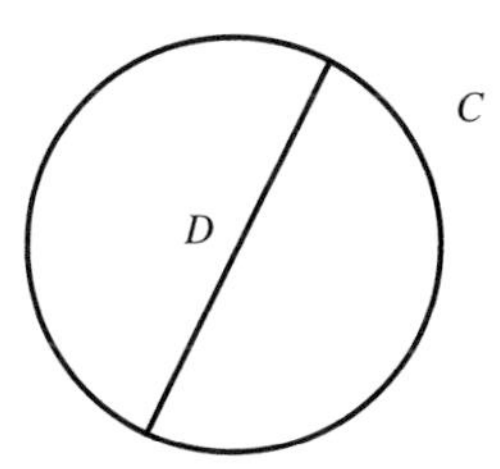

Figure 0.1

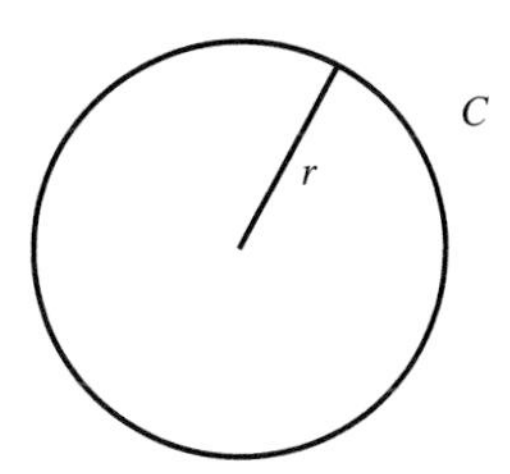

Figure 0.2

EXAMPLE 0.2 Translate into words the formula $C = 2\pi r$, accompanied by the picture given in Figure 0.2.

Solution We identify what the letters in the formula represent: C = circumference of the circle, r = radius of the circle, and π is a constant. We state in words what the algebraic formula means: The circumference of any circle is twice the product of its radius and the number π. The circumference is the length of the boundary of the circle, and the radius is the distance from the center of the circle to its boundary; π is approximately 3.1416. ■

EXAMPLE 0.3 Use the formula $A = \pi r^2$ for the area of a circle to create a two-column table that gives at least three possible radii and corresponding areas.

Solution One possible table is shown below. For each radius the area is calculated by substituting for r in the formula. To get a decimal value for the area, we **approximate** π by 3.1416. For example, the area of a circle of radius 2 cm is calculated as $A = \pi(2)^2 = 4\pi \approx 12.5664$ cm^2.

r	$A = \pi(r)^2$
1	$1\pi \approx 3.1416$
2	$4\pi \approx 12.5664$
2.5	$6.25\pi \approx 19.635$
3	$9\pi \approx 28.2744$
3.7	$13.69\pi \approx 43.0085$

■

A Companion to Calculus uses all four modes of expression—words, pictures, numbers, and symbolic formulations. It is important to be able to communicate in all these modes and to be able to move easily from one mode to another.

Chapter 0 Exercises

1. Each of the following pictures has an algebraic formula attached to it in which letters are used as symbols to represent quantity. Do the following for each:

(i) Identify in words what the pictures represent and what the symbols represent. Label the pictures with symbols in the appropriate positions, adding any necessary lines.

(ii) Write a sentence using words only that explains the algebraic formula.

a. $A = l \cdot w$

b. $A = b \cdot h$

c. $A = \pi r^2$

d. $V = l \cdot w \cdot h$

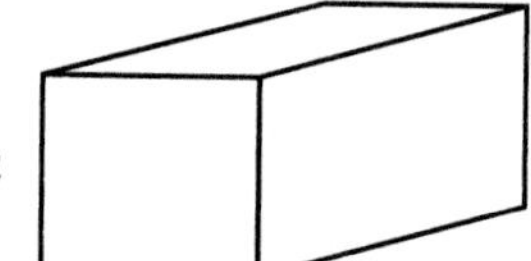

e. $V = \pi r^2 h$

f. $V = \frac{1}{3}\pi r^2 h$

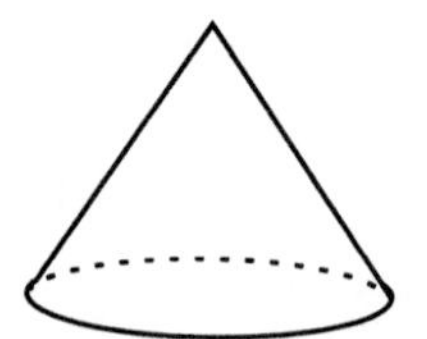

g. $V = s^3$

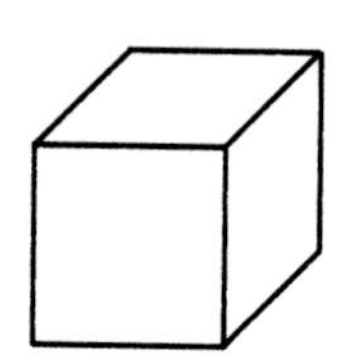

h. $c^2 = a^2 + b^2$

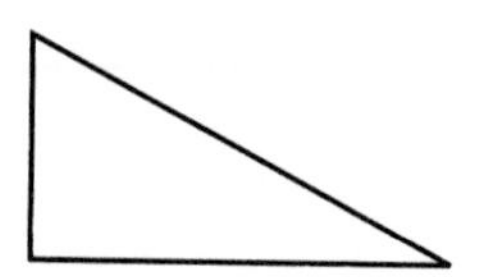

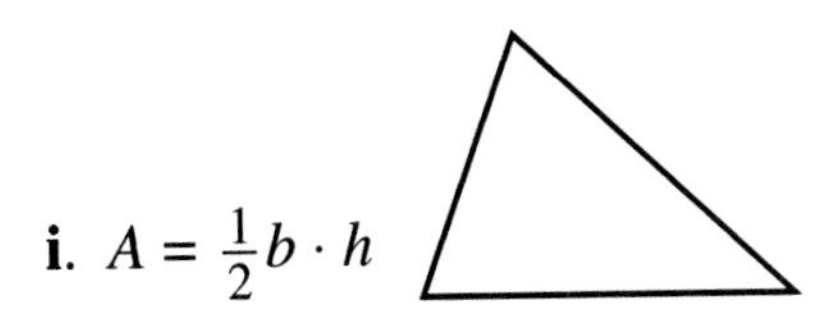

i. $A = \frac{1}{2}b \cdot h$

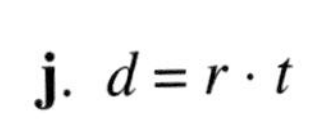

j. $d = r \cdot t$

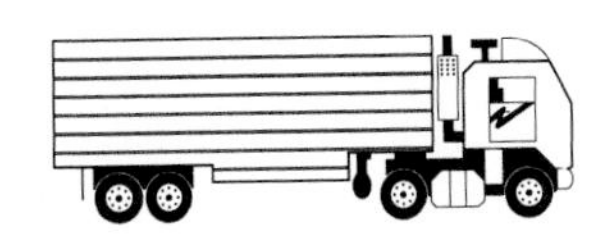

k. $F = \frac{9}{5}C + 32$

32°F 0°C

l. $A = \frac{r^2}{2}\,\theta$

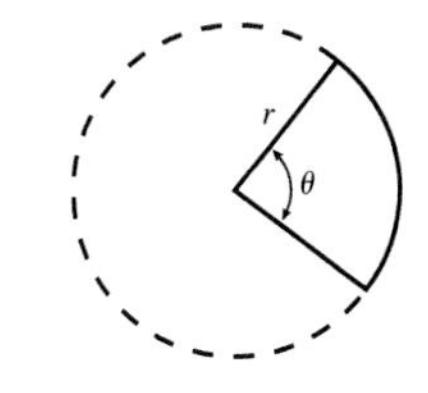

m. $V = \frac{4}{3}\pi r^3$

2. Translate the following descriptive sentences into symbolic notation. Make a picture of the situation and label its parts with appropriate symbols.

> **a.** The length of the perimeter of a rectangle is the sum of the lengths of the four sides.
>
> **b.** The surface area of a rectangular box is the sum of the areas of the six faces.
>
> **c.** The length of fence needed to enclose a semicircular field is the sum of one-half the length of the circumference of the circle plus the length of the diameter of the circle.

3. Use the formula $F = \frac{9}{5}C + 32$ to produce a two-column table that converts these benchmark temperatures from the Celsius scale to the Fahrenheit scale: $0°C$ (water freezes), $100°C$ (water boils), $20°C$ (room temperature), and $37°C$ (normal body temperature).

4. Give two additional examples of symbols used to represent operations in mathematics.

5. Give two additional examples of symbols used to signify relations in mathematics.

6. Give two additional examples of symbols used for abbreviations in mathematics.

CHAPTER 1

Cartesian Coordinates

Horizontal and vertical grids of lines are used to specify locations of cells in spreadsheets and sites on maps; a letter or a number is assigned to each interval between a pair of grid lines. To locate a point on a map, you look it up in the table that gives its position, or coordinates. (Such a table is often on the back of the map.) You then proceed to determine the rectangle identified by those coordinates. Your intended destination is within that rectangle. For example, on the diagram of a map below, the location of your destination, indicated by •, is identified in the table by its coordinates B3. To locate the point, you proceed along the horizontal to the interval identified by B and along the vertical to the interval identified by 3 (see Figure 1.1). The horizontal and vertical intervals identify two strips that intersect in the rectangle that contains the location of the point.

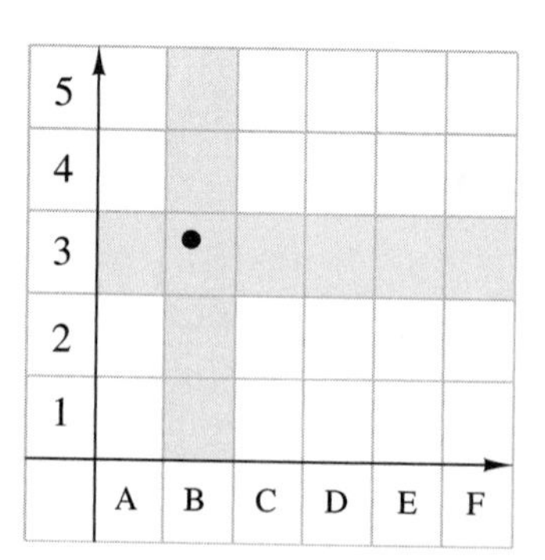

Figure 1.1

To pinpoint a single point in the plane, rather than find an approximate location, we can use a similar system of horizontal and vertical grid lines. To do this, we replace the intervals $A, B, C, \ldots$ on the horizontal grid and the intervals $1, 2, 3, \ldots$ on the vertical grid by individual points on two perpendicular lines.

1-A THE CARTESIAN COORDINATE PLANE

A *Cartesian coordinate system* for a plane consists of two intersecting perpendicular lines, one horizontal and one vertical, called *coordinate axes*. The point of intersection of these two axes is called the *origin* of the coordinate system. Each point on a coordinate axis is assigned a real number called the *coordinate* of the point. (For information on the real numbers, see the Appendix.) The absolute value of the coordinate represents the distance from the point to the origin, and the sign of the coordinate represents the direction from the origin. By convention, the origin, which lies on both axes, is assigned the number 0. On the horizontal axis, points to the right of the origin represent positive numbers and those to the left of the origin represent negative numbers. On the vertical axis, points above the origin represent positive numbers and those below the origin represent negative numbers.

To locate numbers on the axes, a unit of measure is needed. This unit of measure can represent any quantity; for example, it can be a measure of weight, temperature, or dollars. The length of 1 unit of measure can differ on the two axes, but for some applications it is assumed that the vertical and horizontal axes have the same scale. We will indicate when this is the case. Figure 1.2 shows a horizontal axis where 1 unit of measure has length 1 centimeter (cm). To locate the point on this axis that represents the number +2.3, we proceed to the right of the origin a distance of 2.3 cm. Similarly, to locate the point that represents −1.2, we proceed to the left of the origin by a distance of 1.2 units of measure. Note that when a number is positive, it is usually written without a + sign. By convention, no sign indicates a number is positive.

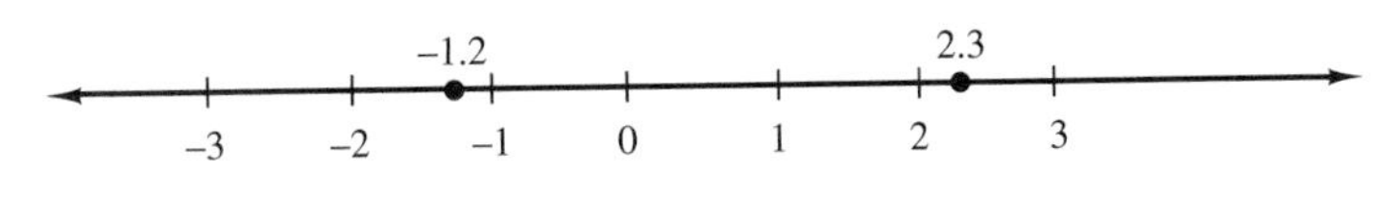

Figure 1.2

A plane with a Cartesian coordinate system is called a *Cartesian coordinate plane* (or just a *coordinate plane*). Each point in any Cartesian coordinate plane is identified by an ordered pair of real numbers, called the *coordinates of the point*. The origin is identified by the coordinates (0, 0). Any ordered pair of real numbers identifies a unique point in the plane, and any point in the plane uniquely determines its corresponding coordinates. To find the position of a point in this coordinate plane, we proceed as we did previously when we located the position of a destination on the map.

For example, the coordinates (–2.5, 4.68) identify a point P as indicated in Figure 1.3. The coordinate –2.5 is located on the horizontal axis, and the coordinate 4.68 is located on the vertical axis. Perpendicular cross hairs parallel to the axes and through these coordinates on the axes locate the point P as the point where the cross hairs intersect. It is common to use the notation $P(-2.5, 4.68)$ to mean the point P with coordinates (–2.5, 4.68).

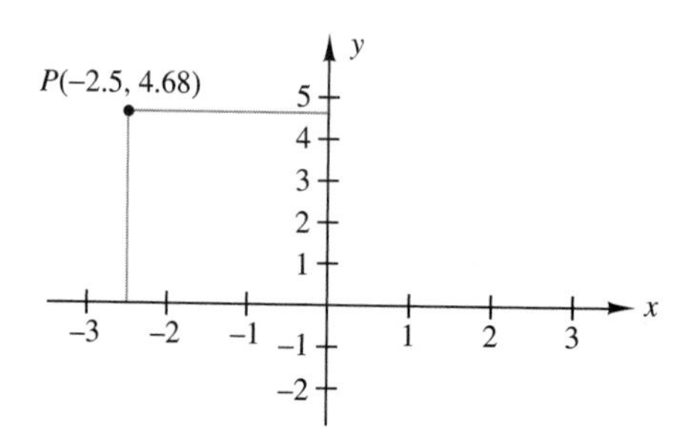

Figure 1.3

By convention, the horizontal axis is often called the *x-axis* and the vertical axis the *y-axis.* With these conventions, the first of the ordered pair of numbers that gives the position of a point is known as the *x-coordinate,* and the second number is known as the *y-coordinate.* The origin, with coordinates (0, 0), is sometimes referred to by the letter O. Points to the right of the origin O (to the right of the y-axis) are assigned positive x-coordinates, and those to the left of O (to the left of the y-axis) are assigned negative x-coordinates. Similarly, points above the origin (above the x-axis) are given positive y-coordinates, and those below O (below the x-axis) are given negative y-coordinates.

Coordinate axes divide the plane into *quadrants*, which by convention are numbered I, II, III, IV as shown in Figure 1.4.

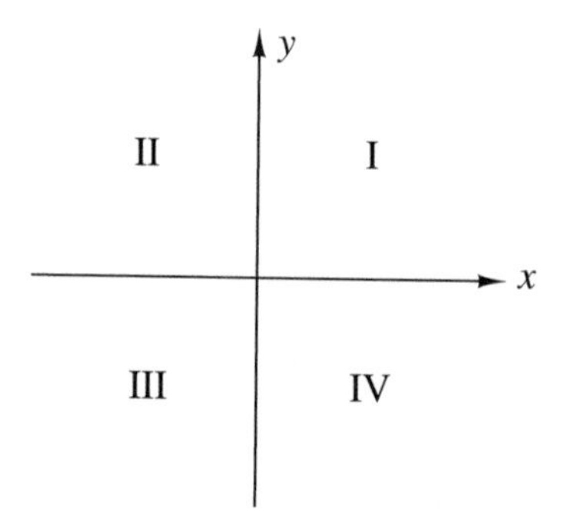

Figure 1.4

EXAMPLE 1.1

a. Consider three points A, B, and C in a Cartesian coordinate plane. The coordinates of these three points are given as (1, 2), (–1, 2), and (0, 4), respectively. (A point and its coordinates are frequently given together, as in $A(1, 2)$.) Draw a Cartesian system and show the positions of the three points (this is called *plotting points A, B, and C*).

b. Describe the position(s) of the point(s) whose first coordinate is 2.

Solution

a. The positions of A, B, and C are shown in Figure 1.5. A lies in quadrant I because both coordinates are positive, B lies in quadrant II, and C lies on the y-axis.

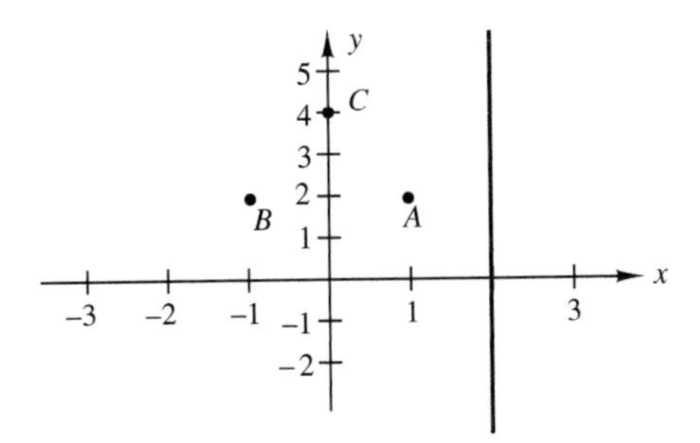

Figure 1.5

b. There are infinitely many points whose first coordinate (x-coordinate) is 2. These are all the points on the line parallel to the y-axis that are at a distance of 2 units to the right of the y-axis as indicated in Figure 1.5. ■

EXAMPLE 1.2 Look at the map of the Washington, D.C., area on the next page (Figure 1.6). Coordinate axes are drawn on the map and units of measure are given at the border of the map in order to answer the following:

a. What are the coordinates of the location of the Smithsonian Institute?

b. What are the coordinates of the location of the Lincoln Memorial?

c. Which Washington, D.C., landmark has coordinates (–6.5, 7.6)?

d. Which Washington, D.C., landmark has coordinates (1.25, 2.95)?

Solution

a. We first locate the site of interest and then proceed to measure how far from the origin we have to travel on the x-axis and on the y-axis to reach the intended destination. To reach the Smithsonian Institute, we proceed from the origin (the White House) in a positive direction (east) along the x-axis a distance of approximately

1 unit, and then from the origin along the y-axis in the negative direction (south) a distance of approximately 1.8 units. This means that the institute has coordinates $(1, -1.8)$ in this Cartesian plane (i.e., the coordinatized map).

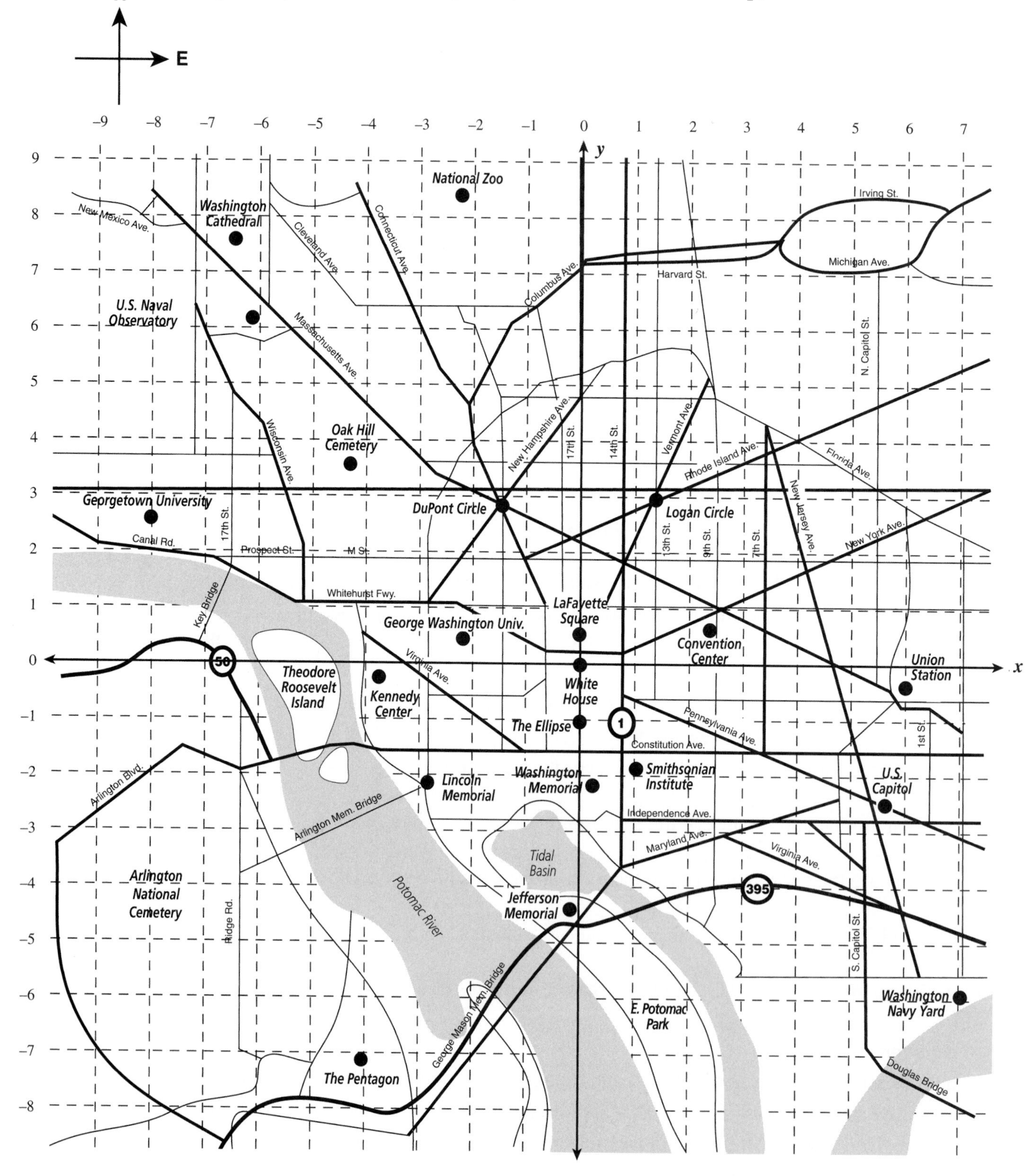

Figure 1.6 The Washington, D.C., Area

b. The Lincoln Memorial has coordinates $(-2.8, -2.1)$. The negative x-coordinate implies that we travel westward rather than eastward from the origin, since we have adopted the convention that the easterly direction is positive and the westerly direction is negative. Similarly, the northern direction is positive and the southern direction is negative.

c. Since $x = -6.5$, we move 6.5 units to the left (west) of the origin. Since $y = 7.6$, we then move 7.6 units up (north) and find we are at the Washington Cathedral.

d. The point with coordinates (1.25, 2.95) lies in quadrant I and corresponds to the location of Logan Circle. ■

Exercises 1-A

---

1. Consider the following four points: $A(-1.2, -2.5)$, $B(-2.7, 3.5)$, $C(1.5, -2.5)$, and $D(1.3, 2.5)$.

a. Show the positions of these points in a Cartesian coordinate plane.

b. In which quadrant does each of the points fall?

c. What can you say about the signs of all points that are in the first quadrant?

d. What can you say about the signs of all points that are in the third quadrant?

2. **a**. Give three points in the plane that have x-coordinate -1.2.

b. Determine how many points there are in the plane with x-coordinate -1.2, describe their location in words, and sketch these points in a Cartesian coordinate plane.

c. Describe all points whose y-coordinate is 2.5, and sketch these points in a Cartesian coordinate plane.

3. The points with coordinates $(-1, -3)$, $(4, -3)$, and $(4, 5)$ are three vertices of a rectangle. Find the fourth vertex and sketch the rectangle.

4. For each of the following landmarks on the Washington, D.C., map in Figure 1.6, identify the quadrant in which the landmark lies, and give your best estimate for the coordinates of the point at the center of the landmark.

a. The White House

b. Dupont Circle

c. The Convention Center

d. Union Station

e. The Jefferson Memorial

f. The National Zoo

g. The Washington Memorial

5. Identify which landmarks on the Washington, D.C., map have the following coordinates:

a. (−4.1, −7.1) **b.** (5.6, −2.5) **c.** (2.3, 0.6) **d.** (0, −1)

6. Consider the spreadsheet below.

a. What is the number in cell C4?

b. Give the labels for all cells that contain the number 14.

	A	B	C	D	E
1	1	2	3	4	5
2	11	12	13	14	15
3	21	14	23	24	25
4	14	32	33	34	35

1-B GRAPHS

When you open your morning paper, you see an abundance of tables and graphs. Graphs allow us to comprehend, with a single picture, a great deal of numerical information as well as trends in the data pictured. Many graphs feature horizontal and vertical axes, each representing a variable quantity. By examining the shape of the graph, we can tell how two quantities vary simultaneously. For example, the graph in Figure 1.7 shows the change in the annual unemployment rate (not seasonally adjusted) for Wisconsin for the years 1990 through 2003. We can see that the unemployment rate generally declined from 1991 through 1999 and then increased from 1999 through 2003.

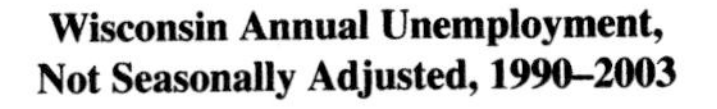

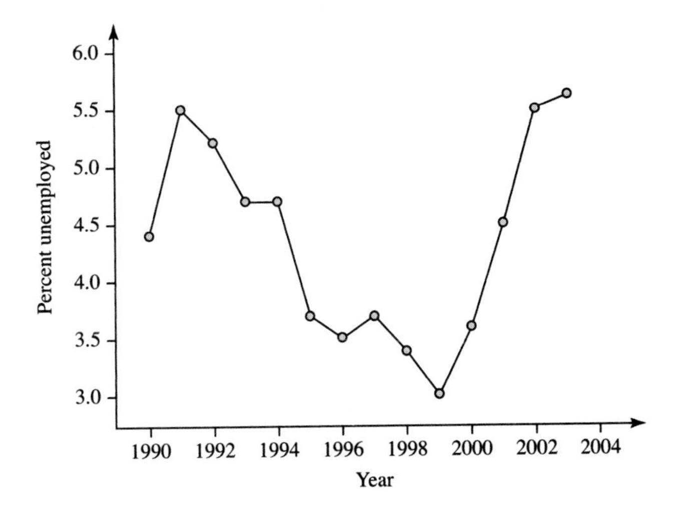

Figure 1.7

Data Source: http://www.dwd.state.wi.us/lmi/xls/urwi/xls

To make this graph, the unemployment data was recorded and organized in a table like this:

Year	90	91	92	93	94	95	96	97	98	99	00	01	02	03
Rate	4.4	5.5	5.2	4.7	4.7	3.7	3.5	3.7	3.4	3.0	3.6	4.5	5.5	5.6

A scatterplot was made by plotting points to represent each ordered pair in this table, using the year as the first coordinate and the rate as the second coordinate. The graph in Figure 1.7 was drawn by connecting consecutive dots in Figure 1.8 with line segments.

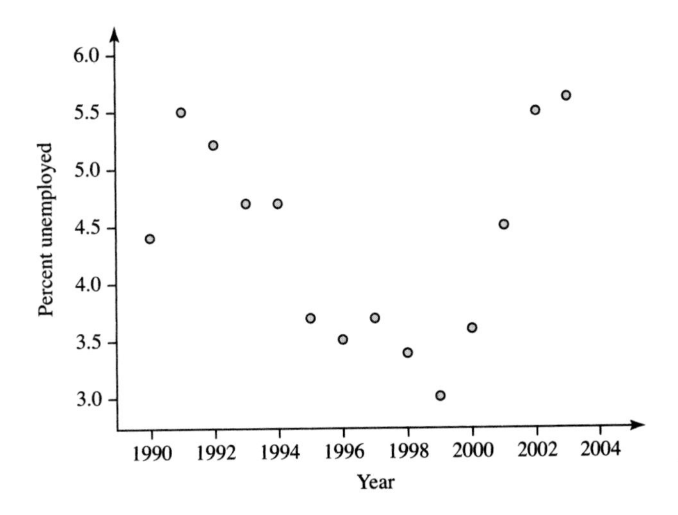

Figure 1.8

Any collection of ordered pairs of real numbers determines a corresponding collection of points in a Cartesian coordinate plane; the points have the ordered pairs as their coordinates. The points are called the *graph* of the set of ordered pairs.

Most graphs studied in calculus are graphs of *equations*. An *equation* is a mathematical statement asserting that two expressions are equal. For example, the statement $2x + 6$ is not an equation because it does not claim that two expressions are equal. The following are equations: $2x + 6 = 3$, $x^2 - 3 = 1$, $x^2 + 2 = 1$, and $2x + 6 = y$. The letter x appearing in the first, second, and third equations and the letters x and y appearing in the fourth equation represent *unknowns* called *variables*; they represent numbers with values that are not yet determined. A *solution* to an equation with one unknown is a number that when substituted for the unknown in the equation makes a true statement. For example, the equation $2x + 6 = 3$ has as a solution $x = -\frac{3}{2}$, since $2\left(-\frac{3}{2}\right) + 6 = 3$ is true; there is no other solution. The equation $x^2 - 3 = 1$ has $x = 2$ and also $x = -2$ as solutions. The equation $x^2 + 2 = 1$ has **no** real solution because $x^2 + 2 \geq 2$ for all real numbers x.

Do not confuse the words "solve" and "simplify." We *solve* equations and *simplify* expressions to derive equivalent expressions. Equations always have an equal sign. Only equations can be solved.

A solution to an equation that involves two unknowns is an ordered pair of numbers that when substituted for the unknowns makes a true statement. The equation $2x + 6 = y$ is true when $x = -1$ and $y = 4$, so the ordered pair $(-1, 4)$ is a solution. Another solution to this equation is $(1.8, 9.6)$; another is $(0, 6)$. In fact, we cannot list all solutions to this equation; it has an *infinite* number of solutions.

Given any equation in two variables, say x and y, the set of ordered pairs of real numbers for which the equation is true is the *solution set* of the equation. The *graph of the equation* is the graph of this solution set. The ordered pairs $(-1, 4)$, $(1.8, 9.6)$, and $(0, 6)$ all are in the solution set of the equation $2x + 6 = y$. If we let $x = a$ and $y = 2a + 6$, no matter what value a has, the ordered pair $(a, 2a + 6)$ is in the solution set of the equation $2x + 6 = y$. The graph of this equation is obtained by plotting all points in the solution set of the equation.

EXAMPLE 1.3 Consider the following equations as equations in two variables, x and y, and draw the graphs of the equations.

a. $x - y = 0$ **b.** $y = -3$

Solution

a. This equation is true when $x = y$, so the solution set is the set of all points whose x-coordinates and y-coordinates are equal. Ordered pairs of real numbers in the solution set are of the form (a, a). The graph of the equation is the line in Figure 1.9.

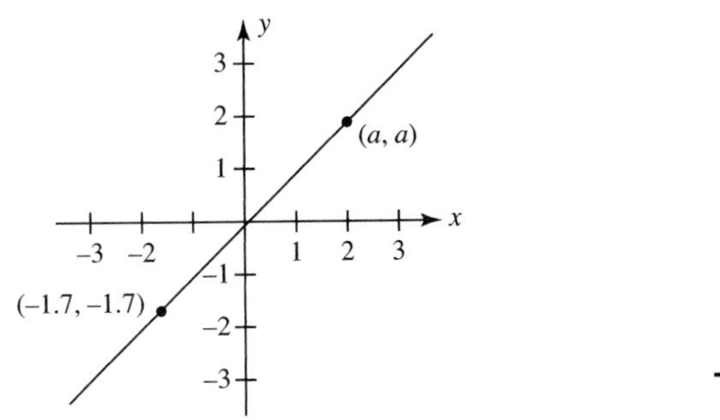

Figure 1.9 $x - y = 0$

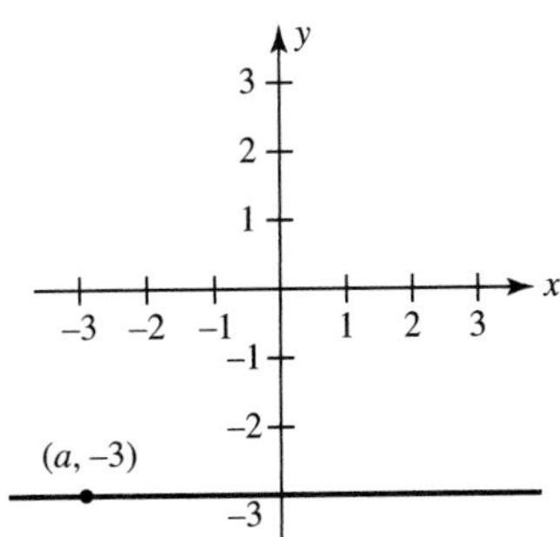

Figure 1.10 $y = -3$

b. Although only the variable y appears in this equation, since we are told it is an equation that involves x and y, we interpret the fact that the variable x does not appear in the equation to mean that there are no conditions or restrictions on x values. Therefore, this graph is the set of all points whose y-coordinate is -3; that is, the set of points with coordinates of the form $(a, -3)$. It is the line parallel to the x-axis and 3 units below it as shown in Figure 1.10. ■

The *intercepts* of the graph of an equation that relates x and y tell us where the graph intersects the axes. The x-intercepts are values of x that satisfy the equation when $y = 0$. The y-intercepts are values of y that satisfy the equation when $x = 0$. The equation $2x + 6 = y$ has -3 as an x-intercept, since when $y = 0$, $2x + 6 = 0$; therefore, $x = -3$. This equation has y-intercept 6, since substituting 0 for x in the equation gives $(2)(0) + 6 = y$; therefore, $y = 6$. The graph intersects the x-axis at $x = -3$ and the y-axis at $y = 6$.

EXAMPLE 1.4 A flower shop bought a new delivery van for \$22,000. The van was depreciated linearly using the function $V = 22{,}000 - 4000t$, where V represents the value of the van and t the age of the van in years. Find the two intercepts for the graph of the function, and interpret the meaning of each intercept with respect to this situation.

Solution The horizontal intercept occurs when $V = 0$. We solve $0 = 22{,}000 - 4000t$ for t and find: $-22{,}000 = -4000t$, or $5.5 = t$. This means that the value of the van is \$0 after $5\frac{1}{2}$ years. The vertical intercept occurs when $t = 0$. At $t = 0$, $V = 22{,}000$. This means that the initial value of the van at an age of 0 years (brand new) is \$22,000. ■

Exercises 1-B

1. Consider the following equations as equations in two variables x and y and draw the graphs of the equations.

a. $x + y = 0$ **b**. $x = 5$ **c**. $y = -2.8$

2. Find the x-intercepts and y-intercepts of the following equations that relate x and y.

a. $3.7y - 2.1x = 5.5$ **b**. $y^2 - 2x = 9$

3. After each equation below, the coordinates of several points are given. For each point, determine whether or not it is a solution of the equation.

a. $x + y^2 = 1$ $(0, -1), (-1, 0), (-1, \sqrt{2})$

b. $x = 1 + 3y$ $(0, 0), (1, 0), (2, \frac{1}{3})$

c. $x - 2y = 0$ $(2, 1), (1, 2), (2a, a)$

d. $y(2 - x) = 2$ $(1, 0), (0, 1), (1, 1), (3, -2)$

4. The rates of unemployment (as a percent) in the city of Baltimore for the years 1990 through 2003 are given in the table below. Plot the values on a coordinate grid, and make a graph like the one shown in Figure 1.7.

Year	90	91	92	93	94	95	96	97	98	99	00	01	02	03
Rate	8.1	10.1	11	10.5	8.8	8.5	8.3	9.6	9.2	7.3	8.1	7.9	8.0	8.6

Data Source: http://www.dllr.state.md.us/lmi/laus/region19902003.htm

1-C LINES AND THEIR EQUATIONS

A *line* is a geometric concept; we learn in geometry that any two distinct points determine a unique line. Any line joining two points in the Cartesian plane has infinitely many points. Each of these points is identified by an ordered pair of numbers, its coordinates. Using coordinate geometry, we can think of a line as the graph of an equation. Since many phenomena can be modeled by linear equations, we want to learn how to express and recognize these equations.

Through any given point in the plane, there are infinitely many lines—one point is not enough to determine a line. A vertical line is a special line parallel to the

y-axis through a given point. Every other line has a unique *slope*; this describes its inclination to the horizontal axis. If we know the slope of a line through a given point, the line is completely determined. Using coordinate geometry, we can easily define the slope of a line. Choose any two distinct points $P(x_1, y_1)$ and $Q(x_2, y_2)$ on the line. The slope of the line is defined as the ratio:

$$\text{slope} = \frac{y_1 - y_2}{x_1 - x_2}.$$

Other ways to express slope are

$$\text{slope} = \frac{\Delta y}{\Delta x} = \frac{\text{change in } y\text{-coordinates}}{\text{change in } x\text{-coordinates}}.$$

The symbol Δ (the Greek letter delta) denotes difference; Δy is a difference of y-coordinates and Δx a difference of corresponding x-coordinates.

> ☞ When calculating a slope, be sure to subtract y-coordinates and x-coordinates of points in the same order.

Several facts about the slope of a line should be noted. All follow from the definition.

1. *The slope of a line does not depend on the choice of the two points used to calculate it.* (Properties of similar triangles assure this.) The following table gives x- and y-coordinates of points that all lie on the same line. The corresponding points are plotted in Figure 1.11.

x	0	1	-3	$\frac{3}{2}$	$-\frac{1}{2}$
y	$\frac{3}{2}$	$-\frac{7}{2}$	$\frac{33}{2}$	-6	4

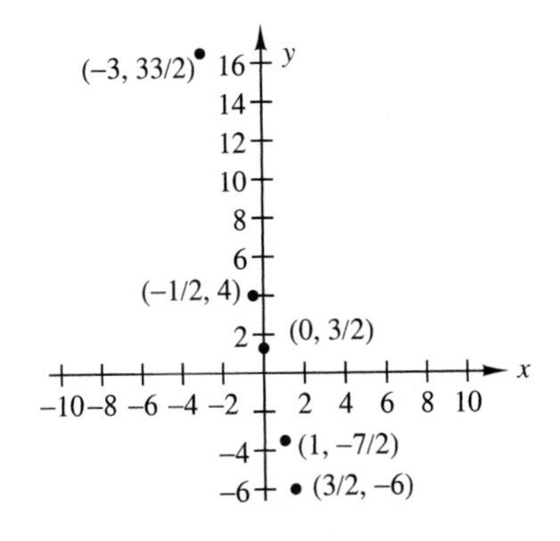

Figure 1.11

If we take the points $\left(0, \frac{3}{2}\right)$ and $\left(1, -\frac{7}{2}\right)$ to compute the slope of the line, then

$$\text{slope} = \frac{\frac{3}{2}-\left(-\frac{7}{2}\right)}{0-1} = \frac{\frac{10}{2}}{-1} = -5\,.$$

If we choose two other points, say $\left(-3, \frac{33}{2}\right)$ and $\left(-\frac{1}{2}, 4\right)$, to compute the slope, then

$$\text{slope} = \frac{\frac{33}{2}-4}{-3-\left(-\frac{1}{2}\right)} = \frac{\frac{33}{2}-\frac{8}{2}}{-\frac{6}{2}+\frac{1}{2}} = \frac{\frac{25}{2}}{-\frac{5}{2}} = -5\,.$$

Note that if we choose these same points, but in the opposite order, we still get the same slope:

$$\text{slope} = \frac{4-\frac{33}{2}}{-\frac{1}{2}-(-3)} = \frac{-\frac{25}{2}}{\frac{5}{2}} = -5\,.$$

2. *The slope of any horizontal line is* 0. Since all y-coordinates on a horizontal line are the same, the difference $y_1 - y_2$, the numerator of the slope ratio, equals 0.

3. *The slope of any vertical line is* **undefined**. Since all x-coordinates of points on a vertical line are the same, the difference $x_1 - x_2$ equals 0, so the quotient that defines slope has 0 in the denominator and cannot be computed.

4. *The slope gives the amount of change in the variable y (the vertical change) that occurs when the variable x is increased by* 1 *unit (horizontal change of* 1 *unit).* This simply says that when $\Delta x = 1$, the slope equals Δy.

EXAMPLE 1.5 Jason bought a 3-year-old car for \$14,400. When the car was 5 years old, he sold it for \$9,900. Assume the depreciation was linear.

a. Find two ordered pairs that show the value of the car at different ages.

b. Use the ordered pairs to find the slope of the line that shows the depreciation of the value of the car over time. What is the meaning of the slope in this instance?

Solution

a. The two ordered pairs are (3, 14,400) and (5, 9,900), which indicate the value of the car at 3 years old and 5 years old, respectively.

b. The slope of the line through these two points is:

$$m = \frac{y_1 - y_2}{x_1 - x_2} = \frac{14{,}400 - 9{,}900}{3-5} = \frac{4{,}500}{-2} = -2{,}250.$$

The value of the car decreased at the rate of \$2,250 per year. (The negative sign for the slope shows decrease.) ■

The slope of a line tells us the steepness of its inclination. Let's compare a line with slope 2 and a line with slope 5. In the first line, when x is increased by 1 unit, y increases by 2 units; in the second line, when x is increased by 1 unit, y increases by 5 units. This means that the second line has a steeper inclination than the first. See Figure 1.12.

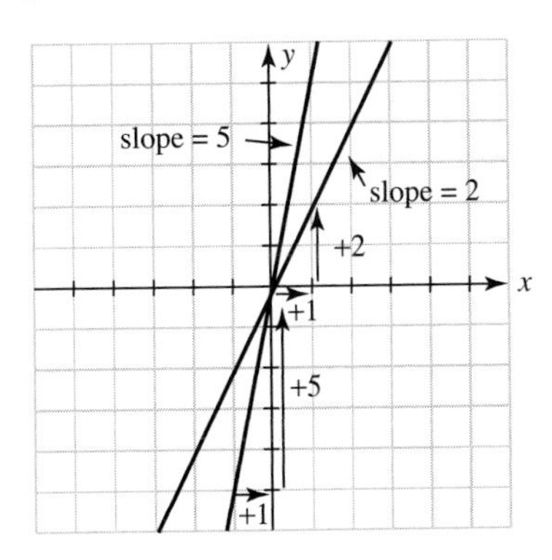

Figure 1.12

Figure 1.13

The sign of the slope tells us which way the line is inclined. Compare two lines in Figure 1.13, one with slope +3 and one with slope −3. In the first line, when x is increased by 1 unit, y increases by 3 units. Therefore, as x is moved 1 unit to the right, y moves up 3 units. The line inclines up to the right. In the second line, when x is increased by 1 unit, y decreases by 3 units. Therefore, as x is moved 1 unit to the right, y moves down 3 units. The line declines down to the right.

5. *Two lines that are parallel (i.e., that never meet) have equal slopes.* The ratio $\frac{\Delta y}{\Delta x}$ is the same for both lines, because as x is increased by 1 unit, the change in y is the same for both lines.

The most common symbol used to denote the slope of a line is the letter m. If the length of 1 unit on each axis is the same, the line through the origin that divides quadrants I and III into two equal pieces has slope $m = +1$. Similarly, the line through the origin that bisects quadrants II and IV has $m = -1$. Other lines through the origin are shown on the graph in Figure 1.14 with their slopes noted. Notice the steepness of the lines relative to each other.

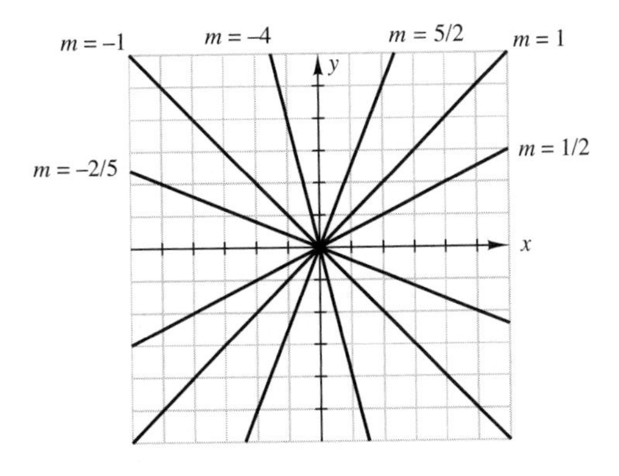

Figure 1.14

■

EXAMPLE 1.6

a. Graph the line through the points (–1, 0) and (2, –2).

b. Find the slope of this line.

c. Give the coordinates of two other points on this line.

Solution

a.

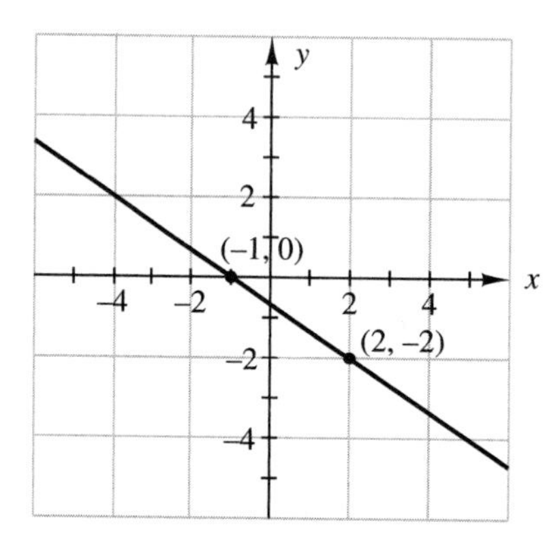

Figure 1.15

b. Slope $= \dfrac{0-(-2)}{-1-2} = \dfrac{2}{-3} = -\dfrac{2}{3}$.

c. Since slope $= -\dfrac{2}{3} = \dfrac{\Delta y}{\Delta x}$, we can choose any Δy and Δx whose ratio is $-\dfrac{2}{3}$ and add these changes to the coordinates of any point on the line to get another point on the line. One choice is $\Delta x = 3$ and $\Delta y = -2$. Add these to the coordinates of (2, –2): (2 + 3, –2 – 2) = (5, –4), which is another point on the graph. Another choice is $\Delta x = -3$ and $\Delta y = 2$. Add these to the coordinates of (–1, 0): (–1 – 3, 0 + 2) = (–4, 2), which is on the graph. See Figure 1.16. ■

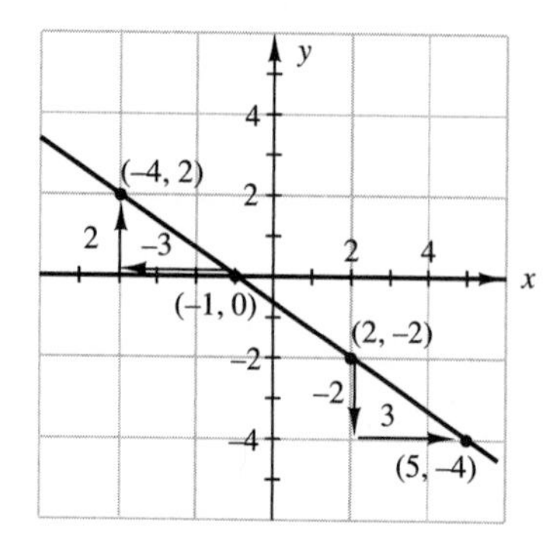

Figure 1.16

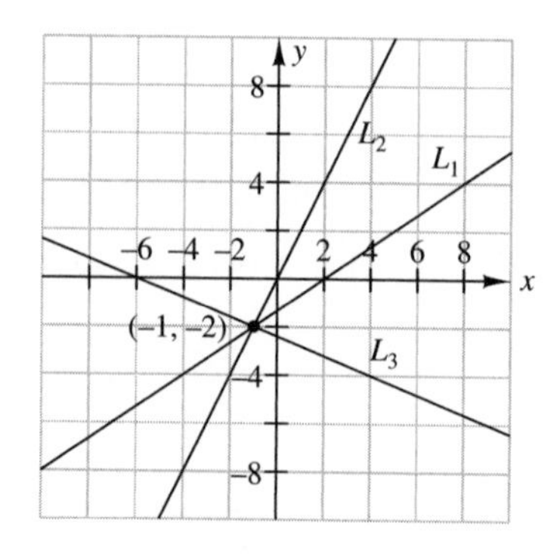

Figure 1.17 ■

EXAMPLE 1.7 The graph in Figure 1.17 contains several lines that pass through the point with coordinates (–1, –2). Estimate the slopes of these lines.

Solution Consider line L_1, and pick two points on the line—for example, (2, 0) and (8, 4). Draw a right triangle whose hypotenuse joins the points, as shown in Figure 1.18, and mark Δx and Δy (with arrows showing the direction of the change in x and y) as the horizontal and vertical legs of the triangle, respectively. Estimate the ratio of the legs, $\frac{\Delta y}{\Delta x}$. This is the slope m of the line. For L_1, it appears that the slope is $\frac{\Delta y}{\Delta x} = \frac{4}{6} = \frac{2}{3}$. For the line L_2, we choose points (1, 2) and (4, 8) to estimate the slope as $m = \frac{\Delta y}{\Delta x} = \frac{6}{3} = 2$. The line L_3 declines as x is moved to the right, so its slope is negative and is estimated to be $m = \frac{-2}{5} = -\frac{2}{5}$. ■

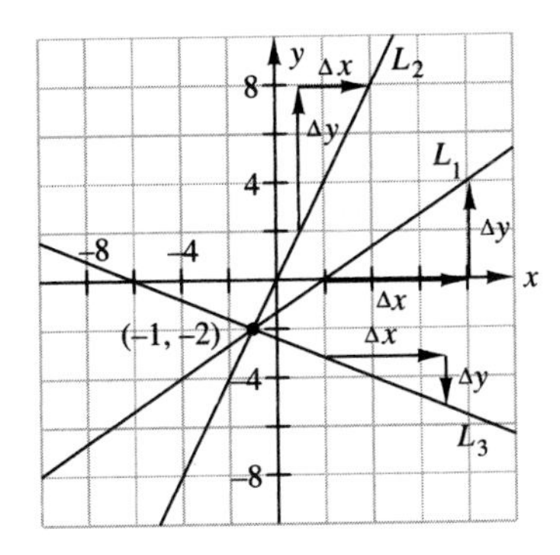

Figure 1.18

The equation of a vertical line is of the form $x = c$; this states that all x-coordinates of points in its solution set equal a single constant c. The equation of a nonvertical line gives a relationship between the x- and y-coordinates of any point on the line. There are many different forms in which the equation of a single nonvertical line may be expressed.

To find the equation of a line, we must have one point $P(x_1, y_1)$ on the line and one other piece of determining information. This second piece of information may be a second point (from which the slope can be calculated), or it may be the value of the slope. Once the slope is known, the fact that **any** two points on the line determine its slope allows an equation to be written. If m is the slope of the line and $Q(x, y)$ is any point on the line, then the equation $m = \frac{y - y_1}{x - x_1}$ must be true. This equation of the line through $P(x_1, y_1)$ with slope m can also be written in the following form:

Point-slope form of the equation of a line: $y - y_1 = m(x - x_1)$.

When the given point on the line is on the y-axis, its coordinates are $(0, b)$; the value b is the y-intercept of the line. Substituting these coordinates for x_1 and y_1 in the above equation gives $y - b = m(x - 0)$, which yields the following form:

Slope-intercept form of the equation of a line: $y = mx + b$.

Each of these equations is an example of the general form of a *linear equation*, $ax + cy + d = 0$, where the coefficients a and c are not both zero. Every linear equation can be put in one of the following two forms:

$$\boldsymbol{x = k} \quad \text{or} \quad \boldsymbol{y = mx + b}.$$

We can then sketch the graph of the line by noting its slope and y-intercept if the equation is $y = mx + b$ or else sketch the vertical line $x = k$.

EXAMPLE 1.8 Consider the line through the points $P(0, 2)$ and $Q(3, 6)$. Give the equation of the line in point-slope form and in slope-intercept form.

Solution First compute the slope of the line: $m = \frac{2-6}{0-3} = \frac{-4}{-3} = \frac{4}{3}$. If P is taken as the given point on the line, the equation of the line in point-slope form is

$$y - 2 = \tfrac{4}{3}(x - 0), \quad \text{or} \quad y - 2 = \tfrac{4}{3}x.$$

If Q is taken as the given point on the line, the equation of the line in point-slope form is

$$y - 6 = \tfrac{4}{3}(x - 3).$$

The slope-intercept form of the equation is $y = \frac{4}{3}x + 2$. It can be obtained from either of the equations above. Just add 2 to both sides of the first equation, or multiply out the right side of the second equation and add 6 to both sides. ■

EXAMPLE 1.9 Put each of the following linear equations into the form $x = k$ or $y = mx + b$ as appropriate. Identify the slope and y-intercept of the line where possible.

a. $3x + 6y - 5 = 0$ **b.** $4x - 6 = 9$ **c.** $4y + 6 = 3$

Solution

a. Since y appears in the equation, we can solve for y to put the equation into slope-intercept form. We have $6y = -3x + 5$, so $y = -\frac{1}{2}x + \frac{5}{6}$. The slope is $-\frac{1}{2}$, and the y-intercept is $\frac{5}{6}$.

b. Since y does not appear in this equation, it can be put in the form $x = k$. We have $4x = 15$, so $x = \frac{15}{4}$. There is no y-intercept, and the slope is undefined for a vertical line.

c. Solving for y, we get $4y = -3$, so $y = -\frac{3}{4}$. This is in slope-intercept form; the slope is 0, and the y-intercept is $-\frac{3}{4}$. The graph of this equation is a horizontal line. ■

Exercises 1-C

1. For each of the given lines, determine if the point $\left(-\frac{1}{2}, 3\right)$ lies on the line.

a. $2y - 3x = 5$

b. $2.4x + y - 1.8 = 0$

c. $2y - 6 = 0$

d. $2x + 1 = 0$

2. For each of the linear equations in Exercise 1, put the equation in the form $x = k$, $y = k$, or $y = mx + b$. Identify the slope and y-intercept where appropriate and sketch the graph of each equation.

3. Consider the points $(-6, 3)$ and $(1, 2)$.

a. Find the slope of the line through the two given points.

b. Find an equation of the line through the two given points.

c. Give coordinates of two other points on the line through the two given points.

d. Sketch the graph of the line and show your points from part c.

4. Estimate the slope of each of the lines drawn in Figure 1.19.

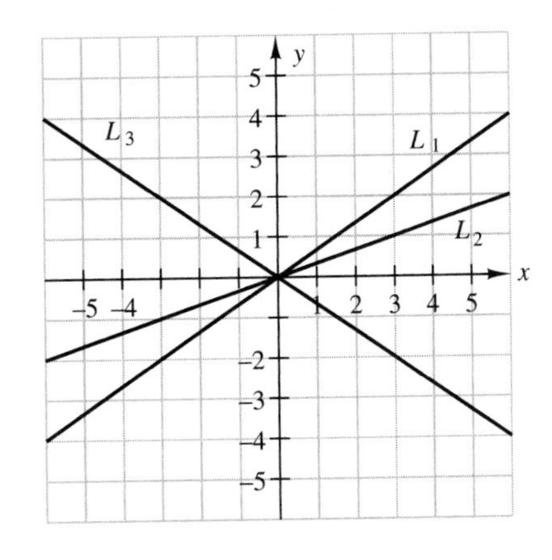

Figure 1.19

5. When the price of a large coffee was \$1.50, Carl's Coffee Corner sold an average of 1200 cups per week. When Carl raised the price to \$2.50, he sold an average of 900 cups per week. Assume the relationship between the demand (number sold) and the price is linear.

a. Find two ordered pairs (p, d) that show the demand d for large coffees at different prices p.

b. Use the ordered pairs to find the slope of the line that relates price and demand. What is the meaning of the slope in this situation?

c. Find an equation for demand in terms of price. Express the equation in point-slope form and in slope-intercept form.

1-D PARALLEL AND INTERSECTING LINES

Any two distinct lines in a plane either intersect at a single point or never intersect. In the latter case, we say that the two lines are *parallel.* Distinct vertical lines are parallel; nonvertical lines are parallel if they have the same slopes. Lines that intersect must have unequal slopes. If two lines intersect, there is a single point that is a solution to the equations for both lines. To find where the two lines intersect, we solve their equations simultaneously.

EXAMPLE 1.10 Two pairs of linear equations are given below. For each pair, find the point of intersection of the two lines, if it exists.

a. $y - x - 1 = 0, \quad y = 2x + 1$ **b**. $y + 5x = 2, \quad 2y + 10x + 5 = 0$

Solution

a. We first find the slope of each line to decide if the given equations represent parallel or intersecting lines. We can rewrite the first equation as $y = x + 1$, so the slope of the line is 1. From the second equation we read the slope of that line as 2. Since the lines have different slopes, they intersect. Assume that the point of intersection of the two lines is (a, b). Since this point lies on both lines, its coordinates must satisfy both equations. So

$$(1) \quad b - a - 1 = 0 \qquad \text{and} \qquad (2) \quad b = 2a + 1.$$

We can substitute for b in equation (1) by using equation (2). Equation (1) then becomes $(2a + 1) - a - 1 = 0$, which simplifies to $a = 0$. To find b, we substitute $a = 0$ in either equation; this gives $b = 1$. So the point of intersection of the two lines is (0, 1). Graphing the two lines on a single coordinate system confirms our answer.

b. We first find the slope of each line. We rewrite the first equation as $y = -5x + 2$. The slope of this line is −5. Now we solve for y in the second equation: $y = \frac{-10x-5}{2} = -5x - \frac{5}{2}$. This line has slope −5 also. The two lines have equal slopes, so they are parallel or coincide. Since their y-intercepts are different, they are parallel and therefore do not have any point of intersection. ■

The Appendix shows a second method for finding the point of intersection of two lines.

Two intersecting lines that are perpendicular (orthogonal) to each other have a special property. A horizontal line and a vertical line are perpendicular to each other. It can be proved that two nonvertical lines with slopes m_1 and m_2 satisfy the equation $m_1m_2 = -1$ if and only if they are perpendicular. Figure 1.20 shows two parallel lines with another line perpendicular to them; note how their slopes are related.

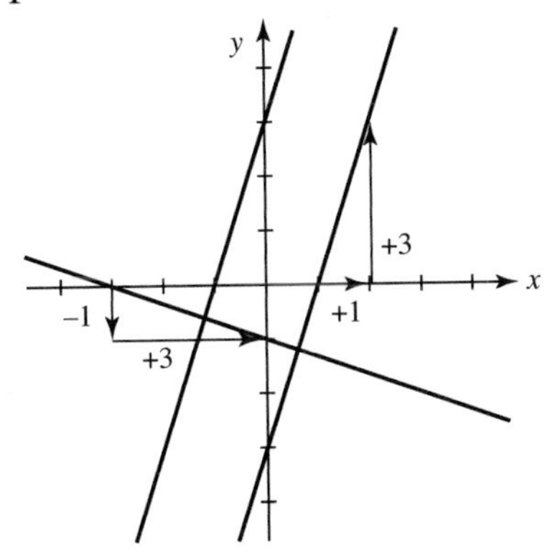

Figure 1.20

> ☞ When perpendicular lines are graphed, the lines will not appear to be perpendicular unless you use a square grid—a grid in which the units on each axis are identical.

EXAMPLE 1.11 Test whether the given pairs of equations below represent two perpendicular lines, and sketch the graphs of the lines.

a. $2y - 5x - 16 = 0$, $5y = -2x + 7$ **b.** $x - 3.2y = 1$, $4x + 1 = 3y$

Solution

a. First, we express the two equations in slope-intercept form to determine the slopes of the given lines. We rewrite the first equation as $2y = 5x + 16$, and then solve for y: $y = \frac{5}{2}x + 8$. The slope of this line is $m_1 = \frac{5}{2}$. The second equation becomes $y = -\frac{2}{5}x + \frac{7}{5}$, so the slope of this line is $m_2 = -\frac{2}{5}$. Since $m_1m_2 = \left(\frac{5}{2}\right)\left(-\frac{2}{5}\right) = -1$, the two lines are perpendicular. Their graphs are shown in Figure 1.21a.

b. The first equation can be rewritten as $y = \frac{1}{3.2}x - \frac{1}{3.2} = 0.3125x - 0.3125$. So $m_1 = 0.3125$. For the second line, $y = \frac{4}{3}x + \frac{1}{3}$ and $m_2 = \frac{4}{3}$. Since the product $m_1m_2 \neq -1$, these lines are not perpendicular. See Figure 1.21b for their graphs.

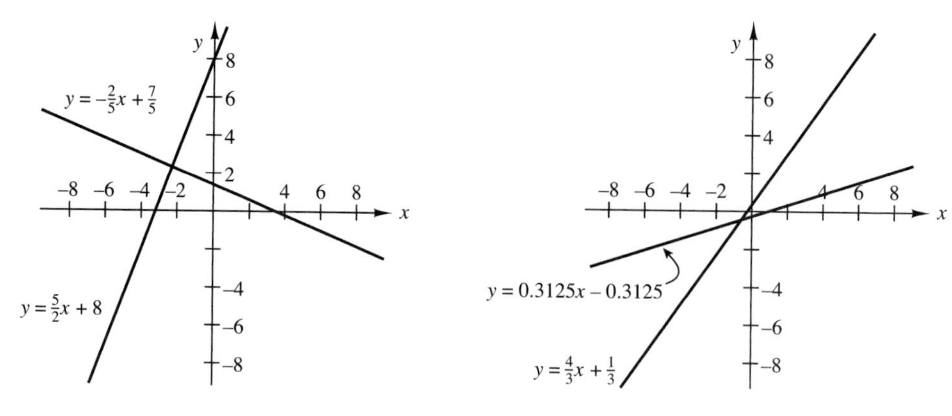

Figure 1.21a

Figure 1.21b ■

EXAMPLE 1.12 Trains A and B travel on adjacent parallel tracks. Train A is 50 miles ahead of train B, and both trains travel at a constant speed of 60 mph in the same direction. At noon, train B passes a mile marker.

a. How far apart are the two trains at 1:00 P.M.? At 2:00 P.M.?

b. Write equations that express the distances of the two trains from the mile marker at t hours past noon.

c. Graph the two equations. What do you observe?

Solution

a. Between noon and 1:00 P.M., each train traveled 60 miles farther, so they are still 50 miles apart. After 2 hours, since each train continues to travel at the same speed, they are still 50 miles apart.

b. Since train B passes the mile marker at noon, its distance to the marker t hours later is given by the equation $d_B = 60t$. Since train A is always 50 miles ahead of train B, its distance to the mile marker is given by the equation $d_A = 60t + 50$. Initially (at noon), train A is 50 miles from the mile marker, so 50 is the y-intercept of the graph.

c. The graphs of the two equations are shown in Figure 1.22. We can observe that the graphs are lines that are parallel (both lines have slope 60); this shows that the distance between the two trains is constant. Since time is nonnegative, the lines are graphed for only $t \geq 0$.

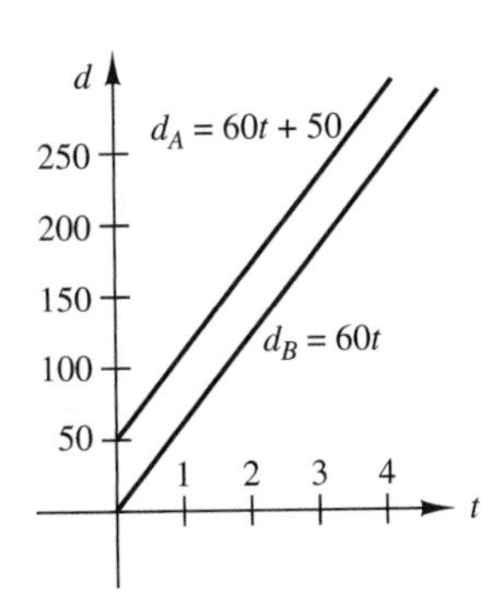

Figure 1.22

■

Exercises 1-D

1. Consider the line with the equation $y = 2x$.

a. Give equations of two different lines that are parallel to the given line.

b. Give equations of two different lines that are perpendicular to the given line.

2. Consider the four points $A(-a, -a)$, $B(-a, a)$, $C(a, a)$, $D(a, -a)$, where a is positive.

a. Give the equation of the line through A and C, and give the equation of the line through B and D.

b. Are the two lines in part a perpendicular?

c. If the area of the figure $ABCD$ is 6.25, what is the value of a?

3. **a.** Give the equations of each of the lines L_1, L_2, and L_3 sketched in Figure 1.23.

b. Give the coordinates of points of intersection of L_1 and L_2 and of L_1 and L_3.

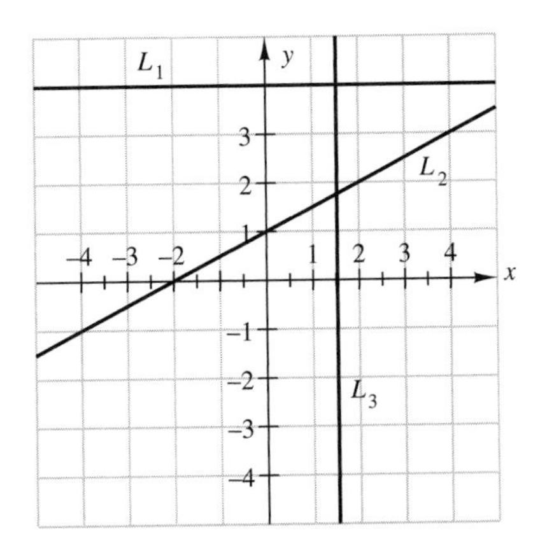

Figure 1.23

4. Consider the line with equation $3y = 5x - 7$.

a. Give the coordinates of two points on the given line.

b. Find the slope of the given line.

c. Find the slope of a line perpendicular to the given line.

1-E DISTANCE BETWEEN TWO POINTS

The distance between any two points on a coordinate line or in a coordinate plane is just the length of the line segment that joins them. In particular, if we have two points on a coordinate axis as shown below, the distance between them is measured by the absolute value of the difference of their coordinates x_1 and x_2. If these are the coordinates of the points P_1 and P_2, respectively, then the distance between P_1 and P_2, denoted by $|P_1P_2|$, is the absolute value of $x_1 - x_2$, denoted by $|x_1 - x_2|$. If, in the figure, $x_1 = -1.3$ and $x_2 = 2.7$, then the distance between them is $|-1.3 - 2.7| = |-4| = 4$.

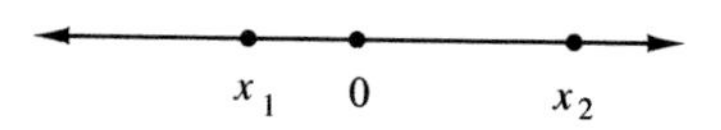

When two points P_1 and P_2 are given in a coordinate plane, either they are on a line parallel to one of the coordinate axes or they lie on a slanted line. In the second case, they are two vertices of a right triangle such as in Figure 1.24.

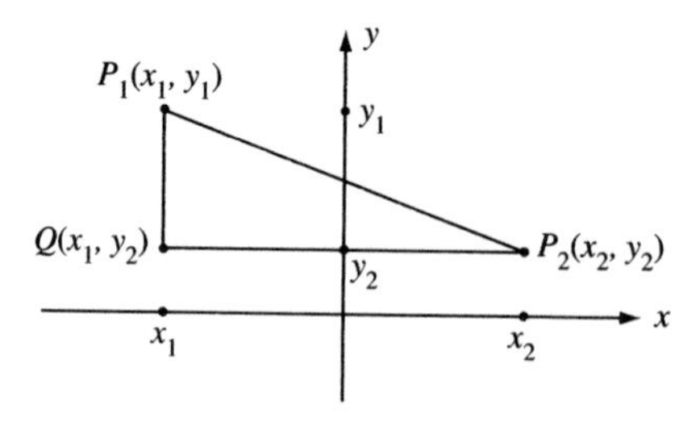

Figure 1.24

The distance between points on lines parallel to the x- or y-axis is the same as the distance between their coordinates on the parallel axis. So, $|P_1Q| = |y_1 - y_2|$ and $|P_2Q| = |x_1 - x_2|$.

The distance between P_1 and P_2 can now be found by using the Pythagorean theorem:

$$\begin{aligned}|P_1P_2|^2 &= |P_2P_1|^2 = |P_1Q|^2 + |P_2Q|^2 \\ &= |y_1 - y_2|^2 + |x_1 - x_2|^2 \\ &= (y_1 - y_2)^2 + (x_1 - x_2)^2.\end{aligned}$$

Thus, the distance between P_1 and P_2 is given by the distance formula:

$$|P_1P_2| = \sqrt{(y_1 - y_2)^2 + (x_1 - x_2)^2} = \sqrt{(x_2 - x_1)^2 + (y_2 - y_1)^2}.$$

This formula is actually true for all points P_1, P_2 in the plane—see what happens when you apply it to the coordinates of P_1 and Q or to P_2 and Q in Figure 1.24.

The distance formula is based on the assumption that all coordinates are given in the same unit of measure. The measured distance between two points in a coordinate plane will equal the calculated distance using the distance formula only when both axes have the same scale.

EXAMPLE 1.13 Use the distance formula to determine the distance between the points $P_1(-2, 3)$ and $P_2(-4, -5)$.

Solution $|P_1P_2| = \sqrt{(-4-(-2))^2 + (-5-3)^2} = \sqrt{4+64} = \sqrt{68} \approx 8.25$ ■

EXAMPLE 1.14 Use a ruler to measure (in centimeters) the most direct distance between the Smithsonian Institute and the Lincoln Memorial on the map of the Washington, D.C., area in Figure 1.6. Then use the distance formula to compute this distance.

Solution The distance on the map measured with a ruler is about 3.8 cm. (Note that the coordinate scale on the map is 1 unit = 1 cm.) Now we use the distance formula. In Example 1.2 we found that the Smithsonian Institute and the Lincoln Memorial had coordinates $(1, -1.8)$ and $(-2.8, -2.1)$, respectively, where these coordinates represent directed distance in centimeters. The distance between these points is

$$\begin{aligned}D &= \sqrt{(1-(-2.8))^2 + (-1.8-(-2.1))^2} \\ &= \sqrt{14.44 + 0.09} \approx 3.81 \text{ cm.}\end{aligned}$$

This answer is reasonably close to the measured distance. ■

Exercises 1-E

1. Consider the three points $A(-1, -5)$, $B(0, -2)$, and $C(2, 4)$.

a. Find the distances $|AB|$, $|BC|$, and $|AC|$.

b. Show that $|AB| + |BC| = |AC|$ and explain why this is true.

2. Find the distance between the two points $(2.7, -5.1)$ and $(-3.3, -4.9)$.

3. Consider the four points $A(-1, -1)$, $B(-1, 1)$, $C(1, 1)$, and $D(1, -1)$.

a. Graph these points in a coordinate plane.

b. Compute the distances $|AB|$, $|BC|$, $|CD|$, $|DA|$, $|AC|$, and $|BD|$.

c. What is the shape of $ABCD$? Justify your answer.

4. State the distance formula in words only. Begin with "To find the distance between two points in a Cartesian coordinate plane, ..."

1-F THE CIRCLE

A *circle* is the set of points equidistant from a fixed point, known as the *center* of the circle. This constant distance from the fixed point is known as the *radius of the circle*. To derive the equation of a circle, we use the definition of a circle and the distance formula.

EXAMPLE 1.15 Find an equation that must be satisfied by any point (x, y) on the circle whose center is the point $(3, -2)$ and whose radius is 5.

Solution The distance from any point (x, y) on the circle to the center $(3, -2)$ is the radius 5. Therefore, $\sqrt{(x-3)^2 + (y-(-2))^2} = 5$. If we square both sides of this equation, it can be rewritten without the square root sign as

$$(x - 3)^2 + (y + 2)^2 = 5^2.$$

This is an equation of the circle with center $(3, -2)$ and radius 5. ■

EXAMPLE 1.16 Find an equation of the circle with center at $(0, 0)$ and radius r, where r is a positive number.

Solution We want to write an equation stating that the distance from the point (x, y) on the circle to $(0, 0)$ is r. Thus, $\sqrt{(x-0)^2+(y-0)^2}=r$. Squaring both sides, we obtain the equation $(x-0)^2+(y-0)^2=r^2$, which simplifies to $x^2+y^2=r^2$. ■

The general equation of a circle with center at the point (h, k) and radius $r > 0$ is:

$$(x-h)^2+(y-k)^2=r^2.$$

This equation is obtained in the manner shown in Example 1.15. The distance from a point (x, y) on the circle to the center (h, k) is set equal to the radius r, and then both sides of the equation are squared. Any equation that can be put into this form by algebraic manipulation (for example, by completing the square) must be the equation of a circle. Note that the radius r must be a positive number, since the distance between two distinct points is always positive. The process of completing the square is discussed in the Appendix.

EXAMPLE 1.17 A small island nation in the Pacific claims fishing rights within a 12-mile radius from the center of the island.

a. Where should you put the origin of your Cartesian coordinate system to most easily draw the map showing the 12-mile circle?

b. Find an equation to describe the boundary of this region.

c. If a boat from another nation is fishing 6 miles east and 10 miles north of the center of the island, is it violating the island nation's fishing laws?

Solution Draw a simple diagram, like Figure 1.25. Diagrams are always helpful in solving problems.

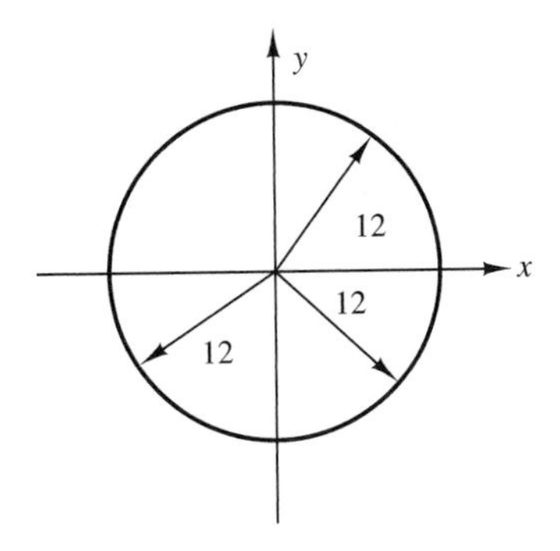

Figure 1.25

a. The origin of the coordinate system should be placed at the center of the island to describe the circle conveniently.

b. The boundary of the region corresponds to a circle centered at the origin (center of the island) with a 12-mile radius. Therefore, the corresponding equation is

$$x^2 + y^2 = 12^2 = 144.$$

c. The distance of the ship from the center of the island (at (0, 0)) is obtained from the distance formula: $d^2 = (6 - 0)^2 + (10 - 0)^2 = 36 + 100 = 136$. Since $d^2 = 136 < 144$, the ship is violating the island's fishing law. See Figure 1.26.

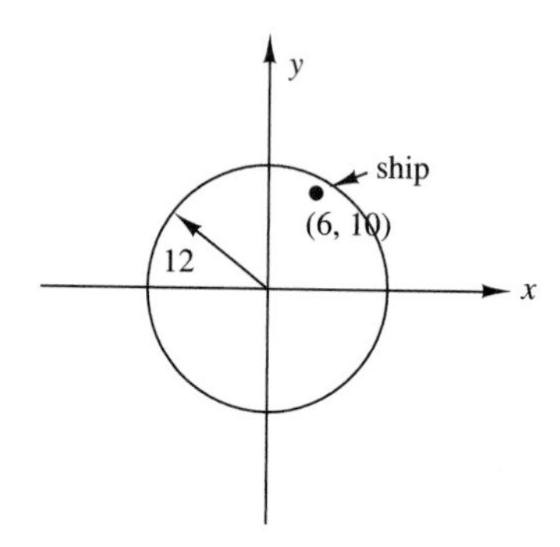

Figure 1.26 ■

Exercises 1-F

1. **a**. Graph the circle with center at the origin and radius $\sqrt{5}$ and give the equation of the circle.

b. Verify that the point (−1, 2) is on the circle.

c. Draw the radius of the circle from (0, 0) to (−1, 2).

d. Find the slope of the line through (0, 0) and (−1, 2).

e. Find the equation of the line tangent to the circle at the point (−1, 2). (Recall that a tangent to a circle is perpendicular to a radius of the circle at the point of tangency.)

2. Consider the circle with center at the origin and radius of 5.

a. Determine the distances between the origin and each of the points $P_1(3, 4)$, $P_2(4, 3)$, $P_3(2, 2)$, and $P_4(4, 5)$.

b. For each of the points given in part a, determine if it is inside the circle, outside the circle, or on the circle.

c. Use the points in part a that fall on the circle, along with the two points $A(0, 5)$ and $B(5, 0)$, to approximate the length of the circular arc from A to B.

(Use the sum of the lengths of the line segments AP_1, P_1P_2, and P_2B, to approximate the length.)

3. **a.** Give the equation of a circle with center (–5, 4) and radius $r = 1$, and graph the circle.

b. Describe the portions of the circle where the tangent lines will have positive slope.

c. Describe the portions of the circle where the tangent lines will have negative slope.

d. At what points on the circle will the tangent lines be horizontal? At what points will they be vertical?

4. Consider the circle in Figure 1.27.

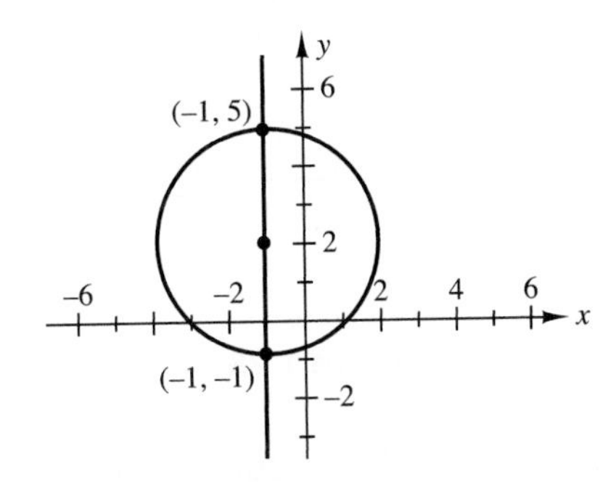

Figure 1.27

a. Find the coordinates of the center of the circle.

b. Find the radius of the circle.

c. Find an equation of the circle.

5. For each of the following equations of circles, find the radius and coordinates of the center. Sketch a graph of each circle.

a. $(x + 4)^2 + (y - \frac{1}{2})^2 = 3$ **b.** $x^2 + y^2 = \pi$

Chapter 1 Exercises

1. Write the equations and sketch the corresponding graphs of the following lines:

a. The horizontal line passing through (–1, 2)

b. The vertical line passing through (–1, 2)

c. The line through $(-1, 3)$ and $(0, 3)$

d. The line through the origin with slope $-\frac{2}{5}$

e. The line through the points $(-1, -2.5)$ and $(2, 5)$

2. What is the relationship between the two lines in parts d and e of Exercise 1 above?

3. For each of the lines shown on the graph in Figure 1.28, write an equation of the line.

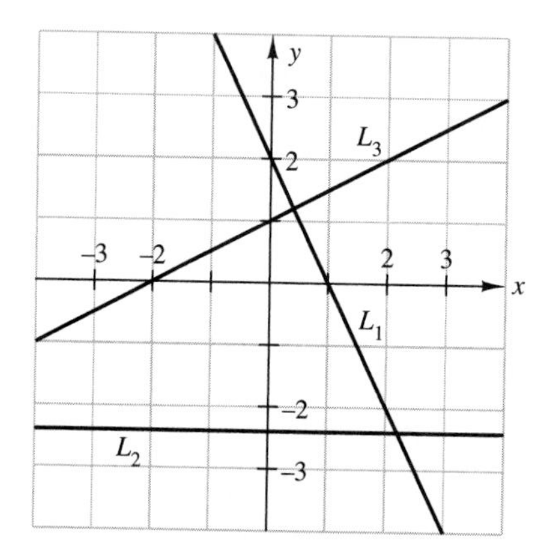

Figure 1.28

4. **a.** Sketch three different lines whose equations have the form $y = mx - 3$. (Choose three different values for m and graph each equation.)

b. Give the slope and y-intercept of each of your lines in part a.

c. What do the three lines in part a have in common?

5. **a.** Sketch three different lines whose equations have the form $y = 2x + b$. (Choose three different values for b and graph each equation.)

b. What are the slope and y-intercept of each of your lines in part a?

c. What do the three lines in part a have in common?

6. Use slopes to determine if the point $(-1, 2)$ lies on the line through the points $(2, 0)$ and $\left(\frac{1}{2}, -1\right)$.

7. Find the slope and intercepts of each of the lines below:

a. $3x - 4y = 9$ **b.** $y = -3$ **c.** $2x = 3y + 1$

8. Draw two lines through the point $(-0.7, 2)$ such that the two lines are perpendicular to each other. Give the slopes of the two lines.

9. Which of the two lines is more steep, one with slope $\frac{3}{2}$ or one with a slope $\frac{2}{3}$? Draw a picture and justify your answer.

10. a. Draw three lines with slope 0.

b. What is the slope of a line that is perpendicular to these?

c. Give the equations of the three lines in part a and an equation of a line perpendicular to them.

11. Consider the rectangle with vertices at the points (2, 6), (8, 6), (2, 3), and (8, 3).

a. Write equations of the two diagonals.

b. Are the diagonals perpendicular? Justify your answer.

12. Consider a circle with center at the origin and radius 5.

a. Determine the distances between the origin and each of the points (–5, 0), (0, –5), (0, 5), (3, 4), (1, 2), (6, 8).

b. For each of the given points, determine if it is inside the circle, outside it, or on the circle.

13. For each, find an equation of the circle with the given conditions:

a. Center (–1, –3), radius $\sqrt{2}$ **b.** Center (–3, 0), radius 5

c. Center (4, –1) and passing through the point (–2, 2)

14. Trains A and B are traveling on adjacent parallel tracks. At noon, train A is 50 miles ahead of train B, as train B passes a grain elevator. Train A travels at a constant speed of 40 mph; train B travels at a constant speed of 50 mph in the same direction.

a. How far apart are the two trains at 1:00 P.M.? At 2:00 P.M.?

b. Write equations that express the distances of the two trains from the grain elevator t hours after noon.

c. What is the y-intercept for each equation in part b? Explain the meaning of each intercept in this context.

d. Graph the two equations. What do you observe?

e. What is the slope of each line? Explain the meaning of each slope in this context.

f. Will train B catch up with train A? If the answer is yes, determine when.

15. A rectangular plot of land is such that its width is half its length.

a. Denote the length by x, and express the perimeter y in terms of x.

b. Sketch the graph of the equation from part a.

c. Explain in words what the graph looks like and why it looks like this.

16. Consider the points $P\left(-1, \frac{5}{2}\right)$, $Q(0, 1)$, and $R(2, -2)$.

a. Plot the points on a Cartesian coordinate system.

b. Verify that the given points lie on the line with equation $y = -\frac{3}{2}x + 1$ and graph the line.

c. Compute the distance from the origin O to each of the given points.

d. Which of the given points is closest to O?

e. Illustrate on the graph drawn in part a each of those distances computed in part c. Does your answer to d make sense?

17. a. Write the distance from O to an arbitrary point (x, y) on the line with equation $y = -\frac{3}{2}x + 1$. Express the distance in terms of x.

b. Use the formula obtained in part a to find the distance from O to the points on the line with x-coordinate: $x = \frac{1}{3}$, $x = \frac{1}{6}$, $x = 0$, $x = -1$, and $x = 1$. Which is the shortest of those distances?

c. How would you find the shortest distance from O to the given line?

18. Consider the line with equation $y = -\frac{3}{2}x + 1$.

a. Write an equation of the line that contains the origin O and that is perpendicular to the given line. Graph both lines.

b. Find the coordinates of the point of intersection of those two lines.

c. Give the distance between O and the point of intersection. Mark this distance on the graph. This is the shortest distance from O to the given line. Does it look like it in your graph?

CHAPTER 2

Functions

When one quantity is described as a function of another, it means that the first quantity depends in some way on the second. The nature of this dependence can be given by an equation, specified in a table, represented by a graph, or described in words. For example, the income tax a person pays depends on the person's taxable income, thus income tax is a function of the taxable income. Number of years of experience on a job has a direct bearing on the salary received, so salary is a function of the number of years of experience. The area of a circle depends on its radius, so we say that area of a circle is a function of its radius. The area of a square depends on the length of its side, so area of a square is a function of the length of its side. Volcanologists have found that it is possible to forecast an eruption of a volcano weeks in advance from the field measurements inserted into a single basic equation. Thus the time of eruption has been found to be a function of geochemical changes measured at the site of the volcano. (See David Berreby, "Barry Versus the Volcano," *Discover,* Vol. 12, No. 6, June 1991.)

When one variable quantity depends on a second, the first is called the *dependent* variable and the second the *independent* variable. In beginning calculus courses, the concept of "function" is usually restricted to cases where the value of the dependent variable is *completely* determined by the value of the independent variable. Such functions are called functions of a *single variable.*

In the examples above, the income tax that a person pays is not a function of the single variable of taxable income, since there are other factors considered in calculating the tax (e.g., marital status). This means that taxable income does not completely determine the tax; two taxpayers with exactly the same taxable income may pay different amounts of income tax. For example, in 2003 a married person with a taxable income of \$46,000 paid \$6,204 in taxes, whereas a single person with the same taxable income paid \$8,316. On the other hand, the area of a square is completely determined by the length of its side. It is not possible to find two squares with the same side length and different areas. Therefore, the area of a square is a function of the single variable—namely, the length of its side.

From now on we will use the word "function" only for functions of a single variable. A *function* can then be thought of as a rule that assigns to each value of the independent variable a **unique** value of the dependent variable. For the functions that we will investigate and analyze, the variables will be real numbers.

To better understand this concept, it is helpful to think of a function as a machine that transforms quantities: When one value of the independent variable is fed into the machine, the unique corresponding value of the dependent variable comes out.

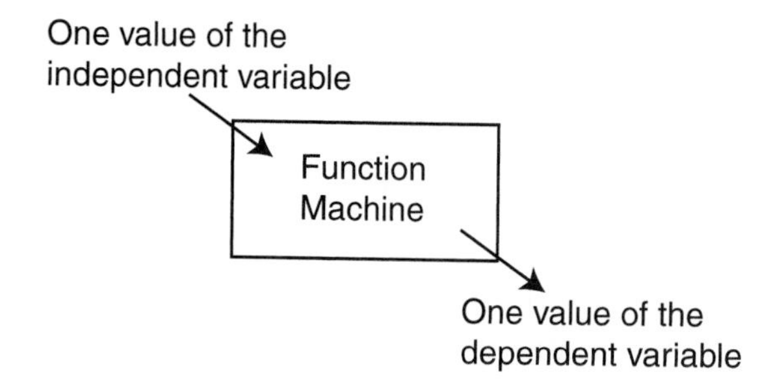

Note that the "2003 income tax machine" would not be a function machine, since when a taxable income of \$46,000 is fed into the machine, several values for income tax come out:

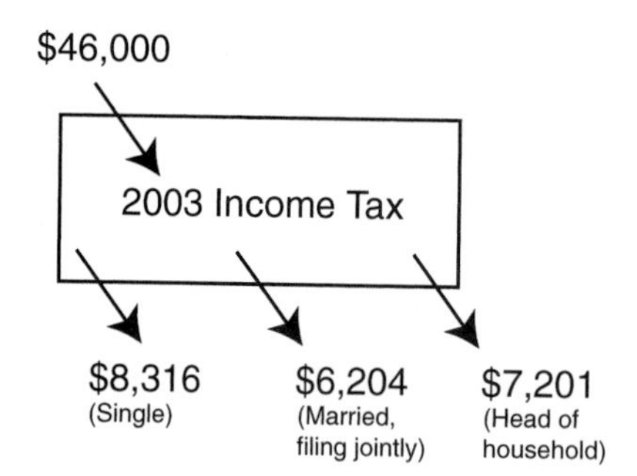

The machine that takes as input the length of a square's side and gives as output the square's area is a function machine. If x denotes the length of a side of the square, the area is given by the equation $A = x^2$. Here the independent variable is x and the dependent variable is A. To each value of x, this function assigns the square of that number, x^2, as the (unique) value of the dependent variable A. This machine can then be represented as follows:

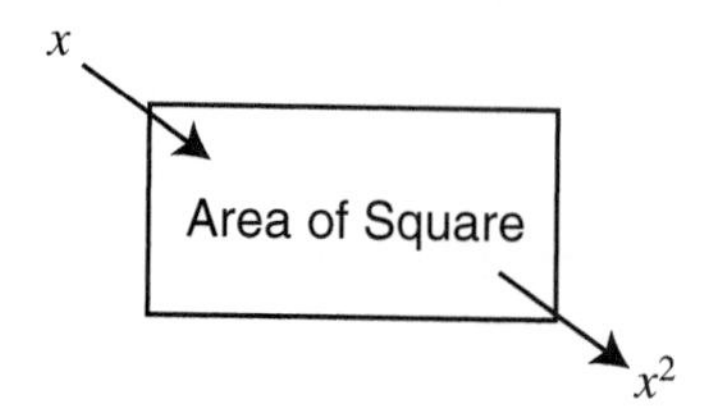

For example, when the number 3 is fed into the machine, the outcome is $3^2 = 9$. If the number 4.5 is put into the machine, the value that comes out is $(4.5)^2 = 20.25$. (The function machine performs the calculations, but you must keep track of the units of measure.)

2-A FUNCTION NOTATION

Functions are usually denoted by letters such as f, g, h, F, G, H or by a suggestive letter like v (for velocity). If f is a function and x denotes the independent variable, then the unique value of the dependent variable that corresponds to x is denoted by $f(x)$, which is read as "the value of f at x" or simply "f of x" or "f at x."

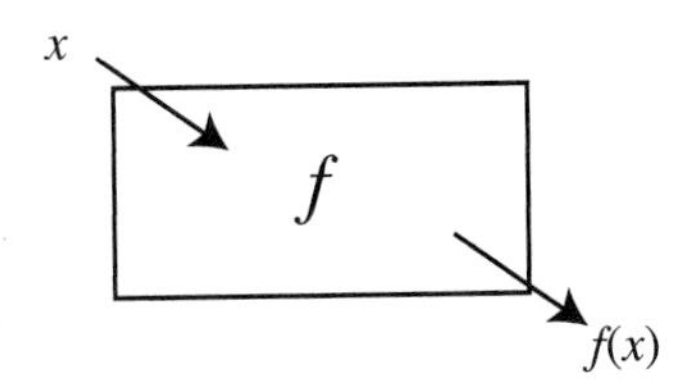

If f is the function that gives the area A of a square in terms of the length of its side, we write $A = f(x) = x^2$. The area function squares the length of one side. In this case, $f(3)$ denotes the area of a square whose side is 3 units long, which is $3^2 = 9$ square units; $f(4.5)$ denotes the area of a square of a side 4.5 units long; $f(4.5) = (4.5)^2 = 20.25$ square units.

Calculus is mainly concerned with the study of functions and how their values change as the independent variable is changed. These functions are most often defined by an equation that gives the function rule as a formula. The formula tells what operations to perform on the independent variable to carry out the rule of the function. The choice of letter to name the variable is immaterial. Thus, the function $f(x) = x^2$ is exactly the same as the function $f(t) = t^2$.

The dummy variable x in $f(x)$ is a placeholder that can be replaced by any constant or algebraic expression. It may help to think of x as a box in which an appropriate expression can be placed. For example, $f(x) = x^2 + 2x + 1$ can be interpreted as

$$f(\square) = \square^2 + 2\square + 1.$$

When the variable in an equation for a function is replaced by a value (a number or algebraic expression), we say that the function is *evaluated* at that value. To evaluate the function $f(x) = x^2 + 2x + 1$ at $x = -3$, we replace each occurrence of x with -3:

$$f(-3) = (-3)^2 + 2(-3) + 1 = 9 - 6 + 1 = 4.$$

We evaluate f at $a + h$ as follows:

$$f(a + h) = (a + h)^2 + 2(a + h) + 1.$$

In the next example, we evaluate a function at several different values.

EXAMPLE 2.1 Suppose a function g is defined by the following equation:

$$g(t) = 3t^2 - \frac{6}{t}.$$

a. Translate the symbolic rule for g into a sentence that uses only words.

b. Express the rule for g as a single quotient.

c. Find $g(4)$. **d**. Find $g(a^2)$. **e**. Find $g(a + 2)$.

f. Some values of the independent variable t are given in the following table. Use a calculator to find the corresponding values of $g(t)$ rounded off to three decimal places, and fill in the table with these values.

t	−3.4	−2.0	−1.2	−0.3	0.5	1.0	1.7
$g(t)$							

g. Is there a value $g(0)$?

Solution

a. The function g assigns to a number three times the square of that number minus the quantity 6 divided by the number.

b. $g(t) = 3t^2 - \frac{6}{t} = \frac{3t^2 - 6}{t}$

To simplify a sum or difference of fractions, find a common denominator.

$$\frac{a}{b} - \frac{c}{d} = \frac{ad - bc}{bd} \quad \text{and} \quad \frac{a}{b} + \frac{c}{d} = \frac{ad + bc}{bd}$$

c. To find $g(4)$, we substitute 4 for t in the expression $g(t) = 3t^2 - \frac{6}{t}$:

$$g(4) = 3(4)^2 - \frac{6}{4} = (3)(16) - \frac{3}{2} = 48 - 1.5 = 46.5.$$

d. We need to substitute a^2 for t in $g(t)$:

$$g(a^2) = 3\left(a^2\right)^2 - \frac{6}{a^2} = 3a^4 - \frac{6}{a^2} = \frac{3a^6 - 6}{a^2}.$$

e. The expression $a + 2$ replaces the variable t in the equation that defines $g(t)$:

$$g(a+2) = 3(a+2)^2 - \frac{6}{a+2} = \frac{3(a+2)^3 - 6}{a+2}.$$

Note that $g(a + 2) \neq g(a) + 2$, since $g(a) + 2 = 3a^2 - \frac{6}{a} + 2$.

Also, $g(a + 2) \neq g(a) + g(2)$, since $g(a) + g(2) = 3a^2 - \frac{6}{a} + 3 \cdot 2^2 - \frac{6}{2} = 3a^2 - \frac{6}{a} + 9$.

f. It is easier (and often more accurate) to calculate values of $g(t)$ in the form of a single quotient, as in part b: $g(-3.4) = \frac{3(-3.4)^3 - 6}{-3.4} = 36.4447\ldots$; similarly, $g(-2) = \frac{3(-2)^3 - 6}{-2} = 15$. This gives us two new entries in the table.

t	−3.4	−2.0	−1.2	−0.3	0.5	1.0	1.7
$g(t)$	36.445	15					

You are asked to complete the remaining calculations (Exercises 2-A, 1).

g. To find the value of g at 0, we would have to divide 6 by 0, but it is not possible to divide by zero. Therefore, for this function g, the independent variable cannot take the value 0, and there is no value $g(0)$. We say $g(0)$ is *undefined.* ■

To find $f(a + h)$, replace every occurrence of the variable in the rule for f by the expression $a + h$.
For almost all functions, $f(a + h) \neq f(a) + h$ and $f(a + h) \neq f(a) + f(h)$.

EXAMPLE 2.2 Given the function $f(x) = x^2 - 4$, use a calculator to complete the following table.

h	$f(3)$	$f(3+h)$	$f(3+h) - f(3)$	$\frac{f(3+h) - f(3)}{h}$
1	5	12	7	7
0.1	5	5.61	0.61	6.1
0.01	5			
0.001	5			
0.0001	5			

Solution When $h = 0.01$, $f(3 + h) = f(3 + 0.01) = f(3.01) = (3.01)^2 - 4 = 5.0601$. This is the third entry in the row for $h = 0.01$. The fourth entry in that row is obtained by subtracting the second from the third entry, which is $f(3+0.01)-f(3)=5.0601-5=0.0601$. The last entry in this row is the number in the third entry divided by $h=0.01$, that is, $\frac{0.0601}{0.01}=6.01$. We add these entries to the table:

h	$f(3)$	$f(3+h)$	$f(3+h)-f(3)$	$\frac{f(3+h)-f(3)}{h}$
1	5	12	7	7
0.1	5	5.61	0.61	6.1
0.01	5	5.0601	0.0601	6.01
0.001	5			
0.0001	5			

You are asked to complete the table in Exercises 2-A, 3. ■

Exercises 2-A

1. Complete the table in Example 2.1, part f.

2. Consider the function $f(t)=\frac{5t-1}{t+2}$.

a. Find each of the following: $f(-0.1)$, $f(0)$, $f\left(\frac{1}{5}\right)$, $f(2)$, and $f(x)$.

b. Is there a value $f(-2)$? Explain.

3. Complete the table in Example 2.2.

4. For the function $G(x)=8x^3-2$, evaluate each of the following: $G(-1)$, $G(1.2)$, $G(a)$, $G(t+1)$, and $G(2+h)$.

2-B DOMAIN AND RANGE OF A FUNCTION

For a given function, possible values for the independent variable may be restricted. If the function is given by a formula, there may be some values for which the formula is undefined. In Example 2.1, the function g is not defined at the number 0 because we can't divide by zero. Also, there are situations where practical considerations lead us to limit the values of the independent variable. For example, if f is the function that gives the area of a square as a function of its side, then we define f only for positive real numbers, since only these can represent the length of a side.

The set of values for which a function is defined is called the *domain* of the function. The function that represents the area of a square as a function of the length of its side has as its domain the set of all positive real numbers, which is the interval $(0, \infty)$.

The domains of most functions studied in calculus are intervals or collections (unions) of intervals. It is usually important to note whether or not the endpoints of the interval are included. Various notations for sets and intervals are used; we now review these.

1 Set Notation for Real Numbers

Curly brackets $\{\}$ are most often used to designate a set; the elements in the set can be described in two different ways:

(1) A list of all elements in the set. This is used for finite sets or well-defined sequences.

For example, $\{0, 1, 2, 3\}$ designates the set that contains just the numbers 0, 1, 2, and 3. The notation $\{\ldots, -2, -1, 0, 1, 2, \ldots\}$ represents the set of integers. The ellipses . . . mean “and so on”; the pattern continues.

(2) Set-builder notation. Unless otherwise specified, these sets contain all real numbers that satisfy the description following the symbol $|$, which means “such that.”

For example, $\{x \mid x \text{ is integer and } 0 \le x \le 3\}$ is read “The set of all real numbers x such that x is an integer and x is between 0 and 3, inclusive.” This is the same as the set $\{0, 1, 2, 3\}$.

The notation $\{x \mid 0 \le x \le 3\}$ designates the set of all real numbers x such that x is between 0 and 3, inclusive. This set contains infinitely many numbers in a continuum, so they cannot be listed. The set can, however, be graphically represented as the segment of a coordinate number line with endpoints 0 and 3 (see the table of interval notation that follows).

2 Interval Notation for Real Numbers

The table that follows summarizes the different ways to describe intervals of real numbers. A note on graphing points: A hollow circle or “hole” on a graph represents a point that is **not** on the graph. A filled-in circle is used to emphasize that the point **is** on the graph.

Graph on the Real Line	Interval Notation	Set of Real Numbers	Set Notation	Description
a b	(a, b)	all x, $a < x < b$	$\{x \mid a < x < b\}$	All real numbers strictly between a and b
a b	$[a, b]$	all x, $a \le x \le b$	$\{x \mid a \le x \le b\}$	All real numbers between a and b, including a and b
a b	$(a, b]$	all x, $a < x \le b$	$\{x \mid a < x \le b\}$	All real numbers between a and b, including b
a b	$[a, b)$	all x, $a \le x < b$	$\{x \mid a \le x < b\}$	All real numbers between a and b, including a
a	(a, ∞)	all x, $a < x$	$\{x \mid a < x\}$	All real numbers strictly greater than a
b	$(-\infty, b)$	all x, $x < b$	$\{x \mid x < b\}$	All real numbers strictly less than b
a	$[a, \infty)$	all x, $a \le x$	$\{x \mid a \le x\}$	All real numbers greater than or equal to a
b	$(-\infty, b]$	all x, $x \le b$	$\{x \mid x \le b\}$	All real numbers less than or equal to b

In interval notation for intervals that extend infinitely in one or both directions, open (rounded) parentheses are always used next to the symbols ∞ and $-\infty$.

When a function is given by an algebraic rule, without explicit mention of the domain, the domain is understood to be the largest set of real numbers for which the operations involved are defined. In other words, a real number c is in the domain of a function f if and only if $f(c)$ is a real number. For example, if $f(x) = \sqrt{x}$, the domain of f is $[0, \infty)$. This is because for $x > 0$, $\sqrt{x}$ means the positive number whose square is x; for $x = 0$, $\sqrt{0} = 0$; and for $x < 0$, $\sqrt{x}$ is not a real number.

When the domain of a function f is not specified, ask the question: For which real numbers c is $f(c)$ a real number?

EXAMPLE 2.3 Determine the domain of each of the following three functions:

a. The function g of Example 2.1

b. The function h defined by $h(x) = \frac{4}{2x+1}$

c. The function $f(x) = \sqrt{x} + 2$

Solution

a. The function $g(t) = 3t^2 - \frac{6}{t}$ involves the operations of multiplication and subtraction, which are always possible for real numbers, and of division, which is possible only when the divisor is different from zero, so the domain of g is the set of all nonzero real numbers. This domain can be written as the set $\{t \mid t \neq 0\}$. It can also be represented as the union of two intervals, $(-\infty, 0) \cup (0, \infty)$. Here the $\cup$ symbol is read "union" and means "together with." A single real number missing from the domain of a function implies that the domain is the union of the two intervals separated by that number.

b. A quotient is undefined when the denominator is zero. We need to exclude from the domain the numbers x for which $2x + 1 = 0$. To solve for x, we first subtract 1 from both sides of the equation, and then divide both sides by 2:

$$2x + 1 = 0 \rightarrow \qquad 2x = -1 \rightarrow \qquad x = -\tfrac{1}{2}.$$

The domain of h is the set $\left\{x \middle| x \neq -\frac{1}{2}\right\} = \left(-\infty, -\frac{1}{2}\right) \cup \left(-\frac{1}{2}, \infty\right)$.

c. Since $\sqrt{x}$ is not a real number when x is negative, the function $f(x) = \sqrt{x} + 2$ is defined only for $x \geq 0$. Therefore, the domain of this function is the interval $[0, \infty)$. ■

 For any expression $\square$, $\sqrt{\square}$ is defined only for $\square \geq 0$.

Another set associated with a function is the *range* of the function, which is the set of all possible values of the dependent variable. For example, if $f(x) = \sqrt{x} + 2$ (as in Example 2.3c), the range of f is the set of all possible numbers of the form $\sqrt{x} + 2$. Since $\sqrt{x} \geq 0$, it follows that $\sqrt{x} + 2 \geq 2$, and therefore the range is the set of all real numbers $y \geq 2$.

Sometimes you are given a function and then asked to determine whether or not a specific number is in its domain or in its range.

EXAMPLE 2.4 Consider the function $f(x) = \sqrt{3x+1} - 5$.

a. Is -1 in the domain of f?

b. Is $-\frac{1}{3}$ in the domain of f?

c. Is -2 in the range of f?

d. Is -8 in the range of f?

Solution

a. We ask: Is $f(-1)$ a real number? Evaluate: $f(-1) = \sqrt{3(-1)+1} - 5 = \sqrt{-3+1} - 5 = \sqrt{-2} - 5$. Since $\sqrt{-2}$ is not a real number, -1 is not in the domain of f.

b. We evaluate: $f\left(-\frac{1}{3}\right) = \sqrt{3\left(-\frac{1}{3}\right)+1} - 5 = \sqrt{-1+1} - 5 = 0 - 5 = -5$. This is a real number, so $-\frac{1}{3}$ is in the domain of f.

c. We ask: Is there a number x for which $f(x) = -2$? This means that we need to solve the equation $\sqrt{3x+1} - 5 = -2$. We add 5 to both sides of the equation; then square both sides of the new equation, and finally solve for x:

$$\sqrt{3x+1} = -2 + 5 = 3$$

$$3x + 1 = 9$$

$$3x = 8 \rightarrow x = \frac{8}{3}.$$

Since $f\left(\frac{8}{3}\right) = -2$, the number -2 is in the range of f.

d. We ask: Does the equation $\sqrt{3x+1} - 5 = -8$ have a solution? When we add 5 to both sides of the equation, we get $\sqrt{3x+1} = -3$, which is never true, since by definition, the square root of every real number is nonnegative. So the number -8 is not in the range of f. ■

We can think of a function as a machine that admits only elements of its domain as inputs and whose outputs are elements of the range. Further discussion on domain and range of a function, as well as additional techniques to find domain and range, can be found in Sections 2-D and 4-C.

Exercises 2-B

1. Use interval notation to describe each set graphed below as an interval or union of intervals. Be careful to use parentheses or square brackets as needed.

a. 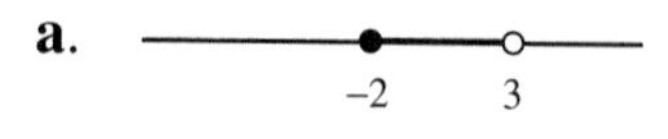

b. 

c. −1 2 4

d. 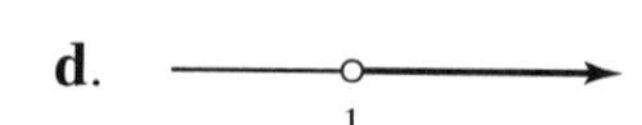

2. Draw and describe as an interval or union of intervals each of the following sets:

a. The set of real numbers x such that $-1 < x \leq 3$

b. The set of real numbers x such that $x < 4$

c. The set of real numbers x such that $x \geq 5.1$

d. The set of real numbers x such that $x \neq 2$ and $x \neq 4$

3. Give the domain of each function:

a. $f(x) = \frac{1}{3x}$ **b**. $g(t) = \frac{-7}{5t-2}$

c. $F(x) = \sqrt{x} - 5$ **d**. $G(x) = 1 + \sqrt{x-2}$

4. **a**. Is the number 0 in the domain of $f(x) = \frac{3}{x-1}$? Explain.

b. Is the number 0 in the range of $f(x) = \frac{3}{x-1}$? Explain.

c. Is the number –2 in the domain of $g(x) = x^2 + 2$? Explain.

d. Is the number –2 in the range of $g(x) = x^2 + 2$? Explain.

5. Give the domain and the range of $F(t) = 3 - \sqrt{t}$.

6. Write a formula for the area A of a circle as a function of its radius x. Give the domain and range of the function.

2-C DIFFERENT WAYS TO REPRESENT FUNCTIONS

Functions that are studied in calculus can be represented in different ways—each way helps us to see how the function behaves. In the discussion that follows, we represent a single function in different ways.

1. *A function is a rule that assigns to each real number in the domain a real number in the range. The rule can be expressed symbolically or in words.*

For example, $f(x) = 2x + 3$ (symbolic rule).
f assigns to each number twice that number plus 3 (descriptive rule).

The equation that gives f as a symbolic rule can be manipulated by algebraic operations, and in this form, calculus techniques can be applied to f.

2. *A function is a set of ordered pairs of real numbers* or a two-column list of real numbers. In each ordered pair, the first coordinate x is in the domain of f and the second coordinate y is the corresponding value $f(x)$ in the range. Note that for most functions, we can write down only a small sample of the complete list of ordered pairs. A short table of ordered pairs for the function $f(x) = 2x + 3$ is given in Figure 2.1a.

Tables like these can be scanned to see patterns of behavior in the function.

3. *A function has a* **graph** *whose points have the ordered pairs* $(x, f(x))$ *as coordinates.*

A portion of the graph of the function $f(x) = 2x + 3$ is shown in Figure 2.1b.

x	$y = 2x + 3$
−1.02	0.96
−0.67	1.66
0	3
0.234	3.468
1	5

Figure 2.1a

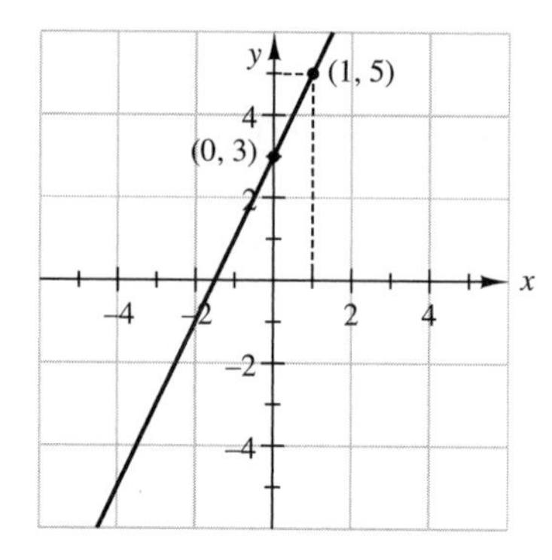

Figure 2.1b Graph of $f(x) = 2x + 3$

From the graph it is often easy to see the general behavior of a function. We can observe whether it has high or low points, when it is increasing or decreasing, and so on.

Sometimes the rule of a function can be inferred from a table of corresponding values, as the following example shows.

EXAMPLE 2.5 Delilah has a New and Almost New shop. Below are wholesale prices she paid for the new clothing and the retail prices she charged for it.

Wholesale	Retail
$10	$17
$12	$19
$9	$16
$26	$33
$14.50	$21.50

a. Delilah used a single rule to determine the retail prices. By comparing the two columns, describe in words the rule that Delilah used to determine the retail prices.

b. Express this rule as a function using symbolic notation.

c. Give this function's domain and range.

d. Draw axes of a Cartesian coordinate system and plot the points (w, r) from the table above, where the first coordinate w is a wholesale price and r is the corresponding retail price, $r = f(w)$.

Solution

a. By comparing the two columns, we see that when the wholesale price is \$10, the corresponding retail price is \$17; when the wholesale price is \$12, the retail price is \$19; and so on. Looking for a pattern, we realize that 17 is 10 plus 7, 19 is 12 plus 7, and so on. We then conclude that Delilah added \$7 to the wholesale price to obtain the retail price.

b. If we use the letter w to represent the wholesale price, r the retail price, and f the function, then $r = f(w) = w + 7$.

c. The domain is the set of wholesale prices of new clothing, and the range is the set of retail prices. If we assume the given list is complete, then

$$\text{domain} = \{10, 12, 9, 26, 14.5\}$$

$$\text{range} = \{17, 19, 16, 33, 21.5\}.$$

d. The points to be plotted are (10, 17), (12, 19), (9, 16), (26, 33), and (14.5, 21.5). The points are plotted in Figure 2.2.

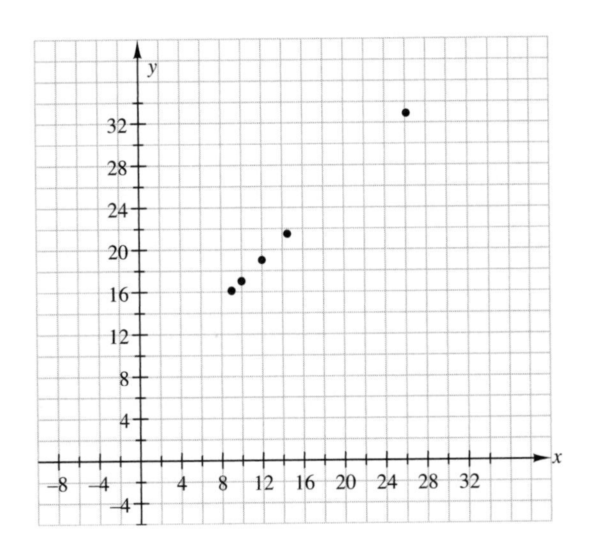

Figure 2.2 ■

A function can be defined by expressing in words the relationship between the dependent and independent variables. Often in applications, information is given in words and a symbolic rule for the function must be derived.

EXAMPLE 2.6 Delilah used a different rule to determine the retail price for furniture in her store. She made a 20% markup; that is, she added 20% of the wholesale price to the wholesale price to obtain the retail price.

a. Express this rule symbolically as a function.

b. Describe the domain and the range of the function.

c. Determine the retail price for each of the wholesale prices: \$75, \$321, and \$890.

Solution

a. As before, let r represent retail price and w wholesale price. Then $r = f(w) = \frac{20}{100}w + w = 0.2w + w = (0.2 + 1)w = 1.2w.$

b. The domain of f is the set of wholesale prices for furniture in the store, and the range is the set of retail prices.

c. $f(75) = (1.2)(75) = \$90$; $f(321) = (1.2)(321) = \$385.20$; $f(890) = (1.2)(890) =$ \$1,068. ■

Exercises 2-C

1. In the Almost New portion of Delilah's shop, used clothing is sold on consignment. When an item sells, Delilah keeps a portion of the selling price to cover her expenses. Below are the selling prices for three items recently sold and how much Delilah kept of the selling price.

Selling Price	*Shop Keeps*
\$10	\$2
\$85	\$17
\$110	\$22

a. On a Cartesian coordinate system, use the values in the table above to plot the points (s, k), where s is the selling price and k is the amount Delilah keeps.

b. Do the plotted points lie on a line? How can you be sure?

c. Describe in words the rule that Delilah uses to determine how much she keeps.

d. Express this rule as a function using symbolic notation.

2. The equation $L = \frac{8}{15}m + 12$ gives the relationship between minutes spent exercising (m) and ounces of liquid beverages (L) that a person should drink if exercising in hot weather.

a. Describe in words the rule that assigns to each number of minutes spent exercising, a recommended amount of liquid the exerciser should consume.

b. Set up a two-column table that shows the values of the function $f(m) = L = \frac{8}{15}m + 12$ for the following values of m: 0, 30, 45, 75 minutes.

c. On a Cartesian system of coordinates, plot the points with coordinates (m, L) for the values in the table in part b.

d. Interpret the slope and y-intercept of this function in the context of exercise and hydration.

2-D THE GRAPH OF A FUNCTION

A graph represents a function visually. The graph of the function f is the graph of the equation $y = f(x)$ on a Cartesian coordinate system. For example, the graph of the function $f(x) = 1 - x^2$ is the graph of the equation $y = 1 - x^2$. Figure 2.3 shows the graph of $y = 1 - x^2$ between $x = -2.5$ and $x = 2.5$. Every point on the graph has coordinates $(x, f(x)) = (x, 1 - x^2)$. The labeled point on the graph has coordinates $x = -1.5$ and $y = f(1.5) = 1 - (1.5)^2 = -1.25$.

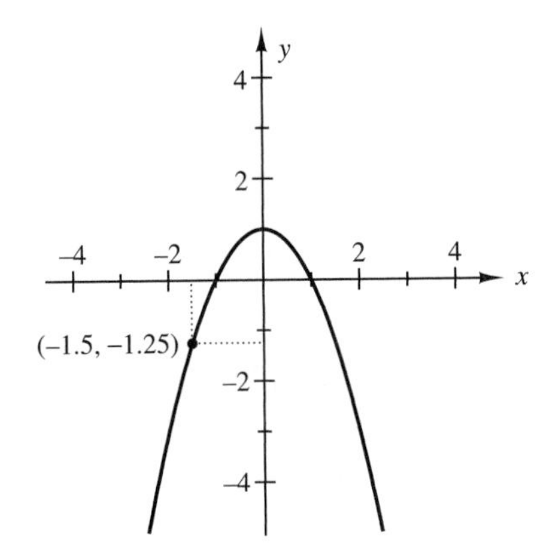

Figure 2.3 $y = f(x) = 1 - x^2$

The graph of a function carries complete information about the function, since it gives all the values of the function. For most functions we can see only a small portion of the whole graph at a time. From the graph in Figure 2.3, we can read the values $f(x)$ for every x in the interval $[-2.5, 2.5]$. (In practice we can read only approximations to those values.) To read the value $f(2)$, for example, we draw the vertical line $x = 2$ and then find its intersection with the graph of f. The y-coordinate of the point of intersection is $f(2)$, which for the function $f(x) = 1 - x^2$ has value -3 (see Figure 2.4).

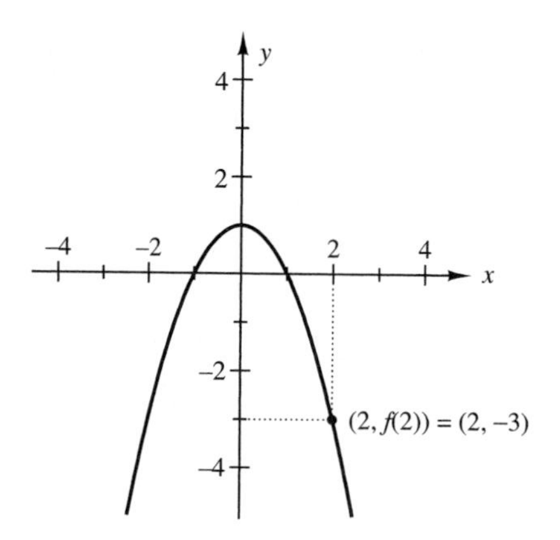

Figure 2.4 $y = f(x) = 1 - x^2$

From the complete graph of a function we can read the domain and the range of the function. The domain is the set of all x-coordinates of points on the graph, and the range is the set of all y-coordinates of points on the graph. To find the domain we project the graph of the function onto the x-axis. To find the range we project the graph onto the y-axis. For example, from the graph of the function $g(x) = \sqrt{4 - x^2}$ shown in Figure 2.5, we determine that the domain of g is the interval [–2, 2] and the range of g is [0, 2].

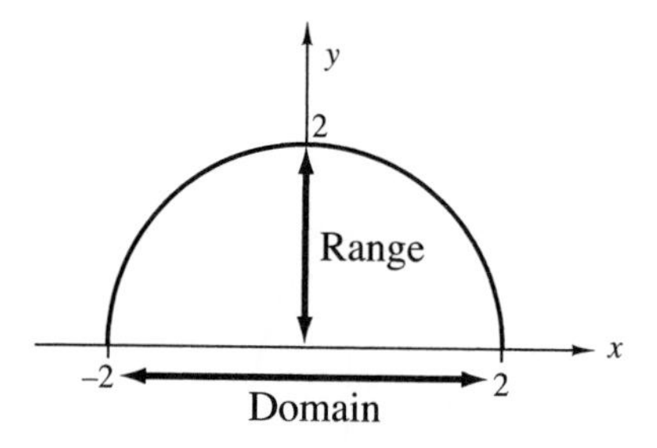

Figure 2.5 $g(x) = \sqrt{4 - x^2}$

EXAMPLE 2.7 The graph in Figure 2.6 represents the temperature y (in degrees Fahrenheit) as a function f of time t (in hours) over a 24-hour period, beginning at noon on a cold winter day.

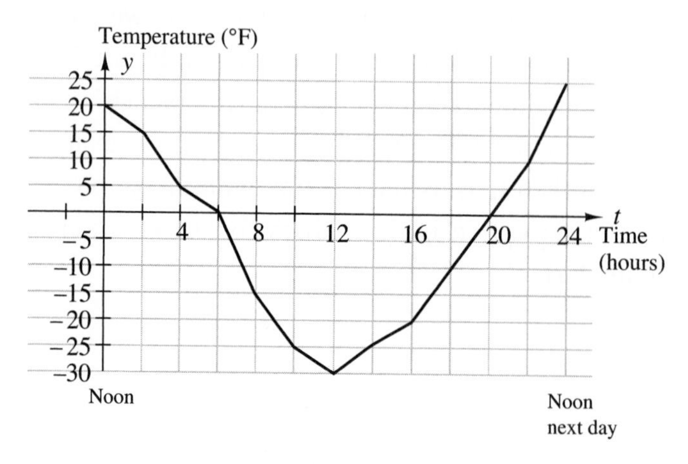

Figure 2.6

a. Give the following values and describe what they represent: $f(0), f(2), f(12)$.

b. Find the values of t for which $f(t) = 0$.

c. At what times was the temperature 10°F? 40°F?

d. Over what time interval was the temperature rising? Falling?

e. At what time was the lowest temperature achieved?

f. At what time was the highest temperature achieved?

g. What are the domain and range of the given function?

Solution

a. To find $f(0)$, we locate the point of the given graph that has 0 as its first coordinate and read its second coordinate: 20. So $f(0) = 20$, which means that the temperature at noon of the first day was 20°F.

In the same way, we read that $f(2) = 15$, which means that the temperature 2 hours later, at 2 P.M., was 15°F.

From the graph, we see that $f(12) = -30$, which means that the temperature at midnight (12 hours after noon) was minus 30°F.

b. Since the values of $f(t)$ are y-coordinates of points on the graph, we look for points that have a second coordinate equal to 0. These are easily found by looking at where the horizontal line through 0 on the y-axis cuts the graph. There are two such points: (6, 0) and (20, 0). The values of t that we are looking for are the first coordinates of these points, $t = 6$ and $t = 20$.

c. To find the times at which the temperature was 10°F, we locate the points of intersection of the graph with the horizontal line $y = 10$ and read the t-coordinates of those points. In doing this, we find two values of t: one is approximately 3 and the other is 22. This means that the temperature was 10°F at approximately 3 P.M. (3 hours after noon) and at 10 A.M. the next day. Since there is no point on the graph 40 units above the horizontal axis, the horizontal line $y = 40$ does not intersect the graph. This means at no time was the temperature 40°F; that is, 40 is not in the range of the function.

d. From the graph, we see that if we move from $t = 0$ to $t = 12$ along the horizontal axis, the corresponding y values (temperature values) of the points on the graph fall, whereas if we continue from $t = 12$ to $t = 24$, the y values rise. So the temperature is falling from noon to midnight and is rising from midnight to noon next day.

e. The lowest point on the graph is (12, –30), so the lowest temperature was at midnight (and it was minus 30°F).

f. The highest point of the graph is (24, 25). So the highest temperature was 25°F at noon the second day.

g. The domain is the set of all values of t for which $f(t)$ is defined or for which there is a point $(t, f(t))$ on the graph. Since the graph contains all the points with coordinates $(t, f(t))$ for t in the 24-hour period, the domain is obtained by collecting all the first coordinates of the points of the graph. This gives us the interval [0, 24] as the domain of f.

The range is the set of all the values $f(t)$. Therefore, to get the range we collect all the second coordinates of the points of the graph. In this case, the range is the interval [–30, 25]. ■

Earlier, when introducing graphs of intervals, we noted a convention that is followed to indicate when a point belongs to or is missing from a graph.

For any printed graph or one drawn by hand, it is a standard convention to indicate that a point is missing by drawing a small hollow circle that represents a "hole." A small filled-in circle represents an isolated point on the graph or emphasizes that a point belongs to the graph.

EXAMPLE 2.8 Answer the following questions about the function F whose graph is given in Figure 2.7.

a. Give each value, or say it does not exist:

(i) $F(-1.5)$ (ii) $F(-1)$ (iii) $F(0)$ (iv) $F(1)$ (v) $F(2)$

b. Give the domain of F. **c**. Give the range of F.

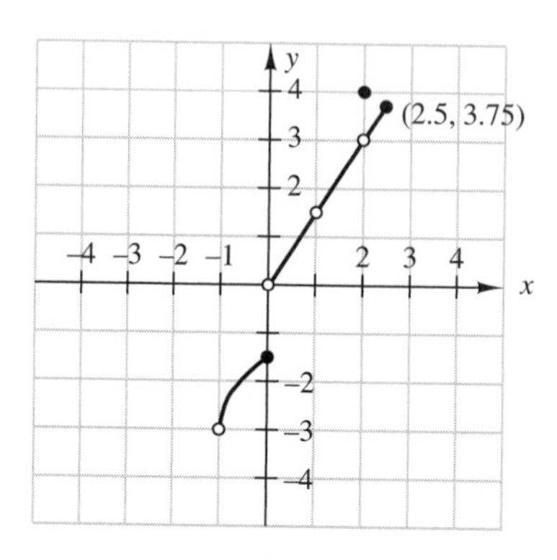

Figure 2.7

Solution

a. (i) The vertical line $x = -1.5$ does not intersect the graph of F; that is, there are no points on the graph with x-coordinate -1.5. Therefore, $F(-1.5)$ does not exist.

(ii) $F(-1)$ is also undefined, since the hollow circle at the point $(-1, -3)$ means the point is not on the graph.

(iii) The point $(0, -1.5)$ belongs to the graph, but the point $(0, 0)$ does not. Thus, $F(0) = -1.5$.

(iv) $F(1)$ does not exist, since there is no point on the line $x = 1$ that belongs to the graph. The point $(1, 1.5)$ represented by the hollow circle is not on the graph.

(v) $F(2) = 4$, since the point $(2, 4)$ is on the graph whereas the point $(2, 3)$ is not.

b. We obtain the domain by taking the points of intersection with the x-axis of all vertical lines drawn from points of the graph. In this manner, we obtain all values of x from -1 to 2.5 with -1 and 1 excluded and 2.5 included; that is, the domain of F is the set $(-1, 1) \cup (1, 2.5]$.

c. To find the range we draw horizontal lines through each point of the graph and take their intersections with the y-axis. From the points of the graph to the left of the y-axis, we obtain all the values of y from -3 to -1.5, with -1.5 included but -3 excluded. From the portion of the graph to the right of the y-axis, we obtain the values of y between 0 and 3.75, excluding $y = 0$, $y = 1.5$, and $y = 3$ and including $y = 3.75$ and the isolated value $y = 4$. Thus, the range of F is the union of four intervals and the single number 4. Symbolically, range of F = $(-3, -1.5] \cup (0, 1.5) \cup (1.5, 3) \cup (3, 3.75] \cup \{4\}$. ■

The points where the graph of a function f crosses the coordinate axes are usually important points. These are the intercepts of the graph of the equation $y = f(x)$, defined in Section 1-B. The y-intercept is the value $f(0)$; it is the y-coordinate of the point of intersection of the graph with the y-axis. An x-intercept is the x-coordinate of a point of intersection of the graph with the x-axis. To find the x-intercepts we need to solve the equation $0 = f(x)$. A function may have several x-intercepts but only one y-intercept. (Why?)

EXAMPLE 2.9 The graph of the function $f(x) = 3 - x^2$ is shown in Figure 2.8.

a. By observing the graph of f, locate the intercepts of f and give their approximate values.

b. Use the formula that defines the function f to find exact values for the intercepts.

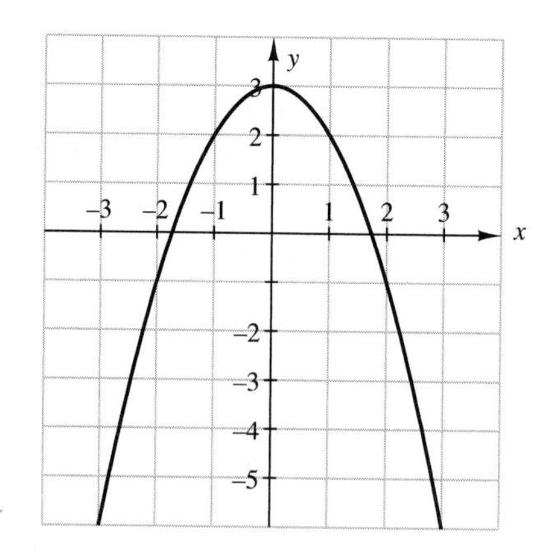

Figure 2.8 $f(x) = 3 - x^2$

Solution

a. To obtain the y-intercept from the graph, we locate the point of intersection of the graph with the y-axis and read its y-coordinate. On this graph we see that the y-intercept is approximately 3. To give the x-intercepts, we locate the points of intersection of the graph with the x-axis and read their x-coordinates. This graph has two x-intercepts and they are approximately −1.7 and 1.7.

b. The y-intercept is the value $f(0) = 3 - 0^2 = 3$. To find the x-intercepts we solve the equation $0 = f(x)$, which in this case is $0 = 3 - x^2$. We add x^2 to both sides to get $x^2 = 3$. The solutions are the two numbers whose square equals 3—namely, $\sqrt{3}$ and $-\sqrt{3}$ (note $\sqrt{3}$ is approximately 1.73). ■

Not every graph (on a system of Cartesian coordinates) represents the graph of a function. Consider, for example, the circle with equation $x^2 + y^2 = 1$ (Figure 2.9).

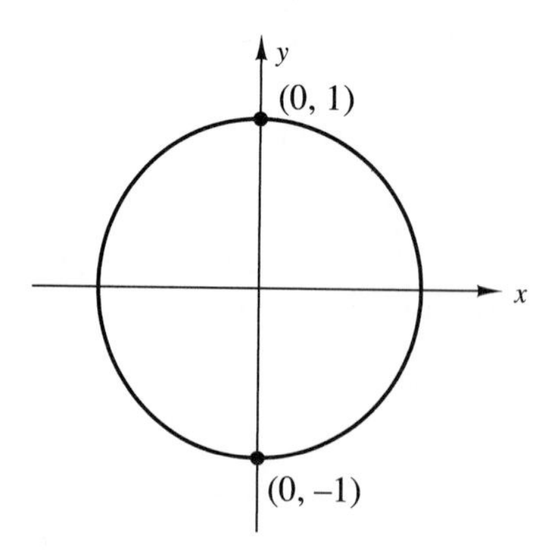

Figure 2.9 $x^2 + y^2 = 1$

To see that this is not the graph of a function, note that two points with the same x-coordinate, such as the points (0, 1) and (0, −1), are both on the graph. The circle cannot be the graph of a function, since this function would then have a rule that assigns two values (–1 and 1) to the number 0.

Since all the points with the same x-coordinate lie on a vertical line, there is a simple test to decide whether or not a graph represents a function.

Vertical line test: If any vertical line intersects a graph at more than one point, then the graph is **not** the graph of a function.

Exercises 2-D

1. An object is thrown upward. The graph in Figure 2.10 represents the object's height y (in feet) above the ground as a function f of time t (in seconds).

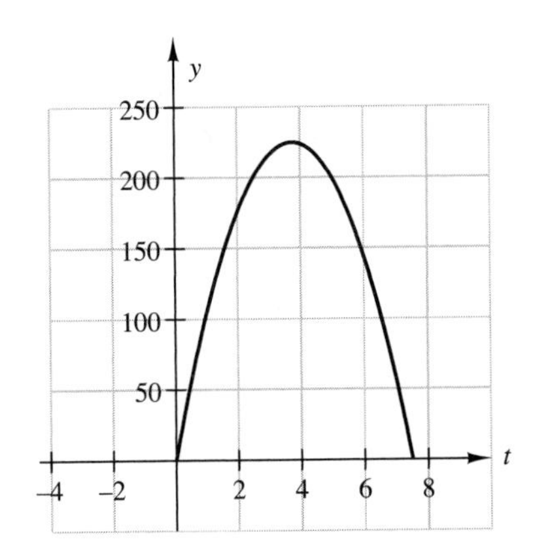

Figure 2.10 $y = f(t)$

a. Estimate the following values from the graph and describe what they represent: $f(0), f(2.2), f(4)$.

b. Give the (approximate) values of t at which $f(t) = 0$.

c. When does the object reach maximum height? How high is the object at that time?

d. Estimate the time(s) when the object is 150 ft above the ground.

e. Is there a time t for which $f(t) = 250$ ft? Explain.

f. Give the domain and the range of the function f.

2. Which of the graphs a–h below are graphs of functions? Give reasons for your answers.

a.

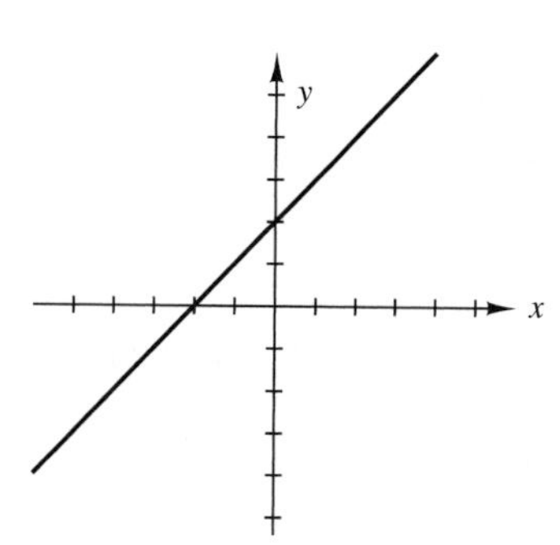

b.

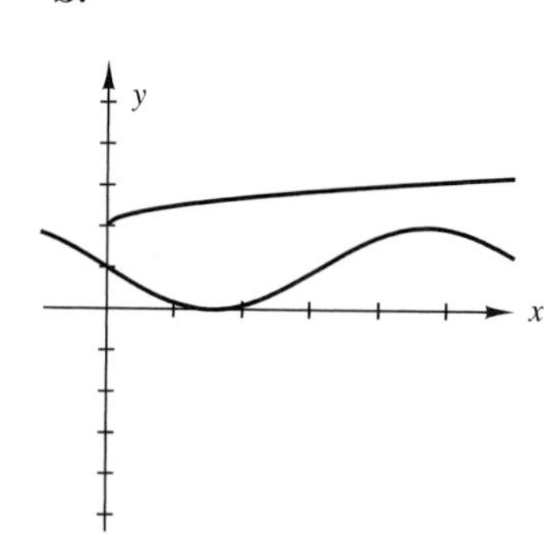

c.

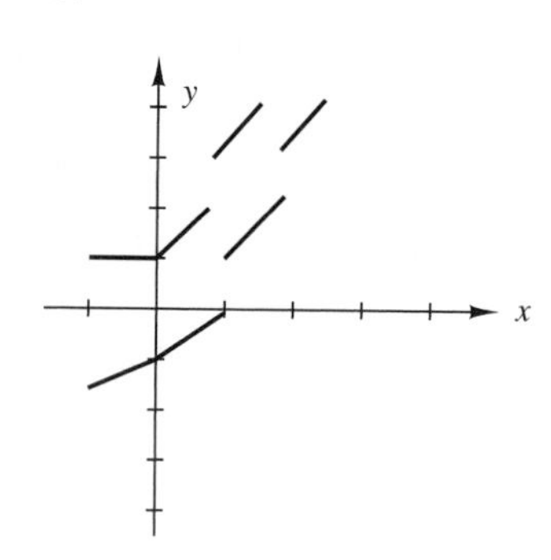

d.

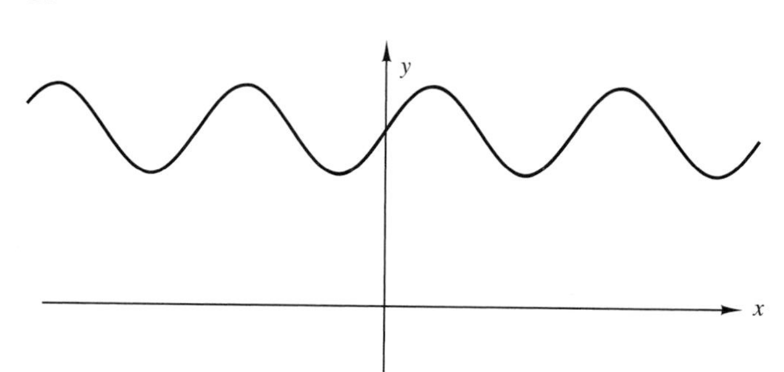

e.

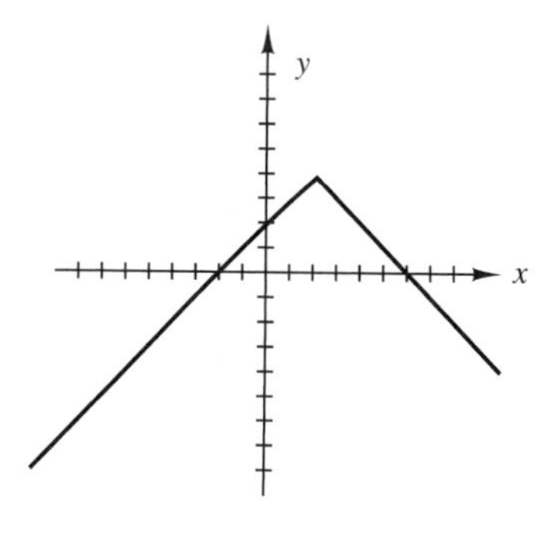

f.

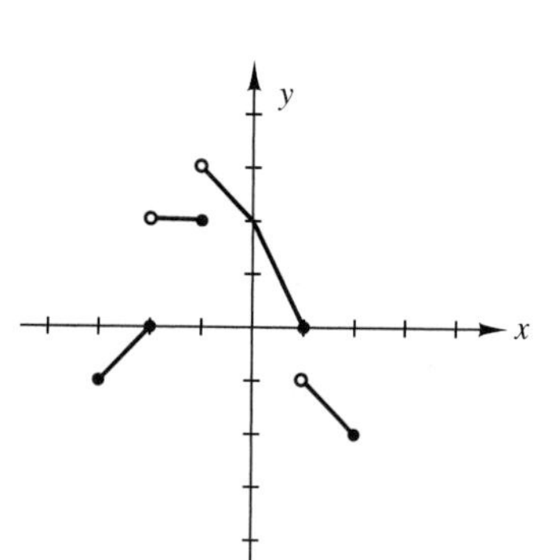

g.

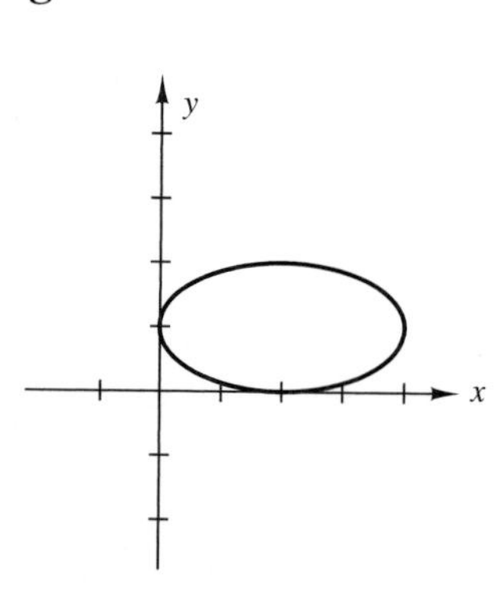

h.

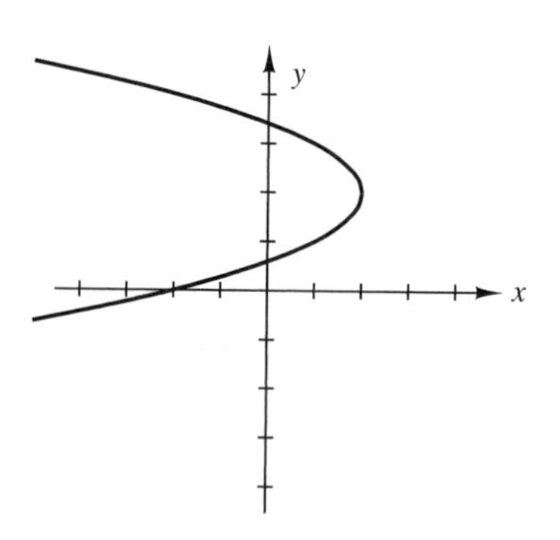

3. Questions a–e refer to the graph of $y = g(x)$ in Figure 2.11.

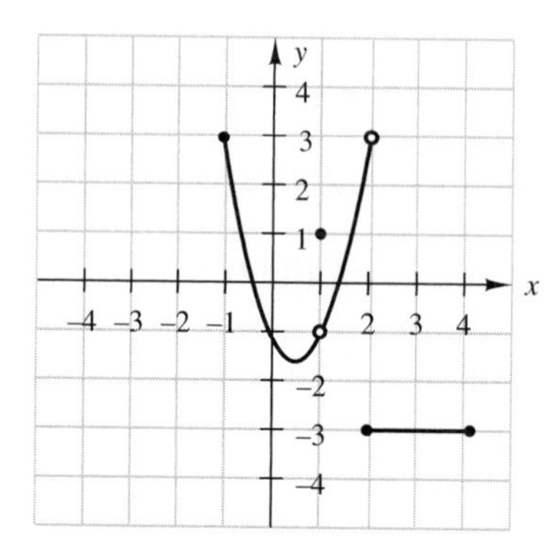

Figure 2.11 $y = g(x)$

a. From the graph of $y = g(x)$, estimate each of the following values or say it does not exist. If it does not exist, explain why.

(i) $g(-1)$ (ii) $g\left(-\frac{1}{2}\right)$ (iii) $g(1)$ (iv) $g(2)$

(v) $g(3.5)$ (vi) The value(s) of a for which $g(a) = 3$

b. Give the domain of the function g.

c. Give the range of the function g.

d. Estimate the value of the y-intercept of the graph of g.

e. Estimate the value(s) of the x-intercepts of the graph of g.

4. Sketch the graph of a function that has no y-intercepts.

5. Explain why a function can have at most one y-intercept.

2-E SPECIAL CLASSES OF FUNCTIONS

1 Linear Functions

A function whose graph is a line is the simplest type of function.

EXAMPLE 2.10 Normal weight, according to the Body Mass Index, is a function of one's height. For adults with height in the range 58 inches to 76 inches, the relationship between height x (in inches) and weight w (in pounds) that gives the cutoff between normal weight and overweight is $w = f(x) = 4.5x - 140$. (The information was taken from the graph "Judging Your Weight" in *The Morning Call*, July 17, 2004.)

a. Find the value $f(64)$ and explain what it means.

b. Graph the function f on its given domain. Give a possible reason for the domain to be restricted to the interval given.

Solution

a. We have $f(64) = 4.5(64) - 140 = 288 - 140 = 148$. This means that a person who is 64 in. tall and weighs more than 148 lb is considered to be overweight.

b. The graph is shown in Figure 2.12. The domain of this function is heights of adults, and most of these heights fall between 4 ft 10 in. (58 in.) and 6 ft 4 in. (76 in.). For heights outside of this prescribed interval, the weights given by the linear function may not be appropriate to define when a person becomes overweight.

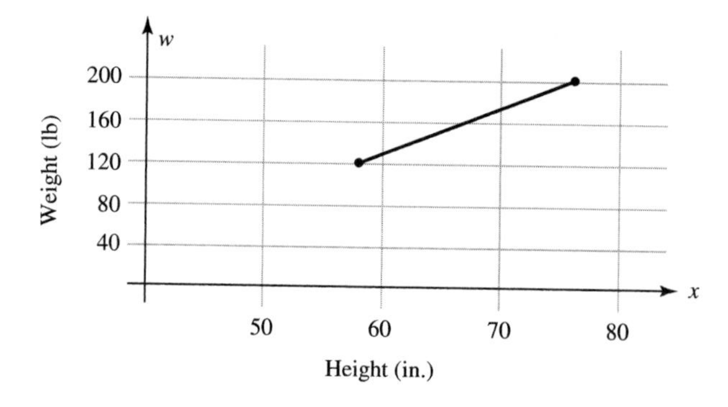

Figure 2.12 $w = 4.5x - 140, \quad 58 \le x \le 76$ ■

The function given in Example 2.10 has a special form and is called a *linear function*. Every linear function can be written $f(x) = mx + b$, where m and b are constants. The graph of this function is the graph of the equation $y = mx + b$, which is the line with slope m and y-intercept b (Section 1-C). Recall that the slope m of a line is the ratio $m = \frac{\Delta y}{\Delta x} = \frac{\text{change in } y}{\text{change in } x} = \frac{y_2 - y_1}{x_2 - x_1}$, where (x_1, y_1) and (x_2, y_2) are any two different points on the line. If a line is the graph of a function f, then $y_1 = f(x_1)$ and $y_2 = f(x_2)$, so that

$$m = \frac{f(x_2) - f(x_1)}{x_2 - x_1}.$$

This shows that the slope of the line represents the change in function values per unit change in x. In Example 2.10, the value of m is 4.5. This means that for each one-inch increment in height, the cutoff between normal weight and overweight increases by 4.5 lb. In the linear function $f(x) = mx + b$, the constant b is the y-intercept of the line, since $f(0) = m \cdot 0 + b = b$.

EXAMPLE 2.11 A manufacturer's total cost consists of a fixed cost of $2,000 (called overhead) plus production costs of $90 per unit. Express the total cost as a function of the number of units produced and draw the graph. What is an appropriate domain for this function?

Solution Since the cost of producing each unit is $90, the cost to produce x units is $90x$. The total cost of producing x units is the fixed overhead plus $90x$: $C(x) = 2000 + 90x$. The graph is shown in Figure 2.13; it is a line.

The domain of $C(x)$ contains only nonnegative numbers (these are numbers of units produced). The graph in Figure 2.13 shows x as a continuous variable, which makes sense if the product is continuous (such as rope measured in yards). But if the product is refrigerators, only integer units make sense. In this case, the graph would consist of individual points that all lie on the line in Figure 2.13.

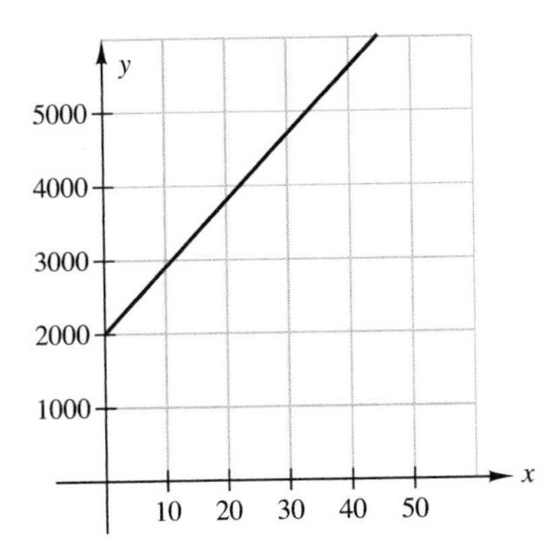

Figure 2.13 $C(x) = 2000 + 90x$ ■

2 Piecewise Linear Functions

Many important functions are not linear; however, their graphs consist of pieces of lines. Such functions are called *piecewise linear.*

EXAMPLE 2.12 To make a telephone call to London, the phone company charges \$1.50 for the first 3 min and \$.15 for each additional minute (or fraction of a minute). Draw a graph and also describe in symbols the function that represents the charges (in dollars) as a function of the length of the call (in minutes). How much does it cost to call for 10 min? $8\frac{1}{2}$ min?

Solution This function cannot be given by a single algebraic expression. Instead, its values are different constants for different intervals of t (time). If the call lasts 3 min or less, the cost is \$1.50; thus, if $0 < t \leq 3$, $f(t) = 1.50$. If the call lasts longer than 3 min but no more than 4 min, the cost is \$1.50 (for the first 3 min) plus \$.15 (for the additional minute or fraction of a minute). We then have $f(t) = 1.50 + 0.15 = 1.65$, if $3 < t \leq 4$. Similarly, if $4 < t \leq 5$, $f(t) = 1.65 + 0.15 = 1.80$. Proceeding in this manner, we obtain the following expression of $f(t)$ for $0 < t \leq 10$:

$$f(t) = \begin{cases} 1.50 & \text{if} \quad 0 < t \leq 3 \\ 1.65 & \text{if} \quad 3 < t \leq 4 \\ 1.80 & \text{if} \quad 4 < t \leq 5 \\ 1.95 & \text{if} \quad 5 < t \leq 6 \\ 2.10 & \text{if} \quad 6 < t \leq 7 \\ 2.25 & \text{if} \quad 7 < t \leq 8 \\ 2.40 & \text{if} \quad 8 < t \leq 9 \\ 2.55 & \text{if} \quad 9 < t \leq 10 \end{cases}$$

Figure 2.14 shows the graph of $y = f(t)$ for t on the interval (0, 10]. This is an example of a *step function.*

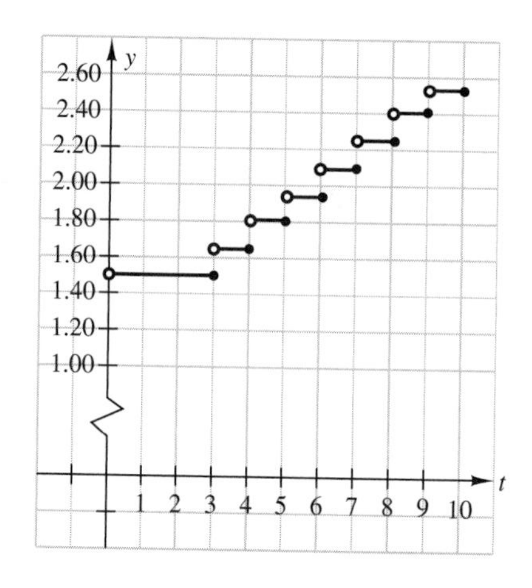

Figure 2.14 $y = f(t)$

To find the cost of a 10-min call, we need to find $f(10)$. For this purpose, we locate the interval in the definition of the function that contains $t = 10$. We see that since $t = 10$ satisfies $9 < t \le 10$, $f(10) = \$2.55$.

The cost of a call $8\frac{1}{2}$ min long is $f(8.5)$. Since 8.5 is in the interval $8 < t \le 9$, we find that $f(8.5) = \$2.40$. ■

EXAMPLE 2.13

a. Give the domain of the function $g(x) = \frac{x^2 - 1}{x - 1}$.

b. Show that for every x in the domain of g, $g(x) = x + 1$.

c. Sketch the graph of g. Is g a linear function? Is it piecewise linear?

Solution

a. The denominator $x - 1$ of the quotient defining $g(x)$ equals 0 only when $x = 1$, so the domain of g consists of all real numbers x except $x = 1$, which in interval notation is $(-\infty, 1) \cup (1, \infty)$.

b. Since $x^2 - 1 = (x - 1)(x + 1)$, we can simplify the expression for $g(x)$:

$$\frac{x^2 - 1}{x - 1} = \frac{(x - 1)(x + 1)}{x - 1} = x + 1, \quad \text{for } x \neq 1.$$

So if $x \neq 1$, then $g(x) = x + 1$ and $g(x)$ is undefined if $x = 1$.

c. The graph of g, shown in Figure 2.15, is the graph of the line $y = x + 1$ with the point (1, 2) missing. Because of this "hole" in its graph, g is not a linear function but is piecewise linear.

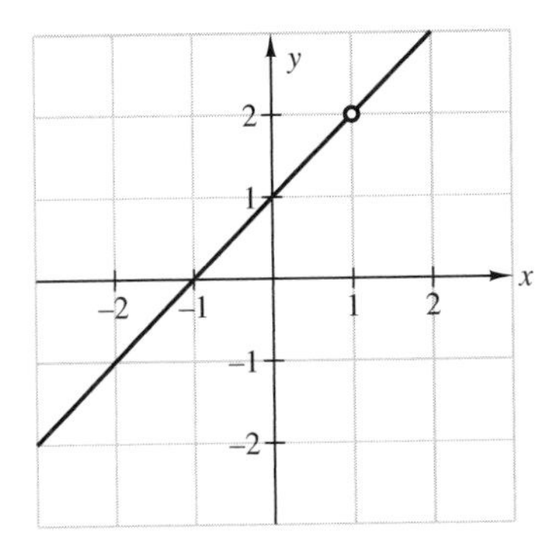

Figure 2.15 $y = \frac{x^2 - 1}{x - 1}$ ■

After simplifying a function formula, be sure to note any numbers that must be excluded from the domain of the simplified rule. (See the solution to Example 2.13b.)

The absolute value function is another example of a piecewise linear function. Recall that $|x|$ is the distance from the number x to 0 on the real coordinate line. Since the numbers x and $-x$ are equidistant from 0 on the coordinate line, $|x| = |-x|$ for all real numbers. When $x \geq 0$, the absolute value of x is just x. (For example, $|1| = 1$, $|2.3| = 2.3$, and $|0| = 0$.) When $x < 0$, its absolute value is $-x$. (For example, $|-2| = -(-2) = 2$, and $|-5.3| = -(-5.3) = 5.3$.)

The absolute value function is then defined as

$$f(x) = |x| = \begin{cases} -x & \text{if } x < 0 \\ x & \text{if } x \geq 0. \end{cases}$$

This shows that when x is negative, the graph of the absolute value function is the part of the line $y = -x$ to the left of the y-axis; when x is positive or zero, the graph is the part of the line $y = x$ on and to the right of the y-axis. Therefore, the graph of $f(x) = |x|$ consists of portions of two different lines. See Figures 2.16, 2.17, and 2.18.

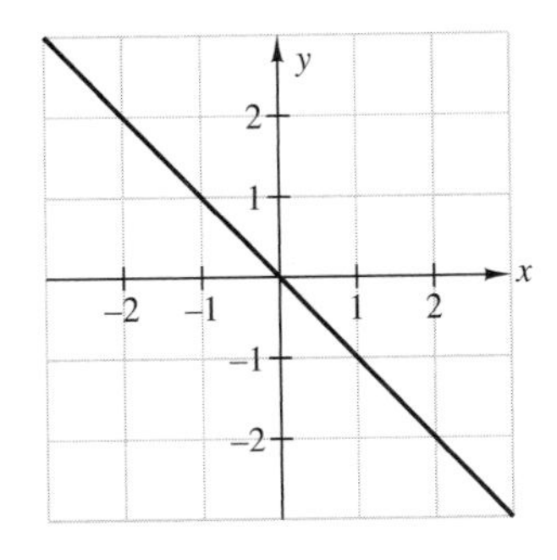

Figure 2.16 $y = -x$

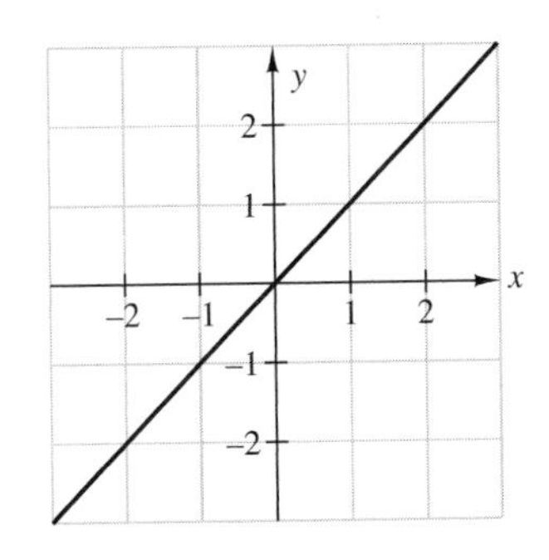

Figure 2.17 $y = x$

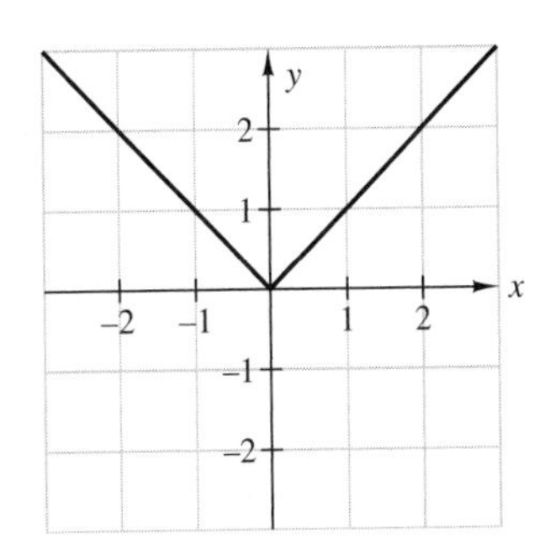

Figure 2.18 $y = |x|$

The graph of the absolute value function makes clear that $|x| = |-x|$ for all real numbers x. The following example shows two functions closely related to the absolute value function. Their graphs have the same "V" shape as the graph of $y = |x|$.

EXAMPLE 2.14 Graph each of the following functions:

a. $g(x) = |x + 3|$ **b.** $h(x) = |x| + 3$

Solution

a. Since $x + 3 < 0$ for $x < -3$ and $x + 3 \geq 0$ for $x \geq -3$, we rewrite $g(x)$ as

$$g(x) = |x+3| = \begin{cases} -(x+3) & \text{if } x < -3 \\ x+3 & \text{if } x \geq -3. \end{cases}$$

The graph of g consists of portions of two different lines:

The line $y = -(x + 3) = -x - 3$ for $x < -3$.

The line $y = x + 3$ for $x \geq -3$. See Figure 2.19.

b. We use the definition of $|x|$ to write

$$h(x) = |x| + 3 = \begin{cases} -x+3 & \text{if } x < 0 \\ x+3 & \text{if } x \geq 0. \end{cases}$$

Thus the graph of h consists of portions of the lines $y = -x + 3$ and $y = x + 3$, as shown in Figure 2.20.

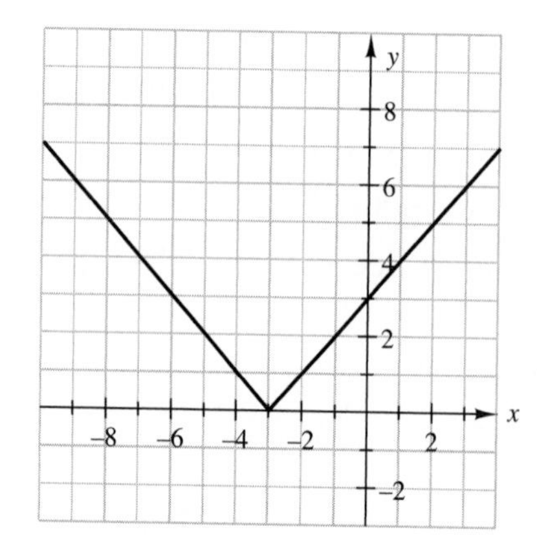

Figure 2.19 $y = |x + 3|$

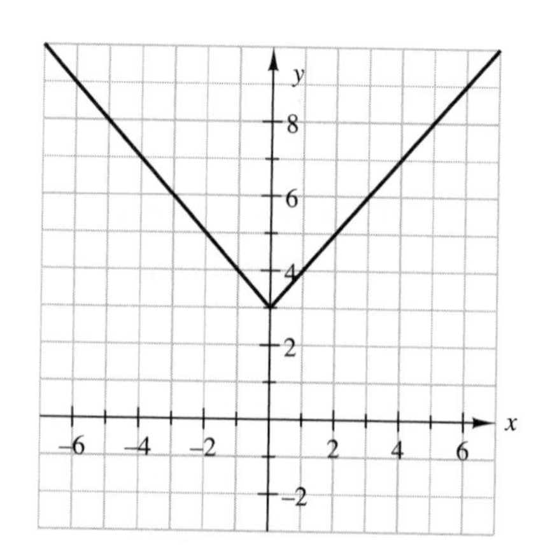

Figure 2.20 $y = |x| + 3$ ■

3 Quadratic Functions

A function given by a formula of the form $f(x) = ax^2 + bx + c$, where a, b, and c are real numbers and $a \neq 0$, is called a *quadratic* function. Its graph is always a parabola. Some specific examples are graphed below.

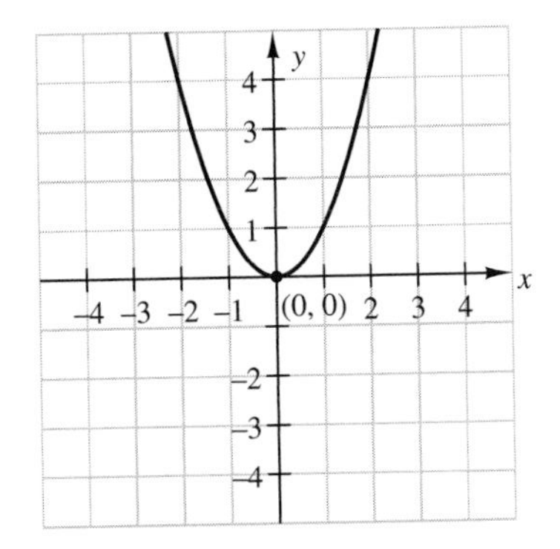

Figure 2.21 $f(x) = x^2$

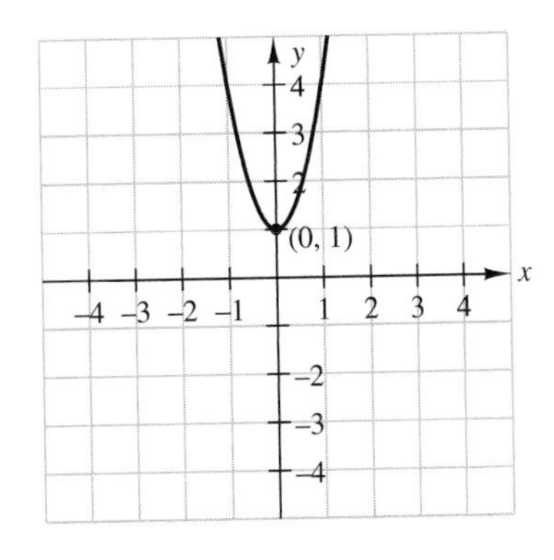

Figure 2.22 $g(x) = 3x^2 + 1$

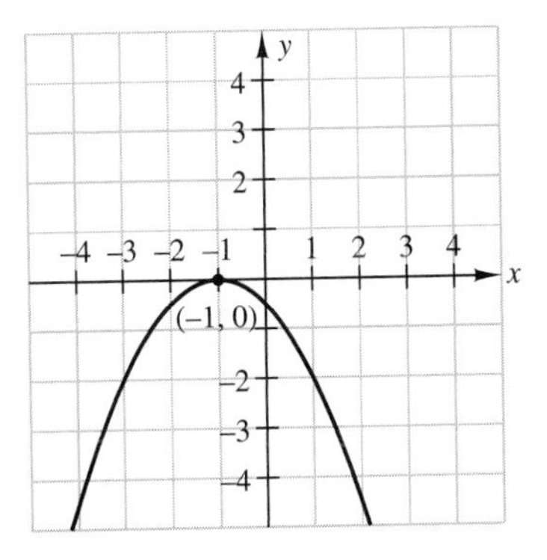

Figure 2.23 $h(x) = -\frac{1}{2}(x+1)^2$

Figure 2.21 shows the graph of the function $f(x) = x^2$, which is of the form $f(x) = ax^2 + bx + c$, where $a = 1$, $b = 0$, and $c = 0$. Figure 2.22 shows the graph of $g(x) = 3x^2 + 1$, which is the case where $a = 3$, $b = 0$, and $c = 1$. Figure 2.23 shows the graph of $h(x) = -\frac{1}{2}(x+1)^2$. To find the values of the constants a, b, c in this case we write

$$\begin{aligned} h(x) &= -\tfrac{1}{2}(x+1)(x+1) \\ &= -\tfrac{1}{2}(x^2 + 2x + 1) \\ &= -\tfrac{1}{2}x^2 - x - \tfrac{1}{2}. \end{aligned}$$

Thus, in this case $a = -\frac{1}{2}$, $b = -1$, and $c = -\frac{1}{2}$.

The points whose coordinates are shown on each graph in Figures 2.21, 2.22, and 2.23 are the *vertices* of those parabolas. They are important because the y-coordinate of the vertex of a parabola gives the smallest value of the function in the first two cases and the largest value in the last case. Vertices are also turning points, since the graph changes from falling to rising or from rising to falling.

The vertex of a parabola is easily found when the defining equation is in the form $y = f(x) = a(x - h)^2 + k$ (where a, k, and h are constants). For this equation, the vertex is (h, k) and represents the lowest point on the graph if $a > 0$ and the highest point if $a < 0$. See Figures 2.24 and 2.25.

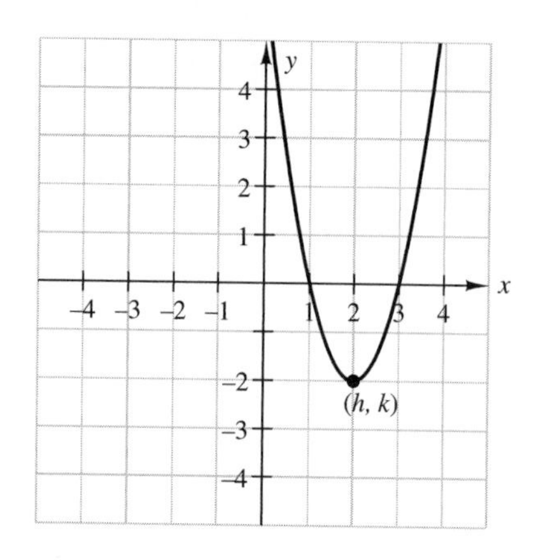

Figure 2.24 $y = a(x - h)^2 + k$, for $a > 0$

Figure 2.25 $y = a(x - h)^2 + k$, for $a < 0$

EXAMPLE 2.15 A ball is dropped from the top of a building. Its height from the ground (in feet) after t seconds is given by $H(t) = -16t^2 + 315$. Graph this function and answer the following questions.

a. How tall is the building?

b. How high is the ball after traveling 3 sec?

c. When does the ball reach the ground?

Solution H is a quadratic function, so its graph (in Figure 2.26) is a parabola (actually a piece of a parabola, since $t \geq 0$ and also the height $H \geq 0$).

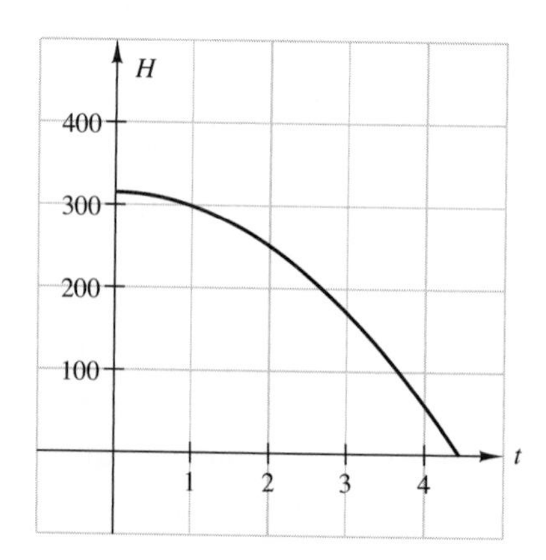

Figure 2.26 Height over time

a. When $t = 0$, the ball is at the top of the building, so $H(0)$ is the height of the building. Since $H(0) = -16(0) + 315 = 315$, the building is 315 ft tall.

b. The height after 3 sec is $H(3) = -16(3^2) + 315 = 171$ ft.

c. The ball reaches the ground at the time t at which the height is 0, so we need to find t so that $0 = H(t) = -16t^2 + 315$. We solve for t in the equation $0 = -16t^2 + 315$. Adding $16t^2$ to both sides, we get $16t^2 = 315$. Dividing both sides by 16, $t^2 = \frac{315}{16}$. So, $t = \pm\sqrt{315/16}$. Since t must be positive, $t = \sqrt{315/16} \approx 4.44$ sec. ■

4 Other Functions

We now discuss graphs of other basic functions.

a. $y = \sqrt{x}$ (square root function)

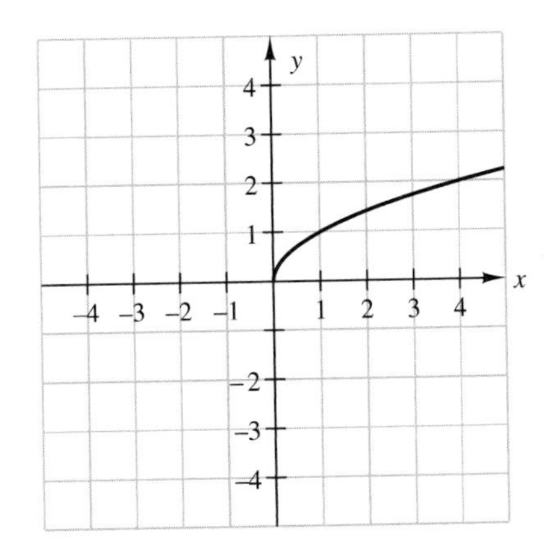

Figure 2.27 $y = \sqrt{x}$ (half-parabola)

This graph is the upper half of the parabola $y^2 = x$. To see this, we solve for y in the equation $y^2 = x$ to get $y = \pm\sqrt{x}$. Then $y = \sqrt{x}$ is the portion of the parabola whose points have nonnegative y-coordinates. Figure 2.28 shows the graph of the parabola $y^2 = x$, and Figure 2.29 shows the graph of the function $y = -\sqrt{x}$, which is the half of that parabola whose points have y-coordinates with $y \leq 0$.

From the graph in Figure 2.27, we see that the domain of the square root function is $[0, \infty)$ and its range is $[0, \infty)$.

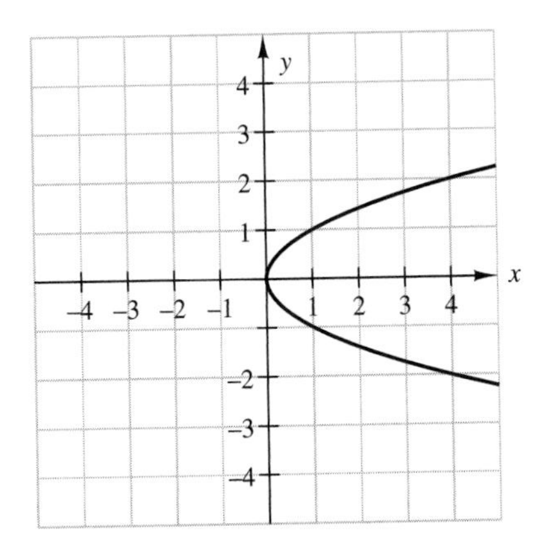

Figure 2.28 $y^2 = x$

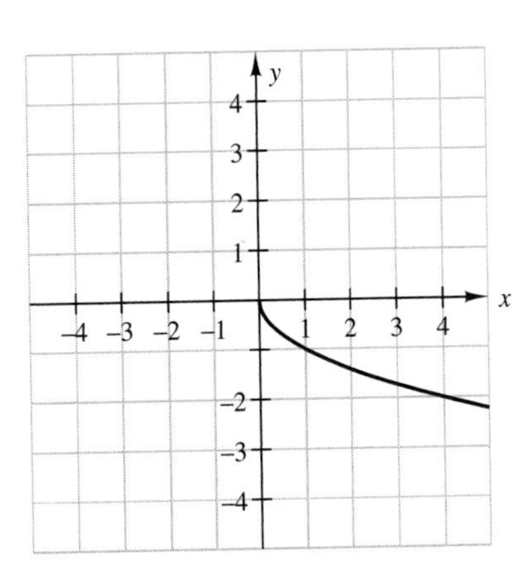

Figure 2.29 $y = -\sqrt{x}$

b. $y = \sqrt{1 - x^2}$

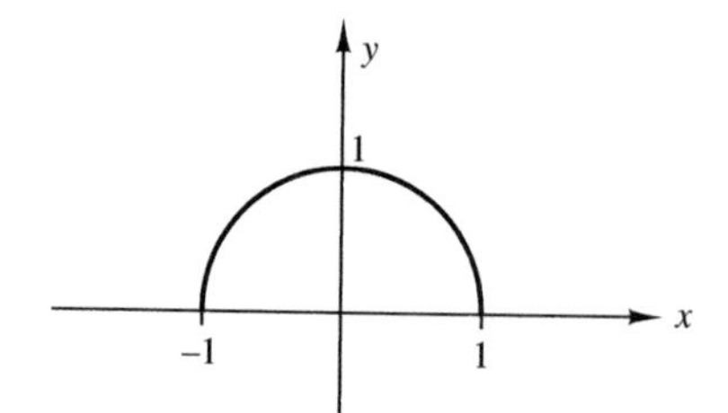

Figure 2.30 $y = \sqrt{1 - x^2}$ (half-circle)

The graph with equation $y = \sqrt{1 - x^2}$ is the upper half of the circle with equation $x^2 + y^2 = 1$. To see this, we solve the equation for y as follows:

$$x^2 + y^2 = 1 \rightarrow \quad y^2 = 1 - x^2 \rightarrow \quad y = \pm\sqrt{1 - x^2}$$

From the graph in Figure 2.30, we see that the domain of the function $y = \sqrt{1 - x^2}$ is $[-1, 1]$ and its range is $[0, 1]$.

c. $y = \frac{1}{x}$ (reciprocal function)

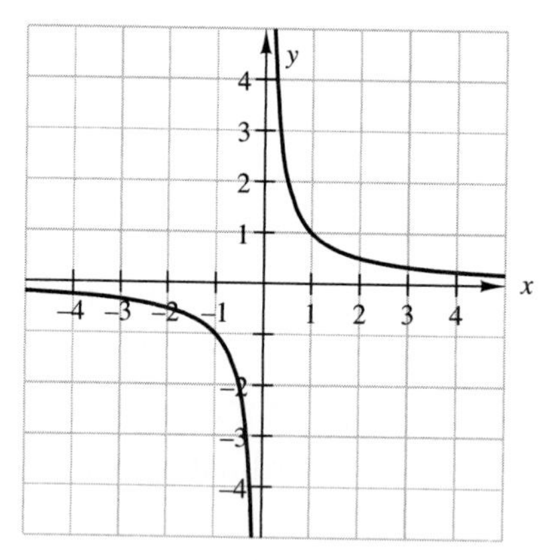

Figure 2.31 $y = \frac{1}{x}$

Note that the domain of this function contains every real number $x \neq 0$. If x is a small positive number (x is close to 0 and to its right), then $\frac{1}{x}$ is a very large positive number. When x is a negative number with small absolute value (x is close to 0 and to its left), then $\frac{1}{x}$ is negative and its absolute value is very large. From the graph, we can see that the range of the reciprocal function consists of every real number except 0. This function is studied in more detail in Chapter 5.

EXAMPLE 2.16 Boyle's law states that for an ideal gas (a fixed quantity kept at a constant temperature), the product of pressure p and volume V is a constant k (determined by the type of gas). This implies that as pressure on the gas is increased, its volume decreases (and vice versa) according to the formula $Vp = k$. Use this information to solve the following problem.

A 3-liter container is filled with gas at a pressure of 1 atm (1 atm is the normal pressure of air at sea level). A piston can compress the gas in the container. Assume that the amount of gas in the container and its temperature are held constant. Express the volume V of gas in the container (in liters) as a function of pressure p (in atm), and sketch the graph of the function V.

Solution From the initial conditions $V = 3$ and $p = 1$, we can calculate k for this gas: $Vp = 3 \cdot 1 = k$. Solving the equation $Vp = 3$ for V gives the equation for the volume function: $V = \frac{3}{p}$. This is a reciprocal function; its graph is shown in Figure 2.32.

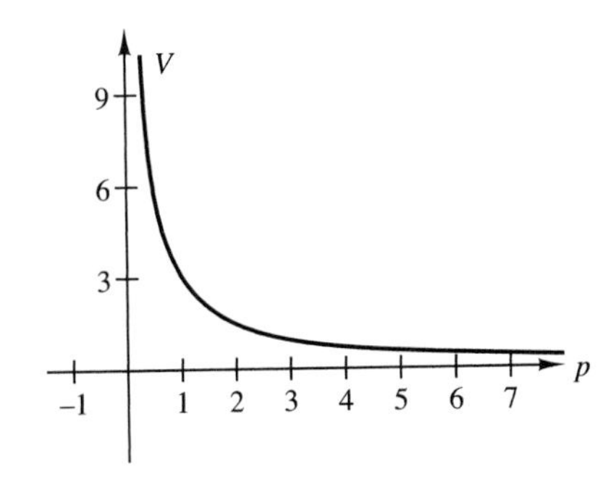

Figure 2.32 $V = \frac{3}{p}$ ■

In addition to piecewise linear functions, there are functions made up of pieces of different kinds of functions. The next example is such a *piecewise-defined function.*

EXAMPLE 2.17 The energy required for walking and for running can be given in terms of the speed of the motion. In both cases, the faster the motion, the more energy required.

If E is the energy (in joules, J) when the speed of motion is x meters per second, $E = f(x)$ is given by

$$E = f(x) = \begin{cases} 155x^2 - 65x + 160 & \text{if } 0 < x < 2.5 \\ 250x + 100 & \text{if } x \geq 2.5. \end{cases}$$

Note that when the speed of walking reaches 2.5 m/sec, the motion changes to running and the pattern changes. Graph the energy function and find the energy required to walk at 1 m/sec and to run at a speed of 3 m/sec.

Solution The graph of this function, shown in Figure 2.33, is part of a parabola for the interval (0, 2.5) and part of a line for the interval [2.5, ∞).

> $f(1) = 250$ (substitute 1 for x in the first formula), so it takes an energy of 250 J to walk at a speed of 1 m/sec.
>
> $f(3) = 850$ (substitute 3 for x in the second formula), so it takes an energy of 850 J to run at a speed of 3 m/sec.

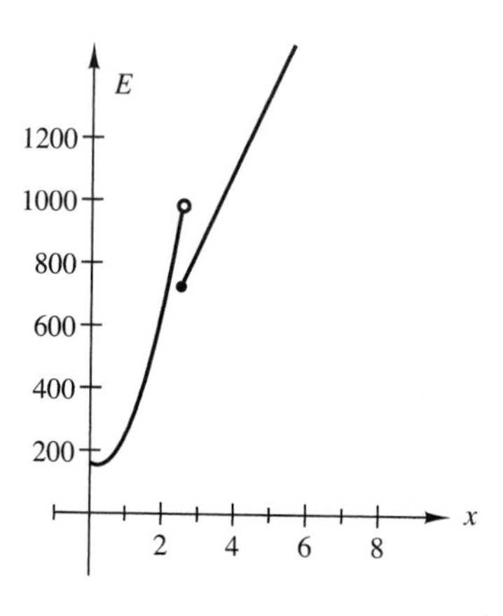

Figure 2.33 ■

Exercises 2-E

1. Which of the following functions are linear? Describe and sketch each graph.

a. $f(x) = -2x + 5$ **b.** $g(x) = -\sqrt{x}$

c. $h(x) = x^2 - 1$ **d.** $H(x) = 4 - x$

2. The following table indicates the dose of a medication a child should receive according to the child's weight. To avoid an overdose, weights are to be rounded down to the previous integer. Thus, the dose for the weight 47.3 lb is 1.5 teaspoons.

Weight (pounds)	*Dose (teaspoon)*
24–35	1
36–47	1.5
48–59	2
60–71	2.5
72–95	3

Let f be the function that assigns to each weight the corresponding dose. Do the following:

a. Find the domain of f.

b. Draw a graph of f.

c. Describe f in symbols using function notation.

d. What is the dose for a child who weighs 46 lb?

3. **a.** Show that the graph of the function $y = \sqrt{4 - x^2}$ is a half-circle. (Start by squaring both sides of the equation.) What is its radius?

b. Sketch the graph of each function:

(i) $f(x) = \sqrt{4 - x^2}$ (ii) $g(x) = -\sqrt{4 - x^2}$

4. Consider the function $f(x) = \begin{cases} -2 & \text{if } x < 0 \\ 3 & \text{if } x = 0 \\ x^2 & \text{if } x > 0. \end{cases}$

a. Sketch the graph of f. **b.** Give each value:

(i) $f(-1.3)$ (ii) $f(0)$ (iii) $f\left(\frac{5}{2}\right)$ (iv) $f\left(\sqrt{3}\right)$

5. Let g be the function $g(x) = \frac{x^2 - 1}{x + 1}$.

a. Give the domain of g. **b.** Write the rule for g in simplified form.

c. Sketch the graph of $y = g(x)$.

6. Sketch the graph of each function. If the graph is a parabola, give the coordinates of the vertex.

a. $F(x) = \frac{10}{x}$ **b.** $y = -x^2 + 4$ **c.** $y = \sqrt{3 - x^2}$

d. $G(t) = (t + 2)^2 - 3$ **e.** $h(x) = -2x + 1$

7. Consider the function h defined as follows:

$$h(x) = \begin{cases} x^2 & \text{if } x < -1 \\ x + 1 & \text{if } -1 \le x \le 2 \\ x & \text{if } x > 2. \end{cases}$$

For each of the following values of x, give $h(x)$:

$$x = 2, \quad x = -1.2, \quad x = -\sqrt{3}, \quad x = 3.4, \quad x = \frac{33}{28}, \quad x = -1.$$

2-F TRANSFORMATIONS OF GRAPHS

We are often faced with the problem of graphing an unfamiliar function. It is helpful to be able to recognize from the algebraic expression of the function rule that the shape of the graph is familiar. It is also important to be able to determine a domain (or calculator window) that will show the main features of the graph. If we know the graph of a function $f(x)$, we can produce the graph of any of the following functions: $y = f(x) + c$, $y = f(x) - c$, $y = f(x + c)$, $y = f(x - c)$, $y = -f(x)$, $y = f(-x)$, $y = f(cx)$, and $y = cf(x)$, where c is a positive constant. To obtain the graphs of the new functions from the graph of f, we apply geometric transformations to the graph of f. The transformations involved are *translations* (*shifts*), *reflections* about one of the coordinate axes, and *scaling*.

1 Vertical Shifts

Suppose c is a positive constant. The graph of $y = f(x) + c$ is the graph of $y = f(x)$ shifted upward c units, since the points of those graphs that correspond to the same value of x are $(x, f(x) + c)$ and $(x, f(x))$, respectively. In the same manner we see that the graph of $y = f(x) - c$ is the result of shifting the graph of $y = f(x)$ downward c units.

EXAMPLE 2.18 Graph the functions $g(x) = x^2 + 2$ and $h(x) = x^2 - 3$.

Solution If $f(x) = x^2$, then $g(x) = f(x) + 2$ and $h(x) = f(x) - 3$. The graph of g is the parabola $y = x^2$ shifted upward 2 units, and the graph of h is the parabola $y = x^2$ shifted downward 3 units. The graphs of f, g, and h are shown in Figure 2.34.

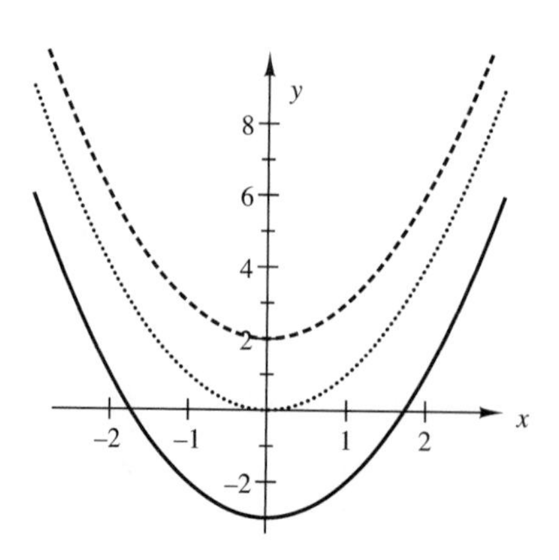

Figure 2.34 $f(x) = x^2$ (dotted)
$g(x) = f(x) + 2 = x^2 + 2$ (dashed)
$h(x) = f(x) - 3 = x^2 - 3$ (solid) ■

2 Horizontal Shifts

If a positive constant c is added to or subtracted from the independent variable in the equation $y = f(x)$, the result is a horizontal shift. The graph of $y = f(x + c)$ is the graph of $y = f(x)$ shifted c units to the left. To see this, let's compare the point $(0, f(0))$ on the graph of $y = f(x)$ with the corresponding point on the graph after a horizontal shift. We ask: For what value of x will the y-coordinate on the graph of $y = f(x + c)$ equal $f(0)$? The answer is $x = -c$, since $f(-c + c) = f(0)$. So $(-c, f(0))$ is the shifted point, and this point is c units to the left of the point $(0, f(0))$. In the same manner, it can be seen that the graph of the function $y = f(x - c)$ is the graph of $y = f(x)$ shifted c units to the right. Figure 2.35 shows the function $f(x) = x^2$ together with the shifted graphs $y = f(x + c)$ and $y = f(x - c)$.

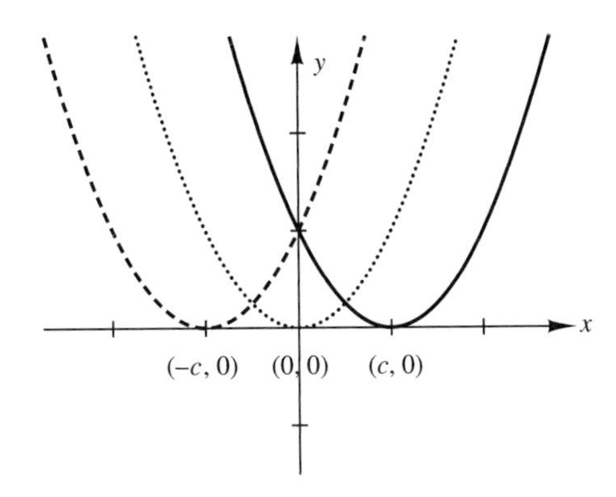

Figure 2.35 $y = f(x) = x^2$ (dotted)
$y = f(x + c) = (x + c)^2$ (dashed)
$y = f(x - c) = (x - c)^2$ (solid)

Note that the graph of $y = f(x + c)$ for c a positive constant is a shift of the graph of $y = f(x)$ to the **left,** whereas the graph of $y = f(x) + c$ is a shift of the graph of $y = f(x)$ **upward.**

EXAMPLE 2.19 Graph these functions:

a. $g(x) = \left|x - \frac{1}{2}\right|$ **b.** $h(x) = \frac{1}{x+1}$

Solution

a. If $f(x) = |x|$, then $g(x) = f\left(x - \frac{1}{2}\right)$, so its graph is the graph of $y = |x|$ shifted to the right $\frac{1}{2}$ unit. Figure 2.36 shows the graph of $y = \left|x - \frac{1}{2}\right|$ in solid lines and the graph of $y = |x|$ in dotted lines. We can verify that our direction of shift is correct by checking the vetex of the "V," which is the x-intercept of $g(x)$. We see that $g(x) = 0$ when $x = \frac{1}{2}$, so the point (0, 0) on the graph of $f(x) = |x|$ is shifted right to the point $\left(\frac{1}{2}, 0\right)$ on the graph of $g(x) = \left|x - \frac{1}{2}\right|$.

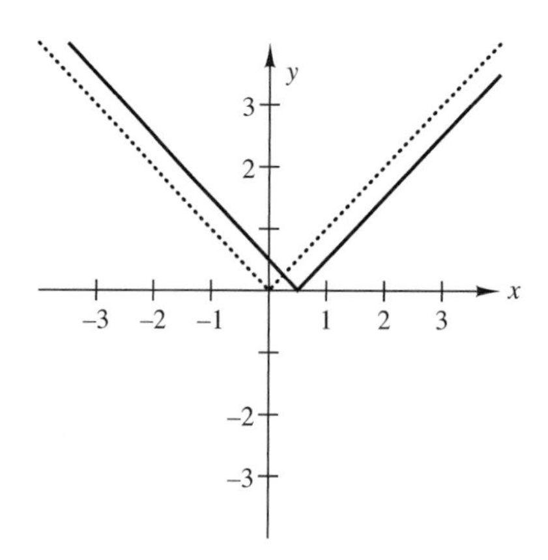

Figure 2.36 $y=\left|x-\frac{1}{2}\right|,\quad y=|x|$ (dotted)

b. Since $h(x)=f(x+1)$ where f is the reciprocal function $f(x)=\frac{1}{x}$, the graph of the function h is the graph of the reciprocal function shifted to the left 1 unit. The graphs of both functions, $y=h(x)=\frac{1}{x+1}$ and $y=\frac{1}{x}$ (dotted), are shown in Figure 2.37. Note that the point (1,1) on the graph of f moves left to become the point (0,1) on the graph of h.

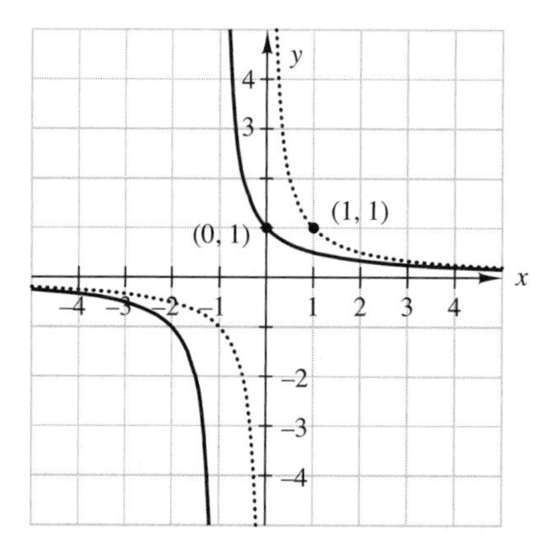

Figure 2.37 $y=\frac{1}{x+1},\quad y=\frac{1}{x}$ (dotted) ■

 When sketching the shifted graph of a function, check the direction of shift by plotting at least one significant point (an intercept, a vertex, or an easily recognizable point).

3 Reflection about the *x*-Axis

The graph of $y=-f(x)$ is obtained from the graph of $y=f(x)$ by reflection about the x-axis. To see this, note that the point $(x,-f(x))$ on the graph of $y=-f(x)$ and the point $(x, f(x))$ on the graph of $y=f(x)$ have the same x-coordinates but opposite y-coordinates. Figure 2.38 shows the graphs of $f(x)=x^2+1$ (dotted) and $g(x)=-(x^2+1)=-x^2-1$.

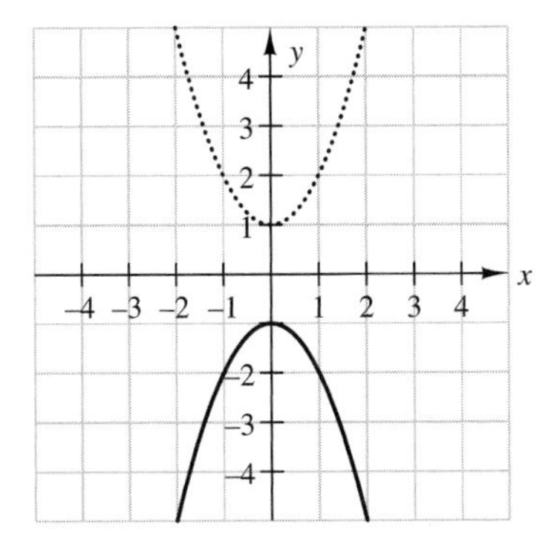

Figure 2.38 $y = x^2 + 1$ (dotted), $y = -(x^2 + 1)$

4 Reflection about the *y*-Axis

The graph of $y = f(-x)$ is the reflection of the graph of $y = f(x)$ about the y-axis. We illustrate this with the functions $f(x) = \sqrt{x}$ and $g(x) = f(-x) = \sqrt{-x}$. Since the square root is defined only for nonnegative numbers, the domain of f is the interval $[0, \infty)$ and the domain of g is the interval $(-\infty, 0]$ (x must be negative, so that $-x$ is positive and $\sqrt{-x}$ is defined). If the point (x, y) is on the graph of $y = \sqrt{x}$, then since $y = \sqrt{-(-x)}$, the point $(-x, y)$ is on the graph of $y = \sqrt{-x}$. This shows that the graph of $y = \sqrt{x}$ reflected about the y-axis is the graph of $y = \sqrt{-x}$. Figure 2.39 shows both graphs.

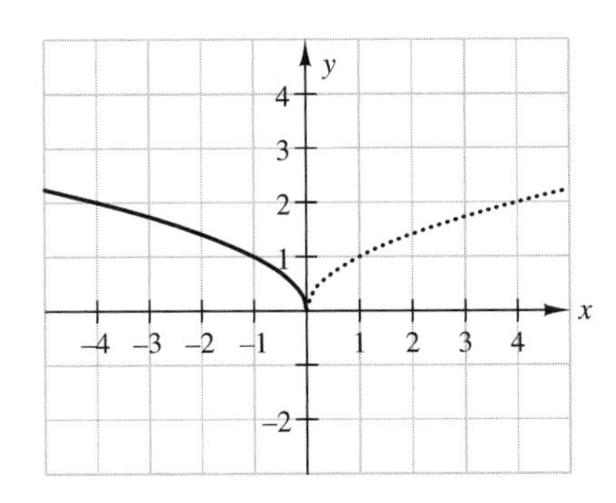

Figure 2.39 $y = \sqrt{x}$ (dotted), $y = \sqrt{-x}$

5 Scaling: Stretching and Compressing

The graph of $g(x) = cf(x)$, where c is a positive constant, is obtained by vertically stretching or compressing the graph of the function f. If $c > 1$, the graph of f is stretched by a factor of c; if $0 < c < 1$, the graph of f is compressed. Figure 2.40 shows the graphs of $f(x) = \sqrt{1 - x^2}$ (dotted), $g(x) = 3\sqrt{1 - x^2}$ (dashed), and $h(x) = \frac{1}{3}\sqrt{1 - x^2}$ (solid). The graph of f is a semicircle as we saw in Section 2-E; the other two graphs are the result of uniformly stretching or compressing the semicircle. Each has the shape of the upper half of an *ellipse*.

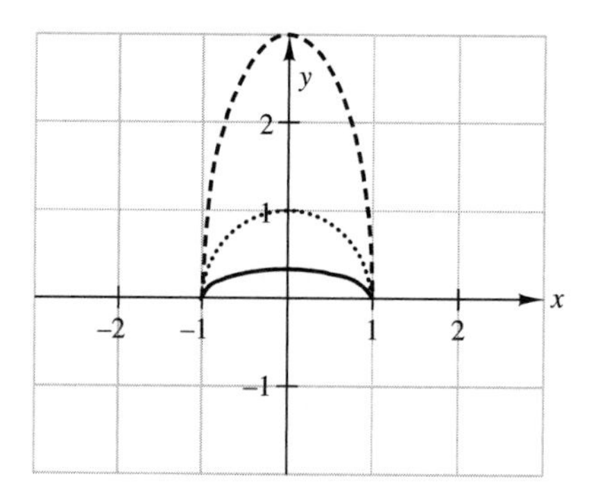

Figure 2.40 $f(x)=\sqrt{1-x^2}$ (dotted)
$g(x)=3\sqrt{1-x^2}$ (dashed)
$h(x)=\frac{1}{3}\sqrt{1-x^2}$ (solid)

The graph of a function $g(x) = f(cx)$ is obtained by horizontally stretching or compressing the graph of f. If $c > 1$, the graph of g is the graph of f compressed horizontally by a factor of c; if $0 < c < 1$, the graph of g is the graph of f stretched horizontally. Figure 2.41 shows the graphs of $f(x)=\sqrt{1-x^2}$, $g(x) = f(2x) = \sqrt{1-(2x)^2}$, and $h(x) = f\left(\frac{1}{2}x\right)=\sqrt{1-\left(\frac{1}{2}x\right)^2}$. As before, the graph of f is a semicircle, and g and h each have half an ellipse as their graph.

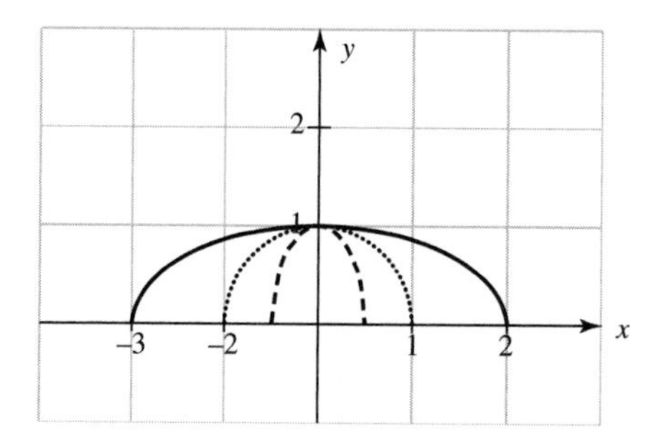

Figure 2.41 $f(x)=\sqrt{1-x^2}$ (dotted)
$g(x)=f(2x)=\sqrt{1-(2x)^2}$ (dashed)
$h(x)=f\left(\frac{1}{2}x\right)=\sqrt{1-\left(\frac{1}{2}x\right)^2}$ (solid)

Multiplying a function $f(x)$ by a constant c > 0 (to produce $cf(x)$) and multiplying the variable x by c (to produce $f(cx)$) have opposite effects. Scaling in a vertical direction ($cf(x)$) stretches the graph when $c > 1$ and compresses it when $c < 1$. However, scaling in a horizontal direction ($f(cx)$) compresses the graph when $c > 1$ and expands it when $c < 1$.

Often more than one transformation needs to be applied to a familiar graph to obtain the graph of a given function. In the next example, we show how to apply several transformations to graphs we already know in order to obtain graphs of new functions. The most important step in doing this is to find the basic shape of the graph by recognizing which simpler function to use.

EXAMPLE 2.20 Sketch the graph of each of the following functions. For each function identify the type of graph (e.g., parabola, semicircle, graph of the reciprocal function). Plot significant points.

a. $y = \sqrt{2x} + 1$ **b**. $y = \frac{3}{2} - 2\sqrt{4 - x^2}$ **c**. $y = |3x + 2|$

Solution

a. The graph of $y = \sqrt{2x} + 1$ is a translation up by 1 unit of the graph of $y = \sqrt{2x}$, which in turn is the result of scaling the graph of $y = \sqrt{x}$. The required graph is a half-parabola. With this information, we can make a small table of values, which allows us to complete the graph shown in Figure 2.42. (Note that the domain contains only values of x for which $2x \geq 0$; therefore, $x \geq 0$.)

x	$y = \sqrt{2x} + 1$
0	1
2	3
$\frac{9}{2}$	4

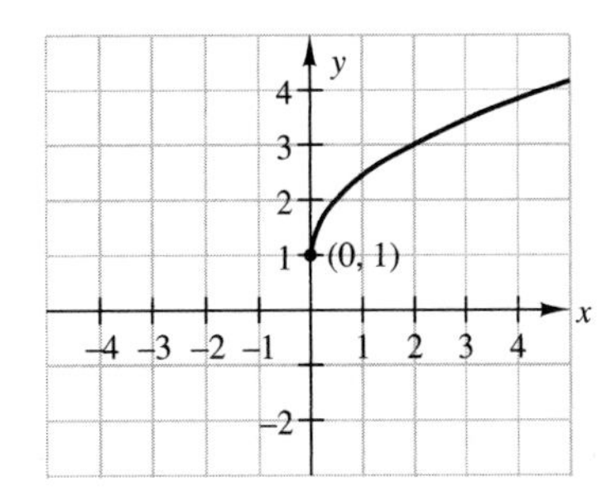

Figure 2.42 $y = \sqrt{2x} + 1$

b. The graph of $y = \frac{3}{2} - 2\sqrt{4 - x^2}$ is the graph of $y = -2\sqrt{4 - x^2}$ shifted upward $\frac{3}{2}$ units. The graph of $y = -2\sqrt{4 - x^2}$ is obtained from the semicircle with equation $y = \sqrt{4 - x^2}$ (which has radius 2 and is centered at the origin) by stretching vertically by a factor of 2 and then reflecting about the x-axis. The required graph is shown in Figure 2.43. It is the lower half of an ellipse.

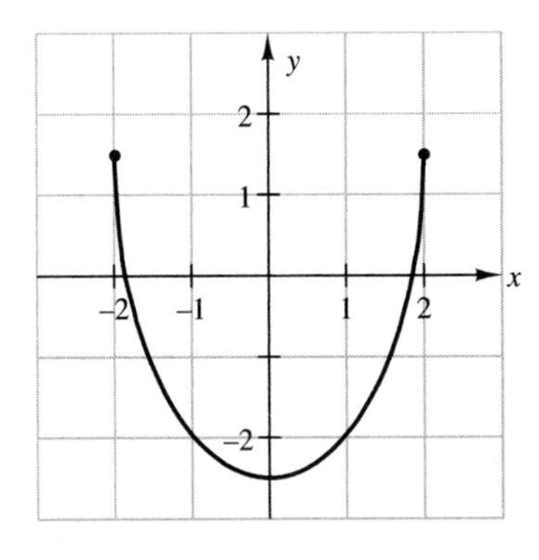

Figure 2.43 $y = \frac{3}{2} - 2\sqrt{4 - x^2}$

c. The graph of $y = |3x + 2|$ is obtained from the graph of $y = |x|$ by scaling and shifting horizontally. Thus, the required graph has a "V" shape. To sketch the graph accurately, we first locate the vertex of the V. This must be where $y = 0$, since $y \geq 0$ for all values of x. We solve the equation $3x + 2 = 0$ and obtain $x = -\frac{2}{3}$. To sketch the graph accurately, we now complete a small table of values, making sure to include values on either side of $x = -\frac{2}{3}$. The graph is shown in Figure 2.44.

x	$y = \|3x + 2\|$
-2	4
$-\frac{2}{3}$	0
0	2
1	5

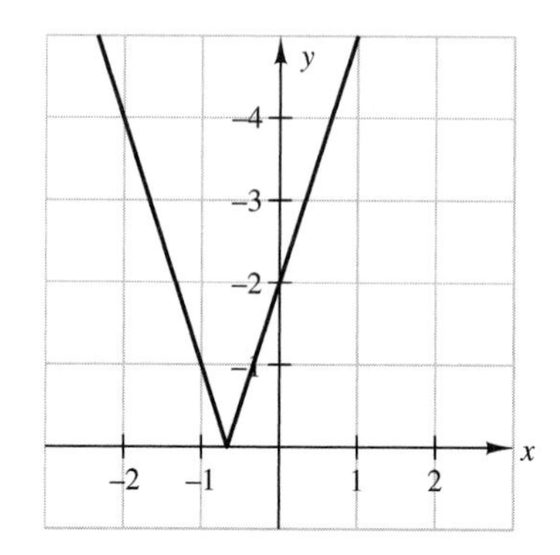

Figure 2.44 $y = |3x + 2|$ ■

Exercises 2-F

1. Six different graphs labeled a–f follow. Each of them is the graph of one of the following functions: $f(x) = 3 - |x|$, $g(x) = |-x + 3|$, $h(x) = |3x + 1|$, $F(x) = -|x - 3|$, $G(x) = 3|x + 1|$, or $H(x) = \left|\frac{x}{3} + 1\right|$. Under each graph, write the corresponding equation for the function.

a.

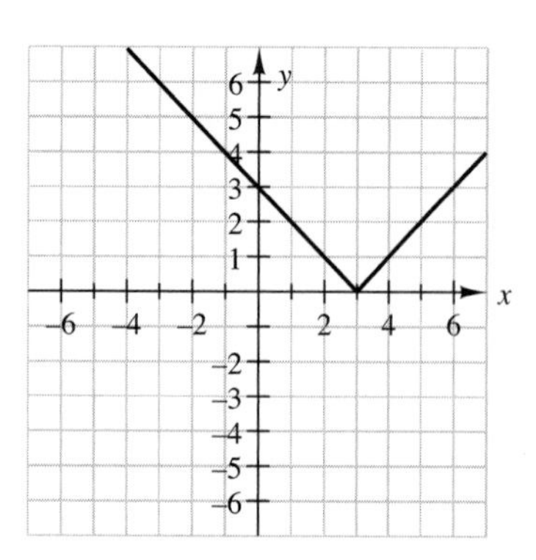

b.

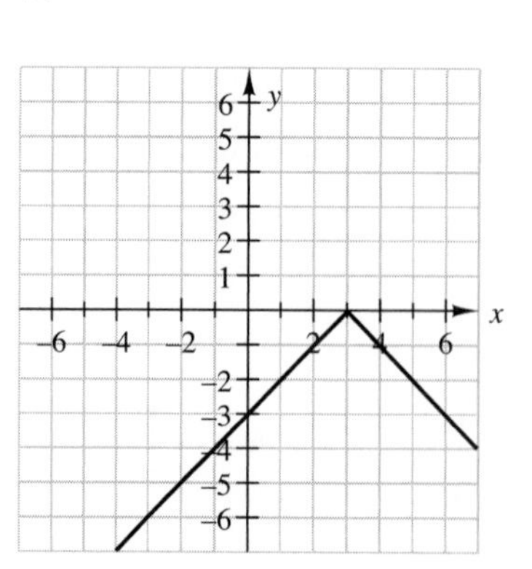

c.

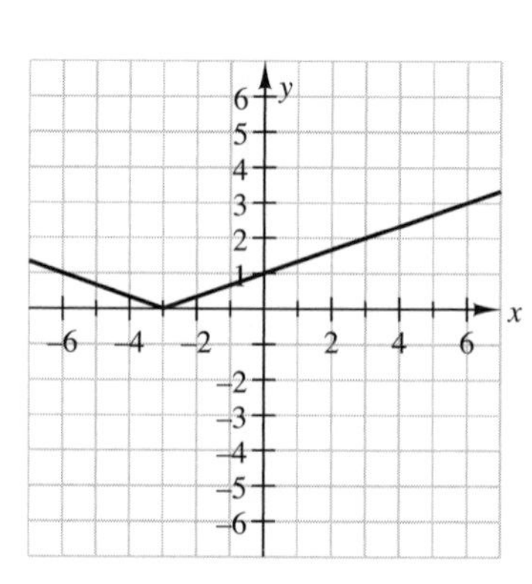

d.

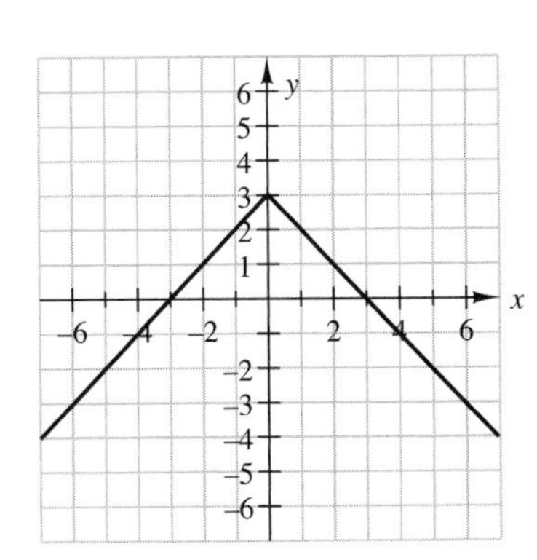

e.

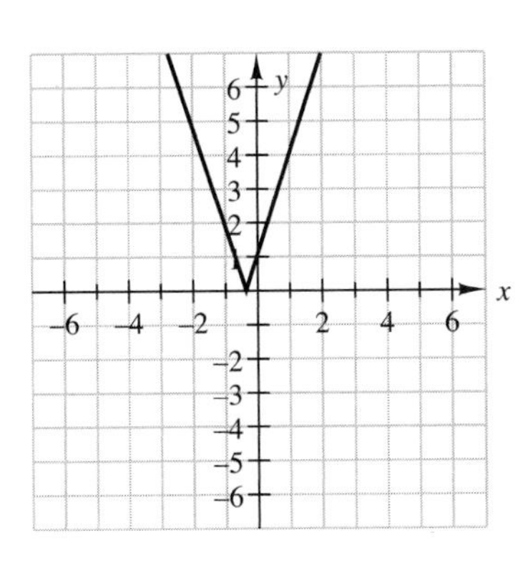

f.

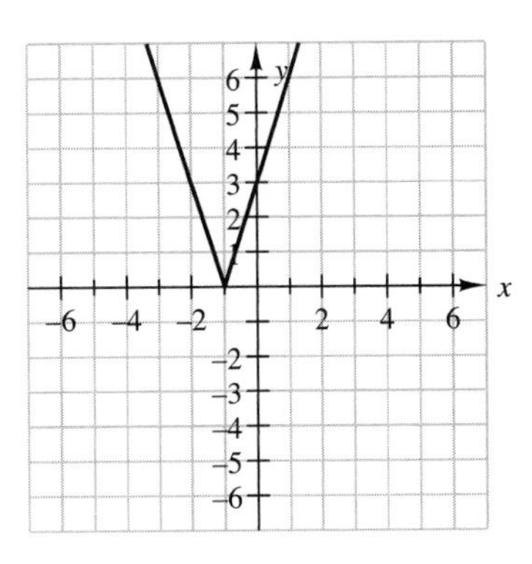

2. Eight functions are defined in a–h below and then six graphs are shown. For each function, identify its graph by number; if the graph is not shown, sketch it yourself.

a. $y=-2(x+1.5)^2+4$

b. $f(x)=4+\sqrt{-x+1.5}$

c. $y=\sqrt{9-4(x+1.5)^2}$

d. $g(x)=|x+1.5|-4$

e. $h(x)=1.5-4x$

f. $y=\dfrac{2}{x-1.5}+4$

g. $y=\dfrac{1}{2x-1.5}+4$

h. $F(x)=4-\sqrt{x+1.5}$

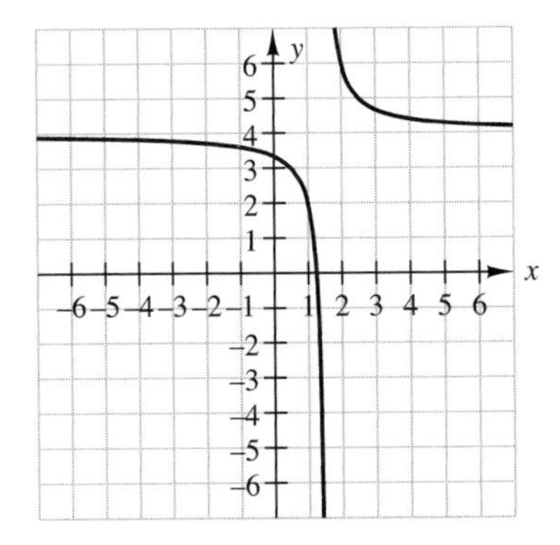

Graph 1

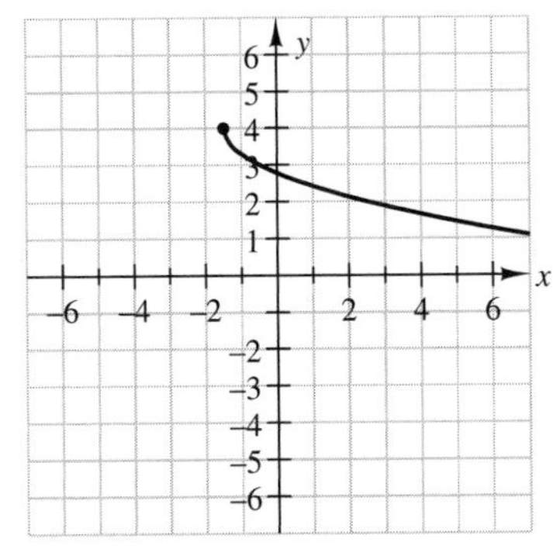

Graph 2

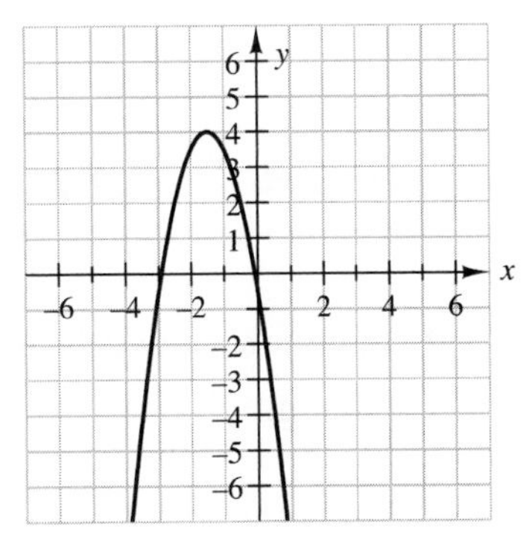

Graph 3

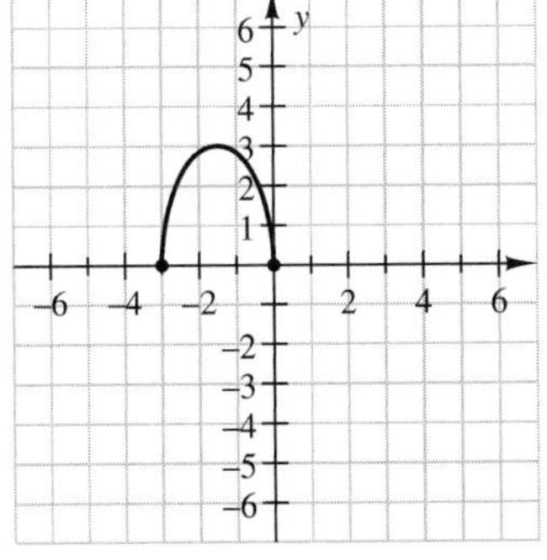

Graph 4

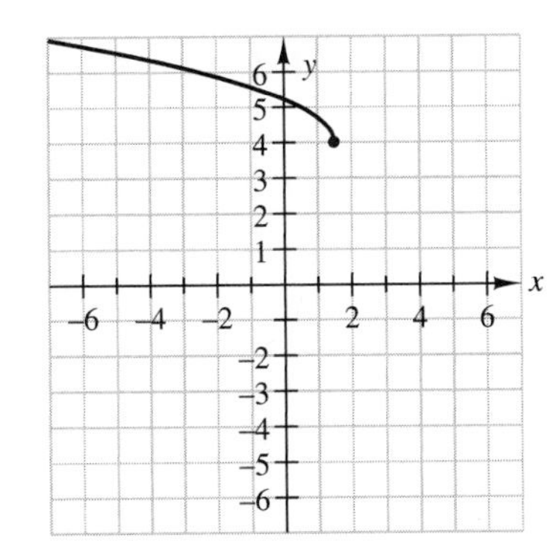

Graph 5

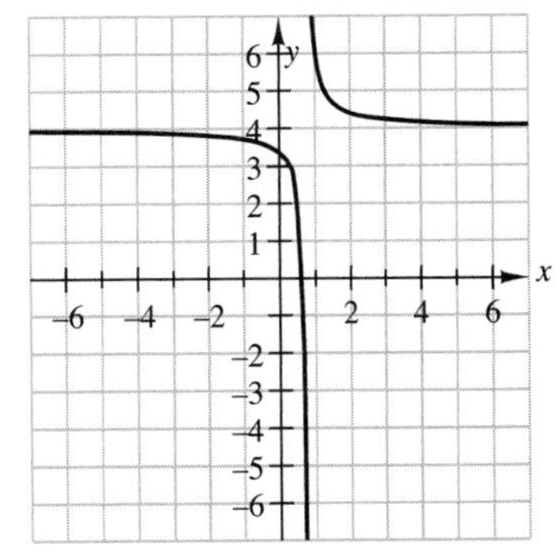

Graph 6

3. Use transformations to sketch the graph of each of the following functions. For each function, identify the type of graph (e.g., parabola, semicircle, graph of the reciprocal function). Plot significant points.

a. $y = -2x^2$ **b**. $y = 2 + \sqrt{-x}$ **c**. $y = 2 + 5x$

d. $y = \sqrt{x+1} + 3$ **e**. $y = 2 - \sqrt{1 - x^2}$ **f**. $y = \frac{1}{3x-1}$

4. The graphs of the following functions can be obtained from the graph of the function $f(x) = x^3$ shown in Figure 2.45. Use transformations to graph each function.

a. $g(x) = (x + 2)^3$ **b**. $h(x) = 2x^3 + 1$ **c**. $F(x) = -\frac{x^3}{2}$

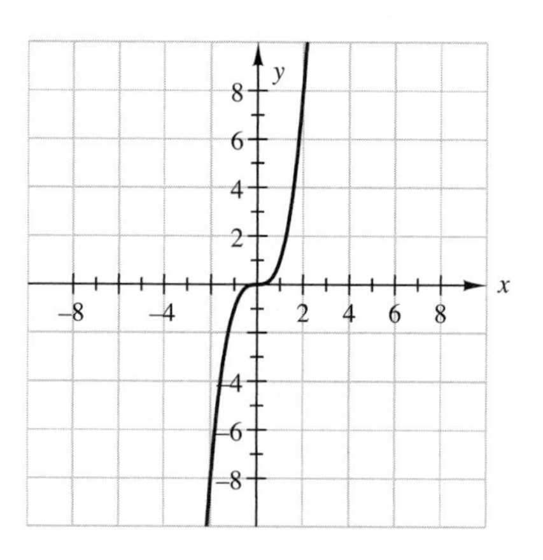

Figure 2.45 $f(x) = x^3$

Chapter 2 Exercises

1. In your own words, explain what is meant by "*s* is a function of *r*." What is meant by "the graph of a function"?

2. Draw the graphs of the following functions. Identify which functions are linear.

a. $f(x) = 2x + 1$ **b**. $g(x) = -2 + 7x$

c. $h(t) = -2 + 7t$ **d**. $k(w) = 7 + w^2$

Is there a difference between the functions in b and c? Explain.

3. Consider the graph in Figure 2.46.

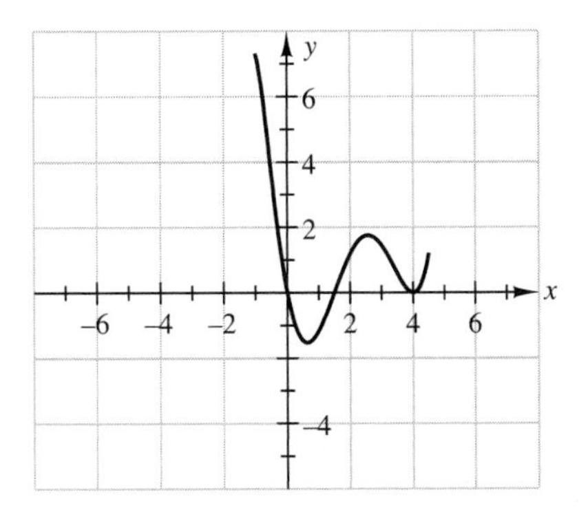

Figure 2.46

a. Explain why it represents the graph of a function.

b. Give the values of the function at $x = 0$, $x = 1.5$, and $x = 4$. What are those values of x called?

4. The graph in Figure 2.47 represents the effect of the market price p of a commodity on the manufacturer's total profit T.

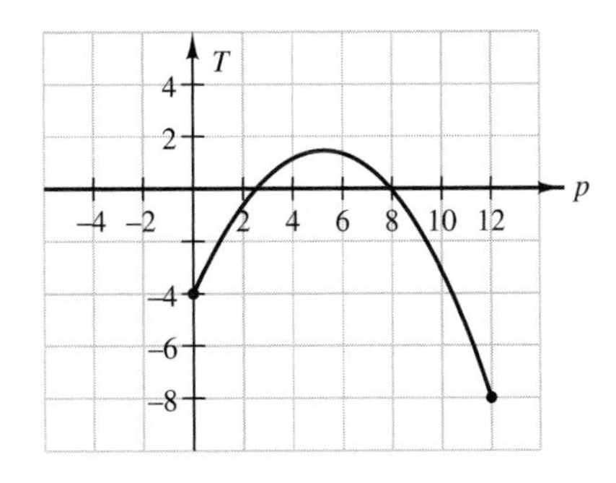

Figure 2.47

a. What is the domain of T? Explain what the graph shows about how T (total profit) is affected by p (market price). Is there an optimal selling price (one that will give the manufacturer maximum profit)? If so, what is it?

b. Which of these could be the formula that represents the function T?

(i) $T = p^2 + 5p - 1$ (ii) $T = -0.2p^2 + 2.1p - 4$ (iii) $T = \frac{4}{3}p - \frac{10}{3}$

5. Look at the graph of the function f in Figure 2.48 and answer questions a–c below.

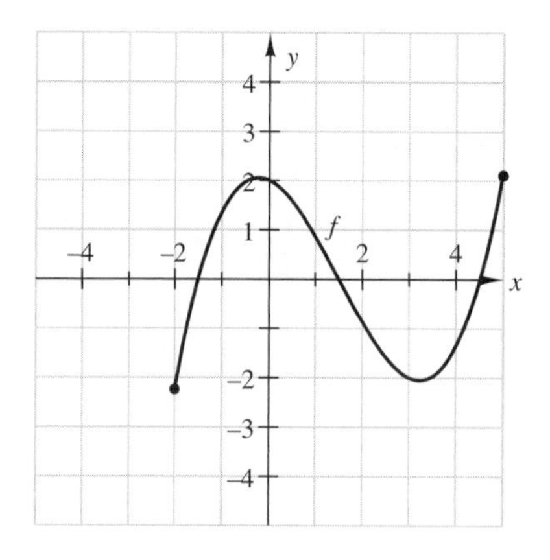

Figure 2.48

a. What are the approximate values of x for which $f(x) = 0$?

b. Give the intervals of values of x for which $f(x)$ is negative.

c. Give the intervals of values of x for which $f(x)$ is positive.

6. Plot the following points:

$$(0, 1), (1, 2), (-2, -1), \left(\frac{1}{3}, \frac{4}{3}\right), (3, 4), (4, 5).$$

a. Draw the graph of a nonlinear function that contains all those points.

b. Is there a linear function that contains all those points? If the answer is yes, give the following:

(i) The graph and the defining equation of such function

(ii) The value of the function at 7.

7. Let $f(x) = \dfrac{8x}{5+x}$.

a. Give each value if it exists. If it does not exist, state why.

$$f(-1),\ f(-0.1),\ f(1.2),\ f\left(\frac{3}{4}\right),\ f(0),\ f(-5),\ f(5)$$

b. Give the domain of f (use interval notation).

c. Is 2 in the range of f? Justify your answer.

d. Is 8 in the range of f? Justify your answer.

8. Consider the graph of $y = f(x)$ in Figure 2.49.

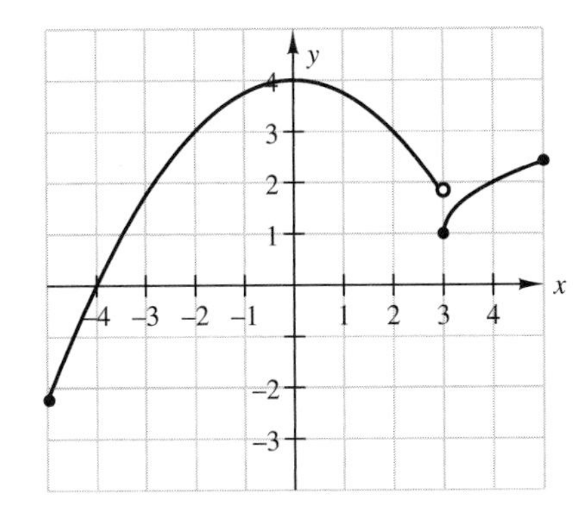

Figure 2.49

From the graph, estimate the following values if they exist. If they do not exist, explain why.

a. $f(3)$ **b**. $f(-3)$ **c**. $f(4)$

d. The values of a for which $0 = f(a)$

e. The values of b for which $3 = f(b)$

f. The values of c for which $5 = f(c)$

9. For each function, identify the type of graph (e.g., line, parabola, semicircle, graph of the reciprocal function) and sketch the graph, plotting significant points. Give the domain and range of each function.

a. $y = 6 - 4x$ **b**. $y = 3x^2 + 2$ **c**. $y = 4 - x^2$

d. $y = |x - 1| - 3$ **e**. $y = \sqrt{1 - x}$ **f**. $y = -\sqrt{x - 1} - 3$

g. $y = 2 + \sqrt{1 - x^2}$ **h**. $y = \frac{1}{x} + 2$

10. If $f(x) = \frac{1}{x^2} - \frac{1}{x}$, find $f\left(\frac{1}{a}\right)$, $f(-a)$, and $-f(a)$.

11. The graph in Figure 2.50 represents a function f.

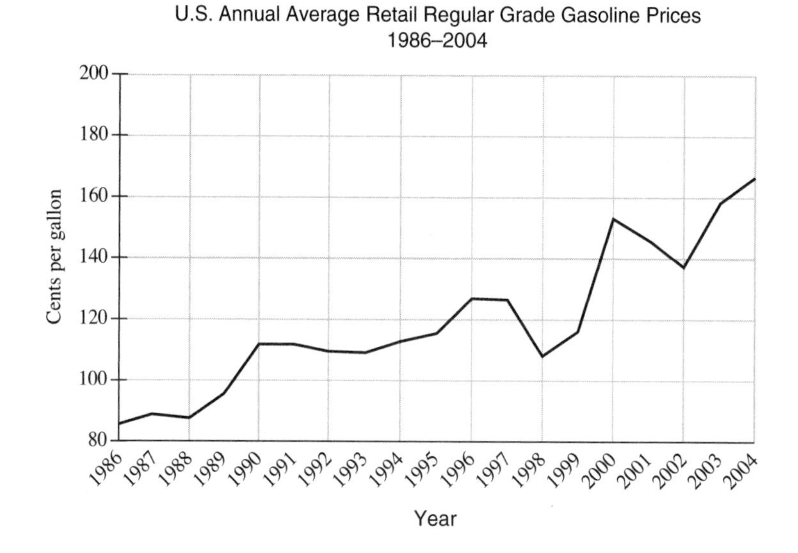

Figure 2.50

Data Source: Report of Attorney General Bill Lockyer (Calilfornia), March 2004

a. Estimate $f(2000)$ and explain what this number represents.

b. Give a complete description of what the graph shows for the period 1986–2004. From the graph, what gasoline price would you predict for 2005?

12. A factory that makes radios estimates that its daily profits from producing x radios per day are $P(x) = 20x - 500$ if it produces at least 25 but not more than 100. If more than 100 radios are produced, the company's fixed costs increase. The company estimates that $P(x) = 20x - 700$ when $100 < x \leq 200$.

a. Graph the function P.

b. What is the company's daily profit if 55 radios are produced?

c. What is the company's daily profit if 155 radios are produced?

13. The local night spot has a cover charge of \$7.50 per person. In addition, each beverage costs \$5.00. Write an equation that gives a person's bill at the night spot as a function of the number of beverages consumed. What would be an appropriate domain for this function?

14. Let $F(x) = \begin{cases} 1-x & \text{if } x \leq -2 \\ 1-x^2 & \text{if } -2 < x \leq 1 \\ 0 & \text{if } x > 1. \end{cases}$

a. Find $F(0)$, $F(-2)$, $F\left(\frac{1}{2}\right)$, $F(1)$, $F(2)$, $F\left(\frac{8}{9}\right)$, $F(\pi)$.

b. Identify values of x where the graph of F has jumps.

15. Write the length of the diagonal d of a square as a function of the length of its side x. Draw and label an appropriate figure.

16. The function f has the rule $f(x) = x^2 + 3x - 5$. Use a calculator to complete the following table.

h	$f(4)$	$f(4+h)$	$f(4+h)-f(4)$	$\frac{f(4+h)-f(4)}{h}$
1	23	35	12	12
0.1	23	24.11	1.11	11.1
0.01	23	23.1101		
0.001	23			
0.0001	23			

Do you notice a pattern? Describe the pattern.

17. An object is dropped from a height of 200 ft. As it falls, its velocity changes. Let v represent the velocity of the object at time t and h the height from the ground at the same time t. The two quantities, v and h, are related by the equation $v = 8\sqrt{200 - h}$. This equation describes v (in feet per second) as a function of h (in feet).

a. Give the domain and the range of the function.

b. Give the object's velocity when its height is 100 ft, 120 ft, 0 ft.

c. What is the object's height when the velocity is 0?

CHAPTER 3

Companion to Limits

3-A COMBINATIONS OF FUNCTIONS

The concept of the limit of a function is one of the most important concepts in calculus. Limits are used to assess the behavior of a function $f(x)$ as x *approaches* a number but does not equal the number. In evaluating limits of functions, we often need to combine or simplify algebraic expressions.

3-A.1 Algebraic Combinations of Functions

To graph the function $F(x) = \sqrt{x-1} + x$, it is helpful to look at it as the sum of two functions; $F(x)$ is the sum of $g(x) = \sqrt{x-1}$ and the linear function $h(x) = x$. This means that for each x-coordinate, we add the y-coordinates from the two corresponding points on the graphs of g and h. The graph of g is a half-parabola and the graph of h is a line; these graphs are shown in Figure 3.1.

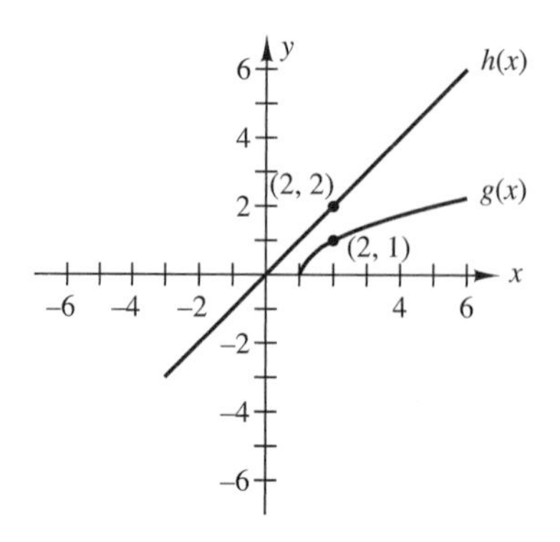

Figure 3.1 $g(x) = \sqrt{x-1}$, $h(x) = x$

For example, to find the y-coordinate of the point $(2, y)$ on the graph of F, we note that the point $(2, 1)$ is on the graph of g and the point $(2, 2)$ is on the graph of h, so the point (2, 3) is on the graph of F. We do this for each value a in the domain of F: The point (a, y) on the graph of F is the point $(a, g(a) + h(a))$. The graph of F is shown in Figure 3.2.

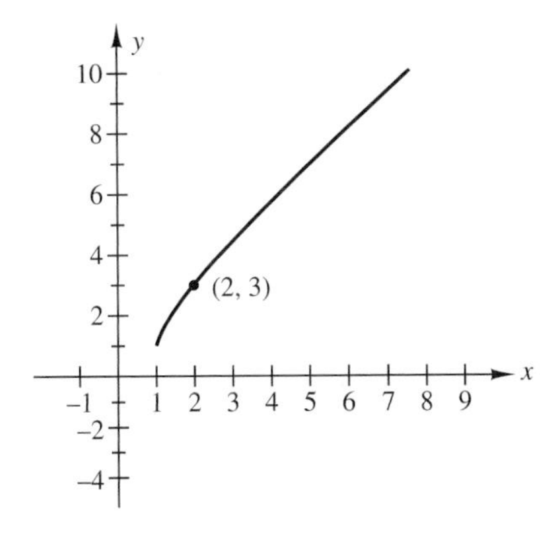

Figure 3.2 $F(x) = g(x) + h(x) = \sqrt{x-1} + x$

New functions can be produced by addition, subtraction, multiplication, and division of other functions. We illustrate this with several examples of functions that occur in business.

Suppose the demand function for a new publication is given by the equation $d = f(p) = 30 - \frac{5}{2}p$, where p is the selling price of the publication (in dollars) and d is the number of copies expected to be sold at that price (in thousands of copies). This type of demand function is typical: It is the *sum* of a constant function $g(p) = 30$ (the maximum sales possible) and a function $h(p) = -\frac{5}{2}p$ that measures how sales drop as price increases.

The marketing department of the publishing company assumes that this function is valid for $2 \leq p \leq 8$. To estimate the total revenue from the new publication as a function of the price per copy, we multiply the number of copies by the price per copy. Thus, total revenue is $R(p) = p(30 - \frac{5}{2}p)$ for $2 \leq p \leq 8$. $R(p)$ is a *product* of two functions.

If $R(x)$ represents the total revenue from the sale of x units of a product and $C(x)$ is the total cost of producing x units, then the total profit $P(x)$ obtained by producing and selling x units is just revenue minus cost:

$$P(x) = R(x) - C(x).$$

The profit function is the *difference* of the revenue and cost functions. The average cost function is $\overline{C}(x) = \frac{C(x)}{x}$; this measures the average cost of production of 1 unit when x units are produced. $\overline{C}(x)$ is a *quotient* of two functions.

Suppose that f and g are any two functions. Then the sum $f + g$, difference $f - g$, product $f \cdot g$ (also denoted fg), and quotient $\frac{f}{g}$ are new functions and are defined as follows for each x in the domain of both f and g:

Sum: $(f + g)(x) = f(x) + g(x)$

Difference: $(f - g)(x) = f(x) - g(x)$

Product: $(f \cdot g)(x) = (fg)(x) = f(x) \cdot g(x)$

Quotient: $\left(\frac{f}{g}\right)(x) = \frac{f(x)}{g(x)}$

Each of these defining equations states the rule for the function. For example, the first rule states that the sum function $f + g$ assigns to a number x the sum of the two numbers $f(x)$ and $g(x)$. For any number $x = a$, if either $f(a)$ or $g(a)$ is undefined, then the sum, difference, product, and quotient of $f(a)$ and $g(a)$ are also undefined. In addition, if $g(a) = 0$, then $\frac{f(a)}{g(a)}$ is also undefined. Domains of combinations of functions are discussed more fully in Section 4-C.

EXAMPLE 3.1 A sports park plans to fence in a track in the shape of a rectangle capped by two semicircles as shown in Figure 3.3. The long side of the rectangle is twice the length of the short side. Write an equation for the amount of fence required to enclose the field as a function of the shorter side of the rectangle.

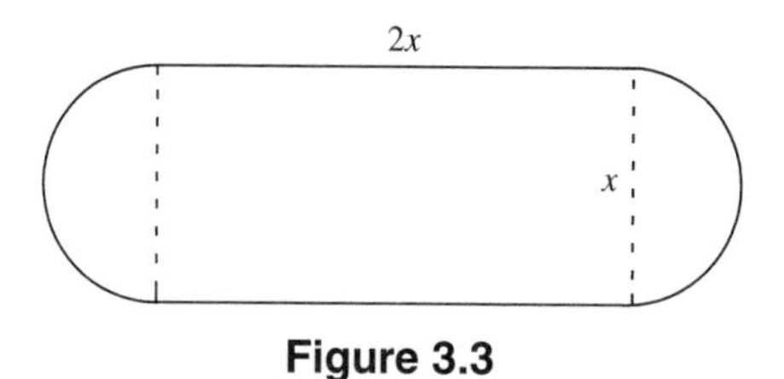

Figure 3.3

Solution We label with an x the part of the picture that is to be the independent variable. (Since we want to express the total fencing as a function of the short side of the rectangle, the short side is labeled x.) The amount of fence needed is the sum of the two longer sides of the rectangle and the perimeter of the two semicircles (which equal a whole circle). The perimeter of the circle is $C = 2\pi r$, where r is the radius of the circle, but $r = \frac{x}{2}$. Thus $C = 2\pi\frac{x}{2} = \pi x$. Therefore, the amount of fence needed is $F = 2 \cdot 2x + C = 4x + \pi x$. ■

EXAMPLE 3.2 The supply function $S(x)$ gives the price at which producers will supply an amount x of a product, and the demand function $D(x)$ gives the price at which consumers demand amount x of a product. The equilibrium point is defined to be the point (x^*, p^*), where x^* satisfies $S(x^*) - D(x^*) = 0$ or $S(x^*) = D(x^*)$ (supply = demand) and p^* is the common value of $S(x^*)$ and $D(x^*)$. If the demand function for a certain power tool is $D(x) = 200 - 3x$ and the supply function is $S(x) = 24 + 5x$, find the equilibrium point.

Solution We look at the difference function:

$$S(x) - D(x) = 24 + 5x - (200 - 3x) = -176 + 8x$$

and set this equal to zero. Solving for x, we get $8x = 176$, so $x = \frac{176}{8} = 22$. Since $S(22) = 24 + 5(22) = 134$, the equilibrium point is (22, 134). This means that according to the given supply and demand functions, when producers supply 22 of the power tools at a price of \$134, consumers will demand exactly 22 tools. The graphs of $S(x)$ and $D(x)$ cross when $x = 22$ (see Figure 3.4). At a price of less than \$134, demand will exceed supply; if the price is more than \$134, supply will exceed demand. Only at a price of \$134 will demand and supply be equal. The graph of $(S - D)(x) = S(x) - D(x)$ is in Figure 3.5.

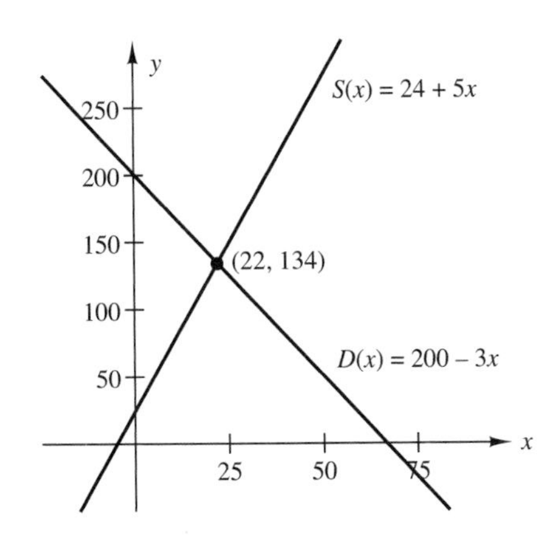

Figure 3.4

Figure 3.5 $(S - D)(x) = -176 + 8x$ ■

EXAMPLE 3.3 Let $f(x) = x^2 - 5x + 6$, $g(x) = x^2 - 4$, $h(x) = \sqrt{x-2}$, and $k(x) = \sqrt{5-x}$. Find each of the following functions and give their domains.

a. $(f + g)(x)$

b. $(g \cdot h)(x)$

c. $(h - k)(x)$

d. $\left(\frac{f}{g}\right)(x)$

Solution

a. $(f + g)(x) = f(x) + g(x) = (x^2 - 5x + 6) + (x^2 - 4) = 2x^2 - 5x + 2$. Both $f(x)$ and $g(x)$ are defined for every real number, so the domain of $f + g$ is $(-\infty, \infty)$.

b. $(g \cdot h)(x) = g(x) \cdot h(x) = (x^2 - 4)\sqrt{x-2} = x^2\sqrt{x-2} - 4\sqrt{x-2}$. The domain of this function is $[2, \infty)$ because h is defined only when $x \geq 2$.

c. $(h-k)(x) = h(x) - k(x) = \sqrt{x-2} - \sqrt{5-x}$. The domain of h is $[2, \infty)$. Since $\sqrt{5-x}$ is defined only when $5 - x \geq 0$—that is, when $5 \geq x$—the domain of k is $(-\infty, 5]$. Thus the domain of $h - k$ consists of all real numbers x that belong to both $[2, \infty)$ and $(-\infty, 5]$. This is the interval $[2, 5]$.

d. $\left(\frac{f}{g}\right)(x) = \frac{f(x)}{g(x)} = \frac{x^2-5x+6}{x^2-4}$. The domain of the quotient function is all real numbers that are in the domain of f and in the domain of g, excluding those values for which $g(x) = 0$. Since $f(x)$ and $g(x)$ are defined for all real numbers, we need to exclude only values of x for which $x^2 - 4 = 0$. Thus, the domain consists of all real numbers except $x = 2$ and $x = -2$. In interval notation this is $(-\infty, -2) \cup (-2, 2) \cup (2, \infty)$. Note that we can simplify the rule for $\frac{f}{g}$:

$$\left(\frac{f}{g}\right)(x) = \frac{x^2-5x+6}{x^2-4} = \frac{(x-2)(x-3)}{(x-2)(x+2)} = \frac{x-3}{x+2}, \quad x \neq 2.$$ ■

EXAMPLE 3.4 Consider the graphs of $y = f(x)$ and $y = g(x)$ shown in Figures 3.6 and 3.7. Estimate the function values indicated in parts a–d below, if they are defined. If they are undefined, explain why.

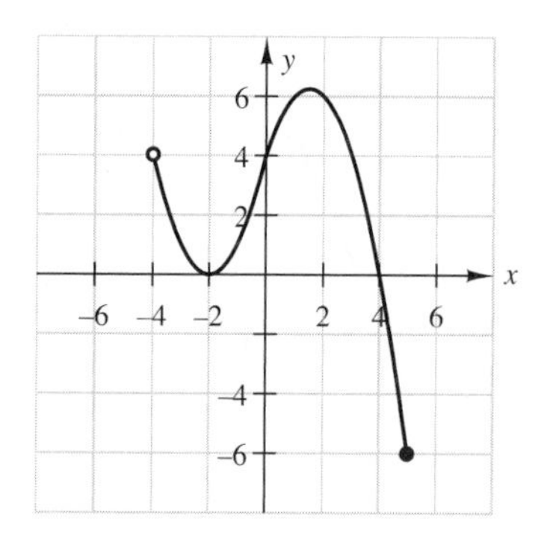

Figure 3.6 $y = f(x)$

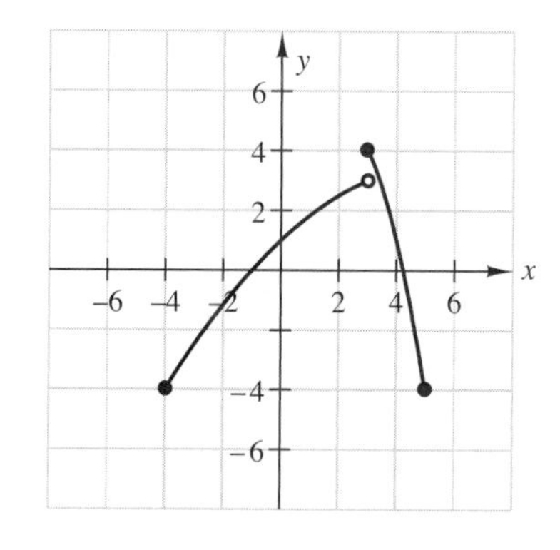

Figure 3.7 $y = g(x)$

a. $(f \cdot g)(0)$ **b.** $\left(\frac{g}{f}\right)(0)$ **c.** $\left(\frac{f}{g}\right)(-1)$ **d.** $(f + g)(3)$

Solution

a. Since $f(0) = 4$ and $g(0) = 1$, $(f \cdot g)(0) = f(0) \cdot g(0) = 4 \cdot 1 = 4$.

b. $\left(\frac{g}{f}\right)(0) = \frac{g(0)}{f(0)} = \frac{1}{4}$.

c. $\left(\frac{f}{g}\right)(-1) = \frac{f(-1)}{g(-1)}$ if $g(-1) \neq 0$. But $g(-1) = 0$, so $\left(\frac{f}{g}\right)(-1)$ is undefined.

d. Since $f(3) = 4$ and $g(3) = 4$, $(f + g)(3) = f(3) + g(3) = 4 + 4 = 8$. ■

3-A.2 Composition of Functions

In the previous section we discussed how to combine functions using the four operations of addition, subtraction, multiplication, and division of functions. In this section we will investigate another operation, called *composition,* which combines two functions to produce a new function. Several important theorems in calculus concern the composition of functions.

Most retail prices are determined by a sequence of markups. For example, a grocer may use the following rule: Double the wholesale cost of cauliflower to get the retail price. This can be expressed as $r(w) = 2w$, where w is the wholesale price and $r(w)$ is the retail price that corresponds to that wholesale price. But the wholesaler sets his price based on how much he has to pay the farmer for the cauliflower. Suppose he adds \$0.15 per head to his cost in setting the wholesale price. This can be expressed as $w(x) = x + 0.15$, where x is the price he pays the farmer per head and $w(x)$ is the corresponding wholesale price. Therefore, if we know that the farmer gets \$.35 a head for cauliflower, we can determine the retail price in two steps. First, we find the wholesale price $w(0.35)$, which is $0.35 + 0.15$, or 0.50. Then we use the result of this calculation to find the retail price $r(0.50)$, which is $2(0.50)$, or 1. Thus the retail price of the cauliflower is \$1 per head.

This two-step operation is called *composition of functions*. The two functions involved are the wholesale price $w(x)$, which is a function of x (the farmer's price), and the retail price $r(w)$, which is a function of the wholesale price w. The *composite* of w and r is the function $r(w(x))$, which is a function of x. The symbolic expression $r(w(x))$ is read as "r of w of x." In some applications a composite function is treated as a two-step process in which the inner function (inside the parentheses) acts first and then the outer function acts on the result of applying the inner function. For this example, this process can be diagrammed as follows:

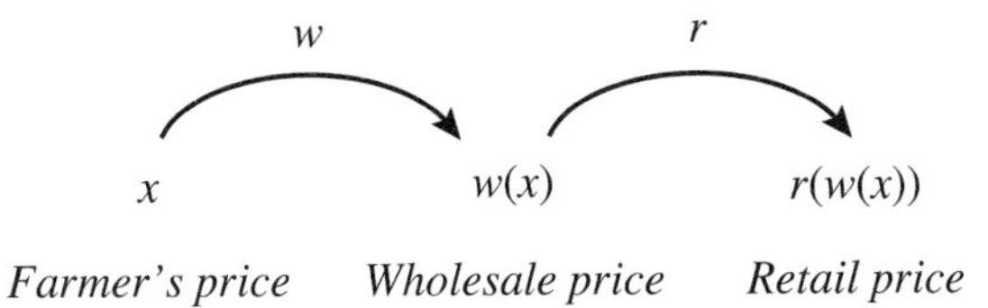

Farmer's price *Wholesale price* *Retail price*

In many cases the two-step composition can be simplified to a single rule. In our example,

$$(1) \quad r(w(x)) = r(x + 0.15) = 2(x + 0.15) = 2x + 0.30$$

is a single rule equivalent to the two-step composition. For example, when $x = 0.35$, $r(w(0.35)) = 2(0.35) + 0.30 = 1.00$, which is the same wholesale price we got before.

If the grocer knows how much the wholesaler is paying the farmer per head, he can compute the new price using equation (1). For example, if the farmer charges \$.75 per head, the retail price is computed as $2(0.75) + 0.30 = \$1.80$.

EXAMPLE 3.5 Compute the following values where the functions r and w are $w(x) = x + 0.15$ and $r(w) = 2w$. Use both the two-step process and the single rule given in equation (1) for the composite function.

a. $r(w(0.65))$ **b.** $r(w(1.04))$

Solution

a. $r(w(0.65)) = r(0.65 + 015) = r(0.80) = 2(0.80) = 1.60$, so the retail price per head is \$1.60. Alternatively, $r(w(0.65)) = 2(0.65) + 0.30 = 1.60$.

b. $r(w(1.04)) = r(1.19) = 2.38$, so the retail price per head is \$2.38. Also, $r(w(1.04)) = 2(1.04) + 0.30 = 2.38$. ■

Since composition is an operation that combines functions, mathematicians have created a special symbol for this operation, $\circ$. The composition of the two functions f and g is denoted $f \circ g$, where $(f \circ g)(x) = f(g(x))$. Note that for a real number c, $(f \circ g)(c)$ is defined only when $g(c)$ is defined ($g(c)$ is a real number) and also $f(g(c))$ is defined. We discuss domains of composite functions in Section 4-C.1. Using the composition symbol, we write our function $r(w(x))$ as $(r \circ w)(x)$.

Note that $f \circ g$ does not mean the product of f and g.
$(f \cdot g)(x) = (fg)(x) = f(x)g(x)$ is the product of f and g.
$(f \circ g)(x) = f(g(x))$ is the composite of f and g.

Although in our example w was a function of x and r was a function of w, we often compose two or more functions of a single variable. To find the simplified expression for the composite function, treat the inner function as a single entity and substitute it for the variable in the outer function.

EXAMPLE 3.6 Let $f(x) = x + 2$ and $g(x) = \dfrac{3}{x-1}$.

a. Find $(f \circ g)(x)$. **b.** Find $(g \circ f)(x)$.

c. Find $(f \circ g)(0)$. **d.** Find $(g \circ f)(0)$.

Solution

a. $(f \circ g)(x) = f(g(x)) = f\left(\dfrac{3}{x-1}\right) = \dfrac{3}{x-1} + 2$

b. $(g \circ f)(x) = g(f(x)) = g(x+2) = \dfrac{3}{(x+2)-1} = \dfrac{3}{x+1}$

We use parts a and b to calculate:

c. $(f \circ g)(0) = \frac{3}{0-1} + 2 = -1$

d. $(g \circ f)(0) = \frac{3}{0+1} = 3$ ■

Exercise 3.6 illustrates that, in general, the order in which two functions are composed makes a difference.

 $f \circ g$ and $g \circ f$ are usually different functions.

EXAMPLE 3.7 Consider the graphs of $y = f(x)$ and $y = g(x)$ shown in Figures 3.6 and 3.7 (which we repeat below). Estimate the values of the composite functions indicated in parts a–d below, if they are defined. If they are undefined, explain why.

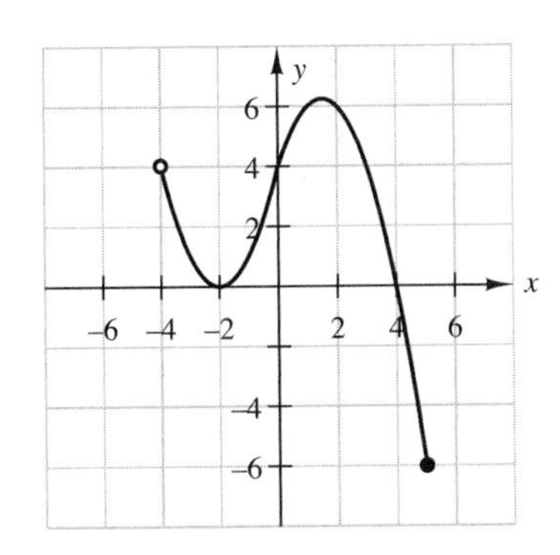

Figure 3.6 $y = f(x)$

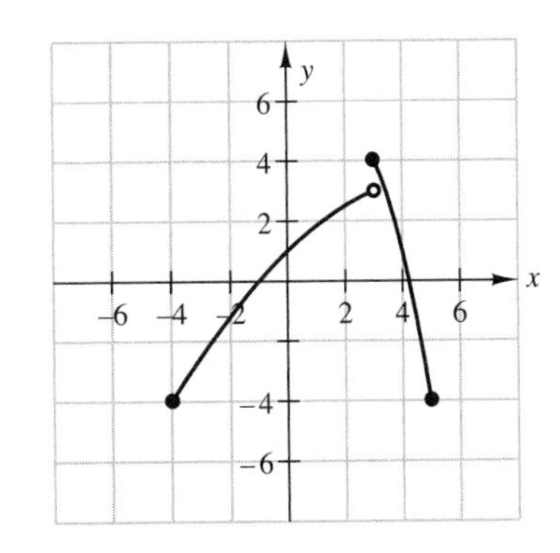

Figure 3.7 $y = g(x)$

a. $(g \circ f)(3)$ **b.** $(f \circ g)(3)$ **c.** $(f \circ f)(5)$ **d.** $(g \circ g)(-1)$ **e.** $(f \circ g)(6)$

Solution

a. Since $f(3) = 4$ and $g(4) = 1$, $(g \circ f)(3) = g(f(3)) = g(4) = 1$.

b. $(f \circ g)(3) = f(g(3)) = f(4) = 0$

c. $(f \circ f)(5) = f(f(5)) = f(-6)$. But -6 is not in the domain of f (there is no point on the graph of f with x-coordinate -6), so $f(-6)$ is not defined. This means that $(f \circ f)(5)$ is not defined.

d. $(g \circ g)(-1) = g(g(-1)) = g(0) = 1$

e. $(f \circ g)(6) = f(g(6))$. But 6 is not in the domain of g, so $g(6)$ is not defined, which means that $(f \circ g)(6)$ is not defined. ■

Exercises 3-A

1. Use the graphs given in Figures 3.6 and 3.7 to estimate the following values if they are defined. If they are undefined, explain why.

a. $(f \cdot g)(-1)$ **b.** $\left(\frac{g}{f}\right)(-1)$ **c.** $(fg)(4)$ **d.** $(g - f)(0)$

2. Use the graphs in Figures 3.6 and 3.7 to estimate the following values, if they are defined. For any that are not defined, explain why.

a. $(f \circ g)(-1)$ **b.** $(g \circ f)(-1)$ **c.** $(f \circ g)(0)$ **d.** $(f \circ g)(5)$ **e.** $(g \circ f)(4)$

3. Let $f(x) = 3x - 4$, $g(x) = 7 - x^2$, and $h(x) = \sqrt{x}$. Find each of the following functions, and give its domain.

a. $f - g$ **b.** $\frac{g}{h}$ **c.** $\frac{f}{g+2}$ **d.** $h \cdot f$ **e.** $\frac{h}{f}$

4. Let f, g, and h be the functions given in Exercise 3. For each of the following composite functions, find its value or explain why it is not defined.

a. $(f \circ g)(-1)$ **b.** $(g \circ f)(-1)$ **c.** $(f \circ h)(4)$ **d.** $(h \circ g)(4)$ **e.** $(g \circ h)(-2)$

5. Let f, g, and h be the functions given in Exercise 3. Give a single rule for each of the following composite functions:

a. $f \circ g$ **b.** $h \circ f$ **c.** $g \circ h$

6. A company makes watches. The material for each watch costs \$3, and weekly fixed cost for the company is \$380. If x watches are produced in a week, the labor cost per watch is $\$16.00 - 0.05x$. The revenue function is $R(x) = 50x - 0.2x^2$.

a. Write the total cost function for a week in which x watches are produced.

b. Write the average cost function per watch for a week in which x watches are produced.

c. Look at the average cost per watch if 10, 100, and 200 watches are produced in a week. Describe in words what these values tell you.

d. Write the profit function for a week in which x watches are produced.

7. Suppose an oil spill is expanding in a circular pattern and the cost of cleaning up the spill depends on the area of the spill.

a. If the radius (in kilometers) of the spill at time t minutes is given by $r(t) = 2\sqrt{t}$ and the cost (in dollars) to clean an area A is given by $C(A) = 100A + 1000$, find the cost if the cleanup starts 25 min after the spill.

b. Find a single rule for the area of the spill as a function of time.

c. Find the area of the spill after 30 min.

d. Find a single rule for the cost of the cleanup as a function of time.

e. Find the cost of the cleanup if it is started 2 hours after the spill.

8. A perfume company has determined that the number of sales x (in thousands of ounces) varies with the amount A spent on advertising (in thousands of dollars) according to the formula $x = 0.1A + 2$.

a. If the company charges \$20 per ounce, its revenue R (in thousands of dollars) is given by $R = 20x$. What is the revenue if \$2,500 is spent on advertising? (Caution: \$2,500 spent on advertising means $A = 2.5$.)

b. Express the revenue R as a function of the amount A spent on advertising.

c. The cost C (in thousands of dollars) for producing and selling the perfume is given by $C = 10 + A + 4x$. What is the cost if \$12,500 is spent on advertising?

d. Express the cost C as a function of the amount A spent on advertising.

e. The profit P (in thousands of dollars) made by the company is found by subtracting the cost from the revenue. What is the profit if the company spends \$15,000 on advertising?

f. Express the profit P as a function of the amount A spent on advertising.

g. What is the profit if \$100,000 is spent on advertising?

3-B ALGEBRAIC SIMPLIFICATION OF FUNCTIONS

To study the behavior of a function, it is often helpful to simplify algebraic expressions. The following examples show some typical cases.

EXAMPLE 3.8 Consider the function $f(x) = \frac{x^3 - 1}{x - 1}$.

a. Give the domain of f.

b. Use a calculator to find the approximate values of f when x takes the following values:

$$x = 0.9, 0.99, 0.999, 0.9999, 1.1, 1.01, 1.001, 1.0001.$$

c. Use the fact that $x^3 - 1 = (x - 1)(x^2 + x + 1)$ to simplify the expression $\frac{x^3 - 1}{x - 1}$.

d. How does the function f differ from the function $g(x) = x^2 + x + 1$? Use a calculator to find the values $g(x)$ when x takes the values listed in part b.

e. Graph (separately) the functions f and g.

Solution

a. Since division by zero is impossible, $x = 1$ must be excluded. So the domain of f is $(-\infty, 1) \cup (1, \infty)$.

b. Although the function is not defined at 1, it is defined for values very close to 1. The required values are given in the following table:

x	0.9	0.99	0.999	0.9999	1	1.0001	1.001	1.01	1.1
$f(x) = \frac{x^3 - 1}{x - 1}$	2.71	2.9701	2.997001	2.9997	undefined	3.0003	3.003001	3.0301	3.31

c. To simplify, we use the factored form of the numerator, $x^3 - 1 = (x - 1)(x^2 + x + 1)$, together with the fact that when $x \neq 1$, $x - 1$ is a nonzero number and $\frac{x - 1}{x - 1} = 1$.

So, for $x \neq 1$, $f(x) = \frac{x^3 - 1}{x - 1} = \frac{(x - 1)(x^2 + x + 1)}{x - 1} = x^2 + x + 1$.

d. The function g is defined for every real number x, whereas f is defined for $x \neq 1$.

x	0.9	0.99	0.999	0.9999	1	1.0001	1.001	1.01	1.1
$g(x) = x^2 + x + 1$	2.71	2.9701	2.997001	2.9997	3	3.0003	3.003001	3.0301	3.31

Note that $f(x) = g(x)$ for $x \neq 1$.

e. The graph of g is a parabola; the graph of f is the same parabola with the point $(1, 3)$ removed, as shown in Figures 3.8 and 3.9.

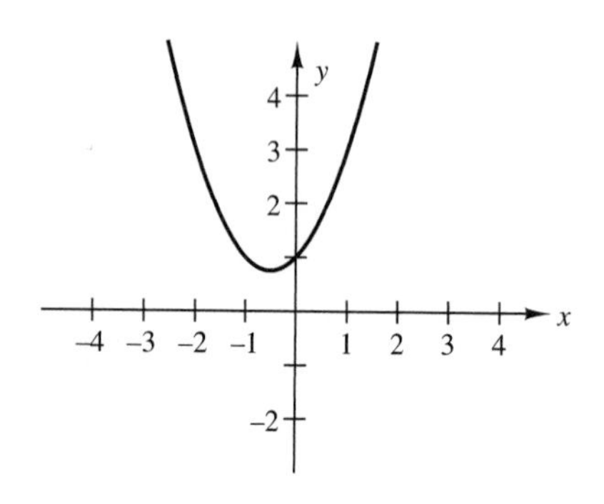

Figure 3.8

Figure 3.9

■

An important concept in calculus is that of the tangent line to the graph of a function. Its slope is obtained from the slopes of *secant lines*. These are lines that contain two points of the graph of the function. The following example illustrates how to obtain an expression for their slopes.

EXAMPLE 3.9

a. Draw the graph of the function $f(x) = x^2$ on the interval [0, 3].

b. On the same graph, draw the secant line that passes through the points (2, 4) and (3, 9) on the graph of f. Give the slope of this line.

c. Let $m(x)$ be the slope of the secant line through the points (2, 4) and (x, x^2) on the graph of f. Find an expression for $m(x)$ and simplify if possible. Is the function m defined for $x = 2$?

d. Draw the secant lines and give the values of $m(x)$ when $x = 1.5$ and $x = 2.5$.

e. Graph the function $y = m(x)$.

Solution

a. and **b.** The graph of f and the secant line through (2, 4) and (3, 9) are shown in Figure 3.10.

The slope of the line through the two points is $\frac{9-4}{3-2} = \frac{5}{1} = 5$.

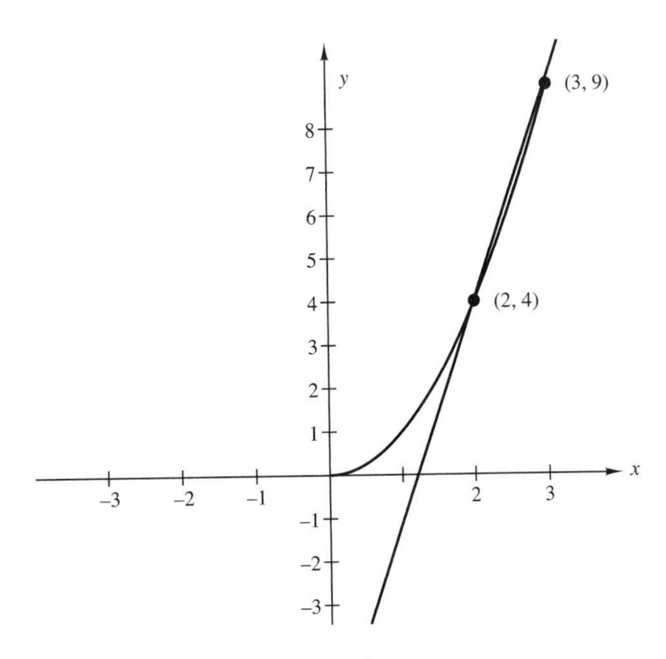

Figure 3.10

c. The slope of the secant line through (2, 4) and (x, x^2) is

$$m(x) = \frac{x^2-4}{x-2} = \frac{\text{difference in } y\text{-coordinates}}{\text{difference in } x\text{-coordinates}}.$$

The function m is not defined at $x = 2$, since the expression for $m(x)$ would then have 0 in the denominator.

When $x \neq 2$, the expression $\frac{x^2-4}{x-2}$ can be simplified by factoring the numerator:

$$\frac{x^2-4}{x-2} = \frac{(x-2)(x+2)}{x-2} = 1 \cdot (x+2) = x+2, \text{ for } x \neq 2.$$

d. If $x = 1.5$, then $y = 2.25$, so the secant line passes through the points (1.5, 2.25) and (2, 4). Its slope is $m(1.5) = 1.5 + 2 = 3.5$. If $x = 2.5$, then $y = 6.25$ and the slope of the secant line is $m(2.5) = 4.5$. The sketches of the two secant lines are left to the reader.

e. The graph of $m(x)$ is the line $y = x + 2$ with the point (2, 4) missing (see Figure 3.11).

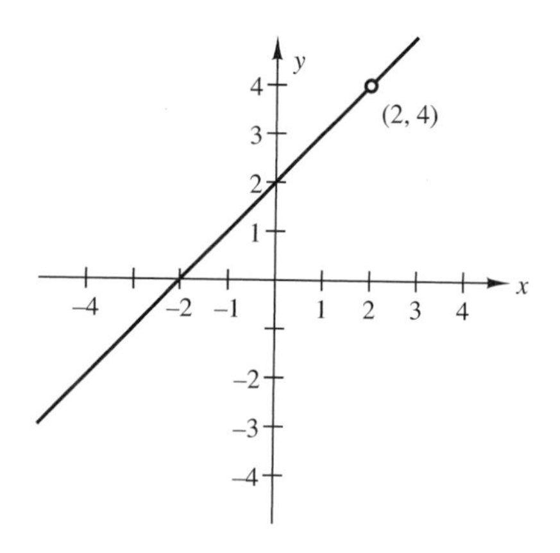

Figure 3.11 $m(x) = \frac{x^2-4}{x-2}$ ■

3-B.1 Quotients of Polynomials (Rational Functions)

The function $f(x) = \frac{x^3-1}{x-1}$ in Example 3.8 is a quotient of the two polynomials, $x^3 - 1$ and $x - 1$. A *polynomial in x* is a function defined by an algebraic expression that involves only the operations of sums and products of real numbers and positive integer powers of x. It can always be written in the form

$$P(x) = a_n x^n + a_{n-1}x^{n-1} + \cdots + a_1 x + a_0$$

where $a_n, a_{n-1}, \ldots, a_1, a_0$ are real numbers and n is a nonnegative integer. A *rational function* is one that can be written as the quotient of two polynomials. Thus $f(x) = \frac{x^3 - 1}{x - 1}$ is a rational function.

When we simplify functions, there are three important facts that form the basis for working with algebraic expressions:

1. Factoring an algebraic expression does not change its value.

2. Multiplying any expression by 1 does not change its value.

3. Adding 0 to any expression does not change its value.

Simplifying quotients often involves using facts 1 and 2. The first step is to factor numerator and denominator to see if there are any common factors. The quotient of two common factors equals 1 (except when the factor equals 0), so we can rewrite the original expression as 1 times a simpler expression. We simplified quotients using this technique in Examples 3.8c and 3.9c.

In some cases we can factor a polynomial directly because we recognize it as the result of a product.

$(x + a)(x + b) = x^2 + (a + b)x + ab$. Use this information when factoring a polynomial of the form $x^2 + cx + d$.

$(x + a)(x - a) = x^2 - a^2$. Use this information when factoring a polynomial of the form $x^2 - c$, when $c > 0$.

EXAMPLE 3.10 Factor each of the following polynomials, if possible:

a. $6x^4 + 3x^2$ **b.** $x^2 + 3x - 10$ **c.** $x^2 - 5$ **d.** $x^2 + 4$

Solution

a. The expression $6x^4 + 3x^2$ has $3x^2$ as a common factor in both terms, so we can write it as $6x^4 + 3x^2 = 3x^2(2x^2 + 1)$.

b. If the polynomial can be factored, it is a product of two factors: $(x + ?)(x + ?)$. In such a case the product of the two unknown numbers should be -10 and the sum of the numbers should be 3. The factors $\pm 2, \pm 5$ of -10 are obvious possibilities; -2 and $+5$ satisfy both conditions, so $x^2 + 3x - 10 = (x - 2)(x + 5)$.

c. The given expression is a difference of squares. The factorization is then $x^2 - 5 = x^2 - \left(\sqrt{5}\right)^2 = \left(x - \sqrt{5}\right)\left(x + \sqrt{5}\right)$.

d. If there are real numbers a and b such that $(x + a)(x + b) = x^2 + 4$, then $a + b = 0$ and $ab = 4$. But then $a = -b$ and also $4 = ab = (-b)b = -b^2$. This is impossible since $-b^2$ is not positive. Therefore $x^2 + 4$ cannot be factored. ■

In general, it is not easy to factor a polynomial into linear factors and in many cases it is impossible. The following theorem is an important tool in factorization. It tells us that if we know a *zero*, or *root*, of the polynomial (a value of x for which the value of the polynomial is 0), then we can factor the polynomial.

Factor Theorem: Let $P(x)$ be a polynomial and r a real number. If $P(r) = 0$, then $x - r$ is a factor of $P(x)$. Also, if $x - r$ is a factor of $P(x)$, then $P(r) = 0$.

One polynomial that has an easily identified root is $P(x) = x^n - a^n$, where a represents some fixed constant. Clearly, $P(a) = 0$ since $P(a) = a^n - a^n = 0$. Thus, by the Factor Theorem, $x - a$ is a factor of $P(x)$.

It can be verified that for any $n > 0$:

$$x^n - a^n = (x - a)(x^{n-1} + ax^{n-2} + a^2x^{n-3} + \cdots + a^{n-2}x + a^{n-1}).$$

When the right side of this equation is multiplied out, all terms except x^n and $-a^n$ cancel. For example, when $x = 4$,

$$(x-a)(x^3+ax^2+a^2x+a^3) = x^4+ax^3+a^2x^2+a^3x-ax^3-a^2x^2-a^3x-a^4 = x^4-a^4.$$

The factorization of $x^n - a^n$ is a key step in finding the derivative of $f(x) = x^n$ at $x = a$.

Our primary concern when simplifying a quotient of polynomials is to identify common factors of the form $x - a$ in the numerator and denominator. For example, consider the quotient $\frac{x^4-2x+1}{x^2+3x-4}$. The denominator factors: $x^2 + 3x - 4 = (x - 1)(x + 4)$. We ask: Is $x - 1$ or $x + 4$ a factor of the numerator? We evaluate the numerator $x^4 - 2x + 1$ at $x = 1$: $1^4 - (2)(1) + 1 = 0$. The Factor Theorem tells us that $x - 1$ is a factor of the numerator. We also evaluate the numerator at $x = -4$: $(-4)^4 - (2)(-4) + 1 = 73$, so $x + 4$ is not a factor of the numerator. To factor $x - 1$ out of the numerator, we can use long division. (Another technique called synthetic division could also be used in this case.)

$$
\begin{array}{r}
x^3 + x^2 + x - 1 \\
x-1 \overline{) x^4 \qquad\qquad - 2x + 1} \\
\underline{x^4 - x^3 \qquad\qquad\qquad} \\
x^3 \qquad - 2x + 1 \\
\underline{x^3 - x^2 \qquad\qquad} \\
x^2 - 2x + 1 \\
\underline{x^2 - x \quad\;\;} \\
-x + 1 \\
\underline{-x + 1} \\
0
\end{array}
$$

So, $(x^4 - 2x + 1) = (x - 1)(x^3 + x^2 + x - 1)$, and we can write

$$\frac{x^4 - 2x + 1}{x^2 + 3x - 4} = \frac{(x-1)(x^3 + x^2 + x - 1)}{(x-1)(x+4)}.$$

Since $\frac{x-1}{x-1} = 1$, the quotient in this factored form can be simplified to $\frac{x^3 + x^2 + x - 1}{x + 4}$, for $x \neq 1$. No further simplification is possible. See the Appendix for additional discussion on when common terms may be cancelled. We discuss other ways to find roots of polynomials in Chapter 4.

EXAMPLE 3.11 Consider the rational function $h(x) = \frac{x^3 - x^2 - 3x + 2}{3x^2 - 6x}$.

a. Give the domain of h.

b. Find any common factors of the numerator and denominator, and write $h(x)$ in simplified form.

Solution

a. The function is not defined at those values of x for which the denominator equals 0. The denominator can be factored as $3x(x - 2)$, which equals 0 when $x = 0$ and when $x = 2$. The domain is then $(-\infty, 0) \cup (0, 2) \cup (2, \infty)$.

b. We first see if $x = 2$ is a root of the numerator by evaluating the numerator at $x = 2$: $2^3 - 2^2 - 6 + 2 = 0$. Since 2 is a root of the numerator, the Factor Theorem tells us that $x - 2$ is a factor. Dividing $x - 2$ into $x^3 - x^2 - 3x + 2$, we find

$$x^3 - x^2 - 3x + 2 = (x - 2)(x^2 + x - 1).$$

Now we are ready to simplify $h(x)$:

$$h(x) = \frac{x^3 - x^2 - 3x + 2}{3x^2 - 6x} = \frac{(x-2)(x^2+x-1)}{3x(x-2)}$$

$$= \frac{(x-2)}{(x-2)} \cdot \frac{(x^2+x-1)}{3x} = \frac{x^2+x-1}{3x}$$

Since $x = 2$ is not in the domain of h but it is in the domain of the simplified expression, we write:

$$h(x) = \frac{x^2+x-1}{3x} \text{ if } x \neq 2.$$

■

3-B.2 Quotients with Radicals

In the previous section we saw that under certain conditions we could simplify a quotient of polynomials by factoring the numerator and denominator. When the numerator or denominator involves a radical, we can change the form of the quotient by *rationalizing* the numerator or denominator—that is, writing the quotient in an equivalent form to remove the radical.

☞ Since $\left(\sqrt{a}+\sqrt{b}\right)\left(\sqrt{a}-\sqrt{b}\right) = a - b$,

$\left(\sqrt{a}-\sqrt{b}\right)$ is rationalized by multiplying by 1 in the form $\frac{\left(\sqrt{a}+\sqrt{b}\right)}{\left(\sqrt{a}+\sqrt{b}\right)}$;

$\left(\sqrt{a}+\sqrt{b}\right)$ is rationalized by multiplying by 1 in the form $\frac{\left(\sqrt{a}-\sqrt{b}\right)}{\left(\sqrt{a}-\sqrt{b}\right)}$.

Note that to rationalize a difference we use a sum, and to rationalize a sum we use a difference.

EXAMPLE 3.12 Let $f(x)=\sqrt{x}$.

a. Give the domain of f.

b. Write an expression for $m(x)$, the slope of the secant line to the graph of $y = f(x)$ through the points $\left(x, \sqrt{x}\right)$ and $\left(9, \sqrt{9}\right) = (9, 3)$.

c. Give the domain of the function m.

d. Write $m(x)$ as a quotient containing no radicals in the numerator, and simplify $m(x)$.

Solution

a. To find the domain of f, we use the fact that the square root is defined for only nonnegative numbers. The domain of f is all real numbers $x \geq 0$.

b. To write the slope of the secant line through the points $(x, \sqrt{x})$ and (9, 3), we use the formula for the slope of the line through two points:

$$m(x) = \text{slope} = \frac{\text{difference in } y\text{-coordinates}}{\text{difference in } x\text{-coordinates}}.$$

Thus, $m(x) = \frac{\sqrt{x}-3}{x-9}$. See Figure 3.12 for the graph of $y = f(x)$ and a secant line.

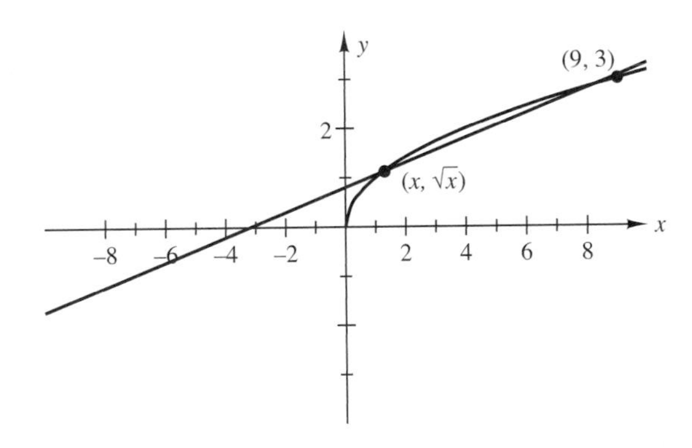

Figure 3.12 $f(x) = \sqrt{x}$ and a secant line

c. To find the domain of m, we note that the expression for $m(x)$ involves two operations that are not possible for all real numbers. The square root function is not defined for negative numbers and a quotient is not defined when a denominator is 0. The domain of m then includes every nonnegative number other than $x = 9$. In interval notation, the domain is described as $[0, 9) \cup (9, \infty)$.

d. To rationalize $\sqrt{x} - 3$, we multiply $m(x)$ by $1 = \frac{\sqrt{x}+3}{\sqrt{x}+3}$. Then,

$$m(x) = \frac{\sqrt{x}-3}{x-9} \cdot \frac{\sqrt{x}+3}{\sqrt{x}+3} = \frac{(\sqrt{x}-3)(\sqrt{x}+3)}{(x-9)(\sqrt{x}+3)} = \frac{(x-9)}{(x-9)(\sqrt{x}+3)} = \frac{1}{\sqrt{x}+3}$$

with this expression for $m(x)$ valid only when $x \neq 9$ and $x \geq 0$. ■

When rationalizing the numerator or the denominator of a quotient, **never** square the whole quotient, since this transforms it into a non-equivalent quotient. For example,

$$\frac{x}{\sqrt{x}} = \frac{x}{\sqrt{x}} \cdot \frac{\sqrt{x}}{\sqrt{x}} = \frac{x\sqrt{x}}{x} = \sqrt{x}\text{, which is **not** equal to}$$

$$\left(\frac{x}{\sqrt{x}}\right)^2 = \frac{x^2}{\left(\sqrt{x}\right)^2} = \frac{x^2}{x} = x.$$

EXAMPLE 3.13 Simplify each of the following quotients by first rationalizing the numerator.

a. $\frac{2-\sqrt{x}}{x-4}$ **b.** $\frac{\sqrt{4+h}-2}{h}$ **c.** $\frac{\sqrt{x}-\sqrt{3}}{x-3}$

Solution

a.

$$\frac{2-\sqrt{x}}{x-4} = \frac{2-\sqrt{x}}{x-4} \cdot \left(\frac{2+\sqrt{x}}{2+\sqrt{x}}\right) = \frac{\left(2-\sqrt{x}\right)\left(2+\sqrt{x}\right)}{(x-4)\left(2+\sqrt{x}\right)} =$$

$$\frac{4-x}{(x-4)\left(2+\sqrt{x}\right)} = \frac{-(x-4)}{(x-4)\left(2+\sqrt{x}\right)} = \frac{-1}{2+\sqrt{x}}, \quad x \neq 4$$

$a - b = -(b - a)$, since $-(b - a) = -b + a = a - b$.

b.

$$\frac{\sqrt{4+h}-2}{h} = \frac{\left(\sqrt{4+h}-2\right)}{h} \cdot \frac{\left(\sqrt{4+h}+2\right)}{\left(\sqrt{4+h}+2\right)} =$$

$$\frac{4+h-4}{h\left(\sqrt{4+h}+2\right)} = \frac{h}{h\left(\sqrt{4+h}+2\right)} = \frac{1}{\sqrt{4+h}+2}, \quad h \neq 0$$

c.

$$\frac{\sqrt{x}-\sqrt{3}}{x-3} = \frac{\left(\sqrt{x}-\sqrt{3}\right)}{x-3} \cdot \frac{\left(\sqrt{x}+\sqrt{3}\right)}{\left(\sqrt{x}+\sqrt{3}\right)} =$$

$$\frac{x-3}{(x-3)\left(\sqrt{x}+\sqrt{3}\right)} = \frac{1}{\sqrt{x}+\sqrt{3}}, \quad x \neq 3$$

■

3-B.3 Complex Fractions

You will sometimes need to simplify a quotient that contains a fraction in the numerator or in the denominator. To simplify in these cases, you may need to find a common denominator, as well as use basic algebra rules for fractions:

1. $\dfrac{\left(\frac{a}{b}\right)}{c} = \dfrac{a}{bc}$

2. $\dfrac{\left(\frac{a}{b}\right)}{\left(\frac{c}{d}\right)} = \dfrac{a}{b} \cdot \dfrac{d}{c} = \dfrac{ad}{bc}$

3. $\dfrac{a}{b} + \dfrac{c}{d} = \dfrac{ad+bc}{bd}$

EXAMPLE 3.14 In a laboratory experiment, a population of 1500 bacteria is placed in a cultured medium and grows in such a way that the number of bacteria after t hours is

$$N(t) = 1500\left(1 + \frac{2t}{30 + t^2}\right).$$

The difference $N(t) - N(0)$ indicates the increase in the number of bacteria during the first t hours. The ratio of that difference to the time elapsed,

$$\frac{N(t) - N(0)}{t}$$

represents the average rate of growth of the bacteria during the first t hours (in number of bacteria per hour).

a. Simplify this quotient.

b. Find the average rate of growth of the bacteria during the first 10 hours.

Solution

a.
$$\frac{N(t)-N(0)}{t} = \frac{1500\left(1+\frac{2t}{30+t^2}\right)-1500}{t} =$$

$$\frac{1500 + \frac{3000t}{30+t^2} - 1500}{t} = \frac{\frac{3000t}{30+t^2}}{t} = \frac{3000t}{\left(30+t^2\right)t} = \frac{3000}{30+t^2}$$

b. We use the simplified form of the quotient $\dfrac{N(t)-N(0)}{t}$ we found in part a, with $t = 10$:

$$\frac{3000}{30+10^2} = \frac{3000}{30+100} = \frac{3000}{130} \approx 23.08 \text{ bacteria per hour.}$$ ■

EXAMPLE 3.15 Simplify each of the following expressions:

a. $\dfrac{\frac{1}{x} - \frac{1}{3}}{x-3}$ **b.** $\dfrac{(h+5)^{-1} - 5^{-1}}{h}$

Solution In both expressions, we find a common denominator for the difference of fractions in the numerator. In solving parts a and b, we show two different techniques of simplification.

a. To "clear" the denominators x and 3, we multiply by $\frac{3x}{3x}$:

$$\frac{\frac{1}{x}-\frac{1}{3}}{x-3} = \frac{3x\left(\frac{1}{x}-\frac{1}{3}\right)}{3x(x-3)} = \frac{3-x}{3x(x-3)} = \frac{-(x-3)}{3x(x-3)} = \frac{-1}{3x}, \quad x \neq 3.$$

b. First we write the terms with negative exponents in fraction form and then use rule 3 for fractions:

$$\frac{(h+5)^{-1}-5^{-1}}{h} = \frac{\frac{1}{h+5}-\frac{1}{5}}{h} = \frac{\frac{5}{5(h+5)}-\frac{(h+5)}{5(h+5)}}{h} = \frac{\frac{5-(h+5)}{5(h+5)}}{h} =$$

$$\frac{5-h-5}{5(h+5)h} = \frac{-h}{5(h+5)h} = \frac{-1}{5(h+5)}, \quad h \neq 0.$$ ■

Always keep parentheses when multiplying by a sum such as $h + 5$.

3-B.4 Quotients with Absolute Values

In simplifying quotients that involve an absolute value, we need to consider two cases. Recall from Chapter 2 that, for **any** expression w,

$$|w| = -w \text{ if } w < 0, \text{ and } |w| = w \text{ if } w \geq 0.$$

Also, $|w| = |-w|$ for all w.

EXAMPLE 3.16 Give the domain and a simplified expression for each of the following functions f and g. Graph each function.

a. $f(x) = \dfrac{|x-3|}{x-3}$ **b.** $g(x) = \dfrac{|-2x+2|}{x-1}$

Solution

a. Since the denominator cannot equal 0, the domain of f contains all numbers except 3. The domain of f is $(-\infty, 3) \cup (3, \infty)$. To find a simplified expression for f, we use the definition of absolute value to write f as a piecewise expression as follows.

If $x - 3 < 0$, then $|x - 3| = -(x - 3)$, so

$$f(x) = \frac{|x-3|}{x-3} = \frac{-(x-3)}{x-3} = -1, \text{ for } x - 3 < 0.$$

If $x - 3 = 0$, then $f(x)$ is undefined.

If $x - 3 > 0$, then $|x - 3| = x - 3$, so

$$f(x) = \frac{|x-3|}{x-3} = \frac{x-3}{x-3} = 1, \text{ for } x - 3 > 0.$$

Thus f can be defined as the step function:

$$f(x) = \begin{cases} -1 & \text{if } x - 3 < 0 \\ 1 & \text{if } x - 3 > 0 \end{cases}$$

or, equivalently, $f(x) = \begin{cases} -1 & \text{if } x < 3 \\ 1 & \text{if } x > 3. \end{cases}$

See Figure 3.13 for the graph of f.

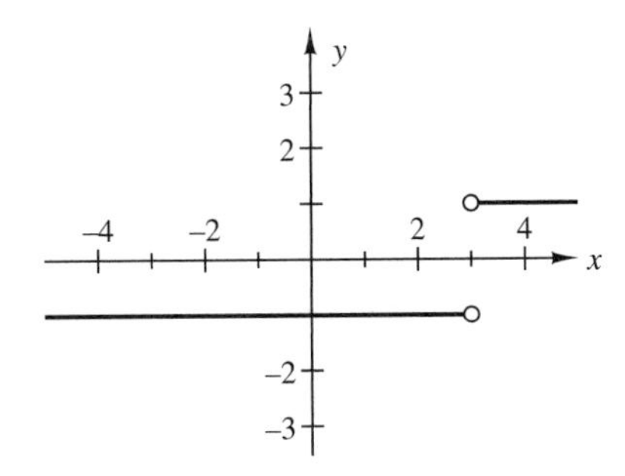

Figure 3.13 $f(x) = \frac{|x-3|}{x-3}$

b. The domain of g is the set of all numbers x such that $x - 1 \neq 0$, so the domain of g is $(-\infty, 1) \cup (1, \infty)$.

If $x < 1$, then $-2x + 2 > 0$ and $\frac{|-2x+2|}{x-1} = \frac{-2x+2}{x-1} = \frac{-2(x-1)}{x-1} = -2.$

If $x > 1$, then $-2x + 2 < 0$ and $\frac{|-2x+2|}{x-1} = \frac{-(-2x+2)}{x-1} = \frac{2x-2}{x-1} = \frac{2(x-1)}{x-1} = 2.$

The function g can be defined as $g(x) = \begin{cases} -2 & \text{if } x < 1 \\ 2 & \text{if } x > 1. \end{cases}$

See Figure 3.14 for the graph of g.

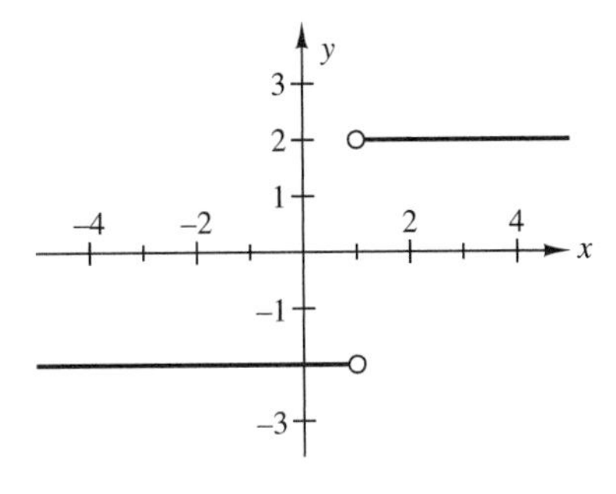

Figure 3.14 $g(x) = \frac{|-2x+2|}{x-1}$

■

Exercises 3-B

1. Simplify the following functions and give the domain of each function.

a. $f(x) = \dfrac{x^2-x-6}{x^2-6x+9}$ **b**. $g(x) = \dfrac{\frac{2}{x+3}-\frac{2}{3}}{x}$

c. $h(x) = \dfrac{\frac{x}{4}-\frac{1}{2}}{x-2}$ **d**. $v(x) = \dfrac{|-3x+2|}{3x-2}$ **e**. $w(x) = \dfrac{x+3}{x^2-9}$

2. **a**. Write an expression $m(x)$ that gives the slope of the secant line through the points $\left(1, \frac{1}{3}\right)$ and $\left(x, \frac{1}{x+2}\right)$ on the graph of the function $f(x) = \frac{1}{x+2}$, for $x > -2$.

b. Simplify the expression $m(x)$ found in part a.

c. Sketch the graph of the function $f(x) = \frac{1}{x+2}$ for $x > -2$, and draw the secant lines through points $\left(1, \frac{1}{3}\right)$ and $\left(x, \frac{1}{x+2}\right)$ for $x = 1.5$ and $x = -0.5$.

3. **a**. Write an expression $n(x)$ that gives the slope of the secant line through the points $\left(2, \frac{1}{4}\right)$ and $\left(x, \frac{1}{x+2}\right)$ on the graph of the function $f(x) = \frac{1}{x+2}$, for $x > -2$.

b. Simplify the expression $n(x)$ found in part a.

4. **a**. Complete the following table using $m(x)$ from Exercise 2 and $n(x)$ from Exercise 3.

x	-1	0	3	5
$m(x)$				
$n(x)$				

b. Explain the values in the table. Are your answers reasonable from the graph?

5. Rationalize the numerator in each of the following expressions and simplify:

a. $\dfrac{\sqrt{x}-\sqrt{2}}{x-2}$ **b**. $\dfrac{\frac{1}{\sqrt{x}}-\frac{1}{5}}{x-25}$ **c**. $\dfrac{\sqrt{3-h}-\sqrt{3}}{h}$

d. $\sqrt{x+1}-\sqrt{x}$ (Note that any expression w can be written as $\frac{w}{1}$, so if w has a square root, we can rationalize the numerator of w by writing it as $\frac{w}{1}$.)

6. Sketch the graph of $y = \sqrt{x}$. Locate the point (1, 1) on the graph, and sketch secant lines through the points (1, 1) and $\left(x, \sqrt{x}\right)$ for several different choices of x. What can you say about the slope of these secant lines?

7. Rationalize the denominator in each of the following expressions:

a. $\frac{x-9}{3-\sqrt{x}}$ **b.** $\frac{5-x}{\sqrt{5}-\sqrt{x}}$ **c.** $\frac{x-4}{-2+\sqrt{x}}$

8. Give the domain and a simplified expression for each of the following functions. Graph each function.

a. $f(x)=\frac{x^2-7x+10}{x-2}$ **b.** $g(x)=\frac{|2-x|}{x-2}$ **c.** $h(x)=\frac{x-2}{x^2-7x+10}$

9. Let $Q(x) = x^4 - 81$. Factor $Q(x)$ into a factor of the form $x - a$ times a polynomial of degree 3.

3-C INEQUALITIES

The limit of a function f at a number c tells about the behavior of the values $f(x)$ when x is "very close to c" but $x \neq c$. What does being very close to c mean? Is the number 1.1 very close to 1? If we use this scale

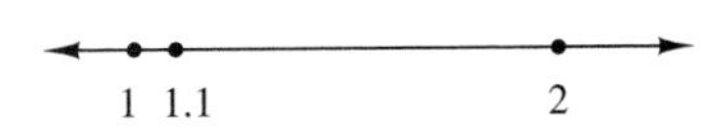

then 1.1 appears to be very close to 1. If instead we work on the following scale

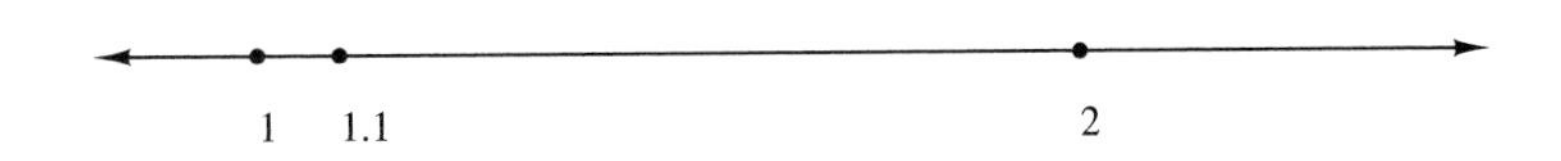

then 1.1 is not so close any more. "Close" is a relative term.

EXAMPLE 3.17 Here is a table of values of a function $f(x)$ for x "close to" 1.

x	$f(x)$
1.01	2.02
1.001	2.002
1.0001	2.0002
1.00001	2.00002
0.99999	1.99998
0.9999	1.9998

a. What would you guess is the value $f(1)$?

b. Can you describe the behavior of the graph of $y = f(x)$ over the interval [0.99999, 1.00001]? Sketch one possible graph for $f(x)$.

Solution

a. We probably would guess that the value of $f(1)$ is 2. That is a good guess, supported by the pattern of values in the table, but any other real number or even "undefined" could also be correct. The table gives no information about the behavior of $f(x)$ over the whole interval $0.99999 < x < 1.00001$. (And an infinite number of values of x lie in that interval.)

b. There are infinitely many possibilities. Figures 3.15a and 3.15b show two.

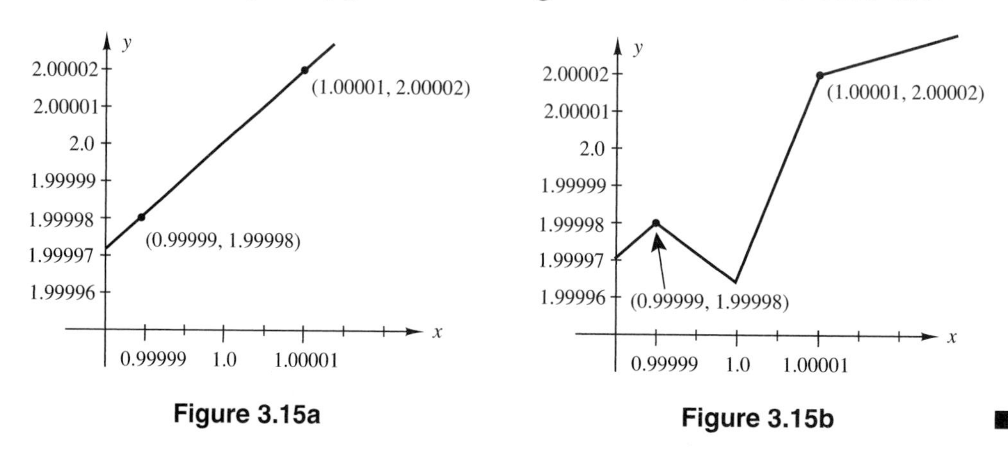

Figure 3.15a **Figure 3.15b** ■

From the table of values in Example 3.17, we are tempted to say that the values of the function f are very close to 2 when x is very close to 1. This statement would be true when f is the function whose graph is in Figure 3.15a. But the values of the function whose graph is in Figure 3.15b are not close to 2 when x is very close to 1. (What number are they close to?) To avoid such ambiguities, we use intervals and inequalities to make more precise what "very close to" means. To say that a certain property is satisfied when x is "very close to" c, we say that the property must be true for **all** values of x within a certain small distance d of c. The set of such numbers x is an interval of length $2d$, with c as its midpoint, and is represented in Figure 3.16.

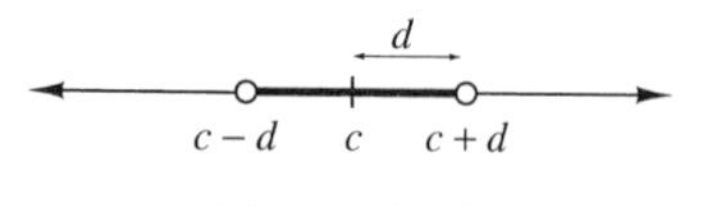

Figure 3.16

 The set of real numbers that are within a distance d of the number $x = c$ form the open interval $(c - d, c + d) = \{\, x \mid c - d < x < c + d \}$.

(Note: Many calculus texts use the Greek letter δ rather than d when describing this set. Both δ and d suggest the word "distance.") Absolute value inequalities are often used to describe the set of all x within a distance d of c and for other sets used in the study of limits. We will discuss them in Section 3-C.2.

3-C.1 Linear Inequalities

To solve an inequality for x means to find all real numbers x that satisfy the inequality. The set of all solutions of an inequality is usually an interval or a union of intervals.

Every inequality can be written in two different ways:
$a < c$ is the same as $c > a$;
$b \leq d$ is the same as $d \geq b$.

We will first solve *linear inequalities*. These are inequalities that can be written in the form $f(x) < g(x)$ or $f(x) \leq g(x)$, where f and g are linear functions. For example,

$3x - 1 < 0.5$, which is the same as $0.5 > 3x - 1$, and

$x + 2 \leq 4x + 1$, which can also be written as $4x + 1 \geq x + 2$.

To solve linear inequalities, we use the same principle as when solving linear equations: Apply the same operations to both sides of the inequality in order to obtain x by itself on one side of the inequality. The following basic algebraic rules are essential. In all expressions, a, b, and c represent real numbers.

1. When we add or subtract the same real number on both sides of an inequality, the inequality remains true.

If $a < b$, then $a + c < b + c$. If $a < b$, then $a - c < b - c$.

If $a > b$, then $a + c > b + c$. If $a > b$, then $a - c > b - c$.

2. If we multiply or divide both sides of an inequality by the same **positive** real number, then the inequality remains true.

If $a < b$ and $c > 0$, then $ac < bc$. If $a < b$ and $c > 0$, then $\frac{a}{c} < \frac{b}{c}$.

If $a > b$ and $c > 0$, then $ac > bc$. If $a > b$ and $c > 0$, then $\frac{a}{c} > \frac{b}{c}$.

3. Multiplying or dividing both sides of an inequality by a **negative** number **reverses** the inequality.

If $a < b$ and $c < 0$, then $ac > bc$. If $a < b$ and $c < 0$, then $\frac{a}{c} > \frac{b}{c}$.

If $a > b$ and $c < 0$, then $ac < bc$. If $a > b$ and $c < 0$, then $\frac{a}{c} < \frac{b}{c}$.

4. If one number is less than a second, and the second is less than a third, then the first is less than the third. (This is called the *transitive property.*)

$$\text{If } a < b \text{ and } b < c, \text{ then } a < c.$$

EXAMPLE 3.18 Carry out each of the following operations to obtain a new equivalent inequality. Give the solution to the inequality in interval notation.

a. Subtract $\frac{5}{2}$ from both sides of the inequality $x + \frac{5}{2} > 6$.

b. Divide the inequality $-4x < 3$ by -4.

c. Multiply the inequality $\frac{x}{3} < -5$ by 3.

Solution

a. We subtract $\frac{5}{2}$ from both sides of the given inequality to obtain $\left(x + \frac{5}{2}\right) - \frac{5}{2} > 6 - \frac{5}{2}$; thus $x > \frac{7}{2}$. In interval notation, the solution is $\left(\frac{7}{2}, \infty\right)$.

b. We divide the inequality $-4x < 3$ by -4 to get $\frac{-4x}{-4} > \frac{3}{-4}$, so $x > -\frac{3}{4}$. In interval notation, the solution is $\left(-\frac{3}{4}, \infty\right)$.

c. We multiply $\frac{x}{3} < -5$ by 3 to get $3 \cdot \frac{x}{3} < (3)(-5)$. The solution is $x < -15$, which is the interval $(-\infty, -15)$. ■

Properties 1–4 for inequalities are also satisfied when the strict inequalities $<$ or $>$ are replaced consistently by $\leq$ or $\geq$, respectively.

EXAMPLE 3.19 Solve the inequality $x - 5 \leq \frac{1}{2} + 3x$, and write the solution using interval notation. Graph the lines that correspond to each side of the inequality, and check your solution graphically.

Solution We multiply both sides by 2 to clear denominators and obtain $2x - 10 \leq 1 + 6x$. Then subtract $6x$ from both sides to get $-4x - 10 \leq 1$; add 10 to both sides to get $-4x \leq 11$. Finally divide by -4 to obtain $x \geq -\frac{11}{4}$. The solution is the interval $\left[-\frac{11}{4}, \infty\right)$.

The solutions to the inequality $x - 5 \leq \frac{1}{2} + 3x$ can be interpreted graphically as the numbers x for which the y-coordinate on the graph of the line $y = x - 5$ is less than or equal to the y-coordinate on the graph of the line $y = \frac{1}{2} + 3x$. This happens when the points on the line $y = x - 5$ (bold line) are equal to or below the points on the line $y = \frac{1}{2} + 3x$. See Figure 3.17 for a graph of the two lines.

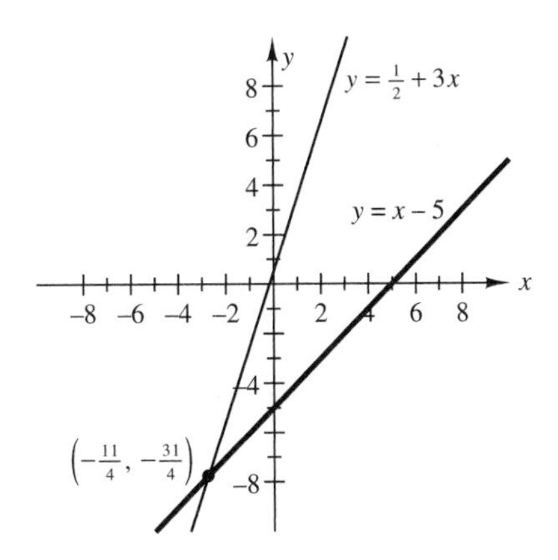

Figure 3.17

The two lines intersect at the point $\left(-\frac{11}{4}, -\frac{31}{4}\right)$. Figure 3.17 shows that points on the line $y = x - 5$ are indeed on or below the line $y = \frac{1}{2} + 3x$ when $x \geq -\frac{11}{4}$. ■

Two separate inequalities $f(x) < g(x)$ and $g(x) < h(x)$ can be combined into a double inequality, $f(x) < g(x) < h(x)$. The rules for inequalities may also be used to solve a double inequality in the same way as for a single inequality.

Both inequalities in a double inequality must have the same direction: $a < b$ and $b < c$ can be written $a < b < c$. The two inequalities $a < b$ and $b > c$ cannot be combined into a double inequality.

The property of combining two inequalities into a double inequality still holds if one or both of the $<$ symbols is replaced by $\leq$.

EXAMPLE 3.20 Solve the double inequality, $-3 \leq 1 - 2x \leq 4$, and write the solution using interval notation.

Solution We actually have two inequalities, $-3 \leq 1 - 2x$ and $1 - 2x \leq 4$, but we can solve both simultaneously in the given form $-3 \leq 1 - 2x \leq 4$. We want to obtain x by itself in the middle of a double inequality. Subtract 1 from the left side, the middle, and the right side of the inequality, $-3 - 1 \leq (1 - 2x) - 1 \leq 4 - 1$, to obtain $-4 \leq -2x \leq 3$. Divide throughout by -2, and reverse the $\leq$ signs: $\frac{-4}{-2} \geq \frac{-2x}{-2} \geq \frac{3}{-2}$. Thus $2 \geq x \geq -\frac{3}{2}$, so the solution is the interval $\left[-\frac{3}{2}, 2\right]$. ■

Exercises 3-C.1

1. Solve the following inequalities. Write solutions using interval notation.

a. $\frac{x}{4} - 5 > x - 1$ **b.** $9x + 1 < 2 - 5x$ **c.** $-4.1 \le 3x + 1 \le 4.1$

d. $0 < 6x - 2.3 < 1$ **e.** $1 < 3 - x < 7$

2. Consider the inequality $2x - 3 > 5x + 7$.

a. Solve the inequality and write the solution using interval notation.

b. Graph the lines that correspond to each side of the inequality and check your solution graphically.

3. Repeat Exercise 2 using the inequality $\frac{3x+1}{2} < 6x - 5$.

4. Solve each of the following double inequalities.

a. $1 < 2x + 5 < 10$ **b.** $6 \ge 7 - 4x \ge 1$

5. Explain why the following double inequality has no solution: $8 < 3x + 5 < -7$.

3-C.2 Absolute Value: Equations and Inequalities

Recall that the distance between b and c on a coordinate line is given by the absolute value of their difference $|b - c| = |c - b|$. In particular, $|b|$ is the distance between b and 0.

To solve equations and inequalities that involve absolute value, we use the definition of absolute value and the following properties:

1. $|ab| = |a||b|$ **2.** $\left|\frac{a}{b}\right| = \frac{|a|}{|b|}$

3. $|-a| = |a|$ **4.** $|a + b| \le |a| + |b|$

EXAMPLE 3.21 Solve each equation and graph its solution on a number line:

a. $|x| = 1.2$ **b.** $|3x + 1| = 6$ **c.** $\left|2 - \frac{x}{3}\right| = 0$

d. $|5x - 7| = -2$ **e.** $|x + 5| = 3$

Solution

a. There are exactly two numbers whose distances from 0 equal 1.2. They are 1.2 and −1.2.

−1.2 0 1.2

b. We give two different methods of solution.

Method 1. Use the definition of absolute value: $|3x + 1| = 6$ is true when $3x + 1 = 6$ or $3x + 1 = -6$. Solve for x in each equation.

$$3x + 1 = 6 \text{ so } 3x = 5. \text{ Thus, } x = \frac{5}{3}.$$

$$3x + 1 = -6 \text{ so } 3x = -7. \text{ Thus, } x = -\frac{7}{3}.$$

The solution is the set of two numbers $\left\{-\frac{7}{3}, \frac{5}{3}\right\}$.

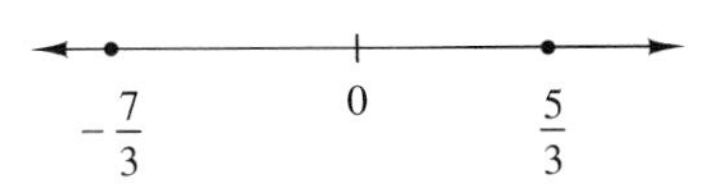

Method 2. $|3x + 1| = 6$, so $\left|3\left(x+\frac{1}{3}\right)\right| = 6$. Thus $3\left|x+\frac{1}{3}\right| = 6$, so $\left|x+\frac{1}{3}\right| = 2$. We can also write this as $\left|x-\left(-\frac{1}{3}\right)\right| = 2$, which says that the distance from x to $-\frac{1}{3}$ is 2. The solutions are then those numbers that are two units away from $-\frac{1}{3}$. These are $-\frac{1}{3} + 2 = \frac{5}{3}$ and $-\frac{1}{3} - 2 = -\frac{7}{3}$.

c. The only number whose absolute value is 0 is 0 itself. So $\left|2-\frac{x}{3}\right| = 0$ is true when $2 - \frac{x}{3} = 0$. Thus $2 = \frac{x}{3}$ so $x = 6$.

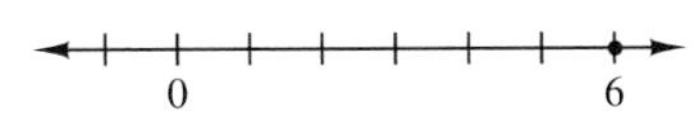

d. Since $|w| \geq 0$ for every real number w, there is no solution to the given equation. The solution set is the *empty set.*

e. $|x + 5|$ is interpreted as the distance between x and −5. Since $|x + 5| = 3$, we want values of x such that the distance between x and −5 is 3. Thus $x = -5 + 3$ or $x = -5 - 3$. The two solutions are $x = -2$ and $x = -8$.

−8 −5 −2 0

■

As we mentioned earlier, absolute value inequalities are very useful in describing sets that are used in the study of limits. The open interval $(c - d, c + d)$ is defined as the set of numbers that satisfy the double inequality $c - d < x < c + d$. (This means that $c - d < x$ **and** $x < c + d$.) This interval is also the set of all numbers x whose distance to c is less than d. Since the distance from x to c is given by $|x - c|$, this is the set $\{x \mid |x - c| < d\}$. We have two different descriptions of the same set. For this reason, we say that $|x - c| < d$ is *equivalent* to $c - d < x < c + d$.

EXAMPLE 3.22 Describe the interval given in Figure 3.18 as follows:

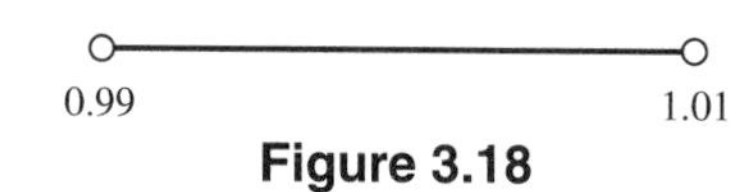

Figure 3.18

a. In a sentence that gives positions of points relative to the number 1

b. Using an absolute value inequality

c. Using a double inequality

d. Using interval notation

Solution

a. The distance from 1 to the left endpoint is $1 - 0.99 = 0.01$ and to the right endpoint is $1.01 - 1 = 0.01$. The set shown in the picture is then the set of all numbers x whose distance to 1 is less than 0.01.

b. $\{x \mid |x - 1| < 0.01\}$ **c**. $\{x \mid 0.99 < x < 1.01\}$ **d**. $(0.99, 1.01)$ ■

If w is any expression involving x, one way to solve the inequality $|w| < d$ is to interpret it as "the distance from w to 0 is less than d." This says that $|w| < d$ is equivalent to $-d < w < d$. This form can be useful in solving inequalities.

EXAMPLE 3.23 Solve each inequality. Graph the solutions on a number line.

a. $|-2x + 5| < 1$ **b**. $|4x - 3| \leq 0.1$

Solution

a. The inequality $|-2x + 5| < 1$ is equivalent to $-1 < -2x + 5 < 1$. To solve, we want to isolate x. Subtract 5 throughout, and then divide by -2 throughout: $-1 < -2x + 5 < 1$ becomes $-6 < -2x < -4$, and this becomes $3 > x > 2$. (The inequalities were reversed because of division by a negative number.) The solution is the interval $(2, 3)$:

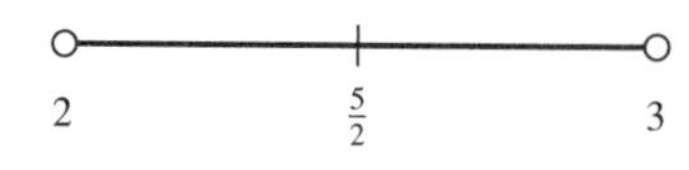

Another approach is to use properties of absolute value to rewrite the inequality as $|x - c| < d$, to be interpreted as the distance from x to c is less than d: $|-2x + 5| < 1$, so $\left|-2\left(x - \frac{5}{2}\right)\right| < 1$. Thus $|-2|\left|\left(x - \frac{5}{2}\right)\right| = 2\left|x - \frac{5}{2}\right| < 1$. This becomes $\left|x - \frac{5}{2}\right| < \frac{1}{2}$. The solution is the set of all real numbers x whose distance to $\frac{5}{2}$ is less than $\frac{1}{2}$, which is the interval $(2, 3)$.

b. The inequality $|4x - 3| \le 0.1$ is equivalent to $-0.1 \le 4x - 3 \le 0.1$. First we add 3 throughout; then divide by 4 throughout to obtain $0.725 \le x \le 0.775$. The solution is the closed interval $[0.725, 0.775]$ shown:

■

Inequalities of the form $|x - c| > d$ have for their solution the set of all numbers x whose distance to c is greater than d. This is the union of two separate intervals: $(-\infty, c - d)$ and $(c + d, \infty)$. This is written as $(-\infty, c - d) \cup (c + d, \infty)$ and graphed as

$c - d \quad c \quad c + d$

One way to solve an inequality of the form $|w| > d$ is to interpret it as "the distance from w to 0 is greater than d," so $|w| > d$ is equivalent to $w > d$ or $w < -d$.

The inequality $|w| > d$ must be written as **two separate** inequalities joined by *or*—that is, $w > d$ **or** $w < -d$. It cannot be written as a double inequality because there are no values of w that satisfy both inequalities simultaneously.

EXAMPLE 3.24 Solve each inequality a–c and graph its solution on a number line:

a. $|x| > 0.5$ **b**. $|3 - x| > 2$ **c**. $|2x - 5| \ge 1$

d. Explain how the solutions to Examples 3.23a and 3.24c are related.

Solution

a. $|x| > 0.5$ means $x > 0.5$ or $x < -0.5$; that is, x is more than 0.5 units away from 0. The solution is $(-\infty, -0.5) \cup (0.5, \infty)$.

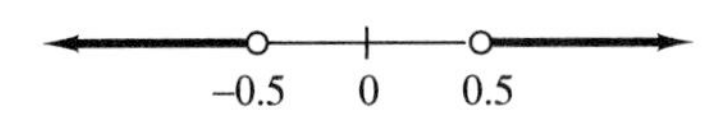

b. We first note that $|3 - x| = |-(x - 3)| = |x - 3|$; thus we can rewrite the given inequality as $|x - 3| > 2$. This has as solutions the numbers whose distances from 3 are greater than 2:

$(-\infty, 1) \cup (5, \infty)$

1 3 5

c. We rewrite the inequality $|2x - 5| \geq 1$ as two inequalities: $2x - 5 \geq 1$ or $2x - 5 \leq -1$. We solve the first inequality: $2x \geq 6$, so $x \geq 3$; then we solve the second inequality: $2x - 5 \leq -1$, so $2x \leq 4$ or $x \leq 2$. The solution is then the set $\{x \mid x \geq 3 \text{ or } x \leq 2\}$:

$(-\infty, 2] \cup [3, \infty)$

2 3

d. The solution of the inequality $|2x - 5| \geq 1$ consists precisely of all the real numbers that are not in the solution set of $|2x - 5| < 1$. The two solution sets have no elements in common, and their union is $(-\infty, \infty)$. ■

When we consider the limit of a function f as x approaches a number c, the actual value $f(c)$ is not important; only the values of f at x close to c are important. So in many cases, the set to consider is not the whole interval $(c - d, c + d)$ but the interval with the point c excluded. This is the union of two intervals $(c - d, c) \cup (c, c + d)$, which is graphed as follows:

$c - d$ c $c + d$

We can describe this set by using absolute value inequalities. To do this, note that c itself is the only number whose distance to c is 0. The above set is then the set of numbers x such that $|x - c| < d$ and $|x - c| \neq 0$. Since $|x - c| \geq 0$, $|x - c| \neq 0$ is equivalent to $|x - c| > 0$. So $(c - d, c) \cup (c, c + d) = \{x \mid 0 < |x - c| < d\}$.

EXAMPLE 3.25 Graph the set $(-2.3, -2) \cup (-2, -1.7)$ on a real line and describe it by using absolute value inequalities.

Solution

-2.3 -2 -1.7

The distance from -2.3 to -2 is $|-2 - (-2.3)| = |-2 + 2.3| = 0.3$. Similar computations show that the distance from -2 to -1.7 is also 0.3. The given set is the set of all points x whose distance from -2 is less than 0.3, excluding the point -2. The given set is then $\{x \mid 0 < |x - (-2)| < 0.3\} = \{x \mid 0 < |x + 2| < 0.3\}$. ■

EXAMPLE 3.26 Write the following sentence in symbolic form using absolute values: If x is within a distance of 0.2 from -1 and $x \neq -1$, then the values of the function $f(x) = 3x + 1$ are within a distance of 0.6 from -2.

Solution The distance from x to -1 is given by $|x - (-1)| = |x + 1|$. So to say that $x \neq -1$ and its distance to -1 is less than 0.2, we write $0 < |x + 1| < 0.2$. To say that the values of the function $f(x) = 3x + 1$ are within a distance of 0.6 from -2 is to say that $|3x + 1 - (-2)| < 0.6$. Thus $|3x + 3| < 0.6$. The whole sentence is: If $0 < |x + 1| < 0.2$, then $|3x + 3| < 0.6$. ■

We could state the sentence given in Example 3.26 in a slightly different way as follows: If x is between -1.2 and -0.8 and $x \neq -1$, then the values of the function $f(x) = 3x + 1$ are between -2.6 and -1.4. We can see from Figure 3.19 that when x is between -1.2 and -0.8, the values of $f(x)$ lie between the horizontal lines $y = -2.6$ and $y = -1.4$.

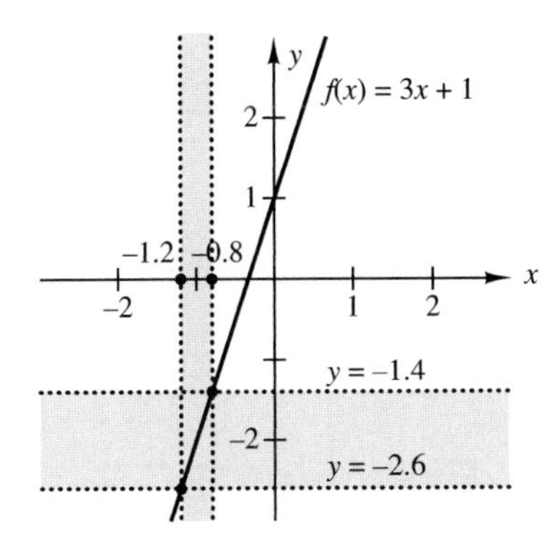

Figure 3.19

EXAMPLE 3.27 The graph of a function f is given in Figure 3.20.

a. Use the graph to estimate a number $\varepsilon > 0$ such that if $|x - 2| < 1$, then $|f(x) - 3| < \varepsilon$.

b. Use the graph to estimate a number $\delta > 0$ such that if $|x + 6| < \delta$, then $|f(x) - 1| < 0.5$.

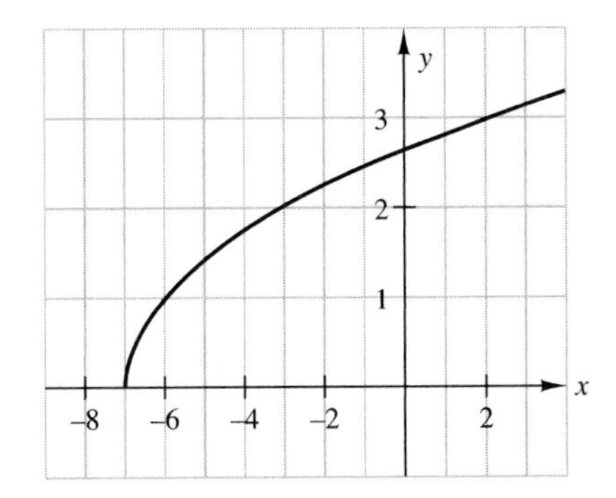

Figure 3.20

(The Greek letter ε (epsilon) represents a distance along the y-axis. The Greek letter δ (delta) represents a distance along the x-axis.)

Solution

a. We want to find a value of ε so that if the distance from x to 2 is less than 1, then the distance from $f(x)$ to 3 will be less than ε. Values of x that are less than 1 unit from the number 2 are in the interval (1, 3). Thus for x between 1 and 3, we need to determine an interval that contains the $f(x)$ values. We estimate from the graph that for x between 1 and 3, the y-coordinates of points on the graph of f lie between 2.8 and 3.2. We can describe the interval $2.8 < f(x) < 3.2$ by the inequality $|f(x) - 3| < 0.2$. Thus we estimate $\varepsilon = 0.2$.

b. Now we want to find a value of δ so that if the distance from x to -6 is less than δ, then the distance from $f(x)$ to 1 will be less than 0.5. Values of $f(x)$ within 0.5 unit of 1 are in the interval (0.5, 1.5). We estimate from the graph that when the y-coordinates of the graph of $f(x)$ are between 0.5 and 1.5, the corresponding x-coordinates of the graph lie between -6.7 and -5. The value -6.7 is 0.7 unit from -6, and the value -5 is 1 unit from -6. We choose the smaller value of 0.7 as our estimate for δ, since if $-6.7 < x < -5.3$, then $f(x)$ lies between 0.5 and 1.5. ■

Exercises 3-C.2

---

1. Solve for x:

a. $\left|\frac{1}{4} - \frac{x}{2}\right| = 1$ **b**. $|3.1x + 1.2| = -1.4$ **c**. $|5.6x + 1.4| = 0$

2. Solve each inequality. Use interval notation to describe the solution set, and graph the set on a real line.

a. $0 < |x - 1| < 0.2$

b. $0 < \left|x + \frac{1}{3}\right| < \frac{1}{4}$

c. $|x + 1.3| \le 0.1$

d. $|x - 4| > -10$

e. $|2x + 4.2| > 2$

f. $|5 - 3x| \ge 7$

g. $|3 - 1.5x| < 0.3$

h. $\left|\frac{x}{2} - 1\right| < -\frac{1}{2}$

3. Express each of the following statements as an English sentence without using the term "absolute value."

a. If $0 < |x - 1| < 0.3$, then $|f(x) - 2| < 0.1$.

b. If $0 < |x + 3| < \frac{1}{5}$, then $|g(x)| > 100$.

4. Consider the interval pictured:

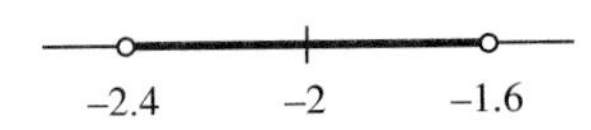

a. Describe the interval in a sentence that gives positions of points relative to the number –2.

b. Describe the interval by a double inequality.

c. Write the interval in interval notation.

d. Give the interval using an absolute value inequality.

5. Consider the set of real numbers whose distance to the number 4.2 is less than 3.

a. Graph this set on a real number line.

b. Write this set in interval notation.

c. Describe the set in an inequality.

6. Consider the graph of the function f given in Figure 3.21. Use the graph to estimate a number ε such that:

a. If $0 < |x - 1| < 0.5$, then $|f(x) - 1.5| < \varepsilon$.

b. If $0 < |x - 2| < \frac{1}{4}$, then $|f(x) - 2| < \varepsilon$.

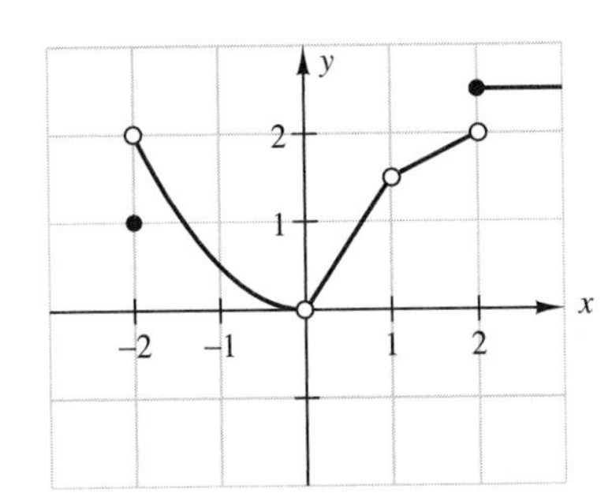

Figure 3.21

7. a. Use the graph in Figure 3.21 to estimate a number δ such that the following is true: If $0 < |x| < \delta$, then $|f(x)| < \frac{1}{2}$.

b. For the same function f is it possible to find a number δ for which the following statement is true? If $|x - 2| < \delta$, then $|f(x) - 2.5| < 0.25$. (If the answer is yes, give the number; if the answer is no, explain why not.)

8. Rewrite each of the following inequalities in the form $|x - c| < d$ or $|x - c| \leq d$, and give the solution set in interval form:

a. $|-5x + 4| < 0.3$ **b.** $|3x + 1.4| \leq 0.7$ **c.** $\left|7 - \frac{x}{3}\right| < \frac{5}{6}$

9. Write each of the following sentences in symbolic form using absolute values.

a. If x is within a distance of $\frac{1}{2}$ from -3, then the values of $g(x)$ are within a distance of $\frac{3}{4}$ from 2.

b. If x is within a distance of 3 from 4.5 and $x \neq 4.5$, then the values of $h(x)$ are within a distance of 1.7 from -5.

3-D IF–THEN STATEMENTS

Many mathematical statements are given in the form, "If P, then Q," where P and Q are statements. You have undoubtedly heard warnings in this form: If you don't do your homework, then you will not do well in this course. In this *conditional statement*, P is "you don't do your homework" and Q is "you will not do well in this course."

Some conditional statements from calculus appear in Exercise 9 of Section 3-C.2. Here are some additional examples.

1. If x is close to a, then $f(x)$ is close to L.

2. If the limit of $f(x)$ as x approaches a exists, then both one-sided limits exist and are equal.

3. If the function f is differentiable at $x = a$, then f is continuous at $x = a$.

In the conditional statement, "If P, then Q," P is called the *hypothesis* and Q is called the *conclusion*. It is important to be able to identify the hypothesis and the conclusion in a conditional statement. When verifying or proving a conditional statement, we need to show that whenever the hypothesis is true, the conclusion is also true. The following if–then statements are examples of results that are useful when working with equations and inequalities.

EXAMPLE 3.28 In each statement, identify the hypothesis and the conclusion.

a. If $ab = 0$, then $a = 0$ or $b = 0$.

b. $ab > 0$ whenever $a > 0$ and $b > 0$.

c. $a < 0$ and $b > 0$ implies that $ab < 0$.

Solution

a. The hypothesis is "$ab = 0$" and the conclusion is "$a = 0$ or $b = 0$."

b. This statement could be rewritten as an if–then statement as follows: If $a > 0$ and $b > 0$, then $ab > 0$. The hypothesis is "$a > 0$ and $b > 0$"; the conclusion is "$ab > 0$."

c. This statement could be written in if–then form as: If $a < 0$ and $b > 0$, then $ab < 0$. So the hypothesis is "$a < 0$ and $b > 0$," and the conclusion is "$ab < 0$." ■

We prove that an if–then statement is true by assuming that the hypothesis is true and then deducing (by a logical argument) that the conclusion is true. To show that the statement "If $ab = 0$, then $a = 0$ or $b = 0$" is true, it must be shown to be true for **all** real numbers a and b.

A general if–then statement about **all** real numbers is false (not true) if there is even **one** real number for which it is false. To show that an if–then statement is **not** true, only one example is needed. Such an example where the hypothesis is true but the conclusion is false is called a *counterexample*. By testing a result with several different values or examples, we can either come up with a counterexample or get evidence that the result is likely to be true.

EXAMPLE 3.29 Decide if the following if–then statements are true or false. If the result is false, modify the hypothesis; that is, include additional conditions so that it is a true if–then statement.

a. If $a < b$, and $c < d$, then $a + c < b + d$. **b**. If $a < b$, then $a^2 < b^2$.

Solution

a. We first test a few specific cases to see if the statement seems plausible. To do this, we pick two numbers a and b that satisfy $a < b$; then pick any other numbers c and d. We first try $a = 2$, $b = 10$, $c = 5$, and $d = 6$. The hypothesis is satisfied since $2 < 10$ and $5 < 6$. Also $2 + 5 < 10 + 6$ so, in this case, the conclusion $a + c < b + d$ is also true. Now we try some other values, including both positive and negative numbers. For example, if $a = 4$, $b = 5$, $c = -7$, and $d = -1$, again we see that $a + c < b + d$. For $a = -3$, $b = -1$, $c = 2$, and $d = 5$, we also see that $a + c < b + d$.

This evidence convinces us that the statement is probably true. The evidence of a few cases does not prove the statement is always true. We can prove the statement using basic properties of inequalities (see Section 3-C.1):

Assume that $a < b$ and $c < d$. Add c to both sides of $a < b$ to get $a + c < b + c$ (by property 1). Similarly, add b to both sides of $c < d$ to get $c + b < d + b$. By the commutative property of addition, $b + c = c + b$ and $d + b = b + d$. By property 4, since $a + c < b + c$ and $b + c < b + d$, we conclude that $a + c < b + d$.

b. If we let $a = 3$ and $b = 7$, we find that the hypothesis is satisfied and so is the conclusion. But if we let $a = -7$ and $b = 3$, we have $a < b$ but $a^2 = 49$ is not less than $b^2 = 9$. This provides a counterexample, and the if–then statement, as given, is false. We can modify the statement in the following way so that it is a true statement. If $a < b$ and $a > 0$, then $a^2 < b^2$. ■

Testing a general statement for several specific values does **not** prove that it is true. To prove a general statement we must use general facts and properties. To disprove a statement—that is, to prove the statement is false—we need to give only one counterexample.

The *converse* of the statement "If P, then Q" is obtained by switching the roles of P and Q. This gives another conditional statement: "If Q, then P." The converse of a true conditional statement may or may not be a true conditional statement. Consider the following two conditional statements from Examples 3.28a and 3.28b:

Statement a: If $ab = 0$, then $a = 0$ or $b = 0$. (True)

Converse: If $a = 0$ or $b = 0$, then $ab = 0$. (True)

Statement b: If $a > 0$ and $b > 0$, then $ab > 0$. (True)

Converse: If $ab > 0$, then $a > 0$ and $b > 0$. (False)

Can you find numbers a and b for which $ab > 0$ and a and b are not both greater than 0?

Biconditional statements are statements of the form "P if and only if Q" or "P iff Q" for short. A statement of this form is called a **bi**conditional statement because it combines the **two** conditional statements "If P, then Q" and its converse "If Q, then P." For the biconditional statement "P if and only if Q" to be true, both the statement "If P, then Q" and its converse "If Q, then P" must be true. Statement a and its converse are both true, so it can be rewritten as a single theorem in if and only if form: $ab = 0$ if and only if $a = 0$ or $b = 0$.

Definitions, although often not stated in if and only if form, are always understood to be if and only if statements. A definition is a name followed by a characterization. If something has that name, then it must also have the characterization; conversely, if something has the given characterization, then it also has the name. For example, a function is defined as a rule that assigns to each value of the independent variable a unique value of the dependent variable. From this definition we know that if a rule has the property that each value of the independent variable is assigned a unique value of the dependent variable, then it is a function. Also, if a rule is a function, then it has this property.

EXAMPLE 3.30 Write each of the following if and only if statements as two separate if–then statements. Then decide if the biconditional statement that combines them is true or not. Explain why or why not.

a. Two nonvertical lines are perpendicular if and only if the product of their slopes is -1.

b. $a > 1$ if and only if $|a| > 1$.

Solution

a. The two conditional statements are

1. If two nonvertical lines are perpendicular, then the product of their slopes is -1.
2. If the product of their slopes is -1, then two nonvertical lines are perpendicular.

Both of the conditional statements 1 and 2 are true, so the biconditional statement is also true.

b. The two conditional statements are

1. If $a > 1$, then $|a| > 1$.
2. If $|a| > 1$, then $a > 1$.

The second if–then statement is not true, since $a = -4$ provides a counterexample: $|-4| = 4 > 1$ but -4 is not greater than 1. Thus the if and only if statement is also not true. ■

An indirect way to say "If $a > 0$ and $b > 0$, then $ab > 0$" is "If $ab \leq 0$, then a and b are not both positive." The second statement is the *contrapositive* of the first. The contrapositive of the conditional statement "If P, then Q" is important because the contrapositive is true exactly when the original conditional statement is true but says it in a different way. The idea of the contrapositive is used in court in concluding that a suspect is innocent if he or she has an alibi. It is often true that "If the suspect committed the crime, then he or she was at the scene of the crime." An alibi is evidence of innocence because it is true that "If he or she was **not** at the scene of the crime, then the suspect did **not** commit the crime." The second statement is the contrapositive of the first, and the two statements are equivalent.

There are statements in mathematics that are more easily proved or more easily applied if the contrapositive form rather than the original statement is used. For example, consider the following statement: "If two nonvertical lines are perpendicular, then the product of their slopes equals -1." We can use the contrapositive to prove lines are **not** perpendicular: "If the product of their slopes is not -1, then two nonvertical lines are not perpendicular."

The *contrapositive* of the conditional statement "If P, then Q" is the conditional statement "If not Q, then not P." The negation or denial of P is given by "not P." In the above court example, P is the statement "the suspect committed the crime," and "not P" is "the suspect did not commit the crime."

The negation of $a < b$ is $a \geq b$ (and not just $a > b$), and the negation of the statement "$a < b$ and $a > 0$" is the statement "$a \geq b$ or $b \leq 0$."

 Care must be taken when negating a statement to be sure that the negation of P covers all cases of the denial of P.

EXAMPLE 3.31 Write the contrapositive of the conditional statement. Which statement do you think is easier to understand?

If $ab = 0$, then $a = 0$ or $b = 0$.

Solution The contrapositive of this statement requires identifying P as "$ab = 0$" and Q as "$a = 0$ or $b = 0$." "Not P" becomes "ab is not 0," and "not Q" is "a is not 0 **and** b is not 0." Thus the contrapositive is "If a is not 0 and b is not 0, then ab is not 0." In this case the contrapositive is easier to understand. It can simply be stated as "The product of two nonzero numbers is also nonzero." ■

EXAMPLE 3.32 Write the contrapositive of each of these conditional statements from calculus.

a. If the limit of $f(x)$ as x approaches a exists, then both one-sided limits exist and are equal.

b. If the function f is differentiable at $x = a$, then f is continuous at $x = a$.

Solution

a. The hypothesis of this statement is "the limit of $f(x)$ as x approaches a exists," and the conclusion is "both one-sided limits exist and are equal." The negation of the hypothesis is "the limit of $f(x)$ as x approaches a does not exist." The negation of the conclusion is "the one-sided limits do not both exist or they are not equal." The contrapositive of the given statement is: "If the one-sided limits do not both exist or they are not equal, then the limit of $f(x)$ as x approaches a does not exist."

b. The contrapositive of the given statement is "If the function f is not continuous at $x = a$, then f is not differentiable at $x = a$." ■

Each of these statements is used in calculus in the form given by its contrapositive, and it is important to remember that the contrapositive says the same thing as the original statement.

Theorems in mathematics and, in particular, in calculus provide rules, tests for conditions, and justification for techniques. To use a particular theorem given as a conditional statement, we must make sure that its hypotheses are true—that is, that they are satisfied—before we can apply the theorem. If the hypotheses are not satisfied in a particular situation, then the theorem cannot be used and no conclusion can be drawn from it.

EXAMPLE 3.33 Consider the following true conditional statement (or theorem): If the function $f(x)$ is of the form $f(x) = mx + b$, with $m \neq 0$, then the graph of $y = f(x)$ crosses the x-axis exactly once.

For each of the following functions $f(x)$, determine if the hypothesis of the statement is satisfied and if the conclusion is satisfied:

a. $f(x) = 3x^2 + 1$ **b.** $f(x) = -5x - 4$ **c.** $f(x) = (x - 2)^3$

Solution

a. The hypothesis is "the function $f(x)$ is of the form $f(x) = mx + b$." The given function is not of the form $f(x) = mx + b$, so the hypothesis is not satisfied and the theorem cannot be used here. Note that the conclusion is also not satisfied.

b. The function $f(x) = -5x - 4$ is of the form $f(x) = mx + b$, so the hypothesis is satisfied. Therefore the conclusion must also be satisfied. The value $x = -\frac{4}{5}$ is the x-coordinate of the only point where the linear function crosses the x-axis.

c. The function $f(x) = (x - 2)^3$ is not of the form required by the hypothesis, so the theorem does not apply. However, the conclusion is true for this function as we can verify by graphing f. ■

Many theorems that are conditional statements are not phrased in the "If P, then Q" form. They may be stated as "Suppose P, then Q" or as "Let P. Then Q." It is important to be able to identify the hypothesis and the conclusion and to determine if the hypothesis is satisfied so that the conclusion may be drawn.

EXAMPLE 3.34 Consider the following theorem about the composition of two linear functions:

Suppose f and g are linear functions. Then $f \circ g$ is a linear function.

For each of the following functions, determine if the theorem applies. If so, what conclusion can be drawn by using the theorem?

a. $f(x) = -2x$, $g(x) = 5x - 9$ **b.** $f(x) = -x + 1$, $g(x) = (3x - 1)^2$

Solution

a. To verify the hypothesis of the theorem we need to see if the functions f and g are linear. Since f and g are given by expressions of the form $y = mx + b$, they are linear. So the hypothesis of the theorem is satisfied and we can apply the theorem to conclude that the function $f \circ g$ is linear. We can also see this by finding $f \circ g$:

$$(f \circ g)(x) = f(g(x)) = f(5x - 9) = -2(5x - 9) = -10x + 18, \text{ which is linear.}$$

b. The function $f(x) = -x + 1$ is linear, but the function $g(x) = (3x - 1)^2 = 9x^2 - 6x + 1$ is not a linear function, since it cannot be written in the form $y = mx + b$.

The hypothesis of the theorem does not hold, so the theorem cannot be applied. ■

Exercises 3-D

1. Determine if the following if–then statements are true or false. If the statement is false, give a counterexample. Then modify the hypothesis so the if–then statement becomes true.

a. If x is a real number, then $\sqrt{x^2} = x$.

b. If $a > 1$, then $|a| > 1$.

c. If $0 < a < b$, then $\sqrt{a} < \sqrt{b}$.

d. If x is a real number, then $-x$ is negative.

2. Write each of the if and only if statements as two if–then statements. Determine if the given biconditional statement is true.

a. $a < b$ if and only if $|a| < |b|$.

b. For $f(x) = x^2 - 4$, $f(x) = 0$ if and only if $x = 2$ or $x = -2$.

3. Write the contrapositive of each of the conditional statements.

a. If two nonvertical lines are parallel, then their slopes are equal.

b. If the slopes of two nonvertical lines are equal, then the lines are parallel.

4. **a.** Write the biconditional statement that combines the two conditional statements in Exercise 3. Is the biconditional statement true or false? Why?

b. Which of the contrapositive statements that you wrote in Exercise 3 can be used to test whether two lines intersect?

5. Consider the following true theorem: If $0 \le c \le d$, then $c^2 \le d^2$. For each of the given values of c and d, determine if the hypothesis is satisfied and if the conclusion is satisfied.

a. $c = -5, d = 6$ **b.** $c = 2.3, d = 4.99$ **c.** $c = 1.7, d = 0$

Chapter 3 Exercises

1. Let $f(x) = x^2$ and $g(x) = \sqrt{2x-3}$.

   (i) For each of the functions in a–d below, give a rule for the function and give its domain.

   (ii) Evaluate each of the functions in a–d at $x = 2$, if possible. If it is not possible, explain why.

   a. $(f+g)(x)$ **b.** $(f \cdot g)(x)$ **c.** $\left(\frac{g}{f}\right)(x)$ **d.** $\frac{f(x)}{g(x)-1}$

2. Let $f(x) = x^2$ and $g(x) = \sqrt{2x-3}$.

   (i) Find each of the following compositions, if they are defined:

   a. $f(g(2))$ **b.** $g(f(2))$ **c.** $(f \circ g)(-4)$

   d. $(g \circ f)(-4)$ **e.** $f(g(a))$ **f.** $(g \circ f)(a+1)$

   (ii) Look at the answers in part i. Is the operation of composition commutative? That is, does $(f \circ g)(a) = (g \circ f)(a)$ for all numbers a?

3. Let f and g be the functions in Exercise 2. Use a graphing utility to graph $(f \circ g)(x)$ and $(g \circ f)(x)$ on the same set of axes to show that the order in which functions are composed makes a difference in the resulting function.

4. Let $g(x) = \sqrt{2x-1}$. Use the formula for g and the graph of f in Figure 3.22 to estimate the following values if they are defined. If they are undefined, explain why.

a. $\left(\frac{f}{g}\right)(2)$ **b.** $(f-g)(0)$ **c.** $(f\cdot g)(5)$ **d.** $\left(\frac{g}{f}\right)(3)$

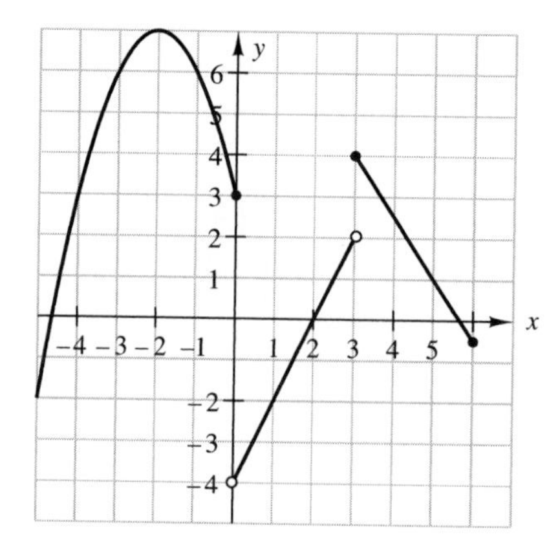

Figure 3.22 $y = f(x)$

5. Let $g(x) = \frac{3}{x-3}$. Use the formula for g and the graph of f in Figure 3.22 to estimate the following values if they are defined. If they are undefined, explain why.

a. $(f\circ g)(0)$ **b.** $(g\circ f)(0)$ **c.** $(g\circ f)(3)$ **d.** $(f\circ g)(2)$

6. Use the graphs of f and g in Figures 3.23 and 3.24 to estimate the values in parts a–g below (when the values are defined) and to answer questions h and i. If a requested function value is not defined, give a reason.

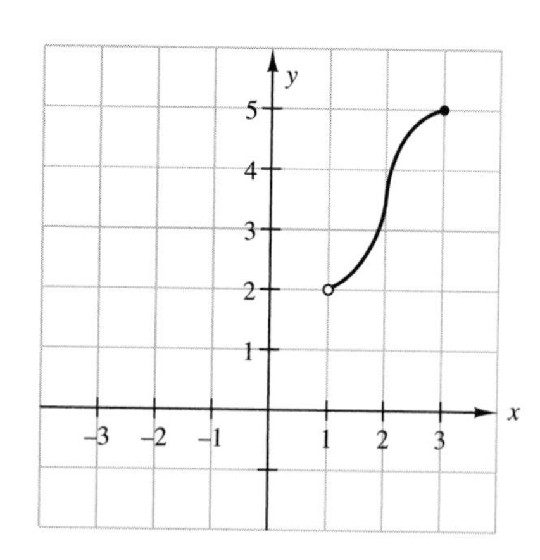

Figure 3.23 $y = f(x)$

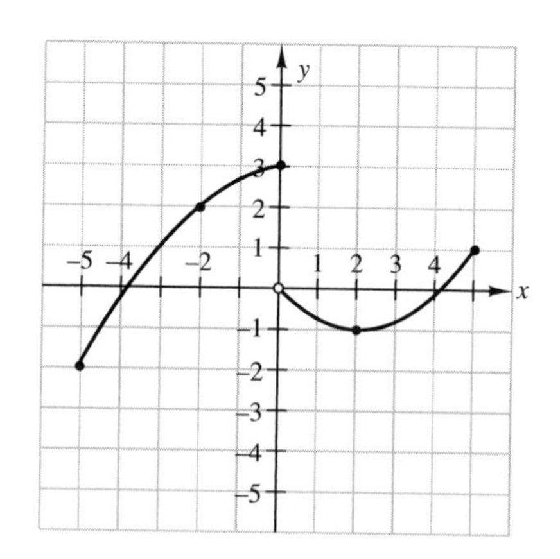

Figure 3.24 $y = g(x)$

a. $g(2) - f(3)$ **b.** $f(g(2))$ **c.** $(f\cdot g)(2)$

d. $g(f(3))$ **e.** $(f\circ g)(0)$ **f.** $(f\circ f)(-2)$

g. $(g\circ g)(-2)$ **h.** What is the domain of f?

i. What is the domain of g?

7. Give the domain of each function and use the techniques of Section 3-B to find a simpler expression for the function.

a. $H(x)=\dfrac{x^4-1}{x-1}$ **b.** $G(x)=\dfrac{\frac{1}{8}-\frac{1}{x^3}}{x-2}$ **c.** $F(x)=\dfrac{3-\sqrt{x+4}}{x-5}$

8. Give the domain and find a simplified expression for each function. Graph each function.

a. $h(x)=\dfrac{x^2+2x-3}{1-x}$ **b.** $g(x)=\dfrac{|x-2|}{3x-6}$ **c.** $f(x)=\dfrac{-x}{3-\sqrt{x+9}}$

9. Consider the interval pictured: 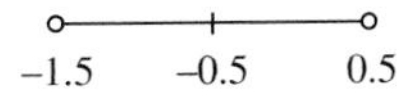

a. Describe the interval in a sentence that gives positions of points relative to the number -0.5.

b. Describe the interval using a double inequality.

c. Write the interval using interval notation.

d. Give the interval using an absolute value inequality.

10. Consider the set pictured: 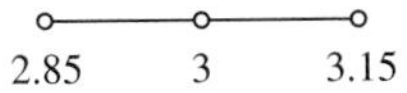

a. Describe the set in a sentence that gives positions of points relative to the number 3.

b. Describe the set using interval notation.

c. Give the set using an absolute value inequality.

11. Consider the set of real numbers whose distance to the number 1.5 is less than 0.8, excluding the number 1.5 itself.

a. Graph this set on a real number line.

b. Write this set in interval notation.

c. Describe the set in terms of inequalities.

12. Solve each of the following inequalities and write the solutions in interval notation. Also graph each solution set on a real line. If the inequality has no solution, explain why.

a. $0<|2x+1|<7$ **b.** $8>|x-\frac{2}{3}|>-2$ **c.** $4>x-4>9$

d. $|3x-8|>1$ **e.** $|4-7x|\le 1.9$ **f.** $|x+\frac{3}{4}|<-\frac{6}{5}$

g. $-9<3-4x<9$

13. Express each of the following statements as an English sentence without using the term "absolute value."

a. If $|x+4| < 0.5$, then $|f(x) - 3| < 0.2$.

b. $|g(x) - 7| < 0.15$ whenever $|x - 2| < 0.6$.

c. If $|x - 5| < \delta$, then $|f(x) + 2| < 2$.

d. If $|x - 1| < 0.3$, then $|f(x) + 4| < \varepsilon$.

14. Consider the statement: $|f(x) - 5| < \varepsilon$ whenever $0 < |x - 2| < \delta$.

a. Write the statement in if–then form.

b. Use your answer to part a to write the contrapositive of this statement.

15. Consider the graph of the function f given in Figure 3.22, which is repeated below. Use the graph to answer parts a and b.

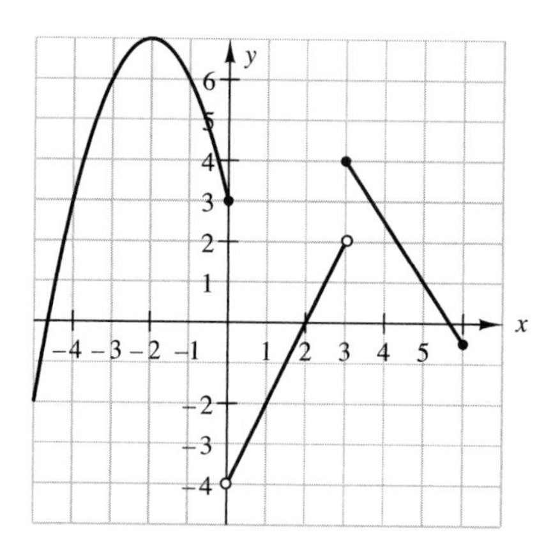

Figure 3.22 $y = f(x)$

a. Estimate a number ε such that, if $0 < |x + 3| < 0.75$, then $|f(x) - 6| < \varepsilon$.

b. Estimate a number δ such that, if $0 < |x - 1| < \delta$, then $|f(x) + 2| < 0.5$.

CHAPTER 4

Companion to Continuous Functions

Suppose you have a laser printer that will print 15 pages per minute and you have p pages to print. If the printer *continues* to print the pages at this rate, you can easily predict the time to complete the job (in minutes) using the function $T(p) = \frac{p}{15}$. This function is shown in Figure 4.1. It is an example of a *continuous function*.

If, however, the printer breaks down after a half-hour and takes a half-hour to fix, your prediction is no longer correct. In this case, the time to complete the job (in minutes) is given by the function

$$S(p) = \begin{cases} \frac{p}{15} & \text{if } p \leq 450 \\ \frac{p}{15} + 30 & \text{if } p > 450. \end{cases}$$

This case is shown in Figure 4.2.

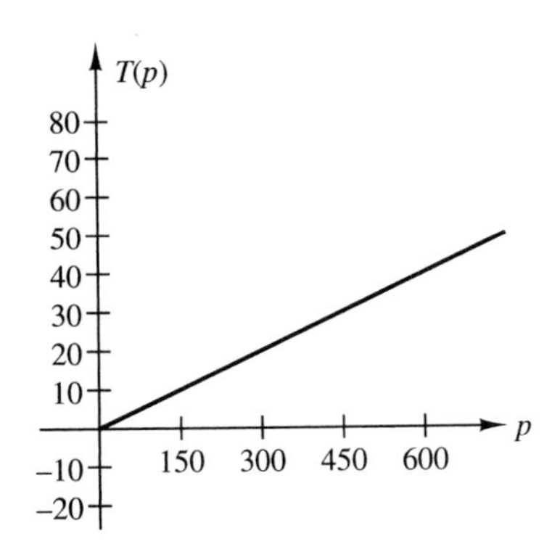

Figure 4.1 $T(p) = \frac{p}{15}$

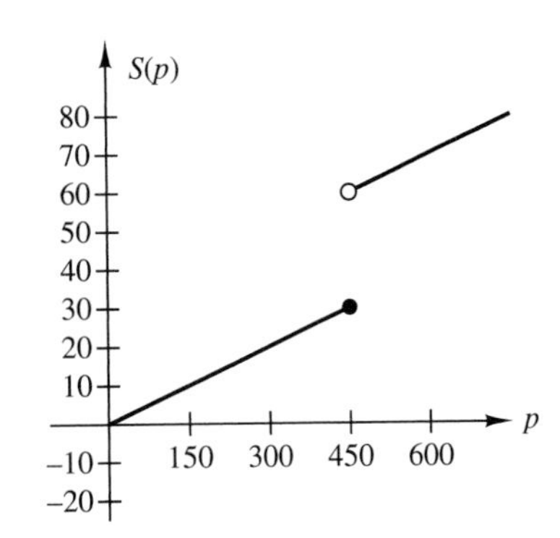

Figure 4.2 $y = S(p)$

Continuous functions are important in mathematics because they have certain useful properties. In calculus, the hypotheses for many theorems (e.g., the Intermediate Value Theorem and the Fundamental Theorem of Calculus) state that the function must be continuous; the conclusion may not hold for discontinuous functions.

Most mathematical functions that model elementary laws of physics are continuous functions. The functions used to describe quantities in business problems are, however, usually discontinuous. For example, the sales tax you pay on a purchase has a graph with jumps because the amount of tax paid must be a whole number of cents.

EXAMPLE 4.1 Suppose a car is traveling at a steady 60 mph on an interstate highway. Suddenly, a deer jumps in front of the car. The driver applies the brakes and swerves to miss the deer. Losing control, the driver crashes into a stone wall at 20 mph and comes to an abrupt stop.

a. Sketch a graph of the distance traveled as a function of time for this scenario.

b. Sketch a graph of the velocity of the car as a function of time for this scenario.

c. Which of these functions is continuous over the interval of time under consideration?

Solution

a. We first consider the car at a point in time we call $t_0 = 0$ when the car is traveling 60 mph, and we measure the distance traveled from time $t = t_0$ on. So at time t_0, the distance traveled is 0. Since the car maintains a constant speed until time $t = t_1$, the straight line with equation $d = 60t$ describes the distance function from $t = t_0$ to $t = t_1$. At time t_1 (point A) the brakes are applied and the distance traveled each second after t_1 decreases until the car hits the wall at time $t = t_2$. At time t_2 (point B) the car stops and the distance traveled does not change—it remains a constant for the remaining time under consideration. Figure 4.3 is the graph of this distance function.

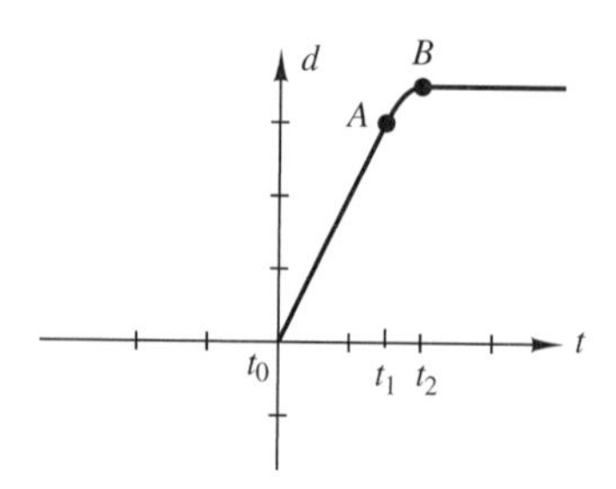

Figure 4.3 Graph of distance function

b. From time $t_0 = 0$ until time $t = t_1$ the car's speed (velocity) is a constant 60 mph. At time t_1 (point A) the brakes are applied and the speed decreases in a linear manner until the car crashes into the wall at time $t = t_2$. The instant before hitting the wall, the car is traveling 20 mph. At time t_2 (point B) the speed of the car is 0 mph;

it has come to a complete stop. For the remaining time under consideration the car remains immobile. See Figure 4.4.

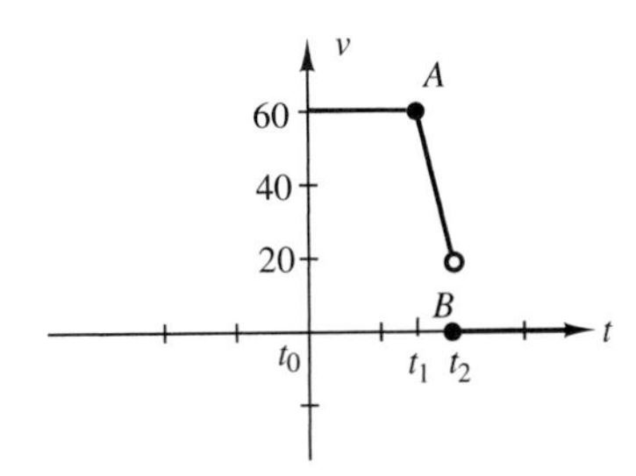

Figure 4.4 Graph of velocity function

c. The graph in Figure 4.3, which shows the distance as a function of time, is continuous. The graph of the velocity function in Figure 4.4 is discontinuous. ■

Visually, a function is *continuous* on an interval if (in theory) you can draw the graph of the function on that interval without lifting your pencil off the paper. This can be accomplished only if there are no holes, gaps, or jumps in the graph of the function. Thus, if a function f is not defined at $x = c$, then f cannot be continuous at $x = c$. A function is *continuous everywhere* if it is continuous on the interval $(-\infty, \infty)$—that is, if it is continuous for all real numbers.

A function can be discontinuous in various ways. It may have a "hole" in its graph at a single value where the function is not defined, as shown in Figure 4.5. It may have a displaced point, as shown in Figure 4.6. In this case, the limit of the function as x approaches c is different from the value $f(c)$. A discontinuous function may have a step or jump in the graph, as seen in Figure 4.2 and in Figure 4.7, which shows a step function similar to those discussed in Chapter 2. A function may also be discontinuous if it has a vertical asymptote; that is, the values of the function increase continuously or decrease continuously without bound near a value $x = c$, as in Figure 4.8.

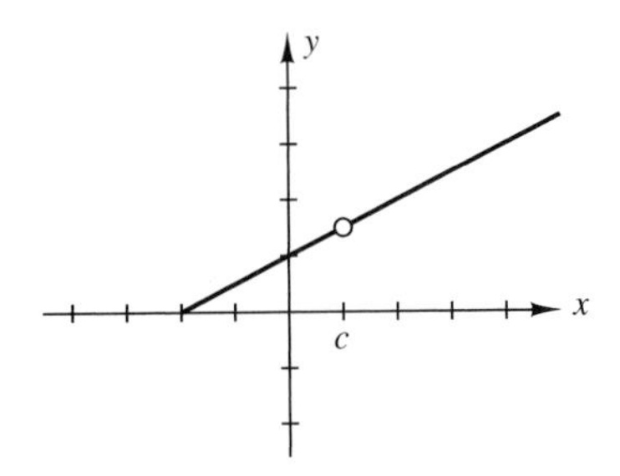

Figure 4.5 $f(c)$ is not defined

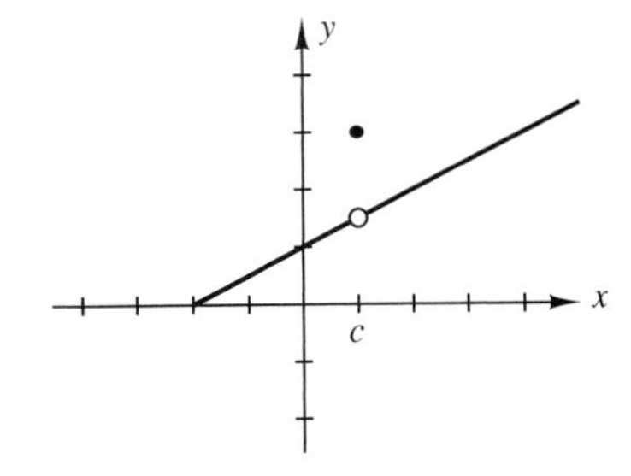

Figure 4.6 f has a displaced point

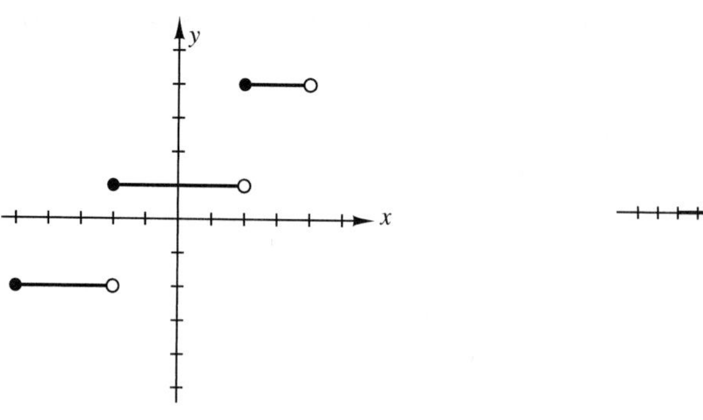

Figure 4.7 *f* has a step or jump

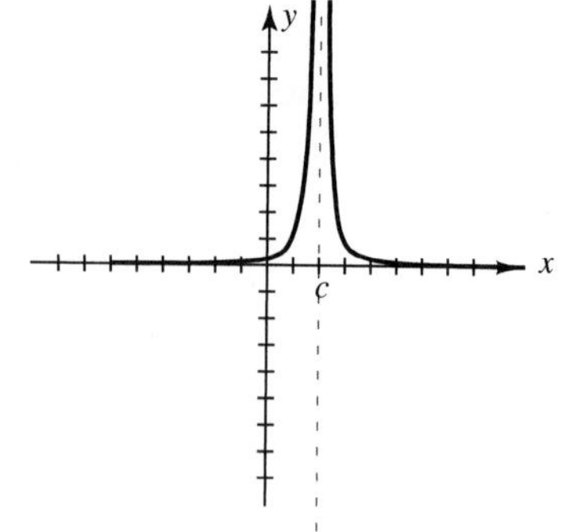

Figure 4.8 *f* has a vertical asymptote

EXAMPLE 4.2 Explain why each of the following functions is discontinuous at $x = 1$:

a. $f(x)=\dfrac{3x^2-3}{x-1}$

b. $g(x)=\dfrac{|x-1|}{x-1}$

c. $h(x)=\begin{cases}\dfrac{|x-1|}{x-1} & \text{if } x\neq 1\\ 2 & \text{if } x=1\end{cases}$

d. $k(x)=\dfrac{3}{x-1}$

Solution

a. The function f is discontinuous because it is not defined at $x = 1$. To sketch the graph, note that the formula for f can be simplified by factoring:

$$f(x)=\frac{3(x-1)(x+1)}{x-1}=3(x+1)=3x+3 \text{ if } x\neq 1.$$

The graph is a line with a hole in it at the value $x = 1$; see Figure 4.9.

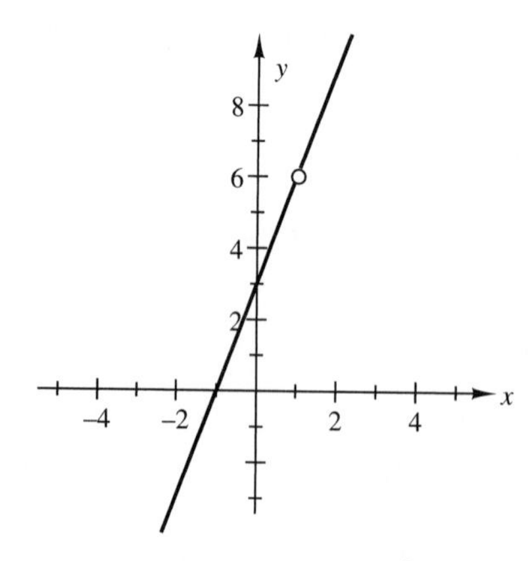

Figure 4.9 $f(x) = \frac{3x^2-3}{x-1}$

b. The function g is discontinuous because it is not defined at $x = 1$. To sketch the graph, we use the definition of absolute value to write $g(x)$ as a piecewise linear function.

$$\text{For } x > 1,\ g(x) = \frac{|x-1|}{x-1} = \frac{x-1}{x-1} = 1.$$

$$\text{For } x < 1,\ g(x) = \frac{|x-1|}{x-1} = \frac{-(x-1)}{x-1} = -1.$$

The graph of $g(x)$ is the step function in Figure 4.10.

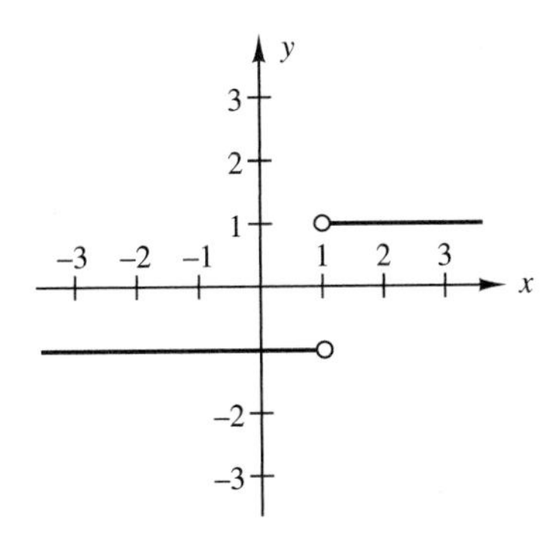

Figure 4.10 $g(x) = \frac{|x-1|}{x-1}$

c. The function h is defined at $x = 1$. It is, however, discontinuous at $x = 1$ because it has a jump at that value. The graph of $h(x)$ is the same as that of $g(x)$ in Figure 4.10 but with the added point (1, 2); see Figure 4.11.

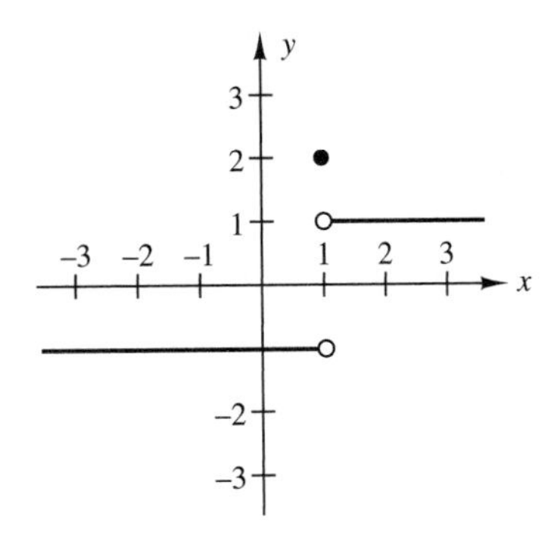

Figure 4.11 $h(x) = g(x)$ if $x \neq 1$; $h(1) = 2$

d. The function k is discontinuous because it is not defined at $x = 1$. It has a vertical asymptote at $x = 1$ because the function values decrease without bound as x approaches 1 from the left and the function values increase without bound as x approaches 1 from the right; see Figure 4.12.

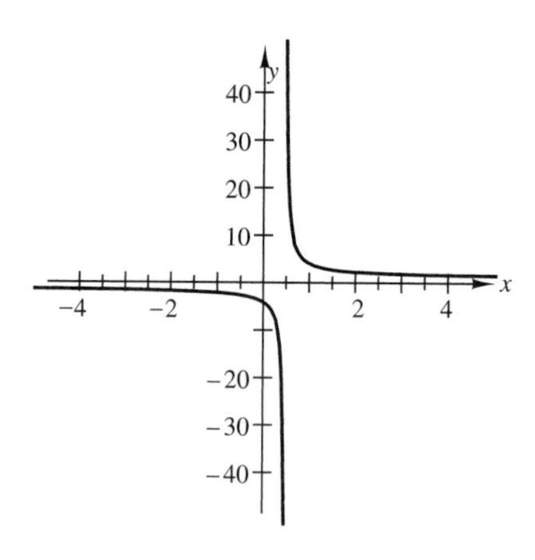

Figure 4.12 $k(x) = \frac{3}{x-1}$ ■

A graphing utility may not show a "hole" in a discontinuous graph. Be sure to identify the domain of a function before graphing it.

4-A POLYNOMIALS

Linear functions studied in Chapter 2 belong to a larger class of functions called polynomial functions. These are the most well-known continuous functions. Recall that a *polynomial function* in x can always be written in the form:

$$f(x) = a_n x^n + a_{n-1}x^{n-1} + \cdots + a_2 x^2 + a_1 x + a_0,\ a_n \neq 0$$

where the coefficients a_i are real numbers and the powers of x are nonnegative integers. The domain of every polynomial function is the set of all real numbers, and every polynomial function is continuous for all real numbers.

The *degree* of the polynomial $f(x)$ written in the standard form above is n, the highest exponent of the variable x appearing in the polynomial. The term $a_n x^n$ is called the *leading term* of the polynomial, and the coefficient a_n is called the *leading coefficient*. The term a_0 is called the *constant term* of the polynomial. Here are some examples of polynomial functions: $f(x) = 2x - \sqrt{3}$, $g(x) = 5.1$, and $h(x) = 5 - 2x^4 - 3x$. The degree of f is 1 (it is a linear polynomial), the degree of g is 0 (it is a constant polynomial), and the degree of h is 4. The leading coefficient of f is 2, of g is 5.1, and of h is –2, since it is the coefficient of x^4.

Polynomial functions are easily evaluated, so mathematicians often use polynomials to approximate functions that are more difficult to evaluate. For example, although you may not be aware of it, your calculator may use a special polynomial called a Taylor polynomial to approximate sin 31° or $\sqrt{2.3}$. In calculus, two important operations on functions are studied—differentiation and integration. It is easy to perform these operations on polynomials. For some other functions these tasks are more difficult. If it is possible to approximate such a function with a polynomial, it may be possible to avoid complicated computations in the solution of a problem. Another nice feature of polynomial functions is that their graphs are "smooth"; that is, there are no "corners" on the graph of a polynomial. (For example, the graph of $f(x) = |x|$ is not smooth, since it has a corner at the point (0, 0).)

EXAMPLE 4.3 Which of the following are polynomial functions? For those that are not polynomials, state why. What are the degree and the leading coefficient of each polynomial?

$f(x) = (x - 3)(2x + 4)$ $\quad$ $g(x) = \sqrt{x} - 2$ $\quad$ $k(x) = |x| + 3$

$h(x) = \frac{x^2 - 1}{x + 1}$ $\quad$ $s(t) = 3t - \frac{2}{t}$ $\quad$ $t(z) = \frac{z^3}{2} - \sqrt{2}z + 1$

Solution Of these functions, only f and t are polynomials. Expanding f, we have $f(x) = (x - 3)(2x + 4) = 2x^2 - 2x - 12$; f has degree 2 and leading coefficient 2. The polynomial t has degree 3 and leading coefficient $\frac{1}{2}$. The functions g and s are not polynomials because not all exponents of the variables are positive integers ($\sqrt{x} = x^{1/2}$ in $g(x)$ and $\frac{2}{t} = 2t^{-1}$ in $s(t)$). Although $h(x)$ can be simplified to $h(x) = x - 1$ for $x \neq -1$, it is not a polynomial because its domain does not contain $x = -1$. The function k is a piecewise linear function and thus the rule for k cannot be put in polynomial form. (See Figure 2.20 in Example 2.13.) ■

Exercises 4-A

1. Sketch the graph of a function that is continuous.

2. Sketch the graph of a function that is not continuous.

3. Explain why the functions below are discontinuous at $x = 2$, and sketch their graphs:

a. $f(x) = \frac{2}{x - 2}$ $\quad$ **b.** $g(x) = \frac{x^2 - 4}{2x - 4}$ $\quad$ **c.** $h(x) = \begin{cases} 2x - 5 & \text{if } x \leq 2 \\ 2x + 5 & \text{if } x > 2 \end{cases}$

4. Use a graphing utility to graph the functions in parts a–c that follow. Where is each function discontinuous? Describe its behavior near the values of x where the function is discontinuous.

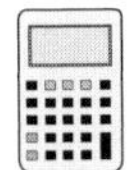

a. $f(x)=\frac{5x}{9-x^2}$ **b.** $g(x)=\frac{2x^2}{x^2-9}$ **c.** $h(x)=\frac{4x}{x^2+9}$

5. Explain why $f(x) = x-\sqrt{2}$ is a polynomial but $g(x) = \sqrt{x}-2$ is not.

6. Explain why $f(x) = \frac{x^2}{3}$ is a polynomial but $g(x) = \frac{3}{x^2}$ is not.

7. Give the degree and leading coefficient of each of the following polynomials:

a. $f(x) = 5x - 3 - \frac{x^3}{5}$ **b.** $g(x) = 3^2 - \sqrt{5}$ **c.** $h(x) = \frac{1}{2}x^3 - \sqrt{7}x^4 + x^2$

8. Consider a linear polynomial $f(x)$.

a. What is the degree of f?

b. What does the leading coefficient describe?

c. What does the constant term describe?

4-B ZEROS OF A FUNCTION

The *zeros*, or *roots*, of a function are values in its domain for which the function equals zero. The zeros of the function are the x-intercepts of the graph of the function, the x-coordinates of points where the graph intersects the x-axis. Note that the zeros of the function $f(x) = x^2 - 4$ are at $x = -2$ and $x = 2$; see the graph in Figure 4.13.

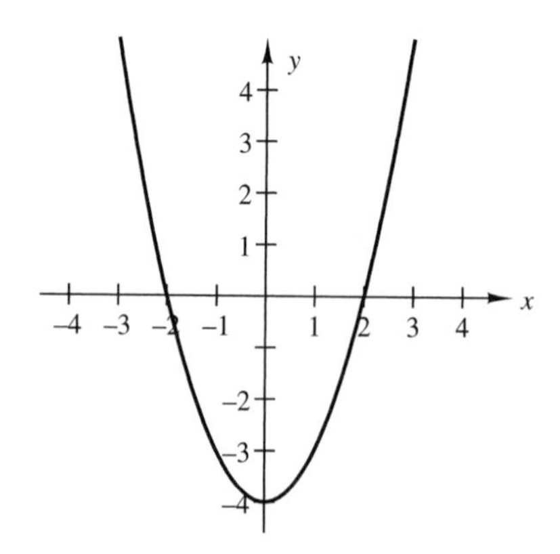

Figure 4.13 $f(x) = x^2 - 4$

4-B.1 Finding Zeros of a Polynomial

Quadratic functions are polynomial functions of degree 2 and can have two zeros (as $f(x) = x^2 - 4$ in Figure 4.13 shows), one zero (e.g., $g(x) = x^2$), or no zeros (e.g.,

$h(x) = x^2 + 1$). We find the zeros of a quadratic polynomial by factoring, if that is possible, or by using the quadratic formula.

To find the zeros of a quadratic polynomial, we put the polynomial in its general form $p(x) = ax^2 + bx + c$, where $a \neq 0$. We first try to factor the function into two linear factors of the form $(jx + k)(mx + n)$, where j, k, m, and n are constants. For example, the quadratic function $p(x) = 8x^2 + 10x + 3$ factors into $(4x + 3)(2x + 1)$. We can then solve for the zeros with $p(x)$ in this factored form because of the following theorem:

Theorem: If $ab = 0$, then either $a = 0$ or $b = 0$. (In other words, if the product of two factors is zero, then one of the factors must equal zero.)

This theorem is illustrated graphically in Figure 4.14, which shows the graphs of the parabola $y = x^2 - x - 6 = (x - 3)(x + 2)$ and the lines $y = x - 3$ and $y = x + 2$. The graph of $x^2 - x - 6$ crosses the x-axis at $x = -2$ and $x = 3$. The graph of the factor $y = x - 3$ crosses the x-axis at $x = 3$; the graph of the second factor $y = x + 2$ crosses the x-axis at $x = -2$. Thus, the zeros of $y = (x - 3)(x + 2)$ are $x = 3$ and $x = -2$.

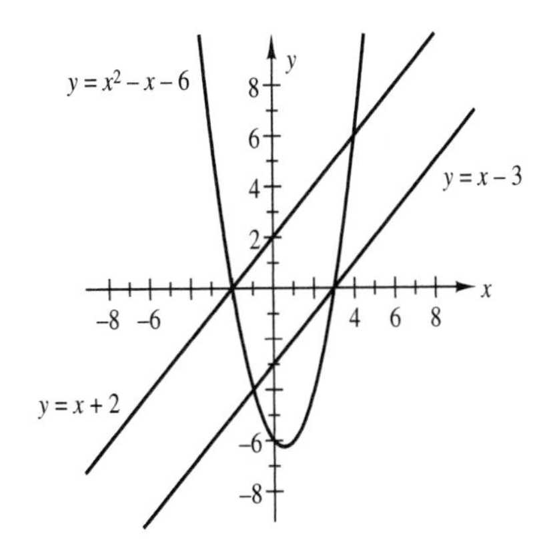

Figure 4.14

If we cannot easily factor the quadratic polynomial, we use the quadratic formula. The *quadratic formula* tells us that the two roots, x_1 and x_2, of the polynomial $p(x) = ax^2 + bx + c$ are

$$x_1 = \frac{-b + \sqrt{b^2 - 4ac}}{2a} \quad \text{and} \quad x_2 = \frac{-b - \sqrt{b^2 - 4ac}}{2a}.$$

When the *discriminant* $d = b^2 - 4ac$ is negative, its square root is not a real number. In this case the graph of the quadratic function (which is a parabola) does not intersect the x-axis. It lies either entirely above or entirely below the x-axis.

The quadratic function $ax^2 + bx + c$ has no real roots when its discriminant $d = b^2 - 4ac$ is negative.

Often the two roots given by the quadratic formula are written in one expression as

$$x = \frac{-b \pm \sqrt{b^2 - 4ac}}{2a}.$$

Since the roots of $p(x) = ax^2 + bx + c$ are x_1 and x_2, we can use the Factor Theorem (in Chapter 3) to write $p(x)$ in the factored form $p(x) = a(x - x_1)(x - x_2)$. (See the Appendix for the derivation of the quadratic formula and further discussion.)

EXAMPLE 4.4 Factor the following quadratic functions to find their zeros:

a. $f(x) = x^2 - 9$ **b.** $g(x) = 2x^2 - 3x + 1$

Solution

a. To find where $f(x) = 0$, we factor $x^2 - 9 = (x - 3)(x + 3)$ and set each factor equal to 0. The equations $x - 3 = 0$ and $x + 3 = 0$ have solutions $x = 3$ and $x = -3$. So the zeros of $f(x) = x^2 - 9$ are 3 and -3.

b. We factor $g(x)$: $2x^2 - 3x + 1 = (2x - 1)(x - 1)$. Now we solve the equations $2x - 1 = 0$ and $x - 1 = 0$. The solutions $x = \frac{1}{2}$, and $x = 1$ are the zeros of $g(x) = 2x^2 - 3x + 1$. ■

To find the zeros of a quadratic function using the quadratic formula, we first put the function in general form $ax^2 + bx + c$. From this form we can determine the values of a, b, and c to substitute into the quadratic formula.

EXAMPLE 4.5 Either factor or use the quadratic formula to find the zeros of the following quadratic functions:

a. $f(x) = 9 - (x + 2)(2x - 1)$ **b.** $g(x) = x^2 - 7x + 10$ **c.** $h(x) = x^2 + x + 2$

Solution

a. We write $f(x)$ in standard form: $f(x) = 9 - (2x^2 + 3x - 2) = -2x^2 - 3x + 11$. Now we can use the quadratic formula with $a = -2$, $b = -3$, $c = 11$:

$$x = \frac{-(-3) \pm \sqrt{(-3)^2 - 4(-2)(11)}}{2(-2)} = \frac{3 \pm \sqrt{9 + 88}}{-4}.$$

So $x_1 = \frac{3 + \sqrt{97}}{-4} \approx -3.21$ and $x_2 = \frac{3 - \sqrt{97}}{-4} \approx 1.71$ are the two zeros of $f(x)$.

b. We factor $g(x)$: $x^2 - 7x + 10 = (x - 5)(x - 2)$. So $x_1 = 5$ and $x_2 = 2$ are the two zeros of $g(x)$. (We also could have used the quadratic formula.)

c. The function $h(x)$ has no zeros, since substitution of $a = 1, b = 1,$ and $c = 2$ into the quadratic formula gives $x = \frac{-1 \pm \sqrt{1^2 - 4 \cdot 1 \cdot 2}}{2} = \frac{-1 \pm \sqrt{-7}}{2}$, which is not a real number. ■

It is usually not easy to find the zeros of a polynomial of higher degree. However, we can always determine the zeros of any polynomial of the form $f(x) = x^n - c$, where c is a constant. This polynomial equals 0 when $x^n = c$, so its zeros are the nth *roots* of the constant c. These are real numbers that, when multiplied by themselves n times, produce the number c. Just as the square root of c is denoted $\sqrt{c}$, the nth root of c is denoted $\sqrt[n]{c}$; it is also denoted by $c^{1/n}$. When n is odd, there is exactly one real zero of the polynomial $x^n - c$, and it is $x = \sqrt[n]{c}$. When n is even and $c \neq 0$, there are two cases: If $c > 0$, this polynomial has the two numbers $x = \pm\sqrt[n]{c}$ as its zeros; if $c < 0$, the polynomial has no zeros.

For example, $f(x) = x^5 + 1$ has only one zero, $x = \sqrt[5]{-1} = -1$, since $(-1)^5 + 1 = -1 + 1 = 0$ and no other number satisfies the equation $x^5 + 1 = 0$. On the other hand, $f(x) = x^4 - 16$ has two zeros, $x = 2$ and $x = -2$. The polynomial $f(x) = x^8 + 5$ has no real zeros, since there is no real number x for which $x^8 + 5 = 0$. (Use a graphing utility to graph these three polynomials and verify the statements about their zeros.)

There is no simple formula like the quadratic formula for finding the zeros of general polynomials of degree 4 or higher. Even factoring cubic polynomials can be very difficult. The Factor Theorem (in Chapter 3) can sometimes help in finding the zeros of such polynomials. We restate it here in a slightly different form:

Factor Theorem: If $P(x)$ is a polynomial, then $P(x) = (x - a)Q(x)$ for some polynomial $Q(x)$ if and only if $P(a) = 0$.

This theorem states the relationship between the zeros of a polynomial and the linear factors of the polynomial. If we know one zero of a polynomial, we can use it to find a second factor ($Q(x)$ in the theorem) of degree less than $P(x)$. We divide $P(x)$ by $x - a$, where a is the zero we know, using either long division or synthetic division to obtain $Q(x)$. If $Q(x)$ is a quadratic polynomial, we can then use the quadratic formula to find other zeros of the polynomial.

EXAMPLE 4.6 Show that $x = 3$ is one of the zeros of $f(x) = 6x^3 - 19x^2 + x + 6$, and use the Factor Theorem to find the other zeros.

Solution We verify that $x = 3$ is a zero of $f(x)$ by substituting $x = 3$ in the polynomial: $f(3) = 6(3)^3 - 19(3)^2 + 3 + 6 = 0$. Thus, $x - 3$ is a factor of $f(x)$. We divide $f(x)$ by $x - 3$ to find a second factor:

$$
\begin{array}{r}
6x^2 - x - 2 \\
x-3\overline{)6x^3 - 19x^2 + x + 6} \\
\underline{6x^3 - 18x^2 } \\
-x^2 + x \\
\underline{-x^2 + 3x } \\
-2x + 6 \\
\underline{-2x + 6}
\end{array}
$$

Now we write $f(x) = (x-3)(6x^2 - x - 2)$. The quadratic polynomial $6x^2 - x - 2$ can be factored, $6x^2 - x - 2 = (2x + 1)(3x - 2)$, so the other zeros of $f(x)$ are $-\frac{1}{2}$ and $\frac{2}{3}$.

■

We can use the Factor Theorem only if we already know a zero of the polynomial. There are several ways to try to find a rational zero of a polynomial. The following theorem provides a technique for determining the only possible rational zeros of a polynomial with integer coefficients. We can use a graphing utility to determine which ones are likely candidates by noting where the graph crosses the x-axis. Long (or synthetic) division or substitution can determine which, if any, of the candidates are rational zeros of the polynomial.

Rational Zeros Theorem: If the rational number $\frac{p}{q}$, in lowest terms, is a zero of a polynomial function with integer coefficients and an integer constant term, then p is an integer factor of the constant term and q is an integer factor of the leading coefficient.

Example 4.7 illustrates how this theorem can be used to find a rational zero (if one exists) for a polynomial.

EXAMPLE 4.7 If possible, find a rational zero of $f(x) = 2x^3 + x^2 - 9$.

Solution The constant term is -9 and the leading coefficient is 2. If $\frac{p}{q}$ is a zero of f, then p must be an integer factor of -9. Therefore, the possible values of p are $\pm 1, \pm 3, \pm 9$. Similarly, q must be one of the integer factors of 2: $\pm 1, \pm 2$. The possible quotients $\frac{p}{q}$ are $\pm 1, \pm 3, \pm 9, \pm\frac{1}{2}, \pm\frac{3}{2}, \pm\frac{9}{2}$.

We can determine if any of these 12 values is a zero of f by substituting each one in turn into $f(x)$. By using a graphing utility to graph f, however, all but one possibility can be eliminated. The polynomial f is graphed in Figure 4.15. From this graph it is obvious that the only zero of f is between 1 and 2; thus the only possible rational zero is $\frac{3}{2}$.

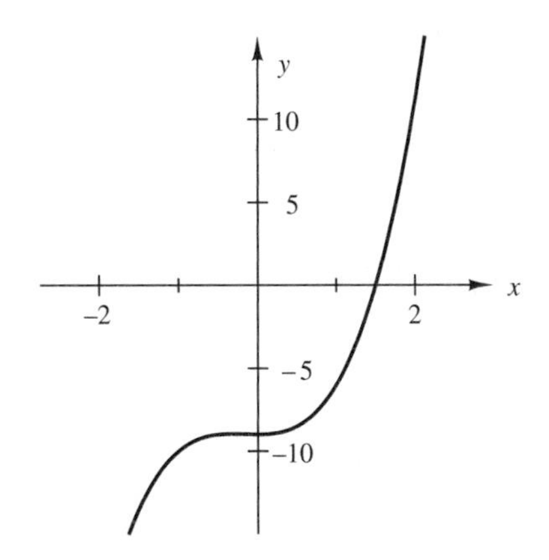

Figure 4.15 $f(x) = 2x^3 + x^2 - 9$

If $f\left(\frac{3}{2}\right) = 0$, then $\frac{3}{2}$ is a zero of the function. When this has been verified, we can divide $2x^3 + x^2 - 9$ by $\left(x - \frac{3}{2}\right)$ to find another factor of the polynomial. We return to this example in Exercise 8 of this section. ■

Exercises 4-B.1

1. Factor each of the following quadratic functions and then find the zeros of the function:

a. $f(x) = 64 - x^2$ **b**. $g(x) = 2x^2 + 3x - 2$ **c**. $h(x) = (x + 2)(x - 3) - (x - 3)$

2. Use the quadratic formula to find the zeros of the following quadratic functions. First, put each function in the form $ax^2 + bx + c$.

a. $f(x) = 5x^2 - 7x + 2$ **b**. $h(x) = 4x^2 - 4x - 2$ **c**. $k(x) = 2x(5 - 3x) - 4$

d. $p(x) = x^2 - \sqrt{12}x + 5$ **e**. $q(x) = x^2 - 4.6x + 5.29$

3. Use a graphing utility to check your answers to Exercises 1 and 2 above.

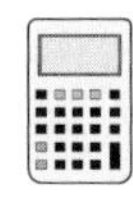

4. If a quadratic function has zeros $x = 4$ and $x = -2$, what are its linear factors?

5. Give a quadratic function that has $x = -\frac{1}{4}$ and $x = 3$ as zeros.

6. If $2 + \sqrt{5}$ is a zero of a quadratic function with integer coefficients and integer constant term, what is the other zero?

7. Find all the zeros of each of the following polynomials.

a. $f(x) = x^3 + 8$ **b.** $g(x) = x^4 - 1$ **c.** $h(x) = x^6 + 64$ **d.** $k(x) = 32x^5 - 1$

8. Verify that $\frac{3}{2}$ is a zero of $f(x) = 2x^3 + x^2 - 9$, and find all other zeros of f.

9. Find all the rational zeros of $f(x) = 2x^3 - 5x^2 - 9x + 18$.

4-B.2 The Approximation of Zeros of Continuous Functions

In many instances it is impractical or unnecessary to find exact values for the zeros of a function. For example, the Rational Zeros Theorem for polynomials is of no use if the polynomial has no rational zeros. In these cases we can use the graph of the function to help us approximate the zeros of the function. Graphing calculators and computer programs with graphing features are especially useful for this task. We can obtain very close approximations to zeros of functions using the trace and zoom features of a graphing utility, or other computational features that display coordinates of points on the graph.

Many computational algorithms have been developed to assist in finding approximations to zeros of continuous functions. One technique is called the *Bisection Method.* To start the Bisection Method, we need to identify an interval that is known to contain a zero. The method produces successively smaller intervals, each of which is contained in the previous interval and contains a zero. For example, if a continuous function f is such that $f(2)$ is negative and $f(3)$ is positive (or vice versa), then the Intermediate Value Theorem guarantees that there must be at least one zero of the function between 2 and 3. Graphs of two such functions are illustrated in Figures 4.16 and 4.17. (There may be several zeros of the function between 2 and 3.)

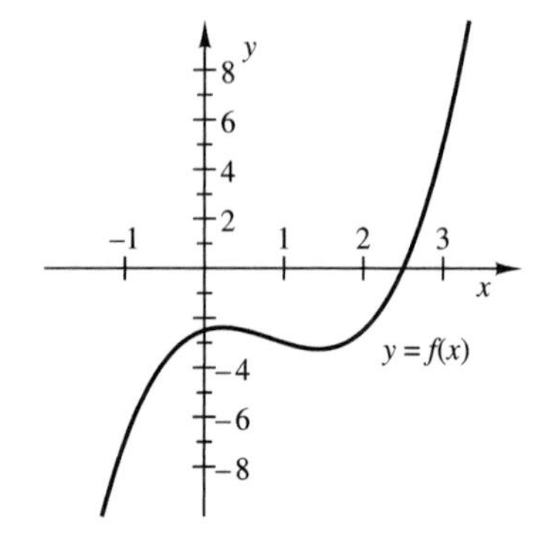

Figure 4.16 f has a zero between $x = 2$ and $x = 3$

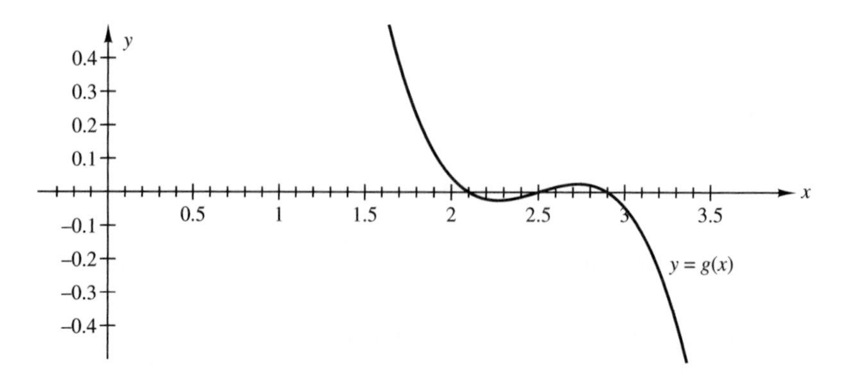

Figure 4.17 g has three zeros between $x = 2$ and $x = 3$

The Bisection Method to Approximate a Zero of a Function *f*

1. Find a pair of numbers a and b such that $f(a)$ and $f(b)$ have opposite signs.

2. Find the midpoint $\frac{a+b}{2}$ of the interval $[a, b]$. Call this midpoint c and evaluate $f(c)$.

3. If $f(a)$ and $f(c)$ have opposite signs, repeat the procedure in step 2 on the interval $[a, c]$; otherwise, repeat the procedure for the interval $[c, b]$. Continue in this manner to narrow the interval to ever smaller intervals, and stop when you attain the desired accuracy of approximation to the value of the zero of $f(x)$. The desired accuracy of approximation may be stipulated to be when the length of the last interval found to contain the zero is less than some previously agreed upon value, such as, for example, 0.001. Alternatively, the desired accuracy may be stipulated to be when the value of $|f(x)|$ is less than some previously agreed upon value, such as, for example, 0.01. A third way to end the approximation process is to specify how many iterations we will use, as in the next example.

EXAMPLE 4.8 Find an interval $[n, n + 1]$ with integer endpoints in which $f(x) = 3x^4 - 2x - 5$ has a zero. Use four iterations of the Bisection Method to estimate the value of this zero. Use values rounded to four decimal places in your calculations.

Solution We evaluate $f(x)$ at several integer values of x and see that $f(1) = -4$ and $f(2) = 39$, so f has a root between 1 and 2. We start with the interval $[1, 2]$. Next, we find the midpoint of this interval, $\frac{1+2}{2} = 1.5$, and evaluate $f(1.5) = 3(1.5)^4 - 2(1.5) - 5 = 7.1875$. This completes the first iteration. Since $f(1) = -4$ and $f(1.5) = 7.1875$, we know f has a zero between 1 and 1.5. Now we apply the Bisection Method to the interval $[1, 1.5]$. The midpoint of this interval is 1.25, and $f(1.25) = 3(1.25)^4 - 2(1.25) - 5 = -0.1758$. This ends the second iteration. Since $f(1.25)$ is negative and $f(1.5)$ is positive, f has a zero in the interval $[1.25, 1.5]$. Repeating the procedure for this interval, we see that $f(1.375) = 3(1.375)^4 - 2(1.375) - 5 = 2.9734$. This completes the third iteration. We now know that f has a root between 1.25 and 1.375, since $f(1.25) = -0.1758$ and $f(1.375) = 2.9734$. For our fourth and final iteration, we find the midpoint of the interval $[1.25, 1.375]$ and evaluate f at that midpoint: $f(1.3125) = 1.2776$, so since $f(1.25)$ is negative and $f(1.3125)$ is positive, we now know f has a root in the interval $[1.25, 1.3125]$. We can choose any point in this interval as our approximation to the zero of f on this interval. At the left endpoint of this interval, $x = 1.25$ and $f(1.25) = -0.1758$. The successive points on the graph of f evaluated in these iterations are shown in Figure 4.18, and the graph of f is shown in Figure 4.19. If we continued the iteration procedure, we would produce a closer approximation to this zero of $f(x)$.

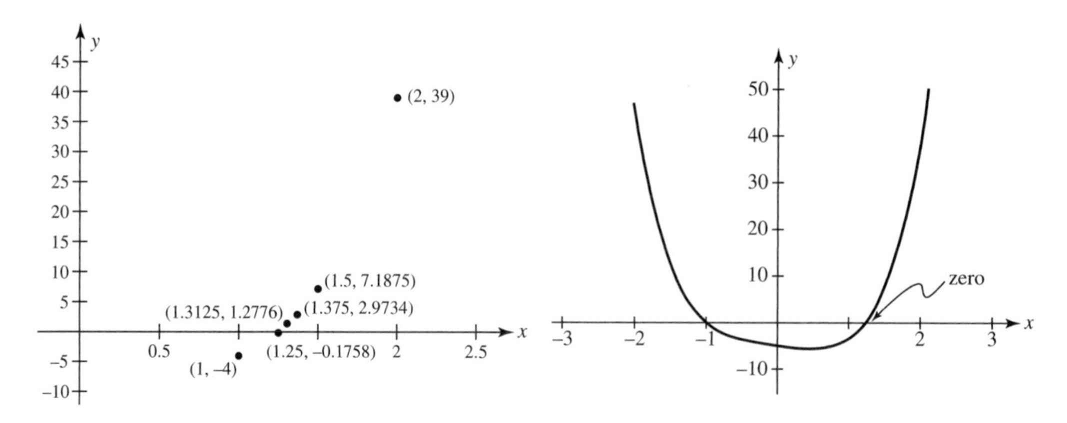

Figure 4.18

Figure 4.19 $f(x) = 3x^4 - 2x - 5$ ■

Exercises 4-B.2

1. Consider the function $f(x) = -18x^3 - 19x^2 + 19x + 10$. Find two consecutive integers x_1 and x_2 such that $f(x_1)$ and $f(x_2)$ have opposite signs.

2. Use three iterations of the Bisection Method to estimate a zero for the function given in Exercise 1. Round to three decimal places in your calculations.

3. Consider the function $f(x) = \frac{2}{x}$. Although $f(-1) = -2$ and $f(1) = 2$, there is no zero for this function between -1 and 1. Explain how this can happen.

4. Consider the function $f(x) = \frac{|x|}{x}$. Although $f(-1) = -1$ and $f(1) = 1$, there is no zero between -1 and 1. Explain why.

5. Use a graphing utility to graph $f(x) = \frac{2x^2 - 6x}{x} + 6$. From the graph, the function appears to have a zero at $x = 0$. Why is this not a zero of the function?

6. Use a graphing utility to graph $f(x) = 2x^4 - 4x^3 - x^2 + 3x - 4$. Then use the zoom and trace capabilities to estimate a zero of the function to the nearest hundredth.

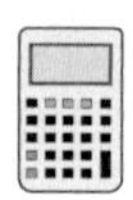

7. The function $f(x) = x^4 - x^3 - x^2 - 9$ has two zeros, one that is positive and one that is negative.

a. Use three iterations of the Bisection Method to approximate the positive zero.

b. Use three iterations of the Bisection Method to approximate the negative zero.

c. Use the trace and zoom features of a graphing utility to estimate the value of each zero to three decimal places.

4-C MORE ON DOMAINS OF FUNCTIONS

In Section 2-B we defined the domain of a function. To determine the domain of a function, it is often necessary to find the zeros of a polynomial. For example, the domain of a rational function excludes the zeros of its polynomial denominator, and the domain of a function that is the square root of a polynomial function includes only those values where the polynomial is nonnegative.

EXAMPLE 4.9 Find the domain of the rational function f defined by

$$f(x)=\frac{6x^2}{5x^2-4x}.$$

Solution The function f is defined everywhere except where the denominator $5x^2-4x$ is 0. We first factor $5x^2-4x=x(5x-4)$; then we set each factor equal to 0: $x=0$ and $5x-4=0$. The zeros of the denominator of f are $x=0$ and $x=\frac{4}{5}$.

The domain of f is all real numbers except 0 and $\frac{4}{5}$, so it is the union of three intervals: $(-\infty, 0)\cup(0, \frac{4}{5})\cup(\frac{4}{5}, \infty)$. ■

EXAMPLE 4.10 Find the domain of g, where $g(x)=\sqrt{x^2-5x+3}$.

(Note that $g(x)$ is the composite $(h\circ k)(x)$, where $h(x)=\sqrt{x}$ and $k(x)=x^2-5x+3$.)

Solution To find the domain of g, we must find the set of real numbers x such that $x^2-5x+3\geq 0$, since the square root of a negative number is not defined.

The graph of $y=x^2-5x+3$ in Figure 4.20 shows where the polynomial is positive; this is the part of the graph above the x-axis. The zeros of the polynomial mark the points at which the function changes from negative to positive or positive to negative. We need to find the zeros of x^2-5x+3. We use the quadratic formula to find

$$x_1=\frac{5+\sqrt{13}}{2}\approx 4.3 \quad\text{and}\quad x_2=\frac{5-\sqrt{13}}{2}\approx 0.7.$$

From the graph in Figure 4.20 we see that $x^2 - 5x + 3 = (x - x_1)(x - x_2) \geq 0$ when $x \leq \frac{5-\sqrt{13}}{2}$ or $x \geq \frac{5+\sqrt{13}}{2}$. Therefore, the domain of g is approximately $(-\infty, 0.7] \cup [4.3, +\infty)$.

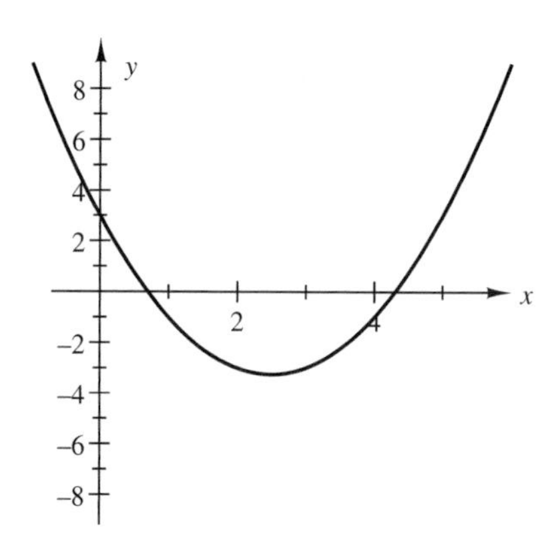

Figure 4.20 $y = k(x) = x^2 - 5x + 3$

Another way to find the domain of the function $k(x) = x^2 - 5x + 3 = (x - 4.3)(x - 0.7)$ is to set up a sign chart for each of the factors. The product of these signs determines a sign chart for $k(x)$ that can be used to determine where $k(x)$ is positive. To form a sign chart for k and use it to find the domain of $g(x) = \sqrt{k(x)}$, we first draw a number line and mark on it the two zeros of k, 0.7 and 4.3. The first factor of $k(x)$ is $x - 4.3$; this is negative when x is less than 4.3 and positive when x is greater than 4.3. We show these signs on the chart, above the number line and next to the factor $x - 4.3$. Similarly, the second factor of $k(x)$ is $x - 0.7$; this is negative when x is less than 0.7 and positive when x is greater than 0.7. We show these signs on the chart next to the factor $x - 0.7$. The product of the two factors, $k(x)$, is positive on the intervals where the two factors are both negative or both positive. We show the sign of $k(x)$ on the chart as well. Such a chart is shown below. The interval graph below the chart shows the domain of $g(x) = \sqrt{k(x)}$. (It is the union of the intervals where $k(x) \geq 0$.)

− − − − − − − − − − − − − 0 + + + sign of $x - 4.3$

− − − − 0 + + + + + + + + + + + + sign of $x - 0.7$

+ + + + 0 − − − − − − − − 0 + + + sign of $k(x) = x^2 - 5x + 3$

0.7 4.3 ■

In general, the following steps can be used to find the domain of a function that is the even root (square root, fourth root, etc.) of a polynomial.

First find the zeros of the polynomial; these divide the x-axis into intervals. Next, set up a sign chart for each of the linear factors associated with the zeros found in the first step and use these to make a sign chart for the original polynomial (the product of all the factors). Finally, identify the domain of the function as the union of the intervals on which the original polynomial is nonnegative.

EXAMPLE 4.11 Find and graph the domain of the following functions:

a. $f(x) = \sqrt{x-3}$ **b.** $g(x) = \sqrt[4]{20+8x-x^2}$ **c.** $h(x) = \sqrt[3]{2-x}$

Solution

a. Since $x-3$ must be nonnegative for $\sqrt{x-3}$ to be defined, we must have $x-3 \geq 0$. Thus, $x \geq 3$ is the domain of $f(x)$. The domain is represented graphically:

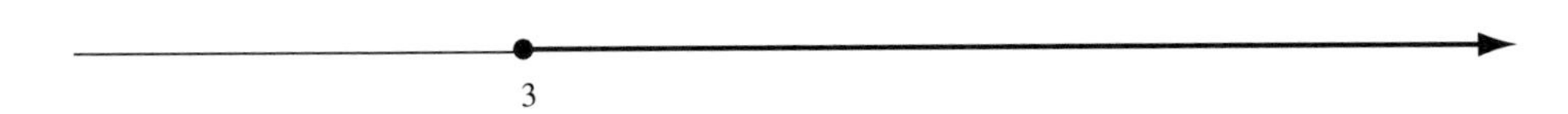

b. We want to find all x for which $20 + 8x - x^2 \geq 0$. By factoring $20 + 8x - x^2 = (10 - x)(2 + x)$, we find the zeros of the polynomial are $x = 10$ and $x = -2$. (We also could use the quadratic formula to find the zeros.) The sign chart is:

+	+	+	+	+	+	+	+	+	+	+	+	+	+	0	–	–	sign of $10 - x$
–	–	–	0	+	+	+	+	+	+	+	+	+	+	+	+	+	sign of $2 + x$
–	–	–	0	+	+	+	+	+	+	+	+	+	+	0	–	–	sign of $(10 - x)(2 + x)$
			–2											10			

The sign chart shows that the domain of $g(x)$ is $[-2, 10]$.

c. $h(x)$ is defined for all values of x, since the cube root of $2 - x$ is defined when $2 - x$ is positive, negative, or zero. The domain of $h(x)$ is $(-\infty, \infty)$. ■

4-C.1 Domains of Composite Functions

To find the domain of a function $f \circ g$ that is the composite of two other functions f and g, we need to consider how we evaluate such a composite function. Recall that $(f \circ g)(a) = f(g(a))$. We first evaluate the inner function g at a to get the value $g(a)$. So a must be in the domain of g. We then evaluate the outer function f at this value $g(a)$. So $g(a)$ must be in the domain of f. This two-step process leads to a description of the domain of the composite function $f \circ g$. The domain of $f \circ g$ is all real numbers a such that a is in the domain of g and the resulting value $g(a)$ is in the domain of f.

> The domain of $f \circ g$ is sometimes smaller than the domain of g. A number a in the domain of g is in the domain of $f \circ g$ only if $f(g(a))$ is defined.

We now illustrate how to use this description of the domain of a composite function when the functions are given in table form, by formulas, or in graphical form.

EXAMPLE 4.12 Consider functions f and g, each with the domain $\{-3, -2, -0.5, 0, 1.5, 3, 4\}$ and whose function values are given in the following table.

x	$f(x)$	$g(x)$
−3	2	3
−2	1.5	4
−0.5	−2	0
0	3	−1
1.5	0	−2
3	2	−3
4	−0.5	−3

Use the table to answer the following questions.

a. Is 0 in the domain of $f \circ g$? If so, find $(f \circ g)(0)$. If not, explain why not.

b. Is -0.5 in the domain of $g \circ f$? If so, find $(g \circ f)(-0.5)$. If not, explain why not.

c. Is 1 in the domain of $g \circ f$? If so, find $(g \circ f)(1)$. If not, explain why not.

Solution

a. To find if 0 is in the domain of $f \circ g$, we first check if 0 is in the domain of g. It is, and $g(0) = -1$. We next check to see that -1 is in the domain of f. It is not in the given domain. Therefore, 0 is not in the domain of $f \circ g$ because $f(g(0)) = f(-1)$ is not defined.

b. We check to see if -0.5 is in the domain of the inner function f of the composite function $g \circ f$. It is, and $f(-0.5) = -2$. Next we check that -2 is in the domain of g. Since $g(-2) = 4$, we can evaluate $(g \circ f)(-0.5) = g(f(-0.5)) = g(-2) = 4$; thus, -0.5 is in the domain of $g \circ f$.

c. The value 1 is not in the domain of $g \circ f$, since 1 is not in the domain of the inner function f. ■

EXAMPLE 4.13 For each pair of functions f and g, find the domain of the composite function $f \circ g$.

a. $f(x) = \dfrac{1}{4 - x^2}$, $g(x) = \sqrt{x}$ **b.** $f(x) = \dfrac{1}{x - 5}$, $g(x) = \dfrac{x+3}{x+1}$

Solution

a. For a real number $x = a$ to be in the domain of $f \circ g$, first a must be in the domain of the inner function g. Since $g(x) = \sqrt{x}$, $a \geq 0$. Second, $g(a) = \sqrt{a}$ must be in the domain of $f(x) = \frac{1}{4 - x^2}$. This domain is all real numbers except +2 and −2, so we rule out all values of a for which $\sqrt{a} = +2$ or $\sqrt{a} = -2$. There are no values of a for which $\sqrt{a} = -2$, and $\sqrt{a} = +2$ for $a = 4$. We must exclude $a = 4$ from the domain of $f \circ g$. Therefore, the domain of $f \circ g$ is all $a \geq 0$ except $a = 4$, or $[0, 4) \cup (4, \infty)$.

b. First, we determine the domain of the inner function g, which is all real numbers $a \neq -1$. Second, $g(a) = \frac{a+3}{a+1}$ must be in the domain of f, so $g(a) \neq 5$. To find which values of a must be excluded, we solve $\frac{a+3}{a+1} = 5$. Multiplying the equation by $a + 1$ gives $a + 3 = 5(a + 1)$, so $a + 3 = 5a + 5$. Thus $4a = -2$, and $a = -\frac{1}{2}$. Therefore, the domain of $f \circ g$ is all real numbers a, $a \neq -1$ and $a \neq -\frac{1}{2}$, or $(-\infty, -1) \cup (-1, -\frac{1}{2}) \cup (-\frac{1}{2}, \infty)$. ■

EXAMPLE 4.14 Consider the functions whose graphs are given in Figures 4.21 and 4.22. Use these figures to answer the following questions.

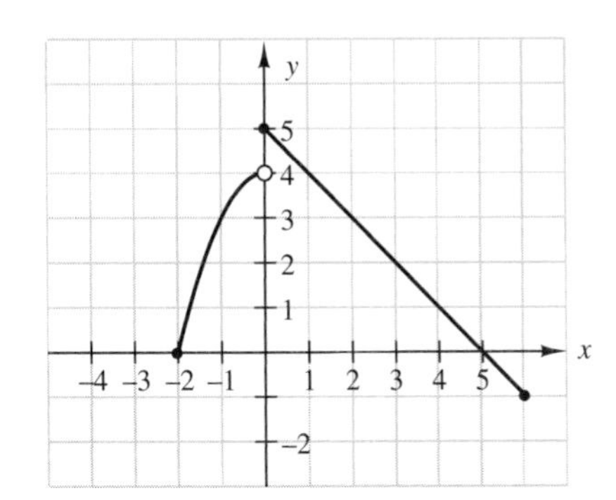

Figure 4.21 $y = f(x)$

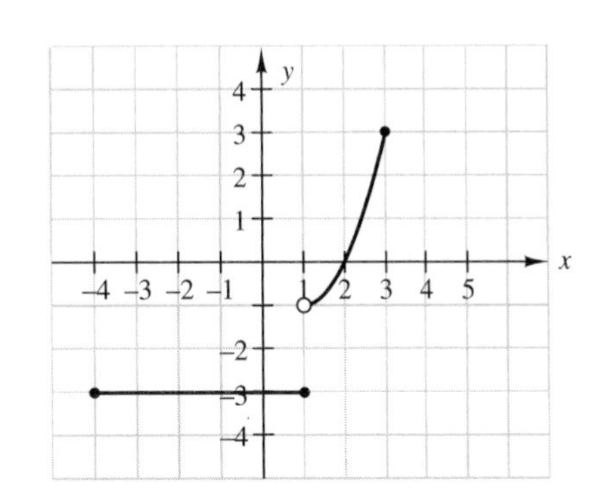

Figure 4.22 $y = g(x)$

a. Is 0 in the domain of $g \circ f$? If it is, find $(g \circ f)(0)$. If not, explain why not.

b. Is −3 in the domain of $g \circ f$? If it is, find $(g \circ f)(-3)$. If not, explain why not.

c. Is 3 in the domain of $g \circ f$? If is it, find $(g \circ f)(3)$. If not, explain why not.

d. Find the domain of $f \circ g$.

Solution

a. From the graph we see that $f(0) = 5$. We now need to evaluate the function g at $f(0)$. But $g(f(0)) = g(5)$ is not defined, so 0 is not in the domain of $g \circ f$.

b. We check the inner function f at −3 and see that $f(-3)$ is not defined, so −3 is not in the domain of $g \circ f$.

c. The inner function is f, and $f(3) = 2$. Thus $(g \circ f)(3) = g(f(3)) = g(2) = 0$, and so 3 is in the domain of $g \circ f$.

d. For $x = a$ to be in the domain of $f \circ g$, first a must be in the domain of the inner function g. From the graph of g, we see that a must be in the interval $[-4, 3]$. Also, the resulting $g(a)$ must be in the domain of the outer function f. We see that $g(a) = -3$ for a in $[-4, 1]$, but -3 is not in the domain of f. Therefore, any a in $[-4, 1]$ is not in the domain of $f \circ g$. For a in $(1, 3]$, $g(a)$ is in $(-1, 3]$ and all real numbers in $(-1, 3]$ are in the domain of f. Therefore, the domain of $f \circ g$ is $(1, 3]$. ■

Exercises 4-C

1. Solve the following inequalities:

a. $3x - 5 > 0$ **b.** $6 - 4x \leq 7$ **c.** $x^2 - 25 > 0$

d. $x^2 + 5x - 6 \geq 0$ **e.** $3x^2 - x - 5 \leq 0$

2. Find the domain of each of the following functions:

a. $f(x) = \dfrac{4}{x^2 + 9}$ **b.** $g(x) = \dfrac{x + 3}{9x^2 + 27x}$

c. $h(x) = \dfrac{5x^2}{x^2 + x - 6}$ **d.** $s(t) = \dfrac{t^2 - 4}{2t^2 - 3t - 5}$

3. Find the domain of each of the following functions:

a. $f(x) = \sqrt{7 - x}$ **b.** $g(x) = \sqrt{x^2 - 4x + 4}$ **c.** $h(x) = \sqrt{25 - 4x^2}$

d. $k(t) = \sqrt[4]{2t^2 + t + 2}$ **e.** $m(x) = \sqrt[3]{x^2 - 12}$

4. Consider the functions f and g, each with domain $\{-5, -4, -1, \frac{1}{4}, \frac{1}{2}, 1, 3, 7\}$, defined as follows: $f(x) = 2x + 1$ and $g(x) = \frac{1}{x}$.

a. Is 1 in the domain of $g \circ f$? If so, find $(g \circ f)(1)$. If not, explain why not.

b. Is -3 in the domain of $f \circ g$? If so, find $(f \circ g)(-3)$. If not, explain why not.

c. Is $\frac{1}{4}$ in the domain of $g \circ f$? If so, find $(g \circ f)\left(\frac{1}{4}\right)$. If not, explain why not.

5. Let $f(x) = \sqrt{x}$ and $g(x) = \dfrac{1}{x - 2}$.

a. Is 0 in the domain of $f \circ g$? If so, find $(f \circ g)(0)$. If not, explain why not.

b. Is 0 in the domain of $g \circ f$? If so, find $(g \circ f)(0)$. If not, explain why not.

c. Find the domain of $f \circ g$.

d. Find the domain of $g \circ f$.

6. Consider the functions whose graphs are given in Figures 4.23 and 4.24. Use these figures to answer the questions that follow.

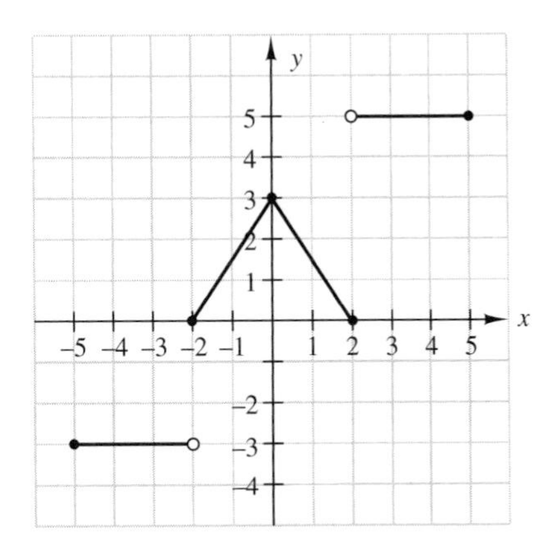

Figure 4.23 $y = f(x)$

Figure 4.24 $y = g(x)$

a. Is -4 in the domain of $g \circ f$? If it is, find $(g \circ f)(-4)$. If not, explain why not.

b. Is 2 in the domain of $g \circ g$? If it is, find $(g \circ g)(2)$. If not, explain why not.

c. Is 0 in the domain of $f \circ g$? If it is, find $(f \circ g)(0)$. If not, explain why not.

d. Find the domain of $g \circ f$.

Chapter 4 Exercises

1. For each of the functions in a–e below that are polynomials, give the degree of the polynomial and its leading coefficient. For those that are not polynomials, give a reason why.

a. $f(x) = \dfrac{2x-3}{3x+4}$ **b.** $g(t) = \dfrac{\sqrt{3}}{2}t^2 - 4t$ **c.** $h(z) = 7$

d. $i(x) = 10^x$ **e.** $j(y) = (3 - 2y)^3$

2. Find the zeros of the following polynomials:

a. $f(x) = 2x - 3$ **b.** $g(x) = x^2 - 7x - 10$

c. $h(t) = 40t - 16t^2$ **d.** $F(y) = 3y^2 - y + 5$

e. $G(z) = 0.125z^3 - 27$ **f.** $C(x) = -5$

3. Use a graphing utility to check your answers to Exercise 2 by noting where the graphs of the functions intersect the x-axis.

4. For each of the functions in Exercise 2, compare the degree of the polynomial with the number of zeros of the function. Is there a pattern? Describe the pattern.

5. Each of the following polynomials has $x = 2$ as one of its zeros. Use the Factor Theorem to find the other real zeros of the polynomial and write the polynomial in factored form.

a. $x^3 - 2x^2 - 9x + 18$ **b.** $x^3 - 2x^2 + 4x - 8$ **c.** $2x^3 - 7x^2 + 5x + 2$

6. How can you prove that $x = 3$ is not a zero of $x^3 + x^2 - 4x - 4$?

7. A polynomial has exactly three zeros, –2, 3, and 4. How can you be certain that $x - 2$ is not a factor of this polynomial?

8. Use three iterations of the Bisection Method to approximate a zero for each of the following functions:

a. $f(x) = x^4 - 2x^2 + 5x - 8$ **b.** $g(x) = x^3 - 2x^2 + 3x - 3$

9. Use a graphing utility with a trace feature to estimate zeros to the nearest hundredth for the two functions in Exercise 8.

10. It is possible for the composition of functions to be several levels deep.
Suppose $f(x) = 3x$, $g(x) = 2x - 5$, and $h(x) = x^2$.
Then $(f \circ g \circ h)(2) = f(g(h(2))) = f(g(4)) = f(3) = 9$.

Evaluate the following compositions of the functions f, g, and h above.

a. $(f \circ g \circ h)(-3)$ **b.** $(g \circ f \circ h)(2)$ **c.** $(h \circ f \circ g)(-1)$

d. $(f \circ h \circ g)\left(\frac{1}{2}\right)$ **e.** $(g \circ h \circ f)(-1.2)$ **f.** $(f \circ f \circ f)(2)$

11. Suppose a city has a population of 100,000 and the population doubles every 30 years. Suppose that immigration into the city adds an additional 50,000 to the population every 30 years.

a. Describe the function $f(x)$ that will give the population after a 30-year period, where x is the current population.

b. Compose the function f with itself (more than once) to answer the following question. If the city can accommodate 2,000,000 residents and the population growth continues at the same rate, in how many years will the population exceed the limits of the city's resources?

12. **a**. Graph the following functions and determine the number of zeros of each:

(i) $f(x) = x^2$ (ii) $f(x) = x^2 - 6x + 9$ (iii) $f(x) = -2x^2 - 3$

(iv) $h(x) = x^2 + 2x + 2$ (v) $g(x) = 4x^2 + 12x + 9$

b. Find the discriminant $d = b^2 - 4ac$ of each function in part a.

c. When the value of the discriminant is 0, how many different real zeros does the function have?

d. When the value of the discriminant is negative, how many real zeros does the function have?

e. If the discriminant is positive, how many different real zeros do you think the function will have? How could you test your conjecture?

13. The graphs of several functions follow. For each graph give the x-values where the function is discontinuous.

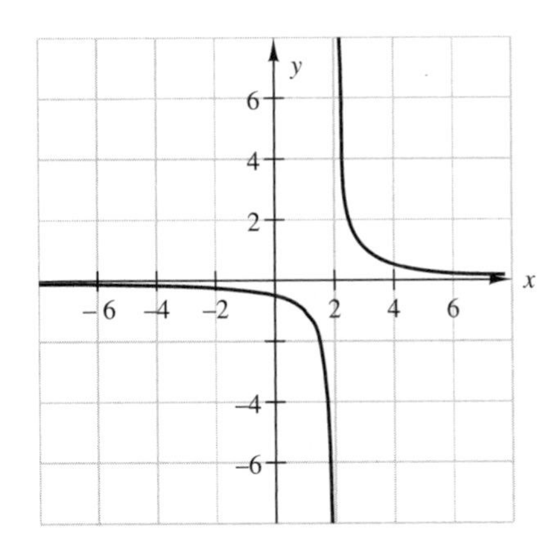

Figure 4.25

Figure 4.26

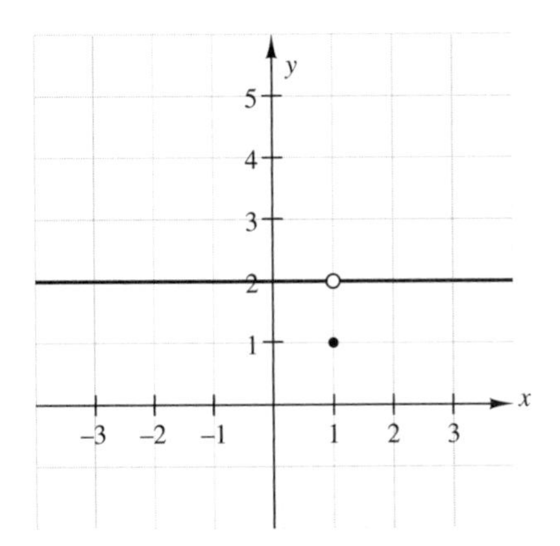

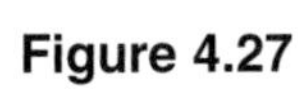
Figure 4.27

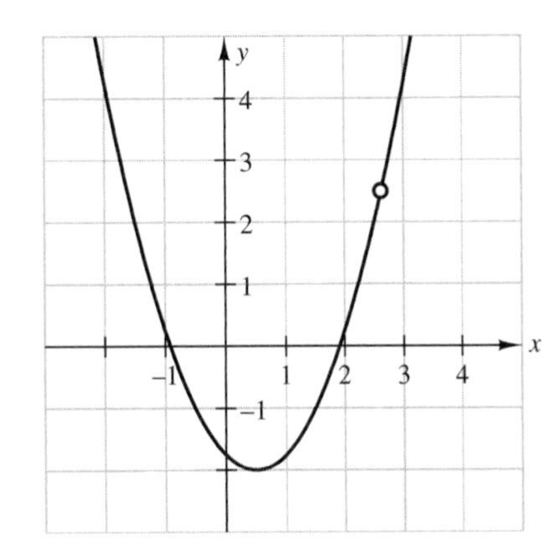

Figure 4.28

14. Find the domain of each of the following functions:

a. $f(x)=\sqrt{x^2+x+3}$ **b.** $g(x)=\sqrt{9x^2-30x+25}$

c. $h(x)=\sqrt[4]{3x-2}$ **d.** $s(t)=\sqrt[6]{2t^2-3t-5}$

e. $v(t)=\sqrt{3t^2-4t-1}$ **f.** $w(x)=\sqrt[3]{5x^3-2x+3}$

15. For each of the following functions, decide if it is continuous for all real numbers. If a function is not continuous everywhere, state the values of x at which it is discontinuous.

a. $f(x)=\dfrac{x-3}{x^2-9}$ **b.** $g(x)=\dfrac{x^2-3}{x^2-9}$

c. $h(x)=\dfrac{x+4}{x+4}$ **d.** $s(t)=\dfrac{t}{t^2+4}$

e. $v(t)=\dfrac{|t-2|}{3t-6}$

16. Use a graphing utility to graph each of the functions in Exercise 15.

a. If the function is discontinuous, describe the type of discontinuity and the x-value where it occurs. (Use answers from Exercise 15.)

b. If a function has a "hole" in its graph, the discontinuity can be removed by defining a piecewise function with the "hole" filled in. Which of the points of discontinuity found in part a are removable in this way?

17. The graphs of f and g are given in Figures 4.29 and 4.30, respectively. Use these graphs to answer the questions that follow.

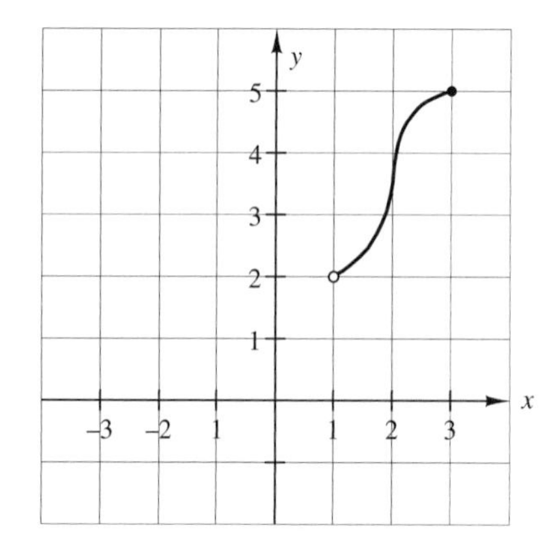

Figure 4.29 $f(x)$

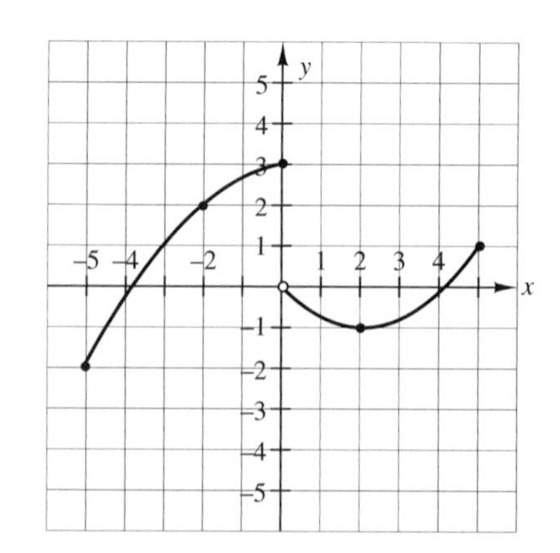

Figure 4.30 $g(x)$

a. Is 1 in the domain of $g\circ f$? If it is, find $(g\circ f)(1)$. If not, explain why not.

b. Is 2 in the domain of $f\circ g$? If it is, find $(f\circ g)(2)$. If not, explain why not.

c. Is 0 in the domain of $f \circ g$? If it is, find $(f \circ g)(0)$. If not, explain why not.

d. Find the domain of $f \circ g$.

e. Find the domain of $g \circ f$.

18. Let $f(x) = \sqrt{x}$, $g(x) = \frac{1}{x+5}$, and $h(x) = \sqrt{x-4}$.

a. Is -3 in the domain of $f \circ g$? If it is, find $(f \circ g)(-3)$. If not, explain why not.

b. Is -3 in the domain of $h \circ g$? If it is, find $(h \circ g)(-3)$. If not, explain why not.

c. Find the domain of $g \circ f$.

d. Find the domain of $f \circ g$.

e. Find the domain of $f \circ h$.

CHAPTER 5

The Role of Infinity

The concept of infinity plays an important role in mathematics. It is used in different contexts and is often associated with the ideas of unlimited, unbounded, or endless. Although its origins can be traced to the Greeks in the sixth century B.C., it was in the seventeenth century with the development of calculus that the mathematical concept of infinity took shape.

The symbol ∞ for infinity was used for the first time in 1655 by the English mathematician John Wallis. He may have taken it from the Roman numeral $\overline{|\infty|}$, which represents the number 100 million (Eli Maor, *To Infinity and Beyond, A Cultural History of the Infinite*, Birkhäuser Boston, Cambridge, MA, 1987). The symbol ∞ seems an appropriate choice since it has the shape of a figure eight curve (technically called a "lemniscate") that can be traveled indefinitely. We have already used this symbol to represent intervals on the real line that are unbounded or unlimited. For example, $(3, \infty)$ denotes the set of all real numbers larger than 3, a set that contains larger and larger numbers and extends forever to the right on the real line.

We should always keep in mind that ∞ represents a concept and **not a real number** and therefore we **cannot** use the rules of operation on this symbol as we do with real numbers.

☞ ∞ is a symbol for a concept. It is not a real number.

In calculus, the concept of infinity in the sense of "unbounded" appears in the study of limits in two different situations:

1. When we analyze the behavior of the values of a function $f(x)$ as x increases without bound, we ask what happens to the values $f(x)$ as x moves farther and farther away to the right on the real line (which is symbolized as $x \to \infty$). We also ask what happens to the values $f(x)$ as x decreases without bound—that is, as x moves farther and farther away to the left on the real line (which is symbolized as $x \to -\infty$).

2. When the values $f(x)$ of a function increase without bound or decrease without bound as x moves closer and closer to a particular number c, we use ∞ to symbolize this behavior.

We illustrate these two situations with the reciprocal function $F(x) = \frac{1}{x}$.

Behavior of $F(x) = \frac{1}{x}$ as x increases without bound

The following table gives values of $F(x)$ for some large values of x:

x	1000	10,000	100,000	1,000,000	10,000,000
$F(x) = \frac{1}{x}$	0.001	0.0001	0.00001	0.000001	0.0000001

Although the table shows just a small sample of values, we could continue this table forever. Taking larger and larger values of x will result in smaller and smaller values of $\frac{1}{x}$. Figure 5.1 shows the graph of $y = \frac{1}{x}$ on an interval (0, 120,000). We say that *as x grows without bound,* $\frac{1}{x}$ *approaches* 0; in other words, *as x approaches infinity,* $\frac{1}{x}$ *approaches* 0. To express this function behavior symbolically, calculus texts use the notation $\lim_{x\to\infty}\frac{1}{x} = 0$, which is read as "the limit of $\frac{1}{x}$ as x approaches infinity equals 0."

Note that x is **never** equal to ∞ and the quotient $\frac{1}{x}$ is **never** equal to 0.

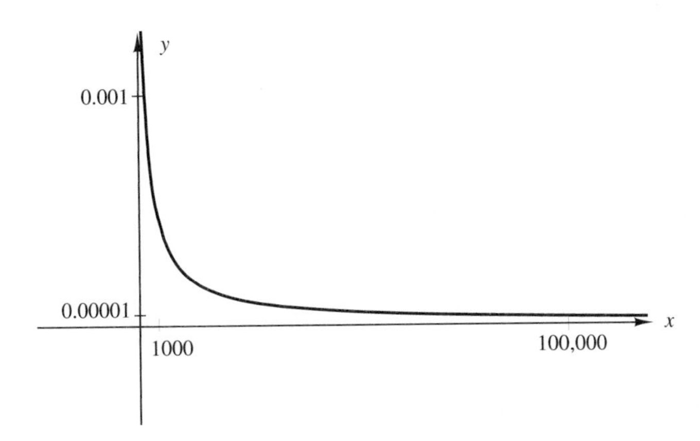

Figure 5.1 $y = \frac{1}{x}$ for $x > 0$

Behavior of $F(x) = \frac{1}{x}$ as x decreases without bound

A similar situation occurs when x takes negative values that are larger and larger in absolute value, as in the following table:

x	-1000	$-10{,}000$	$-100{,}000$	$-1{,}000{,}000$	$-10{,}000{,}000$
$F(x) = \frac{1}{x}$	-0.001	-0.0001	-0.00001	-0.000001	-0.0000001

We say that *as x decreases without bound (or approaches negative infinity),* $\frac{1}{x}$ *approaches* 0. Using limit notation, we write $\lim_{x \to \infty} \frac{1}{x} = 0$. See Figure 5.2.

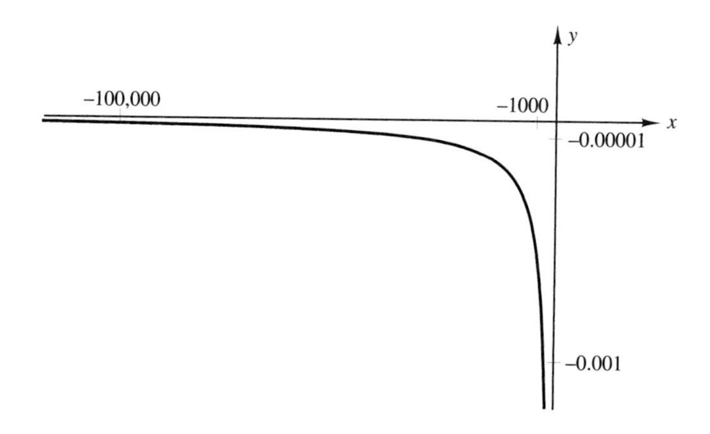

Figure 5.2 $y = \frac{1}{x}$ for $x < 0$

Behavior of $F(x) = \frac{1}{x}$ where $y = F(x)$ increases without bound

The function $F(x) = \frac{1}{x}$ is not defined at $x = 0$, but it is defined for all nonzero values of x. The following table gives values of $F(x)$ for several positive values of x that are close to 0:

x	0.001	0.0001	0.00001	0.000001	0.0000001
$F(x) = \frac{1}{x}$	1000	$10{,}000$	$100{,}000$	$1{,}000{,}000$	$10{,}000{,}000$

We see from the table that as x takes on smaller and smaller positive values, the values of $\frac{1}{x}$ grow larger and larger. We say that *as x approaches* 0 *through positive values (or x approaches* 0 *from the right),* $\frac{1}{x}$ *grows without bound, or approaches infinity.* The notation $x \to 0^+$ or $x \downarrow 0$ indicates that x approaches 0 from the right—that is, through values above or larger than 0. Using limit notation, we write $\lim_{x \to 0^+} \frac{1}{x} = \infty$ or $\lim_{x \downarrow 0} \frac{1}{x} = \infty$. The graph of $y = \frac{1}{x}$ for $-1.5 < x < 1.5$ is sketched in Figure 5.3.

Behavior of $F(x)=\frac{1}{x}$ where $y = F(x)$ decreases without bound

If we take negative values of x, the values of $\frac{1}{x}$ will be negative. The following table shows values of $F(x)$ for some negative values of x that are close to 0:

x	−0.001	−0.0001	−0.00001	−0.000001	−0.0000001
$F(x)=\frac{1}{x}$	−1000	−10,000	−100,000	−1,000,000	−10,000,000

The values of $F(x)$ in the table are negative and very large in absolute value. In this case, we say that *as x approaches* 0 *through negative values* (*or x approaches 0 from the left*), $\frac{1}{x}$ *decreases without bound, or approaches negative infinity.*

In limit notation, we write $\lim_{x\to 0^-}\frac{1}{x}=-\infty$ or $\lim_{x\uparrow 0}\frac{1}{x}=-\infty$. See Figure 5.3.

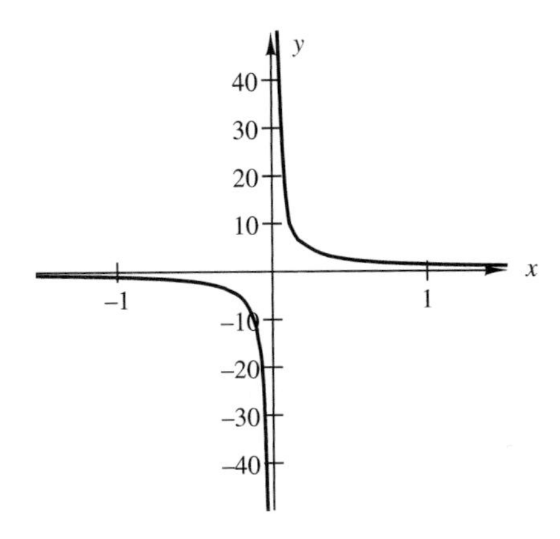

Figure 5.3 $y=\frac{1}{x}$ for $-1.5 < x < 1.5$

Recall that ∞ and $-\infty$ are not real numbers and $\lim_{x\to c^-} f(x)=\infty$ means f increases without bound. There is no x such that $f(x) = \infty$.

5-A GRAPHICAL INTERPRETATION

5-A.1 Horizontal Asymptotes

In a picture of the graph of a function, we often see only a portion of the graph for a subset of the domain. Many functions are defined on intervals of the form $[a, \infty)$, $(-\infty, b]$, or $(-\infty, \infty)$. Often we want to know what happens to the graph beyond the picture we can see. What is the behavior of the graph as x increases without bound or as x decreases without bound?

EXAMPLE 5.1 The potency P (in milligrams) of vitamin C tablets is a function of the time t they have been stored. The graph in Figure 5.4 shows this relationship.

Explain what the graph shows and why it seems reasonable.

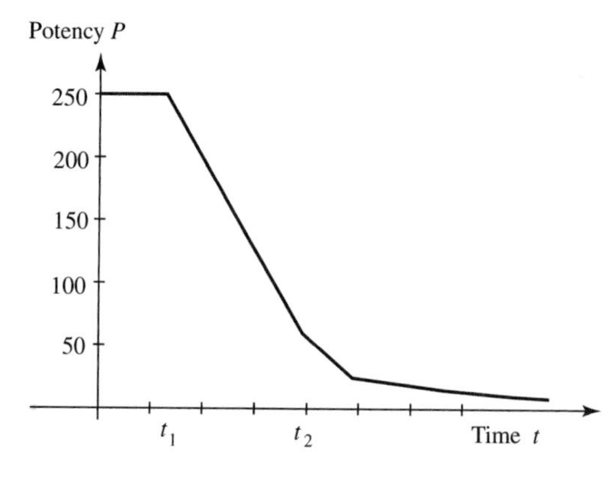

Figure 5.4

Solution The graph shows that the tablets have an initial potency of 250 milligrams (mg), and it remains constant at 250 until time $t = t_1$. From time t_1 to time t_2, the potency declines in a linear manner to approximately 60 mg. After time t_2, the potency continues to decrease, but at a slower pace than before. The potency is actually never 0, but it is very small for large values of t. The larger t is, the closer the value of P is to 0. In other words, storing the vitamins for a long time results in tablets with very little potency.

Since the potency gets arbitrarily close to zero as time increases without bound, we say the line with equation $P = 0$ is a *horizontal asymptote* of the potency function.

■

EXAMPLE 5.2 The number of items N a typical assembly line worker can process in a day is a function of the number of days t he or she has been working on the line. In the graph in Figure 5.5, the vertical axis shows the number of items processed, and the horizontal axis shows the number of days on the job.

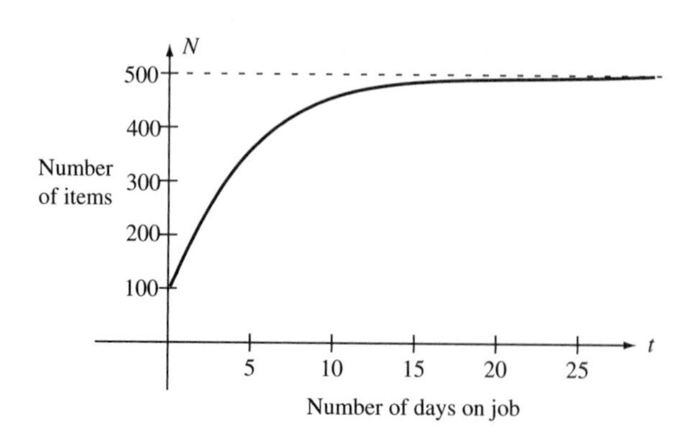

Figure 5.5

a. The point (0, 100) is on the graph. Interpret what this means in this context.

b. Describe what happens as t increases.

c. When can a worker be considered to be experienced?

Solution

a. The point (0, 100) on the graph indicates that a typical worker can process 100 items the first day on the job.

b. As t increases, the worker can process more and more items per day. Reading some values from the graph, we observe, for example, that after 5 days on the job the worker processes 360 items in a day, after 10 days this number is approximately 465, and after 15 days this number is about 490. In the beginning, the number of items processed increases quite rapidly, but after approximately 2 weeks on the job, the worker is very close to maximum efficiency. The worker's level of production after that time is almost constant. After a long time in the job, the average worker will process close to 500 items. As t increases, the curve flattens out and is very close to the horizontal line with equation $N = 500$.

c. We consider the worker experienced after 15 days, since it is approximately then that the number of items processed by the worker almost levels off, or remains almost constant. ■

EXAMPLE 5.3 Draw several possible examples of graphs of functions whose y-values get closer and closer to a constant as x grows larger and larger, and explain the behavior of the graphs.

Solution Three possible graphs are shown in Figures 5.6, 5.7, and 5.8.

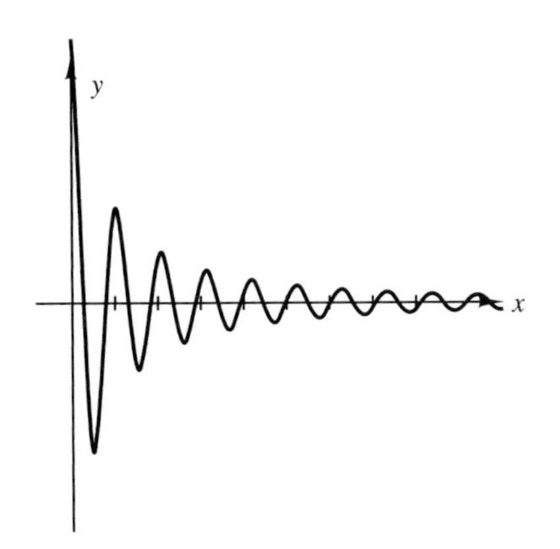

Figure 5.6 $y = f(x)$

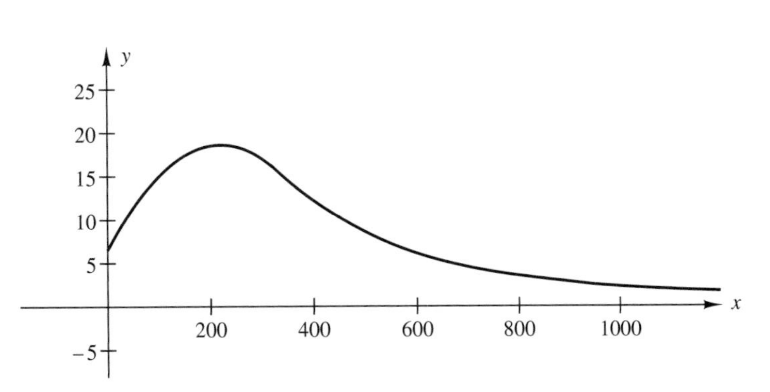

Figure 5.7 $y = g(x)$

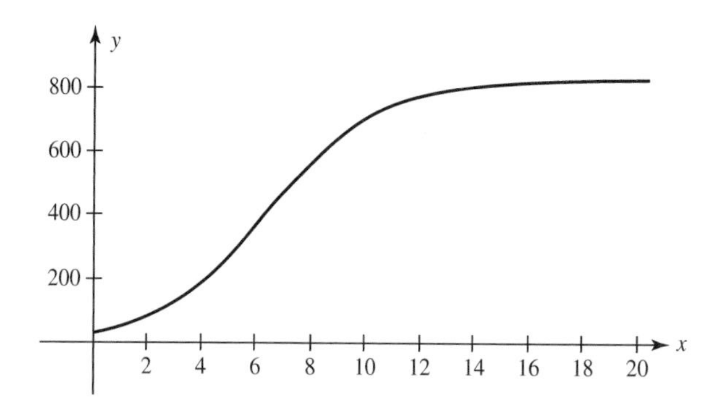

Figure 5.8 $y = h(x)$

The graphs of f and g show the y-values get closer and closer to zero as x grows large without bound. The graph of the function in Figure 5.6 crosses the x-axis repeatedly. For very large values of x, $f(x)$ is sometimes positive, sometimes zero, and sometimes negative. The graph of the function wobbles, or oscillates, as it seems to collapse to the x-axis. The graph might model the height of a hanging spring above its natural length over time after it is compressed and released. The graph in Figure 5.7 represents a function for which the values $g(x)$ are always positive for large values of x. The points on the graph of g get closer and closer to the line $y = 0$, but they never touch the line. The graph might model the amount of drug in the bloodstream x minutes after taking a medicine. The graph in Figure 5.8 is an example of an "S" curve in which the function values grow slowly at the start, then grow rapidly for a short time, and finally level off over time. The graph might model the number of American bison in a reserve during the years after the reserve was started with 32 bison. As x grows larger, the graph approaches a horizontal line $y = 832$, which indicates that the carrying capacity of the reserve is 832 bison.

On the graphs in Figures 5.6 and 5.7, the points on the graph of the function are close to the line $y = 0$ when x is very large. The larger x is, the closer the points on the graph are to $(x, 0)$. In these two cases, we say that as $x \to \infty$ (which means x gets larger and larger without bound), $f(x) \to 0$ (which means the values of f get closer and closer to 0). Using limit notation, we write $\lim_{x \to \infty} f(x) = 0$, or $f(x) \to 0$ as $x \to \infty$. We say the line $y = 0$ is a *horizontal asymptote* of the graph. The graph shown in Figure 5.8 has a horizontal asymptote of $y = 832$ because $\lim_{x \to \infty} f(x) = 832$. ■

Definition: The line $y = b$ is a horizontal asymptote of the graph of $y = f(x)$ if $f(x) \to b$ as $x \to \infty$ or $f(x) \to b$ as $x \to -\infty$.

5-A.2 Vertical Asymptotes

Sometimes the graph of a function is cut off by the upper or lower edge of the rectangular window in which the graph is drawn. We want to know if the graph goes up (or down) forever, in which case no enlargement of the window or scaling of the graph can capture its vertical span.

EXAMPLE 5.4 Three functions and their graphs are given in Figures 5.9, 5.10, and 5.11. None of these functions is defined when $x = 1$, but the functions are defined for x arbitrarily close to 1. In each case, explain what happens to the values of the function when x is close to 1.

a.

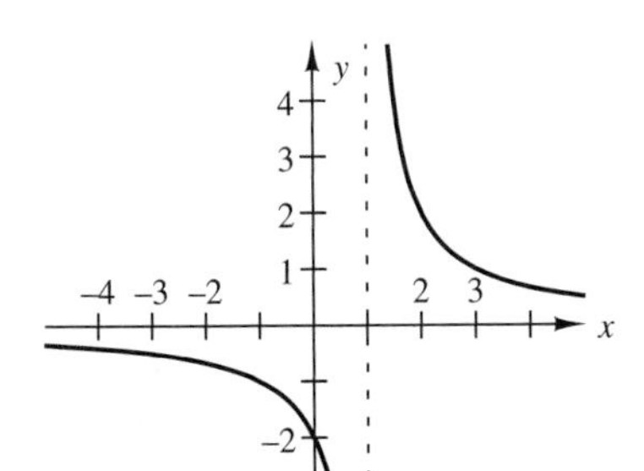

Figure 5.9 $f(x)=\frac{2}{x-1}$

b.

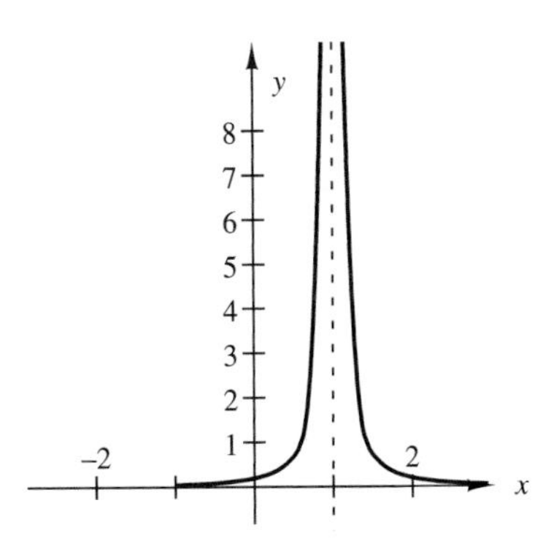

Figure 5.10 $g(x)=\frac{1}{5(x-1)^2}$

c.

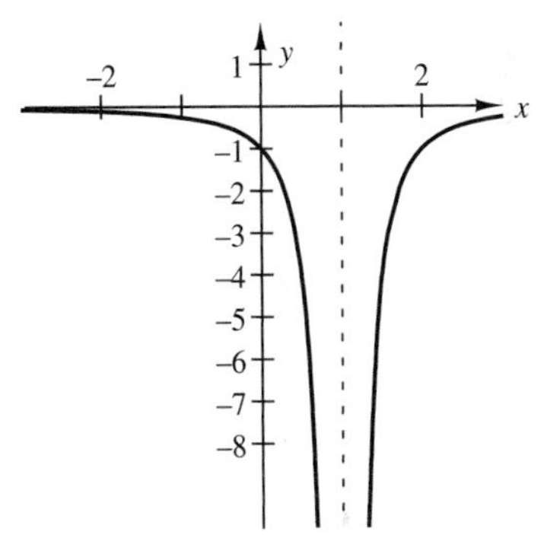

Figure 5.11 $h(x)=-\frac{1}{(x-1)^2}$

Solution

a. The values $f(x)$ increase without bound when x is larger than 1 and is moving toward 1. The values $f(x)$ decrease without bound when x is less than 1 and is moving toward 1. In the first case, we write: As $x \to 1^+$, $f(x) \to \infty$, which is read, *as x approaches* 1 *from the right, f(x) increases without bound.* We can also write $\lim_{x\to 1^+} f(x) = \infty$ and say *the limit of f(x), as x approaches* 1 *from the right, is infinity.*

In the second case, we write: As $x \to 1^-$, $f(x) \to -\infty$, which is read, *as x approaches* 1 *from the left, f(x) decreases without bound.* We can also write $\lim_{x\to 1^-} f(x) = -\infty$ and say *the limit of f(x), as x approaches* 1 *from the left, is negative infinity.*

b. The values $g(x)$ increase without bound as x moves closer to 1, whether x is larger or smaller than 1. We express this as follows: As $x \to 1$, $g(x) \to \infty$ (*as x approaches* 1, *g(x) increases without bound*). We can also write $\lim_{x\to 1} g(x) = \infty$ (*the limit of g(x), as x approaches* 1, *is infinity*).

c. The values $h(x)$ decrease without bound as x moves closer to 1. We write: As $x \to 1$, $h(x) \to -\infty$, and say *as x approaches* 1, *h(x) decreases without bound.* We can also write $\lim_{x \to 1} h(x) = -\infty$ and say *the limit of h(x), as x approaches* 1, *is negative infinity.* ■

In all three cases in Example 5.4, as x approaches 1, the graph of the function approaches the graph of the vertical line $x = 1$. In such cases, the line $x = 1$ is a *vertical asymptote* of the graph of the function.

Definition: The line $x = c$ is a vertical asymptote of the graph of $y = f(x)$ if at least one of the following is true: $f(x) \to \infty$ as $x \to c^-$, $f(x) \to -\infty$ as $x \to c^-$, $f(x) \to \infty$ as $x \to c^+$, or $f(x) \to -\infty$ as $x \to c^+$.

When using a graphing utility to graph a function that has $x = c$ as a vertical asymptote, you may see what looks like a vertical line $x = c$. This line is *not* part of the graph of the function.

EXAMPLE 5.5 The density d of a body is the ratio of its mass m to its volume V: $d = \frac{m}{V}$. When a star is very old, the volume decreases while the mass remains constant.

a. For m fixed, draw a graph that represents density as a function of volume.

b. Does the graph have a vertical asymptote? What happens to the density as the volume decreases?

Solution

a. The density function is the product of a positive constant m and the reciprocal function $\frac{1}{V}$. Figure 5.12 shows the graph of $d = \frac{100}{V}$, the density function when $m = 100$.

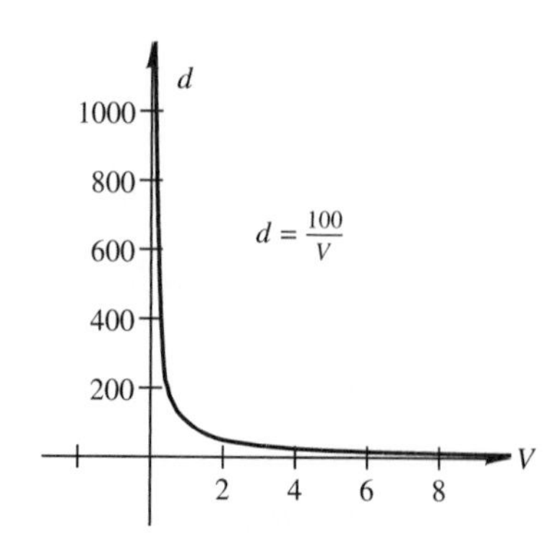

Figure 5.12

b. When the volume gets very small, the density gets very large. The line $V = 0$ is a vertical asymptote. (Theoretically, a large star can become so dense that its gravitational attraction is large enough to not allow light to escape; it has become a "black hole.") ■

Exercises 5-A

1. Draw graphs of three different functions such that the y-values of the functions get arbitrarily close to zero as the x-values decrease without bound.

2. After t weeks of an outbreak of a flu epidemic, the number of people who have caught the disease is $y = g(t)$. The graph of g is given in Figure 5.13.

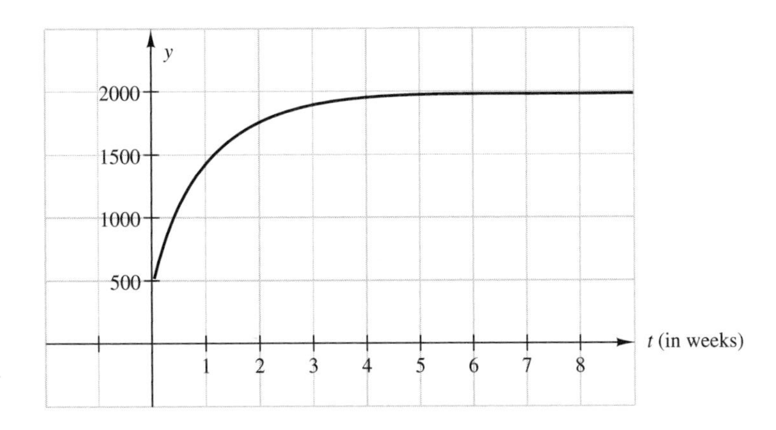

Figure 5.13

a. Approximately how many people caught this flu by the end of the first week?

b. Describe what happens as t increases.

c. Approximately how many people have caught the flu after 2 weeks? 3 weeks? 6 weeks?

d. What is the horizontal asymptote of this graph? Explain the meaning of the asymptote in this context.

e. After how many weeks of the outbreak could you say that the epidemic is over?

3. Is it possible for a function to have more than two vertical asymptotes? If the answer is yes, sketch the graph of such a function. If the answer is no, explain why.

4. Is it possible for a function to have two horizontal asymptotes? If the answer is yes, sketch the graph of such a function. If the answer is no, explain why.

5. Is it possible for a function to have more than two horizontal asymptotes? If the answer is yes, sketch the graph of such a function. If the answer is no, explain why.

6. The graphs of two functions g and h are given in Figure 5.14.

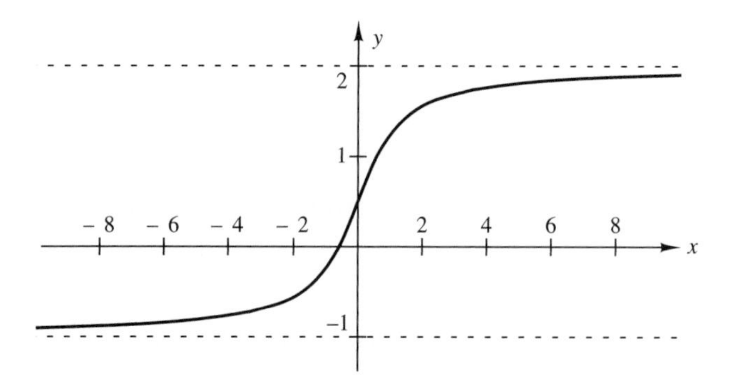

Figure 5.14a $y = g(x)$

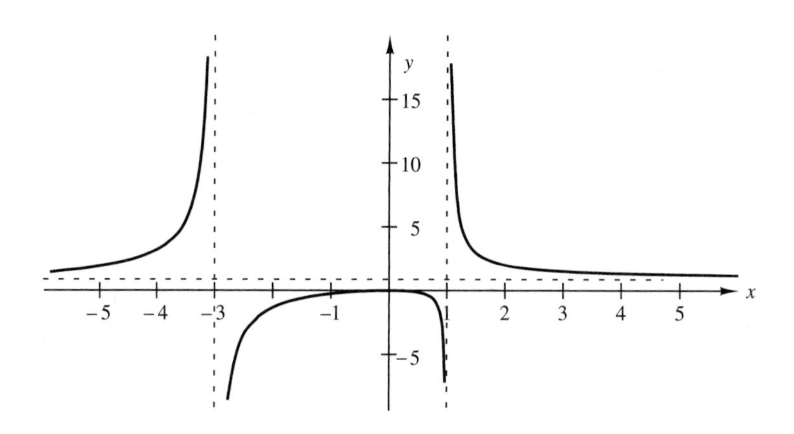

Figure 5.14b $y = h(x)$

Refer to those graphs and complete the following sentences:

a. As $x \to \infty$, $g(x) \to$ _______.

b. As $x \to -\infty$, $g(x) \to$ _______.

c. As $x \to \infty$, $h(x) \to$ _______.

d. As $x \to -\infty$, $h(x) \to$ _______.

e. As $x \to -3^+$, $h(x) \to$ _______.

f. As $x \to -3^-$, $h(x) \to$ _______.

g. As $x \to 1^+$, $h(x) \to$ _______.

h. As $x \to 1^-$, $h(x) \to$ _______.

7. For each function g and h graphed in Figure 5.14, give equations of:

a. All vertical asymptotes of the function (if any)

b. All horizontal asymptotes of the function (if any)

5-B ALGEBRAIC MANIPULATIONS

5-B.1 Finding Horizontal Asymptotes

To find horizontal asymptotes, we ask if the values of a function approach a fixed constant as $x \to \infty$ or $x \to -\infty$. We have seen that the function $y = \frac{1}{x}$ has $y = 0$ as a horizontal asymptote because as $x \to \infty$, $\frac{1}{x} \to 0$. As x gets larger and larger, $\frac{1}{x}$ gets smaller and smaller. It is also true that as $x \to -\infty$, $\frac{1}{x} \to 0$, so the graph of the function $y = \frac{1}{x}$ approaches the horizontal asymptote $y = 0$ as $x \to -\infty$.

In fact, the line $y = 0$ is a horizontal asymptote of any function of the form $y = \frac{1}{x^r}$, where r is a positive rational number, because as x increases without bound, x^r increases without bound and therefore $\frac{1}{x^r} \to 0$ as $x \to \infty$. When x is negative, x^r may not be defined (e.g., for $r = \frac{3}{4}$ and $x = -2$, $x^r = \left(\sqrt[4]{-2}\right)^3$ is not defined). If x^r is defined when x is negative, then as $x \to -\infty$, $x^r \to \infty$ in some cases and $x^r \to -\infty$ in other cases, but in all cases $\frac{1}{x^r} \to 0$.

For $r = 1$, two parts of the graph of $y = \frac{1}{x^r}$ are shown in Figures 5.1 and 5.2. For $r = \frac{1}{2}$, $r = 2$, and $r = 3$, the graphs of $y = \frac{1}{x^r}$ are shown in Figures 5.15, 5.16, and 5.17, respectively.

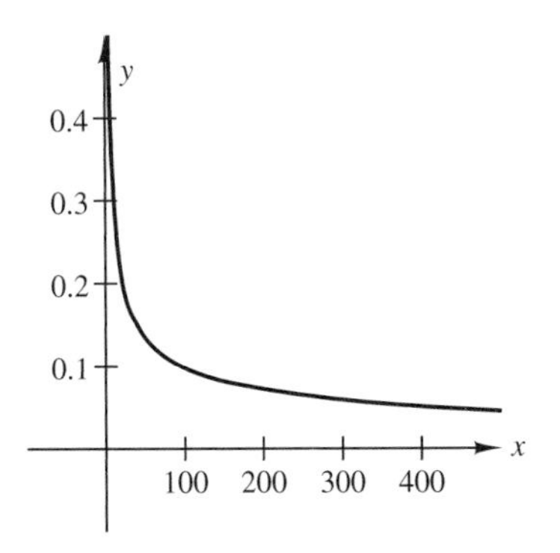

Figure 5.15 $y = \frac{1}{x^{1/2}},\ x > 1$

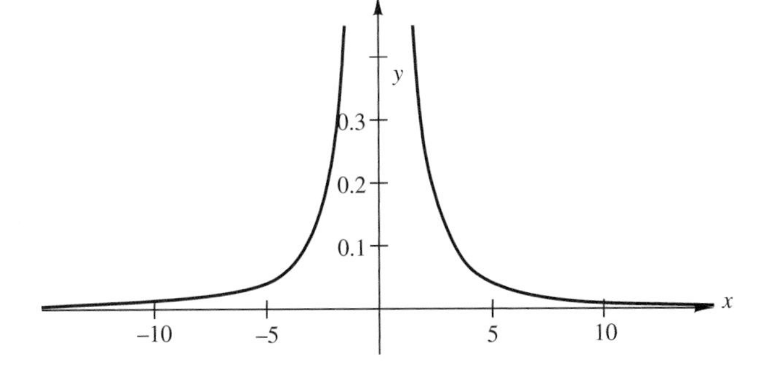

Figure 5.16 $y = \frac{1}{x^2}$

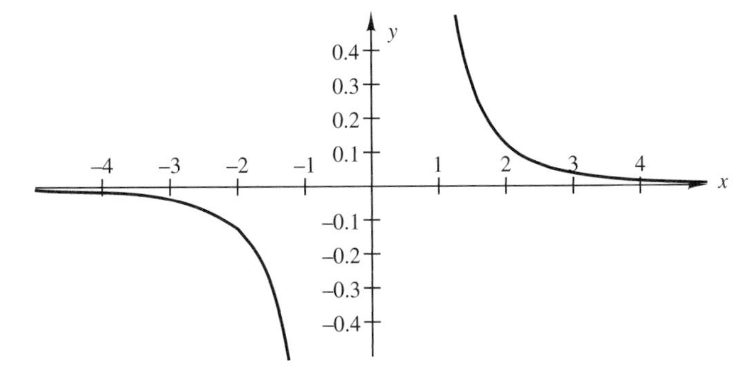

Figure 5.17 $y = \frac{1}{x^3}$

> ☞ Although $\lim_{x \to \infty} \frac{1}{x^r} = 0$, when $c \neq 0$, $\lim_{x \to c} \frac{1}{x^r} \neq 0$.

EXAMPLE 5.6 Analyze the behavior of each function below as $x \to \infty$ and as $x \to -\infty$. Give the equations of any horizontal asymptotes.

a. $g(x) = 3 + \frac{1}{x}$ **b.** $h(x) = \frac{1}{x^2} - 2$ **c.** $F(x) = \dfrac{3 + \frac{1}{x}}{\frac{1}{x^2} - 2}$

Solution

a. As $x \to \infty$ and as $x \to -\infty$, $\frac{1}{x} \to 0$, so $3 + \frac{1}{x} \to 3$.

The line $y = 3$ is a horizontal asymptote of the function g. See Figure 5.18.

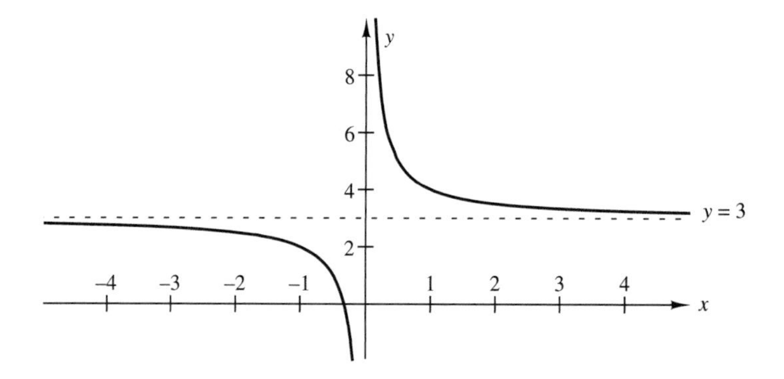

Figure 5.18 $g(x) = 3 + \frac{1}{x}$

b. As $x \to \infty$ and as $x \to -\infty$, $\frac{1}{x^2} \to 0$, so $\frac{1}{x^2} - 2 \to -2$. The line $y = -2$ is a horizontal asymptote of the function h. See Figure 5.19.

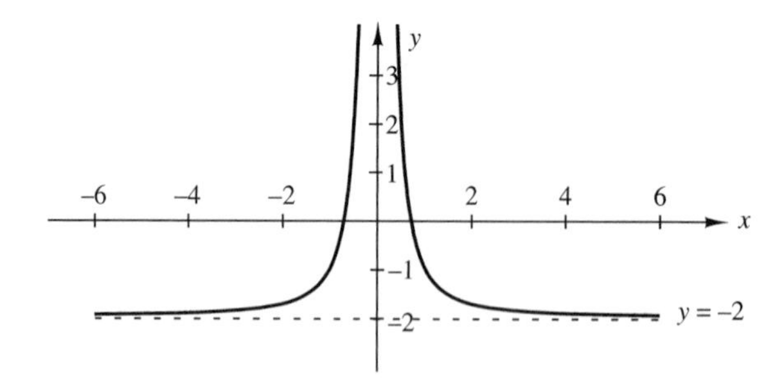

Figure 5.19 $h(x) = \frac{1}{x^2} - 2$

c. Since the function F is the quotient $\frac{g}{h}$, we see that as $x \to \infty$ and also as $x \to -\infty$, $F(x)$ approaches $\frac{3}{-2} = -\frac{3}{2}$. The line $y = -\frac{3}{2}$ is a horizontal asymptote of F. Also, note that $F(0)$ is undefined. See Figure 5.20.

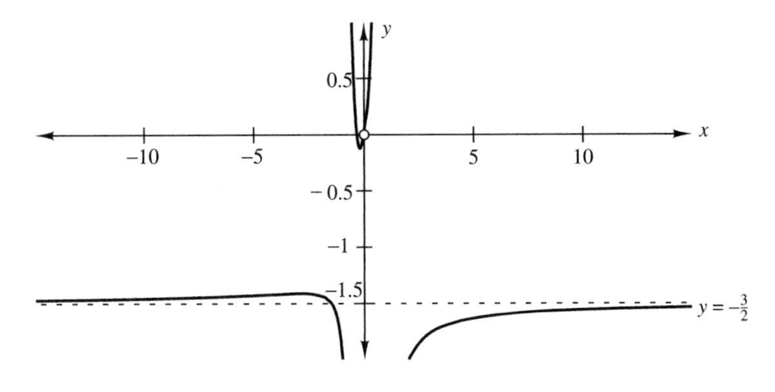

Figure 5.20 $F(x)=\dfrac{3+\frac{1}{x}}{\frac{1}{x^2}-2}$ ■

Note that for $x \neq 0$, the function $F(x) = \dfrac{3+\frac{1}{x}}{\frac{1}{x^2}-2}$ also can be expressed as $\dfrac{3x^2+x}{1-2x^2}$,

since $F(x)=\dfrac{\left(3+\frac{1}{x}\right)}{\left(\frac{1}{x^2}-2\right)}\cdot\dfrac{\left(x^2\right)}{\left(x^2\right)}=\dfrac{3x^2+x}{1-2x^2}$.

But when F is given in this last form, it is not easy to guess what happens to the values of $F(x)$ when x increases without bound or when x decreases without bound. It is when the function is expressed in terms of $\frac{1}{x}$ and $\frac{1}{x^2}=\left(\frac{1}{x}\right)^2$ that we can easily answer those questions.

In general, to analyze the behavior of a function as $x \to \infty$ or as $x \to -\infty$, we often express the function in terms of $\frac{1}{x}$. When the function is a quotient of polynomials, we can divide the numerator and denominator by the highest power of x that appears in the expression.

EXAMPLE 5.7

a. Write $f(x)=\dfrac{x^2-x+1}{1-2x^2}$ in terms of $\frac{1}{x}$ by dividing the numerator and denominator of f by x^2.

b. Use the expression obtained in part a to analyze the behavior of the values $f(x)$ when $x \to \infty$. Does the graph of f have any horizontal asymptotes?

Solution

a. $$\frac{\left(\frac{x^2-x+1}{x^2}\right)}{\left(\frac{1-2x^2}{x^2}\right)}=\frac{\left(\frac{x^2}{x^2}-\frac{x}{x^2}+\frac{1}{x^2}\right)}{\left(\frac{1}{x^2}-\frac{2x^2}{x^2}\right)}=\frac{1-\frac{1}{x}+\frac{1}{x^2}}{\frac{1}{x^2}-2}.$$

b. As $x \to \infty$, $\frac{1}{x} \to 0$ and $\frac{1}{x^2} \to 0$. For this reason, values of the numerator of $f(x)$ get very close to 1 and values of the denominator get close to −2. So, as $x \to \infty$,

$f(x) \to \frac{1}{-2} = -\frac{1}{2}$. The line $y = -\frac{1}{2}$ is a horizontal asymptote. The graph of $y = f(x)$ is shown in Figure 5.21.

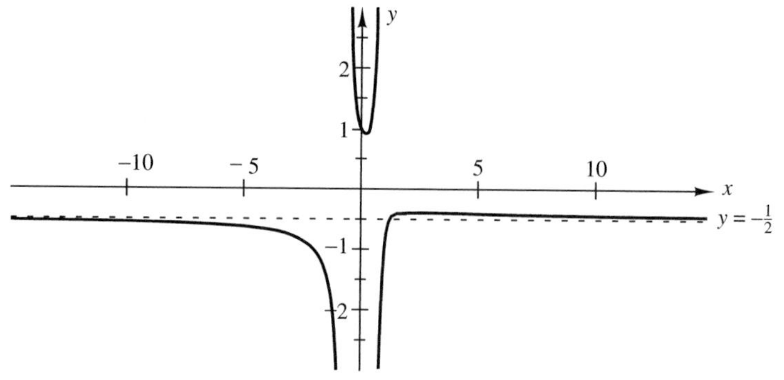

Figure 5.21 $f(x) = \frac{x^2 - x + 1}{1 - 2x^2}$ ■

To solve the next problem, the following facts are useful:

$$\text{For } x \geq 0, \quad x = \sqrt{x^2}.$$

$$\text{For } x \leq 0, \quad x = -\sqrt{x^2}.$$

$$\frac{\sqrt{a}}{\sqrt{b}} = \sqrt{\frac{a}{b}}$$

Sometimes we want to divide an expression such as $\sqrt{p(x)}$ by x. In this case, it is best to write $x = \sqrt{x^2}$ when $x > 0$, so we can rewrite the radicand (expression under the radical) as a quotient. Thus, $\frac{\sqrt{p(x)}}{x} = \frac{\sqrt{p(x)}}{\sqrt{x^2}} = \sqrt{\frac{p(x)}{x^2}}$. When $x < 0$, write $x = -\sqrt{x^2}$. Then $\frac{\sqrt{p(x)}}{x} = \frac{\sqrt{p(x)}}{-\sqrt{x^2}} = -\sqrt{\frac{p(x)}{x^2}}$.

EXAMPLE 5.8

a. Consider the function $g(x) = \frac{\sqrt{5x^2 - x + 3}}{x+1}$. Divide the numerator and denominator by x to write $g(x)$ in terms of $\frac{1}{x}$ for each of the two cases:

(i) x is positive.

(ii) x is negative.

b. Explain what happens to the values $g(x)$ as $x \to \infty$ and as $x \to -\infty$.

Solution

a. (i) $\frac{\sqrt{5x^2 - x + 3}}{x+1} = \frac{\left(\frac{\sqrt{5x^2 - x + 3}}{x}\right)}{\left(\frac{x+1}{x}\right)}$

To divide the numerator by x, we write $x = \sqrt{x^2}$, since in this case, $x > 0$:

$$\frac{\left(\frac{\sqrt{5x^2-x+3}}{x}\right)}{\left(\frac{x+1}{x}\right)}=\frac{\left(\frac{\sqrt{5x^2-x+3}}{\sqrt{x^2}}\right)}{\left(\frac{x+1}{x}\right)}=\frac{\sqrt{\frac{5x^2-x+3}{x^2}}}{\left(\frac{x+1}{x}\right)}=\frac{\sqrt{5\frac{x^2}{x^2}-\frac{x}{x^2}+\frac{3}{x^2}}}{\left(\frac{x}{x}+\frac{1}{x}\right)}=\frac{\sqrt{5-\frac{1}{x}+\frac{3}{x^2}}}{\left(1+\frac{1}{x}\right)}.$$

(ii) We proceed as above but now since x is negative, $x=-\sqrt{x^2}$:

$$\frac{\left(\frac{\sqrt{5x^2-x+3}}{x}\right)}{\left(\frac{x+1}{x}\right)}=\frac{\left(\frac{\sqrt{5x^2-x+3}}{-\sqrt{x^2}}\right)}{\left(\frac{x+1}{x}\right)}=\frac{-\sqrt{\frac{5x^2-x+3}{x^2}}}{\left(\frac{x+1}{x}\right)}=\frac{-\sqrt{5\frac{x^2}{x^2}-\frac{x}{x^2}+\frac{3}{x^2}}}{\left(\frac{x}{x}+\frac{1}{x}\right)}=\frac{-\sqrt{5-\frac{1}{x}+\frac{3}{x^2}}}{\left(1+\frac{1}{x}\right)}.$$

b. In the expression obtained in (i), we use the fact that when $x \to \infty$, $\frac{1}{x} \to 0$ and $\frac{3}{x^2} \to 0$; thus the values $g(x)$ are very close to $\frac{\sqrt{5}}{1}=\sqrt{5}$. In the expression obtained in (ii), valid for negative values of x, we have $\frac{1}{x} \to 0$ and $\frac{3}{x^2} \to 0$ as $x \to -\infty$, and so $g(x) \to \frac{-\sqrt{5}}{1}=-\sqrt{5}$. The function g has two horizontal asymptotes: $y=\sqrt{5}$ and $y=-\sqrt{5}$. See Figure 5.22.

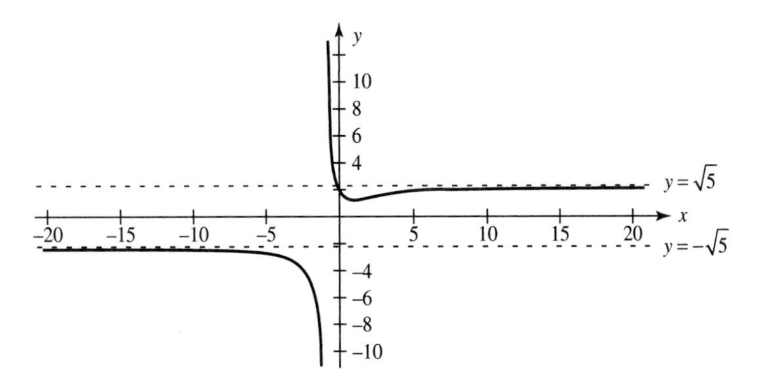

Figure 5.22 $g(x)=\frac{\sqrt{5x^2-x+3}}{x+1}$ ■

In the following example, we illustrate some ways to decide the behavior of a function as $x \to \infty$ when the function is a difference of two quantities, both of which grow arbitrarily large as $x \to \infty$. Although tempting, we cannot say that $\infty - \infty = 0$.

If a function is a difference of two expressions in which at least one is of the form $\sqrt{p(x)}$, we can eliminate the square roots in the difference by multiplying by the sum of the two expressions (this uses the algebraic fact that $(a-b)(a+b) = a^2-b^2$). To guarantee that we do not change the function, we multiply by 1 in the form of this sum divided by itself. Parts b and c of the next example illustrate the procedure.

EXAMPLE 5.9 Analyze the behavior of each of the following functions as $x \to \infty$:

a. $f(x)=x^2-x$ **b**. $g(x)=\sqrt{1+x}-\sqrt{x}$ **c**. $h(x)=\sqrt{x^2+x-1}-x$

Solution

a. We recognize f as a parabola that opens up, so f increases without bound as $x \to \infty$. We can also factor: $f(x) = x^2 - x = x(x - 1)$. As $x \to \infty$, $(x - 1) \to \infty$ also. Since the product of two large positive integers is a larger number, $f(x) \to \infty$.

b. Since $g(x)$ is a difference of two radicals, we multiply $g(x)$ by $\frac{\sqrt{1+x}+\sqrt{x}}{\sqrt{1+x}+\sqrt{x}}$. This will give us an expression for $g(x)$ that is a quotient without the difference that causes the $\infty - \infty$ form:

$$g(x) = \frac{\left(\sqrt{1+x}-\sqrt{x}\right)}{1} \cdot \frac{\left(\sqrt{1+x}+\sqrt{x}\right)}{\left(\sqrt{1+x}+\sqrt{x}\right)} = \frac{1+x-x}{\left(\sqrt{1+x}+\sqrt{x}\right)} = \frac{1}{\sqrt{1+x}+\sqrt{x}}.$$

As $x \to \infty$, $\sqrt{1+x}$ and $\sqrt{x}$ both grow without bound and so does the sum $\sqrt{1+x}+\sqrt{x}$ (the sum of two very large numbers is larger than each of the numbers). So as $x \to \infty$, $g(x) \to 0$. (The function is undefined for $x < 0$.) The graph of g is shown in Figure 5.23.

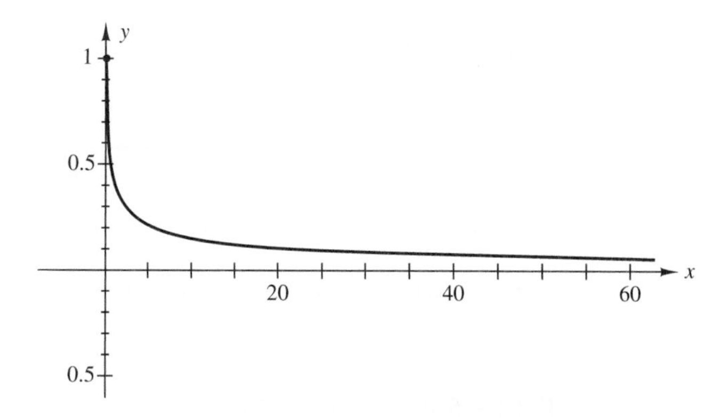

Figure 5.23 $g(x) = \sqrt{1+x} - \sqrt{x}$

c. The function h is also a difference that involves a radical, so:

$$h(x) = \frac{\left(\sqrt{x^2+x-1}-x\right)}{1} \cdot \frac{\left(\sqrt{x^2+x-1}+x\right)}{\left(\sqrt{x^2+x-1}+x\right)} = \frac{x^2+x-1-x^2}{\left(\sqrt{x^2+x-1}+x\right)} = \frac{x-1}{\sqrt{x^2+x-1}+x}.$$

Now, we can divide numerator and denominator by x (which is the highest power of x that occurs in the expression), writing $x = \sqrt{x^2}$ for division in the denominator (since $x > 0$):

$$h(x) = \frac{\frac{x-1}{x}}{\frac{\left(\sqrt{x^2+x-1}+x\right)}{x}} = \frac{1-\frac{1}{x}}{\frac{\sqrt{x^2+x-1}}{\sqrt{x^2}}+\frac{x}{x}} = \frac{1-\frac{1}{x}}{\sqrt{\frac{x^2+x-1}{x^2}}+1} = \frac{1-\frac{1}{x}}{\sqrt{1+\frac{1}{x}-\frac{1}{x^2}}+1}.$$

As $x \to \infty$, $\frac{1}{x} \to 0$ and $\frac{1}{x^2} \to 0$ and therefore $h(x) \to \frac{1}{\sqrt{1}+1} = \frac{1}{2}$. The graph of h is shown in Figure 5.24.

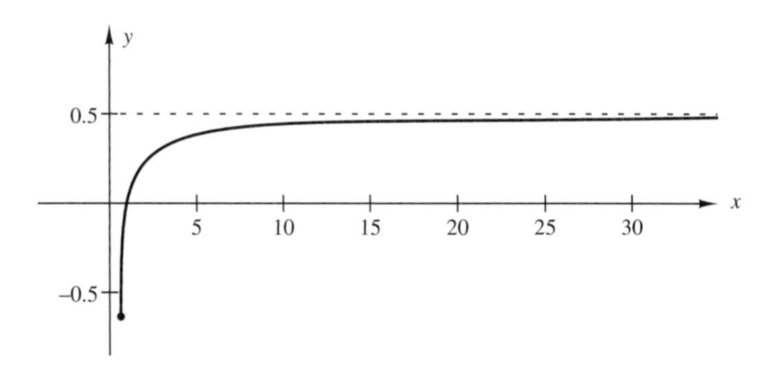

Figure 5.24 $h(x) = \sqrt{x^2 + x - 1} - x$ ■

Exercises 5-B.1

1. For the function $f(x) = \frac{5x^2 - 3x + 1}{x^3 - 2}$, do the following:

a. Divide numerator and denominator by the highest power of x that occurs in the expression.

b. Analyze the behavior of the values of the function as $x \to \infty$.

c. Analyze the behavior of the values of the function as $x \to -\infty$.

d. Give the equations of any horizontal asymptotes of the function.

2. Repeat Exercise 1 for the function $g(x) = \frac{1 - 3x^2}{7x^2 + 1}$.

3. For the function $h(x) = \frac{x^3 - 3x^2 + 2}{4x^3 + x^2 - \sqrt{3}}$, do the following:

a. Analyze the behavior of the values of the function as $x \to 0$.

b. Analyze the behavior of the values of the function as $x \to -2$.

4. Let $F(x) = \frac{\sqrt{x^4 - x}}{4 - x^2}$.

a. Divide numerator and denominator by x^2 to write $F(x)$ in terms of $\frac{1}{x}$. Note that $x^2 = \sqrt{x^4}$ for all x.

b. Analyze the behavior of the values of the function as $x \to \infty$.

c. Write the equations of any horizontal asymptotes.

5. Let $G(x) = \dfrac{\sqrt[3]{x^3 - 2x + 1}}{7x - 2}$.

a. Divide numerator and denominator by x to write $G(x)$ in terms of $\frac{1}{x}$. Note that $x = \sqrt[3]{x^3}$ for all x.

b. Analyze the behavior of the values of the function as $x \to \infty$.

c. Analyze the behavior of the values of the function as $x \to -\infty$.

d. Write the equations of any horizontal asymptotes.

6. Let $f(x) = \dfrac{\sqrt{x^2 - 3}}{2 - x}$.

a. Divide numerator and denominator by x to write $f(x)$ in terms of $\frac{1}{x}$. **Assume x is negative.**

b. Analyze the behavior of the values of the function as $x \to -\infty$.

7. Let $h(x) = \sqrt{x^2 + 3x} - \sqrt{x^2 - 1}$. Analyze the behavior of the values of $h(x)$ when x grows very large. To do this, first write $h(x)$ as $\dfrac{\sqrt{x^2 + 3x} - \sqrt{x^2 - 1}}{1}$ and rationalize the numerator; then divide numerator and denominator by x to write the resulting expression in terms of $\frac{1}{x}$.

8. Use a graphing utility to graph $h(x)$ in Exercise 7 on an interval with left endpoint $x = 1$. Does the graph show the behavior you describe as $x \to \infty$?

5-B.2 Finding Vertical Asymptotes

To determine if the graph of a function f has any vertical asymptotes, we find the values of x near which $f(x)$ grows without bound or decreases without bound. For a function of the form $f(x) = \dfrac{p(x)}{q(x)}$, this may happen when x is a zero of $q(x)$.

EXAMPLE 5.10 The three functions G, H, and K given below are undefined when $x = 2$. For each of these functions, describe what happens to the values of the function when x is close to 2. Give the equations of any vertical asymptotes. Graph each function.

a. $G(x) = \dfrac{x^2 - 4}{x - 2}$ **b**. $H(x) = \dfrac{x + 3}{2 - x}$ **c**. $K(x) = \dfrac{8 - x^3}{(x - 2)^2}$

Solution

a. For $x \neq 2$, $G(x) = \dfrac{x^2 - 4}{x - 2} = \dfrac{(x - 2)(x + 2)}{x - 2} = x + 2$. As $x \to 2$, $G(x) \to 4$. The graph of the function G is the line $y = x + 2$, with the point (2, 4) missing; see Figure 5.25. $G(x)$ does not have any vertical asymptotes.

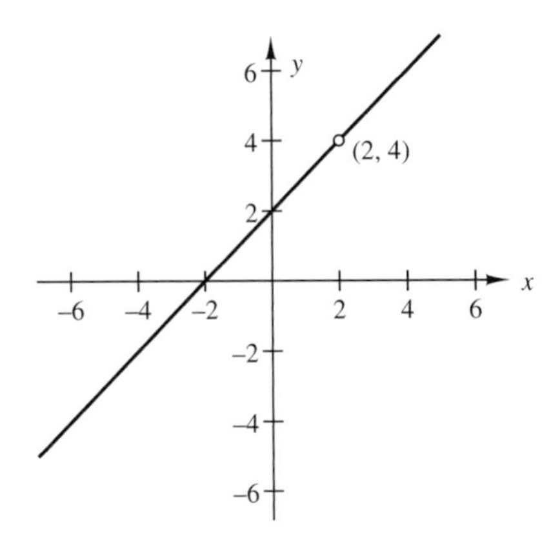

Figure 5.25 $G(x) = \frac{x^2-4}{x-2}$

b. $H(x) = \frac{x+3}{2-x}$ has no common factors in the numerator and denominator, so it cannot be simplified. As $x \to 2$, $(x + 3) \to 5$ and $(2 - x) \to 0$ (through positive values if $x < 2$ and through negative values if $x > 2$). For $x > 2$ and x close to 2, $\frac{x+3}{2-x}$ is negative since it is the quotient of two numbers with opposite signs; therefore, as $x \to 2^+$, $H(x) = \frac{x+3}{2-x} \to -\infty$. On the other hand, when $x < 2$ and is close to 2, $\frac{x+3}{2-x}$ is positive since it is the quotient of two positive numbers; therefore, as $x \to 2^-$, $H(x) = \frac{x+3}{2-x} \to \infty$. The line $x = 2$ is a vertical asymptote. The graph of $y = H(x)$ is shown in Figure 5.26.

c. For $x \neq 2$, $K(x) = \frac{8-x^3}{(x-2)^2} = \frac{-(x-2)(x^2+2x+4)}{(x-2)^2} = \frac{-(x^2+2x+4)}{x-2}$. As $x \to 2$, $-(x^2 + 2x + 4) \to -12$ and $x - 2 \to 0$ (through negative values if $x < 2$ and through positive values if $x > 2$). Therefore, as $x \to 2^-$, $K(x) \to \infty$. On the other hand, as $x \to 2^+$, $K(x) \to -\infty$. The line $x = 2$ is a vertical asymptote. The graph of $y = K(x)$ is shown in Figure 5.27.

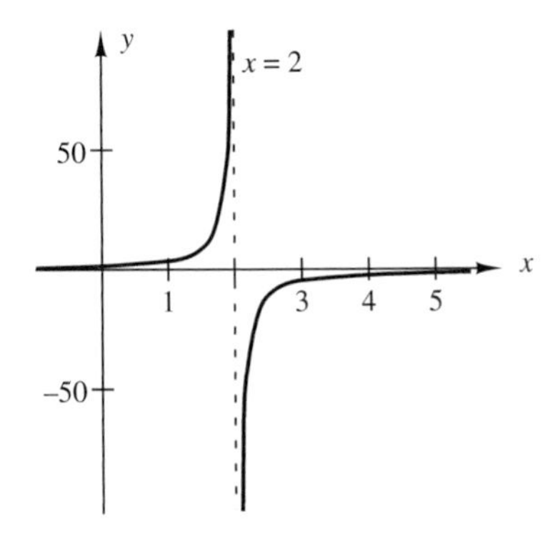

Figure 5.26 $H(x) = \frac{x+3}{2-x}$

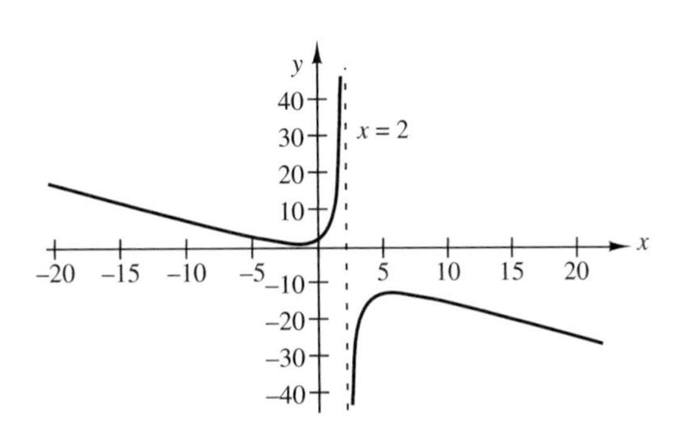

Figure 5.27 $K(x) = \frac{8-x^3}{(x-2)^2}$ ■

EXAMPLE 5.11 Let $f(x)=\dfrac{x^2-2x}{x^2-5x+6}$.

a. Find the values of x at which the denominator is zero.

b. Explain what happens to the values $f(x)$ when x is close to each of the zeros of the denominator.

Solution

a. The denominator $x^2 - 5x + 6$ factors as $(x-2)(x-3)$. It takes the value zero when $x = 2$ and when $x = 3$.

b. When $x = 2$, the numerator is also zero, which means that we can simplify f:

$$f(x)=\frac{x^2-2x}{x^2-5x+6}=\frac{x(x-2)}{(x-2)(x-3)}=\frac{x}{x-3}, \quad \text{for } x \neq 2.$$

Since the sign of f does not change at $x = 2$, we do not need to consider separate cases of $x \to 2^+$ and $x \to 2^-$. For x close to 2, the numerator of $f(x)$ is close to 2 and the denominator is close to $2 - 3 = -1$, so $f(x)$ is close to -2. Thus, $\lim_{x\to 2} f(x) = -2$, but the point $(2, -2)$ is missing from the graph.

For values of x close to 3, we note that $f(x) > 0$ when $x > 3$ and $f(x) < 0$ when $x < 3$. So we need to consider the limits as $x \to 3$ in two parts:

As $x \to 3^+$, x and $x - 3$ are both positive. Since $x \to 3$ and $x - 3 \to 0$, we have $f(x) = \dfrac{x}{x-3} \to \infty$.

As $x \to 3^-$, $f(x) = \dfrac{x}{x-3} \to -\infty$. The line $x = 3$ is a vertical asymptote. The graph of $y = f(x)$ is shown in Figure 5.28.

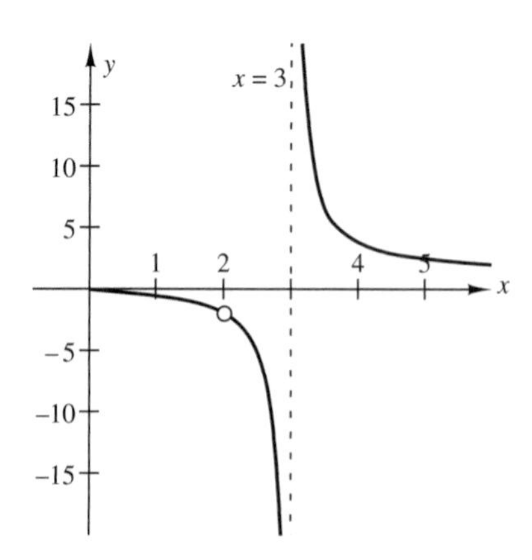

Figure 5.28 $f(x)=\dfrac{x^2-2x}{x^2-5x+6}$ ■

Exercises 5-B.2

1. Let $f(x) = \frac{x - x^2}{x^2 - 1}$.

a. Analyze the behavior of the values $f(x)$ as $x \to 1$.

b. Analyze the behavior of the values $f(x)$ as $x \to -1^-$.

c. Analyze the behavior of the values $f(x)$ as $x \to -1^+$.

d. Is the line $x = 1$ a vertical asymptote of the graph of f? Why?

e. Is the line $x = -1$ a vertical asymptote of the graph of f? Why?

2. Let $h(x) = \frac{x^2 - 3x - 10}{x^2 - 3x - 4}$.

a. Show that when $x = 4$, the denominator is zero but the numerator is not.

b. When x is close to 4 and $x < 4$, state whether (i) the numerator is positive or negative, (ii) the denominator is positive or negative, (iii) $h(x)$ is positive or negative.

c. As $x \to 4^-$, does $h(x) \to \infty$ or $h(x) \to -\infty$?

d. When x is close to 4 and $x > 4$, state whether (i) the numerator is positive or negative, (ii) the denominator is positive or negative, (iii) $h(x)$ is positive or negative.

e. As $x \to 4^+$, does $h(x) \to \infty$ or $h(x) \to -\infty$?

3. Repeat Exercise 2 when $x = 4$ is replaced by $x = -1$.

4. Let $g(x) = \frac{x^2 - x + 1}{x^2 - 3x + 2}$.

a. Find the values of x at which the denominator is zero.

b. Explain what happens to the values of the function when x is close to each of the zeros of the denominator.

c. Give the equations of any vertical asymptotes of the function.

5. Repeat Exercise 4 for the function $H(x) = \frac{x^2 + 2x - 3}{9 - x^2}$.

6. Repeat Exercise 4 for the function $F(x) = \frac{x + \sqrt{5}}{x^3 - 5x}$.

7. Repeat Exercise 4 for the function $G(x) = \frac{2x - 6}{x^2 - 6x + 9}$.

Chapter 5 Exercises

1. A company's advertising expert estimates that the profit P (in thousands of dollars) from selling a new product depends on the advertising expenditure x (in thousands of dollars) according to the graph in Figure 5.29.

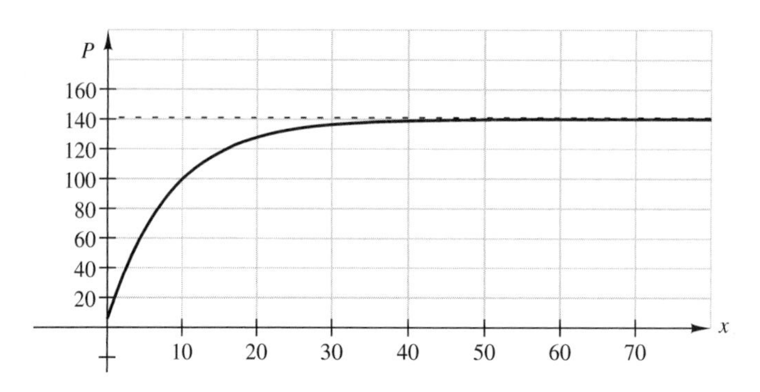

Figure 5.29

a. Approximately what is the profit when the money spent on advertising is

(i) \$10,000 (ii) \$20,000 (iii) \$60,000

b. Explain what happens as $x \to \infty$.

c. Is there a value of x for which the profit is \$200,000? Give the value of x or explain why it does not exist.

d. Is it possible to increase the profit to \$140,000 by increasing the amount spent on advertising? Explain your answer.

e. Give the equations of all asymptotes of the profit function (if any).

2. Refer to the graphs in Figures 5.30, 5.31, and 5.32 to complete statements a–g.

a. As $x \to 3^-, f(x) \to$ _______. **b**. As $x \to 3^+, f(x) \to$ _______.

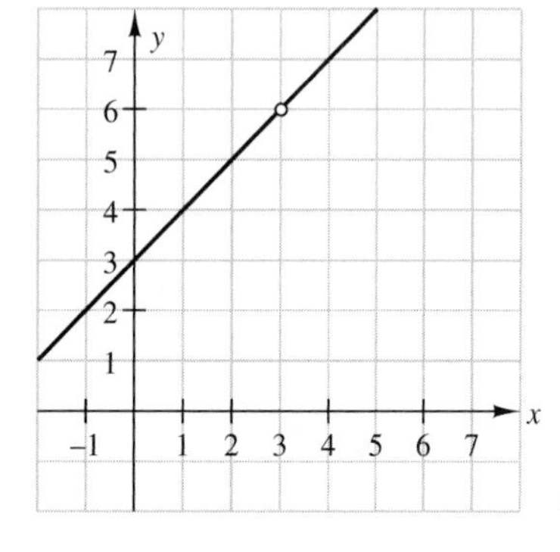

Figure 5.30 $y = f(x)$

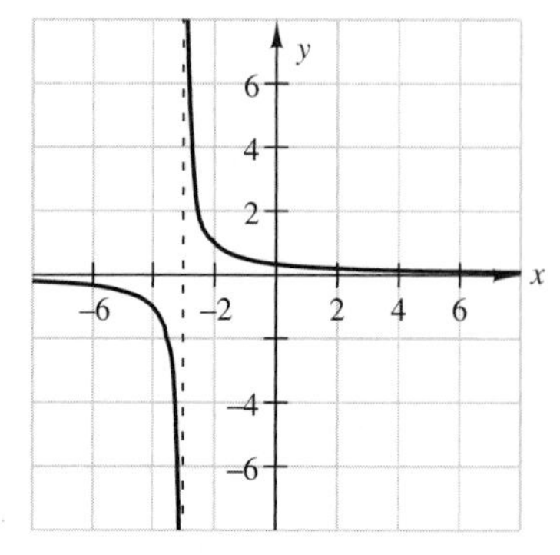

Figure 5.31 $y = g(x)$

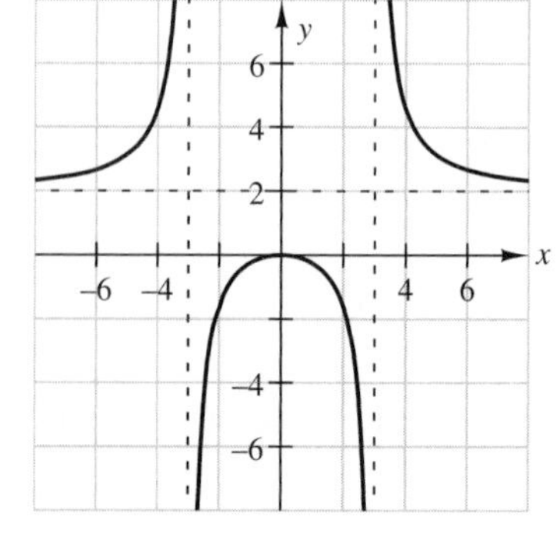

Figure 5.32 $y = h(x)$

c. As $x \to -3^-$, $g(x) \to$ ______.   **d.** As $x \to \infty$, $g(x) \to$ ______.

e. As $x \to -\infty$, $h(x) \to$ ______.   **f.** As $x \to 3^-$, $h(x) \to$ ______.

g. As $x \to 3^+$, $h(x) \to$ ______.

3. Refer to the graphs in Figures 5.30, 5.31, and 5.32. Give the equations of any horizontal asymptotes of the graph of:

a. $y = f(x)$ **b.** $y = g(x)$ **c.** $y = h(x)$

4. Refer to the graphs in Figures 5.30, 5.31, and 5.32. Give the equations of any vertical asymptotes of the graph of:

a. $y = f(x)$ **b.** $y = g(x)$ **c.** $y = h(x)$

5. The functions f, g, and h whose graphs are shown in Figures 5.30, 5.31, and 5.32 can be defined by expressions in this list:

$$\frac{2x^2}{x^2-9},\ x+3,\ \frac{x^2-9}{x-3},\ \frac{x-3}{x^2-9},\ \frac{2x}{x^2-9},\ \frac{1}{x-3},\ \frac{1}{x+3}.$$

Observe the behavior of the graphs and select the correct expression to fill in each blank.

a. $f(x) =$ ________   **b.** $g(x) =$ ________   **c.** $h(x) =$ ________

6. Let $G(x)=\dfrac{3x^2-12}{x^2+2x}$. Questions a–f refer to this function.

a. Analyze the behavior of the values of the function as $x \to \infty$.

b. Analyze the behavior of the values of the function as $x \to -\infty$.

c. Find the values of x at which the denominator is zero.

d. Explain what happens to the values of the function when x is close to each of the zeros of the denominator.

e. Give the equations of the vertical asymptotes (if any).

f. Give the equations of the horizontal asymptotes (if any).

7. Repeat Exercise 6 for the function $F(x)=\dfrac{-x-2}{x^2+5x+6}$.

8. For the function $H(x)=\dfrac{2-x}{x^2-4x+3}$, complete the following statements:

a. As $x \to 3^-$, $H(x) \to$ ______.   **b.** As $x \to 3^+$, $H(x) \to$ ______.

c. As $x \to 1^-$, $H(x) \to$ ______.   **d.** As $x \to 1^+$, $H(x) \to$ ______.

e. As $x \to \infty$, $H(x) \to$ ______.   **f.** As $x \to -\infty$, $H(x) \to$ ______.

9. Let $f(x)=\sqrt{2x^2-1}-\sqrt{2x^2+x}$.

a. Analyze the behavior of the values $f(x)$ as $x \to \infty$. To do this, first write $f(x)$ as $\dfrac{\sqrt{2x^2-1}-\sqrt{2x^2+x}}{1}$ and rationalize the numerator. Then divide numerator and denominator by x to write the resulting expressions in terms of $\frac{1}{x}$.

b. Does the graph of $f(x)$ have any horizontal asymptotes? Explain.

c. Does the graph of $f(x)$ have any vertical asymptotes? Explain.

10. Let $g(x)=\dfrac{3x-5}{\sqrt{6+x^2}}$.

a. Divide numerator and denominator by x to write $g(x)$ in terms of $\frac{1}{x}$. **Assume x is negative.**

b. Analyze the behavior of the values of the function as $x \to -\infty$.

c. Divide numerator and denominator by x to write $g(x)$ in terms of $\frac{1}{x}$. **Assume x is positive.**

d. Analyze the behavior of the values of the function as $x \to \infty$.

e. Does the graph of $g(x)$ have any horizontal asymptotes? Explain.

f. Does the graph of $g(x)$ have any vertical asymptotes? Explain.

11. Let $h(x)=\dfrac{\sqrt[3]{4x-x^3}}{8x-5}$.

a. Divide numerator and denominator by x to write $h(x)$ in terms of $\frac{1}{x}$.

b. Analyze the behavior of the values of the function as $x \to -\infty$.

c. Analyze the behavior of the values of the function as $x \to \infty$.

d. Analyze the behavior of the values of the function as $x \to \frac{5}{8}$.

e. Give the equations of all horizontal asymptotes (if any).

f. Give the equations of all vertical asymptotes (if any).

CHAPTER 6

Problem-Solving and Rates of Change

6-A PROBLEM-SOLVING

Pólya's Approach to Problem-Solving

The mathematician George Pólya spent much of his life advancing the study of problem-solving. In his book, *How To Solve It*, he outlines a four-step approach that can be used to aid in the problem-solving process. These four steps are paraphrased below:

1. Define the problem.
2. Devise a plan for solving the problem.
3. Carry out the plan.
4. Test and evaluate the results.

The following example illustrates each of these steps.

EXAMPLE 6.1 The Ajax Carton Company has decided to make different sizes of boxes by cutting squares of the same size out of each corner of an $8\frac{1}{2} \times 11$-in. piece of cardboard, folding up the resulting flaps, and taping the edges at each corner, as shown in Figure 6.1. How much does the volume V change as the length x of the edge of the cut increases from 1.5 to 1.6 in.?

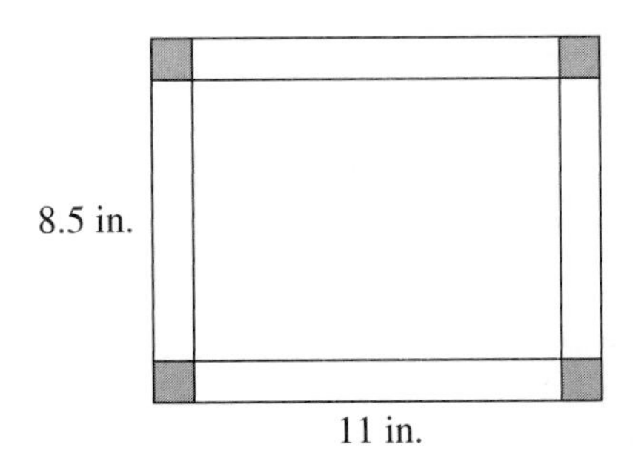

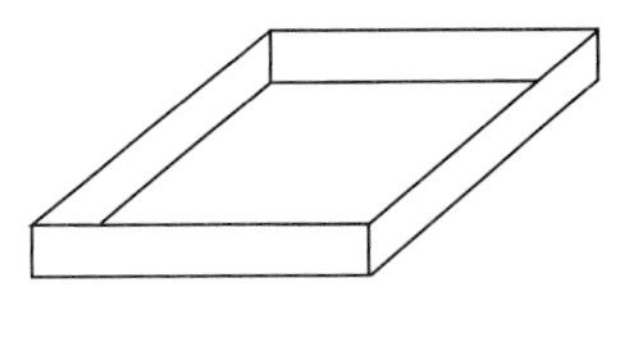

Figure 6.1

Solution

1. Define the problem.

The problem is to find the change in the volume of the open box as the size of the cut changes. To do this we must first express the volume of the box as a function of its length, width, and height, where each of these quantities will vary according to the length x of the side of the square cut from each corner. The independent variable is the length of the side of the squares cut out, and the dependent variable is the volume V of the box.

What are the constraints on the independent variable x? The dimensions of the original sheet of cardboard are $8\frac{1}{2} \times 11$ in., so the size of the cut must be greater than 0 and less than $4\frac{1}{4}$ in. (Why?) We record this constraint as $0 < x < 4.25$. Let's calculate the volume V for a specific instance of the problem—for example, when a 1-in. square is cut from each corner. The length of the resulting box will be $11 - 1 - 1 = 9$ in. The width will be $8\frac{1}{2} - 1 - 1 = 6\frac{1}{2}$ in. The height of the box will be 1 in., exactly the length of the side of the cut-out squares. Since $V = lwh$ is the formula for the volume of a rectangular box, the volume is $V = (9)(6.5)(1) = 58.5$ in.^{3}.

2. Devise a plan.

A *mathematical model* of this problem is an equation that will express the volume V as a function of x, the length of the side of the cut-out squares. To do this we must express the length, width, and height of the box each in terms of x. Since the length is reduced by x inches on each side, the length can be represented by $11 - 2x$. Similarly, the width can be expressed as $8.5 - 2x$. The height of the box is equal to the size of the cut, so the height is x. (See Figure 6.2.) Thus, our plan is to write an equation that defines V in terms of x.

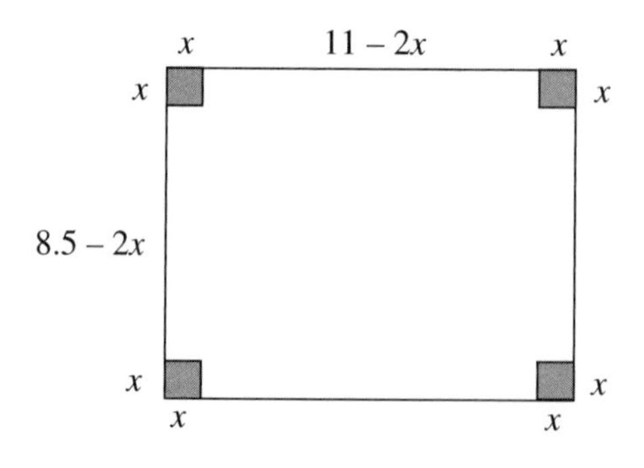

Figure 6.2

3. Carry out the plan.

$V = lwh = (11 - 2x)(8.5 - 2x)(x)$, where $0 < x < 4.25$.

Since we now have an equation for the volume V in terms of the length x of the side of the cut-out square, we can determine the volumes of the boxes when $x = 1.5$ and $x = 1.6$.

For $x = 1.5$: $V = (11 - 2(1.5))(8.5 - 2(1.5))(1.5) = (8)(5.5)(1.5) = 66 \text{ in.}^3$

For $x = 1.6$: $V = (11 - 2(1.6))(8.5 - 2(1.6))(1.6) = (7.8)(5.3)(1.6) = 66.144 \text{ in.}^3$

Thus the change in volume is $66.144 - 66 = 0.144 \text{ in.}^3$.

4. Test and evaluate the results.

Does the answer that you obtained seem reasonable? One way to determine this is to use a graphing utility to graph $y = V(x) = (11 - 2x)(8.5 - 2x)(x)$ and look at what happens to the volume as x changes from 1.5 to 1.6. The graph is shown in Figure 6.3 for a domain of [0, 4.25]. Notice that as the size x of the cut increases from 1.5 to 1.6 in., the volume increases only slightly, as predicted by our solution.

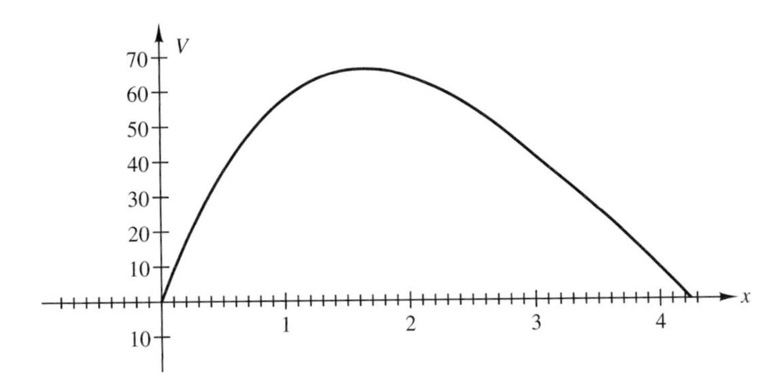

Figure 6.3 $V(x)$ ■

Example 6.1 illustrates several important points. When we set out to solve a problem, it is important to determine the variables in the problem and to name them. We then look for relationships between variables that may be used to limit the number of different variables in the problem. For example, although length, width, and height are all variables in this problem, they can all be expressed in terms of x. In addition, it is useful to list any constraints on variables in the problem.

This example also illustrates several problem-solving strategies. Drawing a picture or organizing given data is often a useful strategy. Another strategy is to solve a particular case (such as we did for a box made using a 1-in. cut) or to solve a simple case of the problem. In our example, the particular case helps us to see that the length and width of the resulting box can be found by subtracting twice the size of the cut from the original length and width. To formulate an equation for volume, we used a known formula for the volume of a rectangular box. In general, to create a mathematical model, we may have to use a pattern, a known formula, or a relation between the variables and constraints discovered from a picture of the situation. Some problem-solving strategies and commonly used formulas are listed below.

Problem-Solving Strategies

1. Draw a picture or graph.
2. Organize the data and look for relationships between variables.
3. Use a table or chart to look for relationships between variables.
4. Use a known formula to state a relationship between variables.
5. Test or model the problem using manipulatives.
6. Look for patterns; make and test conjectures.
7. Guess a solution and check it.
8. Try some examples with numbers. See if these suggest a general solution.
9. Make a similar simpler problem; solve this problem and use a similar tack to try to solve the more complex problem.
10. Look for an analogous problem that you have solved before.
11. Introduce new data or variables to help you see relationships or patterns.
12. Eliminate extraneous or less crucial variables from consideration.
13. Gather additional data and analyze them for patterns and relationships.
14. Brainstorm alternatives with colleagues.
15. Work backward.

Commonly Used Formulas

In formulas 1 through 9, P is the perimeter of the polygon, A is the area of the figure, C is the circumference of the circle, S is the surface area of the solid, and V is the volume.

1. Triangle $P = a + b + c$ $A = \frac{1}{2}bh$

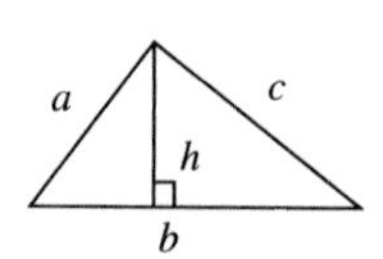

2. Rectangle $P = 2l + 2w$ $A = lw$

3. Parallelogram $P = 2a + 2b$ $A = bh$

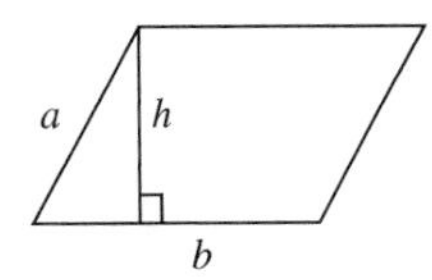

4. Circle $C = 2\pi r = \pi d$ $A = \pi r^2$

5. Trapezoid $P = a + b_1 + b_2 + c$ $A = \frac{1}{2}(b_1 + b_2)h$

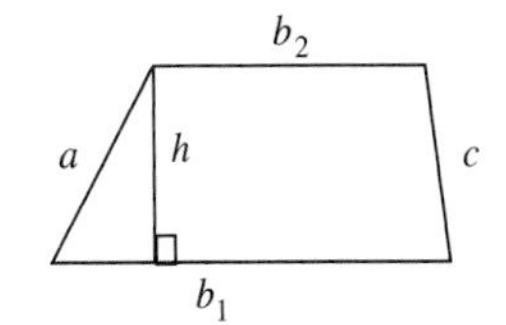

6. Rectangular prism (box) $S = 2lw + 2lh + 2wh$ $V = lwh$

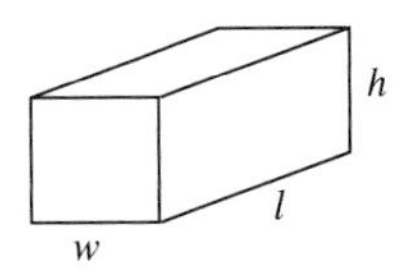

7. Right circular cylinder $S = 2\pi rh + 2\pi r^2$ $V = \pi r^2 h$

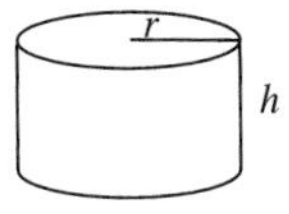

8. Right circular cone $S = \pi r\sqrt{r^2 + h^2}$ $V = \frac{1}{3}\pi r^2 h$

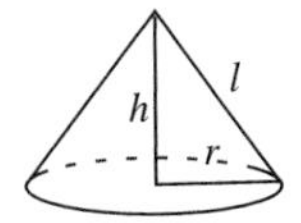

9. Sphere $S = 4\pi r^2$ $V = \frac{4}{3}\pi r^3$

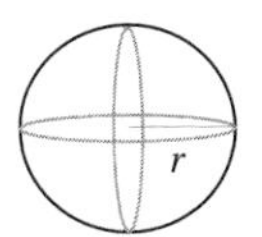

10. Pythagorean Theorem $c^2 = a^2 + b^2$

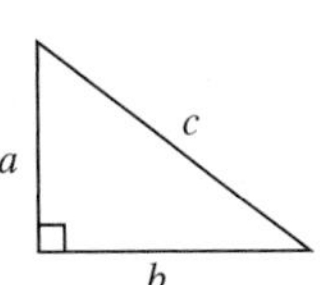

11. Production cost C $C = bx$

where b is the cost per item and x is the number of items produced.

12. Total cost T $T = C + F$

where F is the fixed costs of production and C is production cost.

13. Simple interest I $I = prt$

where p is the principal, r the annual interest rate, and t the time (years).

14. Revenue R $\quad R = px$

where p is the unit price and x is the number sold.

15. Profit P $\quad P = R - T$

where R is the revenue and T is the total cost.

16. Displacement d (net distance) $\quad d = rt$

where r is the average velocity (speed) and t is the time traveled.

Exercises 6-A

Before solving these exercises, you may wish to review the commonly used formulas above.

1. A bus company charges \$20 per person for a bus trip, with a minimum of 25 people to schedule the trip. In an attempt to increase its revenue, the company is offering tours at a discount. Whenever more than 50 people charter a bus, the price per person is reduced by \$.25 for each additional person beyond 50 persons.

a. What is the company's revenue if 25 people make the trip?

b. If 51 people make the trip, the price is reduced to \$19.75 per person. If 52 people make the trip, the price is reduced to \$19.50 per person. What is the price per person if 55 people make the trip?

c. What is the revenue R if 55 people make the trip?

d. What is the price per person if 60 people make the trip?

e. What is the revenue R if 60 people make the trip?

f. Find the difference in revenue between a trip with 55 people and a trip with 60 people.

g. Write a function $R(x)$ that describes the revenue made by the bus company if x people make the trip and x is between 25 and 50, inclusive ($25 \leq x \leq 50$).

h. Write a function $p(x)$, the price per person, when x people take the trip and x is greater than 50 ($x > 50$).

i. Write a function $R(x)$ for the revenue made by the bus company when $x > 50$.

j. Use the results of parts g and i to write a piecewise-defined function $R(x)$ for the revenue made by the bus company when x people make the trip.

2. The Bradley Pencil Company gives bulk discounts. If a firm buys more than \$500 (list price) in supplies, it receives a 10% discount on the price. If a firm buys more than \$1,000 (list price) in supplies, it receives a 15% discount.

a. If a firm orders \$450 worth of supplies, how much will it have to pay?

b. If a firm orders \$550 worth of supplies, how much will it have to pay?

c. If a firm orders \$1,000 worth of supplies, how much will it have to pay?

d. If a firm orders \$1,001 worth of supplies, how much will it have to pay?

e. Sketch a graph that shows how much a firm pays for an order of x dollars (list price). (You will need to consider separately the intervals $0 \leq x \leq 500$, $500 < x \leq 1000$, and $x > 1000$.)

f. For what dollar amounts is it less expensive for a firm to purchase additional supplies to make the total list price more than \$500? More than \$1,000?

g. If your firm needs \$850 worth of supplies, how much more would it cost to purchase \$1,001 worth of supplies to get the larger discount?

3. John has a rectangular yard that is 30 ft by 34 ft. He would like to add a sidewalk along the outside of two adjacent sides of the yard. The contractor quotes a price of \$20 per square foot to put in the sidewalk.

a. Draw a picture of the situation that shows the yard and adjoining sidewalk.

b. If the sidewalk is only 1 ft wide, how many square feet of sidewalk would he have?

c. What is the cost of a 1-ft-wide sidewalk?

d. Write a function $A(x)$ for the area (the number of square feet) of sidewalk installed if the sidewalk is x ft wide.

e. Write a function $C(x)$ for the cost of installing a sidewalk that is x ft wide.

f. Use the function C from part e to find the cost of installing a sidewalk 2 ft wide.

g. If John has \$4,000 to spend, what is the widest he can make the sidewalk (to the nearest tenth of a foot)?

6-B APPLICATIONS

Rates

In 2000, there were 27,987 reported deaths of infants in the United States age one year or younger. The most commonly used index for measuring the risk of dying during the first year of life is the infant mortality rate, which is calculated by dividing the number of infant deaths by the number of live births registered for the same period. The 2000 rate of 6.9 infant deaths per 1,000 live births was the lowest ever recorded up to that time. The drop in infant mortality rate in the U.S. has been steady since 1940: in numbers per 1,000 live births, it was 47 in 1940, 26 in 1960, 12.6 in 1980, and 9.2 in 1990. When one compares mortality rates of white infants and black infants for the same period, the pattern of decline in mortality rates is parallel, but the mortality rate among whites is roughly half that of blacks. Economic differences and quality of prenatal care are a likely reason, but comparative studies have shown that these factors alone do not account for the difference.

Source: U.S. Dept. of Health and Human Services, National Center for Health Statistics, *Monthly Vital Statistics Reports*, *National Vital Statistics Report*, Oct. 9, 2001.

As the above paragraph and examples that follow show, we often use the concept of a rate to make comparative descriptions that measure change:

- The shipping rate is $1.50 per pound.
- Real estate is taxed at a rate of 45 mills ($45 per $1,000) of assessed value.

These examples of rates are ratios of two quantities. The word "per" usually indicates that a rate is being used. *Percent* means "divided by 100." Tax and interest rates are generally expressed as percents. For example,

- A bank advertises an interest rate of 5.1% on savings certificates.
- Sales are taxed at 5%.
- The commission rate on a stock purchase is 1% of the purchase price.

All of the rates exhibited thus far have been *constant* rates. Many rates express how quickly a task can be accomplished:

- The secretary types 80 words per minute.
- The printer can print 20 pages per minute.

- U.S. energy consumption for 2001 was 340 million BTUs per person.
- The new machine packages 400 candy bars per minute.
- The wheel rotates at 1000 revolutions per minute.

These are all *average* rates. At any moment the secretary may be typing slightly faster or slower than 80 words per minute. Many people consume more than 500 million BTUs per year, and others consume less than 200 million BTUs. How do we compute these average rates?

EXAMPLE 6.2

a. In 2002, Greenville County was North Carolina's leading producer of trash per capita, with each resident creating an average of 679 lb of trash each week. What was the average daily rate of trash production for each resident?

b. In a 10-min period, a secretary types 790 words after deductions for mistakes. What is the secretary's average typing rate in words per minute?

c. A company has a fixed "per diem" (daily) expense rate. If an employee is allotted \$1,200 for an eight-day trip, what is the per diem rate?

Solution To compute the rate, we divide the total amount for a time period by the length of that time period.

a. The average daily rate of trash production for each resident is $\frac{679 \text{ lb}}{7 \text{ days}} = 97$ lb per day.

b. The average typing rate is $\frac{790 \text{ words}}{10 \text{ min}} = 79$ words per min.

c. The per diem rate is $\frac{\$1,200}{8 \text{ days}} = \$150$ per day. ■

Average Rates of Change

The Dow Jones Industrial Average (DJIA) started the day on Friday at 9892 and ended the day at 9889. Suppose we are interested in the average change in the DJIA per hour in order to make decisions on investments. On this day, the DJIA decreased by 3 points over a $6\frac{1}{2}$-hour period for an average rate of change of $\frac{3}{6.5} \approx 0.46$ points per hour.

Many of the rates we measure are *average rates of change.*

In common practice the word "average" is often omitted when referring to an average rate of change. In calculus, however, the word "rate" without a qualifier usually means an instantaneous rate of change of a function. For this reason, we will always use the word "average" when we discuss an average rate of change.

EXAMPLE 6.3

a. The population of Gotham City increased from 101,000 to 125,000 last year. What was its average monthly rate of growth?

b. John Smith's travel journal for a recent business trip is shown below. All times are Eastern Standard. His driving distance from New York to Philadelphia was 120 miles and his distance from Philadelphia to Washington, D.C., was 150 miles. What was his average speed from New York to Philadelphia? From Philadelphia to Washington? For the entire trip?

Leave From	Time	Arrive At	Time
New York	10:00 A.M.	Philadelphia	12:30 P.M.
	Eat lunch from 12:30 to 2:00 P.M.		
Philadelphia	2:00 P.M.	Washington	5:00 P.M.

Solution

a. The change in the population was 125,000 – 101,000 or 24,000 over the 12-month period. The average monthly rate of change was $\frac{24{,}000}{12 \text{ months}}$ or 2000/month. This is also called the average monthly *growth rate.*

b. The 120-mile trip from New York to Philadelphia took 2.5 hours, so his average speed (which is average rate of change of distance) was 120 miles/2.5 hours or 48 mph. The 150-mile trip from Philadelphia to Washington took 3 hours for an average speed of 50 mph. Because he stopped for lunch, his average speed for the entire trip was 120 + 150 miles in 7 hours or 270 miles in 7 hours, for an average speed of 39 mph to the nearest mph. ■

EXAMPLE 6.4 Jennifer's fifth-grade class conducted an experiment growing seeds. Her data for the experiment are charted below.

a. What was the average rate of change in the height (average growth rate) of the seedlings from day 0 to day 1? From day 1 to day 5?

b. What was the average growth rate from day 4 to day 10? From day 0 to day 11?

Day	Height(cm)	Day	Height(cm)	Day	Height(cm)
0	0	4	4	8	15
1	0	5	7	9	18
2	1	6	10	10	21
3	2	7	13	11	23

Solution Each average rate of change in height (average growth rate) is

$$\frac{\text{change in height over time interval (in cm)}}{\text{length of time interval (in days)}}.$$

a. Average growth rate (day 0 to 1) = $\frac{0-0}{1-0} = \frac{0}{1}$ = 0 cm/day.

Average growth rate (day 1 to 5) = $\frac{7}{4}$ = 1.75 cm/day.

b. Average growth rate (day 4 to 10) = $\frac{17}{6}$ $\approx$ 2.83 cm/day.

Average growth rate (day 0 to 11) = $\frac{23}{11}$ $\approx$ 2.09 cm/day.

Note that the average growth rate differs when measured over different time intervals. ■

When we look at the different growth rates above it is not obvious whether there is a pattern to the growth rate of the seedlings.

EXAMPLE 6.5 A graph of the data from the table in Example 6.4 may help us see if a pattern exists. This graph has been partially completed in Figure 6.4. The numbers on the horizontal axis represent days, and the numbers on the vertical axis represent height in centimeters.

a. Plot the remaining data points from the chart in Example 6.4.

b. Look at the graph created in part a. Estimate a line that fits the points and draw it. What does the slope of the line measure?

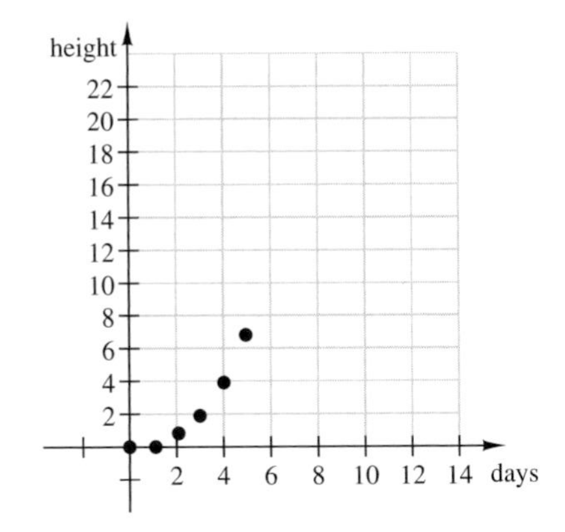

Figure 6.4

Solution The remaining data are plotted in Figure 6.5. The line shown on the graph in Figure 6.5 represents one linear approximation to the seedlings' continuous growth over time; the seedlings' average rate of growth corresponds to the slope of the line. By taking two points on the line, we can find its slope; we choose the points (9, 18) and (7, 13):

$$m = \frac{18 - 13}{9 - 7} = \frac{5}{2}$$

This tells us that the average rate of growth was approximately 2.5 cm per day over the 11-day period. Notice how this method to find the slope is the same as the method we just used to find the average rates of change in Example 6.4.

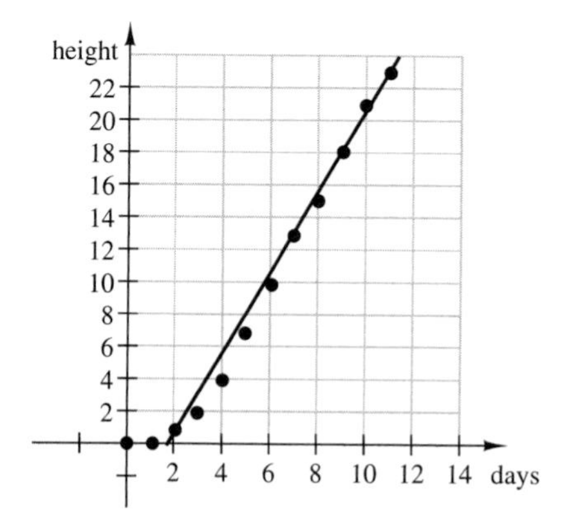

Figure 6.5 ■

The slope of a line is the geometric interpretation of the average rate of change.

EXAMPLE 6.6 [DJIA revisited] A graph of the fluctuations in the DJIA during a single Friday is given in Figure 6.6. Earlier we computed the hourly average rate of change in the DJIA for the whole day; it was 0.46 points per hour. Answer the questions below by referring to the graph in Figure 6.6.

a. Find the average rate of change of the DJIA for each of the 6 hours from 10 A.M. to 4 P.M. by estimating the necessary information from the graph.

b. Are any of these average hourly rates of change close to the average rate of change for the whole day (0.46 point per hour)?

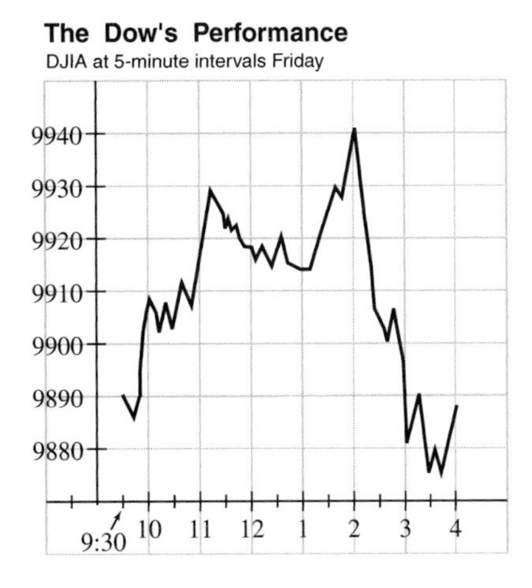

Figure 6.6

Solution

a.

10–11 A.M.:	$\dfrac{9916 \text{ points} - 9908 \text{ points}}{1 \text{ hour}} = 8$ points/hour
11–noon:	3 points/hour
noon–1 P.M.:	–5 points/hour
1–2 P.M.:	27 points/hour
2–3 P.M.:	–60 points/hour
3–4 P.M.:	7 points/hour

b. No hourly rate of change for these six 1-hour periods is close to the average rate of change for the whole day. Although the starting and ending values of the DJIA for the day are almost the same, there are great fluctuations in the graph from hour to hour. ■

Another average rate of change that is often computed is the *average velocity* of a moving object. This is the ratio of the change in position of the object to the length of time in which the change took place.

EXAMPLE 6.7 If a ball is dropped from a 200-ft tower, its height above the ground after t seconds is given by the function $s(t) = 200 - 16t^2$.

a. What is the average velocity of the ball between 0 and 2 sec?

b. Calculate the average velocity of the ball for its whole drop to the ground. ("Whole drop" means from the time it is dropped until it hits the ground.)

Solution

a. We first calculate $s(0)$ and $s(2)$: $s(0) = 200 - 16(0)^2 = 200$ (that is, the ball is 200 ft above the ground at the instant it is dropped) and $s(2) = 200 - 16(2)^2 = 200 - 64 = 136$ (that is, the ball is 136 ft above the ground after 2 sec). These two calculations correspond to the two points (0, 200) and (2, 136) on the graph that shows height as a function of time. To find the average velocity, we find the slope of the line through these two points:

$$\text{Average velocity (during the first 2 sec)} = \text{average rate of change of } s(t) = \frac{s(2)-s(0)}{2-0} = \frac{136-200}{2-0} = \frac{-64}{2} = -32 \text{ ft/sec.}$$

The negative sign indicates that the ball is traveling downward.

b. We first calculate how long it will take the ball to reach the ground (i.e., when $s(t) = 0$) by solving $200 - 16t^2 = 0$: $t^2 = \frac{200}{16} = 12.5$; $t \approx 3.54$ sec.

Then we calculate the average velocity for the time interval from 0 until the time the ball hits the ground:

$$\text{Average velocity} = \frac{s(3.54)-s(0)}{3.54-0} = \frac{0-200}{3.54} \approx -56.5 \text{ ft/sec.}$$ ■

EXAMPLE 6.8 The cost C (in dollars) to produce x bracelets is given by the function $C(x) = 0.05x^2 + 2x + 10$.

What is the average rate of change in the cost of producing a bracelet if production is increased from 10 to 20 bracelets?

Solution $C(10) = 0.05(10)^2 + 2(10) + 10 = 35$ and $C(20) = 70$.

$$\text{Average rate of change} = \frac{C(20)-C(10)}{20-10} = \frac{70-35}{20-10} = \frac{35}{10} = 3.5.$$

Thus, the average rate of change in the cost is $3.50 per bracelet. ■

The previous examples illustrate one method to determine the average rate of change of a function:

If $y = f(t)$, then the average rate of change in y per unit of t from $t = t_1$ to $t = t_2$ is:

$$\textbf{Average rate of change of } f(t) \textbf{ over } [t_1, t_2] = \frac{f(t_2)-f(t_1)}{t_2-t_1}.$$

EXAMPLE 6.9 Suppose in Example 6.7 you want to know the average rate of change in the height of the ball per second for a small increment of time such as the half-second after 2 sec. Then the two times we are considering are 2 sec and $2 + \frac{1}{2}$, or 2.5, sec. Find the average rate of change in the height per second during this half-second period. (This is the average velocity of the ball in the half-second interval after it has dropped for 2 sec.)

Solution Average rate of change $= \dfrac{s(2.5)-s(2)}{0.5} = \dfrac{200-16(2.5)^2-\left(200-16(2)^2\right)}{0.5} = \dfrac{100-136}{0.5} = \dfrac{-36}{0.5} = -72$ ft/sec. ■

Example 6.9 illustrates a second method to find the average rate of change of a function: If $y = f(t)$, then the average rate of change in y per unit of t for the increment h added to $t = t_1$ is as follows:

$$\textbf{Average rate of change of } f(t) \textbf{ between } t_1 \textbf{ and } t_1 + h = \frac{f(t_1+h)-f(t_1)}{h}.$$

The increment h is often denoted Δt, where Δt is interpreted as the *difference in t* or the *change in t*.

Since for an interval $[t_1, t_2]$, the increment h can be taken as $h = t_2 - t_1$, the two definitions for the average rate of change are equivalent; that is,

$$\frac{f(t_1+h)-f(t_1)}{h} = \frac{f\left(t_1+(t_2-t_1)\right)-f(t_1)}{t_2-t_1} = \frac{f(t_2)-f(t_1)}{t_2-t_1}.$$

EXAMPLE 6.10 The total cost C of producing x sneakers is given by the function

$$C(x) = 0.01x^2 - 0.2x + 2000.$$

a. Find the average rate of change in the cost if production is increased from $x = 100$ to $x = 150$.

b. Find the average rate of change in the cost if production is increased by one when $x = 100$.

Solution

a. Average rate of change $= \dfrac{C(150)-C(100)}{50} = \dfrac{2195-2080}{50} = 2.30.$

Therefore, the average rate of change in total cost of production is \$2.30 per sneaker.

b. Average rate of change $= \dfrac{C(100+1)-C(100)}{1} = \dfrac{2081.81-2080}{1} = 1.81$, which says that the average rate of change in cost of production from 100 to 101 sneakers is \$1.81 per sneaker. ■

The average rate of change in cost if production is increased by one at a certain level of production is often called the *marginal cost* at that level. In Example 6.10b it can be thought of as the cost of producing the 101st sneaker. The solution says that the marginal cost at a production level of 100 is \$1.81 per sneaker.

In calculus we are interested in finding out how functions are changing at any point on their graph. To define the *instantaneous rate of change* of a function $f(t)$ at a single value of t, we calculate the average rate of change of $f(t)$ between t and $t + h$, where the incremental change h is very small. If we then let this incremental change h approach 0 and the quotient that defines the average rate of change has a limit, we will have the instantaneous rate of change of $f(t)$ at that point.

This limit, if it exists, is called the *derivative* of the function f at t.

To find a formula for this instantaneous rate of change, we may have to expand or factor polynomials, simplify an algebraic expression, rationalize the numerator or denominator of a quotient, or simplify a complex fraction. The purpose of these algebraic manipulations is to make it possible to take a limit as the increment h in the average rate of change goes to 0. These algebraic techniques were discussed in Section 3-B.

The following example illustrates how such algebraic manipulations can result in a relatively simple expression for the average rate of change. In each, the simplification makes it possible to take the limit of the average rate of change as h goes to 0.

 Remember, to find $f(t + h)$, replace all occurrences of the variable in the rule of the function f by the sum $t + h$.

EXAMPLE 6.11 Find and simplify the average rate of change between t and $t + h$ for each of the following functions. The increment h is always assumed to be nonzero.

a. $f(t) = t^2$ **b.** $f(t) = \sqrt{t-2}$ **c.** $f(t) = \frac{1}{t}$

Solution

a. We use the definition of average rate of change. To evaluate $f(t + h)$, we substitute $(t + h)$ for the variable t in the rule $f(t) = t^2$:

$$\text{Average rate of change} = \frac{f(t+h)-f(t)}{h} = \frac{(t+h)^2-t^2}{h}.$$

Next we expand $(t + h)^2$ and combine like terms in the numerator, and then factor out h and simplify:

$$\frac{t^2+2th+h^2-t^2}{h} = \frac{2th+h^2}{h} = \frac{h(2t+h)}{h} = 2t + h.$$

The average rate of change of $f(t) = t^2$ between t and $t + h$ is $2t + h$.

b. Again, we begin with the definition:

Average rate of change $= \dfrac{f(t+h)-f(t)}{h} = \dfrac{\sqrt{(t+h)-2}-\sqrt{t-2}}{h}$.

We first rationalize the numerator and then simplify:

$$\frac{\sqrt{t+h-2}-\sqrt{t-2}}{h}\cdot\frac{\sqrt{t+h-2}+\sqrt{t-2}}{\sqrt{t+h-2}+\sqrt{t-2}} = \frac{(t+h-2)-(t-2)}{h\left(\sqrt{t+h-2}+\sqrt{t-2}\right)} =$$

$$\frac{h}{h\left(\sqrt{t+h-2}+\sqrt{t-2}\right)} = \frac{1}{\sqrt{t+h-2}+\sqrt{t-2}}.$$

The average rate of change of $f(t) = \sqrt{t-2}$ between t and $t + h$ is $\dfrac{1}{\sqrt{t+h-2}+\sqrt{t-2}}$.

c. Average rate of change $= \dfrac{f(t+h)-f(t)}{h} = \dfrac{\frac{1}{t+h}-\frac{1}{t}}{h}$. To clear the fractions in the numerator, we multiply by $\dfrac{(t+h)t}{(t+h)t}$ and then simplify:

$$\frac{\frac{1}{t+h}-\frac{1}{t}}{h}\cdot\frac{(t+h)t}{(t+h)t} = \frac{t-(t+h)}{h(t+h)t} = \frac{-h}{h(t+h)t} = \frac{-1}{t^2+ht}.$$

The average rate of change of $f(t) = \frac{1}{t}$ between t and $t + h$ is $\dfrac{-1}{t^2+ht}$. ■

Exercises 6-B

1. List at least three constant rates that you have encountered.

2. Look through a newspaper and note all the average rates you see.

3. Express each of the following as a rate:

a. He traveled 120 miles on 4 gal of gas.

b. The cost was $80 for 4 tickets.

c. On the trip, June traveled 240 miles in 6 hours.

4. The table below shows the price per ounce and associated demand for a perfume. Use it to answer the following questions:

a. What is the average rate of change in the demand as the price is changed from $30 to $35?

b. What is the average rate of change in the demand as the price is changed from $25 to $35?

c. What is the average rate of change in the demand as the price is changed from \$30 to \$50?

d. Describe in words what a negative average rate of change in demand means.

Price	Demand (units purchased)
\$25	2400
\$30	2200
\$35	1900
\$40	1550
\$45	1500
\$50	1200

5. $C(x) = 0.1x^2 - x + 1500$ is the cost to produce x watches. Find the following:

a. The average rate of change in cost as x is increased from $x = 100$ to $x = 120$

b. The average rate of change in cost as x is increased by 40 from the current production level of 100

c. The marginal cost at $x = 100$

6. The temperature at time t (in minutes) of a plate that is being heated is given by the function $T(t) = 9t^2 + 10t + 50$. In this problem, you will use a calculator to estimate how rapidly the temperature is changing at $t = 4$ minutes. First, calculate the average rate of change of temperature per minute between t and $t + h$ for each of the following increments h added to $t = 4$:

a. $h = 1$ min **b**. $h = 0.1$ min **c**. $h = 0.01$ min

d. $h = 0.001$ min **e**. $h = 0.0001$ min

Look for a pattern in this sequence of average rates of change. The number that the average rates of change seem to be converging to is called the *instantaneous rate of change* or, simply, the rate of change of the temperature at $t = 4$.

7. The graph in Figure 6.7 shows the changes in the price of gold (adjusted to 100 to start) for a given year. Each interval on the horizontal axis represents 1 month, and numbers on the vertical axis represent price. Use the graph to answer the following questions:

a. Estimate the average monthly rate of change in the price of gold from January 1 to March 31.

b. Estimate the average monthly rate of change in the price of gold from the beginning to the end of the month of August.

c. Estimate the average monthly rate of change in the price of gold from July 1 to December 31.

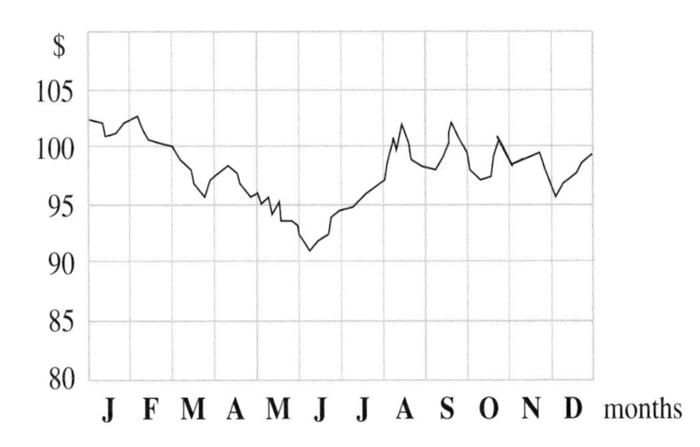

Figure 6.7

8. For each function below, find the average rate of change of f between x and $x + h$ and simplify:

a. $f(x) = 2x - 3$ **b**. $f(x) = x^2 - 3$

c. $f(x) = \sqrt{2x+3}$ **d**. $f(x) = \frac{2}{x-3}$

9. For each function below, find the average rate of change of f from $t = t_1$ to $t = t_2$ and simplify.

a. $f(t) = 5 - 2t$ **b**. $f(t) = 2t^2$

c. $f(t) = \sqrt{5-t}$ **d**. $f(t) = \frac{3}{t+1}$

10. Bob calculated the average rate of change in temperature (in degrees Fahrenheit) for five 2-hour periods and recorded them as follows:

Time Period	Average Rate of Change
9:00 A.M.–11:00 A.M.	2.5°F per hour
11:00 A.M.–1:00 P.M.	0°F per hour
1:00 P.M.–3:00 P.M.	1°F per hour
3:00 P.M.–5:00 P.M.	–1.5°F per hour
5:00 P.M.–7:00 P.M.	–3°F per hour

a. If the temperature at 11:00 A.M. was 35°F, what was the temperature at 1:00 P.M.?

b. If the average rate of change in temperature was recorded each minute for the time period 11:00 A.M. to 1:00 P.M. and it was 0°F each time, what does that tell you about the temperature during this time period?

c. If you know only that the average rate of change in temperature was 0°F per hour from 11:00 A.M. to 1:00 P.M., can you conclude that the temperature was constant over this interval? Justify your answer.

d. If the temperature at 5:00 P.M. was 39°F, what were the temperatures at 3:00 P.M. and at 7:00 P.M.?

e. When the average rate of change in temperature is –3°F per hour, what is happening to the temperature?

11. The graph of the year-to-date rainfall (in inches) in California for a 6-month period is shown in Figure 6.8. For example, one point of the graph has coordinates (J, 1); this says that by the end of January, 1 in. of rain had fallen. Use the graph to answer parts a–d that follow.

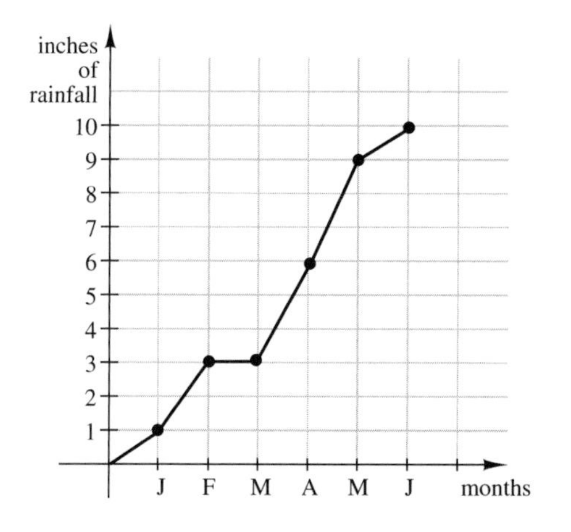

Figure 6.8

a. What is the average rate of change per month in the rainfall for the period from the end of January to the end of March?

b. What is the average rate of change per month in the rainfall for each of the five monthly periods (January to February, February to March, etc.)? What do these numbers represent?

c. What is the average rate of change per month in rainfall for the 6-month period?

d. Are any of the monthly rates of change close to the average rate of change in rainfall for the 6-month period of time?

6-C SECANT AND TANGENT LINES

Rates of change have a geometric interpretation. In every case we have considered, functions vary over a period of time or vary as another quantity (such as production level) changes. In Example 6.5, we saw that for a linear function, the slope of its line gives the average rate of change over any interval. In general, when any function is represented by its graph, we can also represent average rates of change as slopes of lines. A *secant line* of the graph of a function is a line that connects two points on the graph (see Figure 6.9).

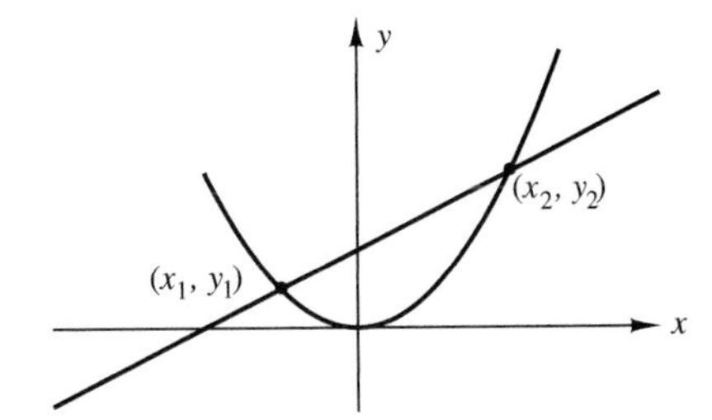

Figure 6.9 A secant line to the graph of a function

The slope m_{sec} of a secant line through the points (x_1, y_1) and (x_2, y_2) on the graph of $y = f(x)$ is defined in the following ways:

$$m_{sec} = \frac{y_2 - y_1}{x_2 - x_1} = \frac{f(x_2) - f(x_1)}{x_2 - x_1} = \frac{f(x_1 + h) - f(x_1)}{h}$$

where $h = x_2 - x_1$.

Notice that these are precisely the definitions used to find the average rates of change of f on the interval $[x_1, x_2]$.

EXAMPLE 6.12 The graph in Figure 6.10 shows the cost function $C(x) = 0.01x^2 - 0.2x + 2000$ given in Example 6.10. A secant line passes through two points on the graph. Find the slope of this secant line, and explain what the slope represents in the context of Example 6.10.

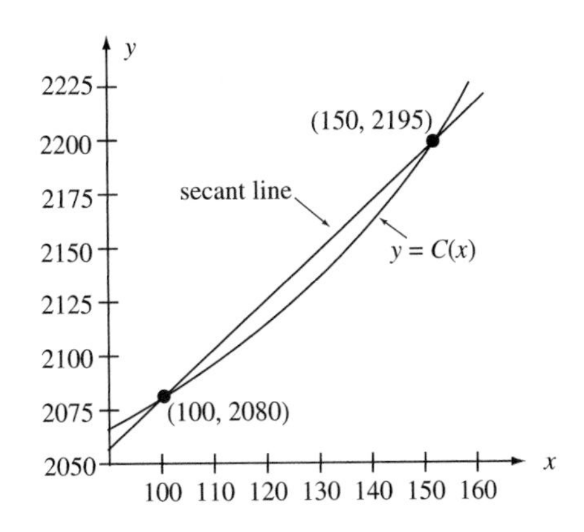

Figure 6.10

Solution The two points on the secant line where it intersects the graph of C are (100, 2080) and (150, 2195); that is, the production cost for 100 sneakers is $2,080 and for 150 sneakers is $2,195. We calculate the slope of the secant line:

$$m_{\text{sec}} = \frac{2195-2080}{150-100} = \frac{115}{50} = 2.30.$$

Earlier, we found that the average rate of change in cost of production over the interval [100, 150] was $2.30 per sneaker, and so the slope of the secant line is equal to the average rate of change in the cost with respect to the change in the number of sneakers produced. ■

In Example 6.10 we also found the marginal cost at a production level of 100 sneakers. If we tried to draw the secant line through the points (100, $C(100)$) and (101, $C(101)$), the points would be so close together that it would be difficult to distinguish between them. For this reason we can approximate this secant line with the tangent line to the graph at the point (100, $C(100)$). The next example shows that this approximation is reasonable. In fact, marginal cost for a particular production level x_1 is generally found by using the slope of the tangent line at $(x_1, C(x_1))$ rather than the slope of the secant line through $(x_1, C(x_1))$ and $(x_1 + 1, C(x_1 + 1))$, since the computations involved in finding the slope of the tangent line are often simpler than those required to find the slope of the secant line.

The *tangent line* to the graph of a function $f(x)$ at a point $(a, f(a))$ is carefully defined in calculus. Intuitively, for a smooth (gently curving) graph, this tangent line will behave like a tangent to a circle; that is, close to the point $(a, f(a))$ it touches the graph only at the point $(a, f(a))$. With a graphing utility, you can zoom in on a point $(a, f(a))$ on a smooth graph, and the graph will appear to approximate a straight line. At this resolution, the tangent line to the graph of f at $(a, f(a))$ is indistinguishable from the graph of f.

EXAMPLE 6.13 The graph of the cost function $C(x) = 0.01x^2 - 0.2x + 2000$ and the tangent line to the graph at $x = 100$ are shown in Figure 6.11. Estimate the slope of this tangent line and compare this number with the marginal cost at $x = 100$ found in Example 6.10.

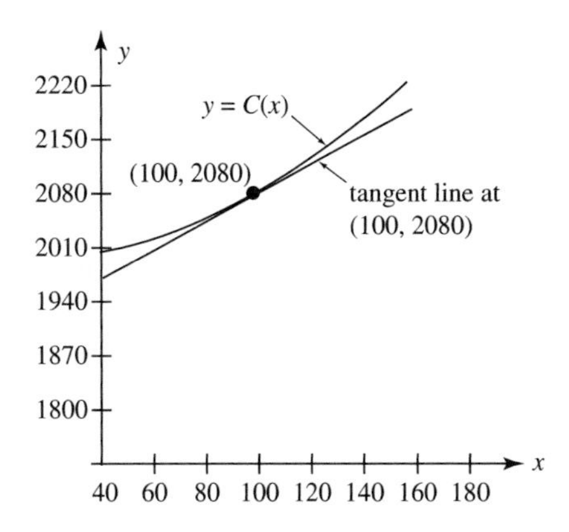

Figure 6.11

Solution The tangent line appears to go through the point (40, 1970) as well as its point of tangency (100, 2080). Thus the slope of the tangent is approximately

$$m_{\tan} = \frac{2080-1970}{100-40} = \frac{110}{60} \approx 1.83.$$

In Example 6.10 we found a marginal cost of \$1.81 at $x = 100$. Therefore, the slope of the tangent line, which is approximately \$1.83 per unit, is a good approximation to this marginal cost. ■

Since the instantaneous rate of change of a function f at a fixed value $x = c$ is the limit of average rates of change of f between c and $c + h$ for small increments h, the tangent line is taken to be the limit of the secant lines through the points $(c, f(c))$ and $(c + h, f(c + h))$ as this second point of intersection with the curve moves closer to the fixed point $(c, f(c))$. Since the slopes of the secant lines correspond to average rates of change of f, the slope of the tangent line at $(c, f(c))$ is the geometric interpretation of the instantaneous rate of change of f at $x = c$.

Exercises 6-C

1. The function that gives the height above the ground (in feet) of a ball t seconds after it has been dropped from a 200-ft tower is $s(t) = 200 - 16t^2$.

a. Sketch the graph of this function over the interval [0, 6].

b. Find the height of the ball from the ground at $t = 0$ sec and after $t = 2$ sec.

c. On the graph, draw the secant line through the two points corresponding to $t = 0$ and $t = 2$.

d. Find the slope of the secant line in part c and compare it with the average velocity found in Example 6.7.

e. Draw (your best guess for) the tangent line to the graph at $t = 2$ sec.

f. Estimate the slope of the tangent line. What is the meaning of this number in the context of the problem?

2. Suppose $C(x) = 0.05x^2 + 2x + 10$ is the cost to produce x bracelets.

a. Sketch the graph of this function over the interval [0, 50].

b. Find the cost to produce 10 bracelets; to produce 20 bracelets.

c. On your graph, draw the secant line through the points corresponding to $x = 10$ and $x = 20$.

d. Find the slope of the secant line in part c and compare it with the average rate of change in the cost found in Example 6.8.

e. Draw (your best guess for) the tangent line to the graph at $x = 20$ bracelets.

f. Estimate the slope of the tangent line. What is the meaning of this number in the context of this problem?

3. Suppose that the temperature T of a metal plate that has been heated for t minutes is given by the function $T(t) = 9t^2 + 10t + 50$.

a. Sketch the graph of this function over the interval [0, 5].

b. Find the temperature at 1 min and 0.5 min later.

c. On your graph, draw the secant line through the points corresponding to $t = 1$ and $t = 1.5$.

d. Find the slope of the secant line in part c. What is the meaning of this number in the context of the problem?

e. Draw (your best guess for) the tangent line to the graph at $t = 1$ min.

f. Estimate the slope of the tangent line. What is the meaning of this number in the context of the problem?

Chapter 6 Exercises

1. Jill has a circular garden with a radius of 10 m (meters). She would like to add a stone walkway of constant width around the outside of the garden. The depth of the sidewalk is to be 0.1 m.

a. Draw a picture of the situation.

b. What is the total area of the garden and a sidewalk that is 2 m wide?

c. What is the area of the garden?

d. What is the area of a 2-m-wide sidewalk?

e. Write an equation for the area A of the sidewalk as a function of the width w of the sidewalk.

f. Write an equation for the volume V of stone used in constructing the sidewalk as a function of its width w.

g. The stone company states that 1 m^3 of the stone to be used weighs 1 ton. Write an equation for the number of tons of stone needed as a function of the width w of the sidewalk.

h. The stone costs $35 per ton. Write an equation for the cost C of the stone to be used as a function of the width w.

i. What is the widest sidewalk Jill can make if she has $1,000 allotted for the stone?

2. The Bond Paper Company charges $30 a case for 20-lb weight $8\frac{1}{2} \times 11$-in. white paper for the first 10 cases. If you buy more than 10 cases, the price per case for the additional cases is reduced by $.10 times the number of additional cases. Thus, the cost of 12 cases is (10)($30) + (2)($29.80), or $359.60.

a. What is the total cost for 20 cases?

b. Write an equation for the total cost C as a function of the number of cases x bought, if $x \geq 10$.

c. What is the average cost per case if you buy 20 cases? (Average cost = $\frac{C(x)}{x}$.)

d. Write an equation for the average cost per case if you buy x cases, if $x \geq 10$.

e. Accounting needs 15 cases of paper, and Marketing needs 20 cases. How much will the company save by combining the orders?

3. Tuition at a state-supported college over a 6-year period was as follows:

Year	Tuition
1990	$9,100
1991	$9,500
1992	$10,000
1993	$10,200
1994	$11,200
1995	$12,300

a. What is the average rate of change in tuition from 1990 to 1991?

b. What is the average rate of change in tuition from 1990 to 1995?

c. In which year was the average rate of change in tuition the largest?

d. The rate of increase in tuition for a given year can be found by dividing the increase for the year by the previous tuition. Express the rate of increase (as a percent) for 1991.

e. Which of the years 1991–1995 had the largest rate of increase?

f. Which rate, the average rate of change in tuition for a given period or the rate of increase as a percent, seems to provide more useful economic information? Justify your response.

4. Express each of the following as an average rate.

a. There were 370 births at the base hospital in 2 months.

b. He completed 9 passes in 16 attempts.

c. Of the 2000 adults surveyed, 300 were unemployed.

5. The graph in Figure 6.12 shows the U.S. federal deficit/surplus (in billions of dollars) for the fiscal years 1982 through 2004.

a. What was the average rate of change in the deficit/surplus from 1982 to 1986?

b. What was the average rate of change in the deficit/surplus from 1986 to 1992?

c. What was the average rate of change in the deficit/surplus from 1992 to 2000?

d. What was the average rate of change in the deficit/surplus from 1996 to 2004?

e. When the deficit is decreasing, is the graph rising or falling?

f. Which year showed the largest rate of increase in the deficit?

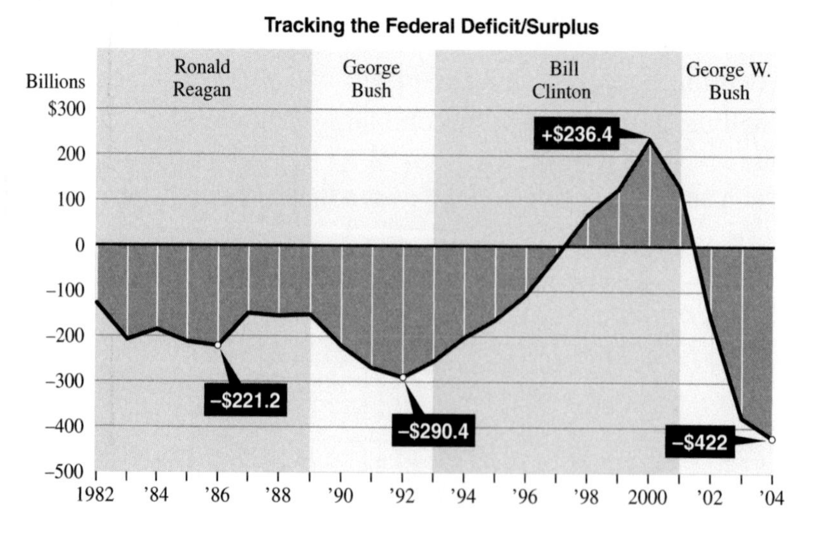

Figure 6.12

Data Source: Congressional Budget Office; White House Graph. *Morning Call,* Sept. 8, 2004

6. $C(x) = 0.005x^2 - 2x + 1200$ is a function that gives the cost to produce x items.

a. Find the average rate of change in the cost of production between 10,000 and $10{,}000 + h$ for each of these increments h: $h = 1000$, $h = 100$, $h = 10$.

b. Estimate the marginal cost at a production level of $x = 10{,}000$.

7. For each function below, find the average rate of change of the function between x and $x + h$ and simplify:

a. $f(x) = 5 - 3x$ **b.** $f(x) = x^2 + 4$

c. $g(x) = \sqrt{x-2}$ **d.** $h(x) = \frac{4}{x+1}$

8. For each of the following functions, find the average rate of change of the function from $t = t_1$ to $t = t_2$ and simplify:

a. $f(x) = 5 - 3t$ **b.** $f(t) = t^2 + 4$

c. $g(x) = \sqrt{t-2}$ **d.** $h(x) = \frac{4}{t+1}$

9. The distance (in meters) a stone falls when dropped from a cliff is a function of time (in seconds) and is given by $s(t) = 4.9t^2$.

a. Sketch the graph of this function over the interval [0, 6].

b. Find the distance the stone dropped from $t = 1$ sec to $t = 3$ sec.

c. On your graph, draw the secant line through the points on the graph corresponding to $t = 1$ and $t = 3$.

d. Find the slope of the secant line drawn in part c.

e. Find the average velocity from $t = 1$ sec to $t = 3$ sec.

f. Draw the tangent line to your graph at $t = 3$ sec.

g. Estimate the slope of the tangent line. What is the meaning of this number in the context of the problem?

CHAPTER 7

Companion to Rules of Differentiation

7-A NEGATIVE AND RATIONAL EXPONENTS

To differentiate functions such as $f(x) = \frac{2}{x^4}$ or $g(x) = \sqrt{x}$, it is often helpful first to express the functions in terms of powers x^r, where r is a rational number (r is the quotient of two integers). For example, we can rewrite $f(x) = \frac{2}{x^4} = 2x^{-4}$ and $g(x) = \sqrt{x} = x^{1/2}$. In this section we review the definitions and basic properties of exponents.

If n is a positive integer, x^n means x is multiplied n times by itself:

$$x^n = \underbrace{x \cdot x \ x \ \cdots \ x}_{n \text{ factors}}.$$

For example,

$$x^5 = x \cdot x \cdot x \cdot x \cdot x \quad \text{and} \quad 10^9 = \underbrace{10 \cdot 10 \ \cdots \ 10 \cdot 10}_{9 \text{ factors}} = 1{,}000{,}000{,}000.$$

If n is a positive integer, $x^{1/n}$ is the nth root of x:

$$x^{1/n} = \sqrt[n]{x}.$$

This means that when $x \geq 0$, $x^{1/n}$ is the nonnegative number whose n^{th} power is x. For example, $27^{1/3} = 3$ since $3^3 = 27$, and $16^{1/4} = 2$ since $2^4 = 16$. When $x < 0$, $x^{1/n}$ is defined only for n odd; it is the number whose n^{th} power is x. For example, $(-27)^{1/3} = -3$ since $(-3)^3 = -27$, but $(-16)^{1/4}$ is not defined since there is no real number c for which $c^4 = -16$.

When $x < 0$ and n is even, $x^{1/n}$ is not a real number; therefore, we say $x^{1/n}$ is not defined. For example, $(-16)^{1/4}$ is not defined. (However, $-16^{1/4} = -(16^{1/4}) = -2$.)

If m and n are positive integers, then $x^{m/n}$ is defined as follows:

$$x^{m/n} = \left(x^{1/n}\right)^m = \left(\sqrt[n]{x}\right)^m$$

The domain of $x^{m/n}$ is restricted to $x \geq 0$ when n is even and is all real numbers when n is odd.

To evalute $x^{m/n}$ we first take an nth root, then an mth power. When c is in the domain of $x^{m/n}$, we can also compute $c^{m/n}$ by first taking the mth power, then the nth root. For example,

$$4^{3/2} = \left(\sqrt{4}\right)^3 = 2^3 = 8; \quad \text{also } 4^{3/2} = \sqrt{4^3} = \sqrt{64} = 8;$$

$$(-8)^{5/3} = \left(\sqrt[3]{-8}\right)^5 = (-2)^5 = -32 \quad \text{and} \quad (-8)^{5/3} = \sqrt[3]{(-8)^5} = \sqrt[3]{-32{,}768} = -32.$$

When both m and n are even, the functions $f(x) = x^{m/n} = \left(\sqrt[n]{x}\right)^m$ and $g(x) = \sqrt[n]{x^m}$ **are not equal** because their domains are different. The domain of f consists of only nonnegative real numbers, whereas the domain of g includes every real number. For example, $\left(\sqrt{-2}\right)^4$ is undefined whereas $\sqrt{(-2)^4} = 4$.

The composite function $f(x) = \left(x^2\right)^{1/2} = \sqrt{x^2}$ is defined for all x and is **not** the same as $g(x) = x$. In fact, $f(x) = \sqrt{x^2} = |x| = \begin{cases} -x & \text{if } x < 0 \\ x & \text{if } x \geq 0. \end{cases}$

In general, when m and n are both even, $\left(x^m\right)^{1/n} = |x|^{m/n}$.

Figures 7.1 and 7.2 show the graphs of $y = \sqrt{x}$ and $y = \sqrt{x^2} = |x|$, respectively.

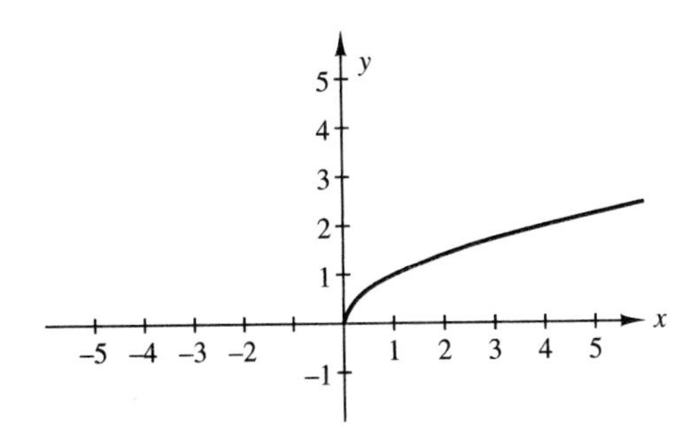

Figure 7.1 $y = \sqrt{x}$

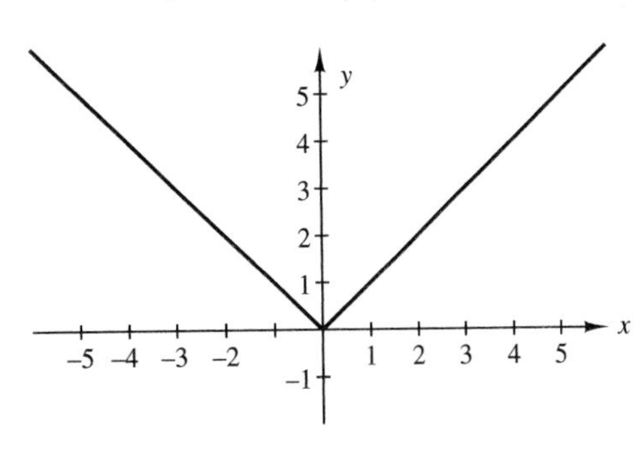

Figure 7.2 $y = \sqrt{x^2} = |x|$

EXAMPLE 7.1 Give the domain of $f(x) = (x+1)^{3/4}$ and find $f(15)$.

Solution Since $(x+1)^{3/4} = \left(\sqrt[4]{x+1}\right)^3$, $f(x)$ is defined only when $x + 1 \geq 0$. So the domain of f is the set of numbers x such that $x \geq -1$, which is the interval $[-1, \infty)$. We evaluate $f(15)$ as follows: $f(15) = (15+1)^{3/4} = (16)^{3/4} = \left(\sqrt[4]{16}\right)^3 = 2^3 = 8$. We could also find this value by taking first the power and then the root: $(16)^{3/4} = \sqrt[4]{16^3} = \sqrt[4]{4096} = 8$. Note that this method can produce large numbers in calculations. ■

An exponent can also be zero or negative. The expression x^0 is defined as follows:

$$x^0 = 1 \text{ for } x \neq 0; \quad \text{the expression } 0^0 \text{ is not defined.}$$

If r is a positive rational number, then x^{-r} is defined as the reciprocal of x^r:

$$x^{-r} = \frac{1}{x^r}, \text{ for } x \neq 0.$$

For example, $x^{-2/3} = \frac{1}{x^{2/3}} = \frac{1}{\left(\sqrt[3]{x}\right)^2}$.

EXAMPLE 7.2 Rewrite each expression so it does not contain negative exponents:

a. $(3x)^{-2}$ **b.** $3x^{-2}$ **c.** $\frac{x^{-2}}{3}$

Solution

a. $(3x)^{-2} = \frac{1}{(3x)^2} = \frac{1}{9x^2}$ **b.** $3x^{-2} = 3\left(\frac{1}{x^2}\right) = \frac{3}{x^2}$

c. $\frac{x^{-2}}{3} = \frac{1}{3}x^{-2} = \left(\frac{1}{3}\right)\left(\frac{1}{x^2}\right) = \frac{1}{3x^2}$ ■

EXAMPLE 7.3 Give the domain of each of the following functions:

a. $g(x) = (4 - 3x)^{-3/5}$ **b.** $h(x) = (1 - 3x)^{-1/6}$ **c.** $f(x) = (x^2 - 1)^{-4}$

Solution

a. $g(x) = (4-3x)^{-3/5} = \frac{1}{(4-3x)^{3/5}} = \frac{1}{\left(\sqrt[5]{4-3x}\right)^3}$. The domain of g consists of all real numbers x such that $4 - 3x \neq 0$; that is, $x \neq \frac{4}{3}$. In interval notation, the domain of g is the set $\left(-\infty, \frac{4}{3}\right) \cup \left(\frac{4}{3}, \infty\right)$.

b. $h(x)=(1-3x)^{-1/6}=\frac{1}{(1-3x)^{1/6}}=\frac{1}{\sqrt[6]{1-3x}}$. Since we are taking an even root of $1-3x$, a number x in the domain of h must satisfy $1-3x\geq 0$; also $1-3x\neq 0$, since the root is in the denominator. So x is in the domain of h only if $1-3x>0$. This is true if $1>3x$, or $\frac{1}{3}>x$, so the domain of h is the interval $\left(-\infty,\frac{1}{3}\right)$.

c. $f(x)=(x^2-1)^{-4}=\frac{1}{(x^2-1)^4}$. The domain consists of all values of x for which $x^2-1\neq 0$. Since the solutions of $x^2-1=0$ are $x=1$ and $x=-1$, the domain of f is $(-\infty,-1)\cup(-1,1)\cup(1,\infty)$. ■

The definitions for exponents lead to the following basic rules of operation with exponents.

Rules of Exponents: Rules of exponents allow functions to be simplified or to be rewritten in a more convenient form to apply rules of differentiation. If r and s are rational numbers, then:

1. $x^r x^s=x^{r+s}$
2. $\frac{x^r}{x^s}=x^{r-s}$
3. $(x^r)^s=x^{rs}$
4. $(xy)^r=x^r y^r$
5. $\left(\frac{x}{y}\right)^r=\frac{x^r}{y^r}$

whenever the expressions on the left sides of the equations are defined.

Although $(ab)^r=a^rb^r$ and $\left(\frac{a}{b}\right)^r=\frac{a^r}{b^r}$, there is no corresponding rule for sums and differences. In general, $(a+b)^r\neq a^r+b^r$ and $(a-b)^r\neq a^r-b^r$. When r is an integer, the distributive law may be used to expand (multiply out) the expressions $(a+b)^r$ and $(a-b)^r$.

The rules of exponents may also be combined with another rule of algebra to rewrite some quotients in a form more convenient for differentiation:

$$\frac{a+b}{c}=\frac{a}{c}+\frac{b}{c}\quad\text{and}\quad\frac{a-b}{c}=\frac{a}{c}-\frac{b}{c}.$$

Although $\frac{a+b}{c}$ may be written as the sum of two quotients, each having denominator c, the quotient $\frac{d}{q+r}$ **cannot** be written as the sum of two separate quotients.

EXAMPLE 7.4 Rewrite each function as a sum where each term in the sum is a constant times a rational power of the independent variable.

a. $F(x) = \frac{3}{x} + 4\sqrt{x^5}$

b. $D(x) = \frac{1}{5x^6} - \frac{x^3}{\sqrt[4]{x}}$

c. $G(t) = \sqrt{t}\left(t^2 + 3\sqrt{t}\right)$

d. $H(z) = \left(3z^{1/3} - z^{2/3}\right)^2$

e. $K(x) = \frac{x^5 - 3\sqrt{x} + \sqrt{2}}{x^6}$

Solution

a. We use rule 3 to write $\left(x^5\right)^{1/2} = x^{5/2}$. Thus, $F(x) = 3\cdot\frac{1}{x} + 4\left(x^5\right)^{1/2} = 3x^{-1} + 4x^{5/2}$.

b. $D(x) = \left(\frac{1}{5}\right)\cdot\left(\frac{1}{x^6}\right) - \frac{x^3}{x^{1/4}} = \frac{1}{5}x^{-6} - x^{(3-1/4)} = \frac{1}{5}x^{-6} - x^{1\,1/4}$. Here we used rule 2 to rewrite $\frac{x^3}{x^{1/4}}$.

c. We use the distributive law and then rule 1: $G(t) = t^{1/2}\left(t^2 + 3t^{1/2}\right) = t^{1/2}t^2 + t^{1/2}3t^{1/2} = t^{1/2+2} + 3t^{1/2+1/2} = t^{5/2} + 3t$.

d.
$$H(z) = \left(3z^{1/3} - z^{2/3}\right)\left(3z^{1/3} - z^{2/3}\right) = \left(3z^{1/3}\right)^2 - 2\left(3z^{1/3}\right)\left(z^{2/3}\right) + \left(z^{2/3}\right)^2$$
$$= 9z^{2/3} - 6z^{1/3}z^{2/3} + z^{4/3} = 9z^{2/3} - 6z + z^{4/3}$$

e. $K(x) = \frac{x^5 - 3\sqrt{x} + \sqrt{2}}{x^6} = \frac{x^5}{x^6} - \frac{3\sqrt{x}}{x^6} + \frac{\sqrt{2}}{x^6} = x^{(5-6)} - 3x^{(1/2-6)} + \sqrt{2}x^{-6} = x^{-1} - 3x^{-1\,1/2} + \sqrt{2}x^{-6}$ ■

Take care in rewriting $\frac{1}{ax^r}$: $\frac{1}{ax^r} = \left(\frac{1}{a}\right)\left(\frac{1}{x^r}\right) = a^{-1}x^{-r} = \frac{1}{a}x^{-r}$. For example, $\frac{1}{3x^2} = \frac{1}{3}x^{-2}$; $\frac{1}{3x^2} \neq 3x^{-2}$ and $\frac{1}{3x^2} \neq (3x)^{-2}$.

Rational exponents occur in a variety of applications; the following example shows one of them.

EXAMPLE 7.5 In a certain community the average daily level of carbon monoxide in the air is estimated to be $c = 0.3p^{2/3}$ parts per million when the population is p thousand. A population study of the same community indicates that t years from now, the population will be $p = \sqrt{3+t}$.

a. Write an expression for the daily level of carbon monoxide t years from now.

b. What will the daily level be 2 years from now?

Solution

a. $c = 0.3p^{2/3} = 0.3\left(\sqrt{3+t}\right)^{2/3} = 0.3\left((3+t)^{1/2}\right)^{2/3} = 0.3(3+t)^{1/3}$

b. When $t = 2$, $c = 0.3(3+2)^{1/3} = 0.3(5)^{1/3} = 0.3\sqrt[3]{5} \approx (0.3)(1.71) = 0.513$. ■

Exercises 7-A

1. If $f(x) = x^{3/5}$, find each of the following values if it exists. If it does not exist, explain why.

a. $f(32)$ **b**. $f(-32)$ **c**. $f\left(\frac{1}{32}\right)$ **d**. $f\left(4^{-5}\right)$ **e**. $f\left(a^{10}\right)$

2. If $g(x) = x^{-5/6}$, find each of the following values if it exists. If it does not exist, explain why.

a. $g(1)$ **b**. $g(-1)$ **c**. $g(0)$ **d**. $g\left(\frac{1}{64}\right)$ **e**. $g(64)$ **f**. $g\left(\frac{1}{a^3}\right)$

3. Give the domain of each function:

a. $F(x) = (2x-5)^{-2/3}$ **b**. $h(t) = (1+4t)^{-3/4}$ **c**. $g(x) = \left(9-x^2\right)^{7/4}$

4. Rewrite each function below as a sum, where each term in the sum is a constant times a power of x.

a. $f(x) = \frac{2}{\sqrt[3]{x}} - \frac{5x+1}{10x^7} - 2\sqrt[5]{x^2}$

b. $g(x) = \frac{3x^3-2x+6}{\sqrt{x}}$

c. $h(x) = \sqrt[3]{x}\left(x^2-2x+1\right)$

d. $F(x) = x^2 + \frac{1}{2x^3} - \frac{2}{3x} - 12$

e. $G(x) = \frac{4\sqrt{x}-2x^3+7}{x^2\sqrt[6]{x}}$

f. $H(x) = \frac{\sqrt{5}}{x^4} + \frac{1}{\sqrt{x}}$

g. $W(x) = \left(2x^{-3/4} + 3x^{3/4}\right)^2$

h. $S(x) = \left(x^{3/2} - \frac{2}{x^2}\right)^2$

5. For each equation below, consider the values of x for which all expressions in the equation are defined. Then decide whether the equation is always true or sometimes false for these values. If sometimes false, find a specific value of x for which the equation is not true.

a. $(5x)^{1/2} = 5\sqrt{x}$

b. $\frac{1}{5x} = 5x^{-1}$

c. $\dfrac{1}{10x^3} = \dfrac{1}{10}x^{-3}$

d. $\dfrac{2}{\sqrt[3]{8x}} = x^{-1/3}$

e. $\dfrac{x^3 + x^2}{\sqrt{x}} = \dfrac{x^3}{x^{1/2}} + \dfrac{x^2}{x^{1/2}} = x^{5/2} + x^{3/2}$

f. $\dfrac{\sqrt{x}+1}{x^3+x^2} = \dfrac{\sqrt{x}+1}{x^3} + \dfrac{\sqrt{x}+1}{x^2} = x^{-5/2} + x^{-3} + x^{-3/2} + x^{-2}$

6. For each equation, determine if it is correct (true for all values of x). If it is not correct, state why, and give a specific value of x for which the equation is false.

a. $\left(x^4 + x^2\right)^{1/2} = x^{4/2} + x^{2/2} = x^2 + x$

b. $\left(x^4 + x^2\right)^{1/2} = \left(x^2\left(x^2+1\right)\right)^{1/2} = x^{2/2}\left(x^2+1\right)^{1/2} = x\left(x^2+1\right)^{1/2}$

c. $\left(x^4 + x^2\right)^{1/2} = \sqrt{x^2\left(x^2+1\right)} = \sqrt{x^2}\sqrt{x^2+1} = |x|\sqrt{x^2+1}$

7-B DECOMPOSITION OF FUNCTIONS

In calculus books several *rules of differentiation* are discussed. They provide methods for finding the derivative of a variety of functions. To apply these rules it is essential to identify a given function as a sum, difference, product, quotient, or composition of simpler functions.

EXAMPLE 7.6 Each of the following functions can be viewed as a sum, difference, product, quotient, or composition of two simpler functions. In each case, say which operation is involved and give the two functions.

a. $F(x) = \left(3x^2 - 1\right)\left(4x^3 - \frac{12}{5}x^2 + 2x - 1\right)$

b. $G(x) = \sqrt{x^3 - 1}$

c. $K(x) = \dfrac{\sqrt{3x-1}}{x^2+4}$

d. $H(x) = \dfrac{4x+7}{\sqrt[3]{x}}$

e. $N(x) = 4x^5 + \sqrt{5x+2}$

Solution

a. F is the product of the functions f and g, where $f(x) = 3x^2 - 1$ and $g(x) = 4x^3 - \frac{12}{5}x^2 + 2x - 1$.

b. G is the composite function $f(g(x))$, where $f(x) = \sqrt{x}$ and $g(x) = x^3 - 1$, since then $f(g(x)) = f(x^3 - 1) = \sqrt{x^3 - 1}$.

c. $K(x) = \frac{f(x)}{g(x)}$, where $f(x) = \sqrt{3x-1}$ and $g(x) = x^2 + 4$.

d. $H(x) = \frac{f(x)}{g(x)}$, where $f(x) = 4x + 7$ and $g(x) = \sqrt[3]{x}$. H can also be interpreted as the sum of two functions by writing $H(x) = \frac{4x+7}{\sqrt[3]{x}} = \frac{4x}{\sqrt[3]{x}} + \frac{7}{\sqrt[3]{x}}$.

e. N is the sum of the functions $f(x) = 4x^5$ and $g(x) = \sqrt{5x+2}$. ■

It is straightforward to find derivatives of functions that are sums of terms of the form cx^r, a constant times a rational power of x. For this reason, to find the derivative of a function, we want to view the function as the result of one or more operations among such functions when possible.

EXAMPLE 7.7 Some of the functions below can be written as sums whose terms are a constant times a power of x. Others are combinations of two functions of that kind. In each case, write the function as a sum whose terms are a constant times a power of x or as a combination of two functions of that type.

a. $F(x) = \frac{x^2}{3} - \frac{x}{\sqrt{2}} + \frac{2}{5x}$

b. $G(x) = \sqrt{x+5}$

c. $K(x) = \frac{4x^2 - 3x}{\sqrt[4]{x}}$

d. $H(x) = \dfrac{\frac{7}{2\sqrt{x}} + 3x^2}{\sqrt{x^5} - \frac{1}{x}}$

Solution

a. $F(x) = \frac{1}{3}x^2 - \frac{1}{\sqrt{2}}x + \frac{2}{5}x^{-1}$ is a sum of terms of the desired form.

b. $G(x) = f(g(x))$, where $g(x) = x + 5$ and $f(x) = \sqrt{x} = x^{1/2}$.

c. $K(x) = \frac{4x^2 - 3x}{\sqrt[4]{x}} = \frac{4x^2}{\sqrt[4]{x}} - 3\frac{x}{\sqrt[4]{x}} = 4x^{(2-1/4)} - 3x^{(1-1/4)} = 4x^{7/4} - 3x^{3/4}$

d. $H(x)$ is the quotient $\frac{f(x)}{g(x)}$, where $f(x) = \frac{7}{2\sqrt{x}} + 3x^2 = \frac{7}{2}x^{-1/2} + 3x^2$ and $g(x) = \sqrt{x^5} - \frac{1}{x} = x^{5/2} - x^{-1}$. ■

There are rules for differentiating a sum, difference, product, quotient, or composition of two functions. The rules for sum and difference are simple: The derivative of a sum (or difference) of two functions is the sum (or difference) of their derivatives. The rules for differentiating a product, quotient, or composition of functions are more complicated. These rules give directions for how to find the derivative of a combination of two functions, provided the derivative of each of the individual functions is known. The rule for the derivative of a product of two functions is called the *product rule.* The rule for the derivative of a quotient of two functions is called the *quotient rule.* The rule for the derivative of a composition of two functions is called the *chain rule.*

Since many functions are obtained by performing more than one operation on simpler functions, it is often necessary to use several steps to decompose a function into sufficiently simple ones.

The rules for derivatives apply to one operation at a time. For example, the product rule gives a formula for the derivative of the product of two functions, provided the derivative of each of the factors is known. To use this rule, it is necessary to view the function as a product of just two functions. But then each of these two, in turn, may have to be viewed as a combination of even simpler functions.

EXAMPLE 7.8 Consider the function $H(x) = \dfrac{\left(\sqrt{x} - \frac{3}{x}\right)^5}{x+5}$.

a. Give two functions, f and g, so that $H(x)$ is the sum, difference, product, quotient, or composition of f and g.

b. Do the same for the functions f and g. Repeat this procedure until the functions obtained are sums, where each term in the sum is a constant times a power of x.

c. Give the sequence of rules to be used to differentiate $H(x)$.

Solution

a. $H(x)$ is the quotient $\frac{f(x)}{g(x)}$, where $f(x) = \left(\sqrt{x} - \frac{3}{x}\right)^5$ and $g(x) = x + 5$.

b. $f(x) = \left(\sqrt{x} - \frac{3}{x}\right)^5 = (k(x))^5$, where $k(x) = \sqrt{x} - \frac{3}{x} = x^{1/2} - 3x^{-1}$. So $f(x)$ can be written $p(k(x))$, where $p(x) = x^5$. The function $g(x) = x + 5$ is already of the required form.

c. To differentiate the function H, first the quotient rule would be used, since $H(x) = \frac{f(x)}{g(x)} = \frac{p(k(x))}{g(x)}$. But to use the quotient rule, the numerator $p(k(x))$ must be differentiated. To do that, the chain rule for composite functions must be used. ■

EXAMPLE 7.9 Repeat parts a, b, and c of Example 7.8 for the function

$$H(x) = \sqrt[3]{\left(2x - x^2\right)^5 + \left(x^3 + x - \frac{1}{\sqrt{5}}\right)(x^2 + 4x - 1)}\,.$$

Solution

a. $H(x)$ is the composite function $\sqrt[3]{f(x)} = (f(x))^{1/3}$, where

$$f(x) = \left(2x - x^2\right)^5 + \left(x^3 + x - \frac{1}{\sqrt{5}}\right)\left(x^2 + 4x - 1\right).$$

If $g(x) = x^{1/3}$, then $H(x) = g(f(x))$.

b. The function f is the sum of the two functions $k(x) = \left(2x - x^2\right)^5$ and $h(x) = \left(x^3 + x - \frac{1}{\sqrt{5}}\right)\left(x^2 + 4x - 1\right)$. These are further decomposed as follows: k is the composite function $k(x) = (z(x))^5$, where $z(x) = 2x - x^2$, and $h(x)$ is the product of $u(x) = x^3 + x - \frac{1}{\sqrt{5}}$ and $v(x) = x^2 + 4x - 1$, which are of the required form.

c. To differentiate the function H, first the chain rule would be used, since $H(x) = (f(x))^{1/3}$. In the first step of the chain rule, the function $f(x)$ must be differentiated. Since $f(x)$ is a sum, each term must be differentiated. The chain rule would be used to differentiate the first term and the product rule to differentiate the second term. ■

Exercises 7-B

1. Describe how each function below is a product, quotient, or composition of two functions f and g. Also, show how these functions f and g can be written as a sum of terms of the form cx^r (a constant times a rational power of x).

a. $F(x) = \sqrt{x+5}$

b. $G(x) = \frac{2x^4 - 1}{\sqrt{x} - 5}$

c. $H(x) = \left(\frac{1}{2\sqrt[5]{x}} + x^3 - \frac{x}{3}\right)\left(7x - \frac{4}{5x}\right)$

d. $B(x) = \sqrt[3]{\left(x^4 - 2x + 1\right)x}$

2. Describe each of the following functions as a sum, difference, product, quotient, or composition of two simpler functions; give the two simpler functions. Repeat this for the two functions obtained, and continue this process until the simpler functions are sums of terms of the form cx^r.

a. $F(x) = \sqrt{x+5}\,\sqrt[3]{x^2 - 1}$

b. $G(x) = \frac{(3x+2)^7}{8-x}$

c. $H(x) = \sqrt{\frac{(x+8)^7}{(3x+2)^3}}$

d. $L(x) = \sqrt[3]{(3x-8)^4} - \frac{x}{x-1}$

e. $N(x) = \dfrac{2 - \frac{1}{x}}{x+1}$ **f.** $p(x) = x\sqrt{x} + \dfrac{1}{x^2\sqrt{x}}$

3. For each function in Exercise 2, give the sequence of rules to be used to differentiate the function.

7-C SIMPLIFYING DERIVATIVES

The derivative of a function can be used to find slopes of tangent lines to the graph of the function and to give information about when the function is increasing or decreasing. It is also used to find the instantaneous rate of change of a function; for example, it can describe a velocity, a growth rate, or a marginal cost. We may want to know when the derivative is zero, positive, or negative, or when it takes on a particular value. The rules of differentiation will often produce derivatives in an algebraic form from which it is difficult to answer these questions. If the derivative can be simplified through algebraic manipulation, these questions can be answered more easily.

EXAMPLE 7.10 A curve has equation $y = x^3 - 3x^2$. The slope of the tangent line to this curve at a point (x, y) is the derivative of the function, which is

$\dfrac{dy}{dx} = 3x^2 - 6x.$

a. Find all values of x where the curve has a horizontal tangent line.

b. Find all points on the curve where the tangent line is parallel to the line with equation $y = 9x + 3$.

c. Find the x-coordinates of all points on the curve where the slope of the tangent line is 2.

Solution

a. The tangent line is horizontal when its slope is zero. We need to find all values of x for which $\dfrac{dy}{dx} = 0$. We solve $3x^2 - 6x = 0$ by factoring: $3x(x-2) = 0$. The solutions are $x = 0$ and $x = 2$.

b. For the tangent line to be parallel to a given line, it must have the same slope as the given line, which is 9. So we need to find all points on the curve for which $\dfrac{dy}{dx} = 9$. To solve $3x^2 - 6x = 9$, we first put the quadratic equation in standard form: $3x^2 - 6x - 9 = 0$. Then factor: $3(x + 1)(x - 3) = 0$. The solutions are $x = -1$ and $x = 3$. These are the x-coordinates of the points on the curve where the tangent lines have slope 9; we find the y-coordinates by substituting these x-values into the equation of the curve $y = x^3 - 3x^2$. When $x = -1$, $y = (-1)^3 - 3(-1)^2 = -4$. When $x = 3$, $y = (3)^3 - 3(3)^2 = 0$. So the points $(-1, -4)$ and $(3, 0)$ are the requested points.

c. We need to find all values of x for which $\frac{dy}{dx} = 2$. First we put $3x^2 - 6x = 2$ in standard form: $3x^2 - 6x - 2 = 0$. This quadratic polynomial does not factor, so we use the quadratic formula to find the solutions of the equation:

$$x = \frac{6 \pm \sqrt{36+24}}{6} = 1 \pm \frac{\sqrt{60}}{6} = 1 \pm \frac{\sqrt{15}}{3}.$$

Thus, $x = 1 + \frac{\sqrt{15}}{3} \approx 2.29$ and $x = 1 - \frac{\sqrt{15}}{3} \approx -0.29$. ■

The product rule of differentiation will produce a sum of terms. To find when this sum equals zero, it is often necessary to find common factors of the terms in the sum.

EXAMPLE 7.11 The function $f(x) = x^3(x-1)^4$ has derivative

$$f'(x) = 4x^3(x-1)^3 + 3x^2(x-1)^4.$$

Find all x-coordinates of points on the graph of the function where the tangent line is horizontal.

Solution We need to solve the equation $f'(x) = 4x^3(x-1)^3 + 3x^2(x-1)^4 = 0$. This equation has common factors x^2 and $(x-1)^3$ in both terms of the sum, so we factor these out and then simplify:

$$x^2(x-1)^3[4x + 3(x-1)] = x^2(x-1)^3(7x-3) = 0.$$

The solutions are $x = 0$, $x = 1$, and $x = \frac{3}{7}$. ■

Often derivatives of functions are sums of terms of the form cx^r, where r is a rational exponent. When some of the exponents are negative, it may be best to express the sum as a single quotient to solve a problem.

EXAMPLE 7.12 The function $y = 4x^{5/2} + 2x^{3/2} - 4\sqrt{x}$ has derivative $y' = 10x^{3/2} + 3x^{1/2} - 2x^{-1/2}$. Find all values of x for which $y' = 0$.

Solution The last term in y' has negative exponent $-\frac{1}{2}$, so we write that term $\frac{-2}{x^{1/2}}$. To express y' as a quotient with common denominator $x^{1/2}$, we multiply each of the first two terms in y' by $1 = \frac{x^{1/2}}{x^{1/2}}$:

$$y' = 10x^{3/2}\frac{x^{1/2}}{x^{1/2}} + 3x^{1/2}\frac{x^{1/2}}{x^{1/2}} - \frac{2}{x^{1/2}} = \frac{10x^2 + 3x - 2}{x^{1/2}}.$$

The derivative y' will equal zero when the numerator of this quotient is zero and the denominator is defined and nonzero. To find the zeros of the numerator $10x^2+3x-2$, we use the quadratic formula: $x=\frac{-3\pm\sqrt{89}}{20}$. One of these solutions is negative and must be eliminated since the denominator $x^{1/2}=\sqrt{x}$ is not defined for negative numbers. So the only value of x for which $y'=0$ is $x=\frac{-3+\sqrt{89}}{20}\approx$ 0.32. ■

The quotient rule of differentiation will produce a quotient in which the numerator is usually a difference of two functions. The next example shows algebraic simplification of a derivative produced using this rule.

EXAMPLE 7.13 The function $F(x)=\frac{2x^3+4x-9}{2x-1}$ has derivative

$$F'(x)=\frac{(2x-1)(6x^2+4)-2(2x^3+4x-9)}{(2x-1)^2}.$$

Simplify $F'(x)$ as much as possible and find the values of x where $F'(x)=0$.

Solution The numerator can be simplified by expanding the products and adding like terms:

$$F'(x)=\frac{12x^3-6x^2+8x-4-4x^3-8x+18}{(2x-1)^2}=\frac{8x^3-6x^2+14}{(2x-1)^2}.$$

To solve $F'(x)=0$, we need to factor the numerator. By inspection, we notice that $x=-1$ is a zero of $8x^3-6x^2+14$, the polynomial in the numerator. So, by the Factor Theorem (see Section 3-B.1), $x+1$ is a factor of $8x^3-6x^2+14$. When we divide $8x^3-6x^2+14$ by $(x+1)$, we obtain $8x^2-14x+14$ as the quotient and 0 as the remainder. Thus $8x^3-6x^2+14=(x+1)(8x^2-14x+14)$. Therefore,

$$F'(x)=\frac{8x^3-6x^2+14}{(2x-1)^2}=\frac{(x+1)(8x^2-14x+14)}{(2x-1)^2}.$$

Since the polynomial $8x^2-14x+14$ has no real roots (verify this), the only value of x for which $F'(x)=0$ is $x=-1$. ■

Often answers to differentiation exercises in calculus texts are given in a simplified form, and it is not possible to know if your solution is correct unless you also simplify your answer.

EXAMPLE 7.14 You differentiated the function $f(x)=\frac{x^2+x}{x^2-1}$ by using the quotient rule and obtained $f'(x)=\frac{(x^2-1)(2x+1)-(x^2+x)2x}{(x^2-1)^2}$. The answer in the back of the book for this derivative is $f'(x)=\frac{-1}{(x-1)^2}$, $x\neq\pm 1$. Is your differentiation correct?

Solution We need to determine whether the derivative you obtained can be simplified to the form $\frac{-1}{(x-1)^2}$. First, we multiply out the numerator of the quotient of your $f'(x)$ and then consolidate terms:

$$(x^2-1)(2x+1)-(x^2+x)2x = 2x^3+x^2-2x-1-2x^3-2x^2$$
$$= -x^2-2x-1\,.$$

So $f'(x) = \frac{-(x^2+2x+1)}{\left(x^2-1\right)^2}$. This still does not match the back-of-the-book answer, so we look for common factors in the numerator and denominator. Both contain the factor $x + 1$:

$$f'(x) = \frac{-(x+1)^2}{\left[(x+1)(x-1)\right]^2} = \frac{-(x+1)^2}{(x+1)^2(x-1)^2} = \frac{-1}{(x-1)^2}\,.$$

valid only for $x \neq \pm 1$ (since f and f' are not defined for $x = \pm 1$).

This shows that your derivative was correct. ■

Exercises 7-C

1. Let $f(x) = \frac{3x+1}{2x-1}$. Then $f'(x) = \frac{-5}{(2x-1)^2}$.

a. Find all values of x where the tangent line to the graph of f is parallel to the line $y = 5 - x$.

b. Find all values of x where the curve $y = f(x)$ has a horizontal tangent line.

2. Let $f(x) = 3x^{2/3} - x^{5/3}$. Then $f'(x) = 2x^{-1/3} - \frac{5}{3}x^{2/3}$. Find all values of x where $f'(x) = 0$.

3. The function $f(x) = (x+1)^3(2x+1)^2$ has derivative

$$f'(x) = 4(x+1)^3(2x+1) + 3(2x+1)^2(x+1)^2\,.$$

Simplify $f'(x)$ and find all values of x for which $f'(x) = 0$.

4. The function $f(x) = (x-2)\sqrt{x+1}$ has derivative

$$f'(x) = \frac{x-2}{2\sqrt{x+1}} + \sqrt{x+1}\,.$$

a. Show that this derivative can be simplified to $f'(x)=\frac{3x}{2\sqrt{x+1}}$.

b. For what value of x does $f'(x) = 0$?

c. For what value of x is f defined but f' is not defined?

Chapter 7 Exercises

1. Let $f(x)=x^{-3/4}$. Find each of the following values if it exists; if it does not exist, explain why.

a. $f(64)$ **b**. $f(-64)$ **c**. $f\left(\frac{1}{16}\right)$ **d**. $f\left(\sqrt[3]{16}\right)$ **e**. $f(4)$

2. If $g(x)=x^{2/3}$, find:

a. $g\left(x^3\right)$ **b**. $g(8x)$ **c**. $g(-x)$ **d**. $g\left(\frac{8}{x^6}\right)$

3. Decide whether each of the following equations is true for all $x > 0$ or sometimes false. If sometimes false, give a specific value of $x > 0$ for which the equation is false.

a. $\frac{2}{\sqrt{7x}}=2(7x)^{-1/2}$ **b**. $\frac{1}{x^2}=x^{-1/2}$ **c**. $\frac{1}{3}x=3x^{-1}$

d. $\frac{x+\sqrt[3]{x}}{x^2}=x^{-1}+x^{-5/3}$ **e**. $\frac{x}{x^2+\sqrt{x}}=x^{-1}+x^{-1/2}$ **f**. $\frac{1}{8x}=\frac{1}{8}x^{-1}$

4. Rewrite each function as a sum where each term is a constant times a power of x.

a. $f(x)=\frac{2x+x^2}{\sqrt[4]{x}}$ **b**. $g(x)=\frac{2}{5x^3}-\frac{\sqrt[3]{x^2}}{7}+x\sqrt{x}$

c. $h(x)=\sqrt[5]{x}\left(x^2+\frac{2}{11}x^5\right)$ **d**. $k(x)=\left(5x^2+\frac{1}{2x^3}+\frac{1}{3}\right)\sqrt[3]{x^7}$

e. $F(x)=(x+1)\left(\frac{\sqrt[3]{7}}{2x}-\frac{1}{\sqrt[5]{x^2}}\right)$ **f**. $G(x)=\left(\frac{\sqrt{3}}{x^2}+2x^{-1}\right)^2$

5. Describe each of the following functions as a sum, difference, product, quotient, or composition of two simpler functions; give the two simpler functions. Repeat this for the two functions obtained and continue this process until the simpler functions are sums whose terms are a constant times a power of x.

a. $F(x) = \dfrac{x^2 + 2x}{-5x + \sqrt{3x}}$

b. $G(x) = \dfrac{\sqrt[7]{2x^3 + 1}}{\sqrt[5]{x - 3}}$

c. $H(x) = x\left(\sqrt{2} + \frac{1}{x^2}\right)^5$

d. $L(x) = \left(\frac{1}{3x} + 8\sqrt[3]{x}\right)\sqrt{x^5 - 2x}$

e. $N(x) = \sqrt[5]{\left(x + \frac{2}{x^5}\right)^2} + \frac{x}{9\sqrt{x}}$

f. $R(x) = \dfrac{-7x + \frac{1}{2}}{x^3 - 4\sqrt{x}} - \frac{6}{5}x^2\sqrt[3]{x}$

6. The derivative of the function $y = \dfrac{(x-3)^4}{x^2 - 2}$ is

$$\frac{dy}{dx} = \frac{4(x-3)^3(x^2 - 2) - 2x(x-3)^4}{\left(x^2 - 2\right)^2}.$$

a. Show that $\frac{dy}{dx}$ can be simplified to $\dfrac{2(x-3)^3(x-1)(x+4)}{\left(x^2-2\right)^2}$.

b. Find all values of x for which $\frac{dy}{dx} = 0$.

7. The function $f(x) = x^3 - 6x^2 + 5$ has derivative $f'(x) = 3x^2 - 12x$.

a. Find all values of x where the curve $y = f(x)$ has a horizontal tangent line.

b. Find all points on the curve $y = f(x)$ where the tangent line is parallel to the line $y = -9x + 7$.

c. Find x-coordinates of all points on the curve where the slope of the tangent line is -13.

8. The function $y = x^{3/2} - 3\sqrt{x}$ has derivative $y' = \frac{3}{2}\sqrt{x} - \frac{3}{2}x^{-1/2}$. Rewrite y' as a single quotient and find the values of x where $y' = 0$.

CHAPTER 8

Review of Trigonometric Functions

The word *trigonometry* comes from the Greek *trigonom* (triangle) and *metry* (measurement). Trigonometry relates an acute angle of a right triangle to ratios of measures of two sides of the triangle. These trigonometric ratios, in turn, can be used to measure indirectly objects or distances that cannot be measured directly. This technique has been used since ancient times for surveying. (See the Appendix for a summary of trigonometric ratios.)

The trigonometric ratios for acute angles in a right triangle can be extended to trigonometric functions defined for an angle of any measure. These trigonometric functions are useful in the study of various rhythmic patterns. Quantities that repeat periodically or cyclically, such as those that appear in the description of sound waves, the swinging of a pendulum, the vibrations of a musical string, and models of economic or population cycles, are represented by trigonometric functions.

8-A ANGLE MEASURES

An *angle* is formed when two rays meet at a point, called the *vertex* of the angle. Angles can be measured in *degrees* or in *radians*, just as length can be measured in inches or centimeters. The degree measure dates back to the Babylonians, who defined the measure of a full revolution as 360 degrees, so that 1 degree is $\frac{1}{360}$ of a complete rotation. In calculus, radians are generally used to measure angles because derivatives of the trigonometric functions are simpler than when degree measure is used. To define a radian, consider a circle of radius 1 unit, called a *unit circle.* Its circumference measures 2π units and is the length of the arc of the circle corresponding to a *central angle* of one complete revolution. (A central angle has its vertex at the center of the circle.) A *radian* is defined as the measure of a central angle that intercepts an arc of length 1 in a unit circle. Figure 8.1 shows an angle of 1 radian and compares it with an angle of 5 degrees.

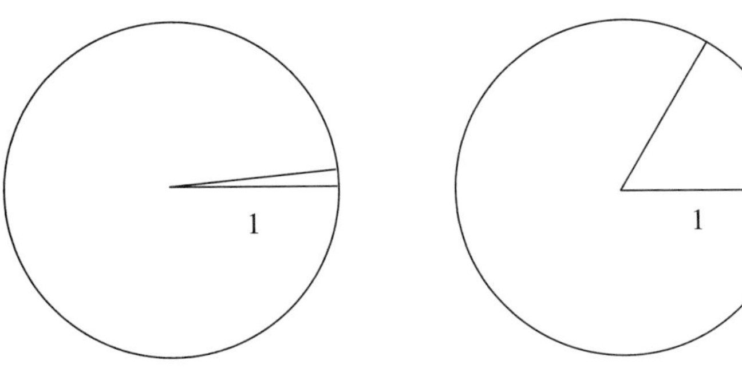

A central angle of 5 degrees in a unit circle; an angle of 1 degree is too small to distinguish

A central angle of 1 radian in a unit circle

Figure 8.1

We measure an angle by identifying one side of the angle as the *initial side* and the other side as the *terminal side* and measuring the directed rotation from its initial side to its terminal side. A *positive angle* is one in which the initial side rotates in a counterclockwise direction to the terminal side. A *negative angle* is one in which the initial side rotates in a clockwise direction to the terminal side (see Figure 8.2).

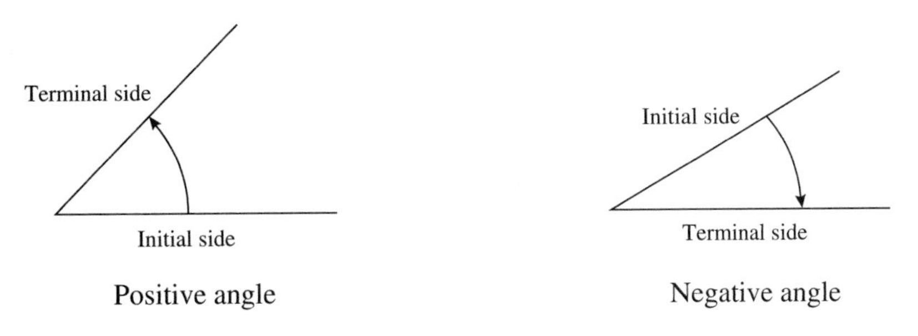

Positive angle

Negative angle

Figure 8.2

Here is how to describe the radian measure of any angle: Let the angle be the central angle of a unit circle, and let the initial side of the angle be a radius of the circle. Sweep out an arc along the circle, moving from the initial side to the terminal side of the angle. The length of the path that is swept out along the unit circle equals the radian measure of the angle (see Figure 8.3).

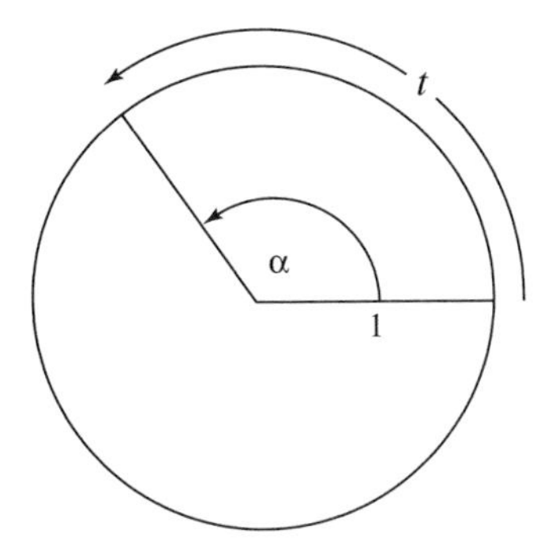

Figure 8.3 Angle α has measure t radians; it sweeps out an arc of t units on the unit circle

When we express angle measure, the notation ° represents degrees; no symbol means the angle is measure in radians. Thus $\theta = 45°$ means θ is an angle of 45 degrees; $\theta = 45$ means θ is an angle of 45 radians (more than seven full revolutions and approximately 2578.3 degrees).

Since an angle of one complete counterclockwise rotation measures 2π radians, an angle of 1 radian is $\frac{1}{2\pi}$ (roughly one-sixth) of a full revolution. The fundamental relationship between the number of radians in a circle and the number of degrees in a circle is

$$1 \text{ full revolution} = 2\pi \text{ radians} = 360 \text{ degrees}.$$

From this relationship we derive the following identities:

$$1 \text{ radian} = \left(\frac{360}{2\pi}\right) \text{ degrees} = \left(\frac{180}{\pi}\right) \text{ degrees} \approx 57.3°$$

$$1 \text{ degree} = \left(\frac{2\pi}{360}\right) \text{ radians} = \left(\frac{\pi}{180}\right) \text{ radians} \approx 0.02 \text{ radian}.$$

Standard Position

Consider a Cartesian coordinate system. An angle is in *standard position* when the vertex of the angle is at the origin and the initial side of the angle is on the positive x-axis. Figure 8.4 shows two angles, one positive and one negative, in standard position.

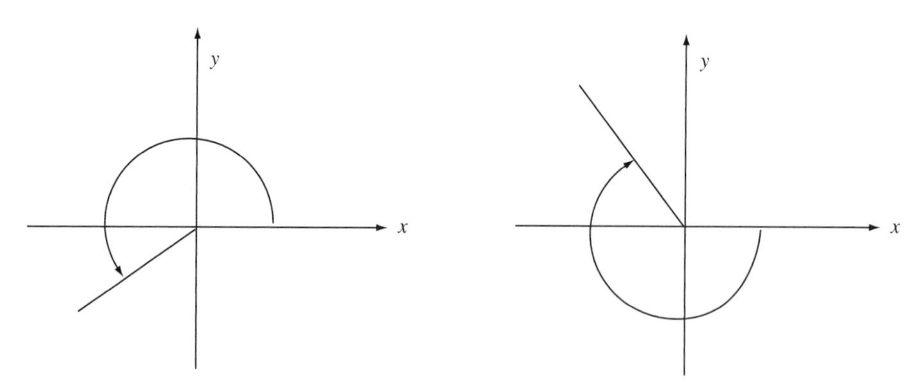

Positive angle in standard position

Negative angle in standard position

Figure 8.4

It is convenient to represent angles in standard position, and to observe relationships between some special angles that are rational multiples of 2π. Terminal sides of angles in standard position that are positive multiples of $\frac{1}{8}$ of a full revolution are shown on the circle diagram in Figure 8.5a. Similarly, terminal sides of angles in standard position that are positive multiples of $\frac{1}{12}$ of a full revolution are shown in Figure 8.5b.

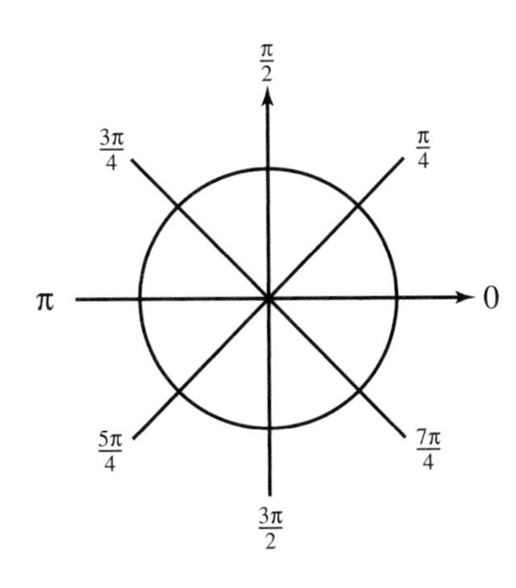

Figure 8.5a

Figure 8.5b

In the next three examples we use the fact that the measure of a half-revolution counterclockwise is π radians, or 180°, and a full revolution is 2π radians, or 360°. This allows us to convert angle measures given in degrees to radians and those given in radians to degrees. It is customary to refer to an angle by its measure.

EXAMPLE 8.1 For each of the angles given below, sketch the angle in standard position and give the measure of the angle in degrees.

a. $\frac{\pi}{4}$ radians **b.** $\frac{7\pi}{3}$ radians **c.** -2 radians

Solution

a. Since π radians is the angle measure of half a circle measured from the positive x-axis in a counterclockwise direction, $\frac{\pi}{4}$ radians is one-fourth of this angle. One-fourth of the half-circle angle in degrees is $\frac{1}{4}(180°) = 45°$. Thus $\frac{\pi}{4}$ radians $= 45°$. Figure 8.6 has a sketch of the angle.

b. One full revolution is $2\pi = \frac{6\pi}{3}$ radians, so an angle of $\frac{7\pi}{3}$ radians is more than one full revolution. Since $\frac{7\pi}{3} = \frac{6\pi}{3} + \frac{\pi}{3}$, this angle will terminate in the first quadrant with the same terminal side as the angle whose measure is $\frac{\pi}{3}$ radians. Figure 8.7 shows a sketch of the angle of $\frac{7\pi}{3}$ radians. An angle of measure $\frac{\pi}{3}$ radians has measure $\frac{1}{3} \times 180° = 60°$. Thus the degree measure of the angle of $\frac{7\pi}{3}$ radians is 360° (one revolution) plus 60°, which is 420°.

c. Since an angle of 1 radian is roughly one-sixth of a full revolution, an angle of -2 radians will be roughly one-third of a full revolution measured from the positive x-axis in a clockwise direction (becase it has negative measure). The angle is sketched in Figure 8.8. The degree measure of the angle of 1 radian is $\frac{180}{\pi}$ degrees, so an angle of -2 radians has measure $-2\left(\frac{180}{\pi}\right) = -\frac{360}{\pi} \approx -114.6°$.

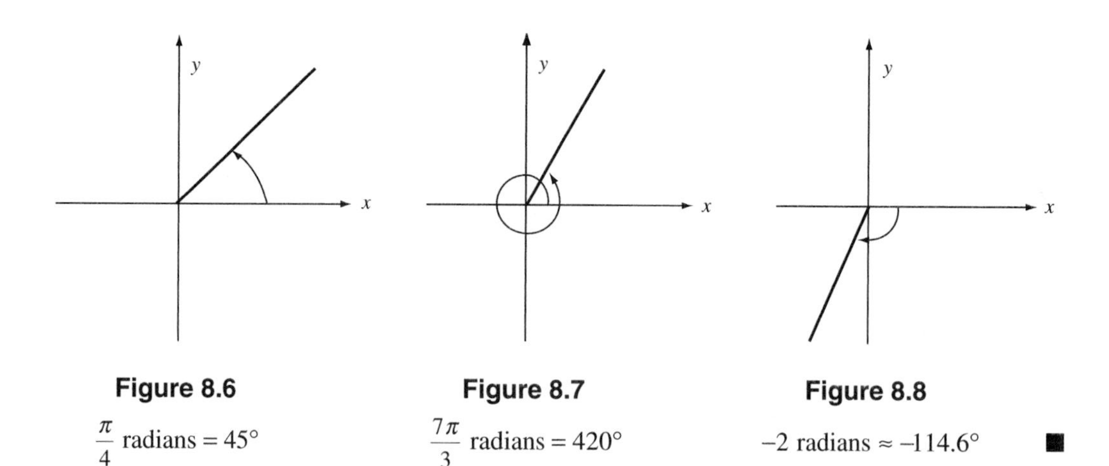

Figure 8.6
$\frac{\pi}{4}$ radians = 45°

Figure 8.7
$\frac{7\pi}{3}$ radians = 420°

Figure 8.8
−2 radians ≈ −114.6° ■

EXAMPLE 8.2 For each of the angles given below, sketch the angle in standard position and give the measure of the angle in radians.

a. −30° **b.** 330° **c.** 270°

Solution

a. An angle of −30° is measured from the positive x-axis in a clockwise direction and is one-sixth of a half-circle. Thus $-30° = -\frac{\pi}{6}$ radians. The angle in standard position is shown in Figure 8.9a.

b. An angle of 330° is measured from the positive x-axis in a counterclockwise direction and sweeps out an arc that stops 30° short of a full 360° revolution. So the terminal side of this angle is the same as that of an angle of −30°. The radian measure is $2\pi - \frac{\pi}{6} = \frac{11\pi}{6}$ radians. The angle in standard position is shown in Figure 8.9b.

c. An angle of 270° has radian measure of $270\left(\frac{\pi}{180}\right) = \frac{3\pi}{2}$ radians, which is three-fourths of a full revolution. When this angle is in standard position, its terminal side lies on the negative y-axis as shown in Figure 8.10.

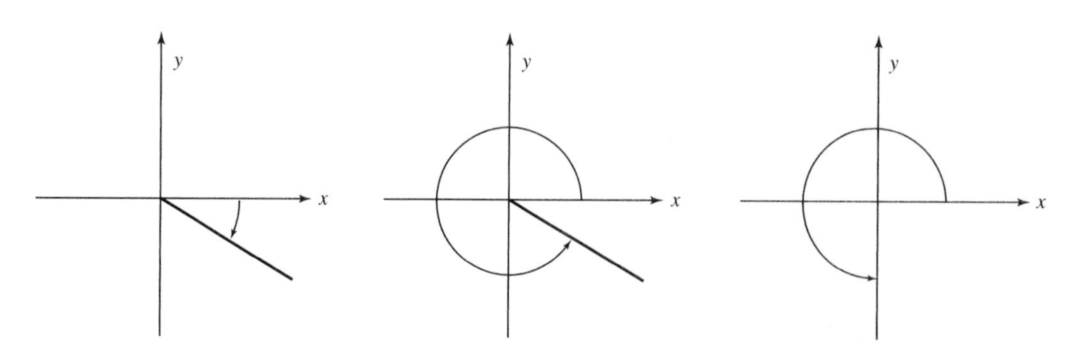

Figure 8.9a
$-30° = -\frac{\pi}{6}$ radians

Figure 8.9b
$330° = \frac{11\pi}{6}$ radians

Figure 8.10
$270° = \frac{3\pi}{2}$ radians ■

EXAMPLE 8.3 Estimate both the radian and degree measures of each angle shown in Figure 8.11.

a.

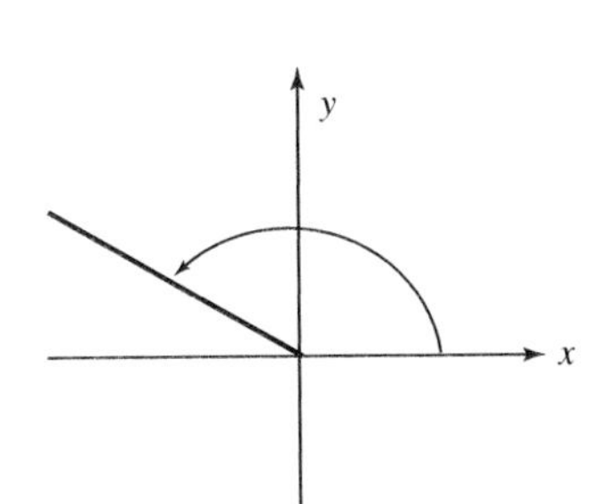

b.

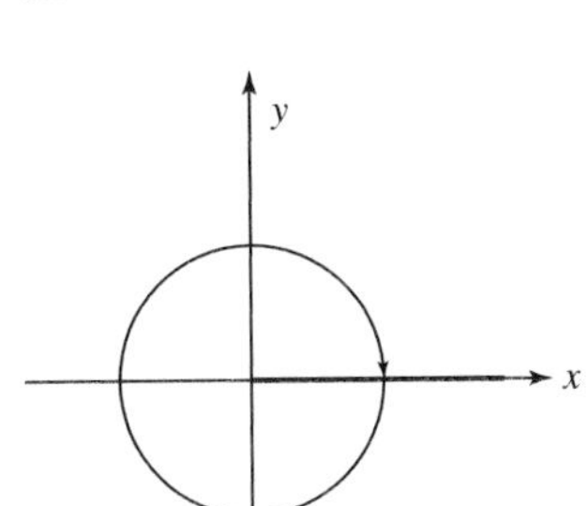

c.

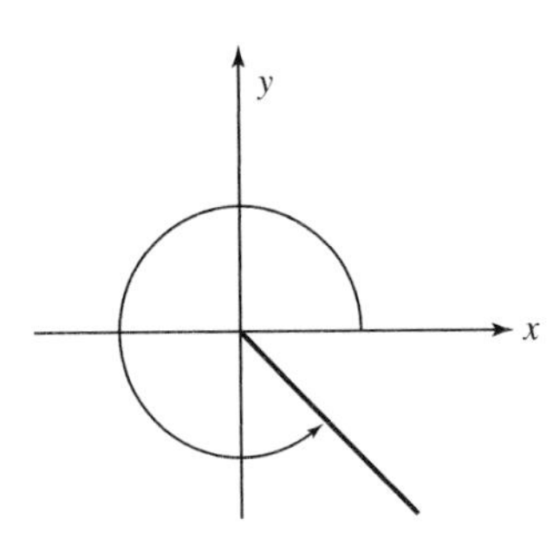

Figure 8.11

Solution

a. This positive angle is in standard position and has terminal side in the second quadrant. The angle covers one quadrant (90° or $\frac{\pi}{2}$ radians) plus approximately two-thirds of the second quadrant (60° or $\frac{\pi}{3}$ radians). So its measure is approximately $90° + 60° = 150°$, or $\frac{\pi}{2} + \frac{\pi}{3} = \frac{5\pi}{6}$ radians. Another way to view this angle is as the half-circle angle minus approximately one-sixth of the half-circle. Thus its measure is $\pi - \frac{\pi}{6} = \frac{5\pi}{6}$ radians, or $180° - 30° = 150°$.

b. This negative angle is in standard position with terminal side on the positive x-axis and represents one revolution in the clockwise direction. It has radian measure -2π and degree measure $-360°$.

c. This positive angle in standard position has terminal side half-way into the fourth quadrant. So its measure is $360° - 45° = 315°$, or $2\pi - \frac{\pi}{4} = \frac{7\pi}{4}$ radians. ■

Note that the terminal side of an angle in standard position does not completely determine the angle. For example, the angle of $\frac{\pi}{2}$ radians in standard position has the same terminal side as the positive angles $\frac{\pi}{2} + 2\pi = \frac{5\pi}{2}$, $\frac{\pi}{2} + 4\pi = \frac{9\pi}{2}$, and $\frac{\pi}{2} + 6\pi = \frac{13\pi}{2}$, as well as the negative angles $\frac{\pi}{2} - 2\pi = -\frac{3\pi}{2}$ and $\frac{\pi}{2} - 4\pi = -\frac{7\pi}{2}$. In fact, given any angle θ, if an angle that is any integer multiple of 2π is added to θ, the new angle will have the same terminal side as θ.

EXAMPLE 8.4

a. Draw an angle in standard position with radian measure between 0 and 2π, with its terminal side passing through the point of the circle shown in Figure 8.12. Estimate both the radian and degree measures of this angle.

b. Draw an angle in standard position with the same terminal side as the one in part a, but with radian measure between -2π and 0. Estimate both the radian and degree measures of this angle.

c. Give both the radian and degree measures of four other angles, two positive and two negative, that have the same terminal side as those in parts a and b.

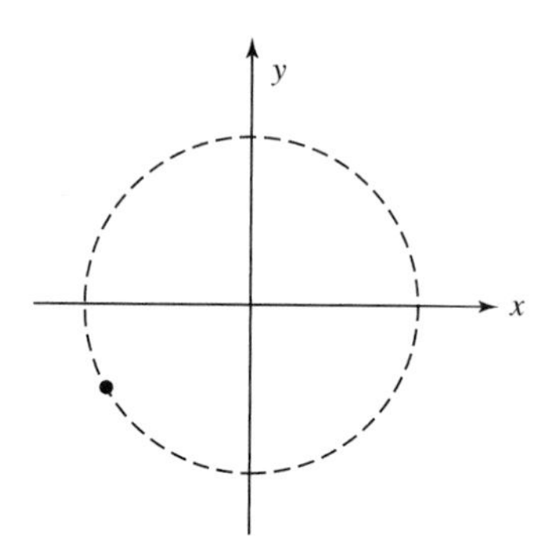

Figure 8.12

Solution

a. We sweep out an angle with measure between 0 and 2π radians by a counterclockwise rotation of at most one revolution. Figure 8.13 shows the angle. This angle covers half a circle plus approximately one-sixth of half a circle, so it measures approximately $\pi + \frac{\pi}{6} = \frac{7\pi}{6}$ radians, which is $180° + 30° = 210°$.

b. We produce a negative angle with radian measure between -2π and 0 by a clockwise rotation of at most one full revolution. Figure 8.14 shows the angle. Since this angle is approximately $\frac{5}{6}$ of a half-circle, it measures $-\frac{5\pi}{6}$ radians, or $\left(-\frac{5}{6}\right)(180°) = -150°$.

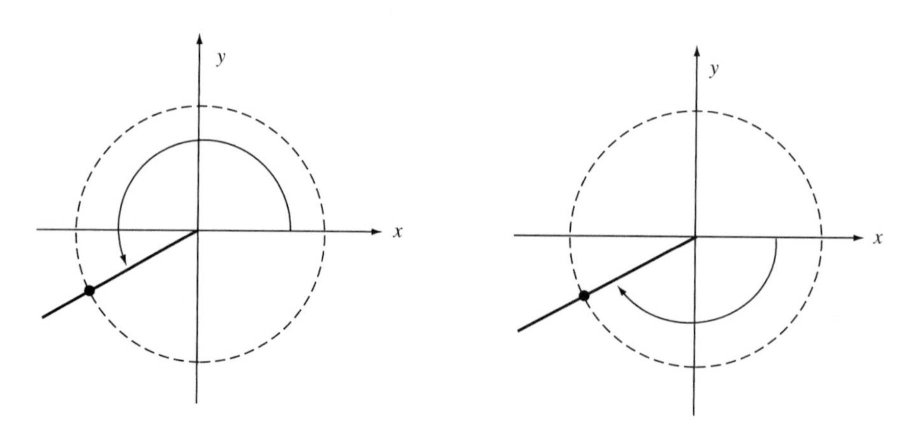

Figure 8.13 **Figure 8.14**

c. There are infinitely many positive angles in standard position with the given terminal side. Two of them are:

$$\frac{7\pi}{6} + 2\pi = \frac{19\pi}{6} \text{ radians (degree measure } 210° + 360° = 570°)$$

$$\frac{7\pi}{6} + 4\pi = \frac{31\pi}{6} \text{ radians (degree measure } 210° + 2(360)° = 930°).$$

There are also infinitely many negative angles in standard position with the given terminal side. Examples of these are:

$$-\frac{5\pi}{6} - 2\pi = -\frac{17\pi}{6} \text{ radians } (-150° - 360° = -510°)$$

$$-\frac{5\pi}{6} - 4\pi = -\frac{29\pi}{6} \text{ radians } (-150° - 720° = -870°).$$

■

Exercises 8-A

1. In Figure 8.15a several points on a circle are given. For each point, draw the terminal side of an angle in standard position that passes through the point. Next to each point, give both the radian and degree measures of the angle, assuming the radian measure is between 0 and 2π.

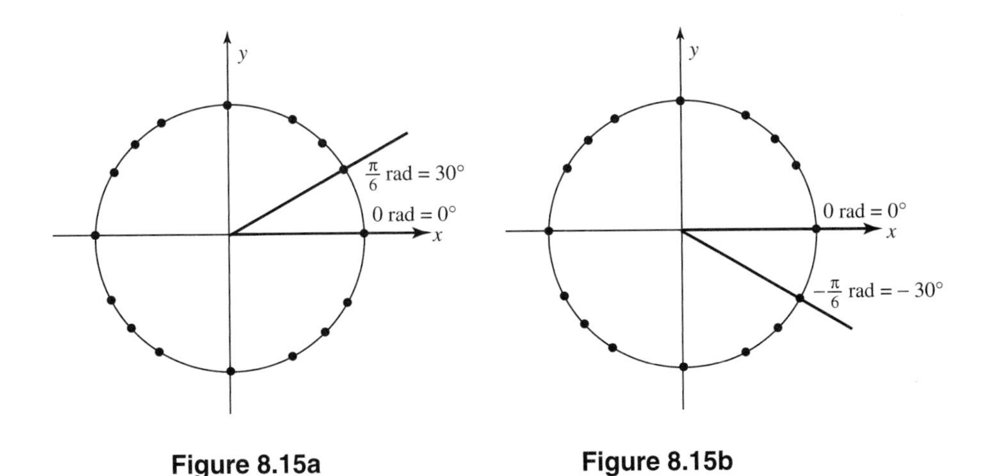

Figure 8.15a **Figure 8.15b**

2. Repeat Exercise 1 for angles with radian measure between -2π and 0. Use Figure 8.15b.

3. For each angle given below, give the radian measures of two positive angles and two negative angles (not counting the given one) in standard position with the same terminal side as the given angle. Draw each of the angles in standard position.

a. $\frac{\pi}{2}$ **b**. $-\frac{\pi}{4}$ **c**. $\frac{2\pi}{3}$ **d**. $-\pi$ **e**. $\frac{7\pi}{6}$

4. Give a decimal approximation of the radian measure of the positive angle in standard position with terminal side on

a. the positive y-axis.

b. the negative x-axis.

c. the negative y-axis.

5. The radian measures of seven different angles in standard position are given below. For each angle, indicate in which quadrant the terminal side lies. (You may wish to use the results of Exercise 4 above.)

a. 2.2 **b.** -3 **c.** 3.4 **d.** $\frac{1}{2}$ **e.** $\sqrt{2}$ **f.** $-\sqrt{5}$ **g.** 14.35

8-B DEFINITION AND EVALUATION OF THE TRIGONOMETRIC FUNCTIONS

There are six different trigonometric functions. Two of them, sine and cosine, assign to the radian measure of each angle θ a real number; the other four trigonometric functions are derived from the sine and cosine. These functions are unlike algebraic functions whose rule of assignment can be expressed as an algebraic formula. Instead, given the radian measure of an angle θ, we must describe in words how to calculate the values that the sine and cosine functions assign to θ. If the angle θ is between 0 and $\frac{\pi}{2}$ radians (an acute angle), the values that the trigonometric functions assign to θ can be expressed as the ratios of lengths of sides of a right triangle. These trigonometric ratios are discussed in the Appendix.

8-B.1 Definitions of the Trigonometric Functions

To describe periodic behavior, the sine and cosine functions must be defined for an angle θ of **any** measure. To do this, we put the angle θ in standard position in a Cartesian coordinate plane and choose any point (x, y) on the terminal side of the angle θ, except $(0, 0)$. Let r denote the distance of the point (x, y) from the origin $(0, 0)$. By the Pythagorean theorem, $r = \sqrt{x^2 + y^2}$ (see Figure 8.16).

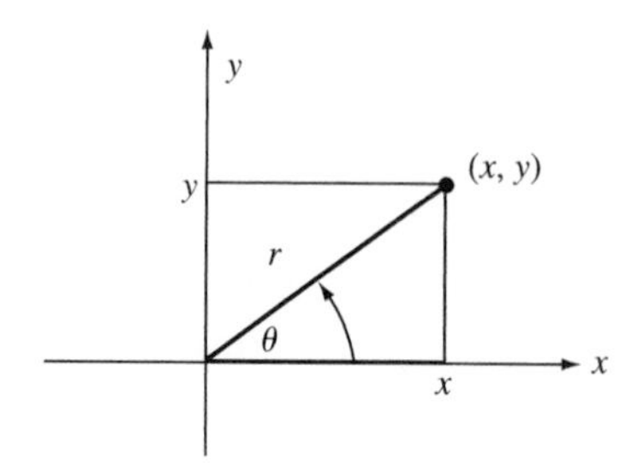

Figure 8.16

The values that the sine and cosine functions assign to θ are given below; the function notation is the standard abbreviated form.

Sine of θ: $\sin(\theta) = \frac{y}{r}$

Cosine of θ: $\cos(\theta) = \frac{x}{r}$

Properties of similar triangles guarantee that these definitions do not depend on the choice of the point (x, y) on the terminal side of θ. (Example 8.6 will illustrate this.)

The other four trigonometric functions are quotients or reciprocals of the sine and cosine functions, so they are not defined for every angle θ. Here are their definitions:

Tangent of θ: $\tan(\theta) = \frac{y}{x}$ (undefined if $x = 0$)

Cotangent of θ: $\cot(\theta) = \frac{x}{y}$ (undefined if $y = 0$)

Secant of θ: $\sec(\theta) = \frac{r}{x}$ (undefined if $x = 0$)

Cosecant of θ: $\csc(\theta) = \frac{r}{y}$ (undefined if $y = 0$)

From these definitions and those of sine and cosine, these functions can be rewritten as follows:

$$\tan(\theta) = \frac{\sin(\theta)}{\cos(\theta)} \qquad \text{(undefined if } \cos(\theta) = 0\text{)}$$

$$\cot(\theta) = \frac{\cos(\theta)}{\sin(\theta)} \qquad \text{(undefined if } \sin(\theta) = 0\text{)}$$

$$\sec(\theta) = \frac{1}{\cos(\theta)} \qquad \text{(undefined if } \cos(\theta) = 0\text{)}$$

$$\csc(\theta) = \frac{1}{\sin(\theta)} \qquad \text{(undefined if } \sin(\theta) = 0\text{)}$$

In many texts, parentheses around the variable are omitted in writing the trigonometric functions; thus, the functions we have just defined are frequently written as $\sin\theta$, $\cos\theta$, $\tan\theta$, $\cot\theta$, $\sec\theta$, and $\csc\theta$. In this text, we will follow this convention; note that parentheses around the variable are used when any ambiguity might exist.

The expressions "sin" and "cos" are meaningless by themselves, without an angle. We always need to state the sine or cosine of an angle, like $\sin\theta$ or $\cos\frac{2\pi}{3}$, or sine or cosine of an expression, like $\sin(2\theta + \pi)$ or $\cos\left(\frac{\theta}{4}\right)$. Also note that $\sin\theta$ is not a product of "sin" times "θ."

EXAMPLE 8.5

a. Find the values of the six trigonometric functions for the angle θ shown in Figure 8.17.

b. Estimate the value of the six trigonometric functions for the angle β shown in Figure 8.18.

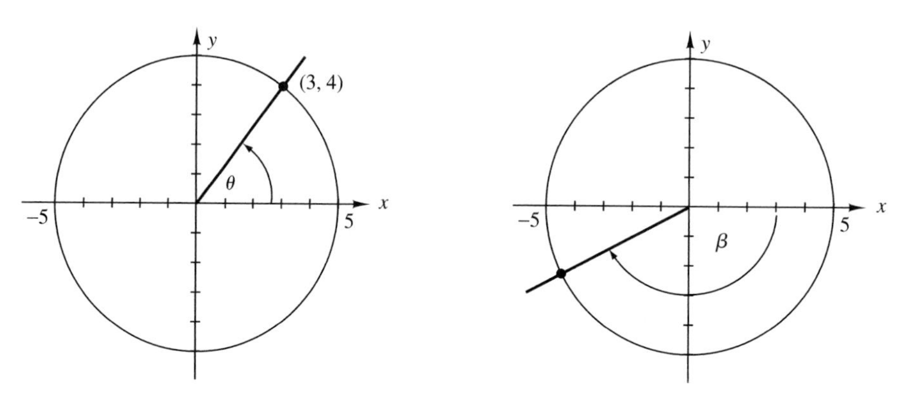

Figure 8.17 Figure 8.18

Solution

a. The point (3, 4) lies on the terminal side of the angle θ in standard position, so we take $x = 3$ and $y = 4$. Then $r = \sqrt{3^2 + 4^2} = \sqrt{25} = 5$ and, therefore, $\sin\theta = \frac{y}{r} = \frac{4}{5}$, $\cos\theta = \frac{x}{r} = \frac{3}{5}$, $\tan\theta = \frac{y}{x} = \frac{4}{3}$, $\cot\theta = \frac{x}{y} = \frac{3}{4}$, $\sec\theta = \frac{r}{x} = \frac{5}{3}$, and $\csc\theta = \frac{r}{y} = \frac{5}{4}$.

b. Consider the point of intersection of the terminal side of the angle β with the given circle. From the graph we estimate the x-coordinate of this point as $x = -4.5$. Since this point lies on a circle with radius 5, we can use the equation of the circle $x^2 + y^2 = 25$ to solve for the y-coordinate:

$$y = -\sqrt{25 - (4.5)^2} = -\sqrt{25 - 20.25} = -\sqrt{4.75} \approx -2.18.$$

The approximate values of the trigonometric functions of β are

$$\sin\beta \approx \frac{-2.18}{5} \approx -0.44,\ \cos\beta \approx \frac{-4.5}{5} = -0.9,\ \tan\beta \approx \frac{-2.18}{-4.5} \approx 0.48,$$

$$\cot\beta \approx \frac{-4.5}{-2.18} \approx 2.06,\ \sec\beta \approx \frac{5}{-4.5} \approx -1.11,\ \text{and } \csc\beta \approx \frac{5}{-2.18} \approx -2.29.$$

Note that $\sin\beta$, $\cos\beta$, $\sec\beta$, and $\csc\beta$ are all negative, since the terminal side of β is in the third quadrant where both x and y are negative. ■

EXAMPLE 8.6 Suppose the terminal side of the angle θ in standard position lies in the second quadrant and coincides with the line $y = -3x$. Draw a picture of the angle θ and find the values of the six trigonometric functions for this angle θ, using the point (x, y) on its terminal side where:

a. $x = -1$ **b.** $x = -\frac{1}{2}$

Solution There are infinitely many possibilities for the angle, two of which are shown in Figure 8.19.

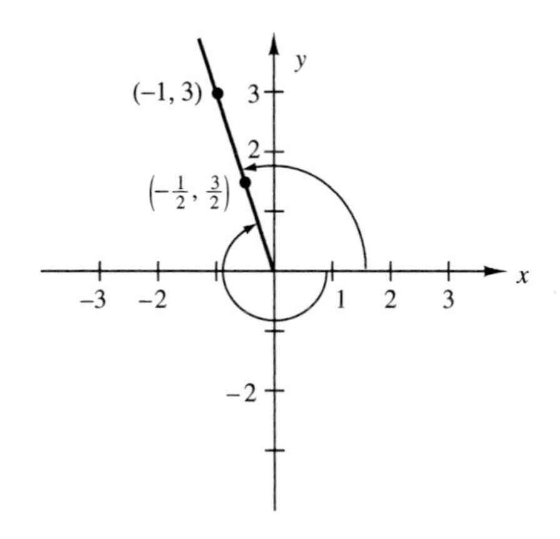

Figure 8.19

a. If (x, y) is on the terminal side of the angle θ and $x = -1$, then $y = -3(-1) = 3$. The distance from $(-1, 3)$ to the origin is $r = \sqrt{(-1)^2 + 3^2} = \sqrt{10}$.

Then $\sin\theta = \frac{3}{\sqrt{10}}$, $\cos\theta = \frac{-1}{\sqrt{10}} = -\frac{1}{\sqrt{10}}$, $\tan\theta = \frac{3}{-1} = -3$,

$\cot\theta = -\frac{1}{3}$, $\sec\theta = \frac{\sqrt{10}}{-1} = -\sqrt{10}$, and $\csc\theta = \frac{\sqrt{10}}{3}$.

b. Since $x = -\frac{1}{2}$ and $y = -3\left(-\frac{1}{2}\right) = \frac{3}{2}$,

$$r = \sqrt{\left(\frac{-1}{2}\right)^2 + \left(\frac{3}{2}\right)^2} = \sqrt{\frac{1}{4} + \frac{9}{4}} = \sqrt{\frac{10}{4}} = \frac{\sqrt{10}}{2}.$$

So $\sin\theta \frac{\left(\frac{3}{2}\right)}{\left(\frac{\sqrt{10}}{2}\right)} = \frac{3}{\sqrt{10}}$, $\cos\theta = \frac{\left(-\frac{1}{2}\right)}{\left(\frac{\sqrt{10}}{2}\right)} = -\frac{1}{\sqrt{10}}$, $\tan\theta = \frac{\left(\frac{3}{2}\right)}{\left(-\frac{1}{2}\right)} = -3$,

$\cot\theta = \frac{\left(-\frac{1}{2}\right)}{\left(\frac{3}{2}\right)} = -\frac{1}{3}$, $\sec\theta = \frac{\left(\frac{\sqrt{10}}{2}\right)}{\left(-\frac{1}{2}\right)} = \frac{\sqrt{10}}{-1} = -\sqrt{10}$, and $\csc\theta = \frac{\left(\frac{\sqrt{10}}{2}\right)}{\left(\frac{3}{2}\right)} = \frac{\sqrt{10}}{3}$. ■

The fact that the answers are the same in parts a and b of Example 8.6 illustrates that any point (x, y) other than $(0, 0)$ on the terminal side of the angle can be chosen to find the values of the trigonometric functions.

When we define the sine and cosine functions, it is frequently convenient to choose a point (x, y) for which $r = \sqrt{x^2 + y^2} = 1$; that is, take a point (x, y) on the unit circle (see Figure 8.20). For (x, y) on the unit circle, the values of the trigonometric functions for the angle θ are:

$$\sin\theta = y \qquad \cos\theta = x$$

$$\tan\theta = \frac{y}{x} \qquad \cot\theta = \frac{x}{y}$$

$$\sec\theta = \frac{1}{x} \qquad \csc\theta = \frac{1}{y}.$$

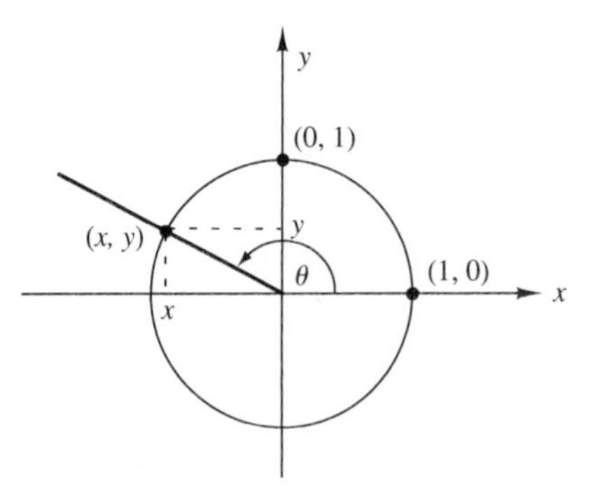

Figure 8.20

In fact, by using the unit circle, we can directly read the values of $\cos\theta$ and $\sin\theta$ for any angle θ simply by looking at the coordinates of the point where the terminal side of the angle intersects the unit circle (see Figure 8.21). The x-coordinate of this point is $\cos\theta$ and the y-coordinate is $\sin\theta$.

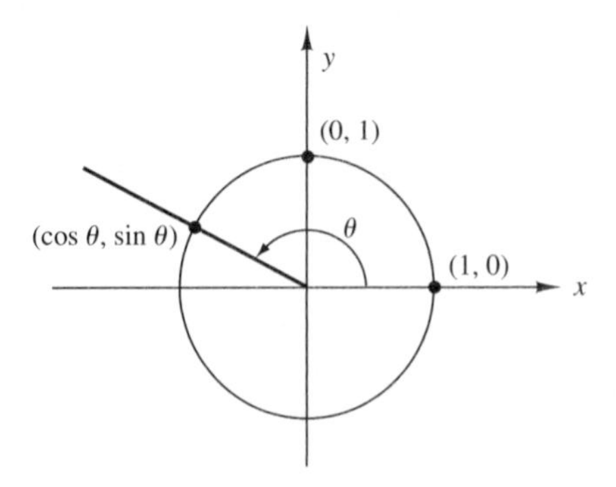

Figure 8.21

Figure 8.22 shows the points where the terminal sides of θ intersect the unit circle and the values of $\sin\theta$ and $\cos\theta$ for $\theta = \frac{\pi}{4}$ and $\theta = \frac{-5\pi}{4}$. Figure 8.23 shows the points where the terminal sides of θ intersect the unit circle and the values of $\sin\theta$ and $\cos\theta$ for $\theta = 0$ and $\theta = \frac{\pi}{2}$. Figure 8.24 shows the points where the terminal sides of θ intersect the unit circle and the values of $\sin\theta$ and $\cos\theta$ for $\theta = \frac{7\pi}{6}$ and $\theta = -\frac{\pi}{6}$.

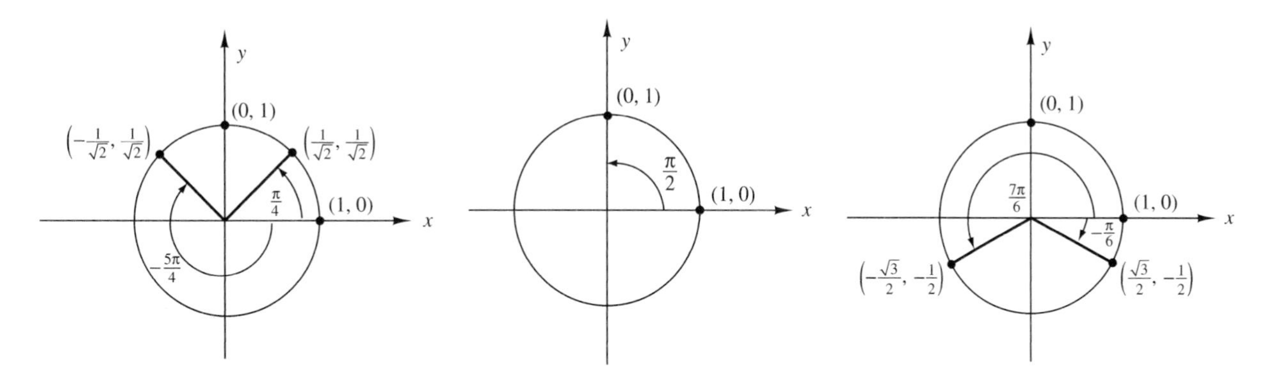

Figure 8.22

$\sin\left(-\frac{5\pi}{4}\right) = \frac{1}{\sqrt{2}}$ $\quad \sin\frac{\pi}{4} = \frac{1}{\sqrt{2}}$

$\cos\left(-\frac{5\pi}{4}\right) = -\frac{1}{\sqrt{2}}$ $\quad \cos\frac{\pi}{4} = \frac{1}{\sqrt{2}}$

Figure 8.23

$\sin 0 = 0$ $\quad \sin\frac{\pi}{2} = 1$

$\cos 0 = 1$ $\quad \cos\frac{\pi}{2} = 0$

Figure 8.24

$\sin\frac{7\pi}{6} = -\frac{1}{2}$ $\quad \sin\left(-\frac{\pi}{6}\right) = -\frac{1}{2}$

$\cos\frac{7\pi}{6} = -\frac{\sqrt{3}}{2}$ $\quad \cos\left(-\frac{\pi}{6}\right) = \frac{\sqrt{3}}{2}$

The values of the trigonometric functions depend only on the terminal side of the angle θ in standard position. For example, angles θ, $\theta + 2\pi$, and $\theta - 4\pi$ all have the same terminal side, so $\sin\theta = \sin(\theta + 2\pi) = \sin(\theta - 4\pi)$.

In general, the sine or cosine of an angle θ is the same as that of any angle obtained by adding to θ any integer multiple of 2π.

Notice that the signs of the trigonometric functions are determined by the quadrant in which the terminal side of the angle lies. For example, if $\frac{3\pi}{2} < \theta < 2\pi$, then the terminal side of θ lies in the fourth quadrant, where $x > 0$ and $y < 0$. Therefore $\cos\theta$ and $\sec\theta$ are positive, whereas $\sin\theta$, $\tan\theta$, $\cot\theta$, and $\csc\theta$ are negative.

8-B.2 Exact Values of the Trigonometric Functions for Some Special Angles

The angles found most frequently in calculus texts are integer multiples of π, $\frac{\pi}{2}$, $\frac{\pi}{3}$, $\frac{\pi}{4}$, and $\frac{\pi}{6}$. You should be able to give the exact values for the trigonometric functions of those angles. For other angles, you will use a calculator to approximate values for the trigonometric functions.

> ☞ When using a calculator, be sure to put it in the correct mode of radians, unless the problem specifies degrees.

Angles in standard position that are integer multiples of $\frac{\pi}{2}$ have terminal sides on one of the coordinate axes, and we can read the values of their sine and cosine as 1, −1, or 0. For integer multiples of $\frac{\pi}{4}$, $\frac{\pi}{3}$, and $\frac{\pi}{6}$ that are outside the interval $\left[0, \frac{\pi}{2}\right]$ and are not multiples of $\frac{\pi}{2}$, we use a *reference angle.* This is the *acute angle* (between 0 and $\frac{\pi}{2}$) whose sides are the terminal side of the given angle and the x-axis.

The two pairs of right triangles shown in Figures 8.25 and 8.26, together with the symmetry of the circle, will help you to remember the values of the trigonometric functions for these special angles. Either of the two triangles in Figure 8.25 can be used to find the values of the trigonometric functions for multiples of $\frac{\pi}{4}$. Note the triangles are similar; the smaller triangle is obtained from the larger by scaling each side by $\frac{1}{\sqrt{2}}$, so that the length of its hypotenuse is 1. Either of the similar triangles in Figure 8.26 can be used to find the values of the trigonometric functions for the angles $\frac{\pi}{3}$ and $\frac{\pi}{6}$.

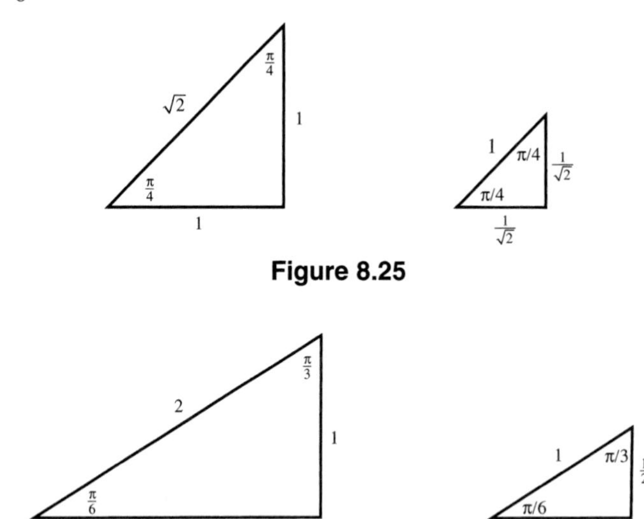

Figure 8.25

Figure 8.26

The following example shows how to find values of the trigonometric functions for angles that are integer multiples of $\frac{\pi}{4}$, $\frac{\pi}{3}$, and $\frac{\pi}{6}$.

EXAMPLE 8.7 Give the values of the six trigonometric functions for each of the following angles.

a. $\theta = \frac{\pi}{4}$ **b.** $\theta = -\frac{\pi}{4}$ **c.** $\theta = \frac{7\pi}{3}$ **d.** $\theta = 5\pi$

Solution

a. From the similar triangles in Figure 8.25, we see that the angle $\frac{\pi}{4}$ in standard position has terminal side through the point $(1, 1)$ and through the point $\left(\frac{1}{\sqrt{2}}, \frac{1}{\sqrt{2}}\right)$ (see Figure 8.27). Since the point $\left(\frac{1}{\sqrt{2}}, \frac{1}{\sqrt{2}}\right)$ lies on the unit circle, we have

$$\sin\frac{\pi}{4} = \frac{1}{\sqrt{2}}, \quad \cos\frac{\pi}{4} = \frac{1}{\sqrt{2}}, \quad \tan\frac{\pi}{4} = \frac{\sin\frac{\pi}{4}}{\cos\frac{\pi}{4}} = \frac{\left(\frac{1}{\sqrt{2}}\right)}{\left(\frac{1}{\sqrt{2}}\right)} = 1,$$

$$\cot\frac{\pi}{4} = \frac{\cos\frac{\pi}{4}}{\sin\frac{\pi}{4}} = 1, \quad \sec\frac{\pi}{4} = \frac{1}{\cos\frac{\pi}{4}} = \frac{1}{\frac{1}{\sqrt{2}}} = \sqrt{2}, \quad \text{and } \csc\frac{\pi}{4} = \frac{1}{\sin\frac{\pi}{4}} = \frac{1}{\frac{1}{\sqrt{2}}} = \sqrt{2}.$$

b. Since the acute angle formed by the x-axis and the terminal side of the angle $-\frac{\pi}{4}$ is the angle $\frac{\pi}{4}$, the reference angle is $\frac{\pi}{4}$. Figure 8.28 shows both angles in standard position. We use the symmetry of the unit circle to get the coordinates of the point $\left(\frac{1}{\sqrt{2}}, -\frac{1}{\sqrt{2}}\right)$ on the terminal side of the angle $-\frac{\pi}{4}$. So we have

$$\sin\left(-\frac{\pi}{4}\right) = -\frac{1}{\sqrt{2}}, \quad \cos\left(-\frac{\pi}{4}\right) = \frac{1}{\sqrt{2}}, \quad \tan\left(-\frac{\pi}{4}\right) = \frac{\left(-\frac{1}{\sqrt{2}}\right)}{\left(\frac{1}{\sqrt{2}}\right)} = -1,$$

$$\cot\left(-\frac{\pi}{4}\right) = \frac{\left(\frac{1}{\sqrt{2}}\right)}{\left(-\frac{1}{\sqrt{2}}\right)} = -1, \quad \sec\left(-\frac{\pi}{4}\right) = \frac{1}{\left(\frac{1}{\sqrt{2}}\right)} = \sqrt{2}, \quad \text{and } \csc\left(-\frac{\pi}{4}\right) = \frac{1}{\left(-\frac{1}{\sqrt{2}}\right)} = -\sqrt{2}.$$

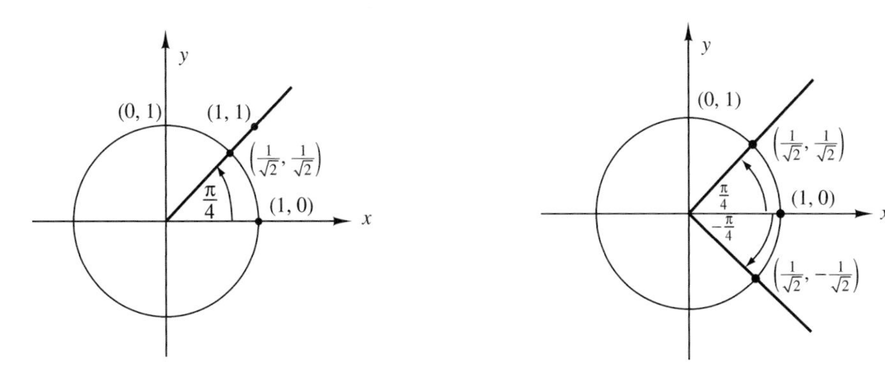

Figure 8.27 **Figure 8.28**

c. Since the terminal side of the angle $\frac{7\pi}{3}$ in standard position coincides with that of $\frac{\pi}{3}$, the reference angle is $\frac{\pi}{3}$. Figure 8.29 shows the angles $\frac{7\pi}{3}$ and $\frac{\pi}{3}$; it also shows one of the triangles of Figure 8.26 placed in such a way that $\frac{\pi}{3}$ is in standard position.

Choosing the point $(x, y) = \left(1, \sqrt{3}\right)$ on the terminal side of the angle, we have $r = 2$ and, therefore,

$$\sin\frac{7\pi}{3} = \sin\frac{\pi}{3} = \frac{\sqrt{3}}{2}, \quad \cos\frac{7\pi}{3} = \cos\frac{\pi}{3} = \frac{1}{2}, \quad \tan\frac{7\pi}{3} = \tan\frac{\pi}{3} = \sqrt{3},$$

$$\cot\frac{7\pi}{3} = \cot\frac{\pi}{3} = \frac{1}{\sqrt{3}}, \quad \sec\frac{7\pi}{3} = \sec\frac{\pi}{3} = 2, \quad \text{and } \csc\frac{7\pi}{3} = \csc\frac{\pi}{3} = \frac{2}{\sqrt{3}}.$$

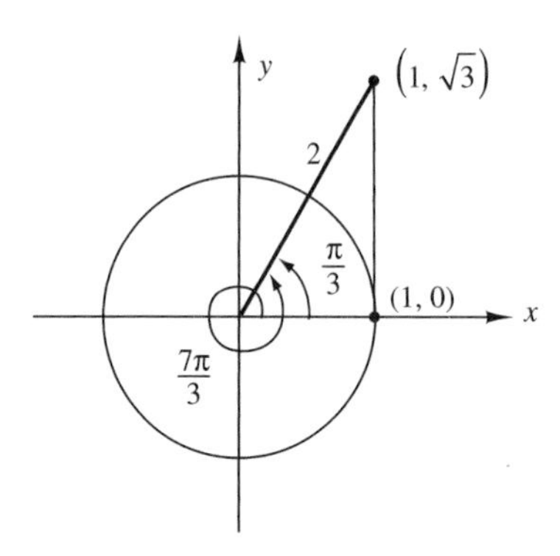

Figure 8.29

d. The terminal side of $\theta = 5\pi$ is the negative x-axis, as shown in Figure 8.30, and cuts the unit circle at $(x, y) = (-1, 0)$. So $\sin 5\pi = 0$, $\cos 5\pi = -1$, $\tan 5\pi = \frac{0}{-1} = 0$, and $\sec 5\pi = \frac{1}{-1} = -1$. Both $\cot 5\pi$ and $\csc 5\pi$ are undefined.

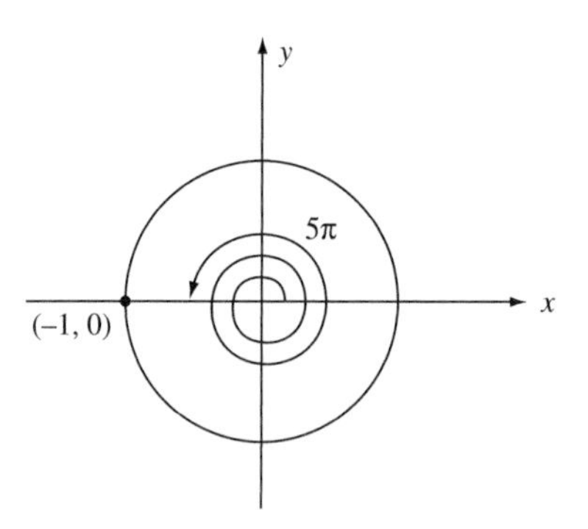

Figure 8.30 ■

In the next example we use the fact that the angles θ and $\theta \pm 2n\pi$, in standard position, have the same terminal side.

EXAMPLE 8.8 For the angle $\theta = \frac{3\pi}{4}$, answer the following questions:

a. Find $\sin\theta$ and $\cos\theta$.

b. Give two other angles θ_1 and θ_2 such that $\theta_1 < 0$ and $\theta_2 > 0$ and $\sin\theta = \sin\theta_1 = \sin\theta_2$ and $\cos\theta = \cos\theta_1 = \cos\theta_2$.

Solution

a. Figure 8.31 shows the angle $\theta = \frac{3\pi}{4}$ in standard position. Its terminal side passes through the point $(-1, 1)$ (the reference angle is $\frac{\pi}{4}$). Thus $\sin \frac{3\pi}{4} = \frac{1}{\sqrt{2}}$ and $\cos \frac{3\pi}{4} = -\frac{1}{\sqrt{2}}$.

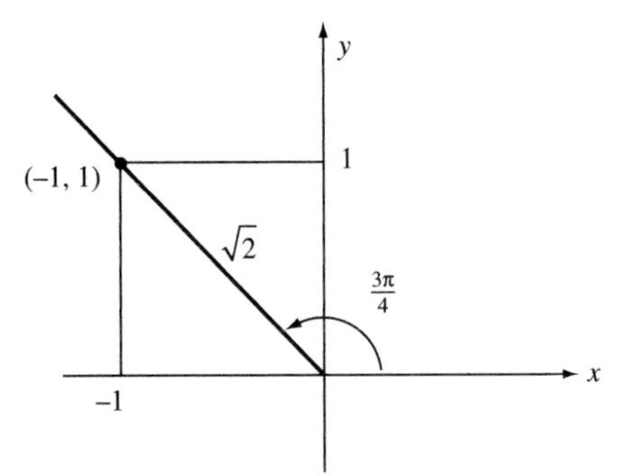

Figure 8.31

b. If θ is increased or decreased by an integer multiple of 2π, the values of sine and cosine of the angle will not change. So we can take $\theta_1 = \frac{3\pi}{4} - 2\pi = -\frac{5\pi}{4}$ and $\theta_2 = \frac{3\pi}{4} + 2\pi = \frac{11\pi}{4}$. ■

Exercises 8-B

1. Several points on the unit circle are given in Figure 8.32. These are points on the terminal sides of angles in standard position that are integer multiples of $\frac{\pi}{6}$, $\frac{\pi}{4}$, $\frac{\pi}{3}$, $\frac{\pi}{2}$, and π. Use the reference angles and symmetry of the circle to fill in the two coordinates of each of these points. Also, label with radian measures the terminal sides of these angles in standard position between 0 and 2π. This circle can serve as a convenient chart for values of trigonometric functions for these special angles.

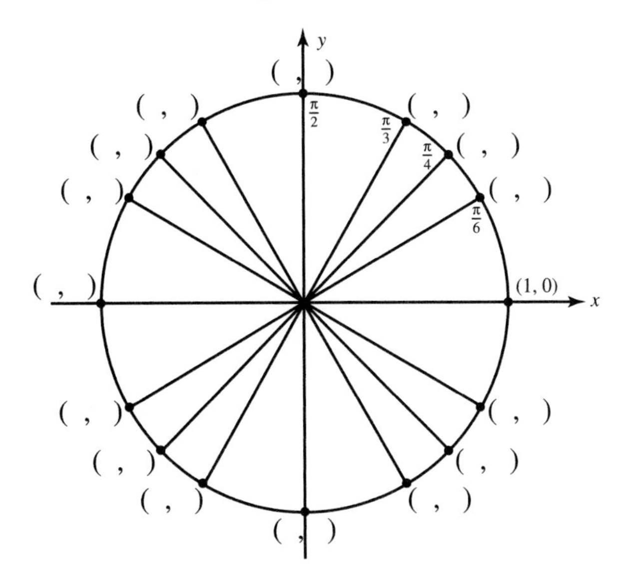

Figure 8.32

2. For the angle $\theta = \frac{5\pi}{6}$, answer the following questions:

a. Find $\sin\theta$ and $\cos\theta$.

b. Give two other angles θ_1 and θ_2 such that $\theta_1 < 0$ and $\theta_2 > 0$ and $\sin\theta = \sin\theta_1 = \sin\theta_2$ and $\cos\theta = \cos\theta_1 = \cos\theta_2$.

3. Answer the two questions of Exercise 2 for the following values of θ:

(i) $\theta = \frac{4\pi}{3}$ (ii) $\theta = -5\pi$ (iii) $\theta = \frac{11\pi}{4}$ (iv) $\theta = -\frac{3\pi}{2}$ (v) $\theta = -\frac{2\pi}{3}$

4. For each of the angles θ pictured in standard position on the unit circle in Figure 8.33, give an estimate of the values of $\sin\theta$ and $\cos\theta$. Explain how you obtained your estimates.

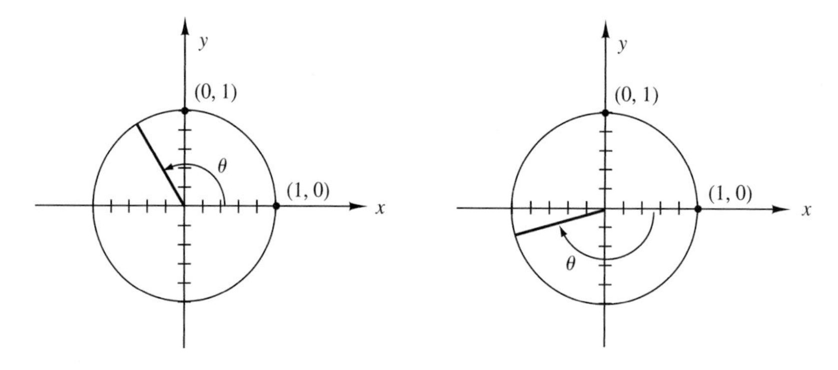

Figure 8.33

5. **a.** If the terminal side of the angle θ in standard position is in the first quadrant, give the sign of the value of each of the six trigonometric functions for θ.

b. Answer the same question for θ with terminal side in

(i) quadrant II (ii) quadrant III (iii) quadrant IV.

6. Give the values of the six trigonometric functions for each of the following angles:

a. $\theta = -7\pi$ **b.** $\theta = \frac{10\pi}{3}$ **c.** $\theta = \frac{11\pi}{4}$ **d.** $\theta = \frac{-17\pi}{6}$

7. Suppose the terminal side of the angle θ in standard position lies in the third quadrant and coincides with the line $y = 4x$. Draw a picture and find the values of the six trigonometric functions for this angle θ.

8-C PROPERTIES AND IDENTITIES FOR THE TRIGONOMETRIC FUNCTIONS

We noted earlier that the trigonometric functions are sometimes called circular functions because of their definitions that use a unit circle. The following examples show how properties of the unit circle lead to some important properties of the trigonometric functions.

EXAMPLE 8.9 Explain why each of the statements is true for every angle θ.

a. $-1 \le \sin\theta \le 1$ **b.** $-1 \le \cos\theta \le 1$

c. $(\sin\theta)^2 + (\cos\theta)^2 = 1$ (This equality is often called the *fundamental identity.*)

Solution

a and **b**. For any angle θ, $\cos\theta = x$ and $\sin\theta = y$, where (x, y) is the point on the terminal side of θ where it intersects the unit circle. Since all coordinates on the unit circle satisfy $-1 \le x \le 1$ and $-1 \le y \le 1$, inequalities for parts a and b follow.

c. The coordinates of any point on the unit circle must satisfy the equation of the circle; therefore, $(\sin\theta)^2 + (\cos\theta)^2 = x^2 + y^2 = 1$. ■

It is customary to write $\sin^2\theta$ to mean $(\sin\theta)^2$ and to use similar notation when multiplying any trigonometric function by itself. For example, $\sin^2\left(\frac{\pi}{3}\right) = \left[\sin\left(\frac{\pi}{3}\right)\right]^2 = \left[\frac{\sqrt{3}}{2}\right]^2 = \frac{3}{4}$. With this notation the equation in part c of Example 8.9 is written

$$\sin^2\theta + \cos^2\theta = 1. \tag{1}$$

Note the difference between $\sin^2\theta$ and $\sin\theta^2$. The first expression, $\sin^2\theta$, means $(\sin\theta)(\sin\theta) = (\sin\theta)^2$, while the second expression, $\sin\theta^2$, means $\sin(\theta^2)$. When in doubt, use extra parentheses.

The equation $\sin^2\theta + \cos^2\theta = 1$ is true for **all** angles θ; for this reason it is called a *trigonometric identity*. Two other identities can be derived from this one; these identities are true for all values of θ for which the functions are defined:

$$\tan^2\theta + 1 = \sec^2\theta \tag{2}$$

$$1 + \cot^2\theta = \csc^2\theta. \tag{3}$$

To obtain the first identity, divide both sides of the equation $\sin^2\theta + \cos^2\theta = 1$ by $\cos^2\theta$. To obtain the second, divide both sides of the equation by $\sin^2\theta$.

EXAMPLE 8.10

a. For an angle θ with $\sin\theta = -\frac{1}{4}$, give the possible values of

(i) $\cos\theta$ (ii) $\tan\theta$ (iii) $\csc\theta$

b. On a unit circle draw an angle θ for which $\sin\theta = -\frac{1}{4}$.

Solution

a. (i) From the identity $\sin^2\theta + \cos^2\theta = 1$, we solve for $\cos\theta$. We first write $\cos^2\theta = 1 - \sin^2\theta$, then take square roots to get

$$\cos\theta = \pm\sqrt{1-\sin^2\theta}.$$

When $\sin\theta = -\frac{1}{4}$, we have

$$\cos\theta = \pm\sqrt{1-\left(-\frac{1}{4}\right)^2} = \pm\sqrt{1-\frac{1}{16}} = \pm\sqrt{\frac{15}{16}} = \pm\frac{\sqrt{15}}{4}.$$

Thus there are two possible values of $\cos\theta$: $\frac{\sqrt{15}}{4}$ and $-\frac{\sqrt{15}}{4}$.

(ii) Since $\tan\theta = \frac{\sin\theta}{\cos\theta}$, there are two possible values of $\tan\theta$:

$$\frac{\left(-\frac{1}{4}\right)}{\left(\frac{\sqrt{15}}{4}\right)} = -\frac{1}{\sqrt{15}} \quad \text{and} \quad \frac{\left(-\frac{1}{4}\right)}{\left(-\frac{\sqrt{15}}{4}\right)} = \frac{1}{\sqrt{15}}.$$

(iii) Since $\csc\theta = \frac{1}{\sin\theta}$, the only possible value of $\csc\theta$ is $\frac{1}{\left(-\frac{1}{4}\right)} = -4$.

b. The graphs in Figure 8.34 show the two angles between 0 and 2π for which $\sin\theta = -\frac{1}{4}$. Any other angle θ for which $\sin\theta = -\frac{1}{4}$ must have its terminal side the same as one of these two angles.

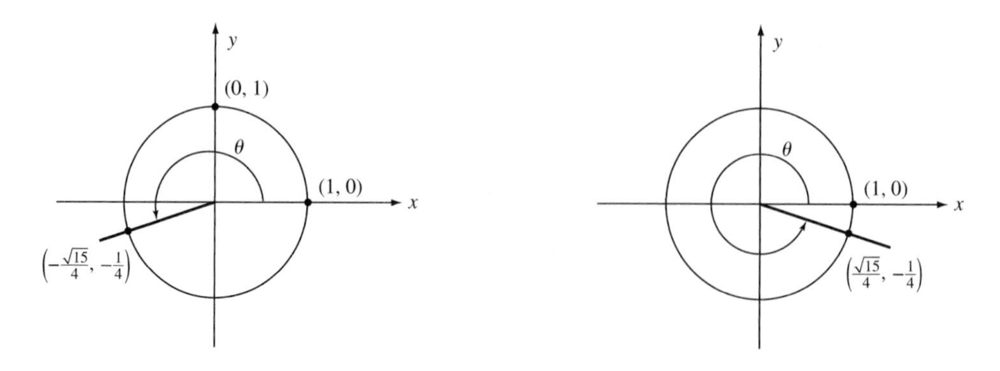

Figure 8.34

■

Other properties of the trigonometric functions can be derived from their definition and the symmetry of points on the circle. For example, the terminal sides of the angles θ and $-\theta$ are symmetric with respect to the x-axis. This means that the points where they intersect the unit circle have the same x-coordinate and their y-coordinates differ only in sign. This gives the identities:

$$\sin(-\theta) = -\sin\theta \quad (4)$$
$$\cos(-\theta) = \cos\theta. \quad (5)$$

See Figure 8.35.

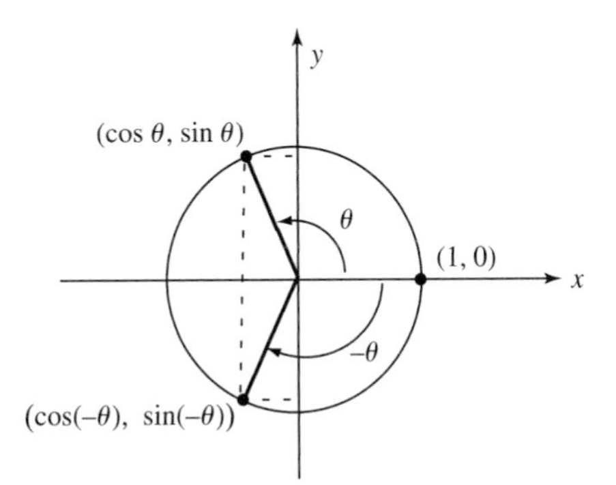

Figure 8.35

Similar identities can be derived for the four other trigonometric functions. For example, identities (4) and (5) imply $\tan(-\theta) = \dfrac{\sin(-\theta)}{\cos(-\theta)} = \dfrac{-\sin\theta}{\cos\theta} = -\tan\theta$. The points of intersection of the terminal sides of the two angles θ and $(\theta + \pi)$ with the unit circle lie on the same line and in opposite quadrants (see Figure 8.36). This means that their x- and y-coordinates have opposite signs and the same absolute value. This gives the following identities:

$$\cos(\theta + \pi) = -\cos\theta \quad (6)$$
$$\sin(\theta + \pi) = -\sin\theta. \quad (7)$$

Similarly, since the angles $(\theta + \pi)$ and $(\theta - \pi)$ have the same terminal side, we have

$$\cos(\theta - \pi) = -\cos\theta \quad (8)$$
$$\sin(\theta - \pi) = -\sin\theta. \quad (9)$$

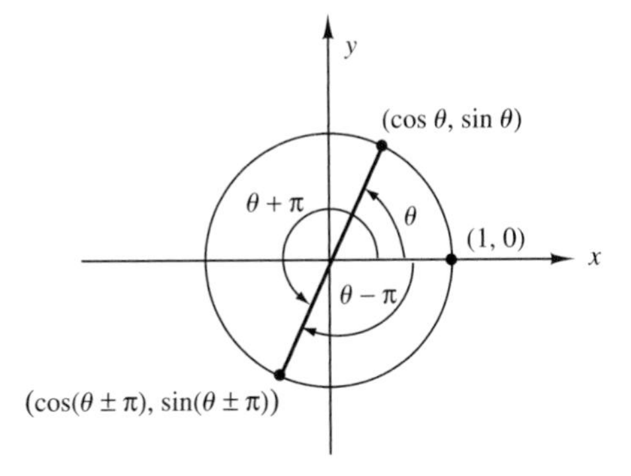

Figure 8.36

The relationship between the values of sine and cosine of an angle θ and the angles $\left(\theta+\frac{\pi}{2}\right)$ and $\left(\theta-\frac{\pi}{2}\right)$ is illustrated in the next example.

EXAMPLE 8.11 Figures 8.37a and b show an angle θ and the angles $\left(\theta+\frac{\pi}{2}\right)$ and $\left(\theta-\frac{\pi}{2}\right)$.

Use the information in the figures to express $\sin\left(\theta+\frac{\pi}{2}\right)$, $\cos\left(\theta+\frac{\pi}{2}\right)$, $\sin\left(\theta-\frac{\pi}{2}\right)$, and $\cos\left(\theta-\frac{\pi}{2}\right)$ in terms of $\sin\theta$ and $\cos\theta$.

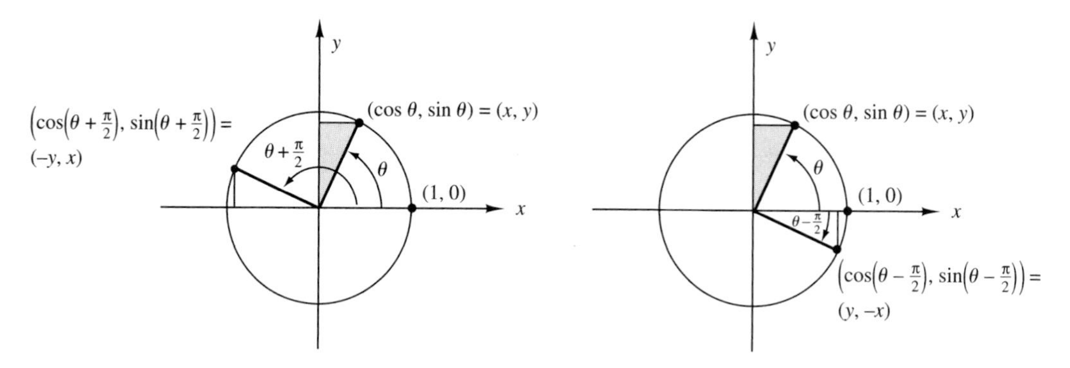

Figure 8.37a **Figure 8.37b**

Solution The terminal side of the angle θ is perpendicular to the terminal sides of the other two angles. We can see from Figure 8.37a that if we rotate the shaded triangle $\frac{\pi}{2}$ radians (that is, in the counterclockwise direction) about (0, 0), we obtain the unshaded triangle shown in the second quadrant. This implies that if (x, y) is the point of intersection of the terminal side of θ with the unit circle, then the point of intersection of the terminal side of $\left(\theta+\frac{\pi}{2}\right)$ with the unit circle has coordinates $(-y, x)$. Since $x=\cos\theta$ and $y=\sin\theta$, we obtain

$$\cos\left(\theta+\frac{\pi}{2}\right)=-\sin\theta \tag{10}$$

$$\sin\left(\theta+\frac{\pi}{2}\right)=\cos\theta. \tag{11}$$

Figure 8.37b shows that if we rotate the shaded triangle $-\frac{\pi}{2}$ radians (that is, in the clockwise direction) about (0, 0), we get the unshaded triangle shown in the fourth quadrant. Thus, the terminal side of $\left(\theta-\frac{\pi}{2}\right)$ intersects the unit circle at the point $(y, -x)$. So,

$$\cos\left(\theta-\frac{\pi}{2}\right)=\sin\theta \tag{12}$$

$$\sin\left(\theta-\frac{\pi}{2}\right)=-\cos\theta. \tag{13}$$

■

We saw earlier that angles θ and $\theta + 2n\pi$, n any integer, have the same terminal side. This gives the identities

$$\cos(\theta + 2n\pi) = \cos\theta \quad (14)$$

$$\sin(\theta + 2n\pi) = \sin\theta \quad (15)$$

where n is any integer. Identities (14) and (15) along with the graphs of the sine and cosine functions show that these functions are *periodic with period* 2π. In general, we have the following definition:

A function f is periodic with period p if for every x in the domain of f, $f(x + p) = f(x)$, and p is the smallest positive number for which this property is true.

EXAMPLE 8.12

a. Show that for every θ in the domain of the tangent function, $\tan(\theta + \pi) = \tan\theta$.

b. Is the tangent function periodic? What can you say about its period?

Solution

a. Write $\tan(\theta + \pi) = \dfrac{\sin(\theta + \pi)}{\cos(\theta + \pi)}$ and use identity (7), $\sin(\theta + \pi) = -\sin\theta$, and identity (6), $\cos(\theta + \pi) = -\cos\theta$, to obtain

$$\tan(\theta + \pi) = \frac{\sin(\theta + \pi)}{\cos(\theta + \pi)} = \frac{-\sin\theta}{-\cos\theta} = \frac{\sin\theta}{\cos\theta} = \tan\theta.$$

b. Since $\tan(\theta + \pi) = \tan\theta$ for all θ in its domain, the tangent function is periodic with period $p \leq \pi$. (We will see in the next section that the period is actually π.) ■

The fact that the sine and cosine functions are periodic implies that all of the trigonometric functions are periodic. Thus, some equations that involve trigonometric functions have infinitely many solutions.

EXAMPLE 8.13 For each of the following equations, find all angles θ that satisfy the equation or explain why there is no solution.

a. $\sin\theta = 0$ **b.** $\sec\theta + \frac{1}{2} = 0$ **c.** $\sin\theta + \cos\theta = 0$

Solution

a. To find angles θ where $\sin\theta = 0$, we look for all points on the unit circle with y-coordinate 0. There are two points: (1, 0) and (−1, 0). (See Figure 8.38.) Thus θ must have its terminal side through one of these points.

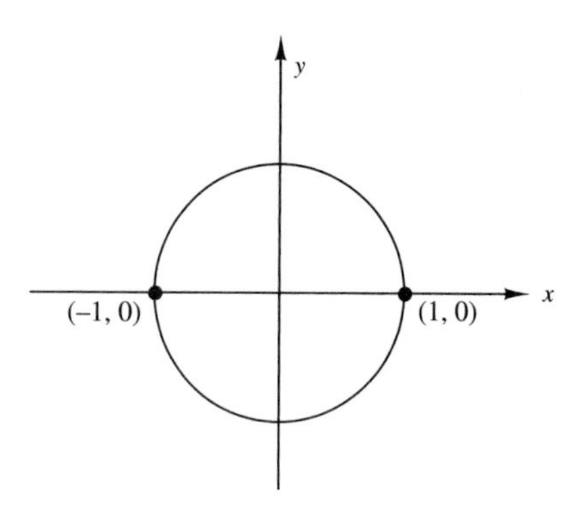

Figure 8.38

The angles θ for which $\sin\theta = 0$ are all integer multiples of π: $0, \pm\pi, \pm2\pi, \pm3\pi, \ldots$.

b. To solve this equation, we first subtract $\frac{1}{2}$ from both sides to write $\sec\theta = -\frac{1}{2}$. We now use the identity $\sec\theta = \frac{1}{\cos\theta}$ to rewrite the equation as $\frac{1}{\cos\theta} = -\frac{1}{2}$, which is equivalent to $\cos\theta = -2$. This equation has no solution, since for every angle θ, $-1 \le \cos\theta \le 1$.

c. We subtract $\cos\theta$ from both sides to get $\sin\theta = -\cos\theta$. To find angles θ for which this is true, we look for all points (x, y) on the unit circle where $y = -x$; that is, we look for the points of intersection of the unit circle and the line $y = -x$. There are two points: $\left(-\frac{1}{\sqrt{2}}, \frac{1}{\sqrt{2}}\right)$, and $\left(\frac{1}{\sqrt{2}}, -\frac{1}{\sqrt{2}}\right)$. (See Figure 8.39.) The angles θ that satisfy the given equation have terminal sides through one of these points. These are

$$\theta = \frac{3\pi}{4},\ \frac{3\pi}{4} \pm 2\pi,\ \frac{3\pi}{4} \pm 4\pi,\ \frac{3\pi}{4} \pm 6\pi,\ \ldots$$

$$\theta = \frac{7\pi}{4},\ \frac{7\pi}{4} \pm 2\pi,\ \frac{7\pi}{4} \pm 4\pi,\ \frac{7\pi}{4} \pm 6\pi, \ldots.$$

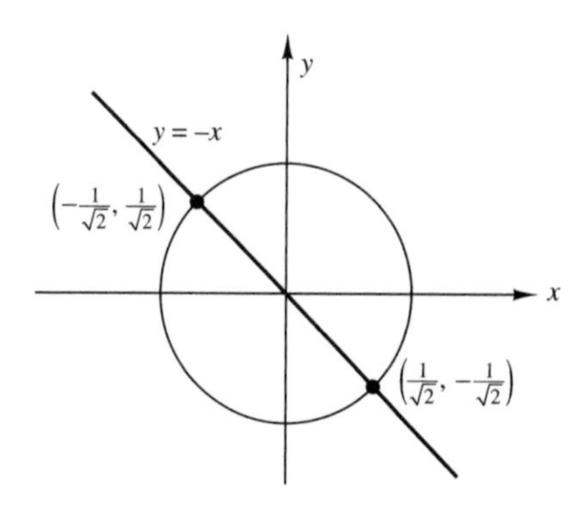

Figure 8.39 ■

Given any angle θ, the symmetry of points on the unit circle leads to the special identities for sine and cosine of $\theta + \frac{\pi}{2}$ and $\theta - \frac{\pi}{2}$ that we have already discussed. In general, the values of sine and cosine of the sum of any two angles θ and ϕ can be obtained from the values of sine and cosine of each angle through the following identities:

$$\sin(\theta + \phi) = \sin\theta\cos\phi + \cos\theta\sin\phi \qquad (16)$$

$$\cos(\theta + \phi) = \cos\theta\cos\phi - \sin\theta\sin\phi. \qquad (17)$$

If we replace ϕ by $-\phi$ in these identities, then by using identities (4) and (5), we obtain the following identities for sine and cosine of the difference of two angles:

$$\sin(\theta - \phi) = \sin\theta\cos\phi - \cos\theta\sin\phi \qquad (18)$$

$$\cos(\theta - \phi) = \cos\theta\cos\phi + \sin\theta\sin\phi. \qquad (19)$$

Identities (18) and (19) allow us to find exact values for sines and cosines of some angles, such as multiples of $\frac{\pi}{12}$.

EXAMPLE 8.14 Write $\frac{\pi}{12} = \frac{\pi}{3} - \frac{\pi}{4}$ and use appropriate identities to find $\sin\frac{\pi}{12}$ and $\cos\frac{\pi}{12}$.

Solution Since $\sin\frac{\pi}{3} = \frac{\sqrt{3}}{2}$, $\cos\frac{\pi}{3} = \frac{1}{2}$, and $\sin\frac{\pi}{4} = \cos\frac{\pi}{4} = \frac{1}{\sqrt{2}}$, we substitute in identities (18) and (19) with $\theta = \frac{\pi}{3}$ and $\phi = \frac{\pi}{4}$ to get

$$\sin\frac{\pi}{12} = \sin\left(\frac{\pi}{3} - \frac{\pi}{4}\right) = \sin\frac{\pi}{3}\cos\frac{\pi}{4} - \cos\frac{\pi}{3}\sin\frac{\pi}{4} = \left(\frac{\sqrt{3}}{2}\right)\left(\frac{1}{\sqrt{2}}\right) - \left(\frac{1}{2}\right)\left(\frac{1}{\sqrt{2}}\right) = \frac{\sqrt{3}-1}{2\sqrt{2}}$$

$$\cos\frac{\pi}{12} = \cos\left(\frac{\pi}{3} - \frac{\pi}{4}\right) = \cos\frac{\pi}{3}\cos\frac{\pi}{4} + \sin\frac{\pi}{3}\sin\frac{\pi}{4} = \left(\frac{1}{2}\right)\left(\frac{1}{\sqrt{2}}\right) + \left(\frac{\sqrt{3}}{2}\right)\left(\frac{1}{\sqrt{2}}\right) = \frac{1+\sqrt{3}}{2\sqrt{2}}.$$ ■

Identities (16) through (19) also provide verification of identities (6) through (15).

EXAMPLE 8.15 Use the identity for $\sin(\theta - \phi)$ to express $\sin\left(\theta + \frac{\pi}{2}\right)$ and $\sin(\theta + \pi)$ in terms of sine and cosine of the angle θ.

Solution With $\phi = \frac{\pi}{2}$ in identity (16), we obtain

$$\sin\left(\theta + \frac{\pi}{2}\right) = \sin\theta\cos\frac{\pi}{2} + \cos\theta\sin\frac{\pi}{2} = (\sin\theta)(0) + (\cos\theta)(1) = \cos\theta.$$

We set $\phi = \pi$ in identity (16) to get

$$\sin(\theta + \pi) = \sin\theta\cos\pi + \cos\theta\sin\pi = (\sin\theta)(-1) + (\cos\theta)(0) = -\sin\theta.$$

(These results verify identities (7) and (11).) ■

In the following example we develop the "double-angle formulas."

EXAMPLE 8.16 Use identities (16) and (17) for $\sin(\theta + \phi)$ and $\cos(\theta + \phi)$ to:

a. develop a formula for $\sin(2\theta)$.

b. develop a formula for $\cos(2\theta)$.

Solution

a. Since $\sin(\theta + \phi) = \sin\theta\cos\phi + \cos\theta\sin\phi$ we can replace each occurrence of ϕ by θ in the identity to obtain:

$$\sin(\theta + \theta) = \sin(2\theta) = \sin\theta\cos\theta + \cos\theta\sin\theta = 2\sin\theta\cos\theta.$$

Thus,

$$\sin(2\theta) = 2\sin\theta\cos\theta. \tag{20}$$

b. Similarly, since $\cos(\theta + \phi) = \cos\theta\cos\phi - \sin\theta\sin\phi$, replacing ϕ by θ gives

$$\cos(\theta + \theta) = \cos(2\theta) = \cos\theta\cos\theta - \sin\theta\sin\theta = \cos^2\theta - \sin^2\theta.$$

Therefore,

$$\cos(2\theta) = \cos^2\theta - \sin^2\theta. \tag{21a}$$ ■

Note that by using identity (1), $\sin^2\theta + \cos^2\theta = 1$, we can get two additional identities for $\cos(2\theta)$:

$$\cos(2\theta) = 2\cos^2\theta - 1 \tag{21b}$$

$$\cos(2\theta) = 1 - 2\sin^2\theta. \tag{21c}$$

The next example shows how we can use forms (21b) and (21c) of the double-angle formula for $\cos(2\theta)$ to develop the "half-angle formulas."

EXAMPLE 8.17 Solve identities (21b) and (21c) for $\cos^2\theta$ and $\sin^2\theta$, respectively, to obtain identities for $\cos^2\theta$ and $\sin^2\theta$ in terms of $\cos(2\theta)$.

Solution From identity (21b), we get $2\cos^2\theta = 1 + \cos(2\theta)$. Dividing both sides by 2 gives

$$\cos^2\theta = \frac{1 + \cos(2\theta)}{2} \tag{22}$$

Starting with identity (21c), we have $2\sin^2\theta = 1 - \cos(2\theta)$. Dividing by 2 gives

$$\sin^2\theta = \frac{1 - \cos(2\theta)}{2}. \tag{23}$$

■

Reference List of Trigonometric Identities

The following identities developed and discussed in this section are collected here for convenience:

1. $\sin^2 \theta + \cos^2 \theta = 1$ — fundamental identity

2. $\tan^2 \theta + 1 = \sec^2 \theta$

3. $1 + \cot^2 \theta = \csc^2 \theta$

4. $\sin(-\theta) = -\sin \theta$

5. $\cos(-\theta) = \cos \theta$

6. $\cos(\theta + \pi) = -\cos \theta$

7. $\sin(\theta + \pi) = -\sin \theta$

8. $\cos(\theta - \pi) = -\cos \theta$

9. $\sin(\theta - \pi) = -\sin \theta$

10. $\cos\left(\theta + \frac{\pi}{2}\right) = -\sin \theta$

11. $\sin\left(\theta + \frac{\pi}{2}\right) = \cos \theta$

12. $\cos\left(\theta - \frac{\pi}{2}\right) = \sin \theta$

13. $\sin\left(\theta - \frac{\pi}{2}\right) = -\cos \theta$

14. $\cos(\theta + 2n\pi) = \cos \theta$

15. $\sin(\theta + 2n\pi) = \sin \theta$

16. $\sin(\theta + \phi) = \sin \theta \cos \phi + \cos \theta \sin \phi$ — sum formula

17. $\cos(\theta + \phi) = \cos \theta \cos \phi - \sin \theta \sin \phi$ — sum formula

18. $\sin(\theta - \phi) = \sin \theta \cos \phi - \cos \theta \sin \phi$ — difference formula

19. $\cos(\theta - \phi) = \cos \theta \cos \phi + \sin \theta \sin \phi$ — difference formula

20. $\sin(2\theta) = 2\sin \theta \cos \theta$ — double-angle formula

21. a. $\cos(2\theta) = \cos^2\theta - \sin^2\theta$ double-angle formula

b. $\cos(2\theta) = 2\cos^2\theta - 1$

c. $\cos(2\theta) = 1 - 2\sin^2\theta$

22. $\cos^2\theta = \dfrac{1+\cos(2\theta)}{2}$ half-angle formula

23. $\sin^2\theta = \dfrac{1-\cos(2\theta)}{2}$ half-angle formula

EXAMPLE 8.18 Use the suggested identities to rewrite each trigonometric function in the requested form.

a. Use the fundamental identity (1) to rewrite $\sin^5\theta$ as an expression of the form $\sin\theta \cdot f(\theta)$, where $f(\theta)$ is a sum of terms of the form $a\cos^n\theta$, a is a constant, and n is a nonnegative integer.

b. Use the half-angle formulas, identities (22) and (23), to rewrite $\sin^4\theta$ as an expression that is a sum of terms of the form $a\cos(n\theta)$, where a is a constant and n is a nonnegative integer.

Solution

a. We can rewrite $\sin^5\theta$ by factoring out one factor of $\sin\theta$:

$$\sin^5\theta = (\sin\theta)\cdot(\sin^4\theta) = (\sin\theta)(\sin^2\theta)(\sin^2\theta).$$

Since we want to obtain terms that contain powers of $\cos\theta$, we can use identity (1) to substitute $1-\cos^2\theta$ for each occurrence of $\sin^2\theta$. Thus, we get

$$\sin^5\theta = (\sin\theta)(\sin^2\theta)(\sin^2\theta) = (\sin\theta)(1-\cos^2\theta)(1-\cos^2\theta) =$$

$$(\sin\theta)(1-2\cos^2\theta+\cos^4\theta).$$

This fits the requested form.

b. We can rewrite $\sin^4\theta$ as $\sin^4\theta = \sin^2\theta\cdot\sin^2\theta$. Then, using the half-angle identity (22), we get

$$\sin^4\theta = \sin^2\theta\cdot\sin^2\theta = \left(\frac{1-\cos(2\theta)}{2}\right)\cdot\left(\frac{1-\cos(2\theta)}{2}\right) =$$

$$\frac{1}{4}(1-\cos(2\theta))\cdot(1-\cos(2\theta)) = \frac{1}{4}(1-2\cos(2\theta))+\cos^2(2\theta)).$$

We're not quite finished because the last term is not in the required form. We can use a half-angle identity again, substituting for $\cos^2(2\theta)$:

$$\sin^4\theta = \frac{1}{4}\left(1 - 2\cos(2\theta) + \frac{1+\cos(4\theta)}{2}\right).$$

We multiply out and add the two constant terms to obtain

$$\sin^4\theta = \frac{3}{8} - \frac{1}{2}\cos(2\theta) + \frac{1}{8}\cos(4\theta),$$

which is the required form. ■

Exercises 8-C

In each of the exercises, remember that all angles are measured in radians.

1. **a.** If $\cos\theta = -\frac{1}{3}$, what are the possible values for $\sin\theta$?

   b. On the unit circle, draw an angle θ in standard position for which $\cos\theta = -\frac{1}{3}$.

2. In each case, find all values of θ for which

   a. $\cos\theta = 0$ **b.** $\cos\theta = 1$ **c.** $\cos\theta = -1$

   d. $\sin\theta = 1$ **e.** $\sin\theta = -1$ **f.** $\tan\theta = 1$

3. Solve each equation for θ. If the equation has no solution, explain why. If the equation has a solution, find all solutions.

   a. $\cos\theta + \frac{\sqrt{3}}{2} = 0$ **b.** $\cos\theta - 1.5 = 0$

   c. $\csc\theta + 1 = 2$ **d.** $\sin\theta - \cos\theta = 0$

   e. $(\sin\theta)^2 - \frac{1}{2} = 0$ **f.** $(\sin\theta)^2 + \frac{1}{2} = 0$

4. Suppose θ is in standard position and has its terminal side in the first quadrant and $\sin\theta = \frac{2}{3}$. Find each of the following values:

   a. $\cos\theta$ **b.** $\sin(-\theta)$ **c.** $\cos(-\theta)$ **d.** $\tan\theta$

   e. $\tan(-\theta)$ **f.** $\sin(\theta+\pi)$ **g.** $\sin\left(\theta+\frac{\pi}{2}\right)$

5. Show that the secant function is periodic with period 2π.

6. If $\cos\theta = -\frac{1}{4}$, $\sin\theta = \frac{\sqrt{15}}{4}$, $\cos\phi = -\frac{\sqrt{7}}{4}$, and $\sin\phi = \frac{3}{4}$, find each of the following values:

   a. $\sin(\theta+\phi)$ **b.** $\sin(\theta-\phi)$ **c.** $\sin\theta + \sin\phi$ **d.** $\sin\theta - \sin\phi$

7. If $\cos\theta = -\frac{1}{4}$, $\sin\theta = \frac{\sqrt{15}}{4}$, $\cos\phi = -\frac{\sqrt{7}}{4}$, and $\sin\phi = \frac{3}{4}$, find each of the following values:

a. $\cos(\theta + \phi)$ **b.** $\cos(\theta - \phi)$ **c.** $\cos\theta - \cos\phi$ **d.** $\cos\theta + \cos\phi$

8. **a.** Verify that $\frac{\pi}{4} + \frac{\pi}{6} = \frac{5\pi}{12}$.

b. Find the exact value of $\cos\frac{5\pi}{12}$.

9. Use the identities for $\sin(2\theta)$ and $\cos(2\theta)$ to develop an identity for $\tan(2\theta)$.

10. **a.** Rewrite $\cos^3\theta$ as a function of the form $\cos\theta \cdot f(\theta)$, where $f(\theta)$ is a sum of terms of the form $a\sin^n\theta$, a is a constant, and n is a non-negative integer.

b. Rewrite $\cos^4\theta$ as an expression that is a sum of terms of the form $a \cdot \cos(n\theta)$, where a is a constant and n is a non-negative integer.

8-D DOMAIN, RANGE, AND GRAPHS OF THE TRIGONOMETRIC FUNCTIONS

The trigonometric functions have angles measured in radians as their domain, and we have written them as functions of the variable θ. In the study of the behavior of these functions using the techniques of calculus, and in many applications, it is more convenient to think of these functions as assigning real numbers to other real numbers and to write them as functions of a real variable t. It is easy to make this transition if we identify a real number t with an angle of radian measure t. With this identification, we can state the following: The domain of the sine and cosine functions is the interval of all real numbers $(-\infty, \infty)$, and

$$\sin t = \text{sine of an angle of } t \text{ radians}$$

$$\cos t = \text{cosine of an angle of } t \text{ radians.}$$

In Example 8.9, we noted that the sine function satisfies $-1 \le \sin t \le 1$. Since $\sin t$ takes on all the values of y-coordinates on the unit circle as the angle of t radians varies, the range of this function is the closed interval $[-1, 1]$. This tells us that the graph of the sine function lies between the horizontal lines $y = -1$ and $y = 1$. We now graph the function $y = \sin t$ over the interval $0 \le t \le 2\pi$.

Given any real number t, $\sin t$ is the y-coordinate of the point on the unit circle that is the terminal side of an angle of t radians in standard position. But $\sin t$ is also the y-coordinate of the graph of the function $y = \sin t$ drawn on a Cartesian coordinate system. If we choose the same unit of measure, we can map out the graph of $y = \sin t$ on a Cartesian coordinate system (with horizontal t-axis) by recording for each t the y-coordinate of the point on the unit circle determined by the angle of t radians in standard position (see Figure 8.40).

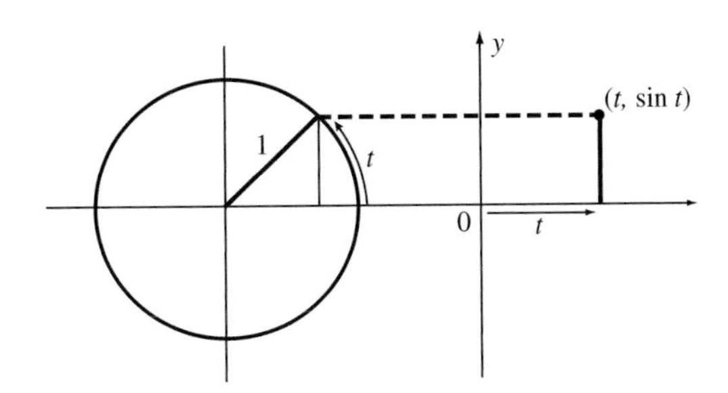

Figure 8.40

As t varies from 0 to 2π, t moves horizontally along the t-axis of the Cartesian coordinate system. On the unit circle, the angle of t radians sweeps around the circle for one full revolution (see Figure 8.41).

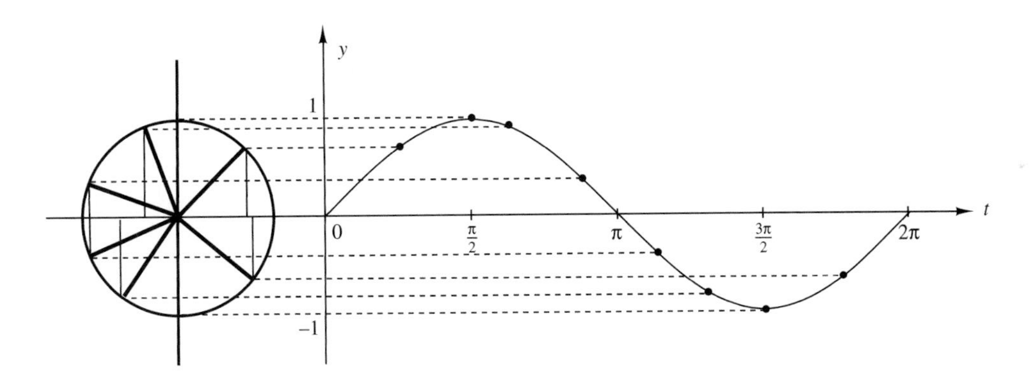

Figure 8.41 $y = \sin t$ for $0 \le t \le 2\pi$

Instead of using the letter t for the independent variable in these functions, we could have used the letter x. That is, $\sin x$ is the sine of an angle of x radians, and $\cos x$ is the cosine of the angle of x radians. (Note that this x represents the radian measure of an angle and is not the coordinate of a point on the unit circle.)

In the following example we examine the graph of the sine function in more detail.

EXAMPLE 8.19

a. Sketch the graph of $y = \sin x$ on the interval $\left[-\frac{7\pi}{2}, \frac{7\pi}{2}\right]$.

b. Identify the intercepts of the graph of $y = \sin x$.

c. Identify where the graph of $y = \sin x$ attains its maximum and minimum values.

Solution

a. Since the sine function is periodic with period 2π, we obtain the complete graph of the sine function by translating the graph on the interval $[0, 2\pi)$ to the right and left repeatedly by 2π units. Figure 8.42 shows the portion of the graph of $y = \sin x$ for the interval $-\frac{7\pi}{2} \le x \le \frac{7\pi}{2}$. (The whole graph extends forever in both directions.)

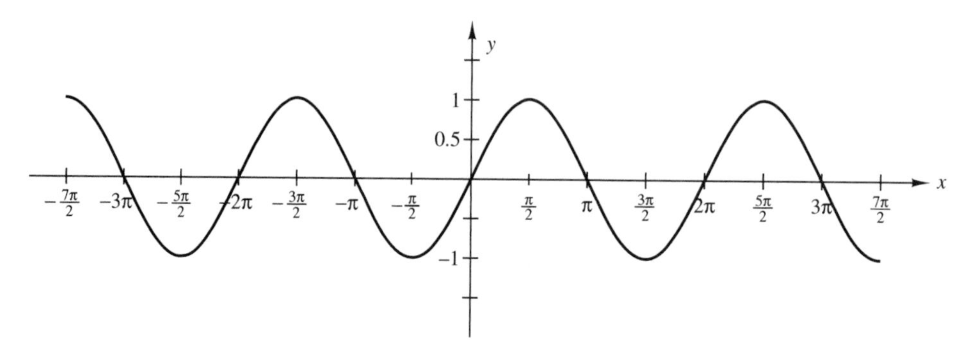

Figure 8.42 $y = \sin x$

b. From Example 8.13a we know that $\sin x = 0$ when $x = 0, \pm\pi, \pm 2\pi, \pm 3\pi, \ldots$, so the x-intercepts of the sine graph are all the integer multiples of π. The y-intercept is 0, since $\sin 0 = 0$.

c. The maximum value of $\sin x$ is 1 and it occurs at $x = \frac{\pi}{2}, -\frac{3\pi}{2}, \frac{5\pi}{2}, -\frac{7\pi}{2}, \ldots$, when the terminal side of the angle of x radians in standard position passes through the point (0, 1) on the unit circle. The minimum value of $\sin x$ is -1; it occurs at $x = -\frac{\pi}{2}, \frac{3\pi}{2}, -\frac{5\pi}{2}, \frac{7\pi}{2}, \ldots$, when the terminal side of the angle x in standard position passes through the point (0, −1) on the vertical axis. ■

Identity (11) (Section 8-C) shows that the graph of the cosine function is a translation of the graph of the sine function. This enables us to quickly analyze the cosine function.

EXAMPLE 8.20

a. Find the domain and range of the cosine function. Then use identity (11), $\sin\left(\theta + \frac{\pi}{2}\right) = \cos\theta$, to sketch the graph of the cosine function on the interval $\left[-\frac{7\pi}{2}, \frac{7\pi}{2}\right]$.

b. Identify the intercepts of the graph of $y = \cos x$.

c. Identify where the graph of $y = \cos x$ attains its maximum and minimum values.

Solution

a. The function $y = \cos x$ has domain $(-\infty, \infty)$ and range $[-1, 1]$ for reasons similar to those given for the sine function. The graph of $y = \cos x$, which is given in

Figure 8.43, has exactly the same shape as the graph of $y = \sin x$, but it has been shifted left by a distance of $\frac{\pi}{2}$ on the x-axis, since $\cos x = \sin\left(x + \frac{\pi}{2}\right)$.

b. The y-intercept of the graph of $y = \cos x$ is $\cos 0 = 1$, so the graph passes through the point (0, 1). The x-intercepts of this function are $x = \pm\frac{\pi}{2}, \pm\frac{3\pi}{2}, \pm\frac{5\pi}{2}, \ldots$ (all odd multiples of $\frac{\pi}{2}$), since $\cos x = 0$ for those values of x.

c. The maximum value of $\cos x$ is 1 and it occurs when $x = 0, \pm 2\pi, \pm 4\pi, \ldots$ (even multiples of π). The minimum value of $\cos x$ is -1 and it occurs when $x = \pm\pi, \pm 3\pi, \pm 5\pi, \ldots$ (odd multiples of π). (You were asked to find these values in Exercise 2 of Section 8-C.)

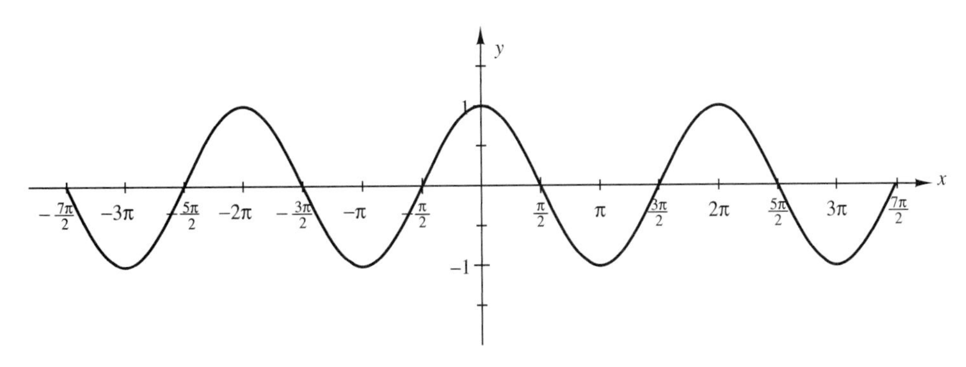

Figure 8.43 $y = \cos x$ ■

We now turn our attention to the tangent function, $y = \tan x$, and use the fact that $\tan x = \dfrac{\sin x}{\cos x}$ to analyze the behavior of the graph.

EXAMPLE 8.21

a. Find the domain of the function $y = \tan x$.

b. Find the vertical asymptotes of the graph of $y = \tan x$, and give the range of the function.

c. Sketch the graph of the function $y = \tan x$ on the interval $[-2\pi, 2\pi]$.

Solution

a. The domain of the tangent function consists of all x for which $\cos x \neq 0$. Since $\cos x = 0$ at $x = \pm\frac{\pi}{2}, \pm\frac{3\pi}{2}, \pm\frac{5\pi}{2}, \ldots$ (which are odd multiples of $\frac{\pi}{2}$), the domain of $y = \tan x$ is the set $\left\{x \mid x \neq \pm\frac{\pi}{2}, \pm\frac{3\pi}{2}, \pm\frac{5\pi}{2}, \ldots\right\}$.

b. As x approaches $\frac{\pi}{2}$ from the left, $\sin x$ increases to 1 while $\cos x$ decreases to 0; thus $\tan x = \dfrac{\sin x}{\cos x}$ increases without bound. This says that the line $x = \frac{\pi}{2}$ is a vertical asymptote. In a similar way, we can see that as x approaches $\frac{\pi}{2}$ from the right, $\tan x$ decreases without bound. Similar behavior occurs at the other odd multiples of $\frac{\pi}{2}$, so the graph has vertical asymptotes $x = k\frac{\pi}{2}$, for k an odd integer. For $x \in \left(0, \frac{\pi}{2}\right)$, $\tan x$ is positive, since both $\sin x$ and $\cos x$ are positive. For $x \in \left(-\frac{\pi}{2}, 0\right)$, $\sin x < 0$ and $\cos x > 0$, so $\tan x$ is negative. The range of this function is $(-\infty, \infty)$; that is, every real number y can be written as $\tan x$ for some value of x.

c. The requested portion of the graph of $y = \tan x$ is shown in Figure 8.44.

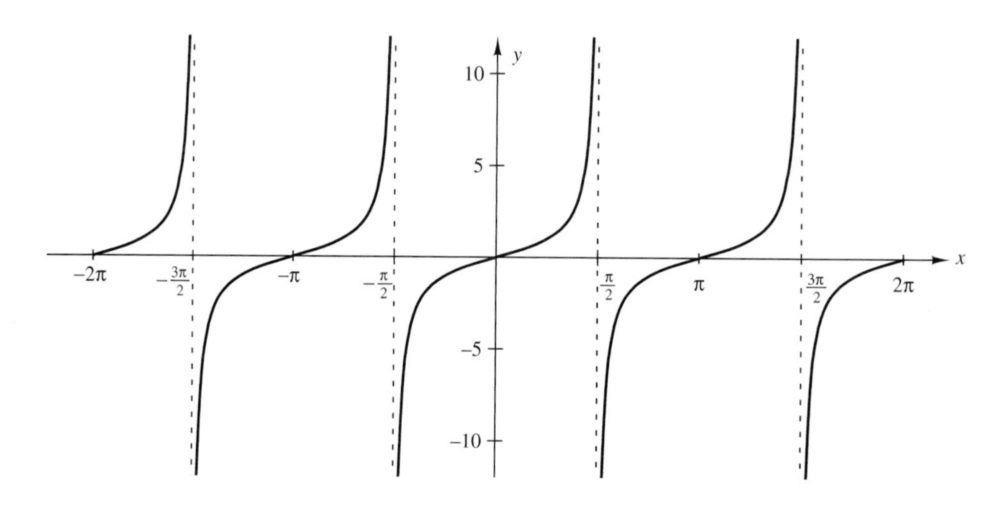

Figure 8.44 $y = \tan x$ ■

Recall that in Example 8.12 we showed that the tangent function is periodic with period less than or equal to π. From the graph we can see that the period is exactly π, since the graph for x in $\left(-\frac{\pi}{2}, \frac{\pi}{2}\right)$ has the same shape as the graph for x in the intervals $\left(\frac{\pi}{2}, \frac{3\pi}{2}\right)$, $\left(\frac{3\pi}{2}, \frac{5\pi}{2}\right)$, and so on.

Knowing the behavior of the sine and cosine functions enables us to analyze the secant and cosecant functions, since they are reciprocals of the sine and cosine functions:

$$\sec x = \frac{1}{\cos x} \qquad \csc x = \frac{1}{\sin x}.$$

EXAMPLE 8.22

a. Find the domain of the secant function.

b. Sketch the graph of the secant function on the interval $[-2\pi, 2\pi]$, and describe its behavior.

Solution

a. The secant function, $y = \sec x = \frac{1}{\cos x}$, has the same domain as the tangent function, since the values of x where $\cos x = 0$ must be excluded. Hence, the domain of $y = \sec x$ is $\left\{x \mid x \neq \pm\frac{\pi}{2}, \pm\frac{3\pi}{2}, \pm\frac{5\pi}{2}, \ldots\right\}$.

b. Because the secant function is the reciprocal of the cosine function, the graph of $y = \cos x$ (shown as a dotted line in Figure 8.45) can be used to graph $y = \sec x$. The highest points on the graph of $y = \cos x$ are the lowest points on the graph of $y = \sec x$ above the x-axis. These are the points where $\cos x = 1$ and $\sec x = \frac{1}{\cos x} = 1$. Also, the lowest points on the graph of $y = \cos x$ are the highest points on the graph of $y = \sec x$ below the x-axis. The x-intercepts of the cosine function determine the vertical asymptotes of $y = \sec x$. For example, as x approaches $\frac{\pi}{2}$ from the left, $\cos x$ decreases to 0, so $\frac{1}{\cos x}$ increases without bound. The line $x = \frac{\pi}{2}$ is a vertical asymptote to the graph of the secant function. As x approaches $\frac{\pi}{2}$ from the right, $\cos x$ increases to 0 ($\cos x$ is negative for $\frac{\pi}{2} < x < \frac{3\pi}{2}$) and, therefore, $\sec x$ decreases without bound. The same situation occurs when x is any odd multiple of $\frac{\pi}{2}$, as the graph in Figure 8.45 shows. The secant function is periodic with period 2π. The values of y for points on the graph of $y = \sec x$ are all greater than or equal to 1 or less than or equal to -1. Thus, the range of $y = \sec x$ is $(-\infty, -1] \cup [1, \infty)$.

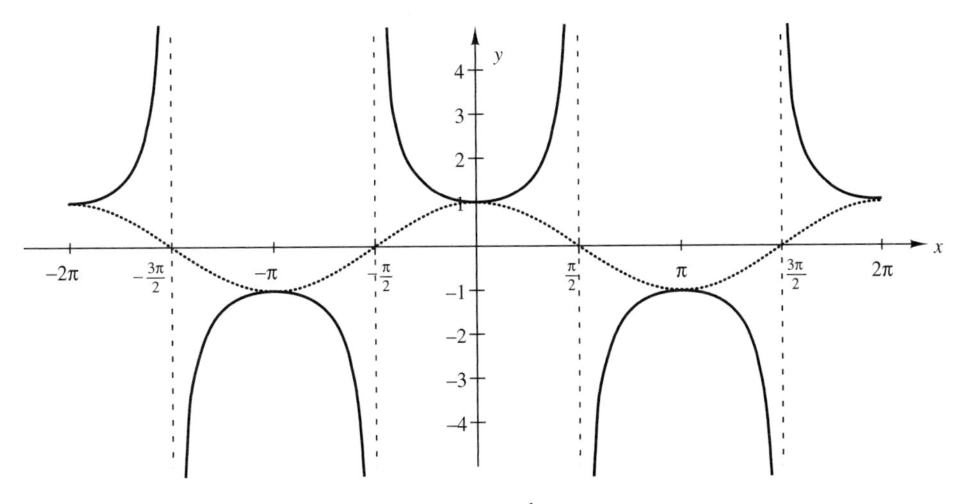

Figure 8.45 $y = \sec x = \frac{1}{\cos x}$, $y = \cos x$ (dotted) ■

The graphs of the cotangent function, $y = \cot x$, and the cosecant function, $y = \csc x$, are similar to those of $y = \tan x$ and $y = \sec x$, respectively; they are investigated in the exercises that follow.

Exercises 8-D

1. Give the intercepts of the graph of $y = \tan x$.

2. Look at the graphs of the sine and cosine functions (Figures 8.42 and 8.43).

a. Give the intervals on which the sine function is increasing. Give the intervals on which the sine function is decreasing.

b. Answer the questions in part a for the cosine function.

3. **a.** From the graph of $y = \sin x$ in Figure 8.42 estimate an approximate value of x in the interval $0 \le x \le \frac{\pi}{2}$ for which $\sin x = 0.25$. Estimate approximate values for the radian and degree measures of the angle for which $\sin x = 0.25$.

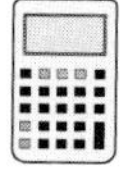

b. Use a graphing utility to graph $\sin x$ in the interval $0 \le x \le \frac{\pi}{2}$. Then use the zoom and trace features to answer the question in part a.

c. Use a calculator to find the value of $\sin x$ for your estimates of x in parts a and b.

d. Find the error of your estimates for x in parts a and b by calculating $\sin x - 0.25$.

4. Answer the following questions for the cotangent function, $y = \cot x$, and its graph:

a. Give the domain.

b. Give all x-intercepts.

c. Give all vertical asymptotes.

d. Graph the function and show its asymptotes.

e. Give the range of the function.

f. Give the period of the function.

5. Answer the questions in Exercise 4 for the cosecant function, $y = \csc x$.

8-E COMBINING FUNCTIONS WITH THE TRIGONOMETRIC FUNCTIONS

In many applications, the trigonometric functions are combined with other functions. In the following example, the cosine function is composed with the function $f(t) = \frac{t}{2}$ and then multiplied by a constant.

EXAMPLE 8.23 An object suspended from a spring is pulled downward from its resting position and then released. The position y (in centimeters) of the object at time t (in seconds) relative to its resting position is given by the equation $y = -12\cos\frac{t}{2}$, where $y = 0$ is taken as the resting position and the positive direction is upward (see Figure 8.46). (This equation is valid for an interval of time $0 \le t \le 20$.)

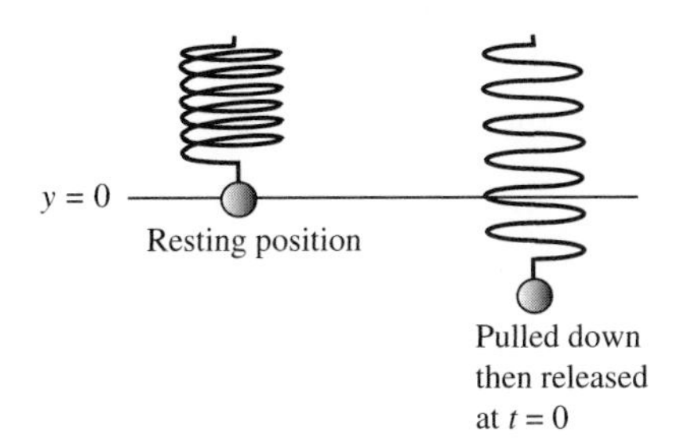

Figure 8.46

a. What is the position of the object at time 0?

b. What is the highest possible position of the object? At what times does the object attain this position?

c. What is the lowest possible position? At what times does the object reach this position?

d. Give all values of t at which the object's position is $y = 0$.

e. Show that the function $y = -12\cos\frac{t}{2}$ is periodic with period 4π.

Solution

a. When $t = 0$, $y = -12\cos 0 = -12$ (since $\cos 0 = 1$), so the object is 12 cm below its resting position.

b. Since $-1 \le \cos\frac{t}{2} \le 1$ for every t, the highest value of $y = -12\cos\frac{t}{2}$ is attained when $\cos\frac{t}{2} = -1$; this value is 12. To find the values of t when this is true, recall from Section 8-D that $\cos x = -1$ when $x = \pm\pi, \pm3\pi, \pm5\pi, \ldots$ (odd multiples of π). Replace x by $\frac{t}{2}$; $\cos\frac{t}{2} = -1$ when $\frac{t}{2} = \pm\pi, \pm3\pi, \pm5\pi, \ldots$, so $t = \pm2\pi, \pm6\pi, \pm10\pi, \ldots$. From these values of t, we choose only those between 0 and 20, which is the given domain of the function $y = -12\cos\frac{t}{2}$; these are $t = 2\pi$ and $t = 6\pi$.

c. The lowest value of $y = -12\cos\frac{t}{2}$ is -12 and occurs when $\cos\frac{t}{2} = 1$. We know from Section 8-D that $\cos\frac{t}{2} = 1$ when $\frac{t}{2} = 0, \pm2\pi, \pm4\pi, \pm6\pi, \ldots$ or $t = 0, \pm4\pi, \pm8\pi, \pm12\pi, \ldots$. Since $0 \le t \le 20$, the only values of t when the object's position is lowest are $t = 0$ and $t = 4\pi$.

d. The position is $y = 0$ when $-12\cos\frac{t}{2} = 0$; that occurs when $\cos\frac{t}{2} = 0$. As seen in Section 8-D, $\cos\frac{t}{2} = 0$ when $\frac{t}{2} = \pm\frac{\pi}{2}, \pm\frac{3\pi}{2}, \pm\frac{5\pi}{2}, \pm\frac{7\pi}{2}, \ldots$, so $t = \pm\pi, \pm 3\pi, \pm 5\pi, \pm 7\pi, \ldots$. Considering the domain of the position function, we see that the object's position is 0 when $t = \pi, 3\pi$, and 5π.

e. We need to first show that the values of the function $y = -12\cos\frac{t}{2}$ at t and at $t + 4\pi$ are the same and then show that $p = 4\pi$ is the smallest number for which $-12\cos\left(\frac{t+p}{2}\right) = -12\cos\left(\frac{t}{2}\right)$. We see that $-12\cos\left(\frac{t+4\pi}{2}\right) = -12\cos\left(\frac{t}{2}+2\pi\right) = -12\cos\frac{t}{2}$. Next we show that the graph of $y = -12\cos\frac{t}{2}$ does not repeat in a period less than 4π. While t ranges from 0 to 4π, $\frac{t}{2}$ ranges from 0 to 2π, which describes exactly one period of $\cos\frac{t}{2}$. ■

The function in Example 8.23 is of the form $y = A\cos(kx)$, with A and k constants; in this example, $A = -12$ and $k = \frac{1}{2}$. Functions of this form and functions of the form $y = A\sin(kx)$ are periodic and occur in the description of oscillating or cyclical phenomena. The graphs of such functions are similar to the graphs of the basic functions $y = \cos x$ and $y = \sin x$ and can be obtained from them. In Section 2-F, we analyzed transformations of the graph of a function f. We recall some of that discussion here and use the same approach to find the graphs of transformations of the cosine function. These techniques can also be applied to transformations of the sine function.

In Section 2-F we learned that the graph of $y = -f(x)$ is a reflection of the graph of $y = f(x)$ across the x-axis. We use this fact to sketch the graph of $y = -\cos x$ by reflecting the graph of $y = \cos x$ across the x-axis.

Figure 8.47 shows the graphs of $y = \cos x$ and $y = -\cos x$.

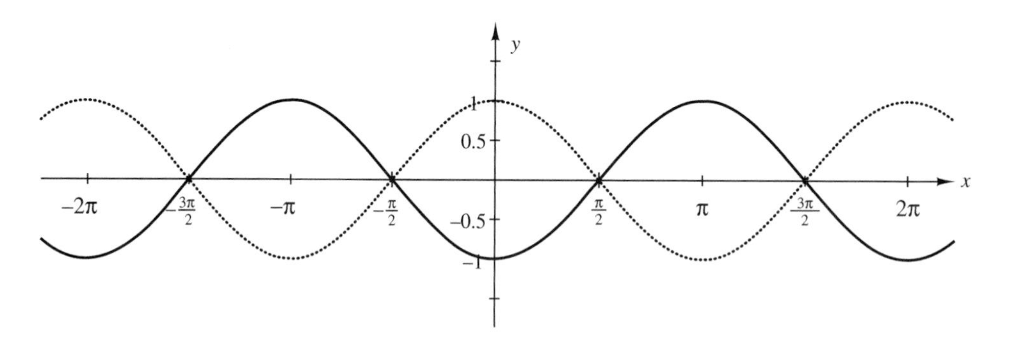

Figure 8.47 $y = -\cos x$, $y = \cos x$ (dotted)

Recall that for positive values of A, the graph of $y = A\,f(x)$ will be a vertical stretch of the graph of $y = f(x)$ when $A > 1$ and a vertical compression of the graph of $y = f(x)$ when $0 < A < 1$. For the periodic functions $y = A\cos x$ and $y = A\sin x$, A determines the maximum and minimum values of these functions. Figure 8.48 shows the graphs of $y = \cos x$ and $y = 3\cos x$. Note that the maximum and minimum points of $y = 3\cos x$ have $y = \pm 3$, and the x-intercepts of $y = 3\cos x$ are the same as those of $y = \cos x$.

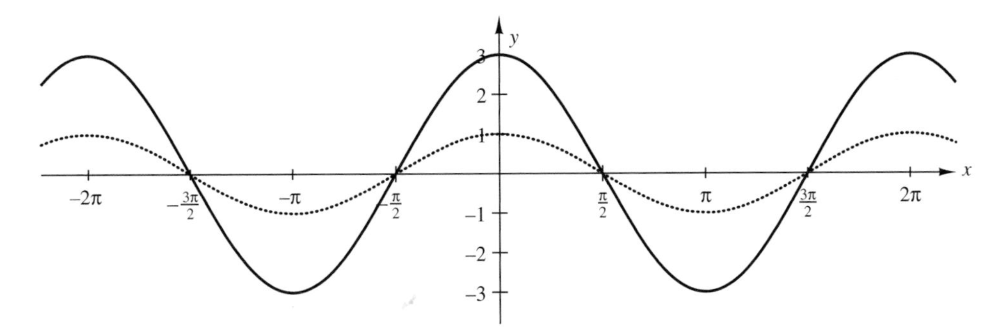

Figure 8.48 $y = 3\cos x$, $y = \cos x$ (dotted)

If k is positive, the graph of $y = \cos kx$ is the graph of $y = \cos x$ stretched or compressed horizontally. If $0 < k < 1$, the graph is stretched; if $k > 1$, the graph is compressed. Figures 8.49 and 8.50 show the graphs of the transformed cosine function when $k = 2$ and $k = \frac{1}{2}$, respectively.

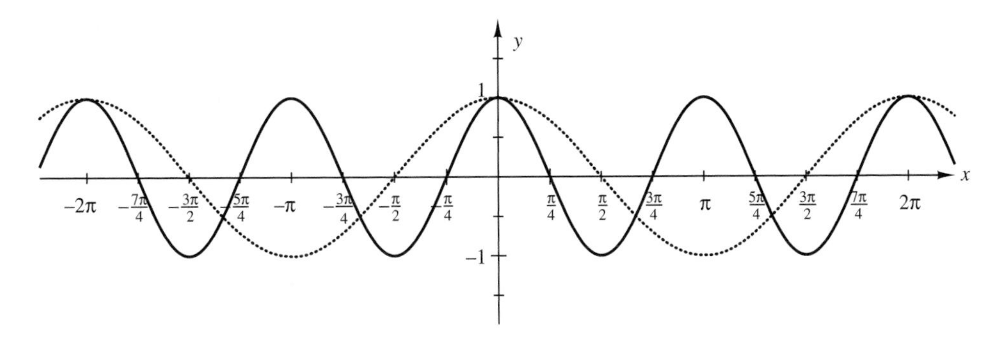

Figure 8.49 $y = \cos 2x$, $y = \cos x$ (dotted)

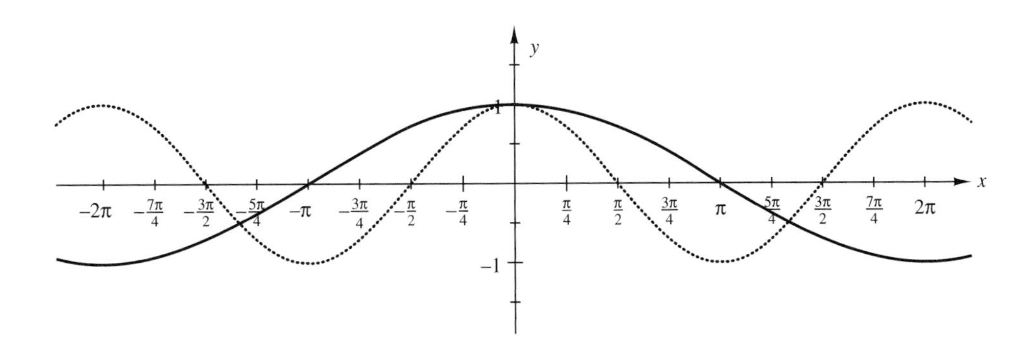

Figure 8.50 $y = \cos \frac{x}{2}$, $y = \cos x$ (dotted)

We are now ready to compare the graphs of $y = -12\cos\frac{x}{2}$ and $y = \cos x$.

EXAMPLE 8.24 The graphs of $y = -12\cos\frac{x}{2}$ and $y = \cos x$ are shown in Figure 8.51. Compare the two graphs, and explain how the graph of $y = -12\cos\frac{x}{2}$ can be obtained from that of $y = \cos x$.

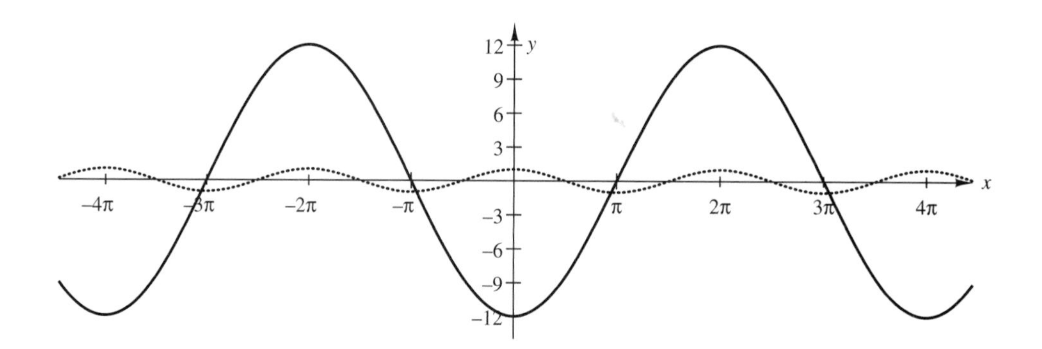

Figure 8.51 $y = -12\cos\frac{x}{2}$, $y = \cos x$ (dotted)

Solution A horizontal stretch by a factor of 2 transforms the graph of $y = \cos x$ into the graph of $y = \cos\frac{x}{2}$. A vertical stretch of this graph by a factor of 12 produces the graph of $y = 12\cos\frac{x}{2}$. Then a reflection across the x-axis gives the graph of $y = -12\cos\frac{x}{2}$. ■

In the graphs of $y = A\cos(kx)$ and $y = A\sin(kx)$, the number A indicates how far the graphs extend vertically. The highest and lowest points on the graphs have y-coordinates $y = |A|$ and $y = -|A|$, respectively. The number $|A|$ is called the *amplitude* of the function. The number $|k|$ determines the period of the function. When the variable x in $\sin x$ and $\cos x$ is multiplied by a number k, the period of the function changes from 2π to $\frac{2\pi}{|k|}$. The number $|k|$ indicates the *frequency* of repetition in the interval $[0, 2\pi]$ of one complete cycle of the graph. Recall that in Example 8.23, $k = \frac{1}{2}$ and we showed that the period was $4\pi = \frac{2\pi}{|k|} = \frac{2\pi}{\frac{1}{2}}$. The frequency is a half-cycle (see Figure 8.50). Figure 8.52 shows the graphs of $y = \cos(kx)$ when $k = 1$, 2, and 3, and $0 \le x \le 2\pi$. For $k = 2$ and 3, the frequency of repetition is increased.

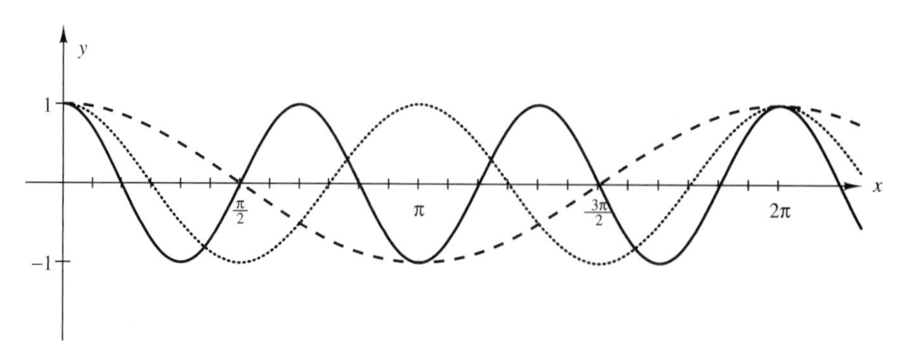

Figure 8.52 $y = \cos x$ (dashed), $y = \cos 2x$ (dotted), $y = \cos 3x$ (solid)

EXAMPLE 8.25 Graph each function and give its amplitude and period.

a. $y = 2 \sin x$ **b.** $y = \sin 2x$ **c.** $y = \dfrac{\sin 4x}{2}$

Solution

a. The graph of $y = 2 \sin x$ is the graph of $y = \sin x$ stretched vertically by a factor of 2 (see Figure 8.53). Since this function is of the form $y = A \sin(kx)$ with $A = 2$ and $k = 1$, it has amplitude 2 and period 2π.

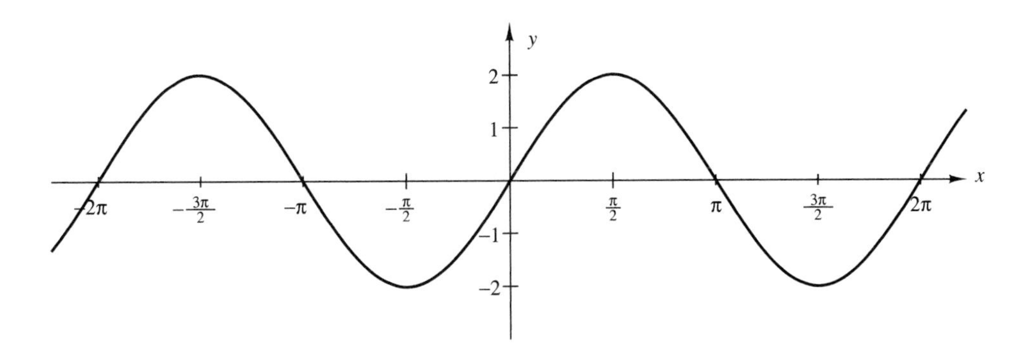

Figure 8.53 $y = 2 \sin x$

b. The graph of $y = \sin 2x$ is obtained from the graph of $y = \sin x$ by a horizontal compression (see Figure 8.54). The given function is of the form $y = A \sin(kx)$ with $A = 1$ and $k = 2$, so its amplitude is 1 and its period is $\dfrac{2\pi}{2} = \pi$.

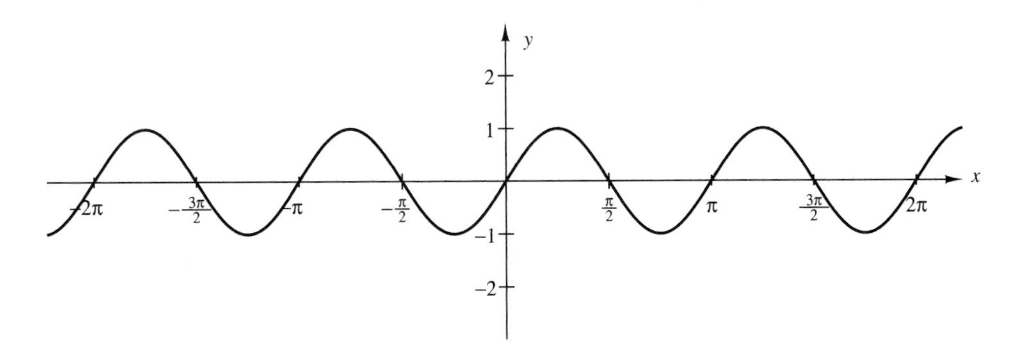

Figure 8.54 $y = \sin 2x$

c. Figure 8.55 shows the graph of $y = \dfrac{\sin 4x}{2} = \dfrac{1}{2} \sin 4x$. This graph is obtained from the graph of $y = \sin x$ by a horizontal compression by a factor of 4 and a vertical compression by a factor of 2. The amplitude of this function is $\dfrac{1}{2}$, and the period is $\dfrac{2\pi}{4} = \dfrac{\pi}{2}$.

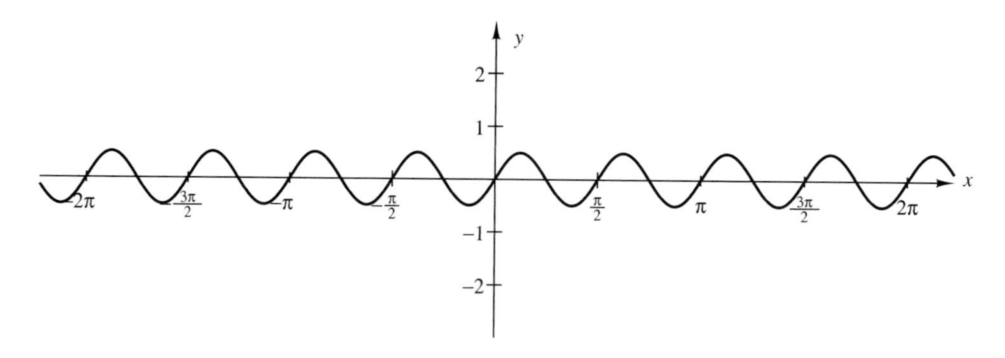

Figure 8.55 $y = \frac{\sin 4x}{2}$ ■

Note that in the expression $\frac{\sin 4x}{2}$, 4 is **not** a factor of the numerator and therefore no cancellation is possible: $\frac{\sin 4x}{2} = \frac{1}{2}\sin 4x$ is **not** the same as sin 2x. Both amplitude and period are different for $\frac{1}{2}\sin 4x$ and sin 2x.

Recall from Section 2-F that the graph of the function $y = f(x + c)$ is obtained from the graph of $y = f(x)$ by a horizontal translation of $-c$ units, and the graph of $y = f(x) + c$ is obtained from the graph of $y = f(x)$ by a vertical translation of c units.

EXAMPLE 8.26 For each function, describe how the graph can be obtained from the graph of $y = \sin x$ and graph the function.

a. $y = \sin(x + \pi)$ **b.** $y = \sin x + \pi$

Solution

a. The graph of $y = \sin(x + \pi)$ is the graph of $y = \sin x$ translated to the left π units (see Figure 8.56). Note that $\sin x = 1$ when $x = \frac{\pi}{2}$, while $\sin(x + \pi) = 1$ when $x = -\frac{\pi}{2}$.

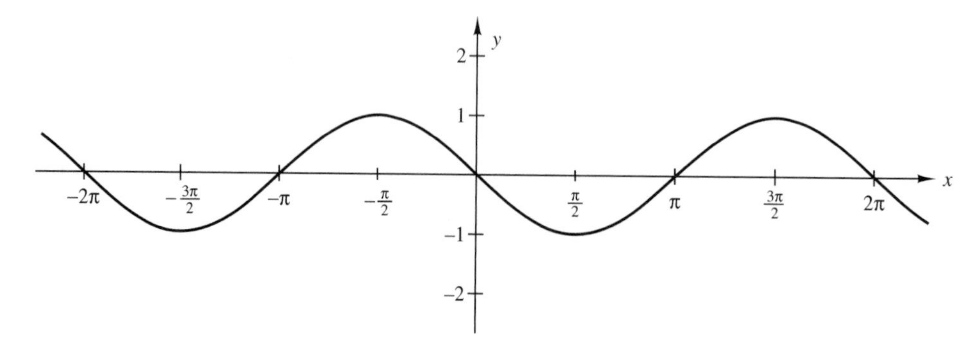

Figure 8.56 $y = \sin(x + \pi)$

b. The graph of $y = \sin x + \pi$ is obtained from the graph of $y = \sin x$ by a vertical shift of π units upward (see Figure 8.57).

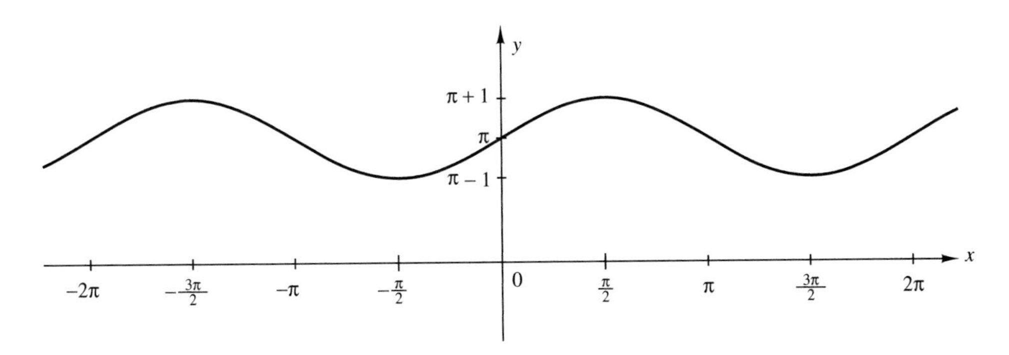

Figure 8.57 $y = \sin x + \pi$ ■

Notice that the graph of $y = \sin(x + \pi)$ in Figure 8.56 is the same as the graph of $y = -\sin x$. This verifies identity (7) in Section 8-C.

In the following examples we investigate functions obtained by combining two or more functions, where at least one is a trigonometric function. When a trigonometric function is composed with another function, parentheses around the inside function are often omitted. The standard convention is that in the absence of parentheses, the trigonometric function is to be evaluated at the first term following it, not including the next trigonometric function. For example, $\tan 2x - 1$ means $\tan(2x) - 1$, and $\sin x \cos x$ means $(\sin x) \cdot (\cos x)$. Before we decompose such functions to use rules of differentiation, it is useful to rewrite the function to show all implied parentheses explicitly.

When you work with trigonometric functions on a computer or calculator, it is often necessary to make all parentheses explicit to evaluate composite functions correctly. For example, $\sin(x + \pi) \neq \sin x + \pi$, as Figures 8.56 and 8.57 illustrate.

The placement of exponents and parentheses in the function formulas is also important for their correct interpretation. As we noted in Section 8-C, $\sin^2 x$ denotes $(\sin x)^2$. This notation applies to any trigonometric function and any exponent. Thus $\tan^3 x = (\tan x)^3$ means the function $\tan x$ is cubed, whereas $\tan x^3 = \tan(x^3)$ means that only the variable x is cubed.

EXAMPLE 8.27 Rewrite each function to show all implied parentheses explicitly. Use a calculator (in radian mode) to evaluate each function at $x = 0$.

a. $f(x) = \sin^3(2x + 5)$ **b.** $g(x) = \sin(2x + 5)^3$

c. $h(x) = \sin 2x^3 + 5$ **d.** $F(x) = \tan 2x - 5\sin(2x - 5)$

e. $G(x) = \tan 2x \sin(2x - 5)$ **f.** $H(x) = \tan(\sin 2x - 5)$

Solution

a. The placement of the exponent 3 indicates that the trigonometric function is cubed; thus $f(x) = \sin^3(2x + 5) = (\sin(2x + 5))^3$. Its value at $x = 0$ is $f(0) = (\sin(2(0) + 5))^3 = (\sin(5))^3 \approx (-0.9589)^3 \approx -0.8817$.

b. The placement of the exponent 3 indicates that only $2x + 5$ is cubed, so $g(x) = \sin(2x + 5)^3 = \sin((2x + 5)^3)$, and $g(0) = \sin((5)^3) = \sin(125) \approx -0.6160$.

c. Since there are no parentheses, only the term immediately after the trigonometric function is assumed to be in parentheses. Thus $h(x) = \sin 2x^3 + 5 = \sin(2x^3) + 5$, and $h(0) = \sin(2(0)^3) + 5 = \sin(2(0)) + 5 = \sin(0) + 5 = 0 + 5 = 5$.

d. The tangent function is to be evaluated at $2x$ only, so $F(x) = \tan(2x) - 5\sin(2x - 5)$, and $F(0) = \tan(0) - 5\sin(0 - 5) = 0 - 5\sin(-5) \approx -5(0.9589) = -4.7945$.

e. The argument of the tangent function (the expression at which it is to be evaluated) is $2x$; $G(x)$ is a product of two functions. Thus $G(x) = \tan(2x) \cdot \sin(2x - 5)$, and $G(0) = \tan(0) \cdot \sin(-5) = 0 \cdot \sin(-5) = 0$.

f. The parentheses indicate that the argument of the tangent function is $\sin 2x - 5$. In this expression, sine is evaluated only at $2x$, so $H(x) = \tan(\sin 2x - 5) = \tan(\sin(2x) - 5)$ and $H(0) = \tan(\sin(2(0)) - 5) = \tan(-5) \approx 3.3805$. ■

EXAMPLE 8.28 Rewrite each function to show all implied parentheses explicitly. Indicate the differentiation rules that would have to be used to find the derivative of each function.

a. $F(x) = \cos^3(4 - \sqrt{x})$ **b.** $G(x) = \sin(\cos 2x)$

c. $H(x) = \sin(4x + 1)\cos 3x$ **d.** $K(x) = \sec x^2$

Solution

a. $F(x) = \cos^3\left(4-\sqrt{x}\right) = \left[\cos\left(4-\sqrt{x}\right)\right]^3$. Thus, $F(x)$ is the composite function $f(g(x))$, where $g(x) = \cos(4 - \sqrt{x})$ and $f(x) = x^3$. Also, $g(x) = h(k(x))$, where $k(x) = 4 - \sqrt{x}$ and $h(x) = \cos x$. To differentiate $F(x) = f(g(x)) = f(h(k(x)))$ we would use the Chain Rule twice—once for the composite function $F(x) = (g(x))^3$ and a second time for the composite function $g(x) = \cos(4 - \sqrt{x})$.

b. $G(x) = \sin(\cos 2x) = \sin(\cos(2x))$. Thus, $G(x)$ is the composite function $f(g(x))$, where $g(x) = \cos(2x)$ and $f(x) = \sin x$. Also, $g(x) = h(k(x))$, where $k(x) = 2x$ and $h(x) = \cos x$. To differentiate G, we would use the Chain Rule twice—once for the composite function $G(x) = \sin(g(x))$ and a second time for the composite function $g(x) = \cos(k(x)) = \cos(2x)$.

c. $H(x) = \sin(4x + 1) \cos 3x = \sin(4x + 1) \cos(3x)$. The function H is the product of two composite functions, $f(x) = \sin(4x + 1)$ and $g(x) = \cos(3x)$. To differentiate H we would use the Product Rule, and in so doing we would use the Chain Rule to find the derivatives of f and g.

d. $K(x) = \sec x^2 = \sec(x^2)$. Thus, $K(x)$ is the composite function $f(g(x))$, where $g(x) = x^2$ and $f(x) = \sec x$. We would use the Chain Rule to find the derivative of $K(x)$. ■

In the next example we find domains of composite functions that involve trigonometric functions.

EXAMPLE 8.29 Give the domain of each of the following functions:

a. $f(x) = \cot\left(x - \frac{\pi}{2}\right)$ **b.** $g(x) = \cos\left(\frac{1}{3x}\right)$

Solution

a. Since $\cot\theta = \dfrac{\cos\theta}{\sin\theta}$, the cotangent is undefined whenever the sine has value 0. Since the sine function is 0 at $n\pi$ for any integer n, the function f is undefined for all x such that $x - \dfrac{\pi}{2} = n\pi$, or

$$x = n\pi + \frac{\pi}{2} = \frac{2n\pi + \pi}{2} = (2n + 1)\frac{\pi}{2}.$$

The domain is all real numbers except odd multiples of $\dfrac{\pi}{2}$.

b. Since the cosine has domain all reals, we need only concern ourselves with the domain of the function $y = \frac{1}{3x}$. Since this function is undefined when $x = 0$, the function g, which is the composite of the cosine with $y = \frac{1}{3x}$, has domain all reals except $x = 0$. ■

Exercises 8-E

1. Graph each function and give its amplitude and period. On the graph, show at least one full period of the function.

a. $y = 3\cos 2x$

b. $y = -\cos\frac{x}{2}$

c. $y = \frac{3\cos 6x}{2}$

d. $y = 8\sin\frac{x}{3}$

2. Without using a calculator, find the exact value of each given function at the indicated value of x.

a. $\sin^3\left(x^2 - 1\right); x = -1$

b. $\cos 3x^2; x = \frac{\sqrt{\pi}}{3}$

c. $\sec(5x)^2;\ x = \frac{2\sqrt{\pi}}{5}$

d. $\tan(\sin x); x = \pi$

e. $\tan\left(\frac{2\pi}{3}x\right)\cos\left(\frac{5\pi}{4}x\right); x = 1$

3. A study of a certain deer population indicates that the population at time t is estimated by the function

$$P(t) = 4500 + 600\sin\frac{\pi}{6}t$$

where t is measured in months and $t = 0$ is taken as March 21, 2001. This equation is expected to be valid until 2005.

a. What was the estimated deer population at the end of the spring (June 21) of 2001?

b. What was the estimated deer population at the end of spring 2002?

c. What was the estimated population on December 21, 2002?

d. Is the function $P(t)$ periodic? What is the period? What does this say about the deer population?

e. When is the population highest? Lowest?

f. Graph the function $P(t)$.

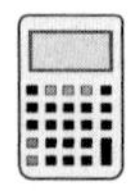

4. Rewrite each function to show explicitly all implied parentheses. Use a calculator to evaluate each function at the given value of x.

a. $\sec(\tan 5x); x = 1$

b. $\sin^3(-\cos x);\ x = \frac{1}{2}$

c. $\cos\left(\sin x^3\right); x = -0.5$

d. $\frac{\tan 3x^2}{6};\ x = \frac{2}{3}$

5. Rewrite each function to show explicitly all implied parentheses. Indicate the differentiation rules that would be used to find the derivative of each function.

a. $F(x) = \tan x^3 \cos\frac{x}{2}$ **b.** $G(x) = \sin(\tan x - 1)$

c. $H(x) = \cos\frac{x^2+7x}{3}$ **d.** $K(x) = 2x + \cos x^5$

e. $R(x) = \cos x^3 - \tan(x+1)$ **f.** $T(x) = \sin^4(x+2)$

6. Each of the functions $f(x)$ and $g(x)$ below has been written carelessly; parentheses need to be inserted to make the rule of each function clear.

a. There is only one correct way to interpret $f(x)$ below. Insert the parentheses to show the intended meaning of

$$f(x) = \sin \cos x.$$

b. There are two ways to insert parentheses in the function $g(x)$ below. The two functions that result have very different definitions. Show the two different ways to insert parentheses and explain how the two resulting functions differ.

$$g(x) = \sin 2 \cos x$$

c. Which of the two functions (with parentheses) that you obtained in part b is represented by the function given in part b (without parentheses)?

7. Find the domain of each of the following functions:

a. $f(x) = \sec\left(x - \frac{\pi}{3}\right)$ **b.** $g(x) = \sin\left(\frac{x}{x-3}\right)$

Chapter 8 Exercises

1. Consider the angles θ, ϕ, and γ, where $\gamma = \theta + \phi$ and the radian measures of θ and ϕ are $\frac{\pi}{3}$ and $\frac{\pi}{4}$, respectively.

a. Find both the radian and the degree measures of γ.

b. Draw a unit circle and the three angles θ, ϕ, and γ in standard position. Use this diagram to give approximate values of $\sin \gamma$ and $\cos \gamma$.

c. Use the sum formulas, identities (16) and (17), for sine and cosine to give the exact values of $\sin \gamma$ and $\cos \gamma$.

2. The terminal side of an angle α in standard position lies in the second quadrant and $\sin \alpha = \frac{1}{5}$.

a. Give the value of

(i) $\cos \alpha$ (ii) $\csc \alpha$ (iii) $\cot \alpha$ (iv) $\tan \alpha$ (v) $\sec \alpha$

b. Draw two angles, one positive and one negative, that could be the angle α.

3. Without using a calculator, give the exact value of each expression.

a. $\sin(72\pi) + 2\cos\left(\frac{3\pi}{4}\right)$ **b.** $\tan\left(\frac{8\pi}{3}\right)$

c. $\sec(3.2)\cos(3.2)$ **d.** $\tan^2\left(\frac{25\pi}{6}\right) - \sec^2\left(\frac{25\pi}{6}\right)$

e. $\sin^2\left(\frac{\sqrt{2}}{3}\right) + \cos^2\left(\frac{\sqrt{2}}{3}\right)$

4. Explain what each of the following identities says about the graph of $y = \sin x$.

a. $\sin(-x) = -\sin x$ **b.** $\sin(x + \pi) = -\sin x$ **c.** $\sin(x - \pi) = -\sin x$

5. Explain what each of the following identities says about the graph of $y = \cos x$.

a. $\cos(-x) = \cos x$ **b.** $\cos(x + \pi) = -\cos x$ **c.** $\cos(x - \pi) = -\cos x$

6. Give the domain of each of the following functions.

a. $f(x) = \tan(3x - 1)$ **b.** $g(x) = \sec(2x)$ **c.** $h(x) = \sin\left(\frac{1}{x^2+1}\right)$

7. Examine the graph of $y = \tan x$ in Figure 8.44 and answer the following questions:

a. Give the intervals where the tangent function is increasing.

b. Give the intervals where the tangent function is decreasing.

c. Use the graph to determine whether each of the following equations is true (for all x in the domain of the functions) or false (for at least one x). Give reasons for your answer.

(i) $\tan(-x) = \tan x$ (ii) $\tan(-x) = -\tan x$ (iii) $\tan(x + 2\pi) = \tan x$

(iv) $\tan(x + \pi) = \tan x$ (v) $\tan\left(x + \frac{\pi}{2}\right) = \tan x$

8. For each expression below, give the exact value of the expression or say it is undefined. If the expression is undefined, explain why.

a. $\sin\left(\frac{31\pi}{4}\right)$ **b.** $\cos\left(\frac{-2\pi}{3}\right)$ **c.** $\tan\left(\frac{-3\pi}{4}\right)$

d. $\csc\left(\frac{5\pi}{6}\right)$ **e.** $\sec\left(-\frac{10\pi}{3}\right)$ **f.** $\csc\left(\frac{5\pi}{2}\right)$

g. $\cot(-5\pi)$ **h.** $\csc^3 \pi$ **i.** $(\cot 3)(\tan 3)$

j. $2 + \sec^2\left(\frac{1}{2}\right) - \tan^2\left(\frac{1}{2}\right)$ **k.** $\sin^2\left(5\sqrt{2}\right) + \cos^2\left(5\sqrt{2}\right)$

9. For each of the following equations, say whether it is true (for every x in the domain of the functions) or false (for at least one x). If false, give a specific value of x for which both sides of the function are defined but for which the equation is false.

a. $\cos(\sin 3x) = \cos(3\sin x)$ **b.** $\cos(\sin 3x) = \cos x \sin 3x$

c. $\tan 5x \cos x^2 = (\tan 5x)\left(\cos\left(x^2\right)\right)$ **d.** $\sec^4 x = (\sec x)^4$

e. $\sec x^4 = (\sec x)^4$ **f.** $\sec x^4 = \sec\left(x^4\right)$

g. $\dfrac{\cos \pi}{\cos x} = \dfrac{\pi}{x}$ **h.** $\dfrac{\cos \pi}{\sin x} = \dfrac{-1}{\sin x}$

i. $\sin x \sec x = \tan x$ **j.** $\sin x \csc x = 1$

10. Choose the expression(s) that represent the function $f(x) = \dfrac{\tan x}{\tan\left(\frac{\pi}{3}\right)}$.

(i) $\dfrac{\tan x}{\sqrt{3}}$ (ii) $\tan\left(\dfrac{3x}{\pi}\right)$ (iii) $\dfrac{3x}{\pi}$ (iv) $\dfrac{x}{\left(\frac{\pi}{3}\right)}$ (v) $\dfrac{\sqrt{3}\tan x}{3}$

11. Rewrite each function to show explicitly all implied parentheses. Indicate the differentiation rules that would be used to find the derivative of each function.

a. $F(x) = \cos x^5$ **b.** $G(x) = \cos^5 x$

c. $H(x) = \cos(\tan 3x) - 1$ **d.** $K(x) = \dfrac{\cos 7x}{\sec(5x-2)}$

12. The graph of a function $y = A \sin kx$ is given in Figure 8.58. Give the values of A and k.

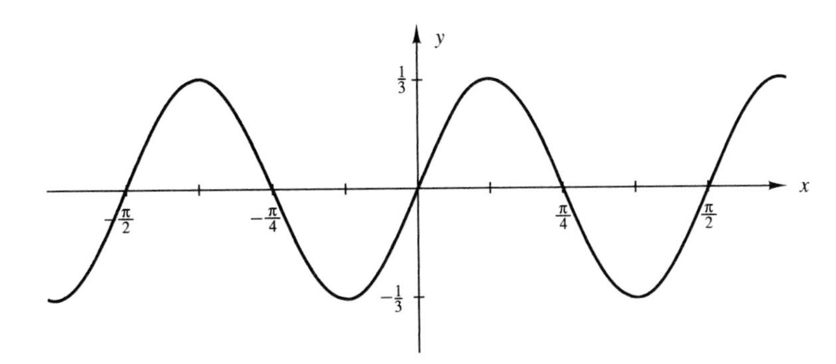

Figure 8.58

13. The graph of a function $y = A \cos kx$ is given in Figure 8.59. Give the values of A and k.

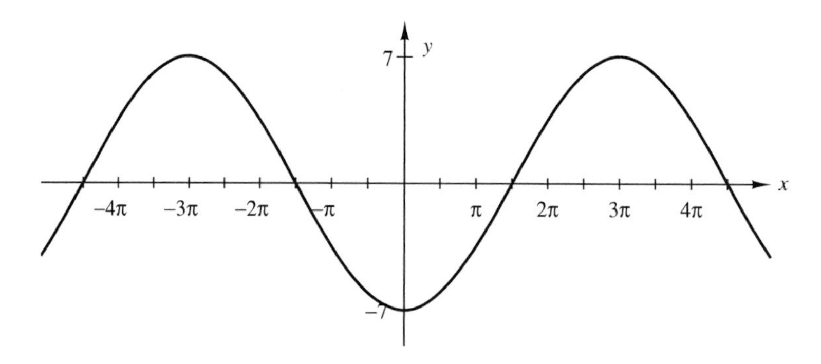

Figure 8.59

14. **a**. On the same Cartesian coordinate system, graph the functions $y = \sin 3x$ and $y = \csc 3x$.

b. Give the domain and the range of the function $y = \sin 3x$.

c. Give the domain and the range of the function $y = \csc 3x$.

15. For each function a–h, identify the number of its graph. Write the equation of the correct function under each graph.

a. $y = \tan 2x$ **b**. $y = 2\tan x$

c. $y = \cot 2x$ **d**. $y = \cos\left(x + \frac{\pi}{4}\right)$

e. $y = \cos x + \frac{\pi}{4}$ **f**. $y = \csc\left(\frac{x}{2}\right)$

g. $y = 2\sec x$ **h**. $y = \sin\left(x + \frac{\pi}{4}\right)$

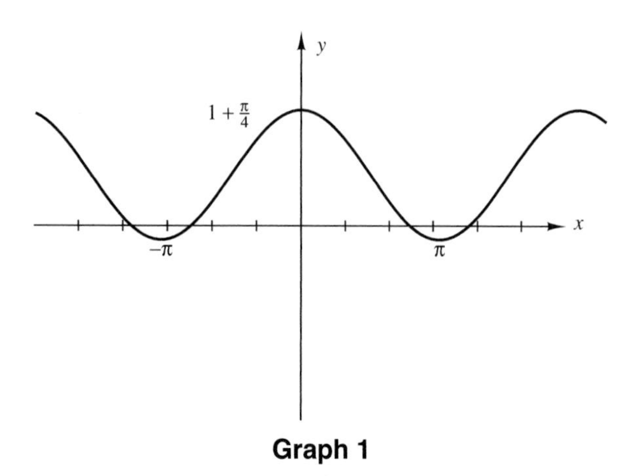

Graph 1

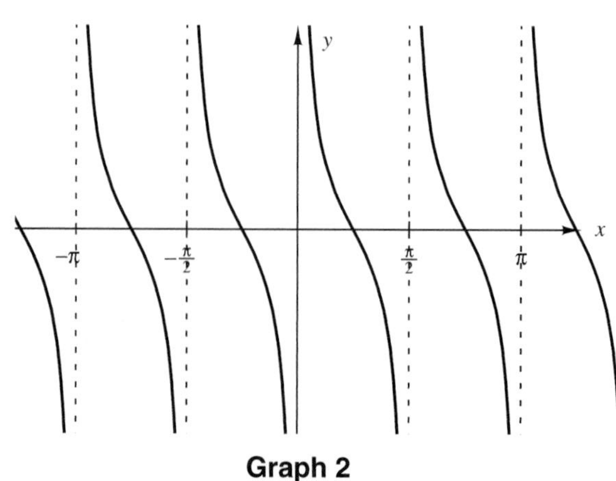

Graph 2

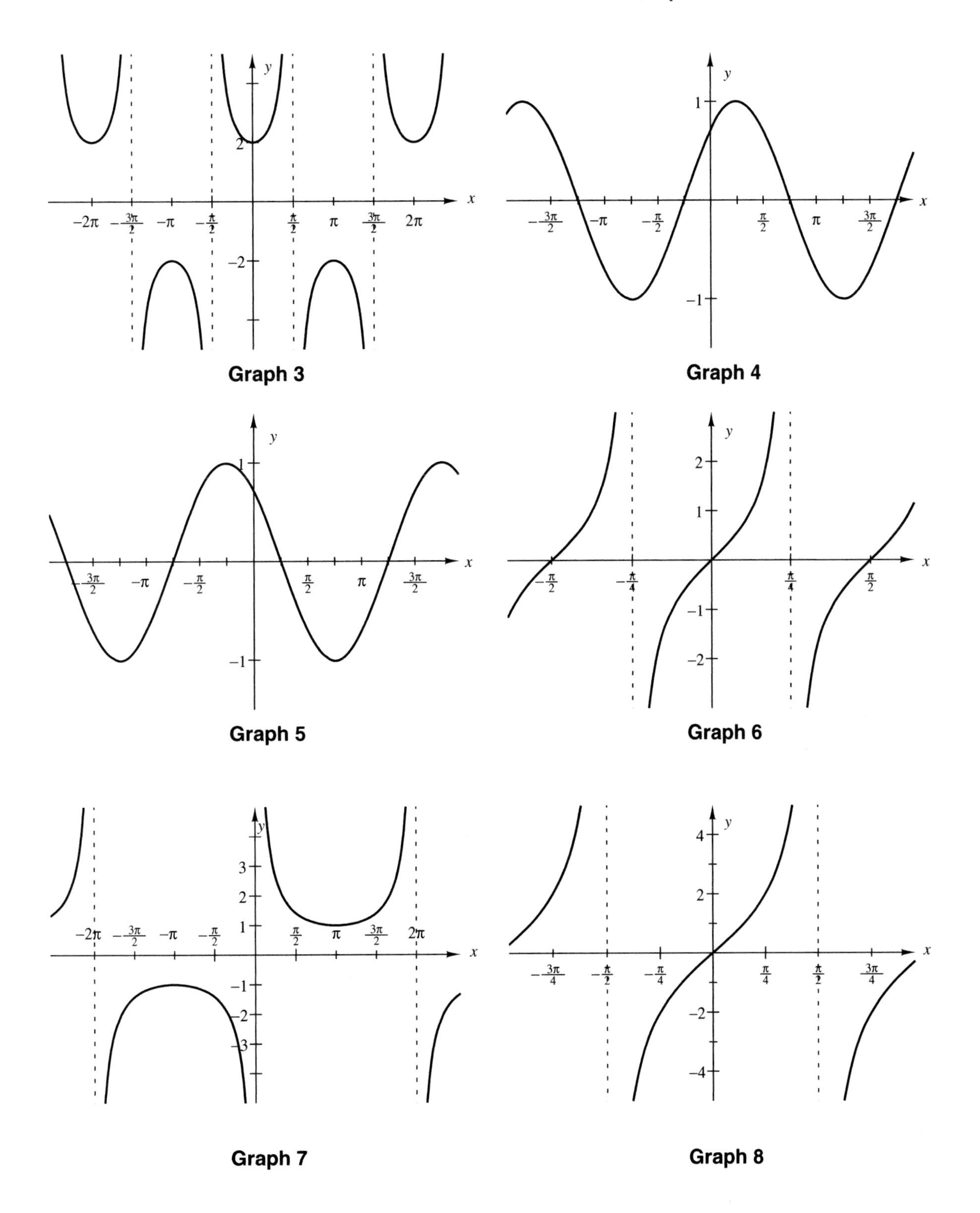

Graph 3

Graph 4

Graph 5

Graph 6

Graph 7

Graph 8

16. Use the sum and difference formulas for sine and cosine, identities (16)–(19), to prove each identity.

a. $\sin(\pi - x) = \sin x$

b. $\cos\left(\frac{\pi}{2} - x\right) = \sin x$

c. $\sin\left(\frac{\pi}{2}+x\right) = \cos x$

d. $\cos(\pi + x) = -\cos x$

17. Combine identities for the sine and cosine functions to obtain identities for each of the following:

a. $\tan(\theta + \pi)$

b. $\tan(\theta - \pi)$

c. $\tan\left(\theta + \frac{\pi}{2}\right)$

d. $\tan\left(\theta - \frac{\pi}{2}\right)$

e. $\tan^2 \theta$

18. Prove the identity $2\csc(2t) = \sec t \cdot \csc t$ by first rewriting $2\csc(2t)$ in terms of $\sin(2t)$. Then use the double-angle formula, identity (20), and finish the proof.

CHAPTER 9

Companion to Implicit Differentiation

9-A IMPLICITLY DEFINED FUNCTIONS

All functions we have considered thus far can be expressed by an equation of the form $y = f(x)$. Such an equation is said to define y *explicitly* as a function of x. An equation in the two variables x and y that is not of this form is said to define y *implicitly* as a function of x. For example, the equation $\sqrt{3y} + x = x^2$ defines y implicitly as a function of x.

EXAMPLE 9.1 Which of the following equations define y implicitly as a function of x?

a. $3x - 4 = 2(y - 4)$ **b**. $\dfrac{x^2}{16} + \dfrac{y^2}{25} = 1$ **c**. $y = 3x^2 - 4x + 2$

Solution The first two equations, a and b, define functions implicitly since they are not written in the form $y = f(x)$. The third equation, c, defines y explicitly as a function of x. ■

There are four possibilities for equations with independent variable x and dependent variable y.

1. *The dependent variable y is defined explicitly as a function of x.* For example, $y = 3x - 7$.

2. *The equation can be solved uniquely for y as a single explicitly defined function of x.* For example, the linear equation $3x + y = 2$ can be rewritten as $y = -3x + 2$, and the quadratic equation $3x^2 - x + y = 0$ can be rewritten as $y = -3x^2 + x$.

3. *The equation cannot be solved for a single explicitly defined function of x, but it can be solved for several different functions y that make the equation true.* For example, the equation $x = y^2$ has as its graph a parabola that opens to the right. This parabola is not the graph of a function, since it fails the vertical line test. However, the parabola equation can be solved for the functions $y = \sqrt{x}$ and $y = -\sqrt{x}$.

4. *The equation defines y implicitly as a function of x, but it is impossible to solve the equation for y using standard algebraic techniques.* In this instance, we will still assume that $y = f(x)$ for some (unknown) function f. The equation $x + y^3 = \sin(xy) - xy^4$ is such an example.

Earlier we noted that the domain for a function of x is the set of all real numbers that, when substituted for x in the function, produce real numbers as function values. This domain will be used for implicitly defined functions as well as explicitly defined functions.

EXAMPLE 9.2 Figure 9.1 shows the graph of the circle with equation $x^2 + y^2 = 9$.

a. Solve $x^2 + y^2 = 9$ for y to find two equations that define y explicitly as a function of x.

b. Verify that the two functions you found in part a satisfy the equation $x^2 + y^2 = 9$.

c. Graph the two functions you found in part a and give the domain and range of each.

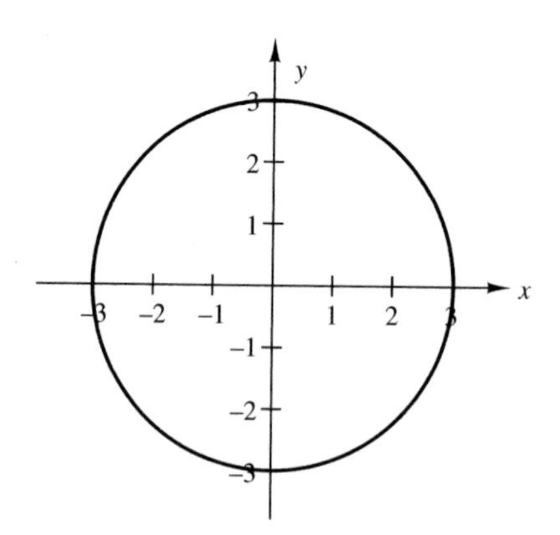

Figure 9.1 $x^2 + y^2 = 9$

Solution

a. We isolate the y^2 term in $x^2 + y^2 = 9$ to obtain $y^2 = 9 - x^2$. Taking the square root of both sides yields the two equations below (since every positive real number has two square roots). Each of these equations defines y explicitly as a function of x:

$$y = \sqrt{9 - x^2}, \quad y = -\sqrt{9 - x^2}.$$

b. To show that the functions $y=\sqrt{9-x^2}$ and $y=-\sqrt{9-x^2}$ satisfy the original equation, we must show that when each is substituted for y, the equation is true. For example, when $y=\sqrt{9-x^2}$, we have $x^2 + y^2 = x^2+\left(\sqrt{9-x^2}\right)^2 = x^2 + 9 - x^2 = 9$. A similar calculation shows that $y=-\sqrt{9-x^2}$ also satisfies the original equation.

c. The graph of $y=\sqrt{9-x^2}$ is the upper half of the circle (see Figure 9.2), and the graph of $y=-\sqrt{9-x^2}$ is the lower half of the circle (see Figure 9.3). Although the domain of each function is [–3, 3] (as can be seen from their graphs or found by solving the inequality $9 - x^2 \geq 0$), their ranges are different. The range of the first function is [0, 3]; the range of the second is [–3, 0].

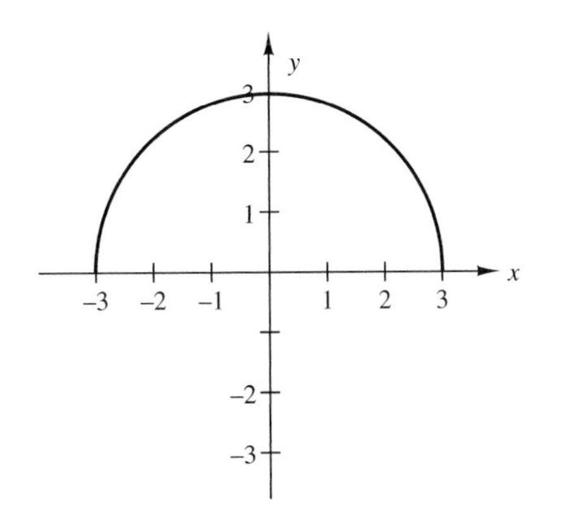

Figure 9.2 $y = \sqrt{9-x^2}$

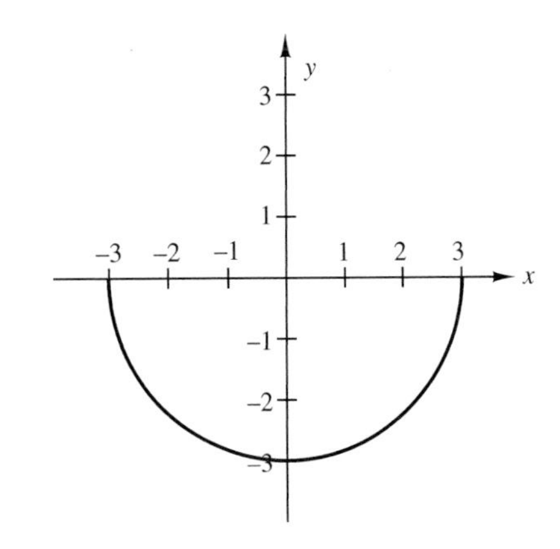

Figure 9.3 $y = -\sqrt{9-x^2}$ ■

In the following example, we solve some additional equations in x and y to find and graph their implicitly defined functions.

EXAMPLE 9.3 For each of the following equations, find one or more explicitly defined functions by solving for y. Graph each explicitly defined function and state its domain and its range.

a. $3x - 4y = 7$ **b**. $y^2 + 1 = x$ **c**. $2x + |y| - 2 = 0$

Solution

a. We recognize this as a line, since it is of the general form $ax + cy = d$. We solve for y to get the linear equation in explicit form: $y=\frac{3}{4}x-\frac{7}{4}$.

The domain and range of this function are each $(-\infty, \infty)$; its graph is in Figure 9.4.

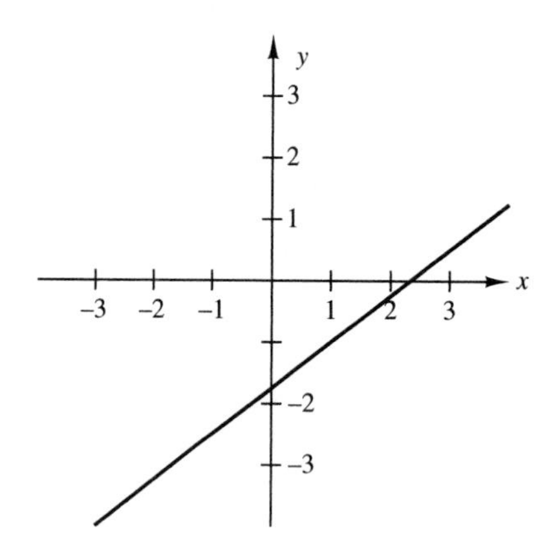

Figure 9.4 $y = \frac{3}{4}x - \frac{7}{4}$

b. We solve for y^2 and then take the square root of each side of the equation $y^2 = x - 1$ to get $y = \pm\sqrt{x-1}$. Then we write this as two separate functions:

$$y = \sqrt{x-1}$$
$$y = -\sqrt{x-1}.$$

For $\sqrt{x-1}$ to be defined, we must have $x - 1 \geq 0$, so the domain is $[1, \infty)$ for both functions. The range of the function $y = \sqrt{x-1}$ is $[0, \infty)$ (see Figure 9.5), and the range of the function $y = -\sqrt{x-1}$ is $(-\infty, 0]$ (see Figure 9.6).

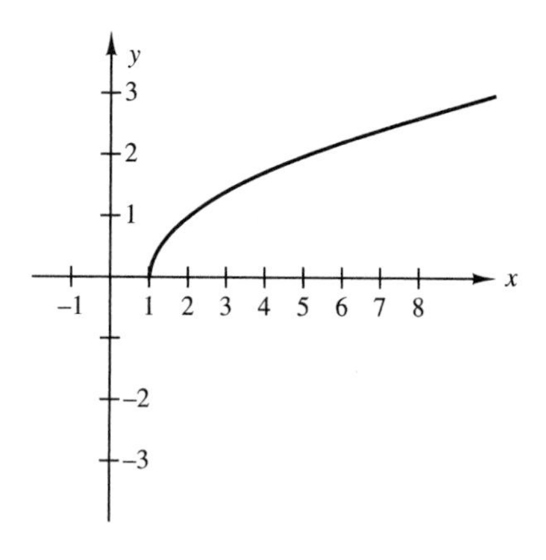

Figure 9.5 $y = \sqrt{x-1}$

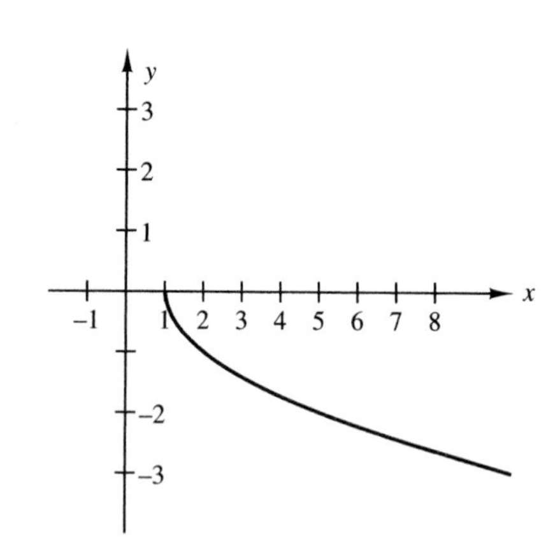

Figure 9.6 $y = -\sqrt{x-1}$

c. First, we solve $2x + |y| - 2 = 0$ for $|y|$: $|y| = -2x + 2$. Now there are two cases to consider: If $y \geq 0$, $|y| = y = -2x + 2$. If $y \leq 0$, $|y| = -y = -2x + 2$, so $y = 2x - 2$.

To find the domain of each function, we put together the facts $|y| \geq 0$ and $|y| = -2x + 2$ to conclude $-2x + 2 \geq 0$. Solve this inequality for x: $-2x + 2 \geq 0$, $-2x \geq -2$, $x \leq 1$. Thus, the domain is $(-\infty, 1]$ for each function. The graph of each of the functions defined by the original equation is a half line (a ray). The range of the first function is given by the condition $y \geq 0$, so the range is $[0, \infty)$ (see Figure 9.7). Similarly, the range of the second function is $(-\infty, 0]$ (see Figure 9.8).

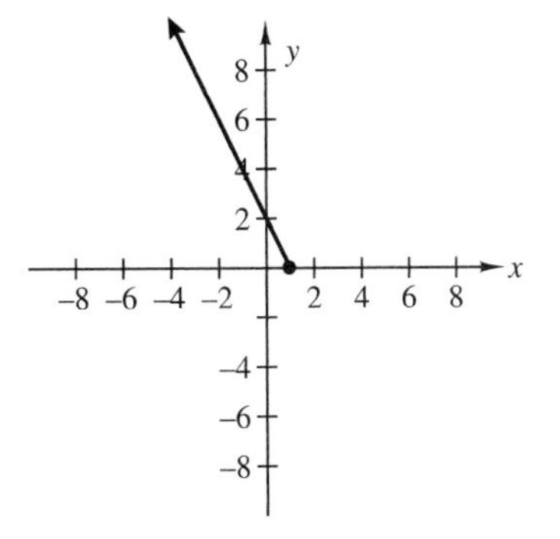

Figure 9.7 $y = -2x + 2,\ y \geq 0$

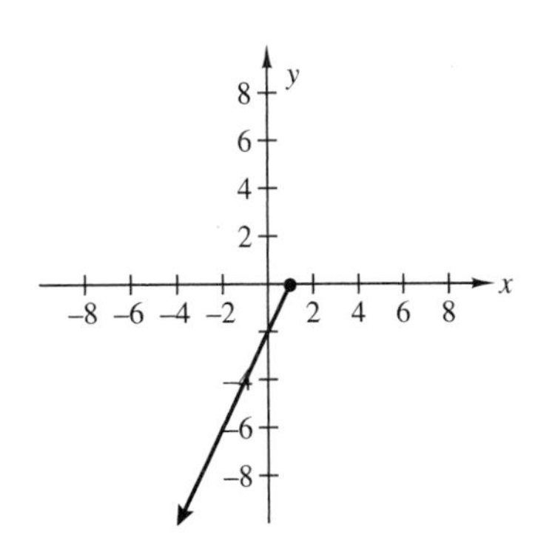

Figure 9.8 $y = 2x - 2,\ y \leq 0$ ■

When the graph of an equation in x and y is given, it is possible to use the vertical line test (see Section 2-D) to decide if it is the graph of a function. When the graph fails the vertical line test, it is possible to produce graphs of functions that are implicitly defined by the equation of the graph, each having the same domain as the original graph.

EXAMPLE 9.4 For each of the graphs in Figures 9.9 and 9.10, do the following:

a. Decide if it is the graph of a function of x.

b. If the graph is not the graph of a function, sketch graphs of two functions that are implicitly defined by the equation of the graph and have the same domain as the original graph.

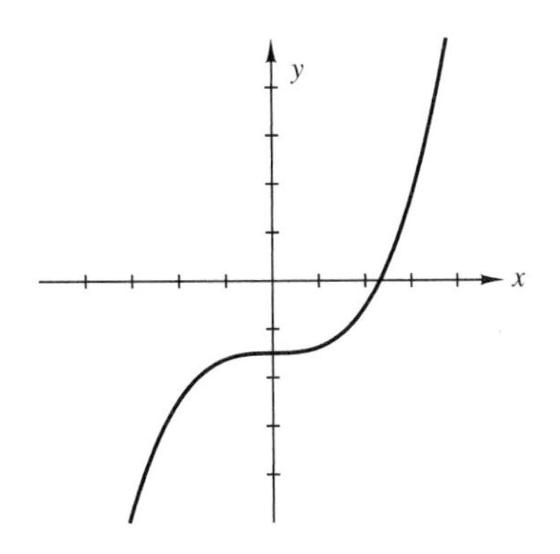

Figure 9.9

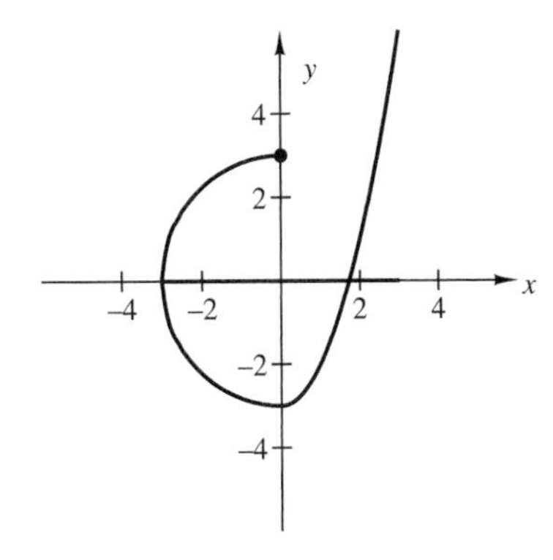

Figure 9.10

Solution

a. The graph in Figure 9.9 is the graph of a function of x since it passes the vertical line test. The graph in Figure 9.10 is not the graph of a function since it fails the vertical line test for the interval $[-3, 0]$.

b. We can split the portion of the graph in Figure 9.10 for the interval $[-3, 0]$ into two pieces, each of which passes the vertical line test, and keep the remaining part of the graph the same. This will produce graphs of two functions (Figures 9.11 and 9.12), each of which has the same domain as the original graph; in addition, each of these is the graph of a function implicitly defined by the equation of the original graph.

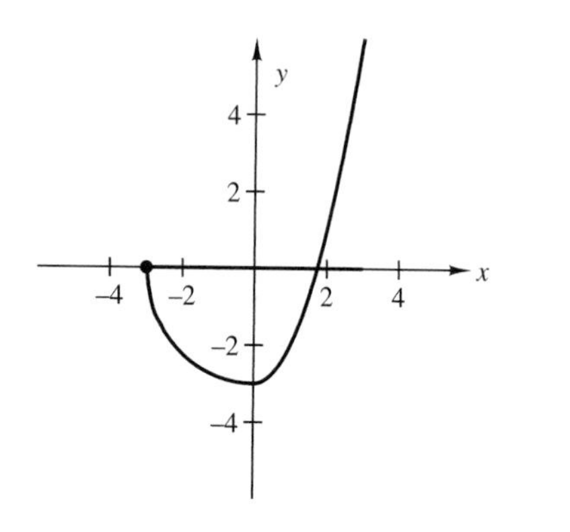

Figure 9.11

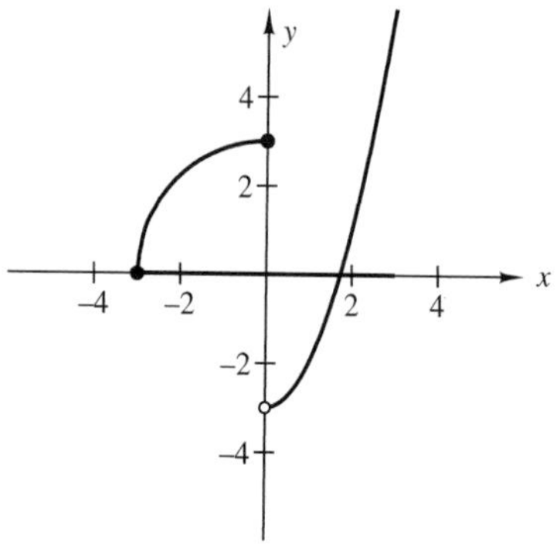

Figure 9.12 ■

Exercises 9-A

1. In each equation below, assume that x is the independent variable and y is the dependent variable. Which of the equations define y implicitly as a function of x?

a. $(x - y)(x + y) = 1$ **b.** $x^2 + y^2 - 4x + 6y - 3 = 0$

c. $x^2 - 4x + 1 = y$ **d.** $xy = -4$

e. $y = 2x - 3y + 1$ **f.** $x^4y + 2x = 1$

g. $x = 2y^3 - 4y + 3$ **h.** $\frac{(x-3)^2}{4} + \frac{(y+1)^2}{9} = 1$

2. For each equation below, find one or more explicitly defined functions $y = f(x)$ with largest possible domains that satisfy the equation. Give the domain of each function. (See Section 4-C to review finding domains.)

a. $x^2 - y^2 = 1$ **b.** $(x - 2)^2 + (y + 3)^2 = 16$ **c.** $xy = -4$

d. $y = 2x - 3y + 1$ **e.** $x = -2(y + 1)^2 + 5$ **f.** $|y + 2| = 3x$

g. $\frac{(x-3)^2}{4} + \frac{(y+1)^2}{9} = 1$

3. For each equation in Exercise 2, use a graphing utility to graph the functions $y = f(x)$ that you found that satisfy the equation.

4. For each of the graphs of equations in Figures 9.13 through 9.16, decide if it is the graph of a function. When it is not the graph of a function, sketch graphs of two functions implicitly defined by the equation of the graph that have the same domain as the original equation.

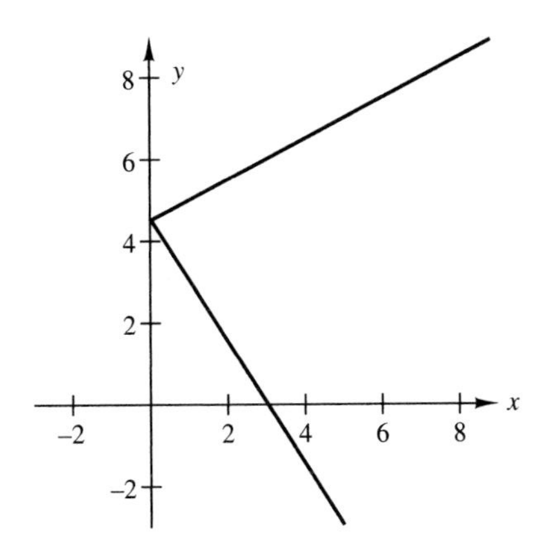

Figure 9.13

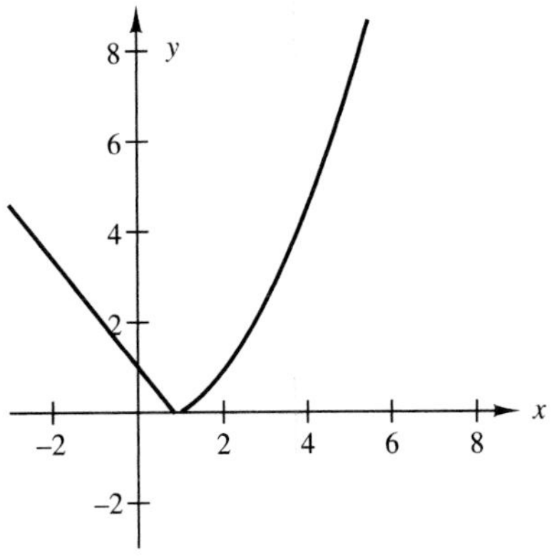

Figure 9.14

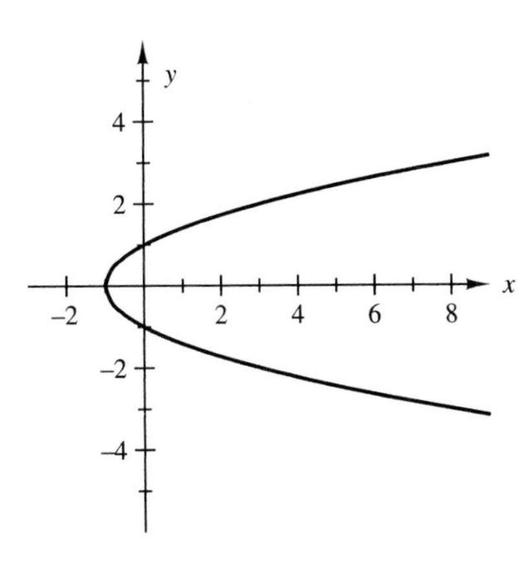

Figure 9.15

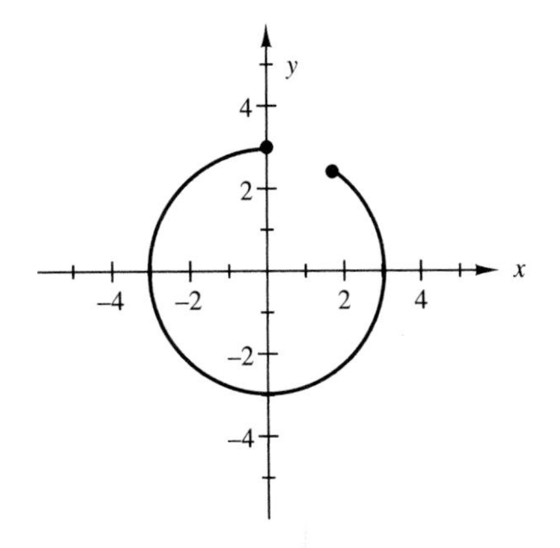

Figure 9.16

9-B SOLVING EQUATIONS THAT CONTAIN $\frac{dy}{dx}$

When y is defined implicitly as a function of x by an equation in x and y, it may be necessary or desirable to find its derivative implicitly. If we solve the implicitly differentiated equation for the derivative $\frac{dy}{dx}$, we have an expression for the derivative in terms of x and y. This expression can then be used to solve problems. For example, if the derivative is interpreted as the slope of the tangent line at the point (x, y), we can find that slope by substituting x and y into the formula for $\frac{dy}{dx}$.

EXAMPLE 9.5 When the equation of the circle $x^2 + y^2 = 25$ is implicitly differentiated, the resulting equation is $2x + 2y\,\frac{dy}{dx} = 0$.

a. Solve this equation for $\frac{dy}{dx}$.

b. Find the slope of the tangent line to the circle at the point (3, 4), and draw this line on a graph of the circle.

c. Look at the graph of the circle. At what points will a tangent line be vertical? Verify your claim using $\frac{dy}{dx}$ found in part a.

d. Use the equation for $\frac{dy}{dx}$ to determine at what points the tangent line will be horizontal.

Solution

a. To solve $2x + 2y\,\frac{dy}{dx} = 0$ for $\frac{dy}{dx}$, treat $\frac{dy}{dx}$ as a single variable that must be isolated. Add $-2x$ to both sides to obtain $2y\,\frac{dy}{dx} = -2x$; then divide both sides by $2y$. (This step is not possible if $y = 0$.) Thus, $\frac{dy}{dx} = \frac{-2x}{2y} = \frac{-x}{y} = -\frac{x}{y}$.

b. To find $\frac{dy}{dx}$ at the point (3, 4) on the graph, we substitute $x = 3$ and $y = 4$ into the equation for $\frac{dy}{dx}$ and find $\frac{dy}{dx} = -\frac{3}{4}$. Thus, the slope of the tangent line at (3, 4) is $-\frac{3}{4}$. This is illustrated in Figure 9.17a.

c. Figure 9.17b shows that the tangent lines to the circle at the points (–5, 0) and (5, 0) will be vertical. The slope of a vertical line is undefined. This agrees with our calculation of $\frac{dy}{dx}$, since at (–5, 0) and (5, 0), $\frac{dy}{dx} = -\frac{x}{y}$ is undefined (we cannot divide by 0).

d. The tangent line is horizontal when $\frac{dy}{dx} = 0$. Since $\frac{dy}{dx} = -\frac{x}{y}$, $\frac{dy}{dx} = 0$ when $x = 0$, which occurs when $y = 5$ or $y = -5$. See Figure 9.17b.

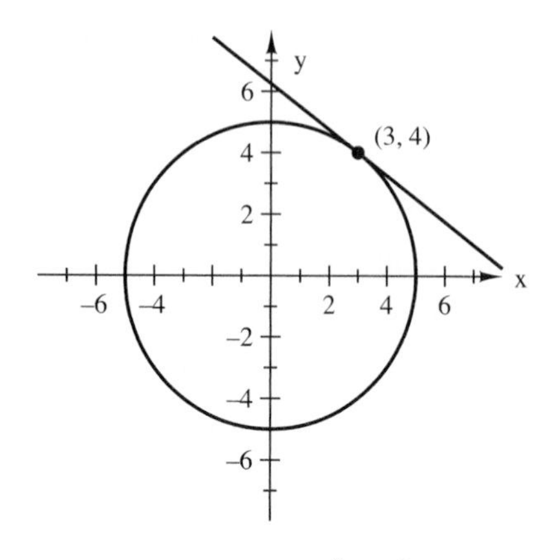

Figure 9.17a $x^2 + y^2 = 25$

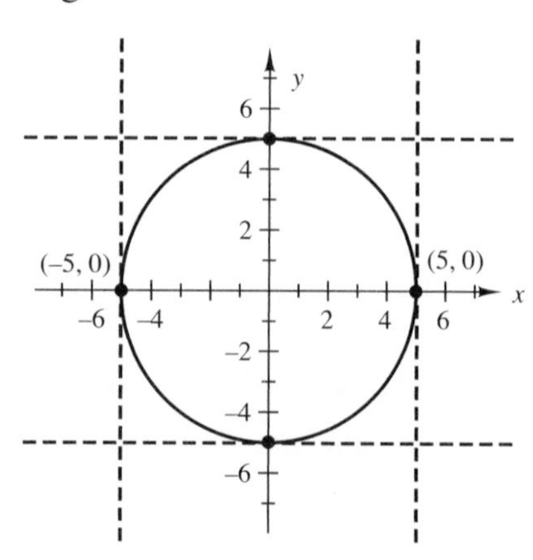

Figure 9.17b $x^2 + y^2 = 25$ ■

The next example gives another illustration of how to solve a linear equation for the derivative of an implicitly defined function y. This derivative is often denoted by y' instead of $\frac{dy}{dx}$. The most important step in the solution is to put all terms containing $\frac{dy}{dx}$ (or y') as a factor on one side and all other terms on the other side. The equations created by implicit differentiation of an equation in x and y have three variables: x, y, and $\frac{dy}{dx}$ (or y'). Since $\frac{dy}{dx}$ (or y') always has an exponent of 1 in these equations, we can solve for $\frac{dy}{dx}$ in the same manner as we solve any linear equation for one of its variables.

EXAMPLE 9.6 Implicit differentiation of the equation $\frac{x}{y} - 2 = \frac{y^2}{3} + 2x$ gives the equation

$$\frac{1}{y} - \frac{x}{y^2}y' = \frac{2y}{3}y' + 2.$$

Solve this equation for y'.

Solution Since we want to solve for y', we isolate all the terms containing y' on one side of the equation: $\frac{1}{y} - 2 = \frac{x}{y^2}y' + \frac{2y}{3}y'$. We then factor out the common factor y': $\frac{1}{y} - 2 = \left(\frac{x}{y^2} + \frac{2y}{3}\right)y'$.

To avoid complex fractions, we multiply each side by the least common denominator, $3y^2$, and simplify:

$$3y^2\left(\frac{1}{y} - 2\right) = 3y^2\left(\frac{x}{y^2} + \frac{2y}{3}\right)y'$$

$$3y - 6y^2 = \left(3x + 2y^3\right)y'.$$

Finally, we divide both sides by $3x + 2y^3$, the coefficient of y':

$$\frac{3y - 6y^2}{3x + 2y^3} = y'.$$

■

Exercises 9-B

1. Each of the following equations was obtained by implicit differentiation of an equation in x and y. Solve each equation for the derivative of y with respect to x (this derivative is denoted by either $\frac{dy}{dx}$ or y').

a. $x - 2y\frac{dy}{dx} = 0$

b. $3 + y' = 5y'$

c. $y\frac{dy}{dx} - 3\frac{dy}{dx} = 2x - 3$

d. $2xyy' + y^2 = \frac{1}{4} + y'$

e. $y\left(x\frac{dy}{dx} - y\right) = 1 - \frac{x}{y}\frac{dy}{dx}$

f. $\frac{x}{y}y' - 1 = 2\left(yy' - xy\right)$

2. The point (1, 2) is on the graph of each of the original equations that was implicitly differentiated to give the equations in Exercise 1. For each equation in Exercise 1, find the slope of the tangent line to the graph of the original equation at the point (1, 2).

3. For each equation in Exercise 1, use your answers from Exercise 2 to find the equation of the tangent line to the graph of the original equation at the point (1, 2).

Chapter 9 Exercises

1. In each of the equations below, assume x is the independent variable and y is the dependent variable. For each, indicate whether the equation defines y implicitly or explicitly as a function of x.

a. $2x^2 - 3y^2 + 3x - 4y - 5 = 0$

b. $y = (x - 5)(3 - 4x)$

c. $y(x - 3) = 4$

d. $y = (3 - 4y)(x - 4)$

e. $x = -4y^2 - y - 7$

f. $(x + 4)^2 - y^2 = 9$

2. In each equation below, solve explicitly for y to obtain one or more functions of x. State the domain and the range for each function you find.

a. $(x + 4)^2 + y^2 = 9$

b. $x^3y = 2y + 3$

c. $(x - 3)(y + 4) = 7$

d. $\frac{(x+1)^2}{4} + \frac{y^2}{16} = 1$

e. $x = -(y - 1)^2 + 4$

3. For the equations in Exercise 2, use a graphing utility to make a separate graph of each of the explicitly defined functions y that you found.

4. Each of the following equations was obtained by implicit differentiation of an equation in x and y. Solve each equation for $\frac{dy}{dx}$ or y'.

a. $y\frac{dy}{dx} - 2x = 3$

b. $x\frac{dy}{dx} - 3 = 2x\frac{dy}{dx}$

c. $x(3 + y') = y'$

d. $\left(\frac{x}{y}+\frac{1}{y^3}\right)\frac{dy}{dx}=\frac{1-y}{2y-1}-\frac{1}{y^3}\frac{dy}{dx}$ **e.** $\frac{y^3}{y-1}(2xy'+1)=x(1+y)y'+y^3$

5. The point (2, −1) is on the graph of each of the original equations that was implicitly differentiated to obtain the equations in Exercise 4. For each equation in Exercise 4, find the slope of the tangent line to the graph of the original equation at the point (2, −1).

6. For each equation in Exercise 4, find the equation of the tangent line to the graph of the original equation at the point (2, −1).

7. For each of the graphs of equations in x and y shown in Figures 9.18 through 9.21, sketch graphs of two or more functions with the same domain that are implicitly defined by the equations of the graphs.

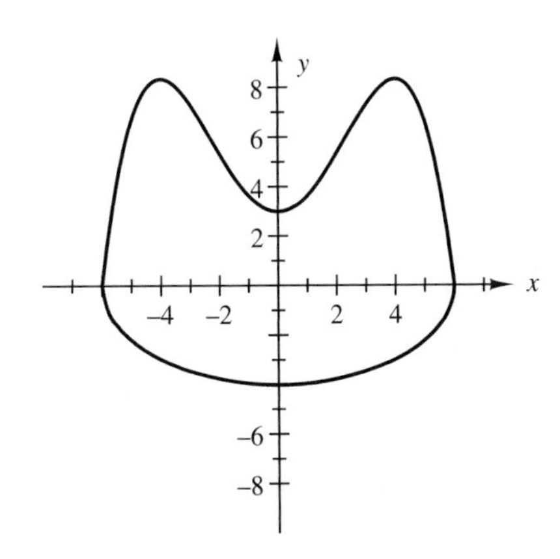

Figure 9.18

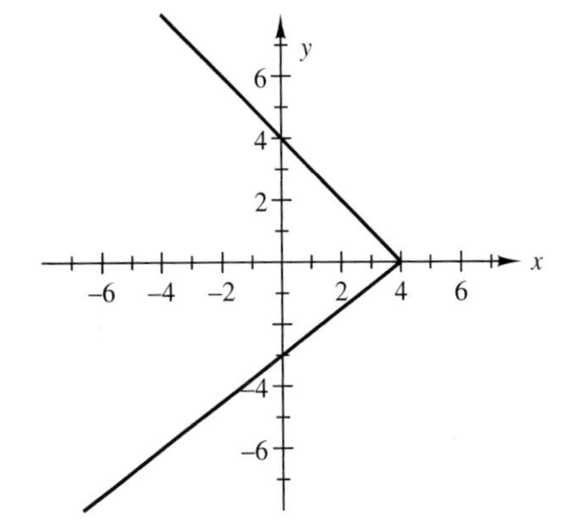

Figure 9.19

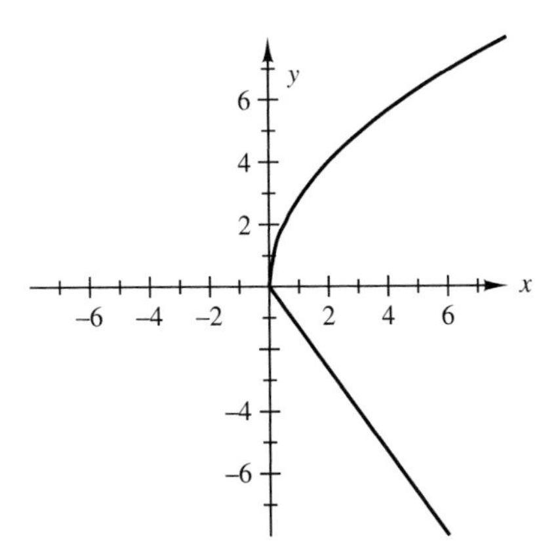

Figure 9.20

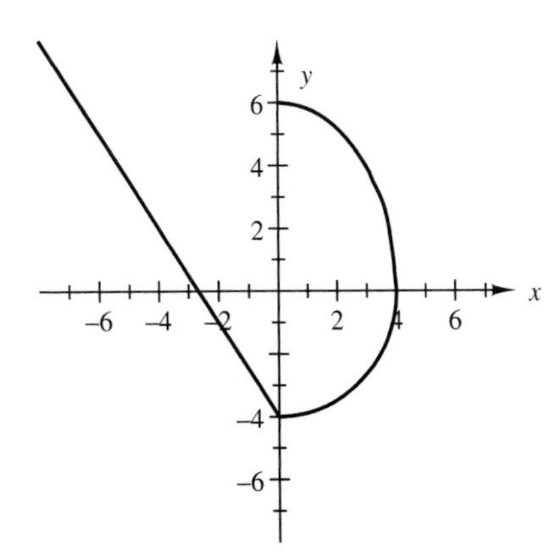

Figure 9.21

CHAPTER 10

Companion to Repeated Differentiation

10-A ITERATION AND PATTERNS IN HIGHER DERIVATIVES

The word "iteration" comes from the Latin word *itero*, which means to perform again or repeat. In mathematics it is used when referring to a process or procedure that is carried out more than once. In Section 4-B.2 we discussed the Bisection Method, which is an iterative procedure used to approximate the zeros of a function. We will now apply iteration to the process of differentiation.

Since the derivative f' of a function f is itself a function, it makes sense to look at the derivative of the function f' (where $f'(x)$ exists). If we apply the definition of derivative to the function $f'(x)$, we obtain the derivative of $f'(x)$, called the *second derivative* of $f(x)$, denoted by $f''(x)$. The second derivative of a function $y = f(x)$ is also denoted by $f^{(2)}(x)$, $\frac{d^2y}{dx^2}$, y'', or $D_x^2 y$. In these notations, the "2" or $''$ indicates that y is twice differentiated as a function of x. For many common functions, the second derivative can be computed by applying rules of differentiation to the function $f'(x)$.

Because the process of differentiation can be carried out on any function, we can look for the derivative of $f''(x)$; this is the *third derivative* of f and is denoted by $f'''(x)$ or $f^{(3)}(x)$. Also we can differentiate the function $f'''(x)$ to find the fourth derivative of $f(x)$, denoted $f^{(4)}(x)$, and so on. (Since the additional primes in the derivative notation become cumbersome and difficult to count, Arabic numbers in parentheses are used to denote higher derivatives.) As long as the derivatives exist, the process of differentiation can be repeated as many times as desired to yield higher-order derivatives of a function, or simply, higher derivatives.

Higher derivatives of polynomial functions have a predictable pattern.

EXAMPLE 10.1 Apply the rules of differentiation to the given function, its derivative, its second derivative, its third derivative, its fourth derivative, and its fifth derivative to find the first six derivatives of $f(x) = 7x^4 - 3x^2 + 5x + \pi$.

Solution The function $f(x)$ is a polynomial function and $f'(x) = 28x^3 - 6x + 5$. The function $f'(x)$ is also a polynomial; its derivative is $f''(x) = 84x^2 - 6$. Similarly, $f'''(x) = 168x$ and $f^{(4)}(x) = 168$. Since $f^{(4)}(x)$ is a constant, its derivative is 0, so $f^{(5)}(x) = 0$ and $f^{(6)}(x) = 0$. In fact, all derivatives higher than the fourth derivative will be zero. ■

Example 10.1 illustrates patterns that hold for higher derivatives of all polynomial functions. First, we observe that every derivative of a polynomial function is a polynomial. Also, the degree of the polynomial is reduced by 1 with each successive higher derivative. In Example 10.1, f has degree 4, f' has degree 3, f'' has degree 2, f''' has degree 1, and $f^{(4)}$ has degree 0; it is a constant function. Since $f^{(4)}$ is a constant, $f^{(5)}$ and all derivatives higher than the fifth will be zero. The same patterns will hold for any polynomial function. If a polynomial function g has degree 57, then $g^{(57)}$ will be a constant, and all derivatives higher than the 57th will be zero.

The next example illustrates a different pattern in higher derivatives.

EXAMPLE 10.2 Find $f^{(10)}(x)$ for $f(x) = (2x + 1)^{10}$.

Solution $f(x) = (2x + 1)^{10}$, so by the Chain Rule, $f'(x) = 10(2x + 1)^9 \cdot 2$. Then:

$$f''(x) = 9 \cdot 10(2x + 1)^8 \cdot 2^2$$

$$f^{(3)}(x) = 8 \cdot 9 \cdot 10(2x + 1)^7 \cdot 2^3$$

$$f^{(4)}(x) = 7 \cdot 8 \cdot 9 \cdot 10(2x + 1)^6 \cdot 2^4$$

$$f^{(5)}(x) = 6 \cdot 7 \cdot 8 \cdot 9 \cdot 10(2x + 1)^5 \cdot 2^5$$

$$f^{(6)}(x) = 5 \cdot 6 \cdot 7 \cdot 8 \cdot 9 \cdot 10(2x + 1)^4 \cdot 2^6$$

$$f^{(7)}(x) = 4 \cdot 5 \cdot 6 \cdot 7 \cdot 8 \cdot 9 \cdot 10(2x + 1)^3 \cdot 2^7$$

$$f^{(8)}(x) = 3 \cdot 4 \cdot 5 \cdot 6 \cdot 7 \cdot 8 \cdot 9 \cdot 10(2x + 1)^2 \cdot 2^8$$

$$f^{(9)}(x) = 2 \cdot 3 \cdot 4 \cdot 5 \cdot 6 \cdot 7 \cdot 8 \cdot 9 \cdot 10(2x + 1) \cdot 2^9$$

$$f^{(10)}(x) = 1 \cdot 2 \cdot 3 \cdot 4 \cdot 5 \cdot 6 \cdot 7 \cdot 8 \cdot 9 \cdot 10 \cdot 2^{10}$$ ■

The product $1 \cdot 2 \cdot 3 \cdot 4 \cdot 5 \cdot 6 \cdot 7 \cdot 8 \cdot 9 \cdot 10$ that appears in $f^{(10)}(x)$ in Example 10.2 could be multiplied out to yield 3,628,800, but we can sometimes observe a pattern by keeping the product in factored form. The product of successive integers from 1 to 10 is called a *factorial* and denoted $10! = 1 \cdot 2 \cdot 3 \cdot 4 \cdot 5 \cdot 6 \cdot 7 \cdot 8 \cdot 9 \cdot 10$. The expression 10! is read as "ten factorial." Using this notation, we can write $f^{(10)}(x) = 10! \cdot 2^{10}$.

The general definition of $n!$ (read "n factorial") for any positive integer n is:

$$n! = 1 \cdot 2 \cdot 3 \cdot 4 \cdots (n-1) \cdot n = n \cdot (n-1) \cdots 3 \cdot 2 \cdot 1.$$

From the expression for $n!$, we see that $n!$ can also be written as $n! = n(n-1)!$ (That is, n factorial can be written as the product of n and $n-1$ factorial.) We define $0! = 1$ so that $n!$ is defined for all nonnegative integers n. Factorial notation simplifies many expressions and equations.

EXAMPLE 10.3 Find a general expression for the nth derivative of $g(x) = \frac{1}{x}$, for n any positive integer.

Solution We use the power rule to find the first six derivatives of $g(x) = \frac{1}{x} = x^{-1}$.

$$g'(x) = -x^{-2} = -1!\, x^{-2}$$

$$g''(x) = +2x^{-3} = +2!\, x^{-3}$$

$$g^{(3)}(x) = -3 \cdot 2x^{-4} = -3!\, x^{-4}$$

$$g^{(4)}(x) = +4 \cdot 3 \cdot 2x^{-5} = +4!\, x^{-5}$$

$$g^{(5)}(x) = -5 \cdot 4 \cdot 3 \cdot 2x^{-6} = -5!\, x^{-6}$$

$$g^{(6)}(x) = +6 \cdot 5 \cdot 4 \cdot 3 \cdot 2x^{-7} = +6!\, x^{-7}$$

Since we want to identify a general pattern for the nth derivative of g, we leave the products in each derivative in factored form and use factorial notation. We observe the following patterns in our list of higher derivatives for g:

1. Each derivative has an opposite sign from the previous derivative. The first derivative has a – sign, the second derivative has a + sign, the third derivative has a – sign, and so on. We note that the nth derivatives with n even have a + sign and the nth derivatives with n odd have a – sign. Since $(-1)^p = 1$ if p is even and $(-1)^p = -1$ if p is odd, we can represent the sign of the nth derivative as $(-1)^n$.

2. If we ignore the signs of the derivatives, the first derivative has coefficient 1!, the second derivative has coefficient 2!, the third derivative has coefficient 3!, and so on. In general, if we ignore the sign, the nth derivative will have coefficient $n!$

3. The power of x in the first derivative is -2, in the second derivative is -3, in the third derivative is -4, and so on. Since the power of x is reduced by 1 for each derivative, this pattern will continue and the power of x in the nth derivative will be $-(n+1)$.

We combine the three observations to write a general expression for the nth derivative of g: $g^{(n)}(x) = (-1)^n\, n!\, x^{-(n+1)}$. In Exercise 4 below, you are asked to verify this formula for $n = 1, 2, 3, 4, 5, 6$. ■

Exercises 10-A

1. Find the eighth derivative of each of the following functions:

a. $f(x) = x^8 - 17x^3 + 6$ **b**. $g(x) = 49x^6 - 86x^5 + 32x^4 - 27x^3 + 31x^2 + 29x - 6$

c. $h(x) = 100x^9$ **d**. $k(x) = (3x - 3)^{10}$

2. Find the first five derivatives (the first, second, third, fourth, and fifth) of each of the following functions:

a. $F(x) = \sqrt{2}x^{10} - 86x$ **b**. $H(x) = \frac{2}{x}$ **c**. $K(x) = x^8 + x^7$

3. For each function in Exercise 2, decide if any of its higher derivatives is zero. In this case, give all the nonzero higher derivatives and the first value of n such that the nth derivative is zero.

4. Verify the formula $g^{(n)}(x) = (-1)^n n! x^{-(n+1)}$ in Example 10.3 for $n = 1, 2, 3, 4, 5, 6$.

5. Find a general expression for the nth derivative of the function $H(x) = \frac{2}{x}$.

6. Recall that the derivative of $\sin x$ is $\cos x$, and the derivative of $\cos x$ is $-\sin x$.

a. Find the first five derivatives of the function $G(x) = \sin x$.

b. Look for a pattern to describe the nth derivative of G for every positive integer n.

c. Use the pattern found in part b to give the 20th and the 102nd derivatives of G.

7. Let $f(x) = \sqrt{2x}$.

a. Find $f^{(6)}(x)$.

b. For each of the following, give a value or explain why it does not exist:

(i) $f^{(6)}(2)$ (ii) $f^{(6)}(0)$

10-B RATE OF CHANGE OF RATE OF CHANGE

The derivative $f'(x)$ of a function $f(x)$ is the (instantaneous) rate of change of $f(x)$ with respect to x. Since f'' is the derivative of f', the second derivative of a function $f(x)$ is interpreted as the (instantaneous) rate of change of $f'(x)$ with respect to x. Thus the second derivative is a *rate of change of a rate of change*. For example, suppose $f(x)$ represents the number of births in a particular city at time x, where x is the number of years since 1960; that is, $x = 0$ represents 1960, $x = 1$ represents 1961, and so on. Then $f'(x)$ is the instantaneous rate of change of $f(x)$ with respect to x and represents birthrate (births per year) at time x. The second derivative, $f''(x)$, is the instantaneous rate of change of $f'(x)$ with respect to x and represents the *rate of change of the birthrate* (births per year, per year). The first derivative tells us whether the number of births is increasing or decreasing; the second derivative tells us whether the **rate** of births is increasing or decreasing. In general, the units of $f'(x)$ are units of f per unit of x; the units of $f''(x)$ are units of f per unit of x, per unit of x, which is sometimes expressed as units of f per unit of x^2.

Since $f'(x)$ represents the slope of the tangent line to the graph of $y = f(x)$, the derivative of $f'(x)$ is positive where $f'(x)$ is increasing. To determine whether f' is increasing on an interval, we sketch some tangent lines to the graph of $y = f(x)$ and see if the slopes are increasing as x increases. Figure 10.1 shows an example of a function defined on the interval [0, 5] with three tangent lines sketched in that interval. Note the slopes of the tangent lines. Since the slopes are increasing as x increases from 0 to 5, f' is increasing, so f'' is positive in the interval (0, 5].

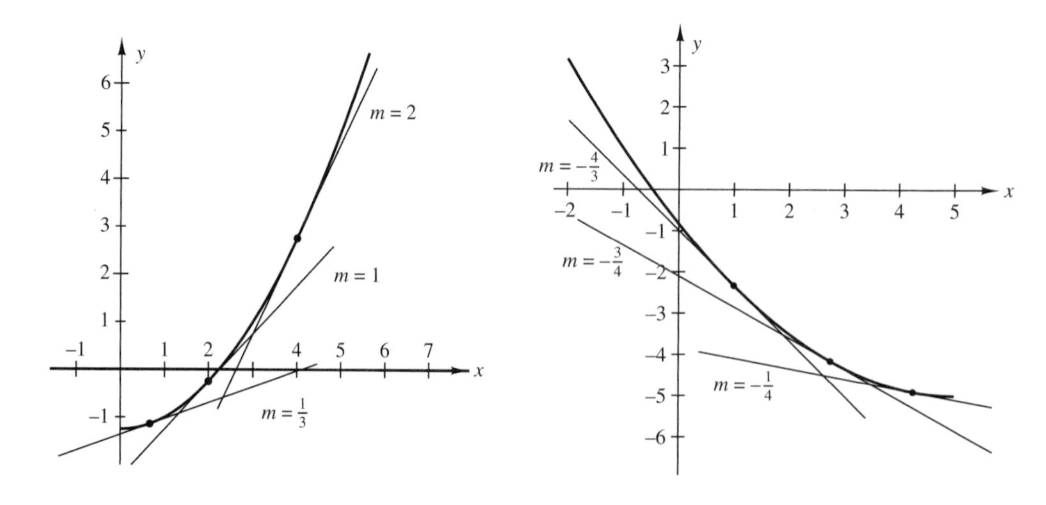

Figure 10.1 $y = f(x),\ f'(x) > 0,\ f''(x) > 0$ **Figure 10.2** $y = g(x),\ g'(x) < 0,\ g''(x) > 0$

The graph in Figure 10.2 shows a function g with the property $g'(x) < 0$ for x in $[-1, 4.5]$. This means that g is decreasing on that interval. However, g' is increasing, since the slopes of tangent lines are increasing in the interval $[-1, 4.5]$; thus $g''(x) > 0$ for x in $[-1, 4.5]$.

In Figures 10.1 and 10.2, each tangent line to the graph of the curve is below the curve except at its point of tangency. Such curves are said to be *concave up*.

The graph in Figure 10.3 illustrates a function h that is increasing on the interval $(-1, 4.5]$ and for which h' is decreasing. Thus $h''(x) < 0$ on the interval $(-1, 4.5]$.

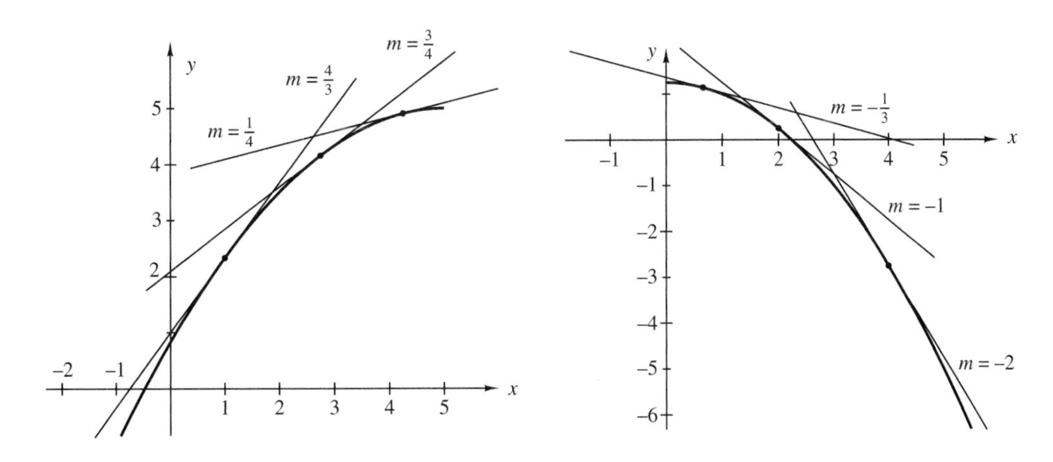

Figure 10.3 $y = h(x), h'(x) > 0, h''(x) < 0$ **Figure 10.4** $y = k(x), k'(x) < 0, k''(x) < 0$

The function k illustrated in Figure 10.4 is decreasing on the interval $(0, 5]$. Since the slopes of tangent lines on $(0, 5]$ are also decreasing, k' is decreasing. Thus $k''(x) < 0$ for x in $(0, 5]$.

In Figures 10.3 and 10.4, each tangent line to the graph of the curve is above the curve except at its point of tangency. Such curves are said to be *concave down*.

The sign of f' on an interval does not predict the sign of f'' on that interval. For example, a function can be increasing and yet have its rate of increase decreasing, as in Figure 10.3.

Looking at concavity and second derivatives can give information about how a rate of change is changing, as Examples 10.4, 10.5, and 10.6 illustrate.

EXAMPLE 10.4 Use the graph of the function $y = f(x)$ given in Figure 10.5 to find where the derivative function f' is increasing and where f' is decreasing.

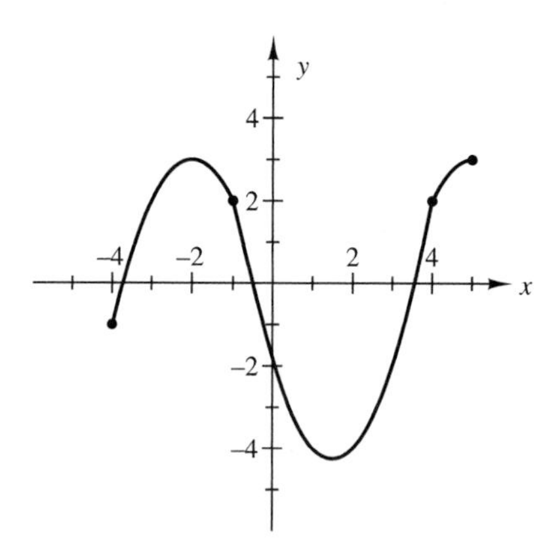

Figure 10.5 $y = f(x)$

Solution Several tangent lines should be sketched on the graph of $y = f(x)$; the behavior of the slopes of these lines can then be observed. We need to identify intervals where slopes of tangent lines are increasing to find where f' is increasing; we identify intervals where slopes of tangent lines are decreasing to find where f' is decreasing. For x in the interval $[-4, -1]$, slopes of tangent lines are decreasing from positive, to 0 (at $x = -2$), to negative. In the interval $[-1, 4]$, slopes of tangent lines are increasing from negative, to 0 $\left(\text{at } x = \frac{3}{2}\right)$, to positive. For x between 4 and 5, slopes of tangent lines are positive and decreasing. Thus $f'(x)$ is increasing for x in $[-1, 4]$ and $f'(x)$ is decreasing for x in $[-4, -1]$, and for x in $[4, 5]$. ■

EXAMPLE 10.5 Each winter at least one variety of influenza spreads throughout our community. Public health officials track the spread of the disease. Let $f(t)$ represent the number of new cases reported on day t. We will pick November 1 (somewhat arbitrarily) as day 0 of the influenza season. The graph of $y = f(t)$ is given in Figure 10.6. Describe the pattern of the disease's spread.

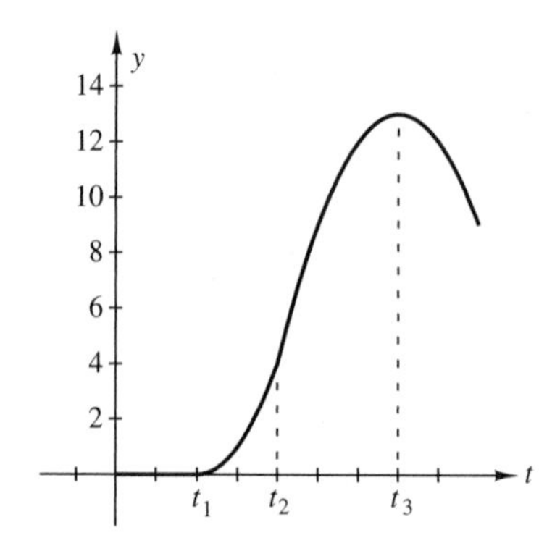

Figure 10.6 $y = f(t)$

Solution Three key days—day t_1, day t_2, and day t_3—are identified on the graph of $y = f(t)$. From day $t = 0$ to day $t = t_1$, there are no identified cases of influenza. On the interval $t = t_1$ to $t = t_2$, the disease starts to spread, slowly at first but then more rapidly. In that interval, the number of people infected is increasing and the rate of increase is also increasing. This can be seen by estimating slopes of tangent lines at points in the interval and observing that the slopes of the tangent lines are positive and getting steeper as the points move to the right within the interval $[t_1, t_2]$. These slopes are the values of the first derivatives $f'(t)$, and in this interval, $f''(t)$ measures the rate of change of $f'(t)$. Since the slopes $f'(t)$ are increasing in value, $f''(t)$ will be positive. On the interval $t = t_2$ to $t = t_3$ the number of new cases reported is still increasing, but the rate of that increase is decreasing. (Look at the slopes of tangent lines at points in this interval to verify this assertion.) Thus after time $t = t_2$ the community can begin to be somewhat hopeful. At time $t = t_3$ the number of new people infected is a maximum, and after time $t = t_3$ the number of new cases decreases each day. ■

EXAMPLE 10.6 A certain electronic toy cost \$85 on December 1, specially priced for the holiday rush. Assume that t months after December 1, for the next 6 months, the price is increasing and will be $P(t)$ dollars, where $P(t) = 85 + t^2 - \frac{1}{9}t^3, 0 \le t \le 6$, t in months. This function models the price for the domain [0, 6].

a. Determine what the price will be on March 1; on May 1.

b. Use a graphing utility to graph the function P.

c. Verify that the price is increasing over the interval $0 \le t \le 6$.

d. For what values of t is the rate of price increase increasing?

e. Find $P'(2)$ and $P''(2)$ and explain what they mean.

f. Find $P'(4)$ and $P''(4)$ and explain what they mean.

Solution

a. March 1 corresponds to $t = 3$, and $P(3)$ is the price on March 1. $P(3) = 85 + 3^2 - \frac{1}{9}3^3 = \$91$. Similarly, May 1 corresponds to $t = 5$ and $P(5) = \$96.11$.

b. See Figure 10.7 for the graph of $y = P(t)$.

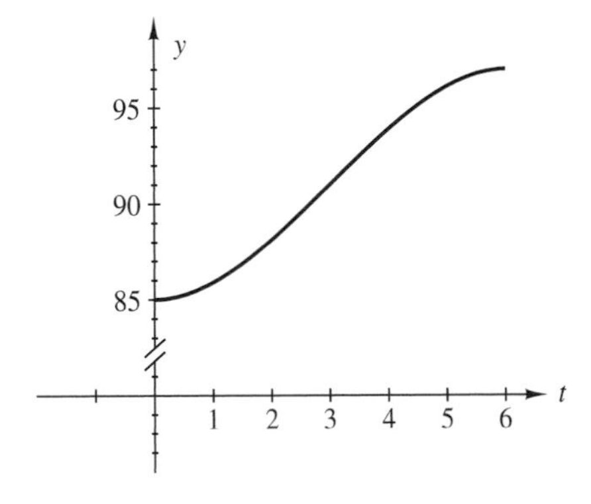

Figure 10.7 $y = P(t)$

c. Although the graph appears to be increasing throughout the interval [0, 6], we use the derivative to confirm this. The price is increasing when the rate of change of price is positive; this occurs when $P'(t) > 0$. $P'(t) = 2t - \frac{1}{3}t^2 = t\left(2-\frac{1}{3}t\right)$. The number lines below give the signs of the two factors t and $\left(2-\frac{1}{3}t\right)$ and of $P'(t)$. $P'(t) > 0$ when both factors have the same sign; thus $P'(t) > 0$ for $0 < t < 6$.

−	−	−	0	+	+	+	+	+	+	+	sign of t
+	+	+	+	+	+	+	0	−	−	−	sign of $2-\frac{1}{3}t$
−	−	−	0	+	+	+	0	−	−	−	sign of $P'(t)=t\left(2-\frac{1}{3}t\right)$
			0				6				

d. The rate of price increase is P', and this function is increasing when $\frac{d(P'(t))}{dt} = P''(t) > 0$. $P''(t) = 2-\frac{2}{3}t$, and $2-\frac{2}{3}t > 0$ when $2 > \frac{2}{3}t$. This occurs when $t < 3$. Thus for $0 < t < 3$, the rate of price increase is increasing. Note also that for $3 < t < 6$, the rate of price increase is decreasing. This means that the rate of price increase, which is the *rate of inflation* of the price of the toy, is decreasing for $3 < t < 6$, even though the price is still increasing in that time interval. We can confirm this behavior by observing the shape of the graph, and noting how the concavity changes at $t = 3$.

e. $P'(2) = 2(2) - \frac{1}{3}(2)^2 \approx 2.67$ and $P''(2) = 2 - \frac{2}{3}(2) \approx 0.67$. This means that when $t = 2$, on February 1, the price is increasing at a rate of \$2.67 per month and the price increase is increasing at the rate of \$.67 a month per month.

f. $P'(4) \approx 2.67$ and $P''(4) \approx -.67$. When $t = 4$, on April Fool's Day, the price is increasing at a rate of \$2.67 per month, but the price increase is decreasing at the rate of \$.67 a month per month. ■

The second derivative has a very different interpretation when we consider the motion of an object in a straight line. Suppose that the position y of the object at time t (relative to some fixed point on the line) is given by $y = f(t)$. Then the (instantaneous) velocity v of the object at any time t is $v = f'(t)$. If y is given in feet and t in seconds, then $f'(t)$ is in feet per second. The *acceleration* of the object at any time t is the instantaneous rate of change of velocity with respect to time. Thus the acceleration a of the object at time t is the derivative of $f'(t)$, which is $f''(t)$. If y is given in feet and t in seconds, then the acceleration $a = f''(t)$ is in feet per second, per second, or feet per second^2.

EXAMPLE 10.7 A train connects terminal A with several other terminals at a large airport, traveling back and forth along a straight track. The graph in Figure 10.8 shows the distance $s(t)$ from the train to terminal A at time t (in minutes) for the interval $0 \le t \le 4$.

a. At what time t is the train at terminal A?

b. During what time interval(s) is the train moving away from terminal A?

c. Is the train's velocity positive or negative during the time interval you found in part b? How do you know?

d. During what time interval(s) is the train's acceleration positive? Negative?

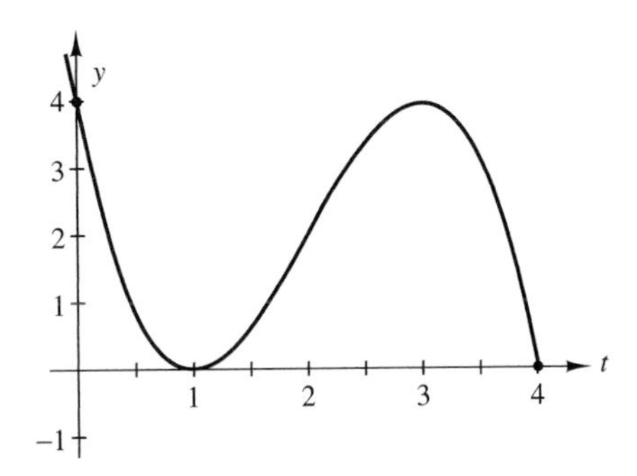

Figure 10.8 $y = s(t)$

Solution

a. The train is at terminal A when $s(t) = 0$, and that occurs when $t = 1$ and $t = 4$.

b. The train is moving away from terminal A when the distance $s(t)$ (the distance from the train to A) is increasing. This occurs in the interval $1 < t < 3$.

c. The train's velocity is positive for $1 < t < 3$. We know this because the velocity is $s'(t)$, which is also the slope of the tangent line to the graph, and in that interval the slopes are all positive.

d. The train's acceleration is positive when $s''(t)$ is positive, and this occurs when the curve is concave up. The interval in which this occurs is $0 < t < 2$. The acceleration is negative when the curve is concave down, and this occurs when $2 < t < 4$. ■

Exercises 10-B

1. The graph in Figure 10.9 shows the number $f(t)$ of people (in hundreds) who become infected at time t (in days) with a fast-spreading stomach flu.

a. On what time intervals is the number of new infections increasing? Decreasing?

b. On what time intervals is the rate of new infections increasing? Decreasing?

c. On what time intervals is the growth rate of new infections positive? Negative?

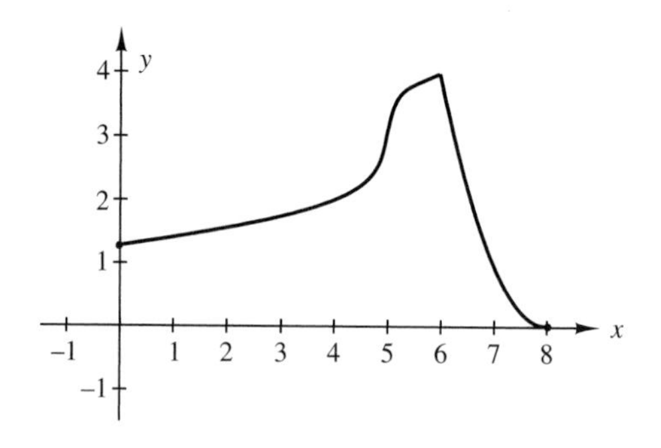

Figure 10.9 $y = f(t)$

2. A train is moving along a straight track. Each of the graphs in Figure 10.10 represents a possible situation with respect to the train's travel. In each case, the graph shows the distance $y = s(t)$ from the train to a signal box at time t. Identify the picture that matches each situation described.

a. The train's acceleration is positive for all t in the interval [0, 10].

b. The train has constant velocity for all t in the interval [0, 10].

c. The train's velocity is steadily increasing throughout the interval [0, 10].

d. The train's acceleration is 0 throughout the interval [0, 10].

e. The train decelerates and then accelerates in the interval [0, 10].

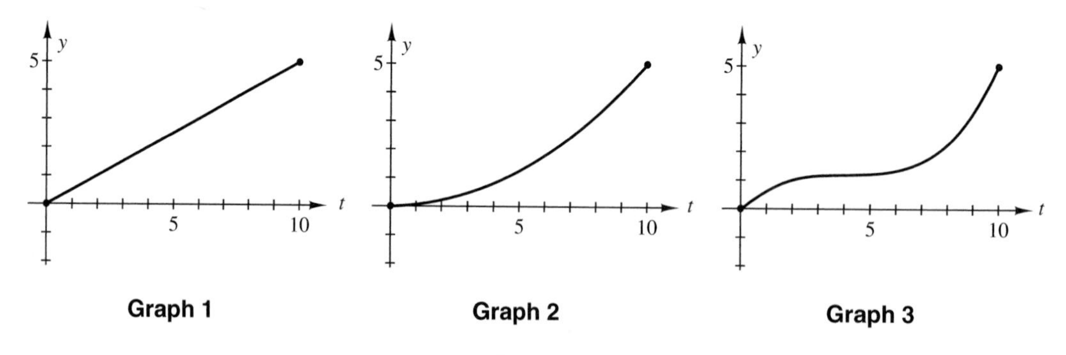

Figure 10.10 Graphs of $y = s(t)$

3. The graph of the derivative f' of a function f is shown in Figure 10.11a, and the graph of the second derivative g'' of a function g is shown in Figure 10.11b. Use these graphs to answer the questions about the functions f and g.

a. Why must the graph of the function f be a line? What is the slope of the line?

b. Suppose that $f(0) = 2$. Sketch the graph of f on the interval $-4 \le x \le 4$.

c. On what interval(s) is the graph of g concave up? On what interval(s) is the graph of g concave down?

d. Suppose that the graph of g is always increasing on the interval $0 \le x \le 10$, and that $g(0) = -1$ and $g(10) = 8$. Sketch a graph that could be the graph of g on the interval $0 \le x \le 10$.

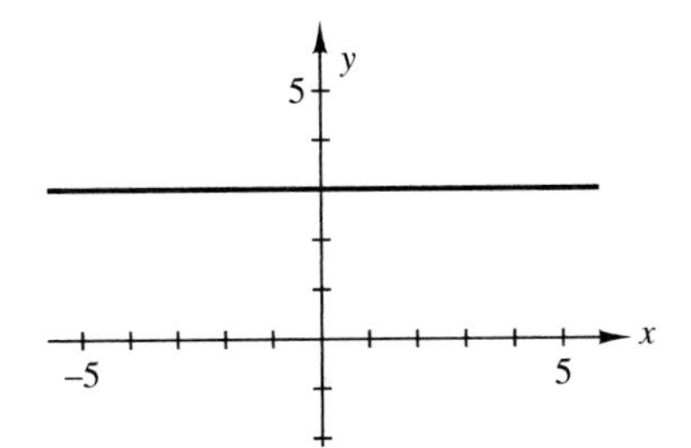

Figure 10.11a Graph of f'

Figure 10.11b Graph of g''

4. A ball is thrown into the air, and its height in feet after t seconds is given by $y = 60t - 16t^2$. Assume ground level corresponds to $y = 0$.

a. Find the time t when the velocity is 0 and explain where the ball is at this time.

b. Find the acceleration when the velocity is 0.

c. Find the time t when the ball hits the ground.

d. Find the acceleration of the ball when it hits the ground.

e. Compare your answers in parts b and d and explain them.

5. In one model, the concentration c (as a percentage) of a certain drug in the bloodstream t hours after it is taken is $c = \dfrac{0.05t}{t^2+9}$, $0 \le t \le 24$.

a. Find the concentration 2 hours after the drug is taken.

b. Find $c'(t)$ and then find the rate of change of the concentration 2 hours after the drug is taken. Is the concentration increasing or decreasing 2 hours after the drug is taken?

c. Use a graphing utility to graph $c'(t)$. Does the graph confirm your answer in part b?

d. Use the graph in part c to determine whether the rate of change of the concentration is increasing or decreasing 2 hours after the drug is taken.

Chapter 10 Exercises

1. Find the seventh derivative of each of the following functions:

a. $F(x) = x^8 - 2x^7 + x^3$ **b.** $G(x) = (4 + x)^5$ **c.** $H(x) = \dfrac{1}{x+2}$

d. $K(x) = (1 + 2x)^{5/2}$ **e.** $R(x) = \sin 3x$

2. Find a pattern in the derivatives found for the functions in Exercise 1, parts c and e, and give a general expression for their n^{th} derivatives.

3. Find the instantaneous rate of change of the slope of the tangent line to the graph of $y = \frac{3}{\sqrt{x}}$ at the point $\left(4, \frac{3}{2}\right)$.

4. The first four derivatives of a function f are given below. Look for a pattern in the derivatives and then answer the questions.

$$f'(x) = (-1)(2)(2x)^{-2}$$

$$f''(x) = (-1)^2(2)^2(2)(2x)^{-3}$$

$$f^{(3)}(x) = (-1)^3(2)^3(2)(3)(2x)^{-4}$$

$$f^{(4)}(x) = (-1)^4(2)^4(2)(3)(4)(2x)^{-5}$$

a. Find $f^{(5)}(x)$. Do not multiply out terms.

b. Give a general formula for the nth derivative $f^{(n)}(x)$.

5. The graph in Figure 10.12 gives an unemployment curve for a certain large city, with time t as the variable on the horizontal axis and $f(t)$ = number of people (in thousands) unemployed on day t.

a. At what time t is the unemployment the highest?

b. On what interval is the unemployment increasing?

c. On what interval is the unemployment decreasing?

d. On what interval is the unemployment rate (rate of change of unemployment) increasing?

e. On what interval is the unemployment rate decreasing?

f. Restate your answers in parts b, c, d, and e in terms of intervals on which $f'(t)$ and $f''(t)$ are either positive or negative.

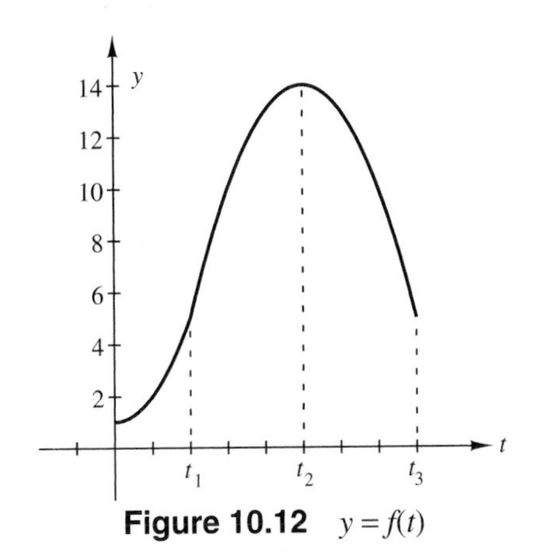

Figure 10.12 $y = f(t)$

6. Each of the graphs in Figures 10.13a–d shows the growth rate of a particular population over time. The function $f(t)$ gives the number of individuals in the population at time t. Figure 10.13a shows a pattern called *logistic growth,* which yields an S-shaped curve. For each graph, determine the intervals on which the rate of change of the population is increasing and the intervals on which it is decreasing.

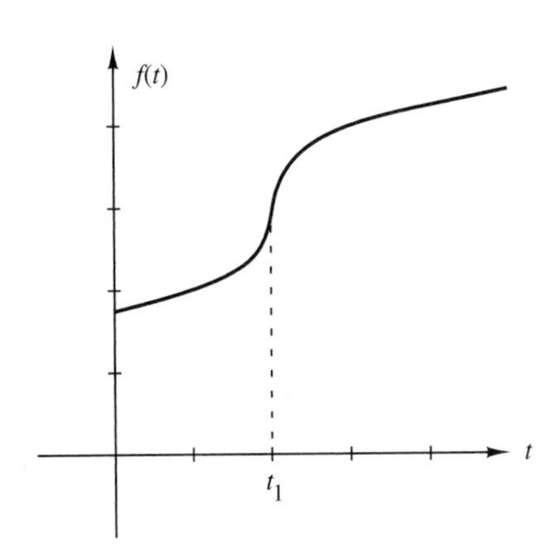

Figure 10.13a

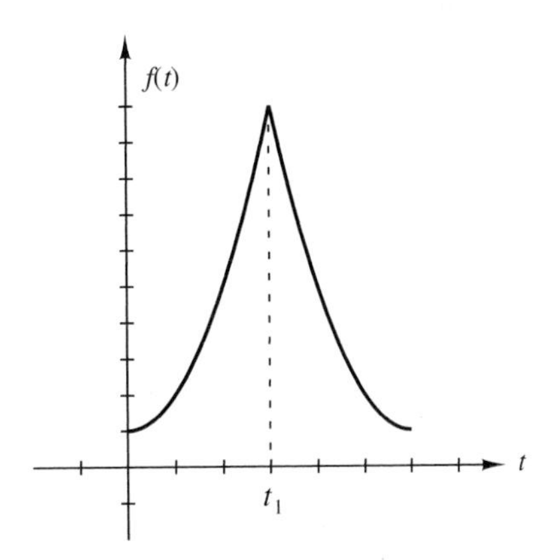

Figure 10.13b

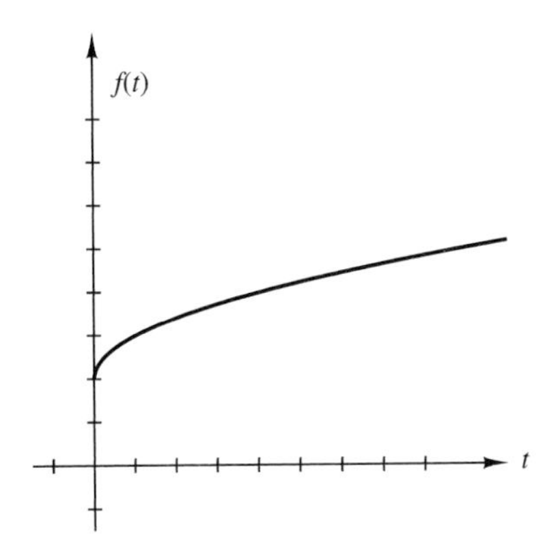

Figure 10.13c

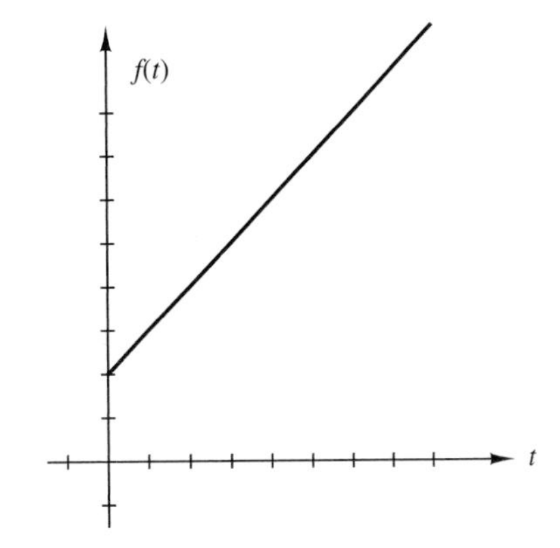

Figure 10.13d

7. Suppose the profit function $P(x)$ for a certain product is given by $P(x) = 20 + 10x + \frac{3}{2}x^2 - \frac{1}{3}x^3$, where x represents the number of lots produced each month.

a. Is the profit increasing or decreasing when four lots are produced each month? How do you know?

b. Is the rate of change of profit per lot increasing or decreasing when four lots are produced each month? Explain your answer.

8. The graphs of the second derivatives f'' and g'' of functions f and g are given in Figures 10.14a and 10.14b. Graphs 1, 2, and 3 show three functions. Two of these graphs are the functions f and g. Match the numbers of the graphs of f and g with the graphs of their second derivatives.

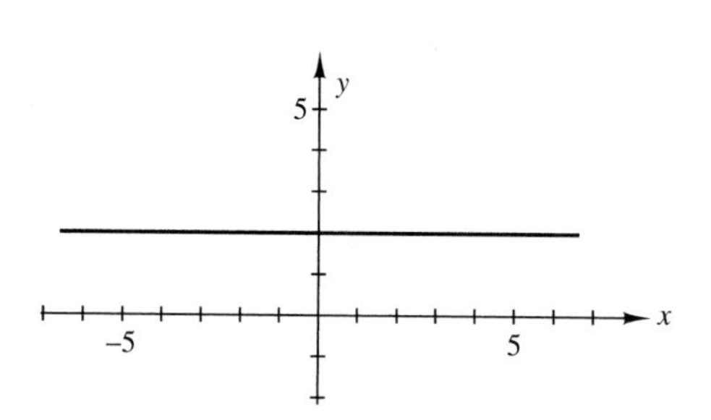

Figure 10.14a Graph of f''

Figure 10.14b Graph of g''

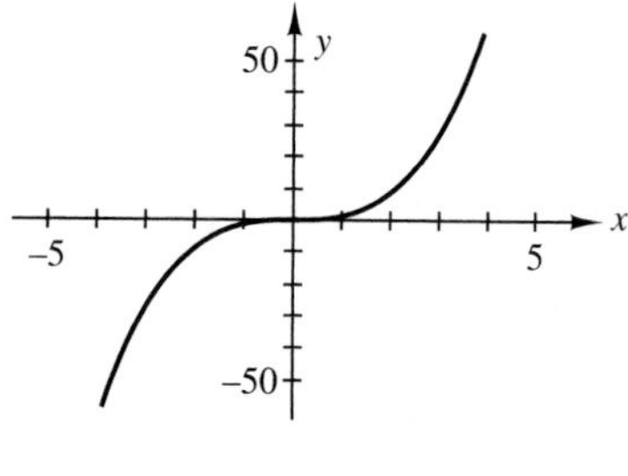

Graph 1

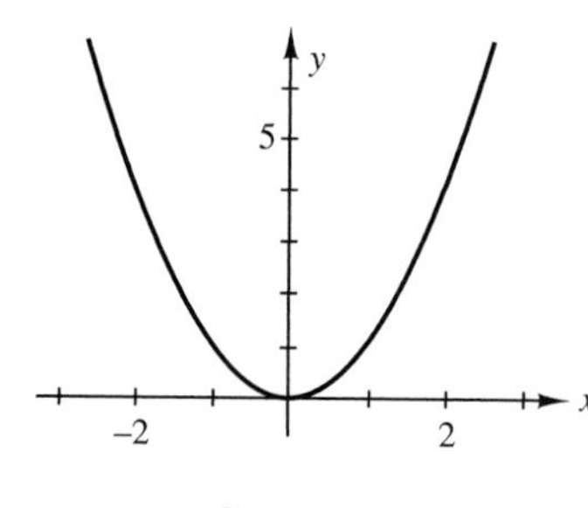

Graph 2

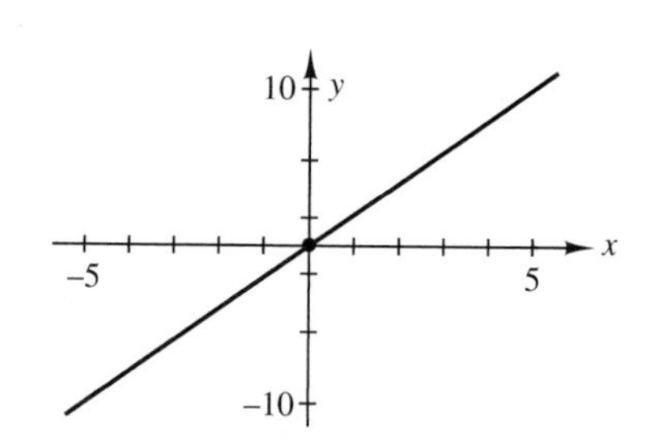

Graph 3

CHAPTER **11**

Companion to Related Rates

In many applied problems, there are two or more variables that are differentiable functions of time—that is, two or more quantities that change as time changes. In Chapter 9 we studied implicit functions and investigated equations in two variables, x and y, where y was considered a function of x. Now we consider situations with two or more variables that all depend on one other variable, usually time. For example, we may know that $x = f(t)$ = bushels of grain in storage at time t, and $y = g(t)$ = total value of stored grain at time t, although the rules for the functions $f(t)$ and $g(t)$ may not be explicitly known. If the price of grain is \$3 per bushel, then x and y are related by the equation $y = 3x$. If we then differentiate both sides with respect to t, we have $\frac{dy}{dt} = 3\frac{dx}{dt}$. This equation says that the rate of change of y with respect to t is 3 times the rate of change of x with respect to t. Since x and y are related and they are differentiable functions of t, their derivatives with respect to t are related as well. These derivatives represent rates of change and are referred to as *related rates.*

11-A SETTING UP EQUATIONS FOR RELATED RATES PROBLEMS

Many problems that involve related rates result from situations that can be represented by a picture. If it is appropriate, we draw a picture to help visualize the relationships between the variables. On the picture, we label with appropriate letters (such as d, s, x, y, or z) any quantities in the problem that can change with time. Any parts of the picture that remain constant for all times t in the domain of the problem are labeled with constants.

EXAMPLE 11.1 A cell-phone tower has height 130 ft and has been assembled on the ground, lying on its side. A motorized device raises the tower until it is in its vertical position.

a. Draw a picture to represent this situation at a time t when the tower is in an intermediate position (neither horizontal nor vertical). Identify and label any constants and any quantities that change as time changes.

b. Find a relationship between the variable quantities and express the relationship as an equation.

Solution

a. The picture in Figure 11.1 represents the situation with a right triangle.

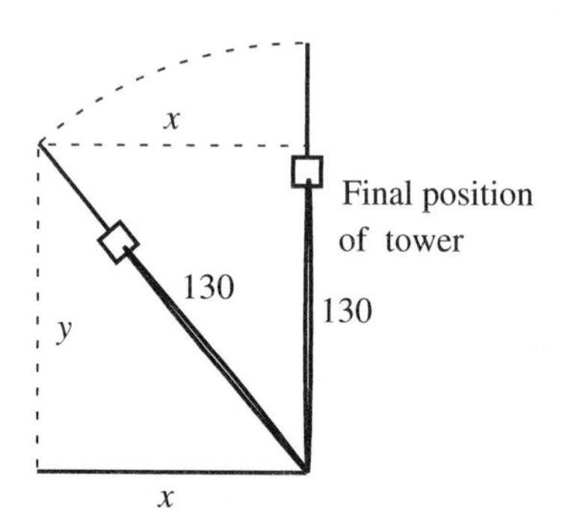

Figure 11.1

The hypotenuse of the triangle represents the tower and remains constant, so it is labeled with length 130. The variable y represents the distance from the top of the tower to the ground at time t, and the variable x represents how far from the vertical position the top of the tower is at time t. When the tower is in its final vertical position, x will equal 0. Both x and y are changing with time.

b. By the Pythagorean theorem, x and y are related by the equation $x^2 + y^2 = 130^2$. ■

As Example 11.1 illustrates, the Pythagorean theorem is appropriate to use in some related rates problems. Other related rates problems may require the formulas for the area of a circle, the volume of a sphere, the volume of a cylinder, or the volume of a cone. (These can be found in Section 6-A.)

Since relationships between parts of similar triangles appear in some related rates problems, it is important to remember that corresponding sides of similar triangles are proportional; that is, they are in the same ratio. For example, suppose we have two similar right triangles A and B as shown in Figure 11.2, with corresponding sides parallel.

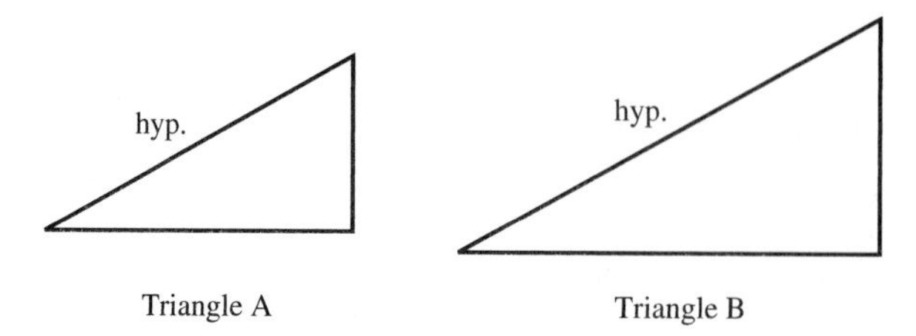

Figure 11.2

One relationship between the sides of triangle A and triangle B can be written as follows:

$$\frac{\text{Horizontal leg of triangle A}}{\text{Vertical leg of triangle A}} = \frac{\text{Horizontal leg of triangle B}}{\text{Vertical leg of triangle B}}.$$

Another way to express the same relationship is as follows:

$$\frac{\text{Vertical leg of triangle A}}{\text{Vertical leg of triangle B}} = \frac{\text{Horizontal leg of triangle A}}{\text{Horizontal leg of triangle B}}.$$

EXAMPLE 11.2 A man walks away from a street light along a straight horizontal sidewalk. The light is at the top of a pole, 18 ft above the ground. If the man is 6 ft tall, find a relationship between x, the distance between the man and the pole, and s, the length of the man's shadow.

Solution A picture of the situation appears in Figure 11.3. The shadow extends from F, the point at the man's feet, to the point B at the tip of the shadow. If H is the point that represents the top of the man's head, then B is on a line from the light at L through H.

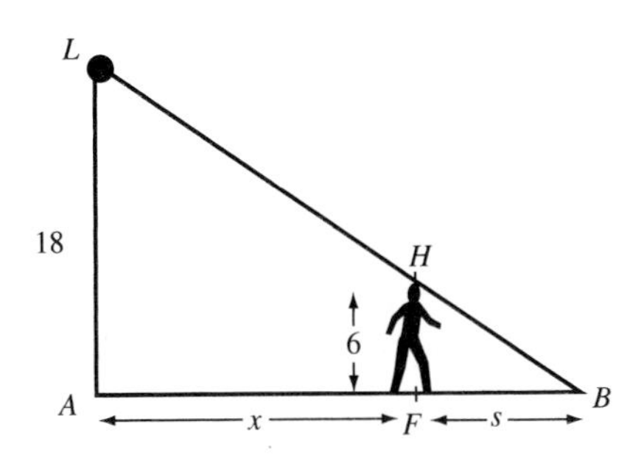

Figure 11.3

Triangle ABL is similar to triangle FBH. We set up a proportion between corresponding legs of these similar right triangles to find an equation that relates x and s: $\frac{s}{6} = \frac{x+s}{18}$. We then multiply both sides of this equation by 18 to get $3s = x + s$, so $2s = x$. This says that the distance from the man to the pole is twice the length of the man's shadow. ■

Some equations involve three variable quantities, but the nature of the relationships allows us to express one of the variable quantities in terms of another. The equation can then be expressed using only two variables, as the next example shows.

EXAMPLE 11.3 Water is poured into a tank in the shape of an inverted right circular cone. The height of the tank is 8 m and its radius at the top is 4 m.

a. Draw a picture to represent this situation, and identify and label all constants and variable quantities.

b. Find an equation that relates the variable quantities, and express this relationship using only two variables.

Solution

a. Let r be the radius (in meters) of the water in the tank at time t, h the height (in meters) of the water in the tank at time t, and V the volume (in cubic meters) of the water in the tank at time t. Figure 11.4 shows this situation. Note that the height, radius, and volume of the tank are constant, but the height, radius, and volume of the water in the tank vary with time; that is, they are functions of time.

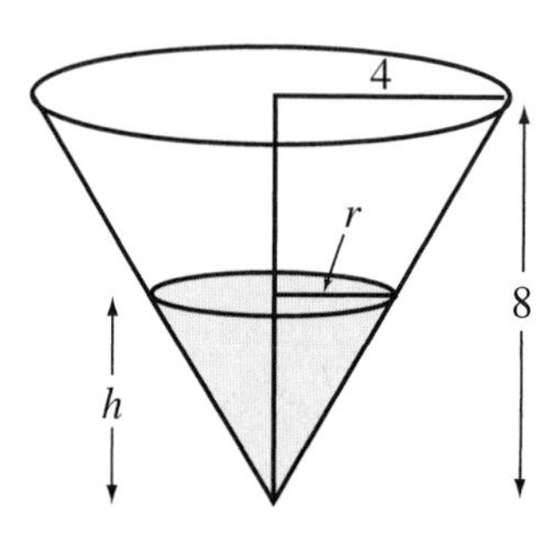

Figure 11.4

b. The equation that relates V, r, and h is the equation for the volume of a right circular cone with base radius r and height h: $V = \frac{1}{3}\pi r^2 h$. (This represents the volume of water in the tank.) We need to express this relationship between V, r, and h using only two variables. The two right triangles that are formed by the radii and heights of the conical tank and the cone of water are similar, so the radii and heights are in the same ratio. Thus, $\frac{r}{4} = \frac{h}{8}$, so $r = \frac{1}{2}h$. If we substitute $\frac{1}{2}h$ for r in the volume equation, then $V = \frac{1}{3}\pi\left(\frac{1}{2}h\right)^2 h = \frac{1}{12}\pi h^3$. This equation relates V and h. If we want an equation that relates V and r, we solve the relationship $\frac{r}{4} = \frac{h}{8}$ for h: $h = 2r$. We substitute $2r$ for h in the volume equation to get $V = \frac{1}{3}\pi r^2 2r = \frac{2}{3}\pi r^3$. We would choose either $V = \frac{1}{12}\pi h^3$ or $V = \frac{2}{3}\pi r^3$ as the equation that relates two variables, depending on what additional information is known and what additional questions need to be answered. ■

Some problems that describe how one quantity is related to another use the phrase "is proportional to" or "is inversely proportional to." If the variable y is *proportional* to the variable x, then x and y are related by an equation $y = kx$, where k is a nonzero constant. In this relationship, a change in x results in a proportional change in y. If x is doubled, then y is also doubled; if x is cut in half, then so is y. The value of k is usually called the *constant of proportionality*, or the *scale factor*. If the variable y is *inversely proportional* to the variable x, then x and y are related by an equation $y = \frac{k}{x}$, where k is a nonzero constant. In this relationship, if, for example, x is doubled, then y is cut in half.

EXAMPLE 11.4 For each of the following statements, identify the variable quantities and write an equation that relates the variables.

a. The velocity of blood flow at a point in an artery is proportional to the difference of the square of the radius of the artery and the square of the distance from the point to the center of the artery.

b. According to Boyle's law for the expansion of gas, if the temperature of the gas is held constant, the number of pounds per unit of pressure is inversely proportional to the number of cubic units of volume of the gas.

Solution

a. Let v be the velocity of blood flow, R the radius of the artery, and d the distance from the point of measurement to the center of the artery. Then $v = k(R^2 - d^2)$, where k is some constant. (The constant k depends on blood pressure, blood cholesterol, the length of the artery, etc.)

b. Let P be the number of pounds per unit of pressure and V the number of cubic units of volume of the gas. Then $P = \frac{c}{V}$, where c is a constant that depends on the type of gas. ■

Exercises 11-A

---

1. An airplane is flying parallel to the ground at an altitude of 3 miles. It flies directly over an observer on the ground and continues flying parallel to the ground. Let d be the distance from the observer to the airplane.

- **a**. Draw a picture that represents the situation.
- **b**. Identify and label all constants and variable quantities.
- **c**. Find a relationship between d and the other variable quantities; express this relationship as an equation.

2. A person on a dock is pulling on a rope attached to a boat at water level. The person's hands are 3 ft above the water level.

- **a**. Draw a picture that represents the situation.
- **b**. Identify and label all constants and quantities that vary.
- **c**. Give an equation that relates the variable quantities.

3. Oil is pumped out of an inverted cone-shaped tank. The radius of the top of the tank is 3 m, and the height of the tank is 10 m.

a. Draw a picture of the tank. Identify and label all constants and variables.

b. Find an equation that relates the volume of oil in the tank and just one other variable.

4. Two cars leave an intersection at the same time. One car travels north and the other travels west. Let d be the distance between the two cars.

a. Draw a picture that represents the situation.

b. Identify and label all constants and variable quantities.

c. Find a relationship between d and the other variable quantities and express this relationship as an equation.

5. The gravitational force of the earth on a rocket traveling in space is proportional to the mass of the rocket divided by the square of the distance from the rocket to the center of the earth. Identify all constants and variable quantities, and write an equation that relates the variables.

6. The number of hours it takes a student group to decorate a gymnasium for an event is inversely proportional to the number of students in the group. Identify all constants and variable quantities, and write an equation that relates them.

11-B PROBLEM-SOLVING STRATEGIES FOR RELATED RATES PROBLEMS

Related rates problems usually contain statements about the rate of change of some quantities that vary over time and ask questions about the rate of change of other variable quantities. In Example 11.1, we could include this additional information: The cell-phone tower is raised so that the distance from the top of the tower to the ground is changing at a constant rate of 10 ft/min. We interpret this rate in terms of the variable y (which represents the distance from the top of the tower to the ground): as the rate of change of y with respect to t is 10 ft/min—that is, $\frac{dy}{dt} = 10$. Note that $\frac{dy}{dt}$ is positive, indicating that y is *increasing* over time.

EXAMPLE 11.5 For each of the following statements, draw a picture, assign a letter name to each quantity that is changing over time, and interpret the statement about rate of change in terms of symbols.

a. A woman leaves her house and walks directly west at a rate of 4 mph.

b. Water is being pumped out of a large cylindrical storage tank at a rate of 5 ft^3/min.

c. One hour after it leaves the station, a train is traveling at 70 mph.

d. The radius of a circular ripple is increasing at a rate of 3 cm/sec.

Solution

a.

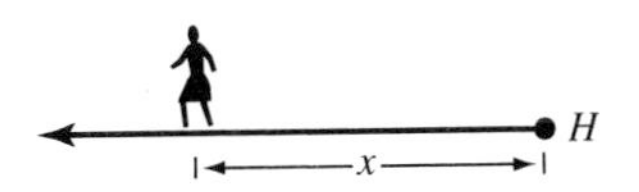

Figure 11.5

Let x be the distance (in miles) that the woman has walked by time t (in hours) (see Figure 11.5). H represents the position of the house. The distance x is a function of time t and is increasing at a rate of 4 mph, so $\frac{dx}{dt} = 4$ mph.

b.

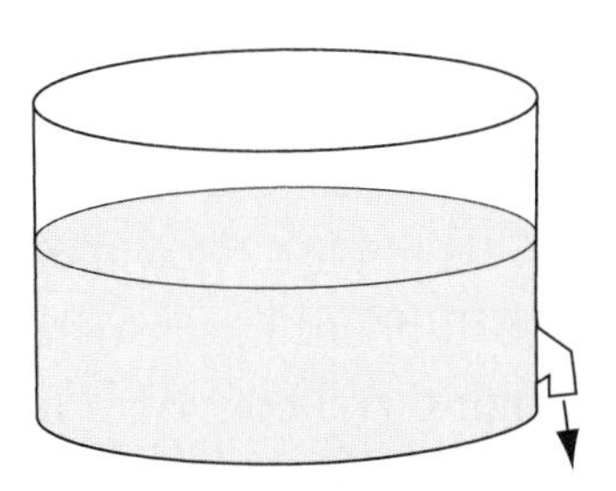

Figure 11.6

Let V be the volume (in cubic feet) of water in the tank at time t (in minutes)(see Figure 11.6). The volume V that is changing is a function of t. Since V is *decreasing* at a rate of 5 ft^3/min, $\frac{dV}{dt} = -5$ ft^3/min.

c.

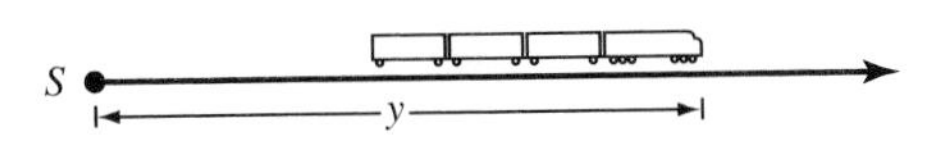

Figure 11.7

Let y be the distance (in miles) the front of the train has traveled by time t (in hours) (see Figure 11.7). S represents the position of the station. Then $\frac{dy}{dt} = 70$ mph when $t = 1$ hour. Note that this rate of change of y with respect to t occurs only at the instant when $t = 1$ hr.

d.

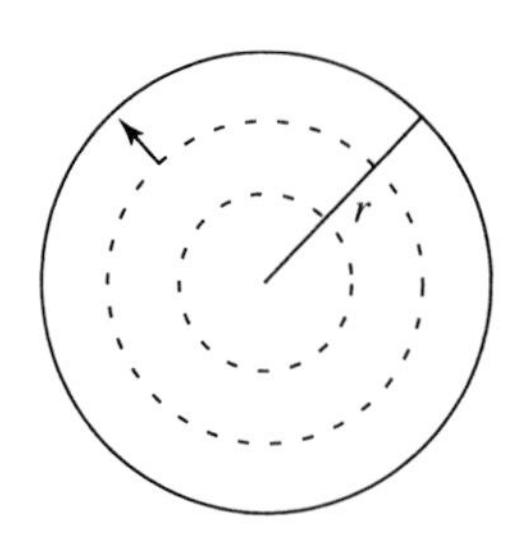

Figure 11.8

Let r be the radius of the circular ripple at time t (see Figure 11.8). Then $\frac{dr}{dt} = 3$ cm/sec. ■

Note that the sign of a rate of change indicates whether a quantity is increasing or decreasing; for example:

$\frac{dy}{dt} = 5$ ft/min means "y is **increasing** at a rate of 5 feet per minute"

$\frac{dy}{dt} = -5$ ft/min means "y is **decreasing** at a rate of 5 feet per minute"

EXAMPLE 11.6 Consider again Example 11.1 with this additional information: A motorized device raises a 130-ft cell-phone tower so that the distance from the top of the tower to the ground is changing at a constant rate of 10 ft/min. Consider the question: What is the rate of change of the distance from the top of the tower to the vertical position when the top of the tower is 120 ft from the ground? (See Figure 11.9.)

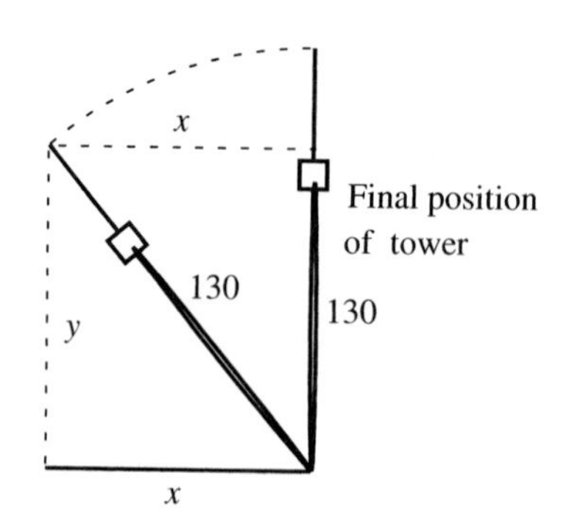

Figure 11.9

a. Represent (with symbols using variables) the rate of change needed to answer this question.

b. Looking at the physical situation, do we expect the rate of change in part a to be positive or negative? Why?

c. Find the distance from the top of the tower to the vertical position when the top of the tower is 120 ft from the ground.

d. Find the rate at which the distance from the top of the tower to the vertical position is changing when the top of the tower is 120 ft from the ground.

Solution

a. We use the same variables as in Example 11.1, so the distance from the top of the tower to the vertical position is x. To answer the question we need to find $\frac{dx}{dt}$, the rate of change of x with respect to t.

b. The distance from the top of the tower to the vertical position is decreasing, so we expect $\frac{dx}{dt}$ to be negative.

c. Recall that y represents the height of the tower from the ground at time t and x and y are related by the equation $x^2 + y^2 = 130^2$. (See Figure 11.9.) We substitute $y = 120$ into this equation and solve for x: $x^2 + 120^2 = 130^2$; then $x^2 = 130^2 - 120^2 = 2500$, so $x = 50$. The top of the tower is 50 ft from the vertical position when the top of the tower is 120 ft from the ground.

d. The rate at which y is increasing is $\frac{dy}{dt} = 10$ ft/min. To find $\frac{dx}{dt}$, the rate of change of the distance from the top of the tower to the vertical position when $y = 120$, we use the Chain Rule to implicitly differentiate the equation $x^2 + y^2 = 130^2$ with respect to t: $2x\frac{dx}{dt} + 2y\frac{dy}{dt} = 0$, or $x\frac{dx}{dt} + y\frac{dy}{dt} = 0$. We want to solve this equation for $\frac{dx}{dt}$ at the instant when $y = 120$. From part c we know that when $y = 120$, $x = 50$. We substitute $x = 50$, $y = 120$, and $\frac{dy}{dt} = 10$ and solve for $\frac{dx}{dt}$: $(50)\frac{dx}{dt} + (120)(10) = 0$, so $\frac{dx}{dt} = -\frac{1200}{50} = -24$ ft/min. The fact that $\frac{dx}{dt}$ is negative confirms our answer in part b. ■

 Recall that the derivative of any constant is always 0. We have seen in part d of Example 11.6 that the derivative of 130^2 is 0.

Note that in Example 11.6, we needed to work with two equations to find the rate at which the distance from the top of the tower to the vertical position is changing when the top of the tower is 120 ft from the ground. We used the original equation $x^2 + y^2 = 130^2$, which relates the variables x and y, to find the value of x when $y = 120$. Then to solve for $\frac{dx}{dt}$ we used the equation obtained when we implicitly differentiated the original equation with respect to t.

To solve a related rate problem that asks for a rate of change at a specific time, we need to **differentiate first** and then evaluate at the specified time. In Example 11.6 we first differentiated the equation $x^2 + y^2 = 130^2$, where x and y are functions of the variable t. Only after we differentiated both sides of this equation did we substitute $y = 120$, $x = 50$, and $\frac{dy}{dt} = 10$ into the resulting equation to find $\frac{dx}{dt}$. (If we evaluate the equation $x^2 + y^2 = 130^2$ at the specified time before differentiating, we lose the relationship between the variables x and y.)

EXAMPLE 11.7 In each of the examples below, x and y are both differentiable functions of the variable t, and x and y are related by the given equation. Find the requested rate of change under the given conditions.

a. $y = x^3 - 2$: Find $\frac{dx}{dt}$ when $x = 2$ and $\frac{dy}{dt} = -3$.

b. $y = \sqrt{x}$: Find $\frac{dy}{dt}$ when $x = 9$ and $\frac{dx}{dt} = 2$.

c. $xy = 24$: Find $\frac{dx}{dt}$ when $y = 3$ and $\frac{dy}{dt} = -4$.

Solution

a. To find $\frac{dx}{dt}$ under the given conditions, we differentiate both sides of the equation $y = x^3 - 2$ with respect to t and obtain $\frac{dy}{dt} = 3x^2\frac{dx}{dt} - 0$. Now we solve for $\frac{dx}{dt}$: $\frac{dx}{dt} = \frac{1}{3x^2}\frac{dy}{dt}$. We then substitute $\frac{dy}{dt} = -3$ and $x = 2$ (as given) and simplify to get $\frac{dx}{dt} = \frac{1}{3(2)^2}(-3) = -\frac{1}{4}$.

b. We differentiate both sides of the equation $y = \sqrt{x}$ with respect to t (using the Chain Rule and keeping in mind that x and y are both functions of t): $\frac{dy}{dt} = \frac{1}{2}x^{-1/2}\frac{dx}{dt} = \frac{1}{2\sqrt{x}}\frac{dx}{dt}$. When $x = 9$ and $\frac{dx}{dt} = 2$, $\frac{dy}{dt} = \frac{1}{2\sqrt{9}}(2) = \frac{1}{3}$.

c. To differentiate the equation $xy = 24$, we need to use the Product Rule. We differentiate both sides of this equation with respect to t to obtain $x\frac{dy}{dt} + y\frac{dx}{dt} = 0$. We then solve for $\frac{dx}{dt}$: $y\frac{dx}{dt} = -x\frac{dy}{dt}$; $\frac{dx}{dt} = -\frac{x}{y}\frac{dy}{dt}$. We need to use the original relationship between x and y to solve for x when $y = 3$. When $y = 3$, since $xy = 24$, $x = \frac{24}{y} = \frac{24}{3} = 8$. Thus when $y = 3$ and $\frac{dy}{dt} = -4$, $\frac{dx}{dt} = -\frac{8}{3}(-4) = \frac{32}{3}$. ■

Exercises 11-B

1. For each of the following statements, draw a picture, assign a letter name to each quantity that is changing over time, and interpret the statement that involves rate of change in terms of symbols.

a. A helium-filled balloon is rising at the rate of 10 ft/sec.

b. Sixty seconds after entering the highway, a police car is speeding to the scene of an accident at 75 mph.

c. Road salt is being dumped into a cone-shaped pile at the rate of 45 ft^3/min.

2. In each of the following, x and y are differentiable functions of t and are related by the given equations. Find the requested rate of change.

a. $y^2 = 5x + 6$: Find $\frac{dx}{dt}$ when $y = 5$ and $\frac{dy}{dt} = -3$.

b. $y^2 = 5x^3 + 9$: Find $\frac{dx}{dt}$ when $y = 8$ and $\frac{dy}{dt} = -3$.

c. $2xy + 5 = 31$: Find $\frac{dy}{dt}$ when $x = 3$ and $\frac{dx}{dt} = 4$.

3. Let A be the area of a circle of radius r, where r is changing at the constant rate of 3 cm/sec.

a. Find $\frac{dA}{dt}$ when $r = 2$ cm.

b. Find $\frac{dA}{dt}$ when $r = 7$ cm.

c. Is $\frac{dA}{dt}$ constant? Explain.

d. Explain why $\frac{dr}{dt}$ can be constant when $\frac{dA}{dt}$ is not constant.

Chapter 11 Exercises

1. Road salt is being dumped into a cone-shaped pile. At all times the height of the pile is equal to the radius of the base.

a. Draw a picture that represents the situation.

b. Identify and label all constants and variable quantities.

c. Find an equation that relates all the variable quantities.

d. Find an equation that expresses the volume as a function of one variable.

2. A man walks along a straight horizontal sidewalk away from a street light. The man is 6 ft tall and the light is at the top of a pole, 18 ft above the ground. (See Example 11.2.)

a. The man is walking at a constant rate of 4 ft/sec. Use a picture of the situation to interpret the statement in symbols.

b. Use a picture of the situation to represent, in symbols, the speed at which the man's shadow is lengthening.

c. Use the information in part a to find how fast the man's shadow is lengthening.

3. Water is poured into a tank in the shape of an inverted right circular cone. The height of the tank is 8 m and its radius at the top is 4 m. (See Example 11.3.)

a. Water is entering the tank at the rate of 3 m^3/min. Use a picture of the situation to interpret this statement in symbols.

b. Use a picture of the situation to represent the rate of change of the height of the water level in symbols.

c. Use the information in part a to find how fast the water level is rising when the water is 5 m deep.

4. An airplane is flying parallel to the ground at an altitude of 3 miles. It flies directly over an observer on the ground and continues flying parallel to the ground. Let d be the distance from the observer to the airplane.

a. The airplane is flying at 350 mph. Use a picture of the situation to interpret this statement in symbols.

b. Use the information in part a to find how fast the distance from the observer to the airplane is increasing when the distance from the observer to the airplane is 6 miles.

5. A person on a dock is pulling on a rope attached to a boat at water level. The person's hands are 3 ft above the water level.

a. The person is pulling at a rate of 30 ft/min. Use a picture of the situation to interpret this statement in symbols.

b. Use the information in part a to find how fast the boat is approaching the dock when it is 10 ft away from the dock.

6. A spherical balloon is being inflated with helium and its volume is increasing at the rate of 6 m^3/min.

a. Interpret this statement in symbols.

b. Find the rate at which the radius of the balloon is increasing when the radius is 5 m.

7. Two cars leave an intersection at the same time. One car travels north at a constant rate of 30 mph; the other car travels west at a constant rate of 20 mph. Let d represent the distance between the two cars.

a. Make a picture of the situation, assigning variables to represent the distances of the cars from the intersection. Then interpret in symbols the statements that involve the speed of the cars.

b. Find how far the car traveling west is from the intersection when the car traveling north is 15 miles from the intersection.

c. Find the rate at which the distance between the cars is increasing when the car traveling north is 15 miles from the intersection.

8. Suppose the cost and revenue functions for a certain product are given by $C = 4000 + 2x$ and $R = 10x - \frac{x^2}{1000}$, where x is the number of units of the product produced in a month (and may change from month to month). The profit function is given by $P = R - C$. If production is increasing at the rate of 700 units per month when production is 2500 units, find and interpret the following in words:

a. The rate of change in cost per month when production is 2500 units

b. The rate of change in revenue per month when production is 2500 units

c. The rate of change in profit per month when production is 2500 units

9. The Body Mass Index indicates how "reasonable" an adult's weight is in relation to the person's height and is given by $B = 703.08\frac{w}{h^2}$, where w is the person's weight (in pounds) and h is the person's height (in inches).

a. Is B directly or inversely proportional to w? Explain.

b. A man 5 ft 8 in. tall is on a diet and is losing 1 lb per week. Find the rate of change of his Body Mass Index when his weight is 180 lb.

c. Is his Body Mass Index increasing or decreasing? How do you know?

10. The demand d for a type of bracelet varies inversely with the price p. Identify all constant and variable quantities, and write an equation that relates them.

CHAPTER 12

Linear Approximations and Differentials

12-A TANGENT LINE APPROXIMATION

Since lines are simple functions, we often approximate a function $y = f(x)$ close to a given point $(c, f(c))$ by a linear function $y = mx + b$ that goes through that point. That is, near a given point we want to approximate the graph of the function $y = f(x)$ by a straight line. For example, the stopping distance of a car depends on a variety of things, including road surface and driving conditions, but as a quick approximation we use the rule of thumb that stopping distance increases by three car lengths for each additional 10 miles per hour of speed. This approximation applies for speeds between 20 and 40 mph. The rule of thumb involves a linear approximation of a more complicated function and provides an estimate of the stopping distance of a car for a restricted interval of speeds.

The linear approximation used for differentiable functions in calculus is the tangent line. Consider the graph in Figure 12.1 with the point (c, $f(c)$) marked. We construct the tangent line to the graph at the point $(c, f(c))$ and use this as the line that approximates $f(x)$ near c. Since the slope of the tangent line to the graph at the point $(c, f(c))$ is $f'(c)$, we find the equation of that line by using the point-slope form $y - f(c) = f'(c)(x - c)$. We can also write the equation of the tangent line to the graph at the point $(c, f(c))$ as $y = f(c) + f'(c)(x - c)$.

In general, the approximation of $f(x)$ by the tangent line will be better for x close to c than for x farther from c, as Figure 12.1 shows. The accuracy of the approximation for a particular value of x is measured by the *error*, which is the absolute value of the difference between $f(x)$ and the y-coordinate on the tangent line corresponding to x.

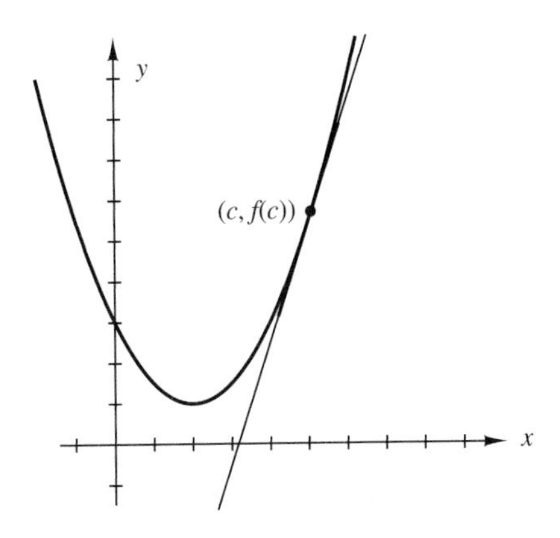

Figure 12.1 $y = f(x)$ and the tangent line $y = f(c) + f'(c)(x - c)$

EXAMPLE 12.1 Let $f(x) = 3 + \sqrt{x+1}$. Then $f'(x) = \frac{1}{2}(x+1)^{-1/2}$.

a. Find the equation of the tangent line to the graph of $y = f(x)$ at the point (3, 5) in slope-intercept form.

b. Find $f(3.2)$, accurate to three decimal places.

c. Compare $f(3.2)$ with the y-value on the tangent line corresponding to $x = 3.2$ by finding the error.

d. Find $f(4.2)$, accurate to three decimal places.

e. Compare $f(4.2)$ with the y-value on the tangent line corresponding to $x = 4.2$ by finding the error.

Solution

a. Since $f'(x) = \frac{1}{2}(x+1)^{-1/2}$, $f'(3) = \frac{1}{2}(3+1)^{-1/2} = \frac{1}{2} \cdot \frac{1}{\sqrt{4}} = \frac{1}{4}$. The line through the point (3, 5) with slope $\frac{1}{4}$ has equation $y - 5 = \frac{1}{4}(x - 3)$. Solving the equation for y, we get the slope-intercept form $y = \frac{1}{4}x + \frac{17}{4}$. Figure 12.2 shows the graph of $y = f(x)$ and the tangent line at the point (3, 5).

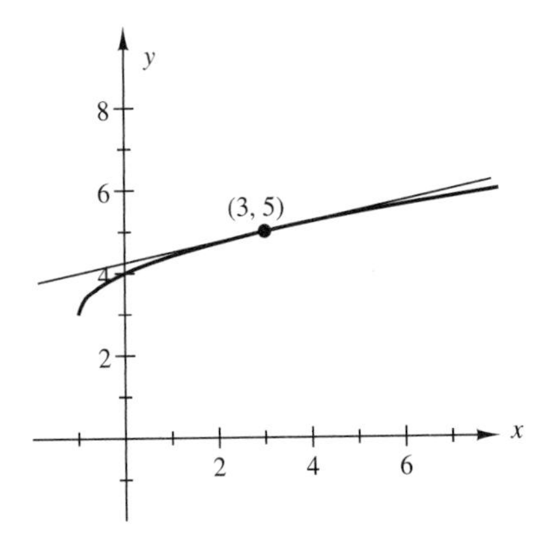

Figure 12.2 $y = f(x) = 3 + \sqrt{x+1}$

b. $f(3.2)=3+\sqrt{3.2+1}=3+\sqrt{4.2}\approx 3+2.0494$, which to three decimal places is 5.049.

c. We substitute $x = 3.2$ in the equation $y=\frac{1}{4}x+\frac{17}{4}$; we find $y=\frac{1}{4}(3.2)+\frac{17}{4}=\frac{20.2}{4}=5.05$. The error introduced by using the linear approximation for $f(3.2)$ is $|5.049-5.05|=|-0.001|=0.001$.

d. $f(4.2)=3+\sqrt{5.2}\approx 5.280$.

e. When $x = 4.2$, the y-value on the tangent line is $y=\frac{1}{4}(4.2)+\frac{17}{4}=\frac{21.2}{4}=5.3$. The error introduced by using the linear approximation for $f(4.2)$ is $|5.280-5.3|=|-0.02|=0.02$. This error is larger than the error found in part c. ■

In Example 12.1, we found the tangent line to the graph of the function $y=f(x)$ at the point $(c, f(c))$ where $c = 3$, and we used that tangent line to approximate the value of the function $f(x)$ at $x = 3.2$ and $x = 4.2$. We can think of the point $x = 3.2$ as $c + \Delta x = 3 + \Delta x$ for an increment $\Delta x = 0.2$, and we can think of the point $x = 4.2$ as $c + \Delta x = 3 + \Delta x$ for an increment $\Delta x = 1.2$. If we wanted to use the tangent line to the graph of $f(x)$ at the point $(3, f(3))$ to approximate the value of $f(2.6)$, we would think of the point $x = 2.6$ as $c + \Delta x = 3 + \Delta x$ for an increment $\Delta x = -0.4$. In each of these cases, Δx represents the directed distance from x to $c = 3$ (Δx is negative if x is less than c and positive if x is greater than c) and gives us a measure of how close x is to c. In general, we want to use the tangent line to the graph of $f(x)$ at the point $(c, f(c))$ to approximate the value $f(c + \Delta x)$ where $|\Delta x|$ is small—that is, for $c + \Delta x$ close to c.

Exercises 12-A

1. Let $f(x)=\sqrt{x^2+1}$.

a. Find the equation of the tangent line to the graph of the function at the point $(0, f(0))$.

b. Approximate $f(-0.05)$ by the value of y on the tangent line for $x=-0.05$.

c. Approximate $f(0.1)$ by the value of y on the tangent line for $x = 0.1$.

d. Find $f(-0.05)$ to three decimal places and use the value obtained in part b to find the error of the approximation.

e. Find $f(0.1)$ to three decimal places and use the value obtained in part c to find the error of the approximation.

2. Let $g(x) = \sin x$.

a. Find the equation of the tangent line to the graph of the function at the point $(0, g(0))$.

b. Approximate $g(-0.05)$ by the value of y on the tangent line for $x = -0.05$.

c. Approximate $g(0.1)$ by the value of y on the tangent line for $x = 0.1$.

d. Find $g(-0.05)$ to five decimal places and use the value obtained in part b to find the error of the approximation.

e. Find $g(0.1)$ to five decimal places and use the value obtained in part c to find the error of the approximation.

3. Consider the three graphs below.

a. On each graph, mark the point $(2, f(2))$ and then draw the tangent line to the graph at the point $(2, f(2))$.

b. Identify for which graph the tangent line to the graph at the point $(2, f(2))$ provides the best approximation to $y = f(x)$ on the interval $(1.5, 2.5)$. Explain why you chose the graph you did.

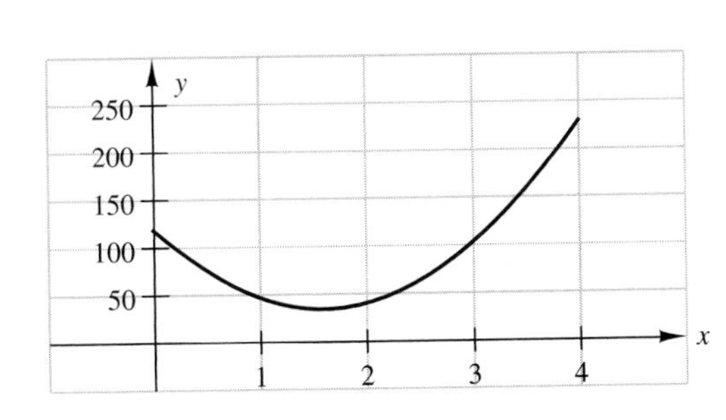

Graph 1

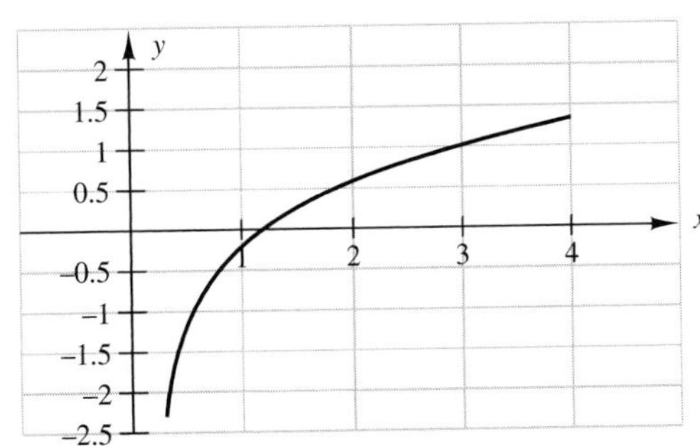

Graph 2

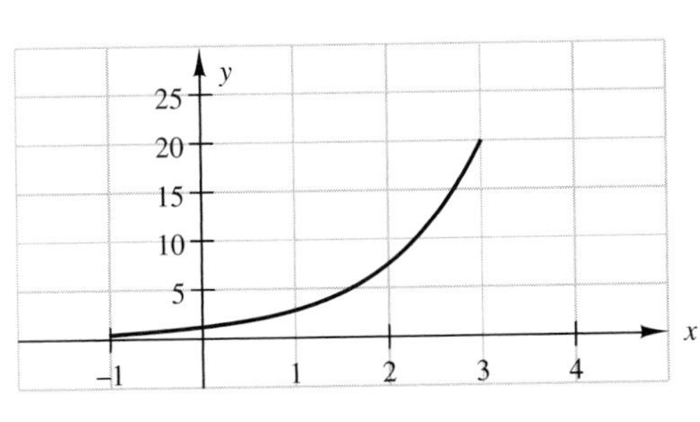

Graph 3

12-B THE DIFFERENTIAL

We saw in the preceding section that the size of Δx makes a difference in the error of a tangent line approximation. In order to measure this error, we want to examine the change in y along the tangent line as x changes from c to $c + \Delta x$. Consider the graph of $y = f(x)$ in Figure 12.3. Two points $(c, f(c))$ and $(c + \Delta x, f(c + \Delta x))$ are marked on the graph. The *increment* Δy is the change in y measured along the curve $y = f(x)$ as x changes from c to $c + \Delta x$. The increment Δy is shown in Figure 12.3 and is defined as $\Delta y = f(c + \Delta x) - f(c)$. Recall from Section 6-C that $\frac{\Delta y}{\Delta x}$ is the slope of the secant line through the points $(c, f(c))$ and $(c + \Delta x, f(c + \Delta x))$.

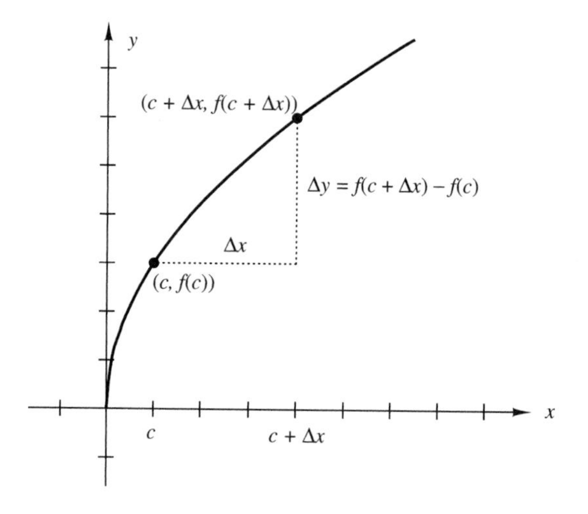

Figure 12.3 $y = f(x)$

Consider the tangent line to the graph of $y = f(x)$ at the point $(c, f(c))$. The *differential of y*, denoted dy, is the change in y measured along the **tangent line** to the graph of $y = f(x)$ at c as x changes from c to $c + \Delta x$. Figure 12.4 shows the graph of $y = f(x)$, the tangent line to the graph at the point $(c, f(c))$, and the differential dy.

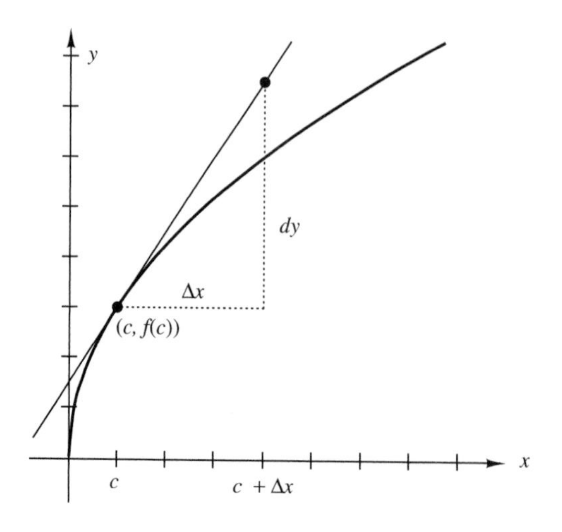

Figure 12.4 $y = f(x)$

Note that both the increment Δy and the differential dy for a given function $y = f(x)$ depend on the choice of the value c and the size of the increment Δx.

Since the tangent line at $(c, f(c))$ can be considered an approximation to the graph of $f(x)$ for x near c, the differential dy can be considered an approximation to the increment Δy. The differential dy is a good approximation for Δy when Δx is small and when the graph of $y = f(x)$ is fairly flat—that is, when the graph of the curve $y = f(x)$ is close to a straight line. (When the graph of the function $y = f(x)$ **is** a line, $\Delta y = dy$ for **all** values of c and Δx.)

EXAMPLE 12.2 Figures 12.5, 12.6, and 12.7 show the graph of a function, a tangent line to the graph, the increment Δy, the differential dy for this tangent line, and the increment Δx. The value $|\Delta x|$ is the same for all three graphs. In which graph is dy the best approximation to Δy? Explain why this is the best approximation.

a. Graph of $y = f(x)$

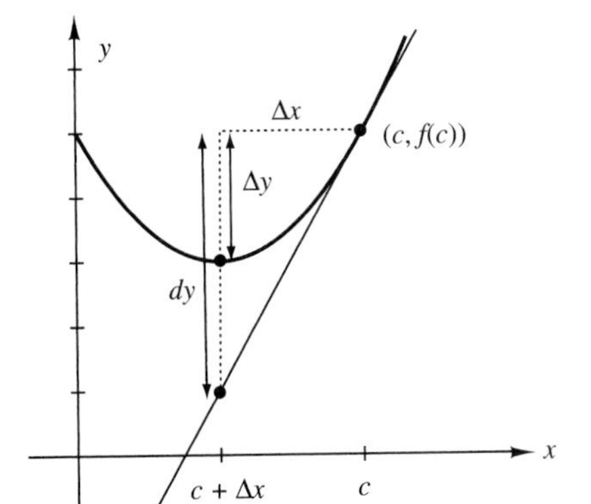

Figure 12.5

b. Graph of $y = g(x)$

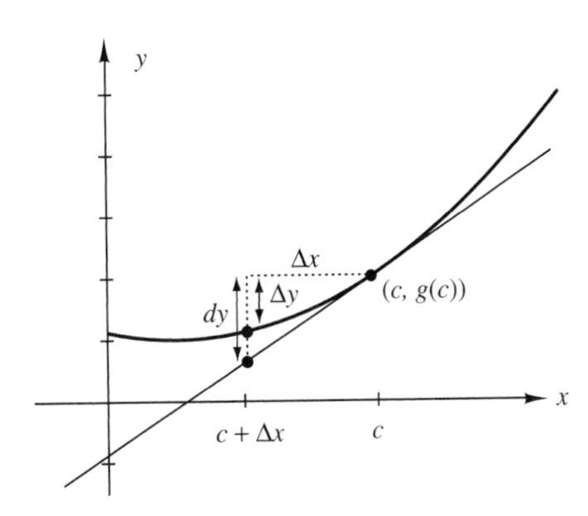

Figure 12.6

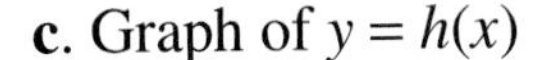
c. Graph of $y = h(x)$

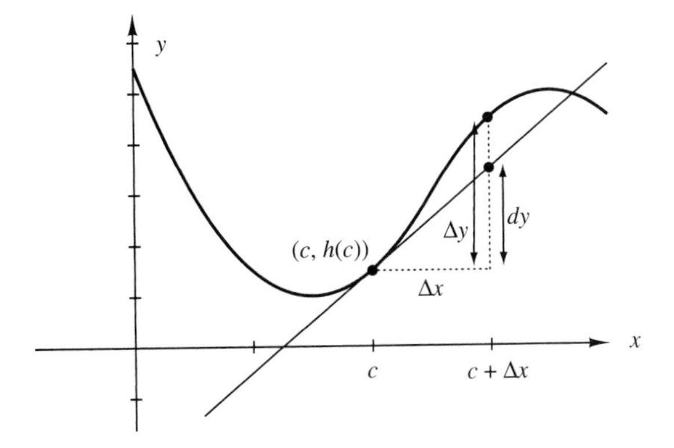

Figure 12.7

Solution The approximation of Δy by dy in part b is the best approximation among the three functions for the given c and Δx, since the function g is fairly flat near the point $(c, g(c))$ and the error $|dy - \Delta y|$ is the smallest. The tangent line to the graph of $y = g(x)$ at the point $(c, g(c))$ is a close approximation to $g(x)$ for x near c. ■

Because dy is the change in y measured along the tangent line, the vertical change dy divided by the horizontal change Δx is the slope $f'(c)$ of the tangent line, as Figure 12.8 shows. Therefore, $\frac{dy}{\Delta x} = f'(c)$ or $dy = f'(c) \cdot \Delta x$. The change in x is sometimes denoted as dx rather than by Δx, so Δx and dx are used interchangeably. In this choice of notation, $\frac{dy}{dx} = f'(c)$, so the derivative notation is given meaning as a quotient.

Thus, we have the following:

$$\Delta x = dx$$

$$\Delta y = f(c + \Delta x) - f(c)$$

$$dy = f'(c) \cdot \Delta x = f'(c) \cdot dx$$

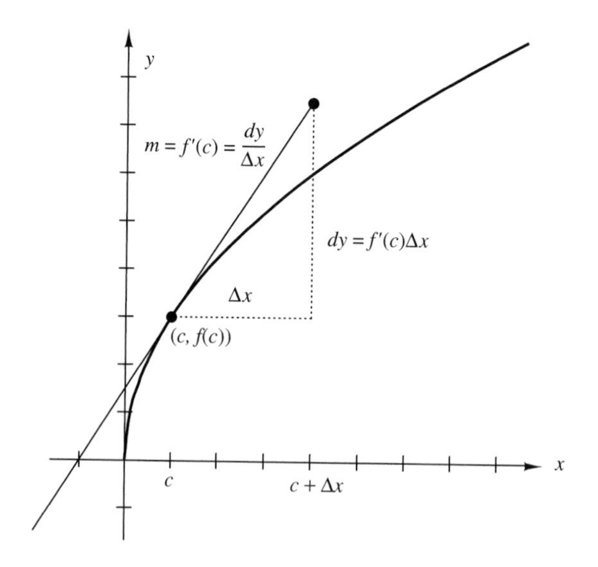

Figure 12.8

EXAMPLE 12.3 Suppose the function $y = p(x) = -0.05x^2 + 2x$ represents the price p (in dollars) of a commodity when there is a supply x (in thousands of units) available. For each given value of $\Delta x = dx$, find dy and Δy for $x = c = 30$ and compare their values. Interpret Δx, Δy, and dy in this context.

a. $\Delta x = dx = 10$ **b.** $\Delta x = dx = 5$ **c.** $\Delta x = dx = 0.5$

Solution In this context, Δy is the change in the price of the commodity when the supply is changed by an amount Δx, and $dy = p'(c)\Delta x$ is an approximation to Δy.

a. $\Delta y = p(30 + \Delta x) - p(30)$
$= p(40) - p(30)$
$= [-0.05(40)^2 + 2(40)] - [-0.05(30)^2 + 2(30)]$
$= [-80 + 80] - [-45 + 60] = -15$

The price will decrease by \$15 if supply is increased by 10 (thousand units). To find dy, we first differentiate $p(x)$: $p'(x) = -0.1x + 2$. Then $dy = p'(c)\Delta x = [-0.1c + 2](10) = [-0.1(30) + 2](10) = -10$. Since dy approximates Δy, the approximate price decrease is \$10 if supply is increased by 10 (thousand units). The error is $|dy - \Delta y| = 5$.

b. $\Delta y = p(30 + 5) - p(30) = p(35) - p(30) = [-0.05(35)^2 + 70] - [15] = -6.25$. The price will decrease by \$6.25 if supply is increased by 5 (thousand units). Next, we compute $dy = p'(c)\Delta x = [-0.1(30) + 2](5) = -5$. The approximate price decrease is \$5 if supply is increased by 5 (thousand units). The error is $|dy - \Delta y| = 1.25$.

c. $\Delta y = p(30.5) - p(30) = 14.4875 - 15 = -0.5125 \approx -0.51$. The price will decrease by \$.51 if supply is increased by 0.5 (thousand units)—that is, by 500 units. Also, $dy = p'(c)\Delta x = [-0.1(30) + 2](0.5) = -0.50$. The approximate price decrease is \$.50 if supply is increased by 0.5 (thousand units). Here the error is $|dy - \Delta y| = 0.01$. ■

The graphs in Figures 12.9a and 12.9b show Δy and dy for parts a and b of Example 12.3. In Figure 12.9c it is impossible to distinguish between the graph of $p(x)$ and the tangent line for x within 0.5 unit of $c = 30$. Note that for smaller Δx, dy better approximates Δy. The closer x is to c, the closer the graph of $y = p(x)$ is to the tangent line to the graph at the point $(c, f(c)) = (30, 15)$.

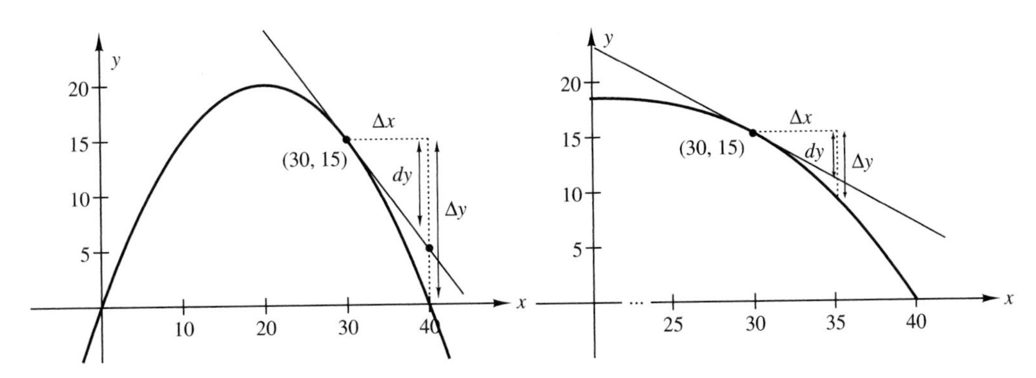

Figure 12.9a **Figure 12.9b**

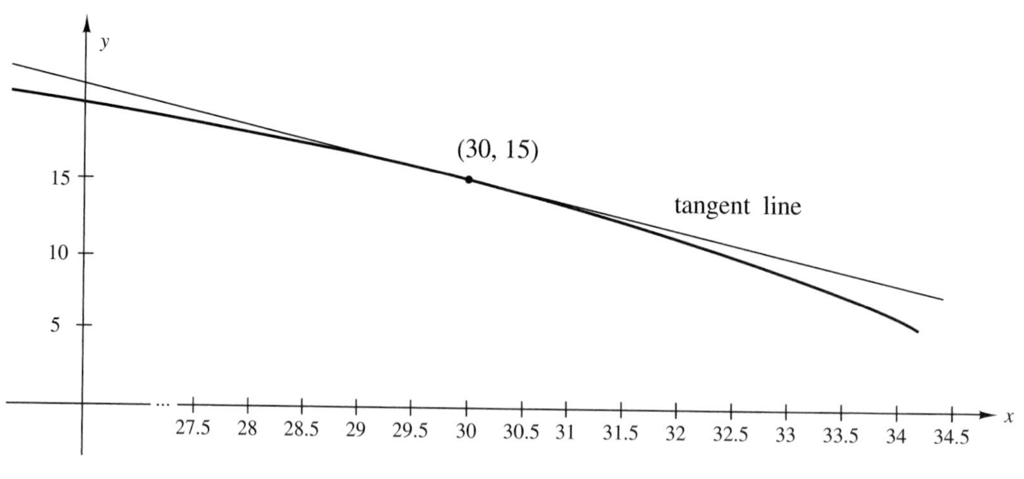

Figure 12.9c

Exercises 12-B

1. In Figures 12.10 and 12.11, use the given value of Δx and the given point on the graph to do the following:

(i) Draw the tangent line to the graph at the given point.

(ii) Sketch the distance Δy.

(iii) Sketch the distance dy.

a. Graph of $y = F(x)$; $\Delta x = 0.5$

b. Graph of $y = G(x)$; $\Delta x = -0.7$.

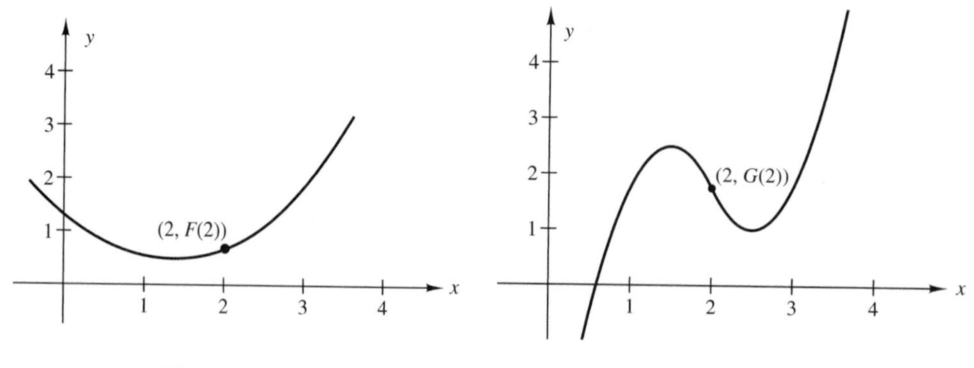

Figure 12.10

Figure 12.11

2. **a**. For each of the graphs in Exercise 1, explain why the differential dy is or is not a good approximation to Δy.

b. For each of the graphs in Exercise 1, estimate a value of Δx such that $|dy - \Delta y| = 0.5$.

3. Suppose the function $y = D(x) = 100 - 2x - 0.01x^2$ represents the demand (in thousands of units) of a commodity when the price is x dollars. For each given value of $\Delta x = dx$, find dy and Δy for $x = \$20$, and find the error of the tangent line approximation. Interpret Δx, Δy, and dy in this context.

a. $\Delta x = dx = 5$ **b.** $\Delta x = dx = 0.5$ **c.** $\Delta x = dx = 0.05$

4. The internal measurement of the edge of a box in the shape of a cube is 20 cm ± 0.02 cm.

a. What is the smallest volume possible for the cube?

b. What is the largest volume possible for the cube?

c. How can this error in the measurement of the volume of the cube be approximated with a differential?

Chapter 12 Exercises

1. Suppose $S(a) = 200a - 0.5a^2$ gives the sales for a cola as a function of the amount a spent on advertising (in thousands of dollars).

a. Use the differential dS to estimate the change in sales that will result from an increase in advertising dollars from \$200,000 to \$210,000.

b. Calculate ΔS, the exact change in sales for the same increase in advertising dollars.

c. Find the error in your approximation from part a.

d. Find an equation of the tangent line to the function S at the point $(a, S(a))$, where $a = 100$.

e. Use the line found in part d to estimate the sales when \$103,000 is spent on advertising.

2. Suppose that $y = g(x)$ represents the volume of a sphere with radius x (see Commonly Used Formulas in Section 6-A).

a. Use the differential dy to estimate the change in the volume of helium in a spherical balloon as the radius increases from 7 to 7.2 in.

b. Find Δy, the exact change in the volume of helium as the radius of the balloon increases from 7 to 7.2 in.

c. Find the error in your approximation from part a.

3. A point is shown on each of the graphs in Figures 12.12, 12.13, 12.14, and 12.15.

(i) Sketch the tangent line to each graph at the given point.

(ii) Sketch dy and Δy on each graph for the point shown, with $\Delta x = 0.5$.

(iii) For each of these points, explain why the differential dy, with $\Delta x = 0.5$, is or is not a good approximation to Δy.

a.

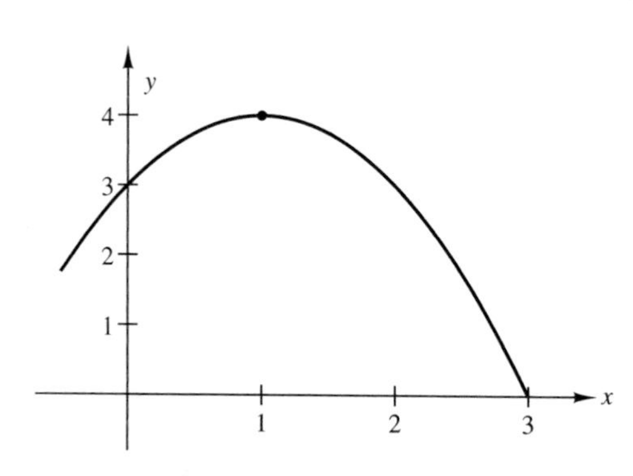

Figure 12.12

b.

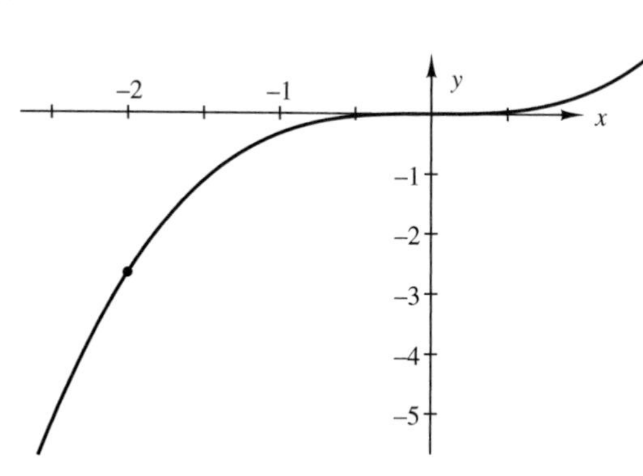

Figure 12.13

c.

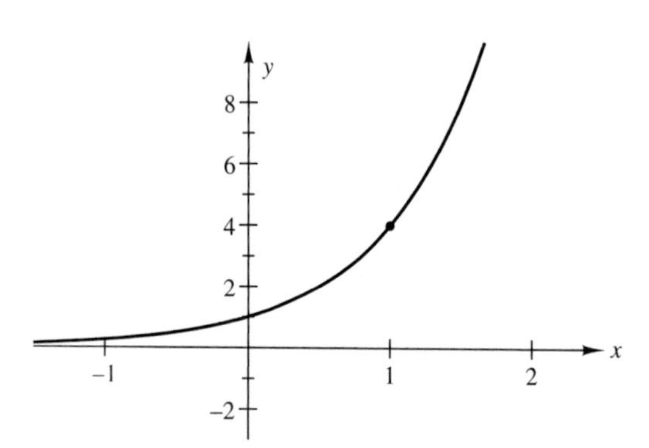

Figure 12.14

d.

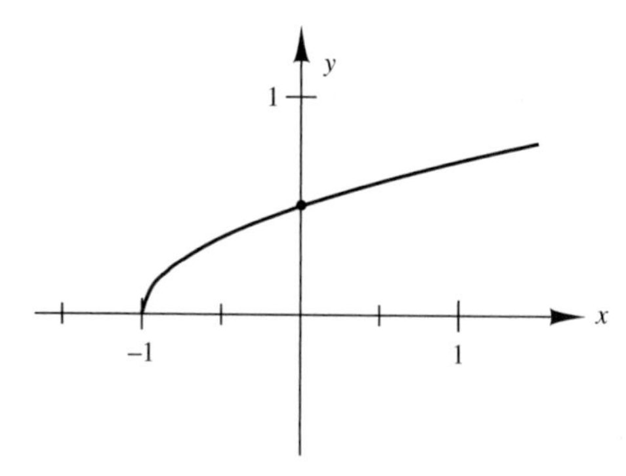

Figure 12.15

4. **a.** For the graph in Figure 12.12, estimate a value of Δx such that $|dy - \Delta y| = 0.5$.

b. For the graph in Figure 12.13, estimate a value of Δx such that $|dy - \Delta y| = 0.5$.

c. For the graph in Figure 12.14, estimate a value of Δx such that $|dy - \Delta y| = 1.0$.

5. Three functions are given below, along with the equation of the tangent line to the function at the point (0, 1). For each, use a graphing utility to graph the function and this tangent line for the window $-0.5 \le x \le 0.5$, $0.3 \le y \le 1.3$. Then answer the questions below.

(i) $f(x) = \sqrt{1+x},\ y = 1 + \frac{x}{2}$ (ii) $f(x) = \sqrt[3]{1+x},\ y = 1 + \frac{x}{3}$

(iii) $f(x) = \sqrt[4]{1+x},\ y = 1 + \frac{x}{4}$

a. For each function, is the tangent line a good approximation to the function for the given interval? Justify your answers.

b. Which of the three tangent line approximations (i, ii, or iii) appears to be best for $f(0.5)$?

c. For each of the three functions, calculate the error introduced by using the tangent approximation to estimate $f(0.5)$. Does this confirm your answer in part b?

CHAPTER 13

Companion to Exponential Functions

Most bacteria reproduce themselves by fission; the cell of a bacterium divides in two. With one particular kind of bacterium, *Escherichia coli,* division takes place every 20 min. This means that it takes 20 min for the number of bacteria to double. If initially there are n bacteria in a colony, then after 20 min there will be $2n$ bacteria. After a second 20-min time period, the $2n$ bacteria have doubled their number and there are $2(2n) = 2^2 n$ bacteria. After three consecutive 20-min time periods, there are $2^3 n$ bacteria. In general, after t consecutive 20-min time periods, there are

$$N = 2^t n$$

bacteria in the colony.

Suppose initially a colony contains 1000 bacteria cells. At the end of 1 hour (three 20-min periods) there will be $2^3(1000) = 8000$ cells. After 2 hours there will be $2^6(1000) = 64{,}000$ cells. After 4 hours there will be $2^{12}(1000) = 4{,}096{,}000$ cells. If the division continues, there will be more than 30 million cells within an additional hour.

In the equation for N, the independent variable t on the right side of the equation appears as an exponent. The kind of growth described above is called *exponential growth*. The function

$$f(t) = 2^t$$

is an example of an *exponential function* because the function consists of a constant raised to a variable power. If we compare the exponential function $f(t) = 2^t$ to a polynomial function like $g(t) = t^2$, we see that the roles of the variable and the constant are interchanged. In the exponential function $f(t) = 2^t$, the constant 2 (called the *base*) is raised to a variable power t, whereas in the polynomial function $g(t) = t^2$, the variable t is raised to the constant power 2.

Because the bacteria cells in the above example divide every 20 min, we considered the expression $N = 2^t n$ for integer values of t (representing 20-min time periods). We now look at exponential functions such as $f(t) = 2^t$, where t takes on all real values.

13-A RULES OF EXPONENTS

We can construct a table of values of $f(t) = 2^t$ for some integer and fraction values of t by using the definition of a^t for integer and fraction values of t. (For example $2^{-3} = \frac{1}{2^3} = \frac{1}{8}$ and $2^{1/2} = \sqrt{2}$. See Section 7-A for further review.) In the table below, decimals are rounded to the nearest thousandth.

t	-3	-2	-1	$-\frac{1}{2}$	0	$\frac{1}{2}$	$\frac{3}{4}$	2	3	4	5
$f(t)=2^t$	$\frac{1}{8}$	$\frac{1}{4}$	$\frac{1}{2}$	0.707	1	1.414	1.682	4	8	16	32

The graph in Figure 13.1 shows a plot of some of the points $(t, f(t))$ from the table. Notice that as t increases, the values of $f(t)$ increase more rapidly. This is typical of exponential growth.

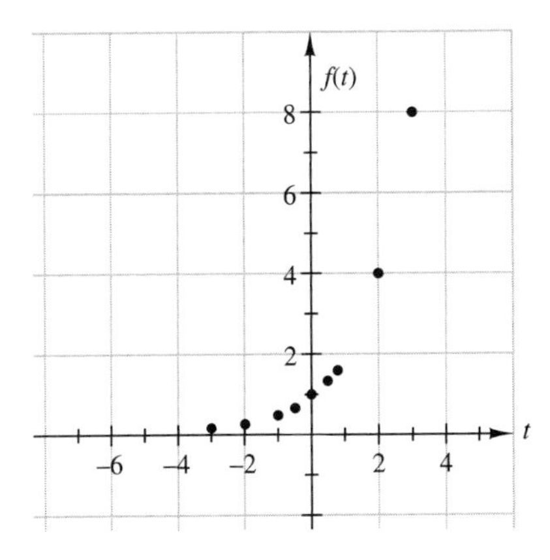

Figure 13.1 Points on the graph of $f(t) = 2^t$

Our discussion of the function $f(t) = 2^t$ illustrates how the function $f(t) = a^t$ is defined in terms of powers and roots when the exponent t is a rational number and the base a is any positive real number. The shape of the graph in Figure 13.1 is strongly suggestive, but there are many missing points. In fact, all points $(t, 2^t)$, where t is an irrational number, are missing. We define $f(t) = 2^t$ at irrational values of t so that f is continuous there. On the picture, this means that we fill in the gaps so that the graph is a smooth curve with no breaks and looks like the graph shown in Figure 13.2.

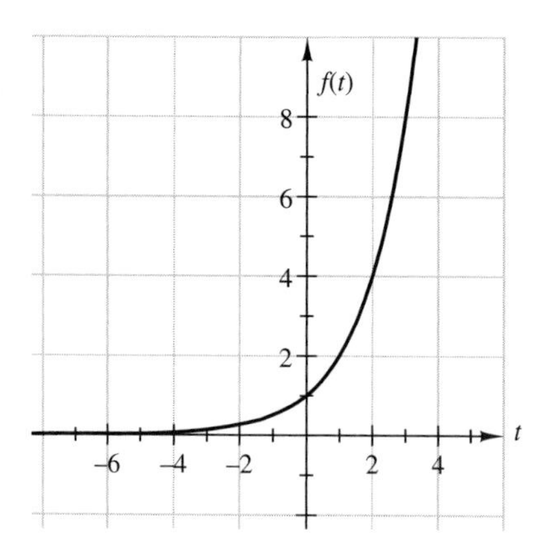

Figure 13.2 $f(t) = 2^t$

Since every irrational number can be approximated to any desired degree of accuracy by rational numbers, values of $f(t) = 2^t$ where t is irrational can be carefully approximated. For example, $2^{\sqrt{2}}$ can be approximated by $2^{1.414} = 2^{1414/1000}$. For a more accurate approximation, $2^{\sqrt{2}}$ can be approximated by $2^{1.4142} = 2^{14,142/10,000}$, and so on.

It is customary to consider only exponential functions of the form $f(t) = a^t$ where $a > 0$ and $a \neq 1$. (There are problems for other values of the base a. If $a = 0$, then $a^t = 0$ for $t > 0$, but a^t is undefined for $t \leq 0$. If $a < 0$, then a^t is not a real number for $t = \frac{1}{4}, \frac{1}{2}, \frac{3}{4}$ and for any other simple fraction with an even denominator. Finally, if $a = 1$, then $1^t = 1$ for all values of t.)

The rules of exponents discussed in Section 7-A continue to hold true for exponential functions $f(t) = a^t$ where t is a real variable. These rules are useful when working with exponential functions, and so we summarize them here for easy reference. In the rules below, a and b are positive constants and x and y are real variables.

1. $a^0 = 1$
2. $a^{-x} = \frac{1}{a^x} = \left(\frac{1}{a}\right)^x$
3. $a^x a^y = a^{x+y}$
4. $\frac{a^x}{a^y} = a^{x-y} = \frac{1}{a^{y-x}}$
5. $\left(a^x\right)^y = a^{xy}$
6. $(ab)^x = a^x b^x$
7. $\left(\frac{a}{b}\right)^x = \frac{a^x}{b^x}$

We can use these rules to express some combinations of exponential functions in the form $a^{u(t)}$, where u is a function of t. (See the Appendix for tips to avoid mistakes when working with exponents.)

EXAMPLE 13.1 Use the rules of exponents to rewrite each of the following as an exponential function of the form $a^{u(t)}$—that is, as a single constant raised to a power that is a function of t.

a. $f(t) = \frac{3^t}{9}$ b. $g(t) = \sqrt{\frac{3^t}{2^t}}$ c. $h(t) = (3^{-t})^t$ d. $k(t) = 3^{-t}3^t$

Solution

a. Since $9 = 3^2$, $f(t)$ can be written as $f(t) = \frac{3^t}{3^2}$. So by rule 4, $f(t) = 3^{t-2}$.

b. By rules 7 and 5, $g(t) = \sqrt{\frac{3^t}{2^t}} = \sqrt{\left(\frac{3}{2}\right)^t} = \left(\left(\frac{3}{2}\right)^t\right)^{1/2} = \left(\frac{3}{2}\right)^{t/2}$.

c. By rule 5, $h(t) = \left(3^{-t}\right)^t = 3^{-t^2}$.

d. By rules 3 and 1, $k(t) = 3^{-t}3^t = 3^{-t+t} = 3^0 = 1$. ■

In some cases we cannot combine exponential functions into a single simpler exponential function. It is important to be aware of which properties of exponents are used when rewriting a function.

EXAMPLE 13.2 In parts a–f two functions are given. If they are equal, give the properties of exponents that show the equality. If they are not equal, find a specific value of t at which the two functions are not equal, and give the values of the functions for that t.

a. $2^t + 2^{t^3} \stackrel{?}{=} 2^{t+t^3}$ b. $4^t + 5^t \stackrel{?}{=} 9^t$ c. $4^t 5^t \stackrel{?}{=} 20^t$

d. $\frac{1}{3} + \frac{1}{5^t} \stackrel{?}{=} \left(3 + 5^t\right)^{-1}$ e. $3^t 3^{t^2} \stackrel{?}{=} 3^{t+t^2}$ f. $\frac{2^{t^3}}{2^t} = 2^{t^2}$

Solution

a. $2^t + 2^{t^3} \neq 2^{t+t^3}$. If we choose $t = 2$, the expression on the left is $2^2 + 2^8 = 260$ and the expression on the right is $2^{10} = 1024$.

b. $4^t + 5^t \neq 9^t$. Choose $t = 0$; then $4^0 + 5^0 = 1 + 1 = 2$, while $9^0 = 1$.

c. $4^t 5^t = (4 \cdot 5)^t = 20^t$ by rule 6 of exponents.

d. $\frac{1}{3} + \frac{1}{5^t} \neq \left(3 + 5^t\right)^{-1}$. Choose $t = 1$; then the function on the left takes the value $\frac{1}{3} + \frac{1}{5} = \frac{8}{15}$ and the function on the right takes the value $(3 + 5)^{-1} = 8^{-1} = \frac{1}{8}$.

e. $3^t 3^{t^2} = 3^{t+t^2}$ by rule 3 of exponents.

f. $\frac{2^{t^3}}{2^t} \neq 2^{t^2}$. Choose $t = 1$; then $\frac{2^{1^3}}{2^1} = \frac{2}{2} = 1$, while $2^{t^2} = 2$. ■

In general, the sum or difference of two exponential expressions cannot be combined to produce a single expression. For example, $3^t + 2^t \neq 5^t$, $3^t - 2^t \neq 1^t$, and $3^{2t} + 3^{5t} \neq 3^{7t}$.

The graph of $f(t) = 2^t$ is shown in Figure 13.2. The graph of **any** exponential function of the form $g(t) = a^t$ where $a > 1$ has a form similar to the graph of $f(t) = 2^t$. Here are several general observations about the graph of $y = a^t$, for $a > 1$:

1. The domain of $y = a^t$, where $a > 1$, is all real numbers.

2. $a^0 = 1$, so the graph goes through the point (0, 1).

3. $a^t > 0$ for all t, so the graph lies in the upper half of the plane (above the t-axis). The range of a^t is $(0, \infty)$.

4. As t decreases, the value of a^t decreases and gets closer and closer to 0. Thus,

$$\lim_{t \to -\infty} a^t = 0 \quad \text{for} \quad a > 1, \text{ so } y = 0 \text{ is a horizontal asymptote.}$$

5. As t increases, the value of a^t increases without bound. Thus, a^t is an increasing function and

$$\lim_{t \to \infty} a^t = \infty \quad \text{for } a > 1.$$

6. The steeper rise of the graph is for larger values of the base a. In general, if $a > b > 1$, then for $t > 0$, $a^t > b^t$ and for $t < 0$, $a^t < b^t$.

The graphs of $y = 2^t$, $y = 3^t$, and $y = 10^t$ are shown in Figure 13.3.

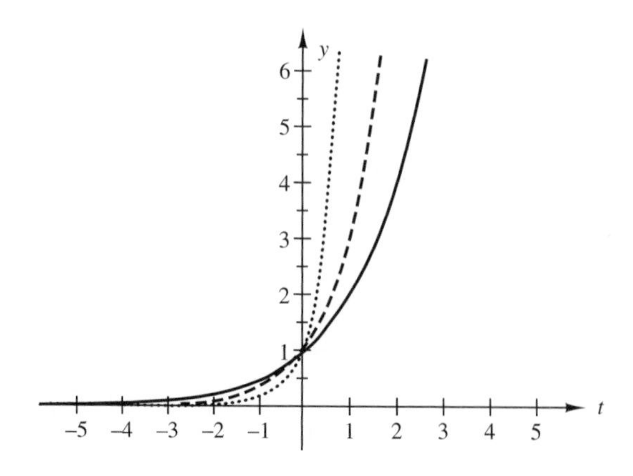

Figure 13.3 $y = 2^t$ (solid); $y = 3^t$ (dashed); $y = 10^t$ (dotted)

When $0 < a < 1$, the behavior of $h(t) = a^t$ can be observed from the example of the function $h(t) = \left(\frac{1}{2}\right)^t$ where $a = \frac{1}{2}$. Note that $\left(\frac{1}{2}\right)^t = \left(2^{-1}\right)^t = 2^{-t}$. Thus there are two ways to write the function $h(t)$:

$$h(t) = \left(\frac{1}{2}\right)^t \quad \text{or} \quad h(t) = 2^{-t}.$$

Several values of the function $h(t) = \left(\frac{1}{2}\right)^t = 2^{-t}$ are charted in the table below.

t	-3	-2	-1	$-\frac{1}{2}$	0	$\frac{1}{2}$	$\frac{3}{4}$	1	2	3
$h(t) = 2^{-t}$	8	4	2	1.414	1	0.707	0.595	0.5	0.25	0.125

The graph of $h(t) = \left(\frac{1}{2}\right)^t = 2^{-t}$ is shown in Figure 13.4.

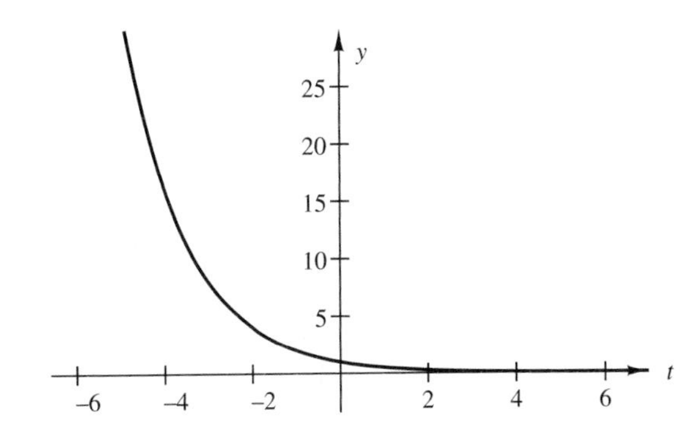

Figure 13.4 $h(t) = \left(\frac{1}{2}\right)^t = 2^{-t}$

We can make some general observations about the graph of a^t for $0 < a < 1$:

1. The domain of $y = a^t$, for $0 < a < 1$, is all real numbers.

2. The graph goes through the point (0, 1).

3. The graph lies in the upper half of the plane (above the t-axis), so $a^t > 0$ for all t.

4. As t decreases, a^t increases without bound:

$$\lim_{t \to -\infty} a^t = \infty \quad \text{for} \quad 0 < a < 1.$$

5. As t increases, a^t decreases and gets closer to 0:

$$\lim_{t \to \infty} a^t = 0 \quad \text{for} \quad 0 < a < 1,$$

so $y = 0$ is a horizontal asymptote.

6. The graphs of $y = a^t$ and $y = \left(\frac{1}{a}\right)^t$ are related by a reflection in the y-axis. This is because if $h(t) = a^t$, then $h(-t) = a^{-t} = \left(\frac{1}{a}\right)^t$, and we learned in Section 2-F that the graph of $y = h(-t)$ is the reflection of the graph of $y = h(t)$ about the y-axis.

Figure 13.5 shows that the graphs of $y = \left(\frac{1}{4}\right)^t$ and $y = 4^t$ are reflections of each other about the y-axis. Notice also that both functions pass through the point (0, 1), both lie above the line $y = 0$, and both have the t-axis as a horizontal asymptote.

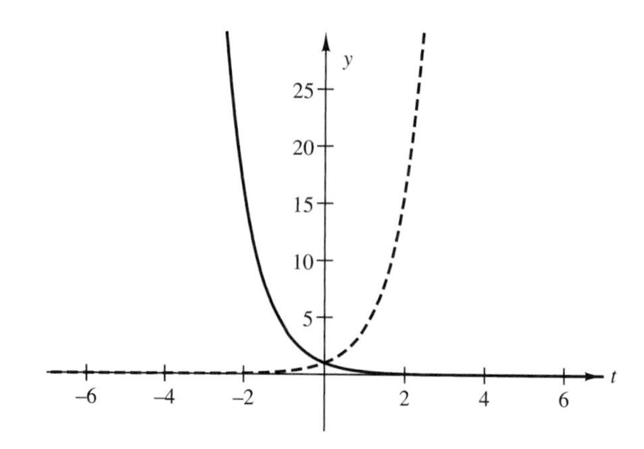

Figure 13.5 $y = \left(\frac{1}{4}\right)^t$ (solid); $y = 4^t$ (dashed)

Compound interest and the effects of inflation can be modeled by exponential functions.

EXAMPLE 13.3 (Compound Interest)

a. Suppose \$1, a one-time deposit, is put into a savings account that earns 5% interest per year compounded annually. The interest is added to the account each year and all the money is left in the account indefinitely. Make a table of values that shows how much money will be in the account after t years for $t = 0, 1, 3, 5, 10, 25, 50$, and 100. Draw a graph of the function.

b. Suppose \$500 is deposited into a savings account that earns 5% interest per year compounded annually, interest is added to the account each year, and all the money is left in the account indefinitely. How much money will be in the account after 10 years? After 25 years?

c. Because of inflation, money loses value (purchasing power). Suppose money loses value at the rate of 6% per year. In other words, today's dollar is worth only \$.94 (in purchasing power) after 1 year. Find the value of \$1 after t years for $t = 0$, 1, 3, 5, 10, 25, 50, and 100 and draw a graph of the function.

Solution

a. If \$1 is deposited into the account at time $t = 0$, after 1 year there will be $\$(1 + 0.05(1)) = \$1.05$ in the account. That amount is then left in the account so that after $t = 2$ years, there will be \$1.05 plus 5% interest on \$1.05: $1.05 + 0.05(1.05)$. We factor out 1.05 to get $1.05(1 + 0.05) = (1.05)^2$ dollars as the amount in the account after $t = 2$ years. That amount is again left in the account so that at the end of $t = 3$ years, there will be 1.05^2 plus interest on 1.05^2. After $t = 3$ years, there will be $1.05^2 + (0.05)1.05^2$. We factor out 1.05^2 to get $1.05^2(1.00 + 0.05) = 1.05^3$ dollars in the account after $t = 3$ years. In general, after t years there will be $f(t) = 1.05^t$ dollars in the account. The following table of values shows how the amount of money in the account grows:

t	0	1	3	5	10	25	50	100
$f(t) = 1.05^t$	1	1.05	1.16	1.28	1.63	3.39	11.47	131.50

Since we are assuming for this problem that interest is compounded annually, the function $f(t)$ is actually a step function. But we will approximate it by the continuous function $f(t) = 1.05^t$. Notice that since $1.05 > 1$, the exponential function $f(t) = 1.05^t$ is an increasing function. See Figure 13.6.

b. In real life, more than \$1 would be deposited into a savings account. To find how much money will be in the account after 10 years if \$500 dollars is deposited initially under the circumstances described above, we multiply by 500 the amount that would be in the account after 10 years if \$1 were deposited initially. Since each \$1 will grow to \$1.63 after 10 years, \$500 will grow to 500(1.63) = \$815. In 25 years, \$500 will grow to 500(3.39) = \$1,695.

c. If inflation causes \$1 to be worth \$(1 – 0.06) = \$.94 after 1 year, then after 2 years, \$1 will be worth \$.94 minus 6% of \$.94, which is $0.94 - (0.06)(0.94) = 0.94(1 - 0.06) = 0.94^2$ dollars. In general, after t years, \$1 will be worth $g(t) = (1 - 0.06)^t = 0.94^t$ dollars. The following table shows how \$1 decreases in value:

t	0	1	3	5	10	25	50	100
$g(t) = 0.94^t$	1	0.94	0.83	0.73	0.54	0.21	0.05	0.002

Notice that since $0 < 0.94 < 1$, the exponential function $g(t) = 0.94^t$ is a decreasing function. Figure 13.7 shows the graph of $y = 0.94^t$ for $t \geq 0$.

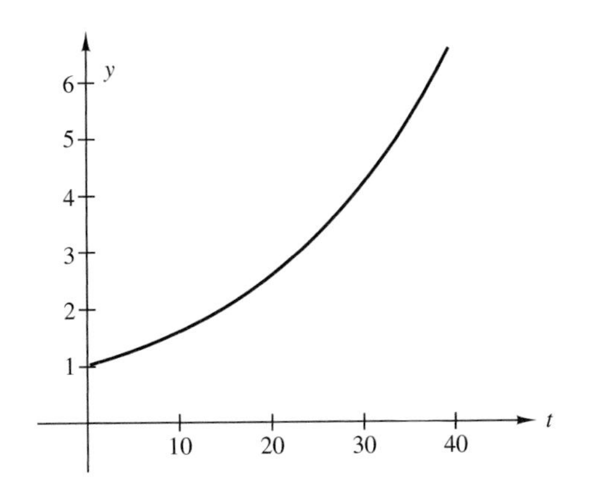

Figure 13.6 $y = 1.05^t, t \geq 0$

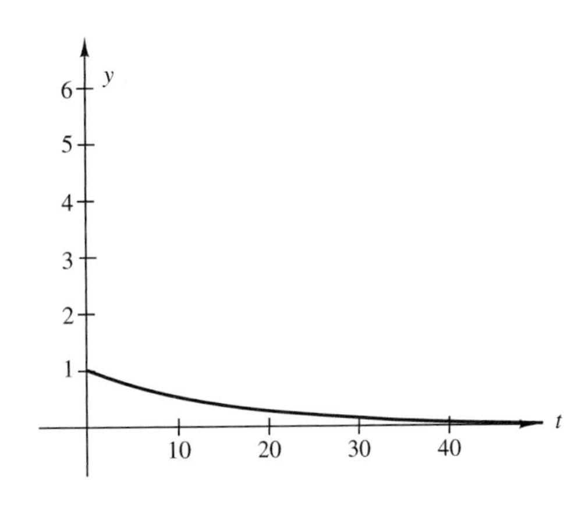

Figure 13.7 $y = 0.94^t, t \geq 0$ ■

The next example shows how to use transformations to sketch graphs that involve exponential functions.

EXAMPLE 13.4 Sketch the graphs of the following functions, using properties of the graph of $y = a^t$ and transformations of graphs discussed in Section 2-F.

a. $y = 3^{t-2}$ **b.** $y = 3^t - 2$ **c.** $y = 3^{-|t|}$

Solution

a. Since $3^0 = 1$, the graph of $y = 3^{t-2}$ goes through the point (2, 1). The graph is the same shape as the graph of $y = 3^t$ but is shifted 2 units to the right. See Figure 13.8.

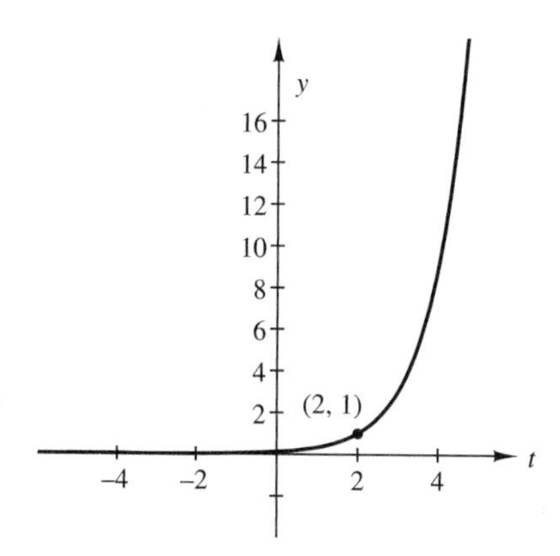

Figure 13.8 $y = 3^{t-2}$

b. The graph of $y = 3^t - 2$ is obtained by shifting the graph of $y = 3^t$ down 2 units. As t decreases, $3^t - 2$ approaches the horizontal asymptote $y = -2$. See Figure 13.9.

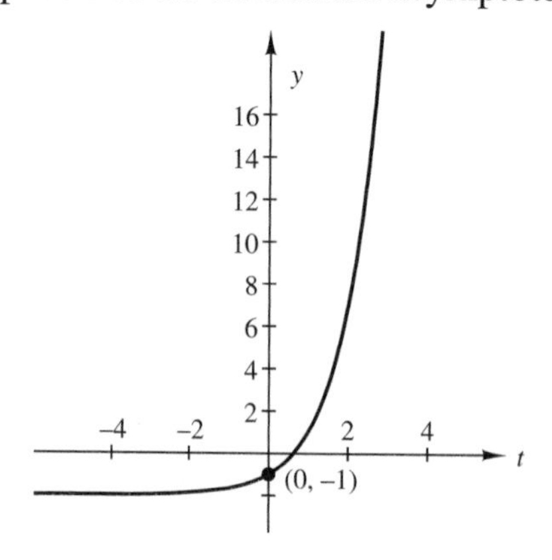

Figure 13.9 $y = 3^t - 2$

c. Since $|t| = t$ if $t \geq 0$, the function $y = 3^{-|t|} = 3^{-t}$ if $t \geq 0$. Thus the portion of the graph to the right of $t = 0$ is exactly that of the graph of $y = 3^{-t} = \left(\frac{1}{3}\right)^t$. Also, $|t| = -t$ if $t < 0$, so the portion of the graph to the left of $t = 0$ is exactly that of the graph of $y = 3^{-(-t)} = 3^t$. See Figure 13.10.

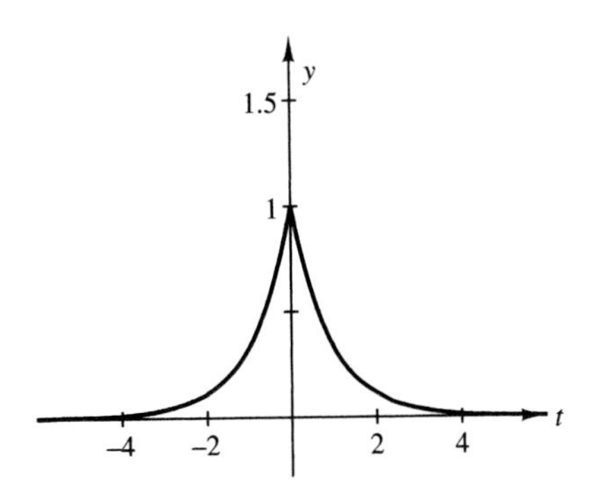

Figure 13.10 $y = 3^{-|t|}$ ■

Exercises 13-A

1. Use what you know about the graphs of exponential functions to match each equation with its graph.

a. $y = 5^x$ **b**. $y = \dfrac{1}{5^x}$ **c**. $y = -3^x$ **d**. $y = 3^{-x}$

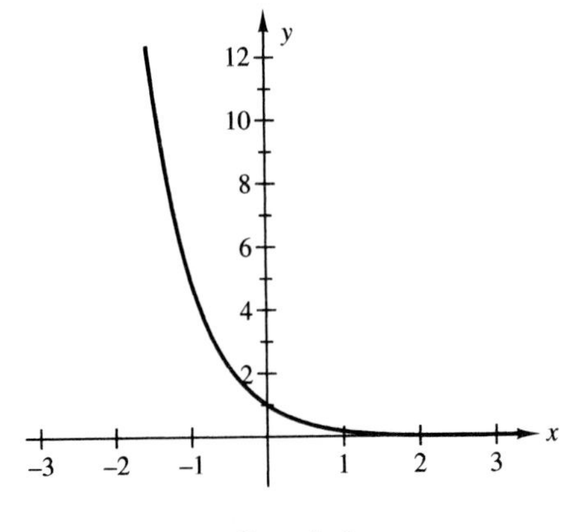

Graph 1

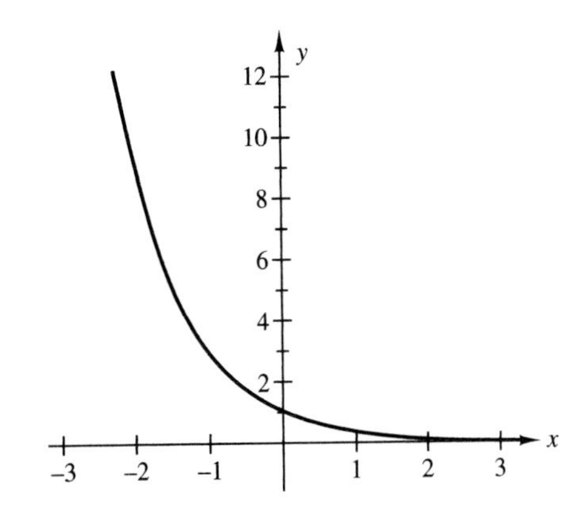

Graph 2

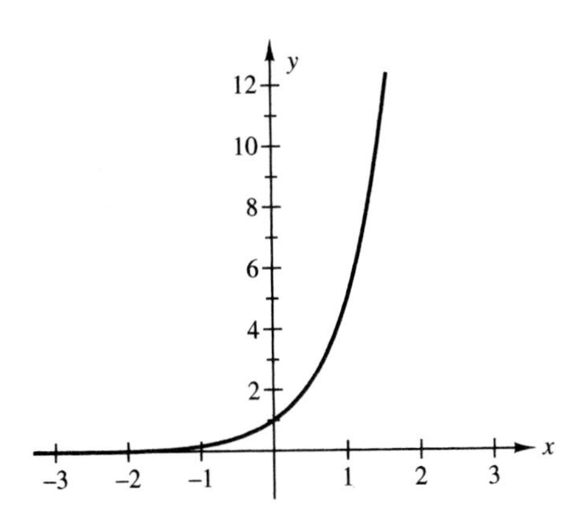

Graph 3

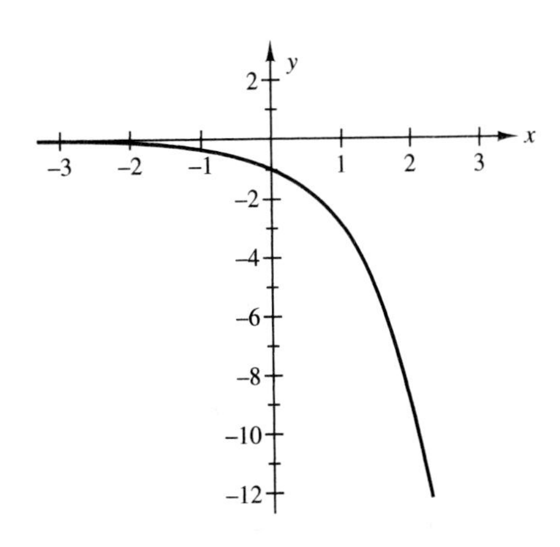

Graph 4

2. The graph of $y = a^x$ is shown below. Sketch the graph of $y = a^{x+2}$.

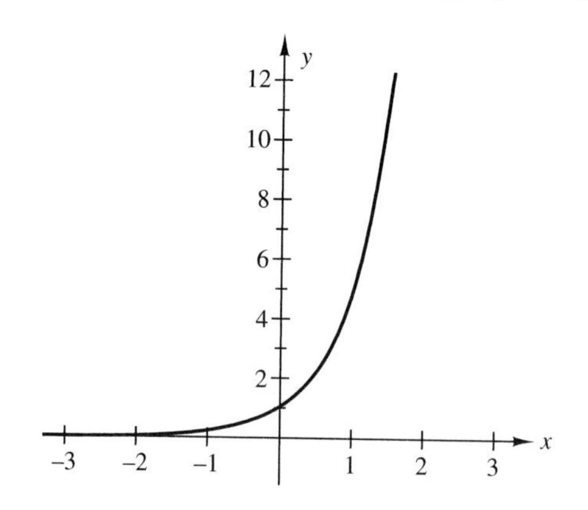

3. The graph of $y = b^x$ is shown in Graph 1 below. What is the equation of the graph shown in Graph 2?

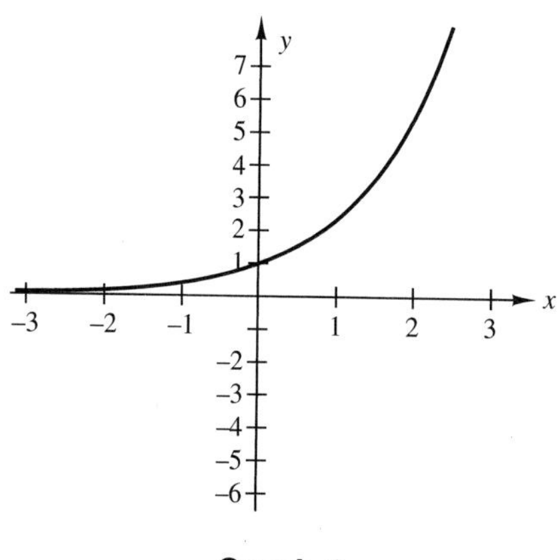

Graph 1

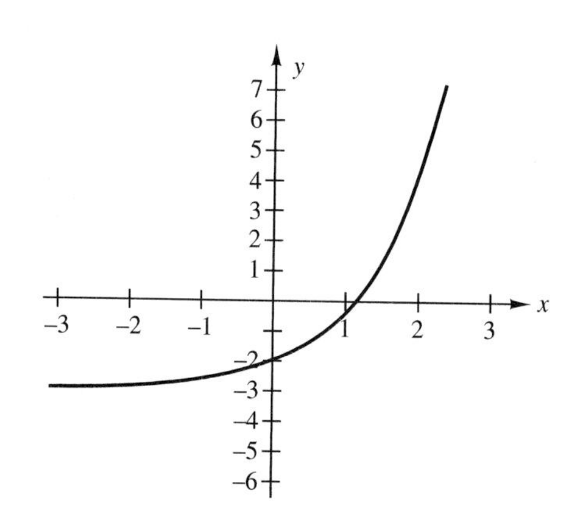

Graph 2

4. Sketch the graph of each of the following functions using properties of transformations discussed in Section 2-F.

a. $f(x) = 2^{1-x}$ **b.** $h(x) = 3^{2x}$ **c.** $g(x) = 5^{-x} + 2$

d. $k(x) = 2^{|x|}$ **e.** $m(x) = (2)(2^x)$

5. Simplify each of the following expressions to the form a^p for constants a and p:

a. $2^3\, 8^6\, 4^{-6}$ **b.** $\sqrt{3}\left(9^{-2}\right)$ **c.** $\frac{8}{27}\sqrt[3]{\frac{2}{3}}$

6. a. In Example 13.3 we showed that $y = 1.05^t$ is a mathematical model for the value of \$1 invested at 5% compounded annually. What is the meaning of $y = 1.10^t$ in this context?

b. Use the window $0 \le x \le 25$, $0 \le y \le 12$ to sketch the graphs of $y = 1.05^x$ and $y = 1.10^x$ on the same set of axes. Use these graphs to compare a 10% growth rate with a 5% growth rate for an investment. What is the value of \$1 after 15 years for each growth rate?

7. **a.** The equation $y = 0.94^t$ gives the real value of \$1 after t years at an annual inflation rate of 6%. What is the meaning of $y = 0.90^t$ in this context?

b. Use the window $0 \le x \le 40$, $0 \le y \le 10$ to sketch the graphs of $y = 0.90^x$ and $y = 0.94^x$ on the same set of axes. Compare the depreciation in the value of \$1 with each inflation rate. What is the real value of \$1 after 15 years for each rate of inflation?

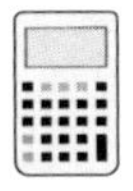

8. Use properties of exponents to simplify each of the following expressions to the form $2^{u(t)}$:

a. $\sqrt{2^{2t}4^{t^2}}$ b. $\dfrac{2^t 4^{t-1}}{8^{t^2}}$ c. $\sqrt{16\sqrt{2^{3t}}}$

9. Decide whether or not each of the following expressions is correct. If an expression is correct, state the rule of exponents used; if it is not correct, explain why.

a. $3^5 + 4^5 = 7^5$ b. $2^{-1/2} = -\dfrac{1}{2^2}$ c. $\dfrac{1}{x+y} = (x+y)^{-1}$

d. $2^3 + 2^4 = 2^7$ e. $\dfrac{1}{x} + \dfrac{1}{y} = \dfrac{1}{x+y}$ f. $2^{-1} = \sqrt{2}$

10. Say whether or not the functions in each given pair are equal. If they are equal, state the rule of exponents you need to show they are equal. If the functions are not equal, find a specific value of t at which the two functions are not equal.

a. $7^{-t} \stackrel{?}{=} -\dfrac{1}{7^t}$ b. $2^t 2^{t^3} \stackrel{?}{=} 2^{t+t^3}$ c. $\dfrac{1}{3^t - 4^t} \stackrel{?}{=} 3^{-t} - 4^{-t}$

d. $\dfrac{1}{2^t + 5^t} \stackrel{?}{=} \left(2^t + 5^t\right)^{-1}$ e. $\dfrac{5^{-t}}{7^{-t}} \stackrel{?}{=} \left(\dfrac{7}{5}\right)^t$ f. $\left(5^t\right)^2 \stackrel{?}{=} 5^{t^2}$ g. $2^t \cdot 2^{t^2} \stackrel{?}{=} 2^{t^3}$

11. If a single deposit of \$500 is made into an account that earns 8% interest compounded annually, how much will be in the account after 10 years if nothing is withdrawn from it?

12. If a certain type of bacteria doubles its number in 30 min, and 1 million bacteria are currently present in a colony, give the number of bacteria present in 1 hour; in 3 hours. (Assume no bacteria die.)

13. To plan for the future value of funds invested today, depreciation due to inflation must be taken into account. If the value of \$1 declines by 4% each year, what will it be worth in 5 years? In 10 years?

13-B THE NATURAL EXPONENTIAL FUNCTION

One of the most frequently occurring exponential functions is the *natural exponential function*. This is the function $y = e^t$, where $\boldsymbol{e}$ denotes an irrational number that occurs frequently in applications. The value of e is approximately 2.71828. Figure 13.11 shows the graphs of three functions: $y = 2^t$, $y = e^t$, and $y = 3^t$. Notice that $2 < e < 3$ and the graph of $y = e^t$ lies between the graphs of the other two functions.

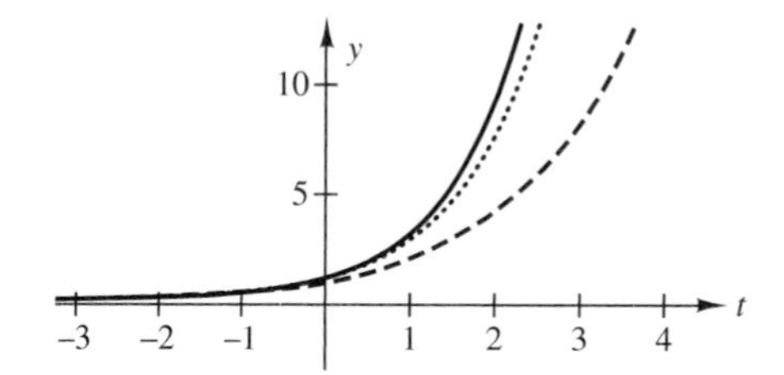

Figure 13.11 $y = e^t$ (dotted); $y = 2^t$ (dashed); $y = 3^t$ (solid)

The number e can be defined in several ways. One way is to consider the expression

$$\left(1+\frac{1}{n}\right)^n.$$

This particular expression applies when we calculate the value of money invested so that interest is compounded n times per year. If \$1 is invested and interest is compounded annually at an interest rate of r per year (in Example 13.3a, r was 0.05), there will be $(1 + r)^t$ dollars in the account after t years. If interest is compounded twice a year, then after t years there will be $\left(1+\frac{r}{2}\right)^{2t}$ dollars in the account. If interest is compounded four times a year, after t years there will be $\left(1+\frac{r}{4}\right)^{4t}$ dollars. In general, if interest is compounded k times a year, the amount after t years will be $\left(1+\frac{r}{k}\right)^{kt}$ dollars. (Some banks advertise that interest is compounded daily, in which case $k = 365$ except in leap years.) The expression $\left(1+\frac{1}{n}\right)^n$ can be interpreted as the value at the end of 1 year of \$1 earning $r = 100\%$ interest compounded n times during the year.

How can we find how much money will be in the account after t years if interest is compounded continuously? This means (in theory) that at every instant, interest is being calculated and added to the account. To derive a function that will give us the amount in the account after t years, we let k become larger and larger in the expression $\left(1+\frac{r}{k}\right)^{kt}$; then we find

$$\lim_{k\to\infty}\left(1+\frac{r}{k}\right)^{kt}.$$

We can relate the expression $\left(1+\frac{r}{k}\right)^{kt}$ to the expression $\left(1+\frac{1}{n}\right)^{n}$ as follows.

If we let $n=\frac{k}{r}$, then we have $\frac{r}{k}=\frac{1}{n}$, and since $k = rn$, $kt = rnt = nrt$. The expression $\left(1+\frac{r}{k}\right)^{kt}$ then becomes $\left(1+\frac{1}{n}\right)^{nrt}$.

As k gets larger, so does n, since n is a multiple of k. Thus, to find how much money will be in the account after t years if interest is compounded continuously, we compute

$$\lim_{n\to\infty}\left(1+\frac{1}{n}\right)^{nrt}=\left[\lim_{n\to\infty}\left(1+\frac{1}{n}\right)^{n}\right]^{rt}.$$

To obtain an expression for this term, we need to look at $\left(1+\frac{1}{n}\right)^{n}$ for larger and larger values of n. The following table shows values of this expression for various values of n:

n	1	5	10	25	50	100	1000	10,000
$\left(1+\frac{1}{n}\right)^{n}$	2	2.4883	2.5937	2.6658	2.6916	2.7048	2.7169	2.7182

Although this table does not prove that the limit of the expression $\left(1+\frac{1}{n}\right)^{n}$ exists, it can be shown that the limit of $\left(1+\frac{1}{n}\right)^{n}$ as n approaches infinity does exist; the limit is the number $\boldsymbol{e}$.

Thus we can write $y = e^{rt}$ as the function that gives the amount after t years if \$1 is invested at a rate r and the interest is compounded continuously. If r is 0.05, then after 5 years the amount will be $e^{(0.05)5} = e^{0.25} \approx \$1.28$. If $r = 1$ and $t = 1$, then $y = e^1 = e$. This gives a *monetary definition of the number e:* If \$1 is invested for 1 year at 100% interest compounded continuously, its value at the end of the year will be e dollars.

Other growth functions can be written using exponential functions of the form $f(t) = Pe^{rt}$ where P is a constant, as shown in the following examples.

EXAMPLE 13.5 The population of bacteria in a certain culture at time t is given by $g(t) = 700e^{0.6t}$, where t is measured in hours. Use a calculator with an exponential key to find the number of bacteria present after:

a. 5 hours **b.** 20 hours

Solution

a. Evaluate $g(5) = 700e^{0.6(5)} = 700e^3 \approx 14{,}060$.

b. After 20 hours, the number of bacteria is $g(20) = 700e^{12} = 113{,}928{,}354$. ■

EXAMPLE 13.6 Sketch the graphs of the following functions and describe their behavior:

a. $y = e^{0.6t}$ **b**. $y = e^{-0.2t}$

Solution The graphs of both functions are shown in Figure 13.12.

a. The function $y = e^{0.6t}$ is an increasing function and behaves like $y = a^t$ for $a > 1$ since $e^{0.6t} = (e^{0.6})^t \approx 1.82^t$.

b. The function $y = e^{-0.2t}$ is a decreasing function since $e^{-0.2t} = (e^{-0.2})^t \approx 0.82^t$.

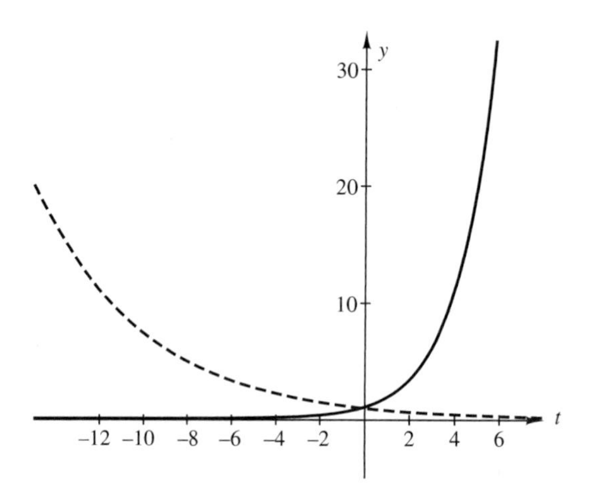

Figure 13.12 $y = e^{0.6t}$ (solid); $y = e^{-0.2t}$ (dashed) ■

On calculator keys and in computer programs, the function $f(t) = e^t$ is sometimes given by "exp" notation: $f(t) = e^t = \exp(t)$. We can think of $\exp(t)$ as "the number e raised to the t power." The notation is particularly useful when expressing composite functions that involve the natural exponential function, such as $g(t) = e^{\sin t} = \exp(\sin t)$ or $h(x) = e^{\sqrt{x}+4} = \exp(\sqrt{x} + 4)$.

EXAMPLE 13.7 Let $f(t) = \exp(t)$ and $g(t) = \sqrt{t}$. Give each of the following functions in simplest form and find the domain of each.

a. $(f \circ g)(t)$ **b**. $(g \circ g)(t)$ **c**. $(g \circ f)(t)$ **d**. $(f \circ f)(t)$

Solution

a. $(f \circ g)(t) = f(g(t)) = f\left(\sqrt{t}\right) = \exp\left(\sqrt{t}\right) = e^{\sqrt{t}}$. Since $\sqrt{t}$ is not defined for $t < 0$, the domain of $f \circ g$ is $[0, \infty)$.

b. $(g \circ g)(t) = g(g(t)) = g\left(\sqrt{t}\right) = \sqrt{\sqrt{t}} = \left(t^{1/2}\right)^{1/2} = t^{1/4}$. As in part a, the domain is $[0, \infty)$.

c. $(g \circ f)(t) = g(f(t)) = g(e^t) = \sqrt{e^t} = (e^t)^{1/2} = e^{t/2}$. Since $e^t > 0$ for all t, the domain of $g \circ f$ is $(-\infty, \infty)$.

d. $(f \circ f)(t) = f(f(t)) = \exp(e^t) = e^{e^t}$. Since the domain of e^t is $(-\infty, \infty)$, the domain of $f \circ f$ is $(-\infty, \infty)$. ■

Exercises 13-B

1. A radioactive substance is decaying. The amount of the substance present at time t is given by the function $f(t) = 2000e^{-0.04t}$, where the amount is given in milligrams and t is measured in decades. Find the amount of the substance present at time

a. $t = 0$ **b.** $t = 5$ **c.** $t = 25$

2. If \$500 is placed in a savings account that earns 8% interest, how much will be in the account in 10 years if nothing is withdrawn from it and:

a. Interest is compounded monthly?

b. Interest is compounded daily in a non-leap year?

c. Interest is compounded continuously?

d. How do the answers to parts a, b, and c compare to each other and with the answer to Exercise 10 in Section 13-A?

3. Use what you know about the graphs of exponential functions to match each equation with its graph.

a. $y = 2^{-x}$ **b.** $y = \frac{1}{4^x}$ **c.** $y = -2^x$ **d.** $y = 4^x$

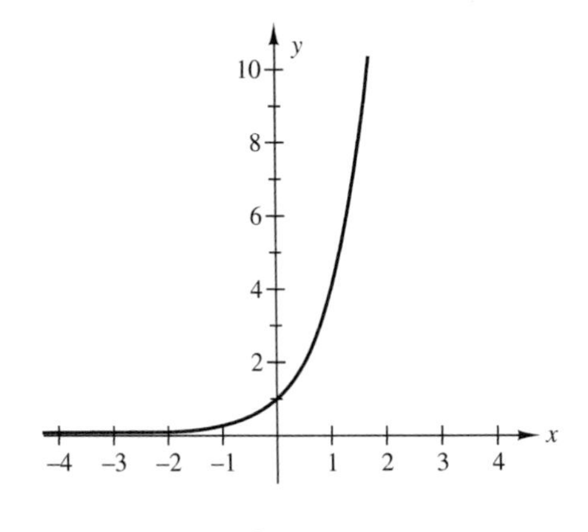

Graph 1

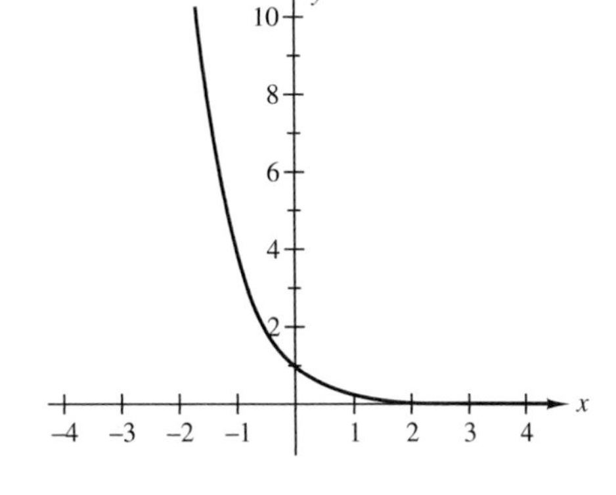

Graph 2

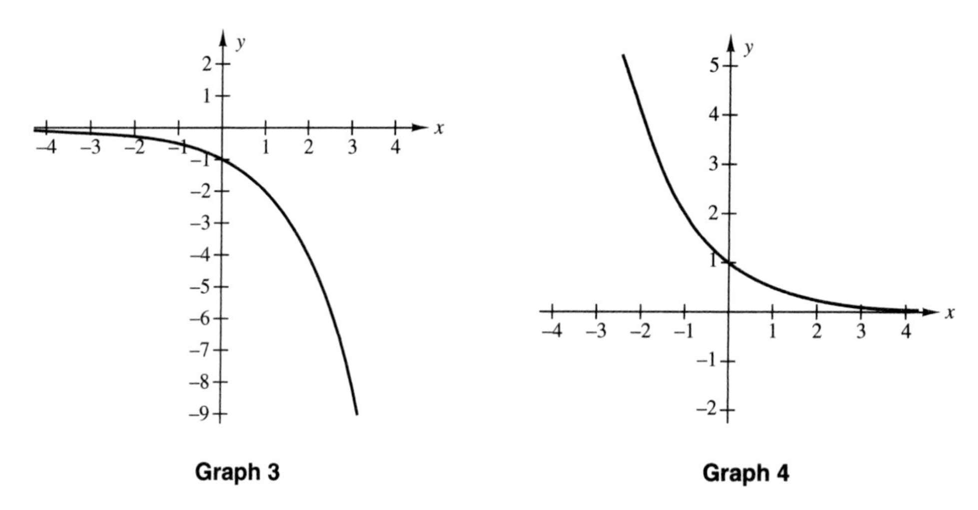

Graph 3 **Graph 4**

4. **a**. The graph of $y = a^x$ is shown below. Sketch the graph of $y = a^{x-3}$.

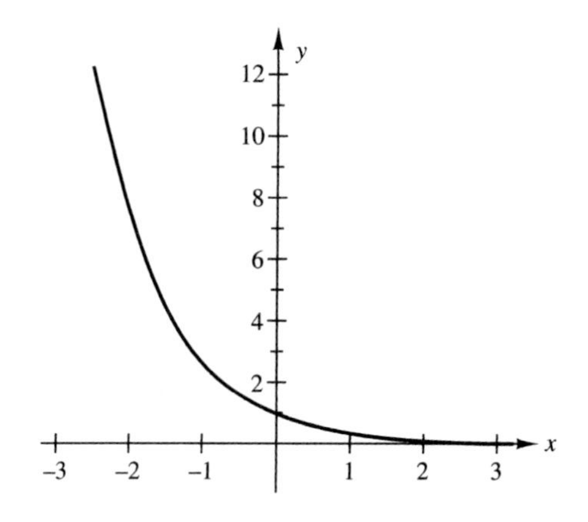

b. The graph of $y = b^x$ is shown in Graph 1. What is the equation of Graph 2?

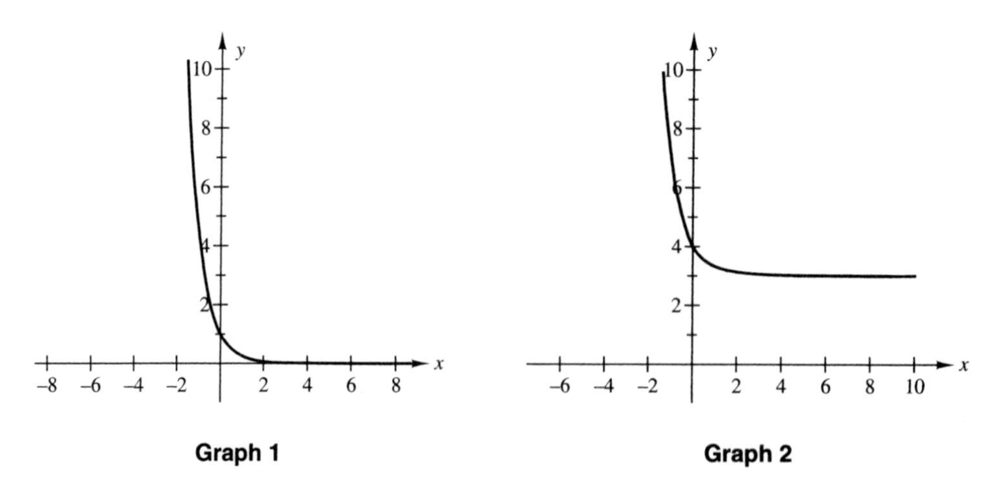

Graph 1 **Graph 2**

5. Use rules of exponents to simplify each of the following expressions to the form $e^{u(t)}$:

a. $\sqrt{e^{3t}}$ **b**. $e^{4t}e^{-5t}e$ **c**. $\dfrac{\left(e^{2t}\right)^e}{e^{3t}}$

6. Let $f(t) = \exp(t)$, $g(t) = \cos t$, and $h(t) = t^2 + 1$. Find, in simplest form:

a. $(f \circ g)(t)$ **b.** $(g \circ f)(t)$ **c.** $(f \circ h)(t)$ **d.** $(h \circ f)(t)$

Chapter 13 Exercises

1. a. Write each of the following functions in the form of x raised to a constant power. Assume $x \geq 0$.

(i) $\sqrt{\sqrt{x^3}}$ (ii) $\sqrt[3]{x}\ \sqrt[4]{x^6}$ (iii) $\sqrt{x\ \sqrt[3]{x^2}}$

b. Which of the functions in part a is defined for $x < 0$? Give the function in simplest form.

2. Simplify each of the following functions to the form $a^{u(t)}$—that is, a single constant raised to a power that is a function of t.

a. $e^t\left(\dfrac{e^{-t}}{e^{3t^2}}\right)$ **b.** $\sqrt[3]{3^t\, 9^{t^2}\, 3^0}$ **c.** $\dfrac{8^t 16^4}{2^t}$

3. Let $f(t) = 2t - 3$, $g(t) = 3\exp(t)$, and $h(t) = \frac{1}{t}$. Give each of the following functions in simplest form and find the domain of each:

a. $(f \circ g)(t)$ **b.** $(g \circ g)(t)$ **c.** $(g \circ h)(t)$

d. $(h \circ g)(t)$ **e.** $(g \circ f)(t)$

4. A worker on an assembly line tests the electronic component that controls the automatic windows for a particular car model. After t hours of experience, the worker can test $N(t)$ components each day, where $N(t) = 150(1 - e^{-0.3t})$.

a. How many components can a worker who has had 10 hours of experience test each day?

b. How many components can a worker who has had 40 hours of experience test each day?

c. Describe what happens to $N(t)$ as t gets larger and larger.

5. Sketch the graph of each of the following functions using properties of transformations:

a. $f(x) = (0.8)^x$ **b.** $g(x) = -3^x$ **c.** $h(x) = 2^{x+1}$

d. $k(x) = 3(5^x)$ **e.** $m(x) = 2 - 4^x$ **f.** $n(x) = \frac{2}{e^x}$

g. $w(x) = 4 + e^x$

6. The graph of each of the functions in parts a–i is one of the six graphs that follow. Write the number of the graph next to the equation of the function.

a. $f(x) = 2^x$ **b**. $g(x) = \sqrt{x}$ **c**. $h(x) = \left(\frac{1}{2}\right)^x$

d. $k(x) = 2x$ **e**. $m(x) = e^2$ **f**. $w(x) = 2^{-x}$

g. $F(x) = x^2$ **h**. $G(x) = x^{1/2}$ **i**. $H(x) = \left(\frac{1}{2}\right)^{-x}$

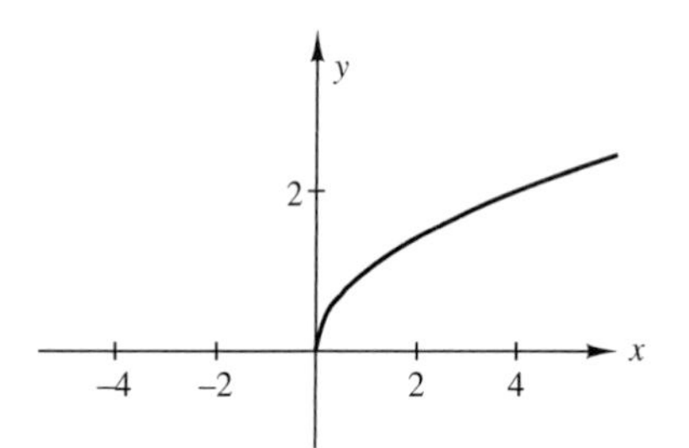

Graph 1

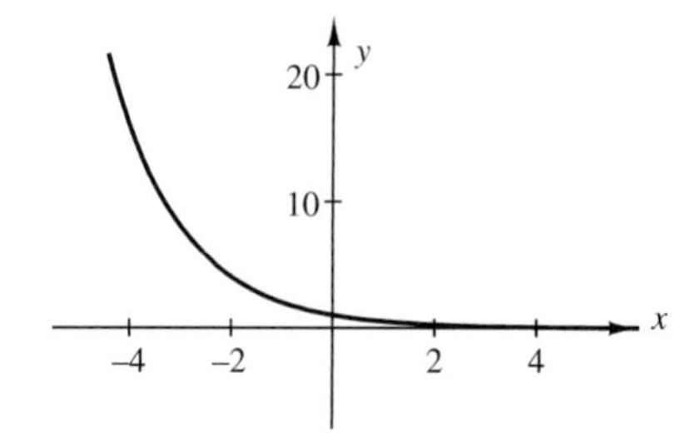

Graph 2

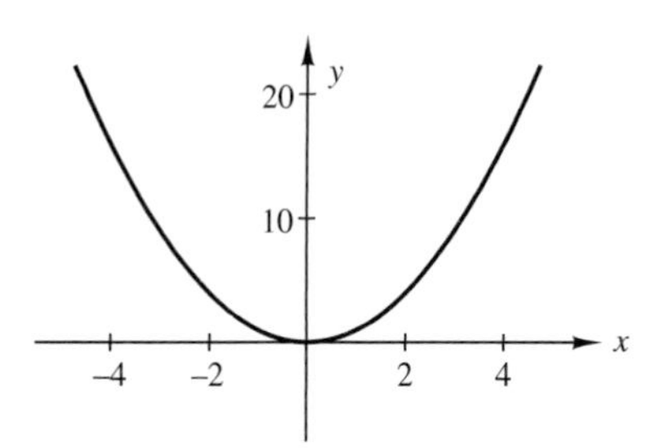

Graph 3

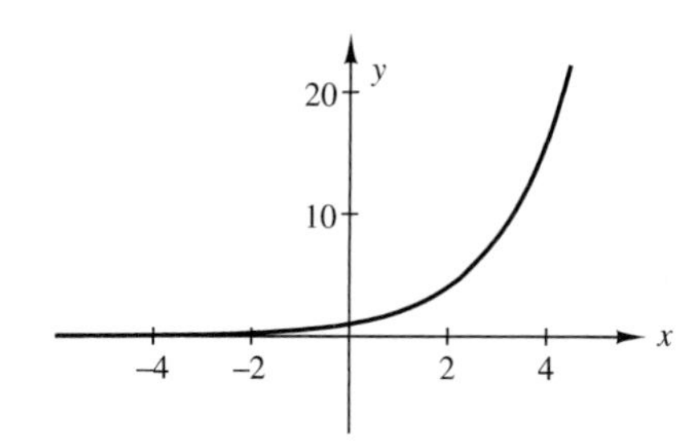

Graph 4

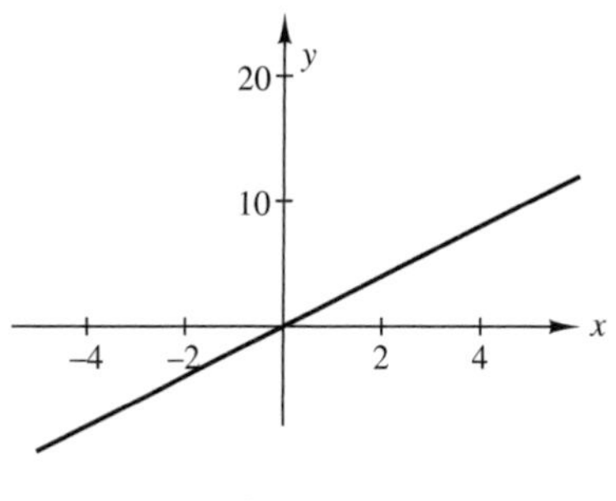

Graph 5

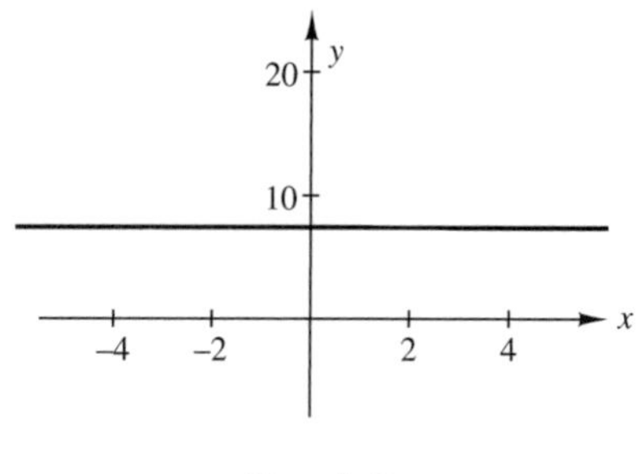

Graph 6

7. If $1,200 is placed in a savings account that earns 4.5% interest and no funds are withdrawn, how much will be in the account after 8 years if:

a. Interest is compounded quarterly?

b. Interest is compounded monthly?

c. Interest is compounded continuously?

8. Decide whether or not each of the following expressions is correct. If it is correct, state the rule of exponents used. If it is not correct, explain why.

a. $2^5 - 2^3 \overset{?}{=} 2^2$ **b**. $\frac{2^5}{2^3} \overset{?}{=} 2^2$ **c**. $\sqrt{x^2} \overset{?}{=} x$

d. $3^{-2} \overset{?}{=} \frac{1}{3^{1/2}}$ **e**. $2^{-3}3^{-2} \overset{?}{=} \frac{1}{72}$ **f**. $x^{-1} + y^{-1} \overset{?}{=} \frac{1}{x+y}$

CHAPTER 14

Companion to Inverse Functions

Each car that is registered in the state of Pennsylvania is assigned a license plate consisting of letters and numbers. (Although most licenses include both letters and numbers, some have only letters. We will still call it a license number.) We can think of this assignment as a *function* from the set of cars registered in the state to the set of licenses. The assignment is a function because each car is assigned only one license number. We can picture the license assignment function f as in Figure 14.1.

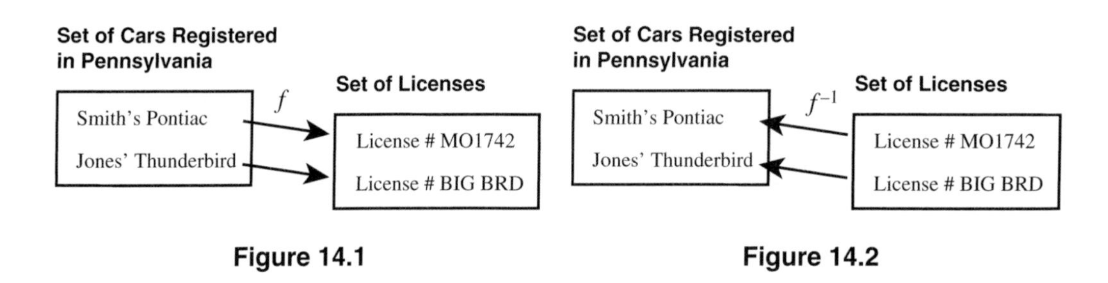

Figure 14.1 **Figure 14.2**

There are many situations in which a license number is known (observed at the scene of a crime, for example) and the car needs to be located. This is a circumstance where we want to map in the reverse direction from the set of licenses to the set of cars; that is, given a license number, we want to identify the car. This function that maps back from the particular license number to the car is called the *inverse function* of the function that maps from the car to the license number. The picture in Figure 14.2 shows the inverse function. When f is used to denote a function, f^{-1} is used to denote its inverse.

14-A ONE-TO-ONE FUNCTIONS

In addition to the fact that no car is given more than one license number, it is also true that no two cars are given the same license number. These two facts taken together mean that the function that assigns a license number to each registered car is a *one-to-one function*. This allows us to identify one and only one license number with each car.

The assignment of ID numbers to students enrolled at a particular college is another example of a one-to-one function. It is a function because each student is given only one number. It is important that it is one-to-one, for if two different students are assigned the same number, serious problems would occur. (Their bills, transcripts, and other records would be confused.)

For example, suppose that Beth, Todd, John, and Mary are all enrolled in the same calculus class, and their student ID numbers have been assigned as shown in Figure 14.3a. At their college, grades are submitted to the registrar by student ID numbers. When the instructor sends their grades to the registrar, the grade sheet looks like the one in Figure 14.3b. How is the registrar supposed to know if Beth got the A and Todd got the B, or if the grades are to be assigned the other way around? From the grade sheet, it is impossible to tell.

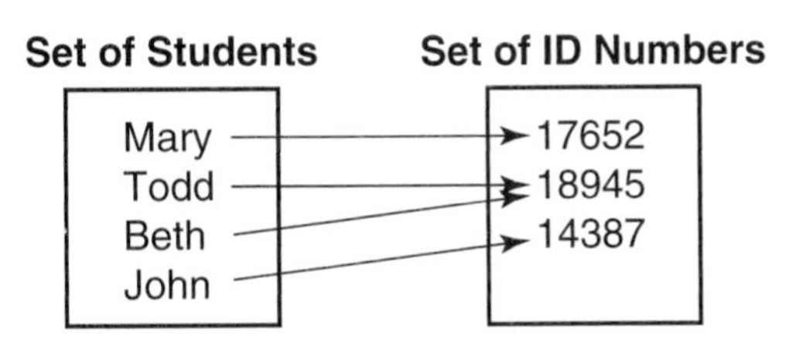

Figure 14.3a
Assignment function that is **not** one-to-one

ID Number	Course Grade
14387	B+
17652	C
18945	A
18945	B

Figure 14.3b

There are two different ways to state the definition of a one-to-one function; statement 2 below is the contrapositive of statement 1.

Definition:

(1) A function is *one-to-one* if, whenever a and b are two values in the domain of f and $a \neq b$, then $f(a) \neq f(b)$.

(2) A function is *one-to-one* if, whenever a and b are two values in the domain of f for which $f(a) = f(b)$, then $a = b$.

Statement 1 says that a function f is one-to-one if (and only if) f maps any two different x-values into different y-values. Statement 2 says that a function f is one-to-one if (and only if), whenever two of its y-values are equal, the x-values that produced them are also equal. Either definition may be used to determine whether or not a function is one-to-one; the given information may determine which definition is more appropriate to use.

EXAMPLE 14.1 Consider two functions, f and g, both defined on the following domain: $\left\{-\frac{5}{2}, -1, -\frac{1}{3}, 0, \frac{1}{3}, 1, \sqrt{2}, \frac{5}{2}, 3.2\right\}$. Let $f(x)$ be the number obtained when x is multiplied by itself and 3 is added to the product, and let $g(x)$ be the number obtained when x is multiplied by $\frac{1}{2}$ and 3 is added to the product. The two tables given in Figure 14.4 show $f(x)$ and $g(x)$ for values of x in the domain. Since for each value of x, the rules for $f(x)$ and $g(x)$ produce unique values, both f and g are functions. Determine whether f and g are one-to-one functions.

x	$-\frac{5}{2}$	-1	$-\frac{1}{3}$	0	$\frac{1}{3}$	1	$\sqrt{2}$	$\frac{5}{2}$	3.2
$f(x)$	$\frac{37}{4}$	4	$\frac{28}{9}$	3	$\frac{28}{9}$	4	5	$\frac{37}{4}$	13.24

x	$-\frac{5}{2}$	-1	$-\frac{1}{3}$	0	$\frac{1}{3}$	1	$\sqrt{2}$	$\frac{5}{2}$	3.2
$g(x)$	$\frac{7}{4}$	$\frac{5}{2}$	$\frac{17}{6}$	3	$\frac{19}{6}$	$\frac{7}{2}$	$\frac{\sqrt{2}+6}{2}$	$\frac{17}{4}$	4.6

Figure 14.4

Solution To determine if f is one-to-one, we want to check if two different values of x give the same value for $f(x)$. If so, then f is not one-to-one. We look at the row of $f(x)$ values and see that $\frac{28}{9}$ appears twice. Both $a = -\frac{1}{3}$ and $b = \frac{1}{3}$ give $f(a) = f(b) = \frac{28}{9}$. This tells us that f is not one-to-one. Since each value of $g(x)$ in the list occurs only once, the function g is one-to-one. ■

 If the domain of a function f is not finite, then checking a list of values $f(x)$ is not sufficient to determine whether or not the function is one-to-one.

Suppose the graph of a function f is given. How can we tell from the picture if f is one-to-one? Recall that we can tell from the graph that f is a function by the vertical line test. If f is a function, each x-value in the domain corresponds to a single y-value in the range. Thus, if we draw a vertical line (a line of the form $x = c$) anywhere on the graph, it will intersect the graph no more than once.

We can tell whether the function f is one-to-one by using the *horizontal line test*. If f is one-to-one, a horizontal line drawn anywhere on the graph will intersect the graph no more than once. If it does intersect the graph in two or more distinct points, then the points have the same y-coordinate but different x-coordinates, which violates the definition of a one-to-one function. In Figure 14.5, the horizontal line $y = 3$ intersects the graph of $y = f(x)$ in two points, so the function is not one-to-one. The picture shows that $f(0) = f(4) = 3$.

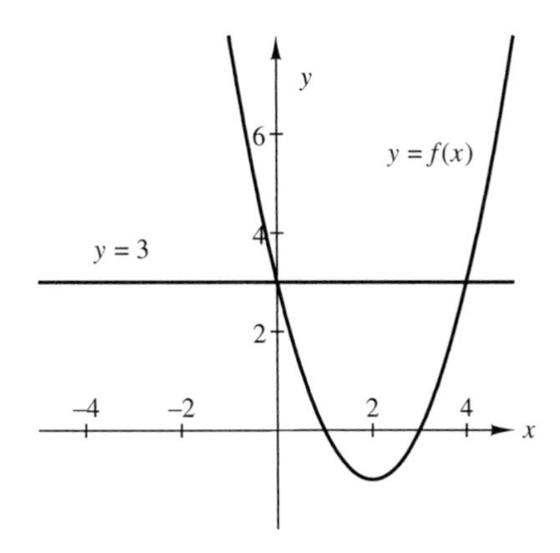

Figure 14.5 A function that is not one-to-one

EXAMPLE 14.2 For each of the graphs given below, determine if it is the graph of a function. If it is the graph of a function, determine if the function is also one-to-one.

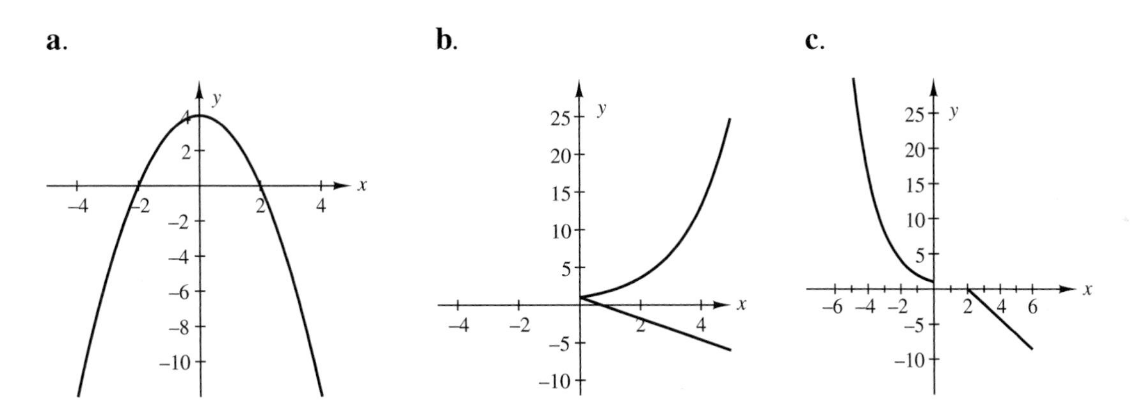

Solution

a. The graph passes the vertical line test, since no vertical line crosses the graph in more than one place. Therefore, it is the graph of a function. However, many horizontal lines (for example, $y = 0$) intersect the graph in two places, so the function is not one-to-one.

b. Since the graph fails the vertical line test (for example, the vertical line $x = 2$ cuts it in two places), it is not the graph of a function.

c. This is the graph of a function since no vertical line intersects the graph in more than one point, and the function is one-to-one since no horizontal line intersects the graph in more than one point. ■

We know from statement 2 in the definition that a function is one-to-one if, when $f(a) = f(b)$, a equals b. We use this property in the following example.

EXAMPLE 14.3 Determine if the following functions are one-to-one:

a. $f(x) = 5x - 4$ **b.** $g(x) = \sqrt{x}$ **c.** $h(x) = x^2 - 3$

Solution

a. We set $f(a) = f(b)$ and show that the only way this could be true is if $a = b$. The equation $f(a) = f(b)$ means $5a - 4 = 5b - 4$. If we add 4 to both sides of this equation, we get $5a = 5b$, so $a = b$. Thus f is one-to-one.

b. For $a \geq 0$ and $b \geq 0$, if $g(a) = g(b)$, we have $\sqrt{a} = \sqrt{b}$. Squaring both sides, we get $a = b$, so g is one-to-one.

c. We set $h(a) = h(b)$ to get $a^2 - 3 = b^2 - 3$; so $a^2 = b^2$. We cannot conclude, however, that $a = b$, since if $a = -2$ and $b = 2$, then $a^2 = b^2$. Therefore, h is not one-to-one. ■

Exercises 14-A

1. In the following two tables, each value of x in the first row is assigned the number y below it. For each table, answer the following questions: Is the assignment $x \to y$ a function (with domain the given values of x)? If so, is the function one-to-one? Give reasons for your answers.

a.

x	$-\frac{3}{2}$	-1.5	$-\frac{1}{7}$	0.2	0.35	1	$\sqrt{2}$	$\frac{5}{3}$	3
y	$\frac{3}{4}$	4	$\frac{2}{9}$	3	$\frac{28}{9}$	4.02	-1.4	$\frac{37}{4}$	1.25

b.

x	$-\frac{3}{2}$	-1.5	$-\frac{1}{7}$	0.2	0.35	1	$\sqrt{2}$	$\frac{5}{3}$	3
y	$\frac{3}{4}$	$\frac{3}{4}$	$\frac{3}{9}$	0.75	$\frac{1}{3}$	-2.02	-1.3	$\frac{37}{4}$	1.25

2. For each of the graphs below, determine if it is the graph of a function. If it is the graph of a function, determine if the function is also one-to-one. Give reasons for your answers.

a.

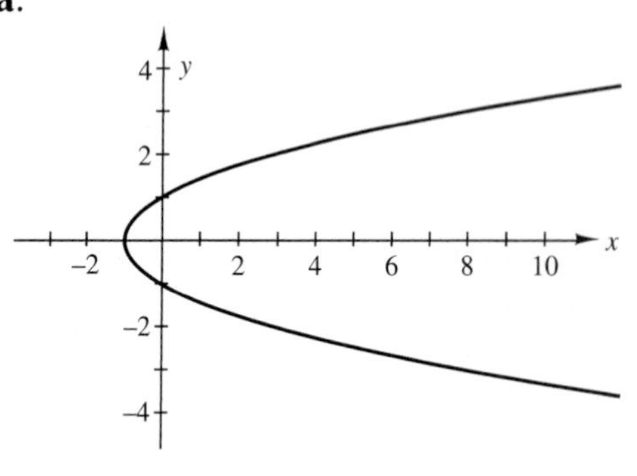

b.

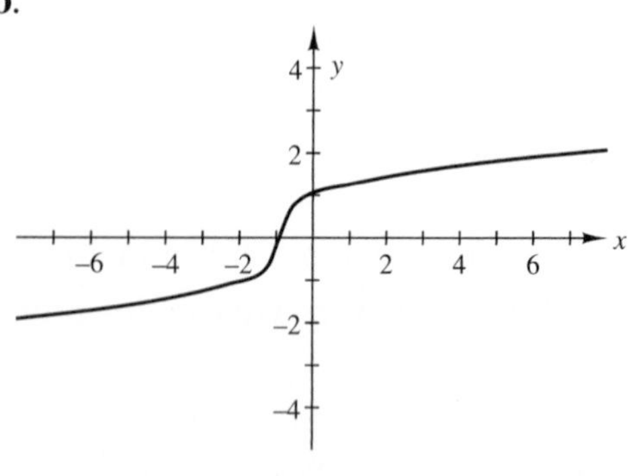

3. Consider the assignment that maps people to Social Security numbers.

a. What would be the practical consequences if this were not a function?

b. What would be the practical consequences if this were not one-to-one?

4. Determine whether or not the following functions are one-to-one.

a. $f(x) = 5 - 7x$ **b.** $g(x) = \frac{3}{x} - 2$ **c.** $h(x) = 3 + \sqrt{4 - x}$

d. $F(x) = \frac{2}{x^2}$ **e.** $G(x) = 6x^3$ **f.** $H(x) = 5 - 2x^2$

5. Suppose that f is a linear function whose graph has nonzero slope. Explain why f is one-to-one.

14-B PROPERTIES OF A FUNCTION AND ITS INVERSE

When a function is one-to-one, each value of x in the domain produces a unique value $f(x)$ in the range, and no other value in the domain produces that same number $f(x)$. This means that the rule that sends $f(x)$ back to x is a function—it is f^{-1}, the inverse function of f.

Consider again the functions f and f^{-1} described in Section 14-A that map from the set of cars to the set of license numbers and from the set of license numbers to the set of cars, respectively. If we apply the function f to the element "Jones' Thunderbird," we get the license number "BIG BRD." If we then apply the function f^{-1} to the element "BIG BRD," we obtain "Jones' Thunderbird." The two functions, when applied one after the other, return us to the element with which we started.

There are other pairs of functions that have this relationship. For example, let f be the function that adds 7 to any number and let g be the function that subtracts 7 from any number. Take any number x and apply the function f. We get $x + 7$. If we then apply the function g to the result, we get $(x + 7) - 7 = x$. We could also have applied the function rules in the opposite order: If we first subtract 7 from x and then add 7 to the result, we get $(x - 7) + 7 = x$. Each function undoes what the other one does; the end result is that the composition of f and g in either order sends x to x.

Definition: The linear function $f(x) = x$ is called the *identity function.*

In the example above, we saw that a composition of the two functions whose rules are "add 7" and "subtract 7" is the same as the identity function. This leads to the following definition of *inverse function*.

Definition: Two functions f and g are *inverses* of each other if $f(g(x)) = x$ for each x in the domain of g and $g(f(x)) = x$ for each x in the domain of f. In other words, $f \circ g$ is the identity function on the domain of g and $g \circ f$ is the identity function on the domain of f.

We denote the inverse of the function f by f^{-1}.

Pay attention to differences in notation. $f^{-1}(x)$ is not the same as $\frac{1}{f(x)}$. $f^{-1}(x)$ denotes the inverse function of f. $[f(x)]^{-1}$ denotes $\frac{1}{f(x)}$, the reciprocal of $f(x)$. On calculators, the "x^{-1}" button means "take the reciprocal."

EXAMPLE 14.4 Determine if the following pairs of functions are inverses of each other.

a. $f(x) = 3x - 2$ and $g(x) = \frac{x+2}{3}$ **b.** $f(x) = \frac{2}{x}$ and $g(x) = \frac{2}{x}$

c. $f(x) = x^2$ and $g(x) = \sqrt{x}$

Solution In each case, we need to see if the composite functions $f \circ g$ and $g \circ f$ are identity functions.

a. The functions f and g each have domain $(-\infty, \infty)$. We have $f(g(x)) = f\left(\frac{x+2}{3}\right) = 3\left(\frac{x+2}{3}\right) - 2 = (x + 2) - 2 = x$ and $g(f(x)) = g(3x - 2) = \frac{(3x-2)+2}{3} = \frac{3x}{3} = x$. Therefore, f and g are inverses of each other.

b. The domain of f and g is all $x \neq 0$. If $x \neq 0$, $f(g(x)) = f\left(\frac{2}{x}\right) = \frac{2}{\frac{2}{x}} = x$. Similarly, $g(f(x)) = x$ for $x \neq 0$. Since the functions f and g are the same, the function $f(x) = \frac{2}{x}$ is its own inverse.

c. For x in the domain of $g(x)$—that is, for $x \geq 0$—$f(g(x)) = f(\sqrt{x}) = (\sqrt{x})^2 = x$. But since the domain of $f(x)$ consists of all real numbers, we also need to have $g(f(x)) = x$ for all real numbers x. If $x < 0$, say $x = -1$, $g(f(-1)) = g((-1)^2) = g(1) = \sqrt{1} = 1 \neq -1$. Therefore, $f(x) = x^2$ and $g(x) = \sqrt{x}$ are not inverses of each other. ■

Note that the function $f(x) = x^2$ in Example 14.4c is not one-to-one, so it does not have an inverse.

Domain and Range of f and f^{-1}

When f is a one-to-one function, f maps from the domain of f to the range of f; that is, the function f takes an x in its domain and maps it to y in its range. The function f^{-1} then maps that y back to x. Thus f^{-1} is defined for all y in the range of f, so the domain of f^{-1} is the range of f. Similarly, the range of f^{-1} is the domain of f. We have the relationships shown below.

$$\text{Domain of } f = \text{Range of } f^{-1}$$

$$\text{Range of } f = \text{Domain of } f^{-1}$$

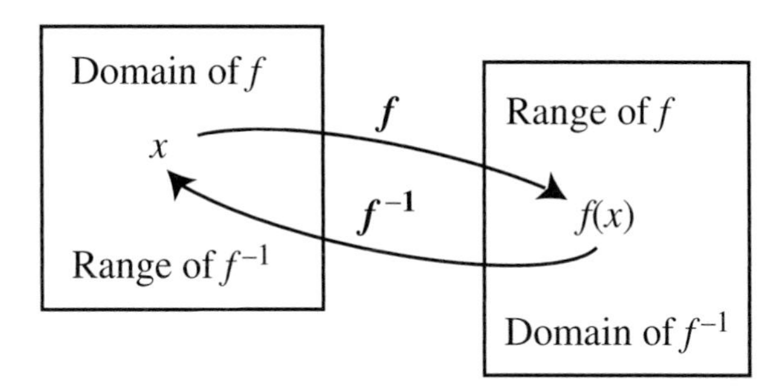

Figure 14.6

Graphs of Functions and Their Inverses

The points (x, y) on the graph of a one-to-one function have a special relationship to points on the graph of the inverse function. We explore that relationship here. Consider the collection of points (x, y): $\{(-2, 1), (-5, -3), (4, 2), (-1, -1)\}$. Look at the new set of points (x, y) obtained from the first set by interchanging the x- and y-coordinates of each of the original pairs: $\{(1, -2), (-3, -5), (2, 4), (-1, -1)\}$. Both sets of points are plotted on the graph in Figure 14.7.

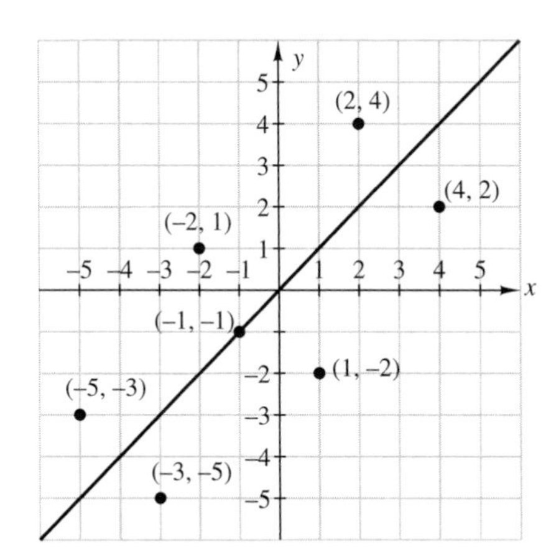

Figure 14.7

Notice that the point $(-1, -1)$ remained unchanged by the switch of coordinates. In fact, any point on the line $y = x$ would remain unchanged. The line $y = x$ is called a *line of reflection symmetry* (or *mirror*) for the two sets of points. If we imagine folding the graph in Figure 14.7 along the line $y = x$, the point $(-2, 1)$ in the first set will land on the point $(1, -2)$ in the second set. Similarly, $(-5, -3)$ will fold onto $(-3, -5)$, and $(4, 2)$ will fold onto $(2, 4)$. In fact, if x- and y-coordinates are interchanged, each point (a, b) will become the corresponding point (b, a) that is its mirror image across the line $y = x$. This fact is illustrated in Figure 14.8.

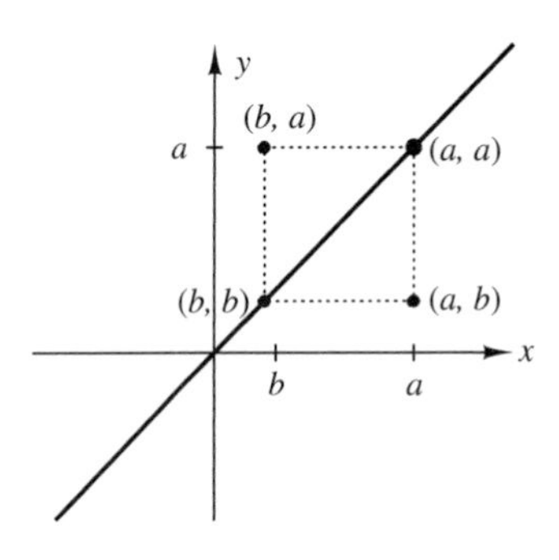

Figure 14.8

EXAMPLE 14.5 Plot each given point. For each, find the coordinates of its mirror image with respect to the line $y = x$, and plot this new point.

a. $(-4, 9)$ **b.** $(6, 6)$ **c.** $(2, -7)$

Solution The point $(-4, 9)$ is shown in Figure 14.9. It mirror image with respect to the line $y = x$ is obtained by interchanging its coordinates. This is the point $(9, -4)$. The point $(6, 6)$ is its own mirror image with respect to the line $y = x$. The reflection of the point $(2, -7)$ about the line $y = x$ is the point $(-7, 2)$.

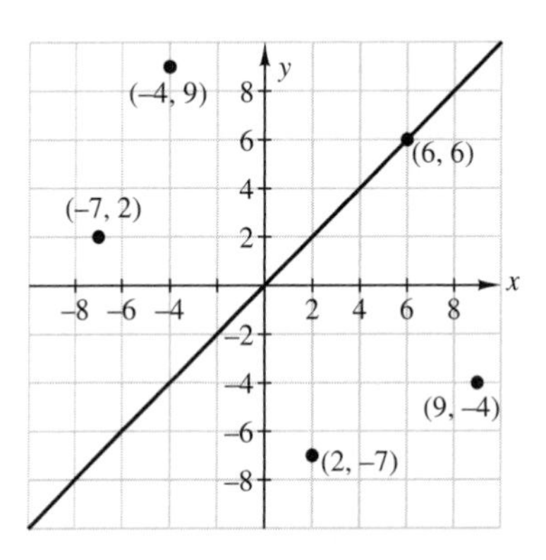

Figure 14.9 ■

Now consider the graph of a one-to-one function $y = f(x)$. For each point $(x, f(x))$, we interchange coordinates and get new points $(f(x), x)$. The set of points $\{(x, f(x))\}$ are mirror images of those in the set $\{(f(x), x)\}$ with respect to the line $y = x$. For example, consider the function $f(x) = x^3 - 4$. Here is a table of some pairs of values $(x, f(x))$ and a second table that gives the interchanged coordinates $(f(x), x)$:

x	$f(x)$
-2	-12
-1	-5
0	-4
$\frac{1}{2}$	$-\frac{31}{8}$
2	4
3	23

$f(x)$	x
-12	-2
-5	-1
-4	0
$-\frac{31}{8}$	$\frac{1}{2}$
4	2
23	3

If we plot these points $(x, f(x)) = (x, x^3 - 4)$ and the corresponding points $(f(x), x)$ on the same coordinate system, plotting the first coordinate along the horizontal axis and the second coordinate along the vertical axis for both sets, we see that the points $(x, f(x))$ are mirror images of the points $(f(x), x)$ with respect to the line $y = x$. See Figure 14.10.

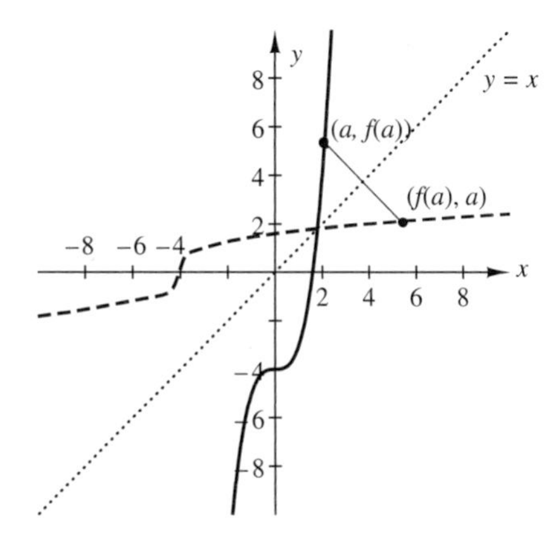

Figure 14.10 The graphs of $(x, f(x))$ (solid) and $(f(x), x)$ (dashed) for $f(x) = x^3 - 4$

What does interchanging coordinates have to do with inverse functions? Look again at the table of values for $(x, f(x))$ and $(f(x), x)$. The function f takes x-values into y-values. The function f^{-1} is defined to be the function that maps the resulting y-value back to x. In the table that follows, f takes $x = -2$ to $y = -12$; then f^{-1} takes -12 back to -2. This relationship holds for any x in the domain of f; f takes x into y, then f^{-1} takes that y back to x. So f^{-1} is the function shown in the second part of the table—the part of the table that was obtained by interchanging coordinates. We can relabel that part of the table as shown below. In the table on the right, x-values are in the domain of f^{-1}.

x	$f(x)$
-2	-12
-1	-5
0	-4
$\frac{1}{2}$	$-\frac{31}{8}$
2	4
3	23

x	$f^{-1}(x)$
-12	-2
-5	-1
-4	0
$-\frac{31}{8}$	$\frac{1}{2}$
4	2
23	3

This example illustrates the following useful fact: The graph of f contains the point (a, b) if and only if the graph of f^{-1} contains the point (b, a). If the graphs of $y = f(x)$ and $y = f^{-1}(x)$ are sketched on the same axes, the graphs are mirror images of each other with respect to the mirror line $y = x$. This relationship means that if we know the graph of a one-to-one function, we can sketch the graph of its inverse.

EXAMPLE 14.6 Sketch the following pairs of inverse functions on the same set of axes:

a. $y = 3x - 2$ and $y = \frac{x+2}{3}$ **b.** $y = \frac{2}{x}$ and $y = \frac{2}{x}$

Solution The graphs are sketched in Figures 14.11 and 14.12, respectively. Notice the symmetry of the graphs with respect to the line $y = x$.

a.

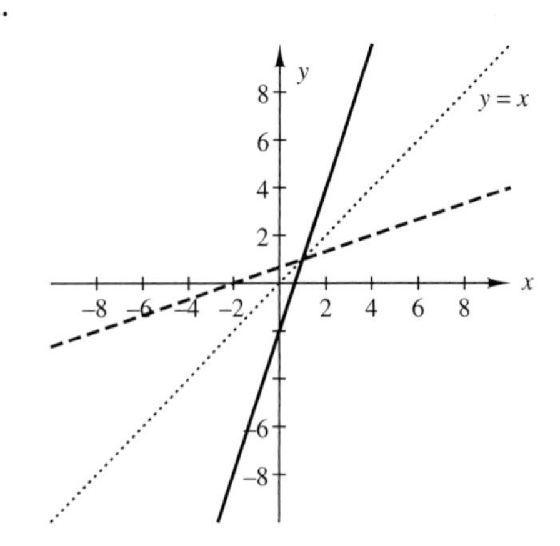

Figure 14.11 $y = 3x - 2$ (solid), $y = \frac{x+2}{3}$ (dashed)

b.

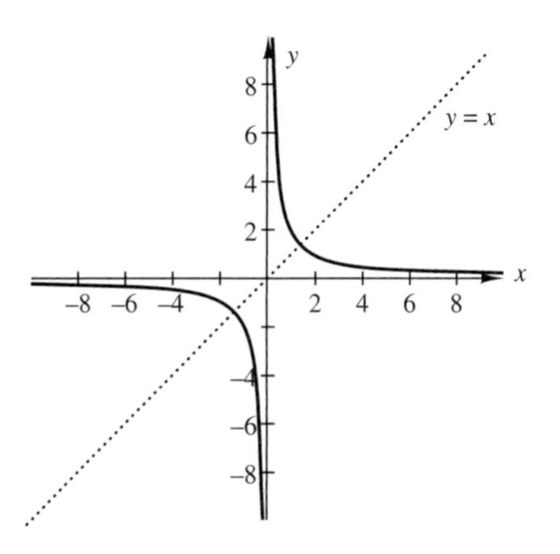

Figure 14.12 $y = \frac{2}{x}$ is its own inverse. ■

EXAMPLE 14.7 For each of the functions whose graphs are shown in Figures 14.13a and 14.13b, decide if the function is one-to-one. If it is, sketch the graph of its inverse on the same set of axes.

a.

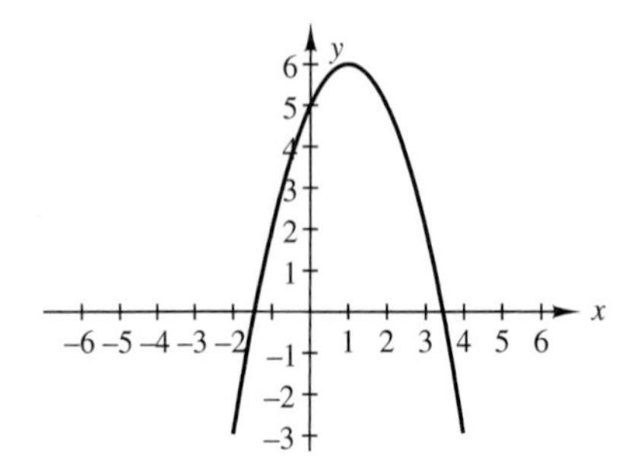

Figure 14.13a $y = f(x)$

b.

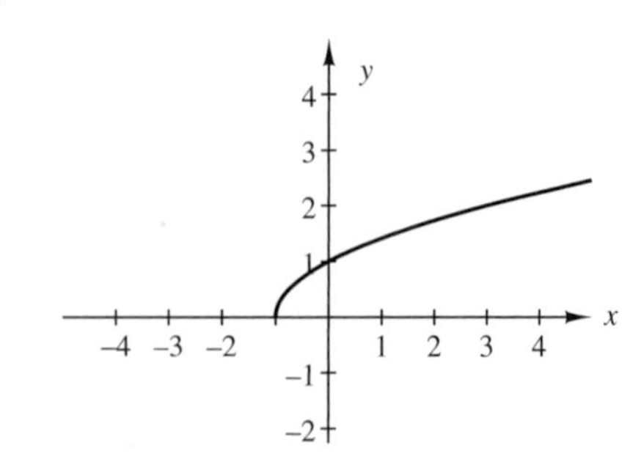

Figure 14.13b $y = g(x)$

Solution

a. The graph of f fails the horizontal line test, so the function is not one-to-one.

b. The graph of g passes the horizontal line test, so it is one-to-one. To sketch the inverse, we need to reflect the graph of g across the line $y = x$; it is helpful to plot a few key points to guide the sketch. For example, $(-1, 0)$ on $g(x)$ becomes $(0, -1)$ on $g^{-1}(x)$, $(0, 1)$ reflects to $(1, 0)$, and the point where the graph of g intersects the line $y = x$ is also on the graph of g^{-1}. The graphs of g and g^{-1} are in Figure 14.13c.

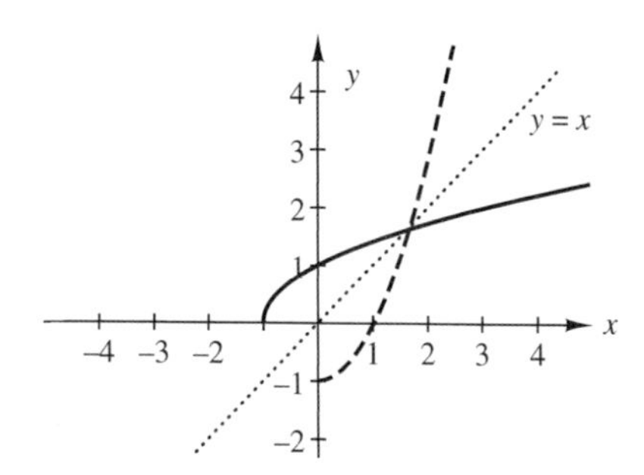

Figure 14.13c $y = g(x)$ (solid), $y = g^{-1}(x)$ (dashed) ■

Exercises 14-B

1. For each table in parts a and b, determine if the table defines a one-to-one function f. If the table does define a one-to-one function, construct a table that gives values of x in the domain of f^{-1} and the corresponding values of $f^{-1}(x)$.

a.

x	$f(x)$
2	–6
0	–13
–1	$\frac{1}{3}$
–24	$\frac{1}{8}$
–44	$\frac{1}{16}$

b.

x	$f(x)$
–4.3	7
–2.2	6
0	16
1	7
3	9

2. For each of the functions in Exercise 1 that are one-to-one, find the following:

a. Domain of f **b.** Range of f

c. Domain of f^{-1} **d.** Range of f^{-1}

3. Plot each given point. For each, give the coordinates of its mirror image with respect to the line $y = x$ and plot this new point.

a. $(-3, 0.65)$ **b.** $(4.2, -5)$ **c.** $(-6, 6)$

d. $(-6, -6)$ **e.** $(2.2, 1.8)$

4. For each of the functions whose graph is sketched in Figures 14.14 and 14.15, decide if the function is one-to-one and give reasons for your decision. If it is, sketch the graph of its inverse on the same set of axes.

a.

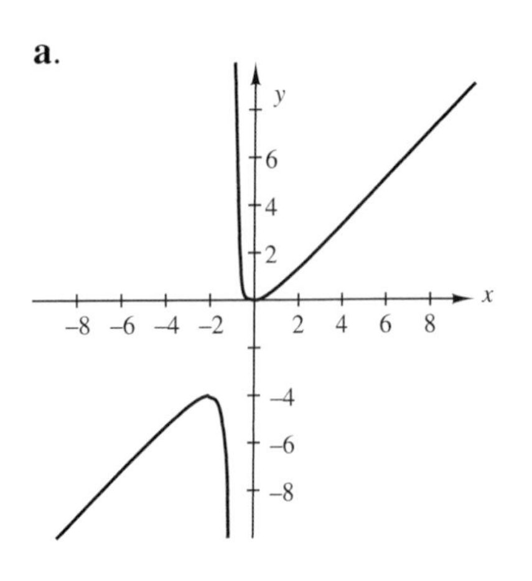

Figure 14.14

b.

Figure 14.15

5. Use the definition of inverse functions to determine if the following pairs of functions are inverses of each other. (See Example 14.4.)

a. $f(x) = \frac{x+1}{3}$ and $g(x) = 3(x-1)$

b. $f(x) = \frac{3}{x+1}$ and $g(x) = \frac{3-x}{x}$

c. $f(x) = \sqrt[3]{x}+1$ and $g(x) = (x-1)^3$

6. For each pair of functions below, use a graphing utility to graph the pair of functions and $y = x$ on the same set of axes in a square window with $-6 \le x \le 6$. ("Square window" means that the scale on the x- and y-axes is the same.)

(i) $y=3x-4,\ y=\frac{x+4}{3}$ (ii) $y = \sqrt[5]{x+3},\ y = (x-3)^5$ (iii) $y=\frac{3}{x-1},\ y=\frac{3+x}{x}$

a. From the graphs, which of the pairs of functions appear to be inverses of each other? Justify your answer.

b. Use the definition of inverse functions to confirm your answers in part a.

14-C FINDING THE INVERSE OF A FUNCTION

To raise money for a special trip, an organization is selling candy bars. The members collect the price of the bars sold plus the required tax from each customer. There is a function that maps from the number of candy bars sold to the total due from the customers. At the end of the sale, the organizers need to send the required tax to the appropriate government office. They wish to find the inverse function, which maps the total collected from customers back to the number of candy bars sold during the candy sale in order to compute the tax due.

14-C.1 Finding the Inverse Function When the Function Is One-to-One

Given a function $f(x)$ in symbolic form, we want to find an explicit representation of its inverse, $f^{-1}(x)$. The following steps show how to find the inverse if it exists.

1. Write the equation $y = f(x)$.

2. Solve the equation $y = f(x)$ for x; that is, rewrite it as an equation of the form $x = g(y)$, if possible. If there is a single function $g(y)$, then the inverse of f exists. If there is not a single function $g(y)$, then f does not have an inverse. (Note that the equation $x = g(y)$ gives the same relationship between x and y as the equation in step 1, so the graphs of these two equations will be identical.)

3. In the equation $x = g(y)$, interchange x and y to get $y = g(x)$. The function $g(x)$ is $f^{-1}(x)$, the inverse of f. (Since x and y have been interchanged, the graph of $y = g(x) = f^{-1}(x)$ will be the reflection of the graph of $y = f(x)$ about the line $y = x$.)

EXAMPLE 14.8 For each given function, find its inverse if it has one. If the inverse exists, check that the inverse satisfies the defintion in Section 14.A. If the inverse does not exist, explain why not.

a. $f(x) = 4x^3 + 1$ **b**. $F(x) = x^2 + 3$

Solution

a. We write the equation $y = 4x^3 + 1$ and solve for x. We first isolate the variable x: $\frac{y-1}{4} = x^3$; then we take the cube root of each side to get $x = \sqrt[3]{\frac{y-1}{4}}$. Interchanging x and y in this equation yields $y = \sqrt[3]{\frac{x-1}{4}}$; so $f^{-1}(x) = \sqrt[3]{\frac{x-1}{4}}$. We now check that $f(f^{-1}(x)) = x$ and $f^{-1}(f(x)) = x$. First, $f(f^{-1}(x)) = f\left(\sqrt[3]{\frac{x-1}{4}}\right) = 4\left(\sqrt[3]{\frac{x-1}{4}}\right)^3 + 1 =$ $4\left(\frac{x-1}{4}\right) + 1 = (x-1) + 1 = x$. Also, $f^{-1}(f(x)) = f^{-1}(4x^3 + 1) = \sqrt[3]{\frac{(4x^3+1)-1}{4}} = \sqrt[3]{\frac{4x^3}{4}} =$ $\sqrt[3]{x^3} = x$.

b. We write the equation $y = x^2 + 3$. In solving for x, we have $y - 3 = x^2$, so $x = \pm\sqrt{y-3}$. Since there is not a single function $x = g(y)$, the inverse of $F(x) = x^2 + 3$

does not exist. Notice also that we would expect that $F(x)$ does not have an inverse since it is not one-to-one; the graph of $F(x)$ is the parabola in Figure 14.16.

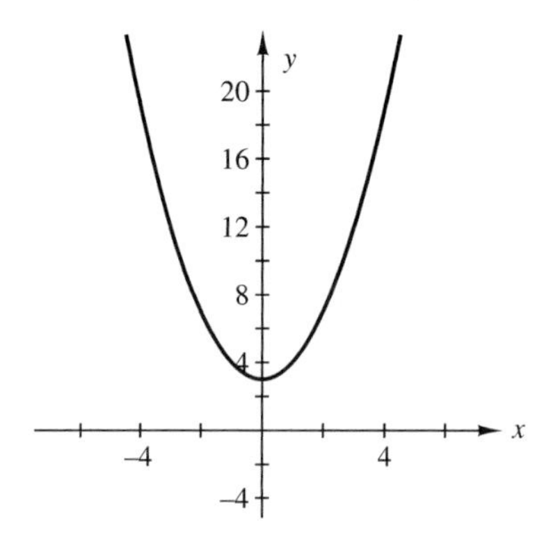

Figure 14.16 $F(x) = x^2 + 3$ ■

14-C.2 Restricting the Domain When the Function Is Not One-to-One

For some functions that are not one-to-one, such as $F(x) = x^2 + 3$ in Example 14.8b, it is possible to restrict the domain (instead of taking the domain to be the set of all real numbers) so that F is one-to-one on the restricted domain. Then we can proceed to find the inverse of the function defined on the restricted domain. When choosing the restricted domain, we try to choose the largest possible one on which the function is one-to-one.

EXAMPLE 14.9 Restrict the domain of the function $F(x) = x^2 + 3$ to be $\{x \mid x \geq 0\}$; then find the inverse of F. Verify that the inverse satisfies the definition of inverse function, and graph F and F^{-1}.

Solution We write $y = x^2 + 3$ and solve for x. Since x is restricted to be nonnegative, $x^2 = y - 3$ has a unique solution: $x = \sqrt{y-3}$. If we interchange x and y, we get $y = \sqrt{x-3}$; so $F^{-1}(x) = \sqrt{x-3}$. We can verify that $F(F^{-1}(x)) = F^{-1}(F(x)) = x$ for $x \geq 0$. The graph of this function $F(x)$, for the domain $x \geq 0$, is the half-parabola shown in Figure 14.17. The graph of its inverse $F^{-1}(x)$ is the reflection of the graph of F across the line $y = x$.

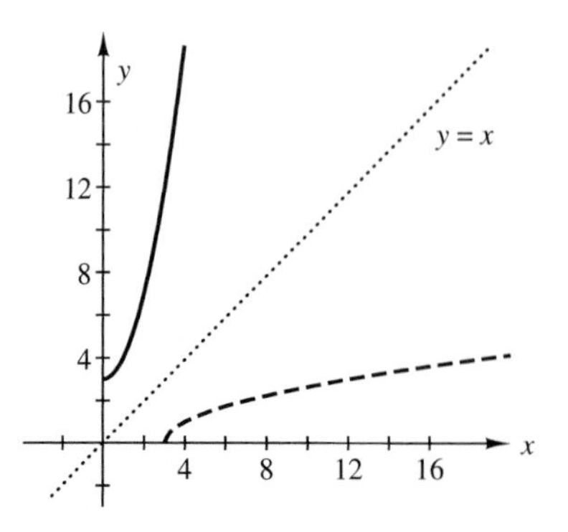

Figure 14.17 $y = F(x) = x^2 + 3$, $x \geq 0$ (solid), and $y = F^{-1}(x) = \sqrt{x-3}$ (dashed) ■

Inverse Trigonometric Functions

None of the trigonometric functions is one-to-one, since each is periodic. Yet, for applications, it is important to be able to define inverses of these functions. For each function, there is a conventional restriction of its domain in order to define an inverse.

Consider the function $f(x) = \sin x$ shown in Figure 14.18. We can see that it is not one-to-one; in fact, if we draw a horizontal line $y = c$ for any c between -1 and 1, the line will cross the graph of $y = \sin x$ infinitely many times. For each y in the range $-1 \le y \le 1$, there are infinitely many real values of x for which $y = \sin x$.

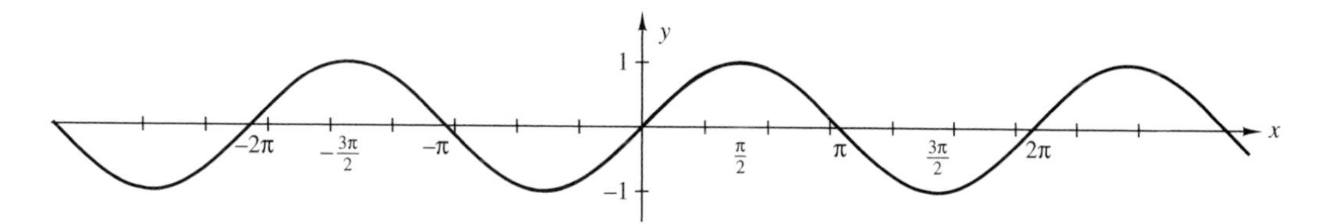

Figure 14.18 $y = \sin x$

We want to restrict the domain of the sine function to some interval in which the graph passes the horizontal line test in order to have a function that has an inverse. We also want to choose an interval on which all values of the range $[-1, 1]$ occur. There are many possible choices for such an interval: $\left[-\frac{\pi}{2}, \frac{\pi}{2}\right]$ or $\left[\frac{\pi}{2}, \frac{3\pi}{2}\right]$ or $\left[\frac{3\pi}{2}, \frac{5\pi}{2}\right]$ and so on. The standard convention is to choose the interval $\left[-\frac{\pi}{2}, \frac{\pi}{2}\right]$. Therefore, we will consider the function $f(x) = \sin x$ restricted to the interval $-\frac{\pi}{2} \le x \le \frac{\pi}{2}$. Its graph is shown in Figure 14.19.

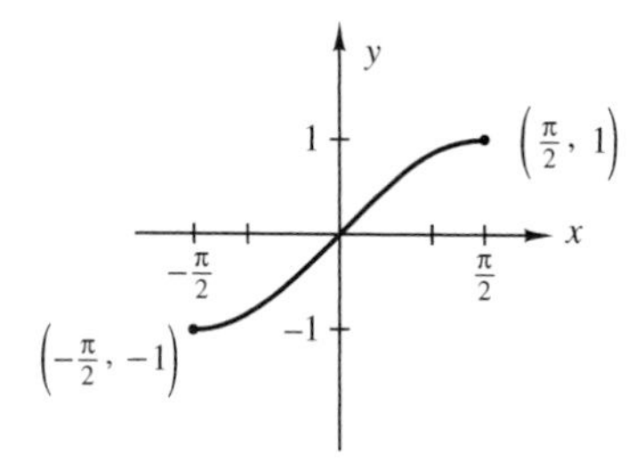

Figure 14.19 $y = \sin x,\ -\frac{\pi}{2} \le x \le \frac{\pi}{2}$

This restricted sine function is one-to-one and so has an inverse, called the *inverse sine function*, denoted $\sin^{-1} x$ or $\arcsin x$. The domain of the inverse sine function is $[-1, 1]$ and its range is $\left[-\frac{\pi}{2}, \frac{\pi}{2}\right]$. Thus we have

$$y = \sin^{-1} x,\ -1 \le x \le 1, \text{ if and only if } x = \sin y,\ -\frac{\pi}{2} \le y \le \frac{\pi}{2}.$$

The graph of $y = \sin^{-1} x$ in Figure 14.20 is obtained by reflecting the graph in Figure 14.19 about the line $y = x$.

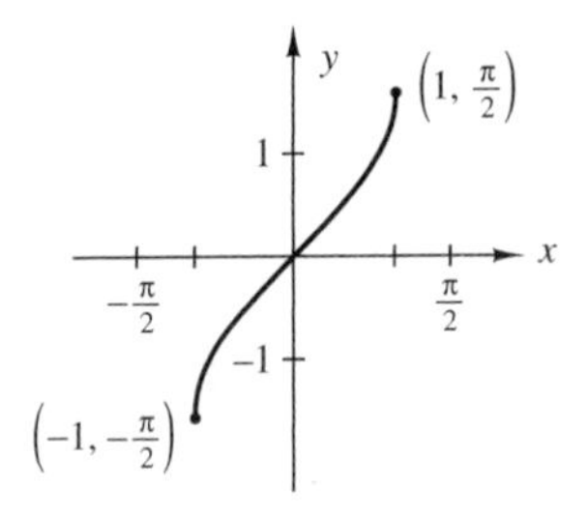

Figure 14.20 $y = \sin^{-1} x$

From the definition of the inverse sine function, $y = \sin^{-1} x$ (or $y = \arcsin x$) means that y is the angle in radian measure between $-\frac{\pi}{2}$ and $\frac{\pi}{2}$ whose sine is x.

> Be aware of differences in notation used with trigonometric functions. The function denoted $\sin^{-1} x$ is the inverse of the restricted sine function. It does **not** mean $(\sin x)^{-1}$, which is $\frac{1}{\sin x}$, the reciprocal of the sine function. The function $\sin^n x$ for n positive means $(\sin x)^n$. For example, $\sin^2 x$ means $(\sin x)^2$.

EXAMPLE 14.10 Find:

a. $\sin^{-1}\left(\frac{\sqrt{3}}{2}\right)$ **b.** $\sin^{-1}\left(-\frac{\sqrt{2}}{2}\right)$ **c.** $\sin^{-1}(-1)$

Solution

a. To find $y = \sin^{-1}\left(\frac{\sqrt{3}}{2}\right)$, we use the definition of the inverse sine function, which says $\sin y = \frac{\sqrt{3}}{2}$ and $-\frac{\pi}{2} \le y \le \frac{\pi}{2}$. Since $y = \frac{\pi}{3}$ is the angle between $-\frac{\pi}{2}$ and $\frac{\pi}{2}$ whose sine is $\frac{\sqrt{3}}{2}$, we conclude $\sin^{-1}\left(\frac{\sqrt{3}}{2}\right) = \frac{\pi}{3}$.

b. To find $y = \sin^{-1}\left(-\frac{\sqrt{2}}{2}\right)$, we need to find the angle y satisfying $\sin y = -\frac{\sqrt{2}}{2}$ and $-\frac{\pi}{2} \le y \le \frac{\pi}{2}$. This angle is $y = -\frac{\pi}{4}$.

c. For $y = \sin^{-1}(-1)$, we need the angle y such that $\sin y = -1$ and $-\frac{\pi}{2} \le y \le \frac{\pi}{2}$. So $y = -\frac{\pi}{2}$. ■

When the domain of a function is restricted to produce an inverse function, the composition of the function and its inverse will be the identity function only when certain conditions are met, as the following example shows.

EXAMPLE 14.11 Evaluate each of the following. From these examples, conjecture when the composite of the sine function and its inverse will be the identity function, and give reasons for your conjecture.

a. $\sin \frac{5\pi}{4}$ **b.** $\sin^{-1}\left(\sin\frac{5\pi}{4}\right)$ **c.** $\sin \frac{\pi}{6}$ **d.** $\sin^{-1}\left(\sin\frac{\pi}{6}\right)$

e. $\sin^{-1}\left(-\frac{1}{2}\right)$ **f.** $\sin\left(\sin^{-1}\left(-\frac{1}{2}\right)\right)$ **g.** $\sin^{-1}\left(\frac{\sqrt{2}}{2}\right)$ **h.** $\sin\left(\sin^{-1}\left(\frac{\sqrt{2}}{2}\right)\right)$

Solution We use the values of $\sin x$ discussed in Chapter 8. When we evaluate expressions with $\sin^{-1} x$, we keep in mind its definition and the fact that the range of this function contains only values in the interval $\left[-\frac{\pi}{2}, \frac{\pi}{2}\right]$.

a. $\sin \frac{5\pi}{4} = -\frac{\sqrt{2}}{2}$ **b.** $\sin^{-1}\left(\sin\frac{5\pi}{4}\right) = \sin^{-1}\left(-\frac{\sqrt{2}}{2}\right) = -\frac{\pi}{4}$

c. $\sin \frac{\pi}{6} = \frac{1}{2}$ **d.** $\sin^{-1}\left(\sin\frac{\pi}{6}\right) = \sin^{-1}\left(\frac{1}{2}\right) = \frac{\pi}{6}$

e. $\sin^{-1}\left(-\frac{1}{2}\right) = -\frac{\pi}{6}$ **f.** $\sin\left(\sin^{-1}\left(-\frac{1}{2}\right)\right) = \sin\left(-\frac{\pi}{6}\right) = -\frac{1}{2}$

g. $\sin^{-1}\left(\frac{\sqrt{2}}{2}\right) = \frac{\pi}{4}$ **h.** $\sin\left(\sin^{-1}\left(\frac{\sqrt{2}}{2}\right)\right) = \sin\left(\frac{\pi}{4}\right) = \frac{\sqrt{2}}{2}$

We can see from parts f and h that the composite function $\sin(\sin^{-1} x)$ seems to be the identity function. The equation $\sin(\sin^{-1} x) = x$ is true for any x in the interval $[-1, 1]$, since $\sin^{-1} x$ is the radian angle in the interval $\left[-\frac{\pi}{2}, \frac{\pi}{2}\right]$ whose sine is x. We can see in part b, for $x = \frac{5\pi}{4}$, that $\sin^{-1}(\sin x) \neq x$; however in part d, for $x = \frac{\pi}{6}$, $\sin^{-1}(\sin x) = x$. So the composite function $\sin^{-1}(\sin x)$ is not always the identity function. It will be the identity function on the interval $\left[-\frac{\pi}{2}, \frac{\pi}{2}\right]$, since by definition, $\sin^{-1}$ is the inverse of the sine function restricted to that interval. Outside of that interval, $\sin^{-1}(\sin x) \neq x$. ■

 When the domain of a function f is restricted in order to define f^{-1}, the composite function $f^{-1} \circ f$ will not always be the identity function. Only when c is in the restricted domain is the equation $(f^{-1} \circ f)(c) = c$ true. For example, $\sin^{-1}\left(\sin\frac{5\pi}{4}\right) \neq \frac{5\pi}{4}$ but $\sin^{-1}\left(\sin\frac{\pi}{6}\right) = \frac{\pi}{6}$.

The process that we have followed to define an inverse for the sine function can be repeated for each of the other five trigonometric functions. For each, we identify an interval on which the function is one-to-one and for which all values in the range occur. Then we specify the inverse as the inverse of the function restricted to that interval.

The tangent function is one-to-one on the interval $\left(-\frac{\pi}{2}, \frac{\pi}{2}\right)$ and takes on every value in its range $(-\infty, \infty)$. Also, $\lim_{x \to \frac{\pi}{2}^+} \tan x = -\infty$ and $\lim_{x \to \frac{\pi}{2}^-} \tan x = \infty$; the lines $x = -\frac{\pi}{2}$ and $x = \frac{\pi}{2}$ are vertical asymptotes. The graph of the tangent function restricted to this interval is in Figure 14.21, with the vertical asymptotes shown as dashed lines. The inverse of the restricted tangent function is denoted $\tan^{-1}x$ or arctan x. The graph of $\tan^{-1}x$ is obtained by reflecting the graph in Figure 14.21 about the line $y = x$. Under this reflection, the vertical asymptotes of the restricted tangent function produce horizontal asymptotes $y = -\frac{\pi}{2}$ and $y = \frac{\pi}{2}$ for the inverse tangent function; that is, $\lim_{x \to -\infty} \tan^{-1} x = -\frac{\pi}{2}$ and $\lim_{x \to \infty} \tan^{-1} x = \frac{\pi}{2}$. These asymptotes are shown as dashed lines in the graph of $\tan^{-1} x$ in Figure 14.22.

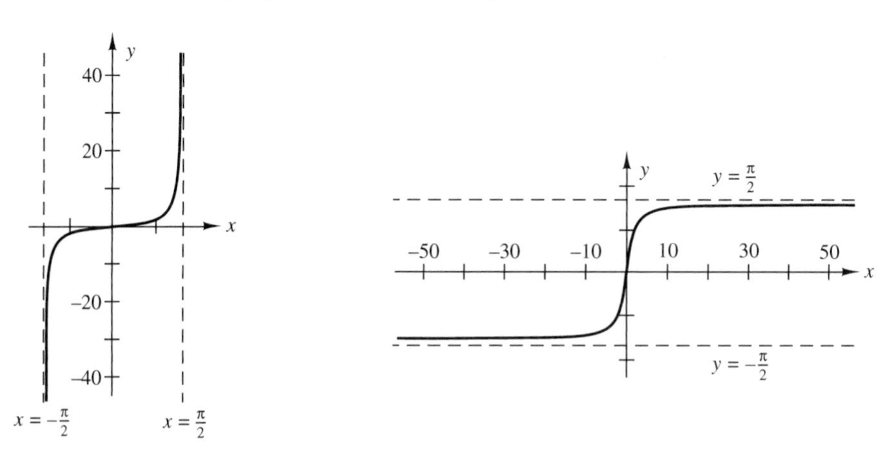

Figure 14.21 $y = \tan x$, $-\frac{\pi}{2} < x < \frac{\pi}{2}$ **Figure 14.22** $y = \tan^{-1} x$

The inverse tangent function has domain all real numbers and range $\left(-\frac{\pi}{2}, \frac{\pi}{2}\right)$; thus we have

$$y = \tan^{-1} x \text{ if and only if } x = \tan y \text{ and } -\frac{\pi}{2} < y < \frac{\pi}{2}.$$

EXAMPLE 14.12 Find:

a. $\tan^{-1}(-1)$ **b.** $\tan^{-1} \sqrt{3}$ **c.** $\tan^{-1} 0$

d. $\tan^{-1}(\tan \pi)$ **e.** $\tan(\tan^{-1} 20)$ **f.** $\tan^{-1}\left(\tan \frac{\pi}{7}\right)$

Solution

a. To find $\tan^{-1}(-1)$, we use the definition of the inverse tangent function. We want the angle between $-\frac{\pi}{2}$ and $\frac{\pi}{2}$ whose tangent is -1. This angle is $-\frac{\pi}{4}$, so $\tan^{-1}(-1) = -\frac{\pi}{4}$.

b. From the definition, we know that $\tan^{-1} \sqrt{3}$ is the angle in the interval $\left(-\frac{\pi}{2}, \frac{\pi}{2}\right)$ whose tangent is $\sqrt{3}$. This angle is $\frac{\pi}{3}$, so $\tan^{-1} \sqrt{3} = \frac{\pi}{3}$.

c. Since 0 is in the interval $\left(-\frac{\pi}{2}, \frac{\pi}{2}\right)$ and $\tan 0 = 0$, it follows that $\tan^{-1} 0 = 0$.

d. Since $\tan \pi = 0$, we can use part c to evaluate this composite function: $\tan^{-1}(\tan \pi) = \tan^{-1} 0 = 0$. We note that since π is not in the interval $\left(-\frac{\pi}{2}, \frac{\pi}{2}\right)$, this composite function does not act as the identity: $\tan^{-1}(\tan \pi) \neq \pi$.

e. To determine $\tan(\tan^{-1} 20)$, we do not need to evaluate $\tan^{-1} 20$. We know that by definition of the inverse tangent function, $y = \tan^{-1} 20$ means that $\tan y = 20$. So $\tan(\tan^{-1} 20) = 20$.

f. To determine $\tan^{-1}\left(\tan \frac{\pi}{7}\right)$, we can use the definition of the inverse tangent function, since $\frac{\pi}{7}$ is between $-\frac{\pi}{2}$ and $\frac{\pi}{2}$. Thus $\tan^{-1}\left(\tan \frac{\pi}{7}\right) = \frac{\pi}{7}$. ■

In the next example, we look at the inverse cosine function.

EXAMPLE 14.13

a. Sketch the graph of the cosine function on the interval $0 \leq x \leq \pi$. How do you know that the function is one-to-one in this interval? What is the range of the cosine function on this interval?

b. Sketch the inverse of the function you sketched in part a. This function is the inverse cosine function, denoted $\cos^{-1} x$, or $\arccos x$. Give the domain and range of this function.

c. Suppose $y = \cos^{-1} x$. For what values of x and what values of y will $x = \cos y$?

d. Find, if possible: $\cos^{-1}(-1)$, $\cos^{-1} 2$, and $\cos^{-1}\left(-\frac{\sqrt{2}}{2}\right)$.

e. For what values of x will the following equations be true?

(i) $\cos(\cos^{-1} x) = x$ (ii) $\cos^{-1}(\cos x) = x$

f. Find: $\cos(\cos^{-1}(0.35))$, $\cos^{-1}\left(\cos\left(-\frac{\pi}{4}\right)\right)$, and $\cos^{-1}\left(\cos \frac{\pi}{3}\right)$.

Solution

a. The graph of the function $y = \cos x$ for x in the interval $[0, \pi]$ is in Figure 14.23. We can see that the graph passes the horizontal line test, so the function is one-to-one. Also, the range of the restricted cosine function is $[-1, 1]$, which is the whole range of the cosine function. The interval $[0, \pi]$ is the conventional one chosen on which to restrict the cosine function in order to define its inverse.

b. The inverse of the function sketched for part a is in Figure 14.24; it is obtained from the graph in Figure 14.23 by reflecting the graph about the line $y = x$. The function $y = \cos^{-1} x$ has domain $[-1, 1]$ and range $[0, \pi]$.

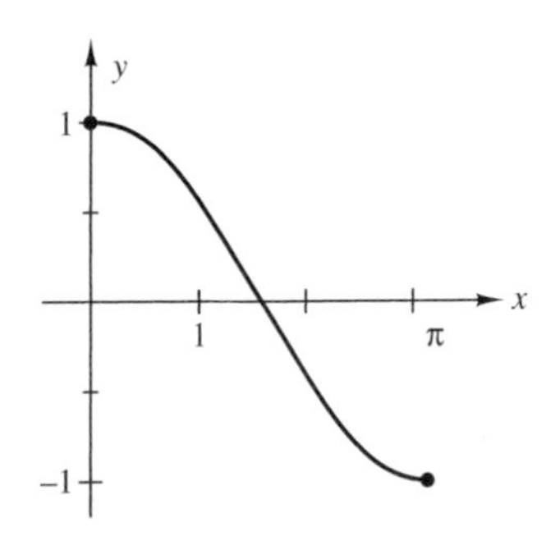

Figure 14.23 $y = \cos x, 0 \le x \le \pi$

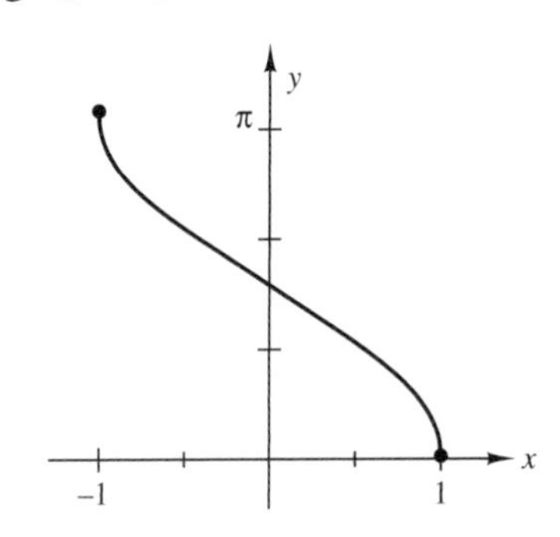

Figure 14.24 $y = \cos^{-1} x$

c. By definition, $\cos^{-1} x$ is the inverse of the cosine function restricted to the interval $[0, \pi]$, so $y = \cos^{-1} x$ means that y is the angle between 0 and π for which $x = \cos y$. Also, x must be between -1 and 1.

d. By definition, $\cos^{-1}(-1)$ is the angle between 0 and π whose cosine is -1. This angle is π; $\cos^{-1}(-1) = \pi$. If $\cos^{-1} 2 = x$, then $\cos x = 2$, which is impossible, since the cosine of an angle is between -1 and 1. Thus $\cos^{-1} 2$ doesn't exist. The angle between 0 and π whose cosine is $-\frac{\sqrt{2}}{2}$ is $\frac{3\pi}{4}$ so $\cos^{-1}\left(-\frac{\sqrt{2}}{2}\right) = \frac{3\pi}{4}$.

e. Equation (i) $\cos(\cos^{-1} x) = x$ will be true for every x between -1 and 1—that is, for every x in the domain of the inverse cosine function. Since the range of the inverse cosine function is the interval $[0, \pi]$, equation (ii) $\cos^{-1}(\cos x) = x$ will be true for every x between 0 and π, and false for all other values of x in the domain of the unrestricted cosine function.

f. We can use part e to evaluate the first and third of these expressions. Since 0.35 is in the interval $[-1, 1]$, we have $\cos(\cos^{-1}(0.35)) = 0.35$. Since $-\frac{\pi}{4}$ is not between 0 and π, we need to first evaluate $\cos\left(-\frac{\pi}{4}\right) = \frac{\sqrt{2}}{2}$. We can now complete the evaluation: $\cos^{-1}\left(\cos\left(-\frac{\pi}{4}\right)\right) = \cos^{-1}\left(\frac{\sqrt{2}}{2}\right) = \frac{\pi}{4}$. Since $\frac{\pi}{3}$ is in the interval $[0, \pi]$, we have $\cos^{-1}\left(\cos\frac{\pi}{3}\right) = \frac{\pi}{3}$. ■

The inverses of the secant, cosecant, and cotangent functions can all be defined by a process similar to that used for the sine, tangent, and cosine. The intervals to which the secant, cosecant, and cotangent functions are restricted in order to define their inverses vary a bit in different calculus texts, and so we do not define those inverse functions here.

In calculus, we encounter functions that are the composition of a trigonometric function and the inverse of a different trigonometric function. In evaluating these composites, we must be careful to be aware of the domains and ranges of the functions involved. Surprisingly, some of these composite functions turn out to have algebraic formulas.

EXAMPLE 14.14

a. Evaluate $\sin(\cos^{-1} 1)$ and $\cos(\sin^{-1} 1)$.

b. Evaluate $\sin\left(\cos^{-1}\frac{1}{2}\right)$ and $\cos\left(\sin^{-1}\frac{1}{2}\right)$.

c. Evaluate $\sin(\cos^{-1} 0)$ and $\cos(\sin^{-1} 0)$.

d. What do you observe about the pairs of values you obtained in parts a, b, and c? Can you show this is always true for any x in the interval $[0, 1]$?

Solution

a. As with all composite functions, we evaluate "from the inside out," using definitions of the inverse functions: $\sin(\cos^{-1} 1) = \sin 0 = 0$. Similarly, $\cos(\sin^{-1} 1) = \cos \frac{\pi}{2} = 0$.

b. Again, we evaluate the inside function first: $\sin\left(\cos^{-1}\frac{1}{2}\right) = \sin \frac{\pi}{3} = \frac{\sqrt{3}}{2}$ and $\cos\left(\sin^{-1}\frac{1}{2}\right) = \cos \frac{\pi}{6} = \frac{\sqrt{3}}{2}$.

c. We have $\sin(\cos^{-1} 0) = \sin \frac{\pi}{2} = 1$ and $\cos(\sin^{-1} 0) = \cos 0 = 1$.

d. In parts a, b, and c, each pair of composite functions has the same value. We want to show that for all x in the interval $[0, 1]$, $\sin(\cos^{-1} x) = \cos(\sin^{-1} x)$. Parts a and c show that this equation is true for the endpoints of the interval, $x = 0$ and $x = 1$. To show that the equation is true for any x in the interval $(0, 1)$, we can use the right triangle definitions of the sine and cosine functions. For each x in the interval $(0, 1)$, the values of $\sin^{-1} x$ and $\cos^{-1} x$ will be in the interval $\left(0, \frac{\pi}{2}\right)$ and so can represent an acute angle in a right triangle. In order to find a simpler expression for $\sin(\cos^{-1} x)$ on the interval $(0, 1)$, we draw a right triangle with hypotenuse 1 and acute angle θ so that $\cos \theta = \frac{x}{1} = x$ (which also means $\cos^{-1} x = \theta$). Then by the Pythagorean theorem, the remaining side has length $\sqrt{1-x^2}$:

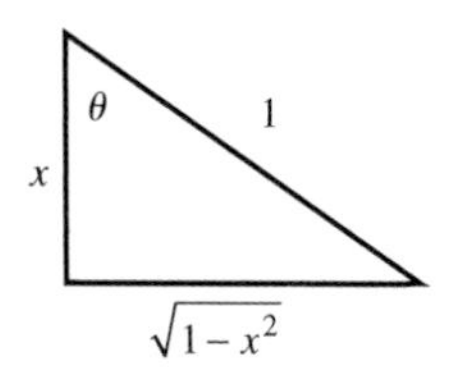

From the picture, we see that $\sin \theta = \frac{\sqrt{1-x^2}}{1} = \sqrt{1-x^2}$. Thus, $\sin(\cos^{-1} x) = \sin \theta = \sqrt{1-x^2}$ for every x in the interval $(0, 1)$. In Exercise 7 of the exercises that follow, you are asked to draw a similar picture to show that $\cos(\sin^{-1} x) = \sqrt{1-x^2}$ for all x between 0 and 1. Thus the two composite functions are equal for all x in the interval [0, 1]. ■

Exercises 14-C

1. Find the inverse of each of the following functions that has an inverse. If the inverse does not exist, explain why not.

a. $f(x) = 9 + 2x$ **b.** $g(x) = 9 + 2x^2$ **c.** $h(x) = 9 + 2x^3$

d. $F(x) = 9 + 2x^4$ **e.** $G(x) = 9 + 2\sqrt{x}$ **f.** $H(x) = 9 + 2\sqrt[3]{x}$

2. For each function in Exercise 1 that has an inverse, use a graphing utility to sketch on the same set of axes the line $y = x$, the function, and the inverse you found. Use a square window and an interval of the domain that includes 0. Verify that your pairs of graphs are mirror images of each other with respect to the line $y = x$.

3. Consider $F(x) = x^2 + 3$ from Examples 14.8 and 14.9. Restrict the domain of F to be $\{x \mid x \leq 0\}$.

a. Sketch the graph of F over the restricted domain.

b. Verify that this restricted F is one-to-one.

c. Find the inverse of F with the restricted domain.

d. Sketch the graph of the inverse of F with the restricted domain.

4. Let $G(x) = x^2 + 4x + 4$. Restrict the domain of G to be $\{x \mid x \geq -2\}$.

a. Sketch the graph of G over the restricted domain.

b. Verify that this restricted G is one-to-one.

c. Find the inverse of G with the restricted domain and sketch its graph.

5. Find the following values, if possible. If not possible, explain why.

a. $\sin^{-1}\left(-\frac{\sqrt{3}}{2}\right)$ **b.** $\cos^{-1}\left(-\frac{\sqrt{3}}{2}\right)$ **c.** $\tan^{-1} 1$

d. $\tan^{-1}\left(-\sqrt{3}\right)$ **e.** $\cos^{-1} \frac{\pi}{2}$ **f.** $\sin^{-1}\left(-\frac{1}{2}\right)$

6. Find the following values, if possible. If not possible, explain why.

a. $\sin(\tan^{-1} 1)$ **b.** $\tan(\sin^{-1} 1)$ **c.** $\cos(\tan^{-1}(-1))$

d. $\tan(\tan^{-1}(-2.4))$ **e.** $\sin^{-1}\left(\sin \frac{3\pi}{2}\right)$ **f.** $\tan^{-1}\left(\sin\left(-\frac{3\pi}{2}\right)\right)$

7. Draw a right triangle with hypotenuse 1 and acute angle θ so that $\sin\theta = \frac{x}{1} = x$. Use this triangle to give an argument similar to the one in part d of Example 14.14 to show that $\cos(\sin^{-1} x) = \sqrt{1-x^2}$ for all x between 0 and 1.

8. a. Use the equations $\cos(\sin^{-1} x) = \sqrt{1-x^2}$ and $\sin(\cos^{-1} x) = \sqrt{1-x^2}$ to find the following values: $\cos(\sin^{-1}(0.2))$ and $\sin\left(\cos^{-1}\left(-\frac{1}{3}\right)\right)$.

b. Use a calculator to evaluate $\cos(\sin^{-1}(0.2))$ and $\sin\left(\cos^{-1}\left(-\frac{1}{3}\right)\right)$. Compare your results with those found in part a.

Chapter 14 Exercises

1. For the functions f and g in Example 14.1, give a general rule for $f(x)$ and for $g(x)$ in symbolic form using the variable x. Check the values of $f(x)$ and $g(x)$ in the table with your symbolic rule.

2. Sketch the graph of each of the following functions. Use the graphs to determine which functions are one-to-one.

a. $f(x) = |x|$ **b.** $g(x) = 5 - 2x$

c. $h(x) = \frac{1}{x}$ **d.** $F(x) = \sqrt{x-1} + 3$

3. For each of the functions in Exercise 2 that is one-to-one, find the symbolic form of the inverse and sketch the graph of the inverse.

4. Determine if the following pairs of functions are inverses of each other:

a. $f(x) = 4 - 3x$ and $g(x) = \frac{4}{3} - \frac{1}{3}x$ **b.** $f(x) = \frac{2}{x}$ and $g(x) = \frac{x}{2}$

c. $f(x) = \frac{1}{x} - 1$ and $g(x) = \frac{1}{x+1}$ **d.** $f(x) = 2x + 1$ and $g(x) = \frac{1}{2}x - 1$

5. Suppose x is the current temperature in degrees Celsius. Then $F(x) = \frac{9}{5}x + 32$ gives the temperature in degrees Fahrenheit. Find the inverse function that converts temperature in degrees Fahrenheit to temperature in degrees Celsius.

6. Suppose x is the area in square inches of the top of a computer desk.

a. Give a function $y = f(x)$ where y is the area of the desktop in square feet.

b. Give the inverse of the function in part a and describe in words what it does.

c. Verify that the functions found in parts a and b are inverses of each other.

7. Let $f(x) = x^3 + 1$. Find:

a. $f(-2)$ **b.** $f^{-1}\left(\frac{9}{8}\right)$ **c.** $f(7)$

d. $f^{-1}(7)$ **e.** $f(f^{-1}(7))$ **f.** $f^{-1}(f(-2))$

8. Let $f(x) = \frac{2}{x+3}$.

a. Show that if $f(a) = f(b)$, then $a = b$.

b. Use your work in part a to determine if f is one-to-one.

9. For each of the following graphs, determine if it is the graph of a function. If so, determine if it is one-to-one. If the function is one-to-one, sketch the graph of its inverse on the same set of axes.

a.

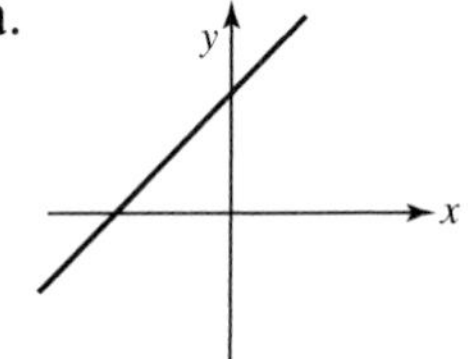

b.

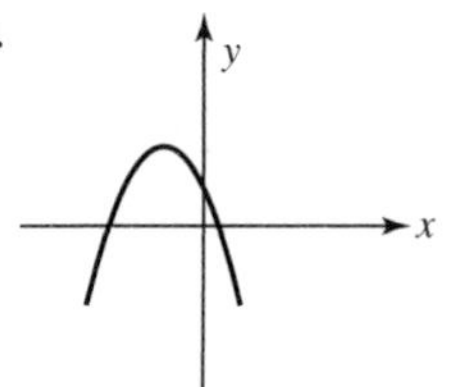

c.

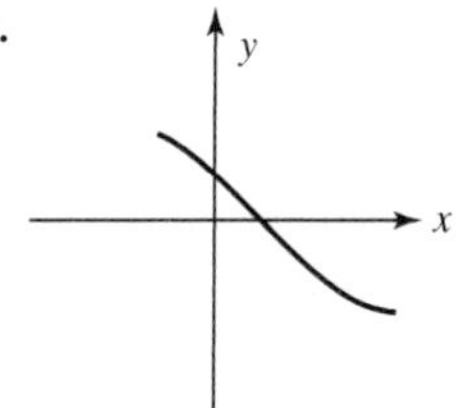

d.

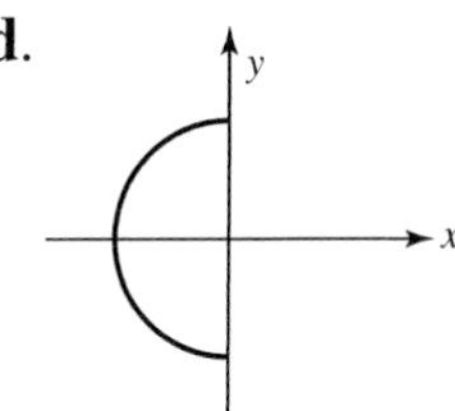

e.

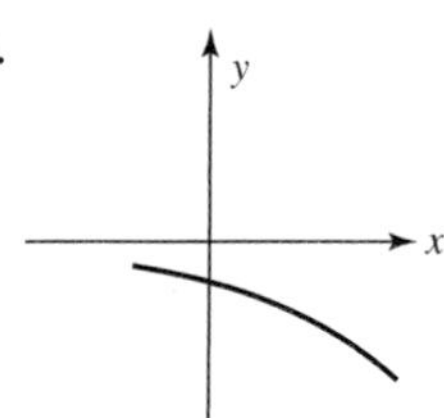

10. Find the following values, if possible. If not possible, explain why.

a. $\sin^{-1}\left(\sin\frac{\pi}{7}\right)$ **b.** $\cos^{-1}\left(\frac{\sqrt{2}}{2}\right)$ **c.** $\tan^{-1}\left(\tan\frac{7\pi}{4}\right)$

d. $\sin\left(\tan^{-1}\left(\frac{\sqrt{3}}{3}\right)\right)$ **e.** $\cos(\cos^{-1}(0.47))$ **f.** $\sin^{-1}\frac{3\pi}{2}$

11. a. In the right triangle below, label one of its acute angles θ so that $\tan\theta = x$. Use the Pythagorean theorem and the definition of $\tan^{-1} x$ to show that each of the following equations is true for all $x \geq 0$: $\sin(\tan^{-1} x) = \frac{x}{\sqrt{1+x^2}}$ and $\cos(\tan^{-1} x) = \frac{1}{\sqrt{1+x^2}}$.

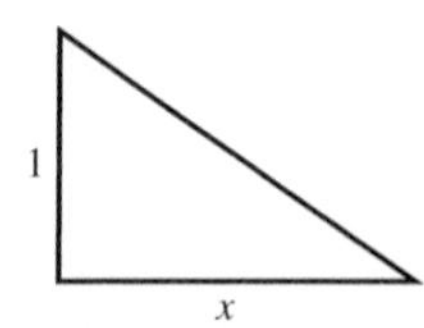

b. Use the equations in part a to find $\cos(\tan^{-1} 3)$ and $\sin(\tan^{-1}(-2))$.

c. Use a calculator to evaluate $\cos(\tan^{-1} 3)$ and $\sin(\tan^{-1}(-2))$, and compare your results with those found in part b.

CHAPTER 15

Companion to Logarithmic Functions

In the 1990 census, the population of a certain city was recorded as 92,800. The 2000 census showed a population of 153,000 for the city, so the city planners have come up with the following equation, using an exponential function, for the population $P(t)$ of the city t years after 1990:

$$P(t) = 92{,}800e^{0.05t}.$$

Note that $t = 0$ corresponds to 1990 and $t = 10$ corresponds to 2000. We can check that the function $P(t)$ agrees with the population given by each census: $P(0) = 92{,}800e^{0} = 92{,}800$ and $P(10) = 92{,}800e^{0.5} \approx 153{,}000$. The city planners have also concluded that the factors that controlled the city's growth from 1990 to 2000 will not change during the foreseeable future.

The hospital in the city is large enough to service a community of 250,000, but if the population increases beyond that value, the hospital will need to be expanded. With the current growth pattern, when should the city planners expect an addition to the hospital will be needed? To answer that question, we want to know the time t at which the population $P(t)$ will reach 250,000. To find this, we set $P(t)$ equal to 250,000 in the equation for $P(t)$: $250{,}000 = 92{,}800e^{0.05t}$.

We need to solve this equation for t to find the time at which the population will equal 250,000. We divide both sides of the equation by 92,800 to get $\frac{250{,}000}{92{,}800} = e^{0.05t}$; this yields $2.694 = e^{0.05t}$.

One way to solve for the unknown t that appears as an exponent on the right side of the equation is to consider the inverse function of an exponential function. (We will complete the solution of this problem in Section 15-C.)

15-A DEFINITION AND PROPERTIES OF LOGARITHMIC FUNCTIONS

The exponential function $f(x) = a^x$, for $a > 1$, is an increasing function, so each horizontal line intersects the graph at most once; this implies the function is one-to-one. See Figure 15.1. For $0 < a < 1$, the function $f(x) = a^x$ is a decreasing function and is also one-to-one. See Figure 15.2.

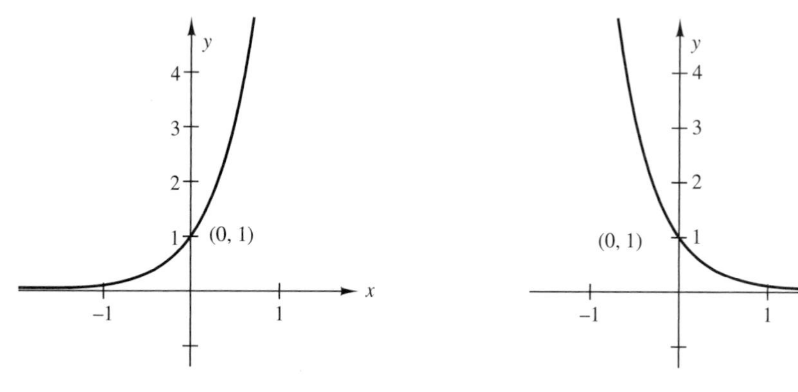

Figure 15.1 $y = a^x\ (a > 1)$ **Figure 15.2** $y = a^x\ (0 < a < 1)$

These graphs show that for $a > 0$, $a \neq 1$, the exponential function $f(x) = a^x$ has an inverse. The inverse is called the *logarithmic function to the base a*. The inverse of $f(x) = a^x$ is denoted by $f^{-1}(x) = \log_a x$ and is read "log to the base a of x."

$$\textit{$\log_a x$ and a^x are inverse functions for $a > 0$, $a \neq 1$.}$$

The expression $\log_a x$ could be written with parentheses as $\log_a(x)$, but usually parentheses are used only when the meaning could be interpreted in more than one way. For example, $\log_a(2x + 1)$, if written without parentheses, could be interpreted as $\log_a(2x) + 1$. If there is any question about the interpretation, insert parentheses to remove ambiguity.

Note that the name "logarithm" does not give any suggestion of its relationship to the exponential function. The word "logarithm" is derived from the Greek words "logos" (ratio) and "arithmos" (number), and was introduced by John Napier (1550–1618), author of the first book on logarithms. Napier defined the logarithm in geometric terms as a ratio of the lengths of two line segments. In today's definition, a logarithm is an exponent.

For any function f with an inverse f^{-1}, we know that $f(f^{-1}(x)) = x$, for x in the range of f, and $f^{-1}(f(x)) = x$, for x in the domain of f.

For every $a > 0$, $a \neq 1$, the exponential function a^x has domain the set of all real numbers and range the set of all positive real numbers. We can express the inverse relationship between the exponential function a^x and its inverse $\log_a x$ as follows:

$$a^{\log_a x} = x \text{ for } x > 0$$

$$\log_a a^x = x \text{ for all real numbers } x.$$

EXAMPLE 15.1 Simplify each of the following expressions:

a. $\log_2 \frac{1}{64}$ **b.** $\log_{10}(0.01)$ **c.** $\log_{10}\left(10^{3x-1}\right)$ **d.** $5^{\log_5(t^2+1)}$

Solution

a. $\log_2 \frac{1}{64} = \log_2 2^{-6} = -6$ since $\log_a a^x = x$ for all real numbers x. Here, the base a is 2 and $x = -6$.

b. $\log_{10}(0.01) = \log_{10}\left(10^{-2}\right) = -2$, since $\log_a a^x = x$ for all real numbers x. Here, the base a is 10 and $x = -2$.

c. $\log_{10}(10^{3x-1}) = 3x - 1$ since $\log_a a^x = x$ for all real numbers x. In this example, the base a is 10 and x is replaced by the expression $3x - 1$.

d. Since $a^{\log_a x} = x$ for $x > 0$, we see that when the base is 5, $5^{\log_5 x} = x$ for $x > 0$. Because $t^2 + 1 > 0$ for all real numbers t, we can replace x by $t^2 + 1$ in the equation $5^{\log_5 x} = x$. Thus, $5^{\log_5(t^2+1)} = t^2 + 1$. ■

From the definition of inverse function, we also know that $y = f^{-1}(x)$ if and only if $x = f(y)$. In terms of the exponential and logarithmic functions, we write this

$$y = \log_a x \text{ if and only if } x = a^y.$$

This gives a way to convert from logarithms to exponents and from exponents to logarithms. For example, for the expressions in Examples 15.1a and b, we have $\log_2 \frac{1}{64} = -6$ if and only if $2^{-6} = \frac{1}{64}$, and $\log_{10}(0.01) = -2$ if and only if $10^{-2} = 0.01$.

The logarithm function with base 10 is called the *common logarithm* because of its ease of use with our base 10 number system. It is often denoted simply as log. Thus $\log x$ means $\log_{10} x$. Common logarithms are used in the Richter scale for measuring earthquake intensity and in measuring intensity of sound in decibels. Most calculators have a special key for the common logarithm.

The natural exponential function $f(x) = e^x = \exp(x)$ arises in many applications and has as its inverse the *natural logarithmic function.* The logarithmic function with base e has a special notation. It is denoted $\ln x$. Thus, $\log_e x = \ln x$, and

$$\ln x \text{ and } e^x \text{ are inverse functions.}$$

Most calculators have a special key for the function $\ln x$.

The fundamental inverse relationships between $\ln x$ and e^x are given by the equations

$$\ln e^x = x, \text{ or } \ln(\exp(x)) = x \text{ for all real numbers } x$$

$$e^{\ln x} = x, \text{ or } \exp(\ln x) = x \text{ for all } x > 0.$$

The inverse relationships between logarithmic and exponential functions allow us to derive properties of logarithms from the rules of exponents (see Section 13-A). For any base $a > 0$, $a \neq 1$, any positive real numbers x and y, and any real number r, the following properties are true. (We list each property twice, once for an arbitrary base a, and once for the natural base e.)

1. $\log_a (xy) = \log_a x + \log_a y$
2. $\log_a \frac{x}{y} = \log_a x - \log_a y$
3. $\log_a(x^r) = r\log_a x$
4. $\log_a a = 1$
5. $\log_a 1 = 0$

1. $\ln (xy) = \ln x + \ln y$
2. $\ln \frac{x}{y} = \ln x - \ln y$
3. $\ln(x^r) = r \ln x$
4. $\ln e = 1$
5. $\ln 1 = 0$

EXAMPLE 15.2 State in words property 1 of logarithms given above.

Solution Property 1 for an arbitrary base a states $\log_a(xy) = \log_a x + \log_a y$. In words this is "The logarithm to the base a of the product of two expressions x and y is equal to the sum of the logarithm to the base a of x plus the logarithm to the base a of y." When the base is e, the equation $\ln(xy) = \ln x + \ln y$ is read as "The natural logarithm of a product xy is the sum of the natural logarithm of x plus the natural logarithm of y. In short, "The logarithm of a product is the sum of the logarithms of the factors." ■

EXAMPLE 15.3 Use properties of exponents and the inverse relationship between logarithms and exponents to show that $\log_2(5 \cdot 7) = \log_2 5 + \log_2 7$. (This is property 1 above for the particular values $x = 5$, $y = 7$, and base $a = 2$.)

Solution Since $5 = 2^{\log_2 5}$ and $7 = 2^{\log_2 7}$, if we multiply the two equations we have $5 \cdot 7 = 2^{\log_2 5} \cdot 2^{\log_2 7}$. We use property 1 of exponents (see Chapter 7) on the right-hand side to conclude $5 \cdot 7 = 2^{\log_2 5 + \log_2 7}$. Then we take log to the base 2 of both sides of this equation to get $\log_2(5 \cdot 7) = \log_2\left(2^{\log_2 5 + \log_2 7}\right) = \log_2 5 + \log_2 7$. ■

If $f(x)$ is a product or quotient, we sometimes want to simplify $\log_a(f(x))$ by using properties 1 and 2 of logarithms.

EXAMPLE 15.4 Use properties of logarithms to write each expression as a sum or difference in simplest form. State which properties you use.

a. $\log_3 81t$ **b**. $\ln \sqrt{x}(x+4)^3$ **c**. $\log_2 \frac{t+1}{2(t-1)}$

Solution

a. We use property 1 to write $\log_3 81t = \log_3 81 + \log_3 t = \log_3 3^4 + \log_3 t$. We can use the inverse relationship $\log_a a^x = x$ to obtain $\log_3 3^4 + \log_3 t = 4 + \log_3 t$.

b. We write $\sqrt{x}$ as $x^{1/2}$ and use property 1 in the first step of the simplification process and property 3 (twice) in the second step: $\ln x^{1/2}(x+4)^3 = \ln x^{1/2} + \ln(x+4)^3 = \frac{1}{2}\ln x + 3\ln(x+4)$.

c. We use property 2 to write $\log_2 \frac{t+1}{2(t-1)} = \log_2(t+1) - \log_2 2(t-1)$. The second term on the right side of the equation can be further expanded by using properties 1 and 4: $\log_2 2(t-1) = \log_2 2 + \log_2(t-1) = 1 + \log_2(t-1)$. Thus, $\log_2 \frac{t+1}{2(t-1)} = \log_2(t+1) - (1 + \log_2(t-1)) = \log_2(t+1) - 1 - \log_2(t-1)$. ■

The logarithm of a simple product or quotient can often be written without parentheses. Parentheses must be inserted when the logarithm is applied to a sum or difference, and whenever there might be ambiguity. For example, $\log_2(3x^2)$ may be written as $\log_2 3x^2$ and $\ln\left(\frac{2x}{x^2+1}\right)$ may be written as $\ln \frac{2x}{x^2+1}$. However, $\ln(x^3+5)$ **cannot** be written without parentheses: $\ln(x^3+5) \neq \ln x^3 + 5$. (Note that $\ln x^3 + 5 = (\ln x^3) + 5$.)

Sometimes we wish to consolidate, rather than expand, an expression that involves logarithms. We can use properties of logarithms to do this.

EXAMPLE 15.5 Use properties of logarithms to write each expression as a single logarithm, if possible. If it is not possible, explain why not.

a. $\log_8 x - 2\log_8 3$ **b**. $4\ln e + 3\ln(x^2+1)$ **c**. $\frac{\log_2 x}{\log_2(x+2)}$

Solution

a. We use property 3 and then property 2 to write $\log_8 x - 2\log_8 3 = \log_8 x - \log_8 3^2 = \log_8 \frac{x}{3^2} = \log_8 \frac{x}{9}$.

b. We use properties 3 and 1 to write: $4\ln e + 3\ln(x^2+1) = \ln e^4 + \ln(x^2+1)^3 = \ln e^4(x^2+1)^3$.

c. Since this expression is the quotient of logarithms, it cannot be written as a single logarithm. ■

In general, the logarithm of a sum or difference cannot easily be expressed as a sum or difference of two logarithms. For example, $\log(100 + 10) = \log 110 \approx 2.04$, but $\log 100 + \log 10 = 2 + 1 = 3$; $\log(100 - 10) = \log 90 \approx 1.95$, but $\log 100 - \log 10 = 2 - 1 = 1$. Also, in general, $\frac{\log_a x}{\log_a y} \neq \log_a\left(\frac{x}{y}\right)$ and $(\log_a x)(\log_a y) \neq \log_a(xy)$.

Parentheses are not always used with exponential and logarithmic functions. To find the derivative of a composite function that involves exponential or logarithmic functions, we insert the implied parentheses to more easily decompose the function. When possible, we use properties of exponential and logarithmic functions to simplify the function before differentiating.

EXAMPLE 15.6 Rewrite each function to show explicitly any implied parentheses and simplify if possible. Indicate the differentiation rules that would be used to find the derivative of each function.

a. $F(x) = e^{x+\ln\sqrt{x}}$ **b**. $G(x) = (\ln x^3)^2$ **c**. $K(x) = \log_2 \sin^2 x$

d. $H(x) = \ln\sqrt{x}\, e^{x+1}$ **e**. $N(x) = \dfrac{\ln 2x^4}{3^{x^2}}$

Solution

a. $F(x) = e^{x+\ln\sqrt{x}} = e^{(x+\ln\sqrt{x})}$. This shows that $F(x)$ is a composite function $f(g(x))$, where $g(x) = x + \ln\sqrt{x}$ and $f(x) = e^x$. To differentiate $F(x)$ in this form we could use the Chain Rule. However, $F(x)$ can be simplified: $F(x) = e^{x+\ln\sqrt{x}} = e^x e^{\ln\sqrt{x}} = e^x\sqrt{x}$, which we can differentiate using the Product Rule.

b. We first insert implied parentheses: $G(x) = (\ln x^3)^2 = (\ln(x^3))^2$. $G(x)$ is a double composite function $G(x) = h(f(g(x)))$, where $g(x) = x^3$, $f(x) = \ln x$, and $h(x) = x^2$. To differentiate $G(x)$, it is easiest to first simplify $\ln(x^3) = 3\ln x$ and then use the Chain Rule to differentiate $G(x) = (3\ln x)^2 = 9(\ln x)^2$.

c. $K(x) = \log_2 \sin^2 x = \log_2((\sin(x))^2)$. Using the properties of logarithms, we can simplify this expression to $K(x) = 2\log_2(\sin(x))$. Since $K(x)$ is the composite function $f(g(x))$, where $g(x) = \sin x$ and $f(x) = 2\log_2 x$, we use the Chain Rule to differentiate $K(x)$.

d. $H(x) = \ln\sqrt{x}\, e^{x+1} = \ln(\sqrt{x}\, e^{(x+1)})$, which can be simplified as follows: $\ln(\sqrt{x}\, e^{(x+1)}) = \ln\sqrt{x} + \ln(e^{(x+1)}) = \frac{1}{2}\ln x + (x+1) = \frac{1}{2}\ln x + x + 1$. In this form, $H(x)$ can be differentiated term by term.

e. $N(x) = \frac{\ln 2x^4}{3^{x^2}} = \frac{\ln(2x^4)}{3^{(x^2)}} = \frac{\ln(2)+4\ln(x)}{3^{(x^2)}}$. $N(x)$ is a quotient so we use the Quotient Rule to find its derivative. Because $3^{(x^2)}$ is a composite function, we use the Chain Rule to find the derivative of this denominator. ■

Exercises 15-A

1. State in words the following properties of logarithms.

a. $\log_a \frac{x}{y} = \log_a x - \log_a y$ **b.** $\log_a(x^r) = r \log_a x$

c. $\log_a a = 1$ **d.** $\log_a 1 = 0$

2. Evaluate the following without using a calculator:

a. $\log_3(3^2) + 1$ **b.** $(\log_3 3)^2 + 1$ **c.** $\log_3 \frac{1}{9}$ **d.** $3^{-\log_3 5}$

3. Use properties of logarithmic and exponential functions to simplify each expression as much as possible. Identify which properties you use.

a. $10^{\log_{10}(x+1)}$ **b.** $e^{x+\ln x}$ **c.** $3 \ln e^{x(x+1)}$

d. $e^{4 \ln x}$ **e.** $5^{3 \log_5(x-2)}$ **f.** $e^{\ln(\sin x)-\ln(\cos x)}$

4. For each equation below, consider the values of x for which all expressions in the equation are defined. Then decide whether the equation is always true or sometimes false for these values. If sometimes false, find a specific value of x for which the equation is not true.

a. $\ln(x) + 1 = \ln(x + 1)$ **b.** $(\ln x)x^2 = \ln(x^3)$

c. $\log(x^3) = 3 \log x$ **d.** $\log_2 x + \log_5 x = \log x$

e. $(\log x)^3 = 3 \log x$ **f.** $\frac{\log_5 x}{\log_5 25} = \log_5\left(\frac{x}{25}\right)$

5. Use properties of logarithms to write each expression as a sum or difference. Simplify as much as possible.

a. $\ln\left(\frac{x^2-1}{x+3}\right)$ **b.** $\log_2 4\sqrt{x+1}$ **c.** $\ln \frac{e^2}{x^4}$

d. $\log_5(5x^2 + 5)^3$ **e.** $\log_3(x^3 + x^2)$ **f.** $\ln e^{x-1}(x - 1)$

6. Use properties of logarithms to write each expression as a single logarithm if possible. If it is not possible, explain why not.

a. $2 \ln x + 5 \ln(x - 1) - \ln(x + 2)$ **b.** $\log_2 x + \log_3(x + 1)$

c. $\frac{1}{2}\log_9 t - \frac{1}{4}\log_9 16$ **d.** $(\log_3 x)[\log_3(x-1)]$

e. $\log_2 x - \log_2(x+1) + 3\log_2 x$ **f.** $\ln(\cos x) - \ln(\sin x)$

7. Rewrite each function to show explicitly any implied parentheses and simplify if possible. Indicate the differentiation rules that would be used to find the derivative of the simplified function.

a. $F(x) = \ln e^{2\cos x}$ **b.** $K(x) = \log_7 \tan^4 x$ **c.** $H(x) = \ln\sqrt{\sin 4x}$

d. $G(x) = \ln x^4 e^{-x}$ **e.** $N(x) = \dfrac{4^{5x-2}}{\ln x^5}$

15-B GRAPHS OF LOGARITHMIC FUNCTIONS

Recall that if a function f has an inverse, then the graph of its inverse, f^{-1}, can be obtained by reflecting the graph of $y = f(x)$ about the line $y = x$. We can use this property to obtain the graph of the logarithmic function to the base a. First we analyze the graph of the natural logarithmic function, which has base e.

EXAMPLE 15.7 Sketch the graphs of the functions $h(x) = e^x$ and $h^{-1}(x) = \ln x$. Give the domain, range, and asymptotes for h and h^{-1}. Describe where each function is increasing, where it is decreasing, where it is concave up, and where it is concave down.

Solution The graph of $h(x) = e^x$ is shown (dashed) in Figure 15.3. The graph of $h^{-1}(x) = \ln x$ is the reflection of the graph of h about the line $y = x$ and is also shown (solid) in Figure 15.3. Note that the graph of $y = e^x$ contains the point $(0, 1)$, so the point $(1, 0)$ is on the graph of $y = \ln x$. The domain of $h(x) = e^x$ is $(-\infty, \infty)$ and its range is $(0, \infty)$.

Since, by properties of inverse functions, the domain of the inverse function h^{-1} is the range of the function h, the domain of $h^{-1}(x) = \ln x$ is $(0, \infty)$. Similarly, the range of h^{-1} is the domain of h. Thus, the range of $\ln x$ is $(-\infty, \infty)$. Since the line $y = 0$ is a horizontal asymptote of $h(x) = e^x$, the line $x = 0$ is a vertical asymptote of $h^{-1}(x) = \ln x$, as Figure 15.3 shows.

From their graphs, we see that the natural exponential function is increasing and concave up on its entire domain $(-\infty, \infty)$, and the natural logarithmic function is increasing and concave down on its entire domain $(0, \infty)$.

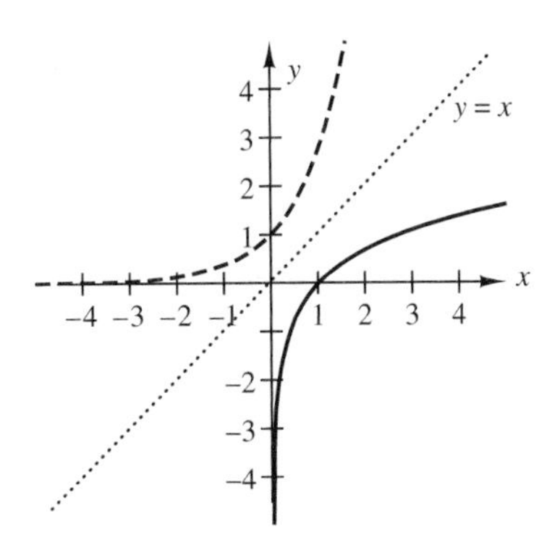

Figure 15.3 $y = e^x$ (dashed), $y = \ln x$ (solid) ■

In the next example we use the same technique to sketch the graphs of the logarithmic functions $\log_2 x$ and $\log_{1/3} x$.

EXAMPLE 15.8 Sketch the graphs of the following pairs of inverse functions and describe the domain, range, and asymptotes of each function.

a. $f(x) = 2^x$ and $f^{-1}(x) = \log_2 x$

b. $g(x) = \left(\frac{1}{3}\right)^x$ and $g^{-1}(x) = \log_{1/3} x$

Solution

a. We first graph the function $f(x) = 2^x$ (discussed in Section 13-A) and then reflect this graph about the line $y = x$ to obtain the graph of its inverse $f^{-1}(x) = \log_2 x$. Figure 15.4 shows both graphs. Since the domain of the exponential function $f(x) = 2^x$ is $(-\infty, \infty)$ and its range is $(0, \infty)$, we conclude that the domain of $\log_2 x$, which is the range of 2^x, is $(0, \infty)$ and the range of $\log_2 x$, which is the domain of 2^x, is $(-\infty, \infty)$. The graph of $f(x) = 2^x$ has only one asymptote, the horizontal line $y = 0$, so the only asymptote of the graph of $f^{-1}(x) = \log_2 x$ is the vertical line $x = 0$.

b. Since the base of $g(x) = \left(\frac{1}{3}\right)^x$ is $\frac{1}{3}$, which is less than 1, g is decreasing on its entire domain $(-\infty, \infty)$, its range is $(0, \infty)$, and its graph contains the point $(0, 1)$ (see Section 13-A). The graph appears in Figure 15.5, together with its reflection about the line $y = x$, which is the graph of $g^{-1}(x) = \log_{1/3} x$. We see that: domain of $\log_{1/3} x$ = range of $\left(\frac{1}{3}\right)^x = (0, \infty)$; also range of $\log_{1/3} x$ = domain of $\left(\frac{1}{3}\right)^x = (-\infty, \infty)$.

Since the horizontal axis, $y = 0$, is the only asymptote of $g(x) = \left(\frac{1}{3}\right)^x$, the vertical axis, $x = 0$, is the only asymptote of $g^{-1}(x) = \log_{1/3} x$. ■

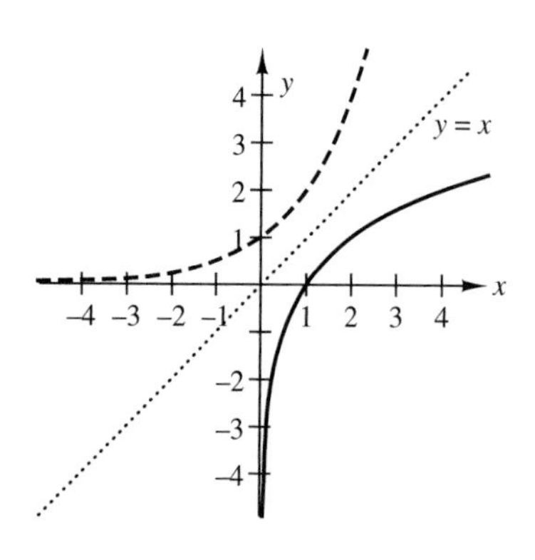

Figure 15.4 $y = 2^x$ (dashed), $y = \log_2 x$ (solid)

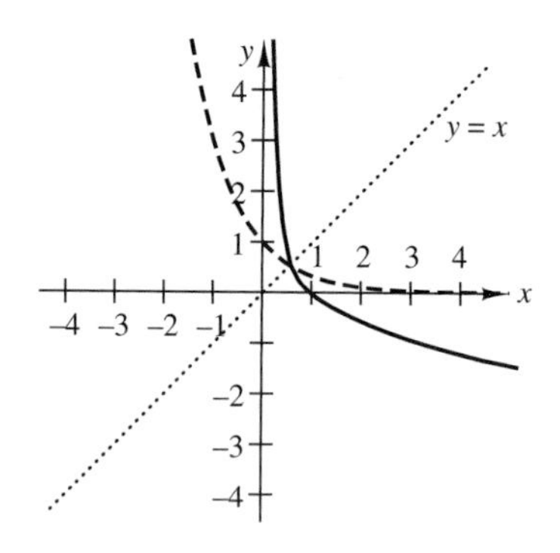

Figure 15.5 $y = \left(\frac{1}{3}\right)^x$ (dashed), $y = \log_{\frac{1}{3}} x$ (solid)

The domain of each of the logarithmic functions shown in Figures 15.3, 15.4, and 15.5 is the interval $(0, \infty)$, and the range of each is $(-\infty, \infty)$. In general, for any real number $a > 0$, and $a \neq 1$,

1. The domain of the function $\log_a x$ is the interval $(0, \infty)$; that is, the graph of $\log_a x$ lies in the right half of the plane.

2. The range of $\log_a x$ is $(-\infty, \infty)$.

3. The point $(1, 0)$ is on the graph of $y = \log_a x$.

Figures 15.3 and 15.4 show that the graphs of the logarithmic functions $y = \ln x$ and $y = \log_2 x$ have similar shapes; any logarithmic function with base $a > 1$ will have a shape similar to these. In particular,

4. When $a > 1$, the function $\log_a x$ is increasing on its entire domain, and $\lim_{x \to \infty} \log_a x = \infty$.

5. When $a > 1$, the values $\log_a x$ decrease without bound as x approaches 0 from the right; that is, $\lim_{x \to 0^+} \log_a x = -\infty$.

6. When $a > 1$, the graph of $y = \log_a x$ is concave down on its entire domain.

The graph of a logarithmic function with positive base $a < 1$ has shape similar to the graph of $y = \log_{1/3} x$ in Figure 15.5. In particular,

7. When $0 < a < 1$, the graph of $y = \log_a x$ is decreasing on its entire domain, and $\lim_{x \to \infty} \log_a x = -\infty$.

8. When $0 < a < 1$, the values $\log_a x$ increase without bound as x approaches 0 from the right; that is, $\lim_{x \to 0^+} \log_a x = \infty$.

9. When $0 < a < 1$, the graph of $y = \log_a x$ is concave up on its entire domain.

Transformations of a logarithmic function will change its graph and may change its domain and its asymptotes.

EXAMPLE 15.9 Give the domains and use transformations of graphs to sketch the graphs of the following functions. Show any asymptotes.

a. $y = \log_3(x - 2)$ **b**. $y = \ln x + 3$ **c**. $y = \ln(1 - x)$

Solution

a. Since the domain of the function $y = \log_a x$ is the set of all real numbers $x > 0$, it follows that the domain of the function $y = \log_3(x - 2)$ is the set of x such that $(x - 2) > 0$, or $x > 2$. The graph of $y = \log_3(x - 2)$ is the graph of $y = \log_3 x$ translated 2 units to the right. It appears in Figure 15.6. Since $x = 0$ is a vertical asymptote of $y = \log_3 x$, the line $x = 2$ is a vertical asymptote of $y = \log_3(x - 2)$.

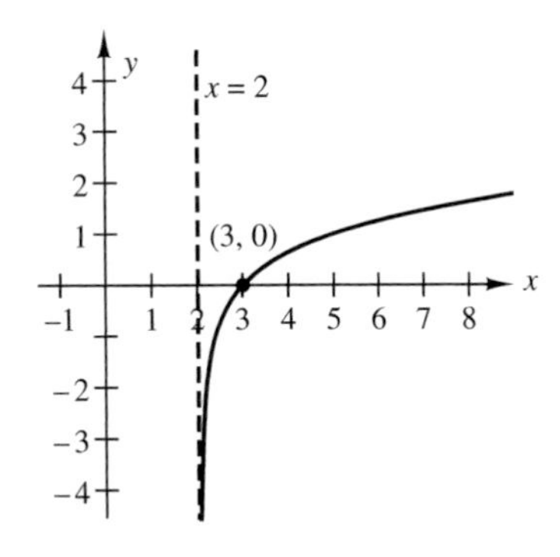

Figure 15.6 $y = \log_3(x - 2)$

b. The domain of $y = \ln x + 3$ is $(0, \infty)$. The graph of $y = \ln x$ is translated up 3 units to yield the graph of $y = \ln x + 3$. The y-axis is a vertical asymptote. See Figure 15.7.

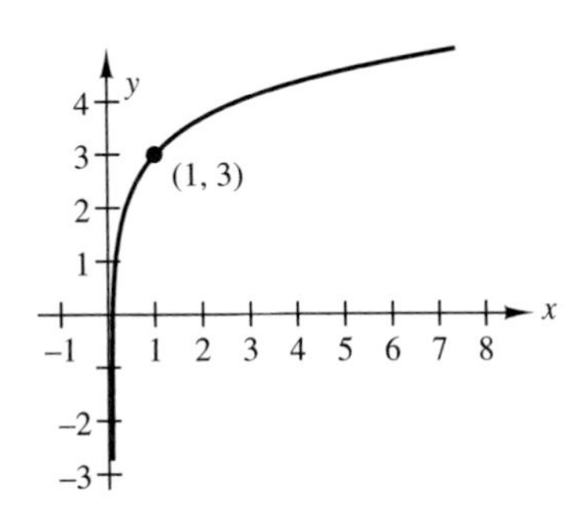

Figure 15.7 $y = \ln x + 3$

c. The function $y = \ln(1 - x)$ is defined when $1 - x > 0$, or $x < 1$; thus the domain of $\ln(1 - x)$ is $(-\infty, 1)$. To obtain the graph of $y = \ln(1 - x)$ by transforming the graph of $y = \ln x$, we can write the given function as $\ln(-(x - 1))$. To obtain its graph we reflect the graph of $y = \ln x$ about the y-axis (this gives the graph of $y = \ln(-x)$) and then translate 1 unit to the right, which gives the graph of $\ln(-(x - 1))$. The line $x = 1$ is a vertical asymptote of the graph of $y = \ln(1 - x)$. See Figure 15.8.

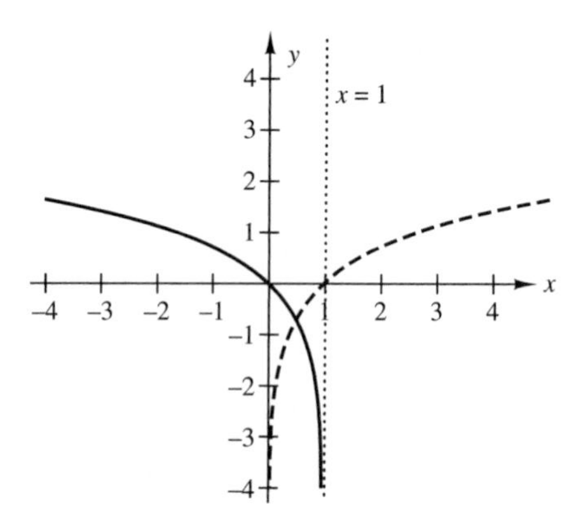

Figure 15.8 $y = \ln(1 - x)$ (solid), $y = \ln x$ (dashed) ■

The domain of $\log_a x$ is $(0, \infty)$; $\log_a x$ is undefined when $x \le 0$. In general, the domain of any function of the form $g(x) = \log_a(f(x))$, $a > 0$, $a \neq 1$, is the set of x-values for which $f(x) > 0$.

EXAMPLE 15.10 Find the domain of each function.

a. $F(x) = \log_2(x^2 + 1)$ **b.** $G(x) = \ln\left(\frac{x+1}{2x-3}\right)$ **c.** $H(x) = \log_{10}(x^2 - 9)$

Solution

a. The domain is the set of all x such that $x^2 + 1 > 0$. Since $x^2 + 1 > 0$ for all x, the domain is all real numbers x.

b. The function $G(x) = \ln\left(\frac{x+1}{2x-3}\right)$ is defined only when $\frac{x+1}{2x-3} > 0$. Note that $2x - 3 = 0$ when $x = \frac{3}{2}$, so $\frac{x+1}{2x-3}$ does not exist (d.n.e.) when $x = \frac{3}{2}$. To solve the inequality $\frac{x+1}{2x-3} > 0$, we use a sign chart:

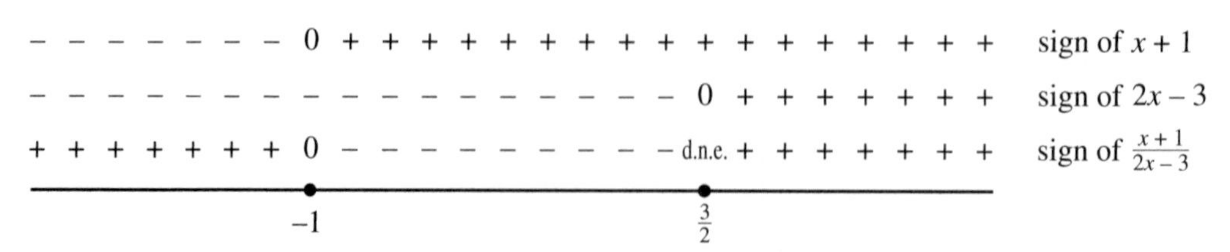

From the chart we see that the quotient $\frac{x+1}{2x-3}$ is positive for $x < -1$ and for $x > \frac{3}{2}$. Thus the domain of the function G is the union of two intervals, $(-\infty, -1) \cup \left(\frac{3}{2}, \infty\right)$.

c. The domain of $H(x) = \log_{10}(x^2 - 9)$ is the set of all real x such that $x^2 > 9$; thus, $x > 3$ or $x < -3$. In interval notation, the domain of H is $(-\infty, -3) \cup (3, \infty)$. ■

Exercises 15-B

1. Find the domain of each function.

a. $f(x) = \ln(x^2)$ **b**. $g(x) = (\ln x)^2$

c. $h(x) = -\ln(-x)$ **d**. $w(x) = \log(5x - 2)$

e. $v(x) = \log_2(x^2 + 3x - 18)$

2. The graphs of two exponential functions, $f(x) = \left(\frac{1}{4}\right)^x$ and $g(x) = 4^x$, are shown in Figures 15.9a and 15.9b.

a. For each of the following evaluate or say why it is undefined:

$f^{-1}(16), f^{-1}(1), f^{-1}(0), g^{-1}(4), g^{-1}(-1), g^{-1}\left(\frac{1}{2}\right).$

b. Use the graph of f to sketch the graph of its inverse f^{-1}. Give the equation of f^{-1} and its domain and range.

c. Use the graph of g to sketch the graph of its inverse g^{-1}. Give the equation of g^{-1} and its domain and range.

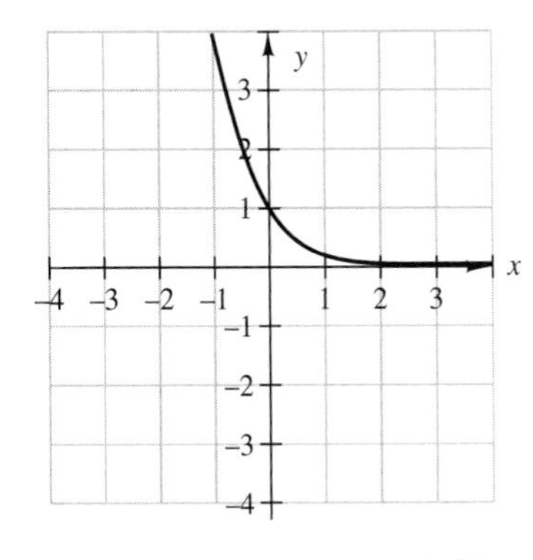

Figure 15.9a $f(x) = \left(\frac{1}{4}\right)^x$

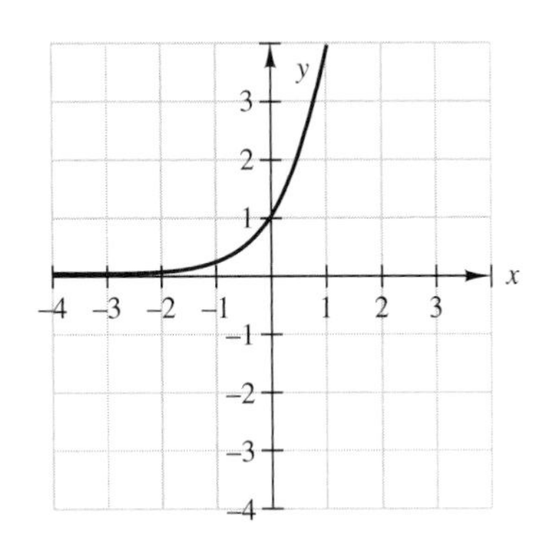

Figure 15.9b $g(x) = 4^x$

3. Use properties of transformations of graphs and the graph of $y = \log_2 x$ (Figure 15.4) to sketch the graphs of the following pairs of functions on the same set of axes:

a. $y = \log_2 x$ and $y = \log_2(-x)$

b. $y = \log_2 x$ and $y = -\log_2(x)$

4. Rewrite the functions in simpler form when possible. Use properties of transformations to sketch the graphs of the following functions:

a. $y = 2^{\log_2 x}$ **b**. $y = \log(x + 3)$ **c**. $y = \log x + 3$

d. $y = 2 - \ln x$ **e**. $y = \log_5 5^x$

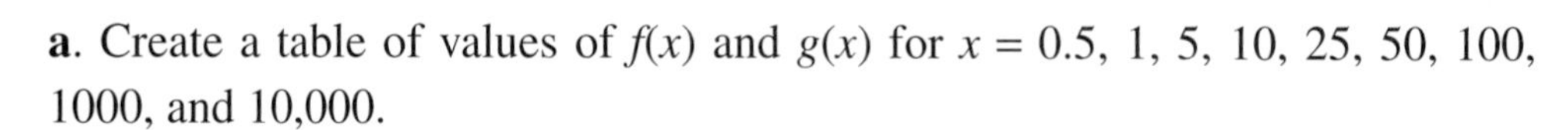

5. Consider the functions $f(x) = \sqrt{x}$ and $g(x) = \ln x$.

a. Create a table of values of $f(x)$ and $g(x)$ for $x = 0.5$, 1, 5, 10, 25, 50, 100, 1000, and 10,000.

b. Sketch the graphs of $y = f(x)$ and $y = g(x)$ on the same set of axes.

c. Describe in words what is happening to values of $f(x)$ and $g(x)$ as x gets larger.

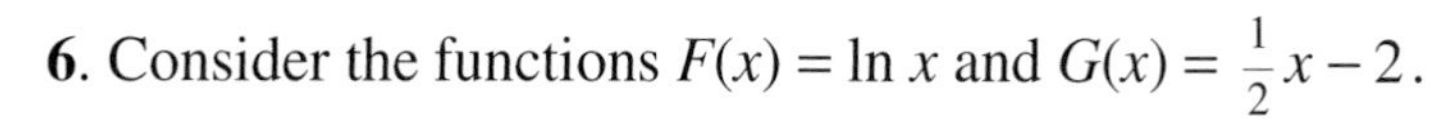

6. Consider the functions $F(x) = \ln x$ and $G(x) = \frac{1}{2}x - 2$.

a. Sketch the graphs of $y = F(x)$ and $y = G(x)$ on the same set of axes.

b. At how many points do the graphs of $y = F(x)$ and $y = G(x)$ intersect?

c. How many solutions are there to the equation $\ln x = \frac{1}{2}x - 2$? Explain how you obtained your answer.

15-C SOLVING EQUATIONS WITH LOGARITHMIC AND EXPONENTIAL FUNCTIONS

Let's return to the example at the beginning of this chapter to find when the population of the city will exceed 250,000. We need to solve the following equation for t:

$$2.694 = e^{0.05t}.$$

To solve this equation and, in general, to solve most equations that involve logarithmic and exponential functions, we make use of the inverse relationships between the two functions:

$$\log_a a^x = x \text{ for all } x, \text{ and } a^{\log_a x} = x \text{ for all } x > 0$$

$$\ln e^x = x \text{ for all } x, \text{ and } e^{\ln x} = x \text{ for all } x > 0.$$

In our example, to solve for the exponent t, we take the natural logarithm of each side:

$$\ln(2.694) = \ln(e^{0.05t}).$$

Since $\ln(e^{0.05t}) = 0.05t$, we have $\ln(2.694) = 0.05t$. To solve for t, we divide each side of the equation by 0.05. To obtain a numerical answer, we use a calculator to find $\ln(2.694) \approx 0.99103$. Thus,

$$t = \frac{\ln(2.694)}{0.05} \approx \frac{0.99103}{0.05} = 19.8 \text{ years.}$$

This means that in 19.8 years from time $t = 0$—that is, 19.8 years from 1990—the population of the city will reach 250,000 and an addition to the hospital will be needed. This will occur in 2009.

The next example illustrates how, when we solve exponential and logarithmic equations, it is sometimes necessary to use the basic properties of exponents and logarithms in addition to their inverse relationship. Also, in general, there are two rules of thumb for solving such equations:

1. To solve an equation for a variable that appears in an exponent, take an appropriate logarithm of both sides of the equation.

2. To solve an equation for a variable x that appears in a composite of the form $\log_a(f(x))$, first apply the base a exponential function to both sides of the equation.

EXAMPLE 15.11 Solve each equation for the variable:

a. $\log_2(1 - x) = 5$ **b.** $3^{x^2+2} = 25$

c. $2^{3x+4} = 5^{x-1}$ **d.** $\ln(t + 1) + \ln(t - 3) = \ln 5$

Solution

a. To solve $\log_2(1 - x) = 5$, we apply the base 2 exponential function to both sides: $2^{\log_2(1-x)} = 2^5$. The base 2 was chosen because $2^{\log_2(1-x)} = 1 - x$. Thus we have $1 - x = 2^5$, so $1 - x = 32$ and $x = -31$.

b. Since we want to solve for x, which appears in the exponent, we want to take a logarithm of each side of the equation, using base 3. We get $\log_3\left(3^{x^2+2}\right) = \log_3 25$, so $x^2 + 2 = \log_3 25$, or $x^2 = \log_3 25 - 2$. Thus $x = \pm\sqrt{\log_3 25 - 2}$. If we want to use a calculator to get a numerical answer, we would use the natural logarithm, since that function is standard on calculators. To solve $3^{x^2+2} = 25$ using the natural logarithm, we begin with $\ln\left(3^{x^2+2}\right) = \ln 25$. Since $\ln\left(3^{x^2+2}\right) = (x^2 + 2)\ln 3$, we have $(x^2 + 2)\ln 3 = \ln 25$, so $(x^2 + 2) = \frac{\ln 25}{\ln 3}$ and $x^2 = \frac{\ln 25}{\ln 3} - 2$. With a calculator we find $\ln 25 \approx 3.2189$ and $\ln 3 \approx 1.0986$, so $x^2 \approx \frac{3.2189}{1.0986} - 2 \approx 0.93$, and the solutions are $x \approx \pm\sqrt{0.93} \approx \pm\, 0.9644$.

c. If we take the natural logarithm of each side of the given equation, we get $\ln(2^{3x+4}) = \ln(5^{x-1})$, so (by property 3 of logarithms) $(3x + 4)\ln 2 = (x - 1)\ln 5$. To solve for x, we multiply out each side of the equation: $3x \ln 2 + 4\ln 2 = x\ln 5 - \ln 5$. To isolate x, we subtract $x \ln 5$ and $4 \ln 2$ from each side of the equation to get $3x \ln 2 - x\ln 5 = -\ln 5 - 4\ln 2$. We factor x out of the left side and -1 out of the right side: $x(3\ln 2 - \ln 5) = -(\ln 5 + 4\ln 2)$. Finally, to solve for x, we divide: $x = -\frac{(\ln 5 + 4\ln 2)}{3\ln 2 - \ln 5}$. We can substitute $\ln 2 \approx 0.6931$ and $\ln 5 \approx 1.6094$ to obtain the numerical answer $x \approx -9.32$.

d. We use property 1 of logarithms to write $\ln(t + 1) + \ln(t - 3) = \ln[(t + 1)(t - 3)]$, so the given equation becomes $\ln[(t + 1)(t - 3)] = \ln 5$. Now we can apply the natural exponential function to both sides: $e^{\ln[(t+1)(t-3)]} = e^{\ln 5}$; this gives $(t + 1)(t - 3) = 5$. To solve this equation, we expand the left side and get $t^2 - 2t - 3 = 5$; then we put the equation in general form $t^2 - 2t - 8 = 0$. This factors as $(t - 4)(t + 2) = 0$. Therefore, $t = 4$ or $t = -2$. The solution $t = -2$ is not a valid solution because $\ln(t + 1)$ for $t = -2$ becomes $\ln(-1)$, which is undefined. The only solution of the original equation is $t = 4$. We note that the extraneous value $t = -2$, which is not a solution of the original equation, was introduced when $\ln(t + 1) + \ln(t - 3)$ was combined and written as $\ln[(t + 1)(t - 3)]$. ■

Extraneous solutions can arise in the solution of equations that involve logarithms. Always check answers to make sure all terms in the original equation are defined for each of the solutions.

EXAMPLE 15.12 Suppose \$1,000 is invested at 6% interest.

a. Find how long it will take to double the investment if interest is compounded quarterly.

b. Find how long it will take to double the investment if interest is compounded continuously.

c. How do the doubling times found in parts a and b compare?

Solution

a. If interest is compounded four times a year, the amount in the account after t years is $A(t) = 1000\left(1 + \frac{0.06}{4}\right)^{4t}$ (see Section 13-B). To find how long it will take to double the \$1,000, we ask: For what t does $A(t) = \$2{,}000$? To solve the equation $2000 = 1000(1 + 0.015)^{4t}$ for t, we divide both sides by 1000 to get $2 = (1.015)^{4t}$. Then we take the natural logarithm of each side of the equation: $\ln 2 = \ln(1.015)^{4t} = (4t)(\ln 1.015)$, so $4t = \frac{\ln 2}{\ln 1.015} \approx 46.56$. Therefore, $t \approx \frac{46.56}{4} = 11.64$ years.

b. If interest is compounded continuously at the rate of 6%, the amount in the account after t years is $A(t) = 1000\,e^{0.06t}$. We set $A(t) = 2000$ and solve for t to find when the amount will double. The equation $2000 = 1000e^{0.06t}$ simplifies to $2 = e^{0.06t}$. We take the natural logarithm of each side to get $\ln 2 = \ln e^{0.06t}$. Thus, $\ln 2 = 0.06t$, or $t = \frac{\ln 2}{0.06} \approx \frac{0.6931}{0.06} \approx 11.55$ years.

c. The difference between the two doubling times is $11.64 - 11.55 = 0.09$, or 9 hundredths of 1 year. This means a difference of $0.09 \cdot 12 = 1.08 \approx 1$ month. So when the interest is compounded continuously, the investment doubles approximately 1 month sooner than if the interest is compounded quarterly. ■

Often exponential functions are used to model a process of decay. For example, carbon-14 dating is used to determine the age of bones. When an animal is alive, the amount of carbon-12, which is not radioactive, and the amount of carbon-14 (denoted ^{14}C), which is radioactive, are equal. When the animal dies, the carbon-14 begins to decay. It takes 5730 years for the amount of carbon-14 to decay to half its initial amount. This number of years is called its *half-life*.

The general function for exponential growth or decay that gives the amount of a substance present at time t is $y = C_0 e^{kt}$, where C_0 is the initial amount (at time $t = 0$) and k is a constant (the rate of decay). To find the function that models given data, we determine C_0 and k.

EXAMPLE 15.13 A certain type of animal bone contains 6 g of carbon-14 when the animal is alive. Carbon-14 is known to start decaying as soon as the animal dies.

a. Find a function that gives the amount of carbon-14 present in the bone t years after the time of death. (Use the fact that the half-life of carbon-14 is 5730 years.)

b. Suppose such a bone is found to have 4 g of carbon-14. How old is the bone?

Solution

a. The initial amount of carbon-14 is 6 g, so $C_0 = 6$. Because the half-life of carbon-14 is 5730 years, we know that when $t = 5730$ years, $y = 3$ g. If we substitute these values in the equation $y = C_0 e^{kt}$, we have $3 = 6e^{k(5730)}$. Dividing by 6 gives $0.5 = e^{5730k}$. To find the value of k, we take the natural logarithm of both sides of this equation: $\ln 0.5 = \ln(e^{5730k}) = 5730k$. Finally, we divide both sides by 5730: $\frac{\ln 0.5}{5730} \approx \frac{-0.6931}{5730} \approx -0.000121 = k$. Thus, the function requested is $y = 6e^{-0.000121t}$.

b. The age of the bone is the time t at which the amount of carbon-14 present is 4 g—that is, $4 = 6e^{-0.000121t}$. To solve for t we divide both sides of the equation by

6: $\frac{4}{6} = e^{-0.000121t}$ and take the natural logarithm of each side: $\ln\left(\frac{4}{6}\right) = \ln e^{-0.000121t} = -0.000121t$. We divide both sides of this equation by -0.000121 to obtain $t = \frac{\ln(4/6)}{-0.000121} \approx \frac{-0.405465}{-0.000121} \approx 3351$. So the bone is approximately 3,350 years old. ■

Exercises 15-C

1. Solve for the variable in each equation.

a. $\log_6(x+1) = 2$ **b**. $\log_9(x^2) = -\frac{1}{2}$ **c**. $\log_2(t^2 + 3t) = 2$

d. $\log_2 t + \log_2(t+3) = 2$ **e**. $\log_3 \frac{1}{x} = \frac{1}{2}$ **f**. $\ln(2z) - \ln z = 3z$

g. $e^{-2x} = 5$ **h**. $2^{x+3} = 4^{x-5}$ **i**. $10^{4x+1} = 3^{x-3}$

j. $\log_2 e^x = 4$

2. The concentration of pollutants in a river is given in grams per liter (g/liter) and is approximately $P(x) = 0.03e^{-2x}$, where x is the number of miles downstream from a paper mill.

a. What is the concentration of pollutants in the river at the mill?

b. If concentrations of 0.01 g/liter or less are safe for water sports, how far downstream do we have to be to pursue water sports in the river safely?

3. In how many years will a savings certificate triple its value if interest at an annual rate of 5% is compounded continuously and no money is withdrawn?

4. An animal bone contained 10 g of carbon-14 when the animal was alive.

a. Find a function for the amount of carbon-14 present in the bone t years after the animal's death.

b. How old is the bone if it now contains 7.5 g of carbon-14?

5. **a**. Use a graphing utility to graph the function found in part a of Exercise 4.

b. Trace along the graph to find the age of the bone when it contains 7.5 g of carbon-14.

c. Compare your answer in part b to the answer obtained in part b of Exercise 4.

Chapter 15 Exercises

1. Simplify each expression.

a. $\left(\frac{1}{2}\right)^{-\log_2 3}$ **b**. $\log_5 25$ **c**. $\log 0.001$

d. $\log_3 3^x$ **e**. $\log_2 \frac{1}{8}$ **f**. $\log 10{,}000$

2. Use the properties of logarithms to write each expression as a sum or difference.

a. $y = \ln\left(\frac{1-x}{1+x}\right)$ **b**. $y = \log_2(2x + 1)^2$ **c**. $y = \ln\frac{(x+2)^2}{(x+1)^3}$

d. $y = \ln\left(x^2 + 3x - 4\right)$ **e**. $y = \log\sqrt{x}\left(2x+1\right)$

3. Use the properties of logarithms to combine each expression into a single term.

a. $2\log_8 x + 3\log_8(x - 1)$ **b**. $\frac{1}{2}\ln(x + 2) + \ln 2 - 3\ln(x + 1)$

c. $\log(2x + 3) - [\log(x + 3) + 2\log 5x]$

4. Use transformations of graphs of functions $y = \log_a x$ to draw the graph of each of the following functions. Give the domain and asymptotes of each function.

a. $y = -\ln x$ **b**. $y = \ln(-x)$

c. $y = \log_2(x + 3)$ **d**. $y = \log_3(1 - x)$

e. $y = \log x - 2$

5. Find the domain of each function.

a. $f(x) = \log_5(-x)$ **b**. $g(x) = \ln(2 - x^2)$

c. $h(x) = \log\left(\frac{5+x}{3-x}\right)$ **d**. $k(t) = \frac{2t}{\log_2(t-1)}$

6. Describe in words how the graphs of logarithmic functions with base less than 1 compare to the graphs of logarithmic functions with base greater than 1. Illustrate your answers with graphs.

7. Solve for the variable in each equation.

a. $e^{3t} = 100$ **b**. $\ln(2x + 1) - \ln x = 2\ln 3$

c. $2^{x+3} = 7^{2x-1}$ **d**. $\log_2 t + \log_2(t + 1) = 1$

e. $\ln(5z - 2) = 3$ **f**. $\log_x 64 = 3$

8. The number of bacteria present in a certain culture at time t (measured in hours) is given by $Q(t) = 2000e^{0.3t}$. Time $t = 0$ corresponds to 8:00 A.M. on a particular day.

a. Find the number of bacteria present at noon the same day.

b. Find when there will be 20,000 bacteria present in the culture.

9. Consider a bank at which interest is compounded continuously. What is the annual interest rate if $1,000 will double in 10 years?

10. A particular car is worth $25,000 in 2005. The worth of the car in dollars is given by $f(t)$, where t is the number of years since 2005 and $f(t) = 25{,}000\,e^{-0.5t}$.

a. How much will the car be worth in 10 years?

b. In how many years will the value of the car be reduced to half of its 2005 value?

11. For each pair of functions, sketch the graph of the exponential function and use it to sketch the graph of the logarithmic function on the same set of axes.

a. $y = \left(\frac{1}{2}\right)^x$ and $y = \log_{1/2} x$ **b**. $y = 5^x$ and $y = \log_5 x$

12. Rewrite each function to show explicitly all implied parentheses. Simplify if possible. Indicate the differentiation rules that would be used to find the derivative of each function.

a. $F(x) = e^{\sin x + \ln x}$ **b**. $G(x) = \ln \frac{x^6}{e^x}$ **c**. $K(x) = \log_3 \cot^2 x$

d. $M(x) = e^{2 \ln x}$ **e**. $N(x) = \frac{3^{-x-2}}{\ln x^4}$

13. Consider the function $f(x) = e^{-x} + 1$.

a. Use transformations of the graph of $y = e^x$ to sketch the graph of $y = f(x)$.

b. Explain how you can tell from the graph of $y = f(x)$ that the function $f(x)$ is one-to-one.

c. Give an equation for $f^{-1}(x)$.

d. Sketch the graphs of $y = f(x)$ and $y = f^{-1}(x)$ on the same set of axes.

14. Consider the function $g(x) = \log(x + 2)$.

a. Use transformations of the graph of $y = \log x$ to sketch the graph of $y = g(x)$.

b. Explain how you can tell from the graph of $y = g(x)$ that the function $g(x)$ is one-to-one.

c. Find a formula for $g^{-1}(x)$.

d. Sketch the graphs of $y = g(x)$ and $y = g^{-1}(x)$ on the same set of axes.

15. Match each function in parts a–f to the number of its graph below without using a graphing utility. Explain how you determined each match.

a. $y = 1 + \log_{10}(x + 1)$ **b**. $y = \sqrt{x} - 1$ **c**. $y = \ln x$

d. $y = \ln \sqrt{x}$ **e**. $y = \left(\frac{1}{x}\right)^2$ **f**. $y = \log_{1/2} x$

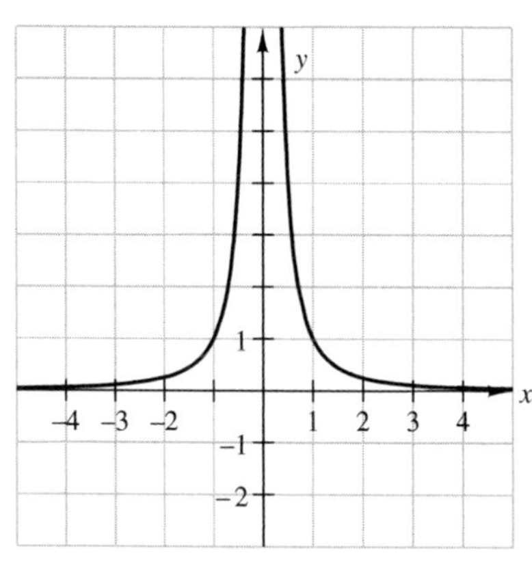

Graph 1

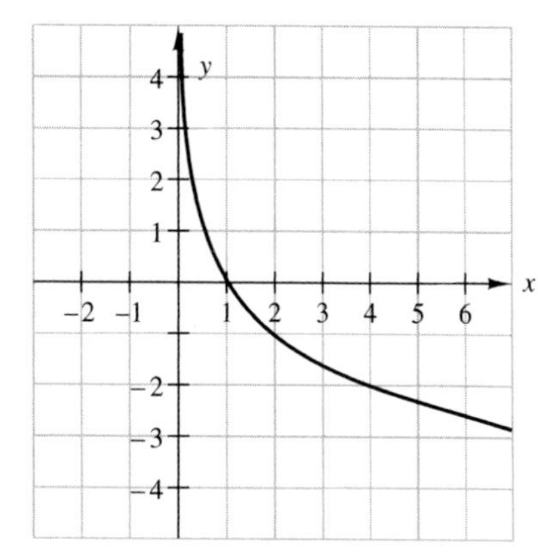

Graph 2

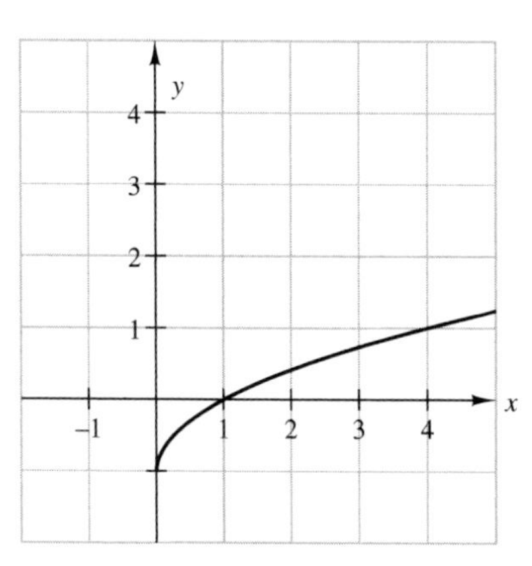

Graph 3

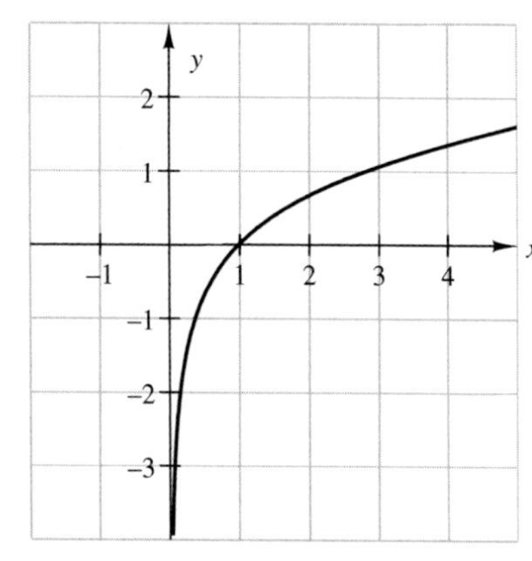

Graph 4

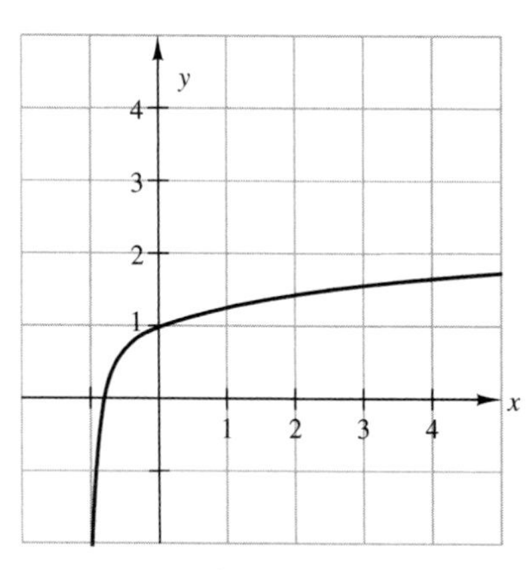

Graph 5

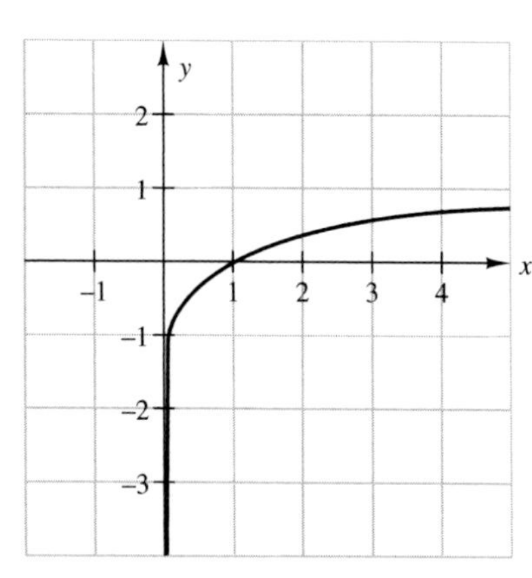

Graph 6

16. An animal bone contained 8 g of carbon-14 when the animal was alive.

a. Find a function for the amount of carbon-14 present in the bone t years after the animal's death.

b. How old is the bone if it has 2 g of carbon-14 left?

17. **a**. Use a graphing utility to graph the function found in part a of Exercise 16.

b. Trace along the graph to find the age of the bone when it contains 2 g of carbon-14.

c. Compare your answer in part b to the answer obtained in part b of Exercise 16.

CHAPTER 16

Companion to Extreme Values of a Function

Many applied problems require finding the maximum or the minimum value of a function. For example, if $P(x)$ is the profit in dollars of producing and selling x items, the producer wants to know how many items the company should produce to maximize the profit. Suppose the profit function is $P(x) = x\left(13 - \frac{x}{400}\right)$. Its graph is given in Figure 16.1.

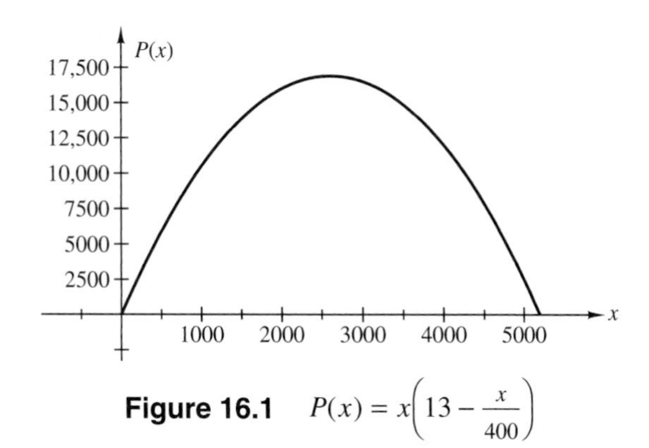

Figure 16.1 $P(x) = x\left(13 - \frac{x}{400}\right)$

From the graph we can see that the highest profit occurs when $x = 2600$ items, and we can compute the maximum profit $P(2600) = \$16{,}900$.

16-A EXTREME VALUES AND CRITICAL VALUES

16-A.1 Absolute and Relative Extreme Values

In most cases we are interested in finding the maximum (largest) or minimum (smallest) value of a function in a specified domain D, often an interval. In the example above, the number of items to be produced cannot be negative, so $x \geq 0$. And since the profit should not be negative, $x \leq 5200$. So $D = [0, 5200]$. Note the domain for the problem is determined by a condition of the problem (profit should not be negative) rather than a mathematical constraint in the function.

Suppose f is a function with domain D and c is in D. If $f(c)$ is the largest among all the values $f(x)$ for x in D, we call $f(c)$ the *maximum value of f on D* (or the *absolute maximum of f on D*) and we say that f has an absolute maximum at c. This maximum value of f on D is the y-coordinate of the highest point of the graph of f, among all points $(x, f(x))$, with x in D.

If d is in D and $f(d)$ is the smallest among all the values $f(x)$ for x in D, we call $f(d)$ the *minimum value of f on D* (or the *absolute minimum of f on D*). In this case we say that f has an absolute minimum at d. On the graph of f, the minimum value of f on D is the y-coordinate of the lowest point among all points $(x, f(x))$, with x in D.

An *extreme value of f* is either a maximum or a minimum value of f on D. When no domain D is explicitly specified for a function, its domain is taken to be the implied domain of the function.

Depending on the function f and the set D, extreme values of f on D may or may not exist, as we see in the following example.

EXAMPLE 16.1 The graph of $y = f(x)$ is given in Figure 16.2. For each of the following intervals, give the values of x at which f has extreme values on that interval. Also give the absolute maximum and the absolute minimum of f on the interval, if they exist.

a. $(-\infty, \infty)$ 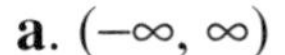 **b**. $(-1, 0)$ **c**. $[-1, 0)$

d. $(0, 1.8)$ **e**. $[0, 2]$

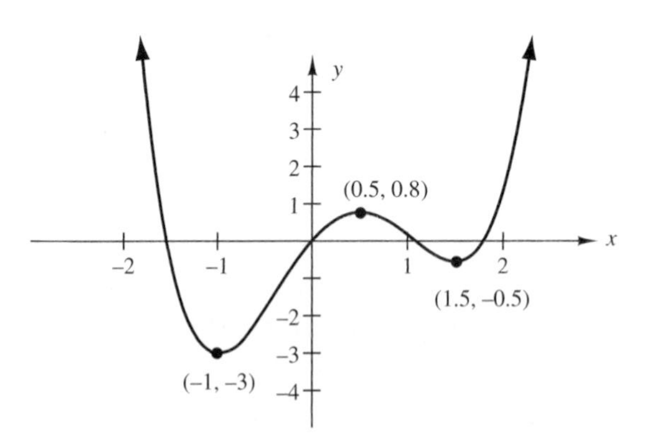

Figure 16.2 $y = f(x)$

Solution

a. Since the values of $f(x)$ grow without bound, there is no maximum value of f on $(-\infty, \infty)$. The lowest point on the graph of the function is $(-1, -3)$, so f has an absolute minimum at $x = -1$ and the minimum value is $-3 = f(-1)$.

b. Since the open interval $(-1, 0)$ does not contain its endpoints, the points $(-1, -3)$ and $(0, 0)$ are not on the graph of f restricted to the interval $(-1, 0)$. The graph of the function over the given interval is highlighted in Figure 16.3.

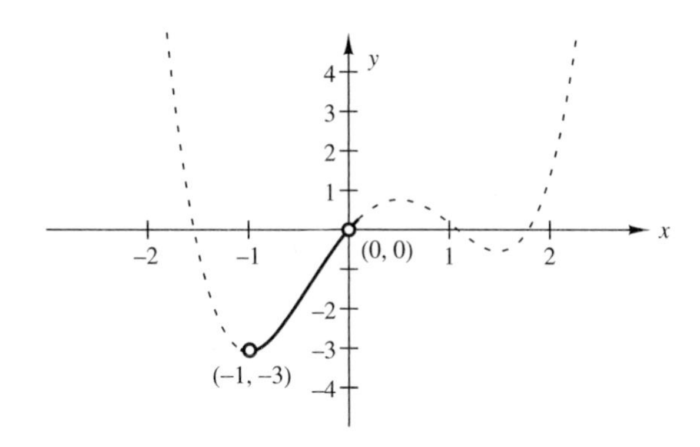

Figure 16.3 $y = f(x)$ on $(-1, 0)$

This graph does not have a highest or a lowest point on the interval $(-1, 0)$. There are no extreme values of f on this interval.

c. The graph of the function f over the interval $[-1, 0)$ is highlighted in Figure 16.4.

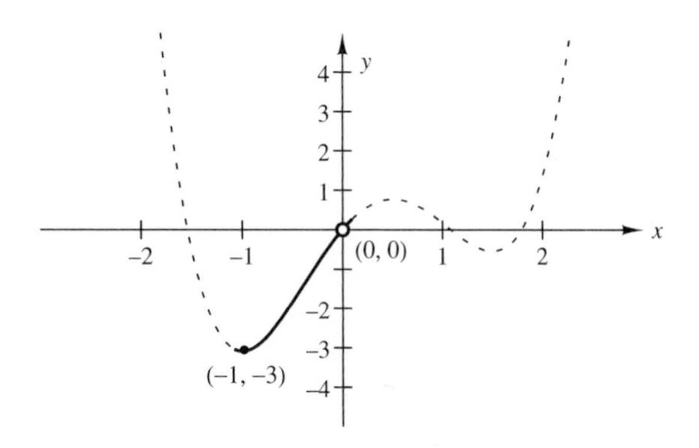

Figure 16.4 $y = f(x)$ on $[-1, 0)$

The lowest point on the graph of the function f on the interval $[-1, 0)$ is the left endpoint $(-1, -3)$, so the minimum value is $-3 = f(-1)$. Since 0 is not in the interval $[-1, 0)$, the point $(0, 0)$ is not on the graph of f restricted to the interval $[-1, 0)$. Therefore, f has no maximum value on the given interval.

d. On the interval $(0, 1.8)$ the highest point is $(0.5, 0.8)$ and the lowest is $(1.5, -0.5)$. See Figure 16.5. So on this interval the function has maximum value at $x = 0.5$, and the maximum is $0.8 = f(0.5)$. The minimum value of f on the interval $(0, 1.8)$ occurs at $x = 1.5$ and is $-0.5 = f(1.5)$.

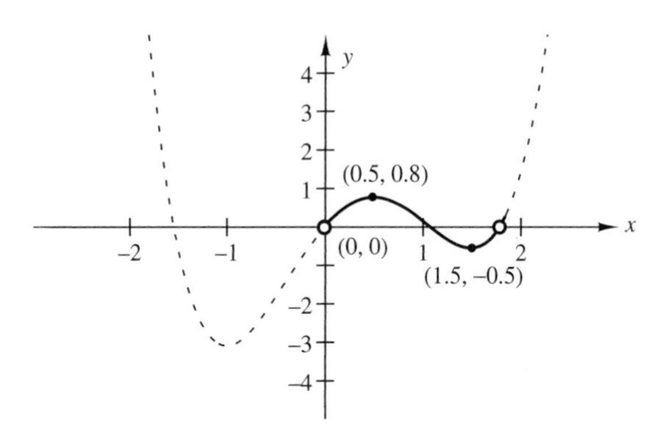

Figure 16.5 $y = f(x)$ on (0, 1.8)

e. The graph of the function over the closed interval [0, 2] has both a highest and a lowest point (see Figure 16.6).

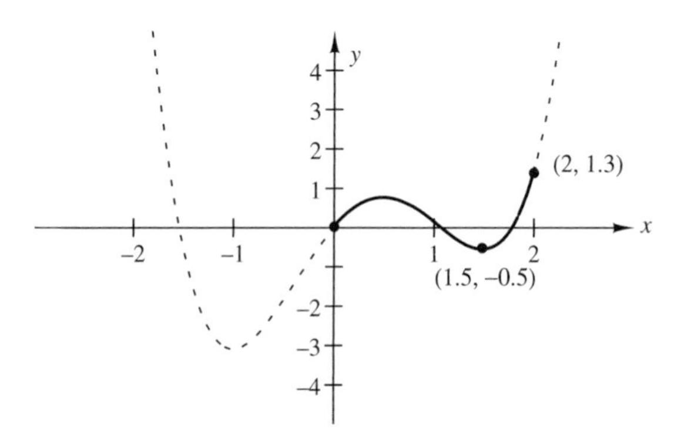

Figure 16.6 $y = f(x)$ on [0, 2]

The maximum value of f on the given interval occurs at the endpoint $x = 2$ and is $f(2) = 1.3$; the minimum occurs at $x = 1.5$ and is $f(1.5) = -0.5$. ■

When looking for the maximum or minimum of a function f on an interval, be sure to pay attention to whether or not the endpoints of the interval are included.

In the example above, the three points (−1, −3), (0.5, 0.8), and (1.5, −0.5) provide important information about the function. They are "turning points"; that is, points where the graph changes from increasing to decreasing or vice versa. Although not all three points are the highest or lowest points on the graph of f, each of them is the highest or the lowest point of the graph of the function over some open interval containing the x-value of the point.

The y-value of each of these "turning points" is called a *relative*, or *local*, *extreme value of the function f.* Since (0.5, 0.8) is the highest point of the function on some open interval containing $x = 0.5$ (for example, the interval (0, 1)), 0.8 is a relative (local) maximum of f and it occurs at $x = 0.5$. This function has two relative minima: -3 (which is also the absolute minimum of f) and -0.5. These relative minima occur at $x = -1$ and $x = 1.5$, respectively.

EXAMPLE 16.2 Consider the function $y = f(x)$ whose graph is given in Figure 16.2. For each of the following intervals, give the values of x at which f has relative extreme values on that interval, and identify any relative maximum or relative minimum values of f. For each interval, compare the relative extrema with the absolute extrema found in Example 16.1 for the same interval.

a. $[-1, 0)$

b. $(0, 1.8)$

c. $[0, 2]$

Solution

a. The graph of the function f over the interval $[-1, 0)$ is highlighted in Figure 16.4. Because f is increasing throughout the interval $[-1, 0)$, it has no relative extrema on that interval. We see that a relative extreme value cannot occur at the endpoint $x = -1$ of the interval $[-1, 0)$ because there is no open interval inside $[-1, 0)$ that contains the endpoint. We note from part c of Example 16.1 that the absolute minimum value of f on $[-1, 0)$ is $f(-1) = -3$ and f has no absolute maximum on $[-1, 0)$.

b. The function f has a relative maximum value on the interval $(0, 1.8)$ at $x = 0.5$ and a relative minimum value at $x = 1.5$. (See Figure 16.5.) The relative maximum value is 0.8, and the relative minimum value is -0.5. From part d of Example 16.1 we see that 0.8 is also the absolute maximum of f on $(0, 1.8)$ and -0.5 is also the absolute minimum of f on $(0, 1.8)$.

c. The function f has a relative maximum value on the closed interval $[0, 2]$ at $x = 0.5$ and a relative minimum value at $x = 1.5$. (See Figure 16.6.) As in part b above, the relative maximum value is 0.8 and the relative minimum value is -0.5. The absolute maximum of f on $[0, 2]$ is 1.3 and occurs at the endpoint $x = 2$, while the absolute and relative minimum values are the same on this interval; both occur at $x = 1.5$. ■

Relative extrema of a function f on an interval can only occur at interior points of the interval; they cannot occur at the endpoints. Absolute extrema of f on an interval can occur at endpoints or at interior points of the interval. A function might have no relative extrema on an interval (if the function is increasing on the interval, for example). A function also might have no absolute extrema on an interval (for example, $f(x) = x$ on the interval $(0, 1)$).

16-A.2 Solving Equations to Find Critical Values

Extreme values of a function are, in theory, easy to spot. The existence of the tangent line and the slope of the tangent line at a point on the graph of a function can tell us whether the point is a possible extreme value for the function.

EXAMPLE 16.3 The graph of the function $y = f(x)$ given in Figure 16.7 shows that f has a relative maximum value at $x = 0$ and a relative minimum value at $x = 1$. What can be said about the tangent line to the graph of f at the points $(0, f(0))$ and $(1, f(1))$? What are the values of the derivatives $f'(0)$ and $f'(1)$?

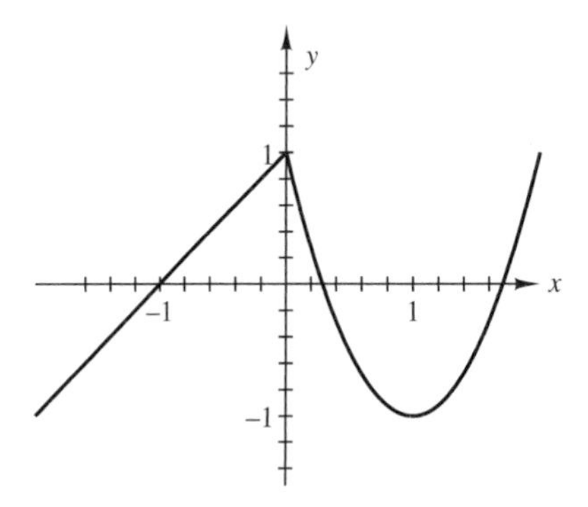

Figure 16.7 $y = f(x)$

Solution Because of the sharp cusp, there is no tangent line at the point $(0, f(0))$; thus, $f'(0)$ is undefined. The tangent line at the point $(1, f(1))$ is horizontal, which means that $f'(1) = 0$. ■

In general, to find extreme values of a function f when we have no picture to guide us, we find all values of x in the domain of f where the tangent line to the graph of f is horizontal and also any values of x in the domain of f where the graph has no tangent line. To find the x-values where the tangent line is horizontal, we need to solve the equation $f'(x) = 0$. To find the values of x where there is no tangent line, we find the values of x where $f(x)$ is defined but $f'(x)$ is undefined.

The *critical values* of f are the values of x in the domain of f where $f'(x) = 0$ or $f'(x)$ does not exist.

To find critical values, it is useful to factor $f'(x)$ whenever possible. When $f'(x)$ consists of a sum of terms that contain quotients or have negative exponents, we first find a common denominator and write $f'(x)$ as a single quotient. If possible, we then factor the numerator and denominator to determine where they are 0. The values of x at which the denominator is 0 are where $f'(x)$ is undefined; the values of x at which the numerator is 0 (and the denominator is not 0) are where $f'(x) = 0$.

EXAMPLE 16.4 In each part, a function and its derivative are given. Determine all critical values of the given function.

a. $f(x) = x^4 - x^2$ $\quad$ $f'(x) = 4x^3 - 2x$

b. $g(x) = \frac{1}{x+1}$ $\quad$ $g'(x) = -(x+1)^{-2}$

c. $h(x) = 5x\sqrt[5]{x-2}$ $\quad$ $h'(x) = 5(x-2)^{1/5} + x(x-2)^{-4/5}$

d. $m(x) = \frac{x+5}{x^2-4}$ $\quad$ $m'(x) = \frac{-x^2-10x-4}{(x^2-4)^2}$

e. $F(x) = 3x + e^{2x}$ $\quad$ $F'(x) = 3 + 2e^{2x}$

f. $G(x) = x + \cos x$ $\quad$ $G'(x) = 1 - \sin x$

Solution

a. Since f and f' are polynomial functions, they are defined for every x. To find the values of x where $f'(x) = 0$, we factor $f'(x)$: $f'(x) = 4x^3 - 2x = 2x(2x^2 - 1) = 2x\left(\sqrt{2}x+1\right)\left(\sqrt{2}x-1\right)$. We now set each factor equal to 0 to obtain $x = 0$, $x = -\frac{1}{\sqrt{2}}$, and $x = \frac{1}{\sqrt{2}}$ as critical values of f.

b. We first write $g'(x) = \frac{-1}{(x+1)^2}$. Since a quotient equals 0 only when the numerator is 0, $g'(x) \neq 0$ for all x. The only value of x where $g'(x)$ is undefined is $x = -1$, since it is a zero of the denominator. But $x = -1$ is not in the domain of g; therefore, it is not a critical value. Thus, the function g has no critical values.

c. We write $(x-2)^{-4/5} = \frac{1}{(x-2)^{4/5}}$ and then put the terms in $h'(x)$ over a common denominator: $h'(x) = 5(x-2)^{1/5} + \frac{x}{(x-2)^{4/5}} = \frac{5(x-2)^{1/5}(x-2)^{4/5} + x}{(x-2)^{4/5}} = \frac{5(x-2)^1 + x}{(x-2)^{4/5}} = \frac{5x-10+x}{(x-2)^{4/5}} = \frac{6x-10}{(x-2)^{4/5}}$. We see that $h'(x) = 0$ when $6x - 10 = 0$, which is when $x = \frac{5}{3}$. The derivative $h'(x)$ is undefined for $x = 2$, whereas the function $h(x)$ is defined for every x, including $x = 2$. So $x = 2$ and $x = \frac{5}{3}$ are the critical values of h.

d. Since $m'(x)$ is a rational function, it is undefined for $x = 2$ and $x = -2$, which are the zeros of the denominator. But the original function $m(x)$ is not defined at those values, so they are not critical values. We now look for the values of x where $m'(x) = 0$. For this we set the numerator equal to 0 and solve for x. We first factor out the constant -1: $-x^2 - 10x - 4 = -(x^2 + 10x + 4) = 0$. Since the resulting quadratic is not easily factored, we use the quadratic formula to find roots of the numerator. These roots are $x = \frac{-10 \pm \sqrt{10^2 - 4 \cdot 4}}{2} = \frac{-10 \pm \sqrt{84}}{2} = \frac{-10 \pm 2\sqrt{21}}{2} = -5 \pm \sqrt{21}$. So the critical values of $m(x)$ are $x = -5 - \sqrt{21} \approx -9.58$ and $x = -5 + \sqrt{21} \approx -0.417$.

e. Since the domain of the exponential function is all real numbers, both F and F' are defined for every x. For this reason, to find critical values we only need to find those values of x where $F'(x) = 0$. But $F'(x) = 3 + 2e^{2x} = 0$ implies $e^{2x} = -\frac{3}{2}$, which has no solution since the values of the exponential function are always positive. So the function F has no critical values.

f. Both G and G' are defined for every x, since the sine and cosine functions have domain $(-\infty, \infty)$. To find the values of x where $G'(x) = 0$, we solve $G'(x) = 1 - \sin x = 0$ or, equivalently, $\sin x = 1$. This is true when $x = \frac{\pi}{2}, \frac{\pi}{2} \pm 2\pi$, $\frac{\pi}{2} \pm 4\pi$, and so on. The critical points of G are all real numbers of the form $x = \frac{\pi}{2} + 2k\pi$, with k an integer. ■

A critical value of a function *f* must be in the domain of *f*. We first find all values of *c* for which either $f'(c) = 0$ or $f'(c)$ does not exist. Then we must check to make sure that $f(c)$ is defined; only if this is true is *c* a critical value of *f*.

Exercises 16-A

1. The graph of $y = f(x)$ in Figure 16.8 has the x-axis as a horizontal asymptote. For each of the following intervals, give the absolute maximum and the absolute minimum of f on the interval, if they exist. If they do not exist, explain why. Also, give the values of x at which each extreme value occurs.

a. $(-\infty, 0)$ **b**. $(-2, 2)$ **c**. $[-2, 0)$ **d**. $[-2, 2]$

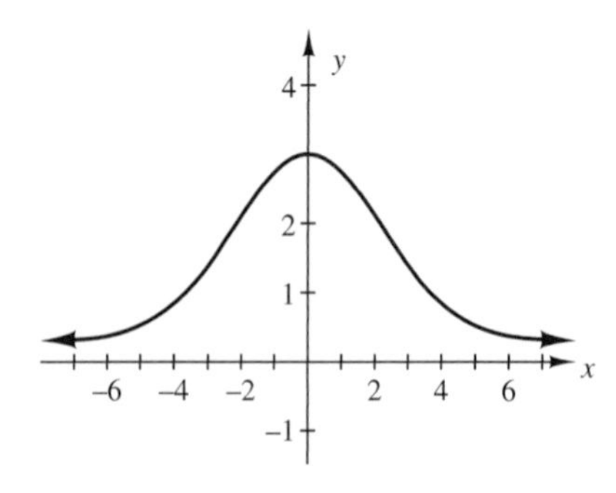

Figure 16.8 $y = f(x)$

2. The graph of $y = h(x)$ is in Figure 16.9; the arrows indicate that the graph continues forever in the direction shown. For each of the following intervals, give the absolute maximum and the absolute minimum of h on the interval, if they exist. If they do not exist, explain why. Also, give the values of x at which each extreme value occurs.

a. $(-\infty, \infty)$ **b**. $[-4, 0]$ **c**. $(-4, 0)$ **d**. $(-4, 3)$

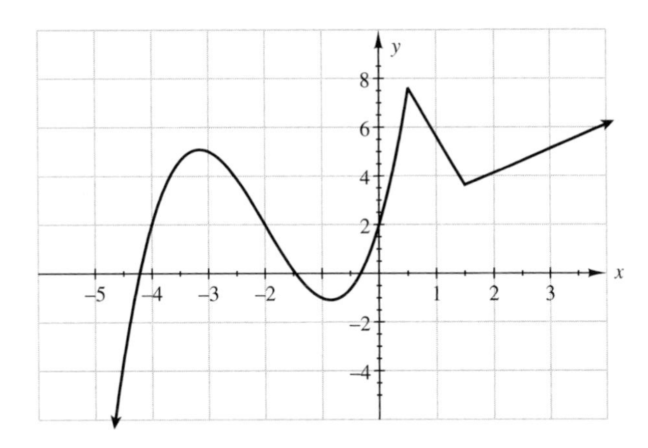

Figure 16.9 $y = h(x)$

3. For the function $y = h(x)$ whose graph is in Figure 16.9:

a. Give all the relative maxima and the values of x at which they occur.

b. Give all the relative minima and the values of x at which they occur.

4. The graph of $y = G(x)$ is in Figure 16.10; dashed lines indicate asymptotes. For each of the following intervals, give the absolute maximum and the absolute minimum of G on the interval, if they exist. If they do not exist, explain why. Also, give the values of x at which each extreme value occurs.

a. $(-\infty, \infty)$ **b**. $[-0.8, 0]$ **c**. $(-2, 2)$ **d**. $(2, \infty)$

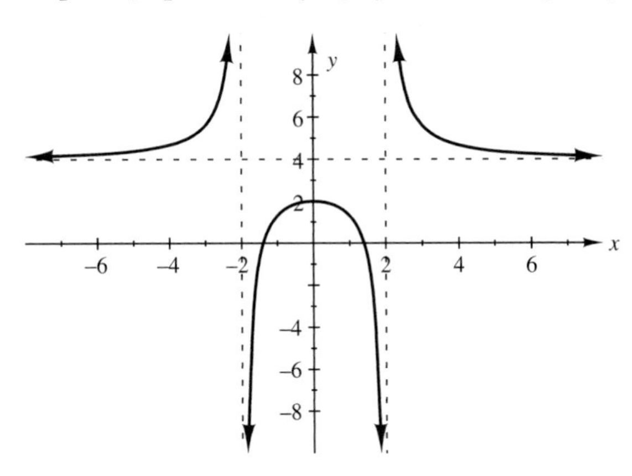

Figure 16.10 $y = G(x)$

5. Find all x-values where the function $G(x)$ in Figure 16.10 has a relative extreme value, and in each case determine whether a relative maximum or a relative minimum occurs.

6. Graph each of the following functions. Then give the absolute maximum and the absolute minimum values of the function on the interval $(-\infty, \infty)$, if they exist. Also give the x-values where these extreme values occur.

a. $f(x) = e^{2x}$ **b**. $y = \log_3 x$ **c**. $y = (x-1)^2 + 2$

7. In each part, a function and its derivative are given. Find all the critical values of the function.

a. $f(x) = x^3 - x^2 - x + 2$ $f'(x) = 3x^2 - 2x - 1$

b. $g(x) = 2x^3 + 2x^2 + 3x$ $g'(x) = 6x^2 + 4x + 3$

c. $h(x) = x^{2/3}(x-5)$ $h'(x) = \frac{5}{3}x^{2/3} - \frac{10}{3}x^{-1/3}$

d. $n(x) = \dfrac{x^3}{x^2 - 1}$ $n'(x) = \dfrac{3x^2(x^2-1) - 2x^4}{(x^2-1)^2}$

e. $k(x) = x\sqrt{x+3}$ $k'(x) = (x+3)^{1/2} + \frac{1}{2}x(x+3)^{-1/2}$

f. $m(x) = x^3 e^{-x}$ $m'(x) = 3x^2 e^{-x} - x^3 e^{-x}$

16-B SETTING UP FUNCTIONS TO SOLVE EXTREME VALUE PROBLEMS

Maximum and minimum problems, when expressed in mathematical form, require us to find the maximum or the minimum value of a function on an interval. This interval is determined by analyzing the constraints or restrictions that apply in the specific situation. To solve extreme value problems it is helpful to use the following steps:

1. Draw a picture where appropriate.

2. Name the variables.

3. Express the quantity that has to be maximized or minimized as a function of a single variable.

4. Find the domain of the function that meets the conditions of the problem.

5. Find the maximum or minimum value of the function on that domain.

In the following examples we carry out steps 1 through 4 to show how to identify the function to be maximized or minimized and the domain for which that function is to be considered.

EXAMPLE 16.5 An efficiency study of the morning shift at a candy factory indicates that a worker's rate of production after t hours on the job is approximately $R(t) = -3t^2 + 17t + 10$ boxes per hour. If a worker arrives at the job at 7:00 A.M., at what time during the morning hours is the worker performing most efficiently? Identify the domain of R that needs to be considered to solve this problem.

Solution Since the morning hours end at 12 noon, which is 5 hours after the starting time, the values of t to be considered are $0 \le t \le 5$. To answer the question, we need to find the time t at which the function $R(t) = -3t^2 + 17t + 10$ has a maximum value on the interval [0, 5]. ■

Sometimes the quantity to be maximized or minimized is easily written in terms of two or more variables and must be rewritten as a function of one variable before we can solve the problem. To do this, we use relations determined by the problem to write all of the variable quantities in terms of a single independent variable.

EXAMPLE 16.6 A farmer has a square plot of land that measures 1 mile on each side. He wishes to fence a rectangular field to enclose an area of 9000 ft^2. The south and west sides of the fence will cost \$3.80 per foot and the other two sides will cost \$4.05 per foot. The farmer wants to know what dimensions of the field will minimize cost. Find a function and its domain that the farmer can use to find the answer.

Solution We start by drawing a picture; then we name the variables.

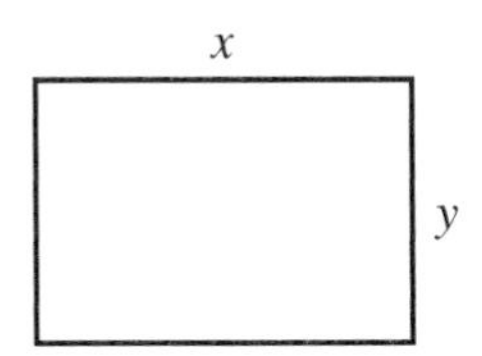

Let x denote the length of the north and south sides of the rectangle, and let y denote the length of the east and west sides, measured in feet. The total length of the south and west sides is $x + y$, so the cost for that part of the fence is $3.8(x + y)$. Similarly, the cost of the rest of the fence is $4.05(x + y)$. The total cost is: $C = 3.8(x + y) + 4.05(x + y) = 7.85(x + y)$.

To write the cost in terms of x only, we notice that the area of the rectangle should be 9000 ft^2; thus, $xy = 9000$. This equation allows us to write y in terms of x: $y = \frac{9000}{x}$. The cost is then $C = C(x) = 7.85\left(x + \frac{9000}{x}\right)$.

The values that x can take are restricted by the fact that x is the length of a side of a rectangle, so $x > 0$, and the fact that $x \le 1$ mile $= 5280$ ft. To complete the solution to this problem, we would find the value of x where the function C has a minimum value on the interval (0, 5280]. The first step in this process is to find all critical values of C in the interval. ■

Many applied maximum and minimum problems deal with volume or surface area. We recall here some of the commonly required geometric formulas. (Some other formulas were given in Section 6-A.) In these formulas, V represents the volume and S represents the surface area.

Geometric Formulas

Cylinder

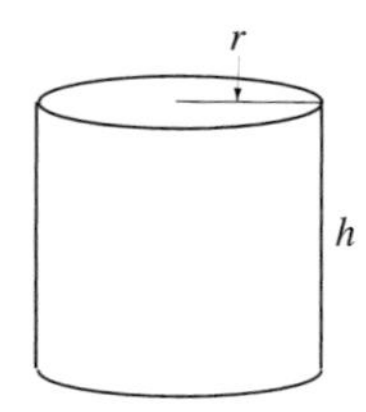

$V = \pi r^2 h$

$S = 2\pi rh$ (no top or bottom)

$S = 2\pi rh + 2\pi r^2$ (both top and bottom)

Cone

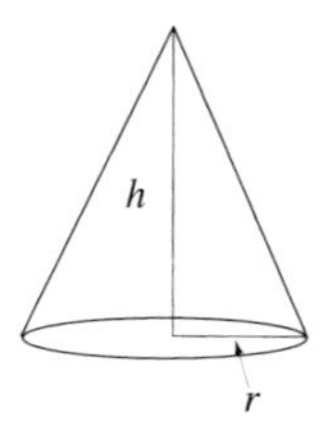

$V = \left(\frac{1}{3}\right)\pi r^2 h$

$S = \pi r\sqrt{r^2 + h^2}$ (open)

$S = \pi r\sqrt{r^2 + h^2} + \pi r^2$ (closed)

Sphere

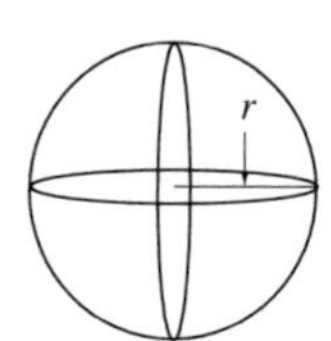

$V = \frac{4}{3}\pi r^3$

$S = 4\pi r^2$

Rectangular parallelepiped (box)

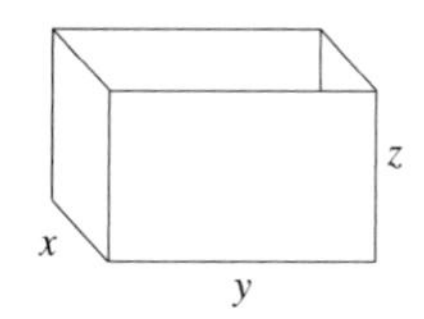

$V = xyz$

$S = 2xy + 2xz + 2yz$ (closed)

In the following example, we have a function with two variables, r and h, where the choice of which variable to write in terms of the other makes a difference; choosing to solve for h in terms of r gives a simpler expression for the function.

EXAMPLE 16.7 Identify a function and its domain that a manufacturer needs to consider to answer the following question: What are the dimensions of a 12-oz soda can designed to be manufactured with the least amount of thin aluminum sheet? Use the fact that 12 fluid ounces is approximately 21.64 in^3. (Ignore overlapping at seams.)

Solution We draw a cylinder to represent the soda can.

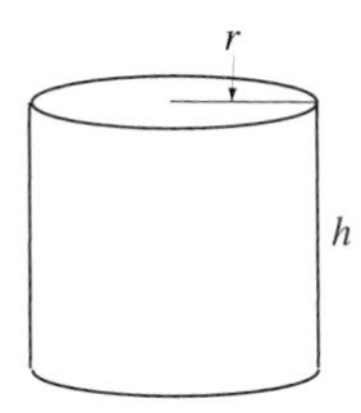

We call r the radius of the base and h the height in inches. Since the function that gives the surface area is to be minimized, we start by writing the formula $S = 2\pi rh + 2\pi r^2$. To write S as a function of only one variable, we use the fact that the volume is 21.64 in^3. Since the volume of the cylinder is $V = \pi r^2 h$, we have the equation $\pi r^2 h = 21.64$. This equation could be solved for either r or h; we choose to solve for h. We divide both sides of the equation by πr^2 to obtain $h = \dfrac{21.64}{\pi r^2}$. Thus, $S = S(r) = 2\pi r\left(\dfrac{21.64}{\pi r^2}\right) + 2\pi r^2 = \dfrac{2(21.64)}{r} + 2\pi r^2 = \dfrac{43.28}{r} + 2\pi r^2$. Since r is the radius, $r > 0$. The manufacturer wants to find the value of r where the function $S(r) = \dfrac{43.28}{r} + 2\pi r^2$ has a minimum on the interval $(0, \infty)$.

Although no upper bound is given on the value of the radius r, practical considerations such as ability to manufacture, reasonable size for display and use, and aesthetics will impose some upper bound. If the solution r that minimizes the value of $S(r)$ falls in the range of reasonableness, it will be accepted by the manufacturer. Otherwise, a more practical but not optimal solution will be used. ■

EXAMPLE 16.8 To extend city water service to a new development, the city plans to install a new pipe. The pipe will carry water from a point A to a point B that lies across a river 100 yd wide and 250 yd downstream from point A. The installation cost is \$30 per yard for the underwater pipe and \$15.50 per yard for the underground pipe. Give the function and its domain that must be considered to minimize installation cost.

Solution We start by drawing a picture.

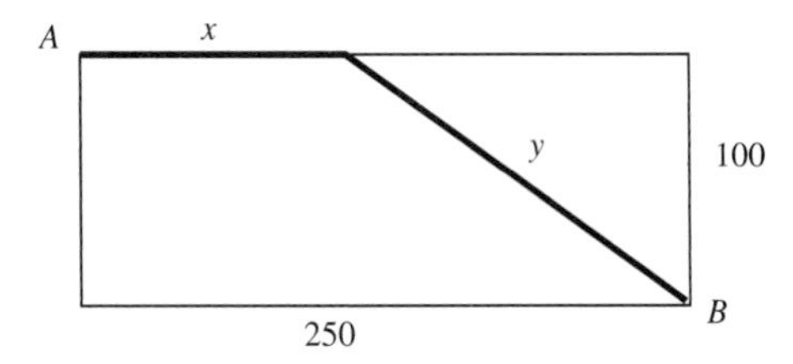

We call x the length of underground pipe and y the length of underwater pipe. The installation cost C is \$15.50x (for underground pipe) plus \$30y (for underwater pipe), so $C = 15.50x + 30y$. We want to write C as a function of only one variable quantity, so we express y in terms of x. To do this we observe from the picture that y is the hypotenuse of a right triangle with sides 100 and $250 - x$. By the

Pythagorean theorem, $y = \sqrt{(250-x)^2 + 100^2}$. Now we can write $C = C(x) = 15.50x + 30\sqrt{(250-x)^2 + 100^2}$. To determine the domain of the function C, we note that x represents the length of underground pipe, so $0 \le x \le 250$. ■

Exercises 16-B

Each of the following problems can be solved by finding a maximum or a minimum value of a function of one variable on an interval. For each problem, give the function and its domain that must be considered to solve the problem.

1. A gardener wants to fence a 45-yd^2 rectangular garden with one side adjoining a neighbor's lot. The fence costs \$4.80 per yard, and the neighbor has agreed to pay for half the cost of the dividing fence. Find the dimensions of the garden that will minimize the gardener's cost. The garden's dimensions cannot exceed her lot size, which is a square 40 yd by 40 yd.

2. A toy manufacturer sells approximately 500 tricycles per month when the price of each tricycle is \$49. The manufacturer estimates that for each \$1 increase in price, he will sell three fewer tricycles a month. The manufacturer's cost to produce each tricycle is \$21. Find the selling price that will maximize the manufacturer's profit.

3. An electric company will run a power line from a plant on one side of a river 500 yd wide to a factory on the other side and 750 yd downstream. The cost of running the line along the shore is \$110 per yard, and the cost of running it under water is \$390 per yard. What is the most economical route over which to run the line?

4. A company that produces fruit juices will order the production of cylindrical cans. Each can is to hold 12 in.^{3} of frozen apple juice. The cost per square inch of constructing the top and bottom is twice the cost per square inch of constructing the cardboard side. What are the dimensions of the least expensive can?

5. The Ajax Carton Company will make boxes by cutting squares of the same size out of an 8.5-by-11-in. piece of cardboard, folding up the resulting flaps and taping the edges at each corner. How long should the sides of the cut-out squares be to obtain a box with maximum volume? (A problem similar to this one was discussed in Section 6-A.)

6. For Exercises 1 through 5 above, use a graphing utility to graph the function that you found on an appropriate domain of the function. Use the graph to estimate the minimum or maximum sought in each exercise.

7. Consider the problem described in Example 16.7.

a. Give the function S for the surface area of the can as a function of the single variable h, the height of the can.

b. Explain why the function $S(r)$ given in the solution of Example 16.7 is a better choice to minimize the surface area of the can.

Chapter 16 Exercises

1. Four functions and their derivatives are given below. Find all the critical values of each function.

a. $f(x) = \sin^2 x$ $\qquad f'(x) = 2 \sin x \cos x$

b. $n(x) = \dfrac{x^2 - 3}{x^2 + 1}$ $\qquad n'(x) = \dfrac{(x^2 + 1)2x - (x^2 - 3)2x}{(x^2 + 1)^2}$

c. $g(x) = x \ln x$ $\qquad g'(x) = 1 + \ln x$

d. $h(x) = 3x^2 \sqrt[3]{x - 4}$ $\qquad h'(x) = 6x(x - 4)^{1/3} + x^2(x - 4)^{-2/3}$

2. A population study indicates that during the next 10 years, the population of a community t years from now will be approximately $P(t) = -t^3 + 10t^2 + 45t + 30$ and its rate of growth will be $P'(t) = -3t^2 + 20t + 45$. Figures 16.11 and 16.12 show the graphs of $P(t)$ and $P'(t)$.

a. Use the graphs in Figure 16.11 and 16.12 to answer the following questions:

(i) During the next 10 years, when will the population be largest?

(ii) During the next 10 years, when will the population be growing most rapidly?

b. Find the critical values of the function $P(t)$.

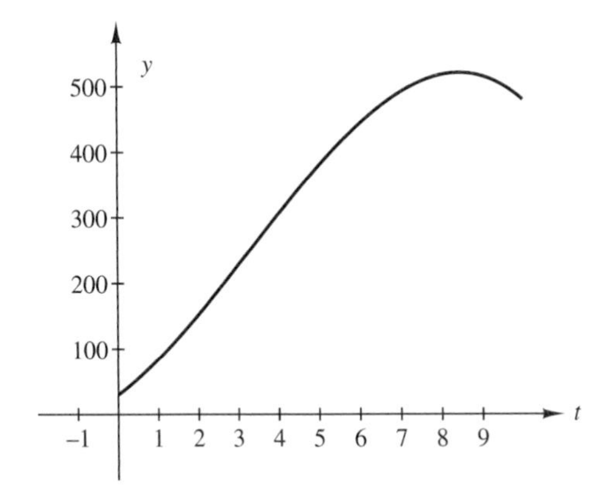

Figure 16.11 $P(t) = -t^3 + 10t^2 + 45t + 30$

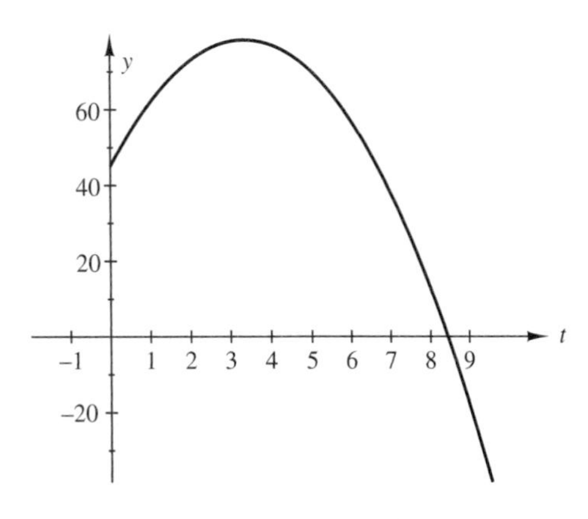

Figure 16.12 $P'(t) = -3t^2 + 20t + 45$

3. Consider $R(t) = -3t^2 + 17t + 10$ of Example 16.5 on the domain [0, 5].

a. Find the critical values of $R(t)$ on [0, 5].

b. Use your answer in part a to find at what time during the morning hours the worker of Example 16.5 will perform most efficiently.

c. Use your answer in part a to find at what time during the morning hours the worker of Example 16.5 will perform least efficiently.

d. Use a graphing utility to graph $R(t)$ on [0, 5]. Explain how your graph confirms your answers to parts a, b, and c.

4. The graph of $y = g(x)$ is in Figure 16.13. For each of the following intervals, give the absolute maximum and the absolute minimum of g on the interval if they exist. If they do not exist, explain why. Also, give the values of x at which each extreme value occurs.

a. $(-\infty, 0)$ **b**. $(-2, 1)$ **c**. $[-2, 0)$ **d**. $\left[-2, \frac{1}{2}\right]$

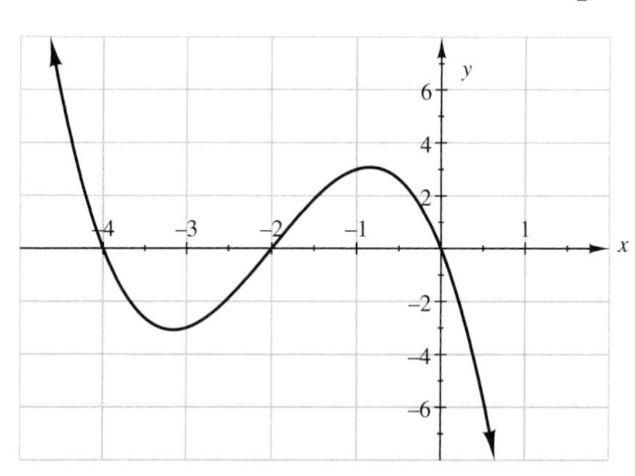

Figure 16.13 $y = g(x)$

5. Give all the relative extreme values of the function $y = g(x)$ in Figure 16.13 and the values of x where each relative extreme value occurs. In each case, say whether the value is a relative maximum or a relative minimum.

6. The number of new jobs each month is an indicator of the health of the labor market. The graph in Figure 16.14 shows the number of new jobs, in thousands, for each month in 2004, adjusted for seasonal variations. (*Data Source:* www.msnbc.msn.com)

a. Identify the longest stretch of months during which the number of new jobs was decreasing in 2004.

b. During what month was the number of new jobs a maximum, and what is that maximum number?

c. During what month was the number of new jobs a minimum, and what is that minimum number?

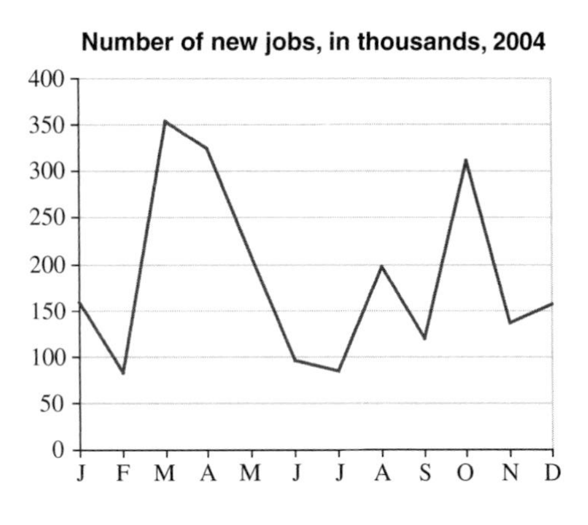

Figure 16.14

7. The greatest period of immigration to the United States was the years 1900–1920. The graph in Figure 16.15 shows the annual number of immigrants for each of the years during this period.

a. Give the years in which the function in the graph has a relative maximum.

b. After what year was the last time (in the period shown) that the number of immigrants started to increase? Does the function have an absolute minimum at this time?

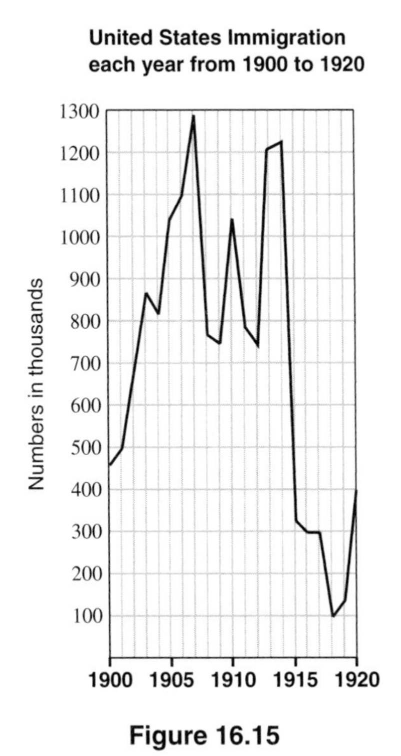

Figure 16.15

Data Source: U.S. Dept. of Justice, *2002 Yearbook of Immigration Statistics* (2003).

8. For each of the descriptions below, sketch the graph of a function $y = f(x)$ on an interval I to provide an example that fits the description:

a. The function f is continuous on the open interval I and has no absolute extrema and no relative extrema on that interval.

b. The function f is continuous on the open interval I, has no absolute extrema on the interval, and has a relative maximum and a relative minimum.

c. The function f is not continuous on the closed interval I and has no absolute maximum or absolute minimum on that interval.

9. A manufacturer needs to construct wooden boxes with square base, no top, and volume of 5 ft^3. Give the function and its domain that the manufacturer can use to find the dimensions of a box made of the least amount of wood.

10. A hotel owner is trying to decide how much to charge per room per day to maximize profit. The hotel has 100 rooms. The owner's records indicate that when the daily rates per room are \$50, the hotel can be filled. Each \$5 increase in the daily rate results in 8 additional vacant rooms. The hotel owner has to take into consideration the fact that when a room is occupied, there is an extra expense of \$6 for servicing that room. Find the function and its domain that the hotel owner needs to consider to make a decision. (Hint: Let x be the number of \$5 increases in the daily rate and write the profit as a function of x.)

CHAPTER 17

Companion to Curve Sketching

Functions and graphs are used by businesses and local, state, and national governments to make decisions. For example, the number of people killed in alcohol-related accidents is a cause for concern and has stimulated many initiatives to reduce the number of such deaths. A reasonably accurate model of the alcohol-related automobile crash deaths between 1982 and 1999 (www.madd.org/stats/fatalities) is given by the function $f(x) = -0.3718x^4 + 21.9933x^3 - 395.6523x^2 + 1880.2042x + 21{,}524.7647$, where x is the number of years since 1982. To evaluate various intervention strategies, we want to know when the number of such deaths was increasing and when the number was decreasing. We also want to know whether the rate of increase or decrease was increasing or decreasing after each intervention. A relative maximum for f means that the number of such deaths started to decrease after this point.

Plotting points can be very helpful when we want to sketch the graph of a function, but it is not enough. We need to know the behavior of the function **between** key points in order to sketch the graph accurately. Calculus provides us with a powerful tool for accurately sketching the graph of a differentiable function. By looking at the first derivative of the function and considering where that derivative is positive and where it is negative, we can determine where the graph of the function is increasing and where it is decreasing. We can use the sign of the second derivative to determine where the graph of the function is concave up (the rate of change is increasing) or concave down (the rate of change is decreasing). By finding critical values of the function and its derivative, we can locate the local extreme values of a function and any points of inflection. By plotting these key points and intercepts and using the additional information to connect the points with curves, we can create a good sketch of the graph of the function. In this chapter, you will learn techniques for finding the key points and determining the graph's behavior between these points.

Consider a graph that contains the following points: $(-5, -5)$, $(-2, 4)$, $(-1, 3)$, $(0, 6)$, and $(3, -3)$; the points are plotted in Figure 17.1.

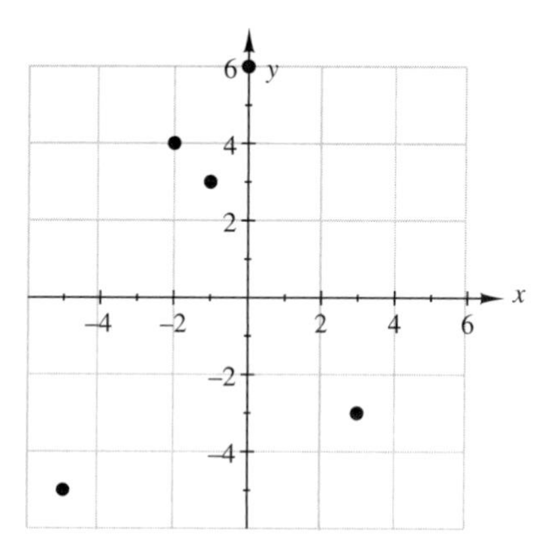

Figure 17.1

Figure 17.2 shows one function whose graph contains these points; Figure 17.3 shows another function whose graph contains these same five points.

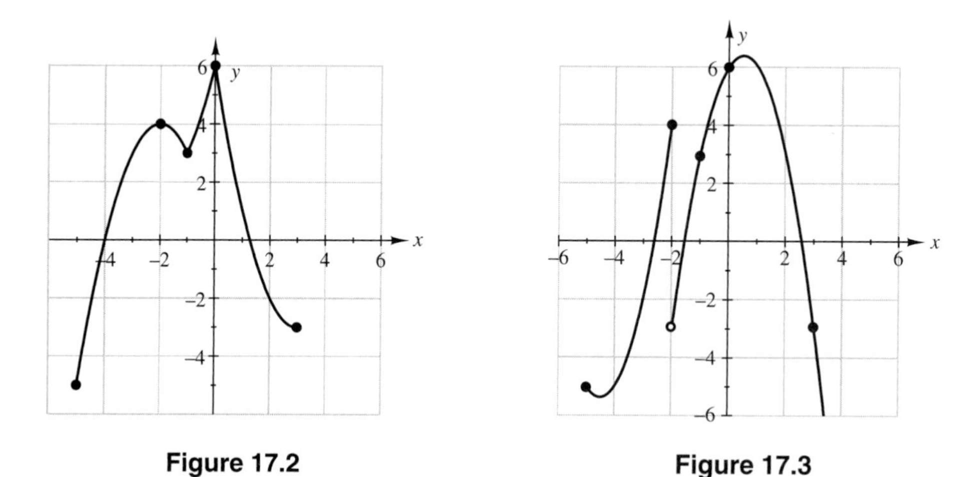

Figure 17.2 **Figure 17.3**

The graphs in Figure 17.2 and 17.3 illustrate the importance of finding and plotting important points and finding the signs of the derivatives between these points if we want to see the essential features of a graph. In Figure 17.2 the x-values of the given points divide the domain $[-5, 3]$ of the function f into the open intervals $(-5, -2)$, $(-2, -1)$, $(-1, 0)$, and $(0, 3)$, and on each of these intervals, the first and second derivatives do not change sign. We see that on the interval $(-5, -2)$, f is increasing (f' is positive) and concave down (f'' is negative); on $(-2, -1)$, f is decreasing and concave down; on $(-1, 0)$, f is increasing and concave up; and on $(0, 3)$, f is decreasing and concave up. Every one of the given points is "special"; they are either relative extrema or endpoints of the domain. In Figure 17.3 the graph contains all of the given points, but not all of them are important points of the graph. Also, not all of the important points are included. The point where there is a relative minimum on the interval $(-5, -2)$ and the point where there is a relative maximum on the interval $(0, 3)$ are not given.

In the sections that follow, we will discuss techniques to find and organize information about the behavior of a function. This will enable us to sketch an accurate graph by first plotting some key points and then connecting them by appropriate curves. These techniques also are useful in ensuring that the graphs drawn with a graphing utility show all essential features of a function.

17-A SOLVING INEQUALITIES

To find the sign of a derivative, we need to solve inequalities of the type $f'(x) > 0$, $f'(x) < 0$, $f''(x) > 0$, and $f''(x) < 0$. The examples that follow illustrate some techniques for solving these inequalities. We use these basic principles:

1. The product or quotient of two positive numbers is positive.

2. The product or quotient of two negative numbers is positive.

3. The product or quotient of a positive number and a negative number is negative.

EXAMPLE 17.1 In a–d below, the derivative of a function is given. For each, find all critical values of the original function and determine for what values of x the derivative is positive and for what values of x the derivative is negative.

a. $f'(x) = (x + 1)(1 - 2x)$ **b.** $g'(x) = \frac{x-3}{x+4}$

c. $h'(x) = \frac{(x-1)(x-5)}{\sqrt[3]{2-x}}$ **d.** $m'(x) = \frac{3x+2}{\sqrt{x}\,(x-4)}$

Solution

a. Since f' is defined everywhere, we need only find where f' equals 0 to find the critical values of f. This occurs where each factor of f' is 0. The first factor, $x + 1$, is 0 when $x = -1$. The second factor, $1 - 2x$, is 0 when $x = \frac{1}{2}$. The two critical values of f are $x = -1$ and $x = \frac{1}{2}$. These critical values divide the domain into the open intervals $(-\infty, -1)$, $\left(-1, \frac{1}{2}\right)$, and $\left(\frac{1}{2}, \infty\right)$. We want to determine the sign of f' on each of these intervals. We can use sign charts for each linear factor to determine the sign of f' on each interval (see Section 4-C). To determine the sign of $x + 1$ to the right of $x = -1$, we can test with the number $x = 0$ and see that $x + 1 = 0 + 1 = 1$ is positive. Therefore, the factor $x + 1$ is positive to the right of -1 and negative to the left of -1. We use $x = 1$, a number to the right of the critical number $\frac{1}{2}$, to determine the sign of the second factor to the right of $\frac{1}{2}$. When $x = 1$, $1 - 2x = 1 - 2(1) = -1$, the second factor is negative to the right of $\frac{1}{2}$ and positive to the left of $\frac{1}{2}$. We make a sign chart for each factor and their product f'.

The product of the two linear functions is positive when both factors have the same sign and negative when they have different signs.

– – – – – – – – – – – 0 + + + + + + + + + +	sign of $x + 1$
+ + + + + + + + + + + + + 0 – – – – – – – –	sign of $1 - 2x$
– – – – – – – – – – – 0 + + 0 – – – – – – – –	sign of $f'(x) = (x + 1)(1 - 2x)$
-1 $\frac{1}{2}$	

From this chart we see that f' is positive on $\left(-1, \frac{1}{2}\right)$ and negative on the intervals $(-\infty, -1)$ and $\left(\frac{1}{2}, \infty\right)$.

b. The critical values of g are the x-values where g' is 0 or undefined. The critical values are $x = 3$, where the numerator of g' is 0, and $x = -4$, where the denominator of g' is 0. We create sign charts and align them one above the other. Where $x + 4 = 0$, $g'(x)$ does not exist, so we write d.n.e.

– – – – – – – – – – – – 0 + + + + + + +	sign of $x - 3$
– – – – – 0 + + + + + + + + + + + + +	sign of $x + 4$
+ + + + + d.n.e. – – – – – – – 0 + + + + + + +	sign of $g'(x) = \frac{x-3}{x+4}$
-4 3	

Thus, g' is positive on the intervals $(-\infty, -4)$ and $(3, \infty)$ and negative on the interval $(-4, 3)$.

c. We find potential critical values of h by determining where the numerator and denominator of h' are 0. These values are $x = 1$, 5, and 2; these values divide the domain into the open intervals $(-\infty, 1)$, $(1, 2)$, $(2, 5)$, and $(5, \infty)$. The denominator $\sqrt[3]{2-x}$ is positive when $2 - x$ is positive and negative when $2 - x$ is negative. Thus we can draw three sign charts, one for each of the three factors of h', and then combine them into a single sign chart. On intervals where there is an even number of negative signs, the sign of the derivative is positive. Otherwise, it is negative.

– – – 0 + + + + + + + + + + + + + +	sign of $x - 1$
– – – – – – – – – – – – – 0 + + + +	sign of $x - 5$
+ + + + + + 0 – – – – – – – – – – –	sign of $\sqrt[3]{2-x}$
+ + + 0 – – d.n.e. + + + + + + 0 – – – –	sign of $h'(x) = \frac{(x-1)(x-5)}{\sqrt[3]{2-x}}$
1 2 5	

From the sign chart, we see that $h'(x) > 0$ on the intervals $(-\infty, 1)$ and $(2, 5)$ and $h'(x) < 0$ on the other two intervals. The derivative $h'(x)$ does not exist at $x = 2$.

d. We set each factor of $m'(x)$ equal to 0 and solve for x to find three possible critical values: $x = -\frac{2}{3}$, 4, and 0. However, because the term $\sqrt{x}$ is in the denominator, $m'(x)$ is defined only for $x > 0$. This means that the only critical value of m is $x = 4$.

The factor $\sqrt{x}$ is always positive and will not affect the sign of the quotient, so we need sign charts for the other two factors to produce the sign chart for m'.

+ + + + + + + + + + + + + + + + + +	sign of $3x + 2$
– – – – – – – – 0 + + + + + + + + +	sign of $x - 4$
d.n.e. – – – – – – – – d.n.e. + + + + + + + + +	sign of $m'(x) = \frac{3x+2}{\sqrt{x}(x-4)}$
0 ———————— 4 ————————	

Thus m' is positive on the interval $(4, \infty)$ and negative on the interval $(0, 4)$. Note that $m'(4)$ does not exist. ■

In Example 17.1, we solved inequalities of the form $f'(x) > 0$ or $f'(x) < 0$, where $f'(x)$ was written as a product of simple factors or a quotient with factored numerator and denominator. When we differentiate a product or quotient function, the derivative obtained is seldom in that form. Some algebraic manipulation is necessary to simplify and factor $f'(x)$. When $f'(x)$ is a sum of terms that are quotients or contain negative exponents, we first find a common denominator and write $f'(x)$ as a single quotient.

EXAMPLE 17.2 In each part a–d, a function and its derivative are given. Find the critical values of each function and the intervals where its derivative is positive and the intervals where its derivative is negative.

a. $f(x) = (x + 1)^{3/2}(2x - 5)^4$; $\quad f'(x) = \frac{3}{2}(x + 1)^{1/2}(2x - 5)^4 + 8(x + 1)^{3/2}(2x - 5)^3$

b. $g(x) = \frac{x^2 - 2}{3x^2 + 5}$; $\quad g'(x) = \frac{(3x^2 + 5)(2x) - (x^2 - 2)(6x)}{\left(3x^2 + 5\right)^2}$

c. $h(x) = 7x^{1/3}(x - 2)$; $\quad h'(x) = \frac{7}{3}x^{-2/3}(x - 2) + 7x^{1/3}$

d. $n(x) = x^2e^{-x}$; $\quad n'(x) = 2xe^{-x} - x^2e^{-x}$

Solution

a. To find the critical values of f we first factor out of $f'(x)$ the common factors of each term. These are $(x + 1)^{1/2}$ and $(2x - 5)^3$, since $(x + 1)$ and $(2x - 5)$ are factors of both terms and $\frac{1}{2}$ and 3 are the smaller exponents in each case. We then simplify:

$$f'(x) = (x+1)^{1/2}(2x-5)^3\left[\frac{3}{2}(2x - 5) + 8(x+1)\right] = (x+1)^{1/2}(2x-5)^3\left(11x + \frac{1}{2}\right).$$

We set each factor equal to 0 and solve for x; we find the possible critical values of f are $x = -1$, $\frac{5}{2}$, and $-\frac{1}{22}$. The function f has domain $[-1, \infty)$, since it contains the factor $(x + 1)^{3/2}$. We are interested in the sign of the derivative on the intervals

$\left(-1, -\frac{1}{22}\right)$, $\left(-\frac{1}{22}, \frac{5}{2}\right)$, and $\left(\frac{5}{2}, \infty\right)$. Since $(x+1)^{1/2}$ is positive on all of these intervals, the sign of f' is determined by the signs of its other two factors. The sign charts are:

− − − − − − − − − − − − − 0 + + + + + +	sign of $(2x-5)^3$ = sign of $2x-5$
− − − 0 + + + + + + + + + + + + + + + +	sign of $11x+\frac{1}{2}$
0 + + 0 − − − − − − − − − − 0 + + + + + +	sign of $f'(x)=(x+1)^{1/2}(2x-5)^3\left(11x+\frac{1}{2}\right)$

$-1 \qquad -\frac{1}{22} \qquad \frac{5}{2}$

Thus f' is positive on the intervals $\left(-1, -\frac{1}{22}\right)$ and $\left(\frac{5}{2}, \infty\right)$ and negative on the interval $\left(-\frac{1}{22}, \frac{5}{2}\right)$.

b. To find the critical values of g we first multiply out and simplify the numerator of $g'(x)$:

$$g'(x)=\frac{(3x^2+5)(2x)-(x^2-2)(6x)}{\left(3x^2+5\right)^2}=\frac{6x^3+10x-(6x^3-12x)}{\left(3x^2+5\right)^2}=\frac{22x}{\left(3x^2+5\right)^2}.$$

Since the denominator is always positive, the sign of the quotient is determined by the sign of its numerator, $22x$. Thus, $g'(x)$ is positive for $x > 0$ and negative for $x < 0$. The critical value is $x = 0$.

c. To find the critical values of h we want to write $h'(x)$ as a single quotient. To do this we write the term that has a negative exponent as a quotient, find a common denominator, and add the two terms. Finally, we factor the numerator:

$$h'(x)=\frac{7}{3x^{2/3}}(x-2)+7x^{1/3}=\frac{7(x-2)+21x^{3/3}}{3x^{2/3}}=\frac{7x-14+21x}{3x^{2/3}}=\frac{28x-14}{3x^{2/3}}=\frac{14(2x-1)}{3x^{2/3}}.$$

The two critical values of h are $x = 0$ and $\frac{1}{2}$. Since $x^{2/3}$ is positive for all $x \neq 0$, the sign of the quotient is determined by the sign of $2x-1$. Thus, h' is positive on $\left(\frac{1}{2}, \infty\right)$ and negative on $(-\infty, 0)$ and $\left(0, \frac{1}{2}\right)$. $h'(0)$ does not exist.

Another way to find the critical values is to use the approach shown in part a—factor out of $h'(x)$ the common factor $x^{-2/3}$:

$$h'(x)=x^{-2/3}\left(\frac{7}{3}(x-2)+7x\right)=x^{-2/3}\left(\frac{7(x-2)+21x}{3}\right)=x^{-2/3}\left(\frac{28x-14}{3}\right)=\left(\frac{14(2x-1)}{3x^{2/3}}\right).$$

d. We factor out of $n'(x)$ the common factor xe^{-x} in order to find critical values of n: $n'(x) = xe^{-x}(2-x)$. Since an exponential function is always positive, the sign of $n'(x)$ is determined by the sign of $x(2-x)$. We draw sign charts for these factors to determine the sign chart for $n'(x)$:

–	–	–	–	0	+	+	+	+	+	+	+	+	+	+	+	+	+	+	+	sign of x
+	+	+	+	+	+	+	+	+	+	+	+	+	0	–	–	–	–	–	–	sign of $2 - x$
–	–	–	–	0	+	+	+	+	+	+	+	+	0	–	–	–	–	–	–	sign of $x(2 - x)$ = sign of $n'(x)$

0 2

We see that n' is positive on $(0, 2)$ and negative on $(-\infty, 0)$ and $(2, \infty)$. ■

Exercises 17-A

1. Consider the following functions and their derivatives. Find the critical values of each function and the intervals where the derivative is positive and the intervals where the derivative is negative.

a. $f(x) = x^3 - x^2 - x + 2;\ f'(x) = 3x^2 - 2x - 1$

b. $g(x) = 2x^3 + 2x;\ g'(x) = 6x^2 + 2$

c. $h(x) = x^{2/3}(x - 5);\ h'(x) = \frac{5}{3}x^{2/3} - \frac{10}{3}x^{-1/3}$

d. $n(x) = \dfrac{x^2 - 3}{x^2 + 1};\ n'(x) = \dfrac{(x^2+1)(2x)-(x^2-3)(2x)}{(x^2+1)^2}$

e. $m(x) = x^3e^{-x};\ m'(x) = 3x^2e^{-x} - x^3e^{-x}$

2. Consider the following second derivatives of the functions given in Exercise 1. For each, determine for what values of x the second derivative is positive and for what values of x the second derivative is negative.

a. $f''(x) = 6x - 2$

b. $g''(x) = 12x$

c. $h''(x) = \frac{10}{9}x^{-1/3} + \frac{10}{9}x^{-4/3}$

d. $n''(x) = \dfrac{8 - 24x^2}{(x^2 + 1)^3}$

e. $m''(x) = e^{-x}(x^3 - 6x^2 + 6x)$

3. Let $f(x) = 5x(x + 1)^{4/5}$. Then $f'(x) = 5(x + 1)^{4/5} + 5x \cdot \frac{4}{5}(x + 1)^{-1/5}$.

a. Find the domain of f.

b. Write $f'(x)$ as a single quotient, in simplest form.

c. Find all critical values of x where $f'(x) = 0$.

d. Find all critical values of x where $f'(x)$ does not exist.

e. For what values of x is $f'(x) > 0$?

f. For what values of x is $f'(x) < 0$?

4. Let $f(x) = \frac{x^2+3x+8}{x-4}$. Then $f'(x) = \frac{(x-4)(2x+3)-(x^2+3x+8)}{(x-4)^2}$. Answer all parts of Exercise 3 for this function and its derivative.

5. Let $f(x) = x^3 + 6x^2 + 12x + 8$. Then $f'(x) = 3x^2 + 12x + 12$. Answer all parts of Exercise 3, except part b, for this function and its derivative.

17-B GRAPHICAL INTERPRETATION

Once we have found critical values and the signs of the first and second derivatives of a function, we can sketch the graph. We use the following principles in sketching the graph of a function f for which f' and f'' exist on an interval I.

1. If $f'(x_1) = 0$, x_1 in I, then f has a horizontal tangent line at the point $(x_1, f(x_1))$.

2. $f'(x) > 0$ on an interval I if and only if f is increasing on I.

3. $f'(x) < 0$ on an interval I if and only if f is decreasing on I.

4. $f''(x) > 0$ on an interval I if and only if f is concave up on I.

5. $f''(x) < 0$ on an interval I if and only if f is concave down on I.

EXAMPLE 17.3 For each of the four functions described below, draw a portion of the graph of a function that is defined and continuous on the interval (–6, 6) and that satisfies the given conditions.

a. $f'(x) > 0$ if $x < 1$; $f'(x) < 0$ if $x > 1$; $f''(x) < 0$ if $-6 < x < 6$.

b. $g'(2) = 0$; $g'(x) > 0$ if $x < 2$; $g''(x) > 0$ if $x > 2$.

c. $h(-1) = 3$; $h'(-1)$ does not exist; $h''(x) > 0$ if $x < -1$; $h''(x) > 0$ if $x > -1$.

d. $k'(x) > 0$ if $x < 1$; $k'(x) > 0$ if $x > 1$; $k''(x) > 0$ if $x < 1$; $k''(x) < 0$ if $x > 1$.

Solution We need to organize the given information, so we create a table. The first column of the table gives the relevant x-intervals. The second, third, and fourth columns give information about the function and its first and second derivatives, respectively. The final column gives a summary of the conclusions we can draw based on the information in that row of the table.

a. Since the sign of the first derivative changes at $x = 1$, we split the interval (–6, 6) into two subintervals: (–6, 1) and (1, 6). Our table follows:

x	f	f'	f''	Summary
$-6 < x < 1$		$+$	$-$	f is increasing; f is concave down
$x = 1$			$-$	$f(1)$ is a relative maximum
$1 < x < 6$		$-$	$-$	f is decreasing; f is concave down

We are not given enough information to fill in all the cells of the table, but we can sketch a graph of a function f that satisfies the given conditions. One such graph is shown in Figure 17.4.

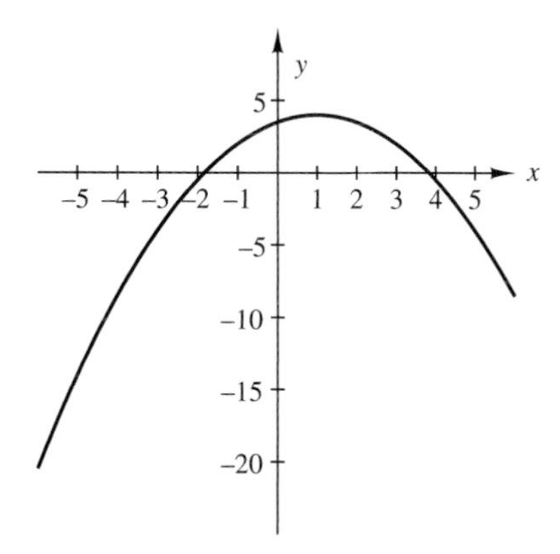

Figure 17.4

b. In this table, we consider the subintervals (–6, 2) and (2, 6) determined by the critical value $x = 2$.

x	g	g'	g''	Summary
$-6 < x < 2$		$+$		g is increasing
$x = 2$		0		g has a horizontal tangent line
$2 < x < 6$			$+$	g is concave up

The graph in Figure 17.5 satisfies the given conditions.

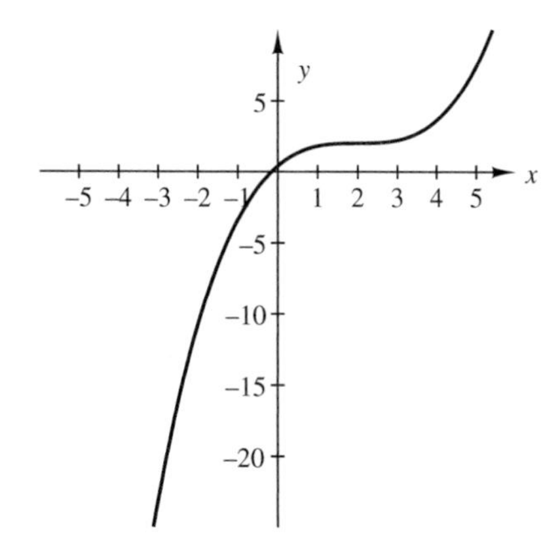

Figure 17.5

c. We organize the given information in this table:

x	h	h'	h''	Summary
$-6 < x < -1$			$+$	h is concave up
$x = -1$	3	d.n.e.		h passes through the point $(-1, 3)$
$-1 < x < 6$			$+$	h is concave up

From the information, we can sketch a graph that satisfies these conditions; see Figure 17.6.

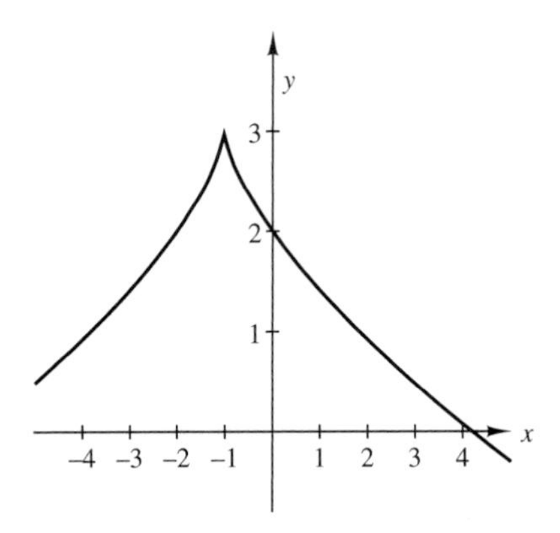

Figure 17.6

d. From the table below, we can sketch the graph of one possible function k (see Figure 17.7).

x	k	k'	k''	Summary
$-6 < x < 1$		$+$	$+$	k is increasing and concave up
$x = 1$				
$1 < x < 6$		$+$	$-$	k is increasing and concave down

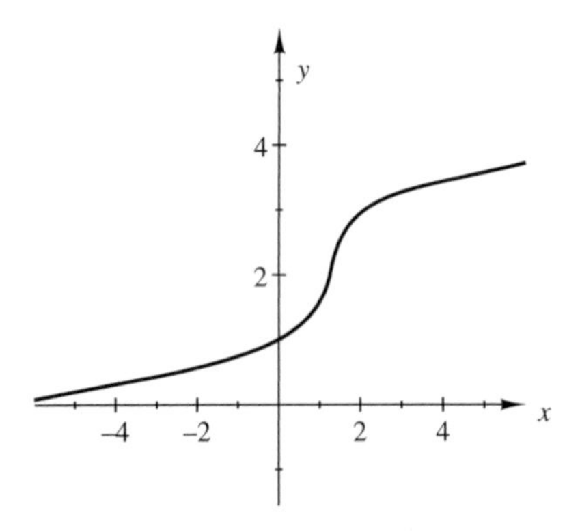

Figure 17.7 ■

In constructing a table of information that shows how the function f and its derivatives f' and f'' behave on subintervals that make up the domain, always list the intervals so that the x-values are in increasing order.

Just as a table of information on the behavior of a function and its derivatives allows us to sketch a possible graph that satisfies specified conditions, we can set up such a table by reading the graph of a given function.

On the graph in Figure 17.7 there is a point—approximately (1.3, 2)—where the function changes from concave up to concave down. A point on a graph that marks a change in concavity is called an *inflection point* of the graph. The x-coordinate of an inflection point will be a value in the domain of the function where the second derivative of the function changes sign.

EXAMPLE 17.4 For each of the following graphs, construct a table that shows all points where the first or second derivative of the function changes sign. Give the signs of the first and second derivatives on the subintervals determined by these points. Describe in words the behavior of the function.

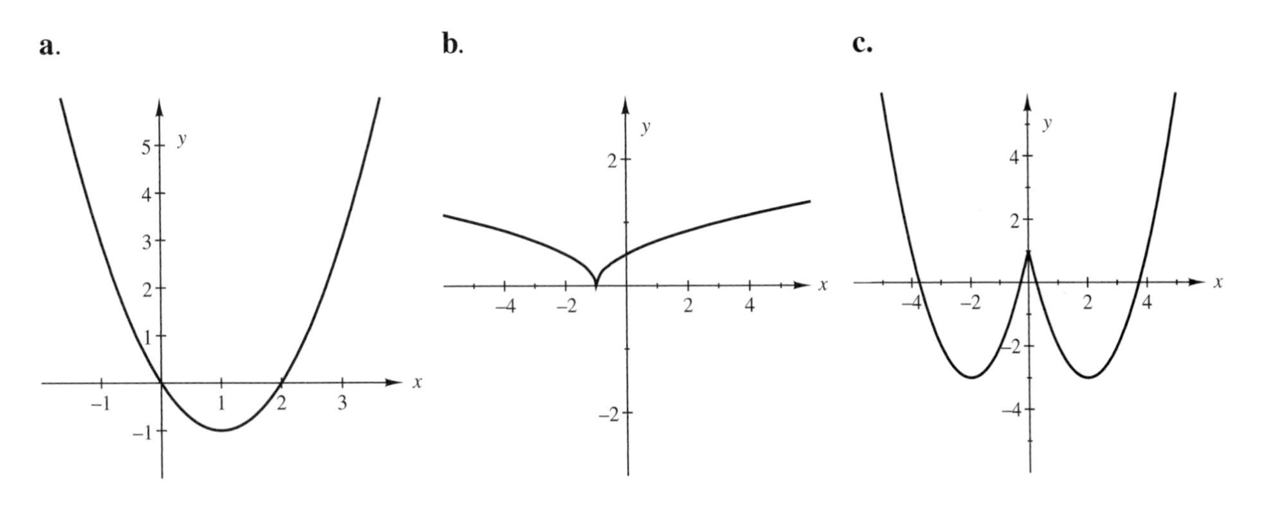

Figure 17.8 $y = f(x)$ **Figure 17.9** $y = g(x)$ **Figure 17.10** $y = h(x)$

Solution

a. We see that the function f graphed in Figure 17.8 is concave up for all values of x shown, so $f''(x) > 0$ for all x in the domain. The function f is decreasing for $x < 1$ and increasing for $x > 1$ and has a horizontal tangent line at $x = 1$. We can construct a table of information as follows:

x	f	f'	f''	Summary
$x < 1$		$-$	$+$	f is decreasing and concave up
$x = 1$	-1	0		f has a horizontal tangent at $(1, f(1))$; $f(1)$ is a minimum
$x > 1$		$+$	$+$	f is increasing and concave up

b. The function g with graph shown in Figure 17.9 is decreasing and concave down for $x < -1$. The function is increasing and concave down for $x > -1$, and the function contains the point $(-1, 0)$. The cusp shows that the derivative does not exist at the point where $x = -1$.

x	g	g'	g''	Summary
$x < -1$		$-$	$-$	g is decreasing and concave down
$x = -1$	0	d.n.e.		$g(-1)$ is a minimum
$x > -1$		$+$	$-$	g is increasing and concave down

c. The graph of the function h shown in Figure 17.10 is concave up for $x < 0$ and concave up for $x > 0$. For $x < -2$, h is decreasing; for $-2 < x < 0$, h is increasing; for $0 < x < 2$, h is decreasing; and for $x > 2$, h is increasing. The function has horizontal tangent lines at $x = -2$ and at $x = 2$. There are absolute minima at these values of x. The cusp at $x = 0$ shows that the derivative of h does not exist at $x = 0$; h has a relative maximum value there.

x	h	h'	h''	Summary
$x < -2$		$-$	$+$	h is decreasing and concave up
$x = -2$	-3	0		$h(-2) = -3$ is a minimum
$-2 < x < 0$		$+$	$+$	h is increasing and concave up
$x = 0$	1	d.n.e.		$h(0) = 1$ is a relative maximum
$0 < x < 2$		$-$	$+$	h is decreasing and concave up
$x = 2$	-3	0		$h(2) = -3$ is a minimum
$x > 2$		$+$	$+$	h is increasing and concave up

■

If we are given the graphs of f' and f'', we can determine the shape of the graph of the original function f. The next example illustrates how we use the graphs of the first and second derivatives to graph a function.

EXAMPLE 17.5 Figures 17.11a and 17.11b show the graphs of the first and second derivatives of a function f that is continuous everywhere.

a. Find the critical values of f.

b. Find the x-values where $f''(x) = 0$ or $f''(x)$ does not exist.

c. Use the x-values found in parts a and b to divide the domain into open intervals, and create a table showing the signs of f' and f'' over the domain of f.

d. Sketch a possible graph of $y = f(x)$.

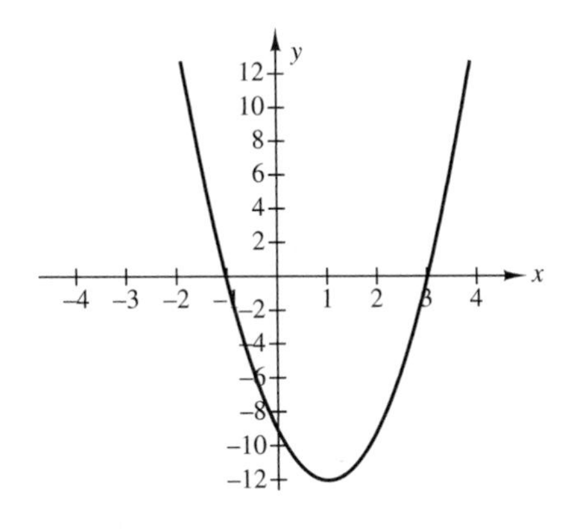

Figure 17.11a $y = f'(x)$

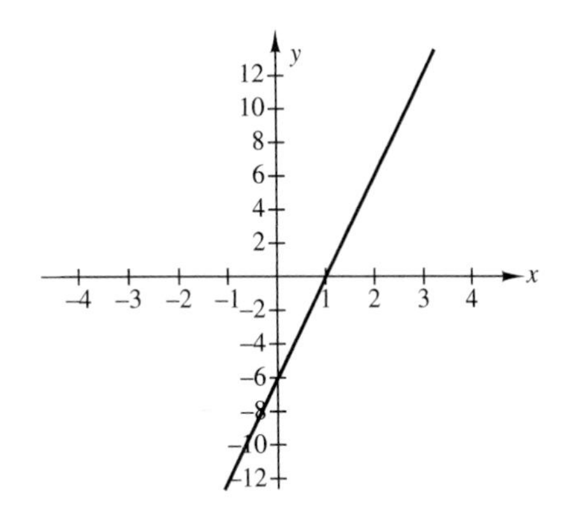

Figure 17.11b $y = f''(x)$

Solution

a. To find critical values of f we look at the graph of f' to find values of x where $f'(x) = 0$ or $f'(x)$ does not exist. Since $f'(x) = 0$ at the x-intercepts of the graph of f' and f' exists everywhere, we see that $x = -1$ and $x = 3$ are the critical values of f.

b. Since f'' exists everywhere, we look for the x-intercepts of the graph of f''. We see that $f''(1) = 0$, and $x = 1$ is the only value for x for which $f''(x) = 0$.

c. The x-values found in parts a and b divide the domain of f into the open intervals $(-\infty, -1)$, $(-1, 1)$, $(1, 3)$, and $(3, \infty)$. We note that f' is positive (the graph is above the x-axis) on the first and last intervals and f' is negative (the graph is below the x-axis) on the other two intervals. We see that f'' is negative on the first two intervals and f'' is positive on the last two intervals. We now organize what we know about the graph of f:

x	f	f'	f''	Summary
$x < -1$		+	–	f is increasing and concave down
$x = -1$		0	–	f has a relative maximum
$-1 < x < 1$		–	–	f is decreasing and concave down
$x = 1$		–	0	f has an inflection point
$1 < x < 3$		–	+	f is decreasing and concave up
$x = 3$		0	+	f has a relative minimum
$x > 3$		+	+	f is increasing and concave up

d. A possible graph of f that illustrates the behavior described in the table is shown in Figure 17.12.

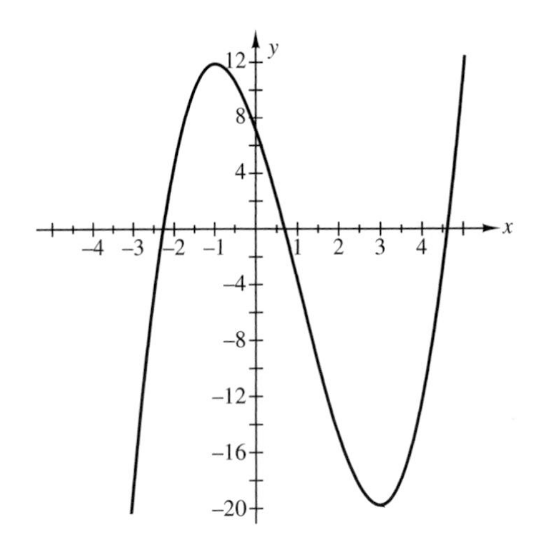

Figure 17.12 $y = f(x)$ ■

Exercises 17-B

1. Answer the following questions for the function $f(x)$ sketched in Figure 17.13.

a. For what values of x does $f'(x) = 0$?

b. For what values of x does $f'(x)$ not exist?

c. Where is $f'(x) > 0$?

d. Where is $f'(x) < 0$?

e. Where is $f''(x) > 0$?

f. Where is $f''(x) < 0$?

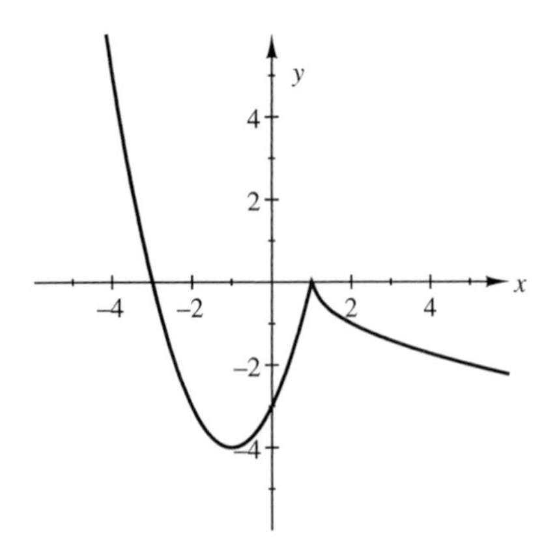

Figure 17.13 $y = f(x)$

2. For each description below, draw a portion of the graph of a function that is defined and continuous on the interval $(-4, 4)$ and that satisfies the given conditions.

a. $f''(x) < 0$ if $x < 2$; $f''(x) > 0$ if $x > 2$; $f'(2) = 0$.

b. $g'(x) > 0$ if $-4 < x < 4$; $g''(x) > 0$ if $-4 < x < 4$; $g(x) > 0$ if $-4 < x < 4$.

c. $h(0) = 0$; $h'(0) = 0$; $h''(x) < 0$ if $-4 < x < 4$.

d. $k'(x) < 0$ if $x < -1$; $k'(x) > 0$ if $-1 < x < 2$; $k'(x) < 0$ if $x > 2$; $k''(x) < 0$ if $x < 0$; $k''(x) > 0$ if $x > 0$.

3. For each of the graphs in Figures 17.14 and 17.15, construct a table that shows all values of x where the first or second derivative of the function changes sign in the interval $(-4, 4)$. Give the signs of the first and second derivatives on the subintervals determined by these x-values. Describe in words the behavior of the function.

a.

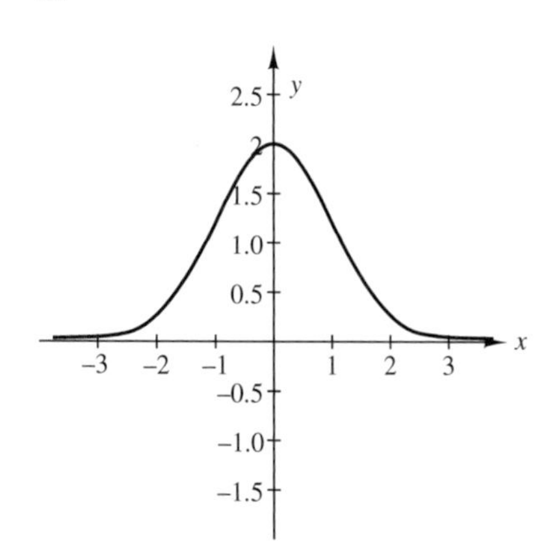

Figure 17.14 $y = f(x)$

b.

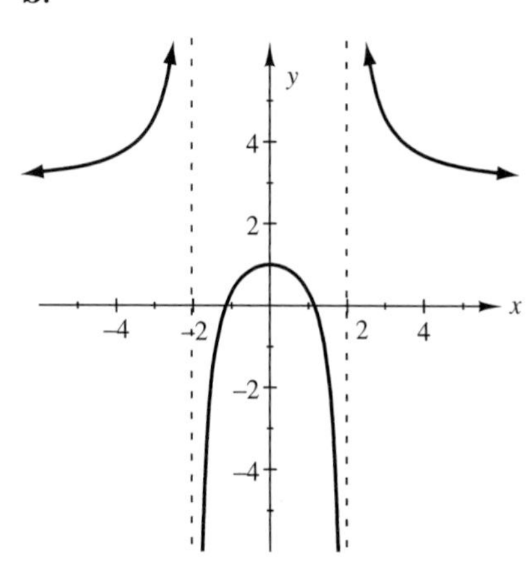

Figure 17.15 $y = g(x)$

4. Draw the graph of a continuous function f with domain $(-\infty, \infty)$ that has exactly one relative maximum and one relative minimum.

a. How many zeros does the function f have? Identify the zeros of f.

b. Identify the relative extrema.

c. Does your graph have any inflection points? If so, identify them.

d. Could a graph with exactly one relative maximum and one relative minimum have more than one inflection point? Explain your answer.

e. For what values of x is $f'(x) > 0$?

f. For what values of x is $f'(x) < 0$?

g. For what values of x is $f''(x) > 0$?

h. For what values of x is $f''(x) < 0$?

5. Figures 17.16a and 17.16b show the graphs of the first and second derivatives of a function f that is continuous everywhere.

a. Find the critical values of f.

b. Find the x-values where $f''(x) = 0$ or $f''(x)$ does not exist.

c. Use the x-values found in parts a and b to divide the domain into open intervals, and create a table showing the signs of f' and f'' over the domain of f.

d. Sketch a possible graph of $y = f(x)$.

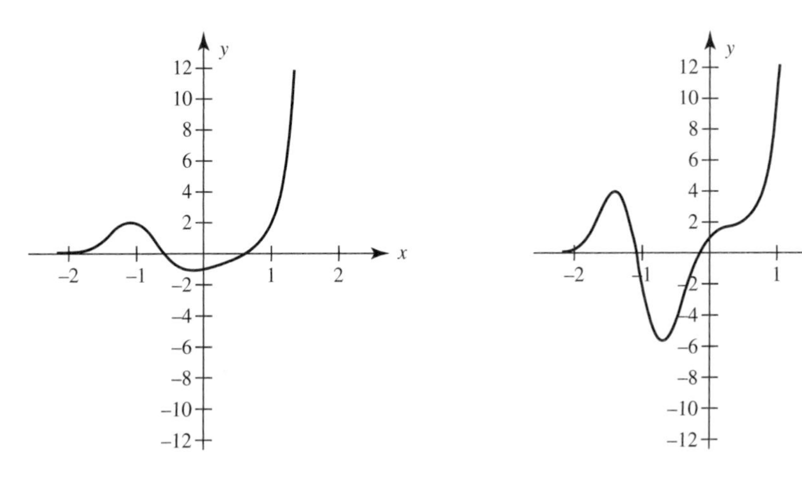

Figure 17.16a $y = f'(x)$ **Figure 17.16b** $y = f''(x)$

17-C PUTTING IT ALL TOGETHER

Often we are given a function in equation form and we wish to sketch an accurate graph, either by hand or using a graphing utility. To do this we construct a table, like the tables given in Example 17.3. First, we need to find the domain of the function f; this tells us what values of x to consider for the x column of the table. In Example 17.3, we were given a domain of (−6, 6), but in many problems we use the implied domain. For example, if a function contains a square root, the expression under the square root must be nonnegative.

To set up the table, we first find the values of x in the domain of f where $f'(x)$ is zero or does not exist and also those values of x where $f''(x)$ is zero or does not exist. In doing this we identify all values of x where either f' or f'' can change sign. These values of x divide the domain into intervals that we list in increasing order in the x column of the table. The following steps help us sketch the graph of a function $y = f(x)$:

1. Find the domain of f.

2. Find the y-intercept, $f(0)$, and any x-intercepts, if convenient. (Sometimes it is very difficult to solve $f(x) = 0$ to find the x-intercepts.)

3. Find any horizontal or vertical asymptotes of f.

4. Find $f'(x)$ and $f''(x)$.

5. Find all critical values of f—that is, values of x in the domain of f where $f'(x) = 0$ or $f'(x)$ does not exist. These are values at which relative extrema may occur.

6. Find all values x in the domain where $f''(x) = 0$ or $f''(x)$ does not exist. These are values at which inflection points may occur.

7. Use the x-values of any vertical asymptotes as well as those found in steps 5 and 6 to partition the domain into subintervals. Set up a table to record the signs of f' and f'' on each subinterval. Evaluate f at the endpoints of these subintervals. Record any information known about f' and f'' at these endpoints.

8. In the table, describe the behavior of the graph in each of the subintervals and at endpoints. Use this information to sketch the graph of $y = f(x)$.

We illustrate these steps in the next example.

EXAMPLE 17.6 Parts a and b below give a function and its first and second derivatives. Follow the steps described above to construct a table that shows the behavior of the function, and use the table to sketch the graph of $y = f(x)$.

a. $f(x) = x^3 - 6x^2 + 10;$ $f'(x) = 3x^2 - 12x;$ $f''(x) = 6x - 12$

b. $f(x) = \dfrac{x^2}{x-1};$ $f'(x) = \dfrac{x^2 - 2x}{(x-1)^2};$ $f''(x) = \dfrac{2}{(x-1)^3}$

Solution

a. Since f is a polynomial function, its domain is the set of all real numbers, and f' and f'' exist for all x in the domain. Also, there are no vertical or horizontal asymptotes. We set $f'(x) = 0$ to get $3x(x - 4) = 0$; so $f'(x) = 0$ when $x = 0$ and $x = 4$. Also, we solve $f''(x) = 6x - 12 = 0$; so $f''(x) = 0$ when $x = 2$. We put these three values where $f'(x) = 0$ or $f''(x) = 0$ in increasing order ($x = 0, x = 2, x = 4$) to divide the domain into four subintervals for the left column of our table of the behavior of f.

We use sign charts to find the sign of $f'(x) = 3x(x - 4)$ on each interval:

```
– – – 0 + + + + +     sign of 3x
– – – – – – – 0 + +   sign of x – 4
+ + + 0 – – 0 + +     sign of 3x(x – 4) = f′(x)
——————•—————•—————
      0     4
```

Finally, we enter the information from the sign charts into our table and write our conclusions for the behavior of the graph of f:

x	f	f'	f''	Summary
$x < 0$		+	−	f is increasing and concave down
$x = 0$	10	0	−	$f(0) = 10$ is a relative maximum
$0 < x < 2$		−	−	f is decreasing and concave down
$x = 2$	−6	−	0	$(2, -6)$ is an inflection point
$2 < x < 4$		−	+	f is decreasing and concave up
$x = 4$	−22	0	+	$f(4) = -22$ is a relative minimum
$x > 4$		+	+	f is increasing and concave up

From the table, we can graph the function $y = f(x)$; the graph is shown in Figure 17.17.

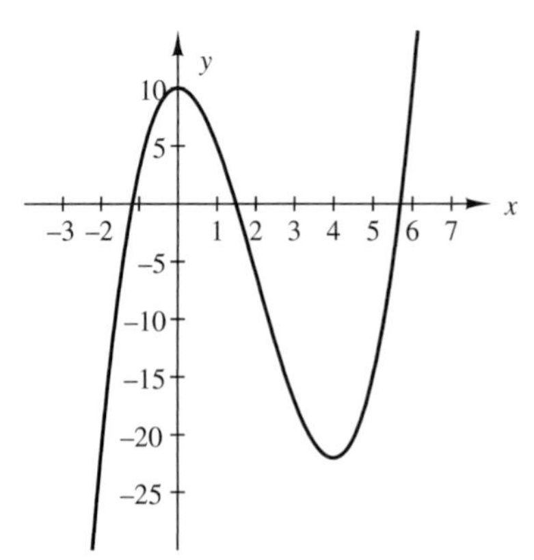

Figure 17.17 $f(x) = x^3 - 6x^2 + 10$

b. The function $f(x) = \frac{x^2}{x-1}$ is a rational function whose domain is the set of all real numbers, excluding $x = 1$. Therefore, $f(1)$, $f'(1)$, and $f''(1)$ do not exist. The line $x = 1$ is a vertical asymptote; there are no horizontal asymptotes. To solve $f'(x) = 0$, we set the numerator $x(x - 2) = 0$; thus $x = 0$ and $x = 2$ are critical values. There are no values of x for which $f''(x) = 0$. The endpoints of subintervals to consider for the table are $x = 0$, $x = 1$, and $x = 2$.

x	f	f'	f''	Summary
$x < 0$		+	−	f is increasing and concave down
$x = 0$	0	0	−	$f(0) = 0$ is a relative maximum
$0 < x < 1$		−	−	f is decreasing and concave down
$x = 1$	d.n.e.	d.n.e.	d.n.e.	$x = 1$ is a vertical asymptote
$1 < x < 2$		−	+	f is decreasing and concave up
$x = 2$	4	0	+	$f(2) = 4$ is a relative minimum
$x > 2$		+	+	f is increasing and concave up

The graph of the function is sketched in Figure 17.18.

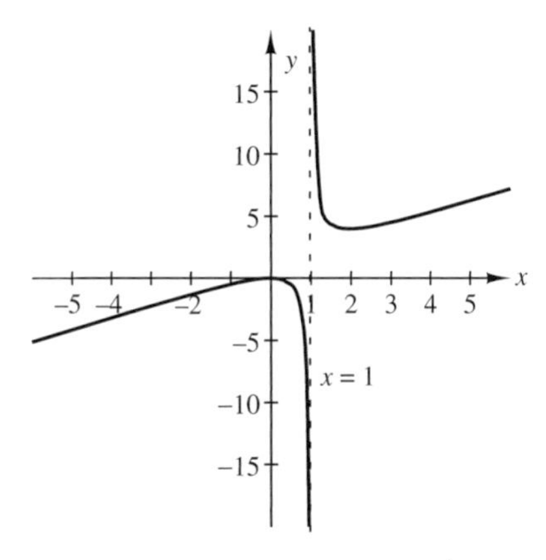

Figure 17.18 $f(x) = \frac{x^2}{x - 1}$ ■

If we are given a formula for a function $y = f(x)$, it is usually easy to produce its graph with a graphing utility. Yet what we see on the screen may show little of the function's important behavior. Such graphs are produced by evaluating a finite number of values of $f(x)$ for x in the specified domain and using curves to connect consecutive points. These connecting curves may not match the actual graph. The techniques we have discussed for graphing functions by hand are also important when using a graphing utility to graph a function. They can help us choose a suitable window to see the overall behavior of the function and also show us where we should zoom in to view important behavior that otherwise would not be seen. The following example illustrates the process.

EXAMPLE 17.7 Consider the function $f(x) = x^3 - 1.2x^2 + 0.24x + 10$. Then $f'(x) = 3x^2 - 2.4x + 0.24$ and $f''(x) = 6x - 2.4$. Find a window $a \le x \le b$, $c \le y \le d$ for a graphing utility that shows all the key features of the graph of f.

Solution Any window we might choose that shows the global behavior of f shows a graph that looks like a translation of the graph of $y = x^3$ with no relative extrema. (See Figure 17.19.) However, if we create a table following the steps outlined in this section, we find that f has a relative maximum, relative minimum, and an inflection point.

x	f	f'	f''	Summary
$x < 0.1172$		+	−	f is increasing and concave down
$x = 0.1172$	10.013	0	−	(0.1172, 10.013) is a relative maximum
$0.1172 < x < 0.4$		−	−	f is decreasing and concave down
$x = 0.4$	9.968	−	0	(0.4, 9.968) is an inflection point
$0.4 < x < 0.6828$		−	+	f is decreasing and concave up
$x = 0.6828$	9.9227	0	+	(0.6828, 9.9227) is a relative minimum
$x > 0.6828$		+	+	f is increasing and concave up

One window that shows key features is $-1 < x < 1.5$, $9.5 < y < 10.5$. See Figure 17.20.

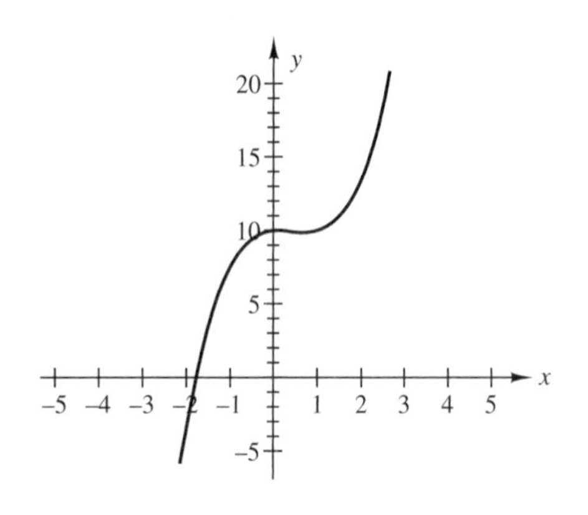

Figure 17.19 $y = f(x)$ with window $-5 < x < 5$, $-5 < y < 20$

Figure 17.20 $y = f(x)$ with window $-1 < x < 1.5$, $-9.5 < y < 10.5$ ■

 A graphing utility may not be able to help you see all important features of the graph of a function. It is best to use a combination of analysis and technology to understand and interpret the graph.

Exercises 17-C

1. Sketch the graph of a function f on the interval $(-6, 6)$ through the points given in Figure 17.1 so that $f'(x)$ is always defined and none of the points is a relative minimum or relative maximum of the function.

2. Let $f(x) = x^3 + 6x^2 + 12x + 8$. Then $f'(x) = 3x^2 + 12x + 12$ and $f''(x) = 6x + 12$. Follow steps 1–8 given in this section to construct a table of information about f, and sketch a graph of f.

3. Let $f(x) = x^2(1 - x)^2$. Then $f'(x) = 2x(1 - x)^2 - 2x^2(1 - x)$ and $f''(x) = 12x^2 - 12x + 2$. Follow steps 1–8 given in this section to construct a table of information about f, and sketch a graph of f.

4. Let $f(x) = \frac{x^2 + 3x + 8}{x - 4}$. Then $f'(x) = \frac{x^2 - 8x - 20}{(x - 4)^2}$ and $f''(x) = \frac{72}{(x - 4)^3}$. Follow steps 1–8 given in this section to construct a table of information about f, and sketch a graph of f.

5. The graph of $y = f(x)$ contains the point $(2, 1)$. f is increasing everywhere and is concave down on the interval $(-\infty, 2)$ and concave up on the interval $(2, \infty)$. Sketch a possible graph of f.

6. The graph of $y = g(x)$ contains the points $(-1, 5)$, $(1, -11)$, and $(3, -27)$. The function g is increasing on the intervals $(-\infty, -1)$ and $(3, \infty)$ and decreasing on $(-1, 3)$. It is concave down on the interval $(-\infty, 1)$ and concave up on the interval $(1, \infty)$. Sketch a possible graph of g.

7. Let $f(x) = x^3 - 7.2x^2 + 17.04x - 5$. Then $f'(x) = 3x^2 - 14.4x + 17.04$ and $f''(x) = 6x - 14.4$. Can you find a single window that shows all the key features of the graph of f? If you can, state the window parameters. If you cannot, use what you learned in this section to explain why it is not possible.

Chapter 17 Exercises

1. For each of the descriptions below, make a table to organize the information given. Draw the graph of a function f defined and continuous on an interval containing $x = 2$ that satisfies the given conditions.

a. $f'(x) < 0$ for $x < 2$; $f'(x) < 0$ for $x > 2$; $f''(x) > 0$ for $x < 2$; $f''(x) < 0$ for $x > 2$.

b. $f(x) < 0$; $f'(x) < 0$; $f''(x) < 0$ for all x.

c. $f(2) = -1$, but $f'(2)$ does not exist; $f''(x) < 0$ for $x < 2$; $f'(x) > 0$ for $x > 2$.

2. For each of the following graphs, identify all intervals where the function is:
(i) Increasing and concave up
(ii) Increasing and concave down
(iii) Decreasing and concave up
(iv) Decreasing and concave down

a.

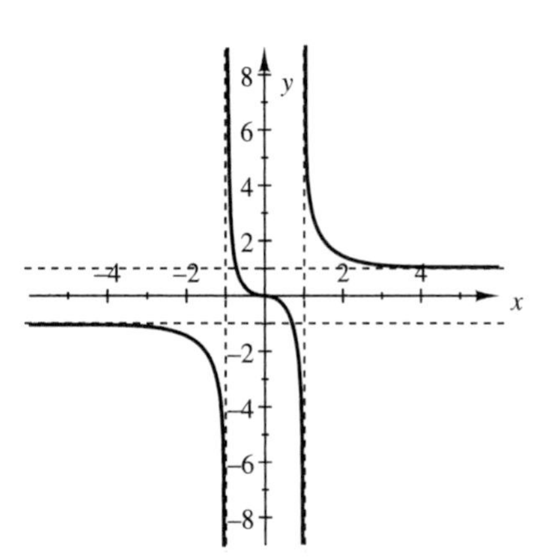

b.

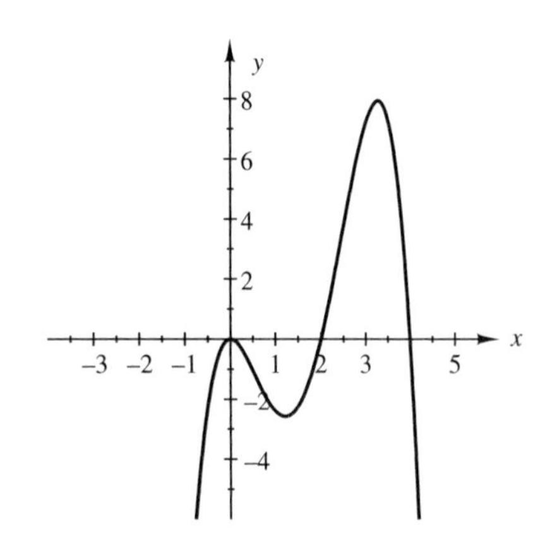

c.

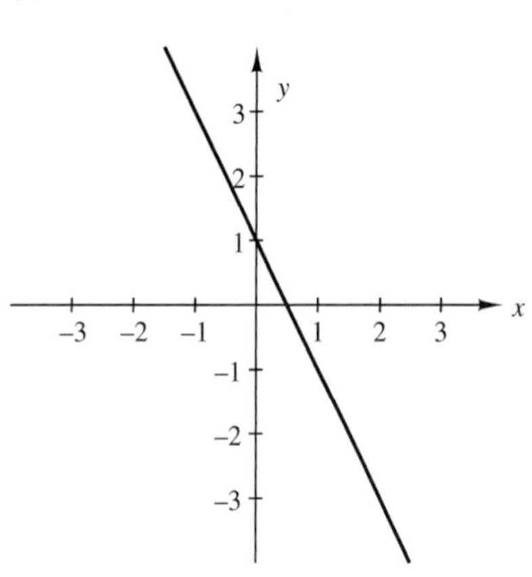

d.

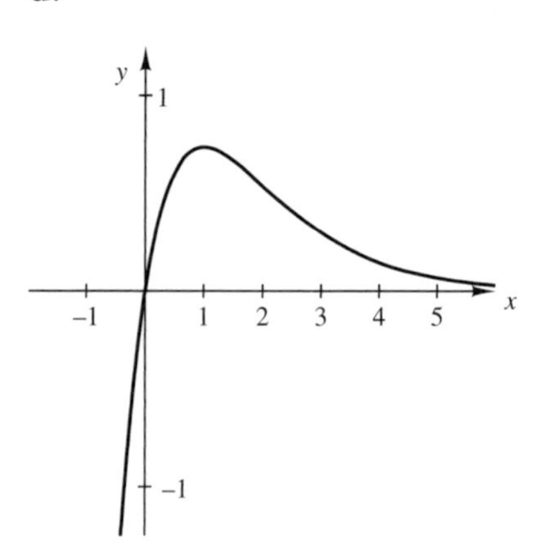

e.

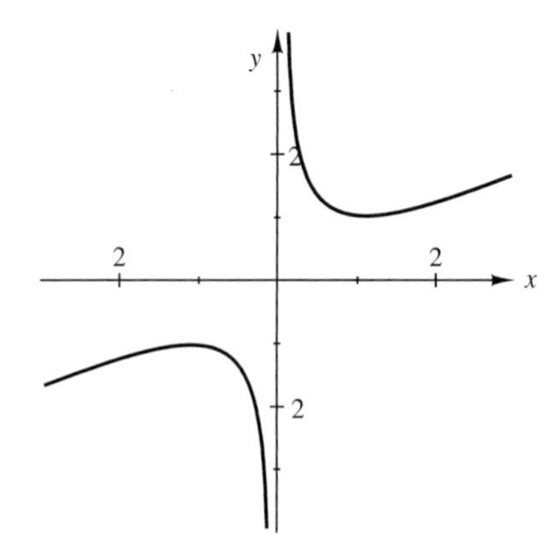

f.

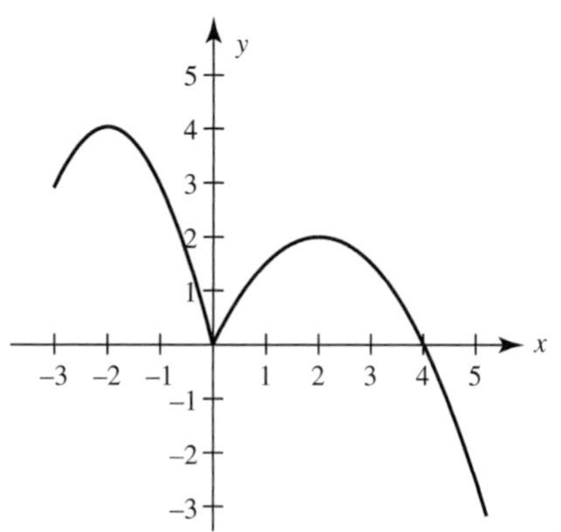

3. For each of the graphs in Exercise 2, construct a table that shows values of x where the first and second derivatives of the function change sign. Give the signs of the first and second derivatives on the intervals between these values.

4. For each of the graphs in Exercise 2, identify all x-intercepts and all y-intercepts.

5. For each of the graphs in Exercise 2, give equations of all vertical and horizontal asymptotes.

6. In a–d below, a function f and its derivatives f' and f'' are given. Use steps 1–8 of Section 17-C to construct a table of information about f, and sketch the graph of f.

a. $f(x) = x^3 - x^2 - x + 2$; $f'(x) = 3x^2 - 2x - 1$; $f''(x) = 6x - 2$

b. $f(x) = 2x^3 + 2x$; $f'(x) = 6x^2 + 2$; $f''(x) = 12x$

c. $f(x) = x^{2/3}(x - 5)$; $f'(x) = \frac{5}{3}x^{2/3} - \frac{10}{3}x^{-1/3}$; $f''(x) = \frac{10}{9}x^{-1/3} + \frac{10}{9}x^{-4/3}$

d. $f(x) = \frac{x^2 - 3}{x^2 + 1}$; $f'(x) = \frac{(x^2+1)(2x) - (x^2-3)(2x)}{\left(x^2+1\right)^2}$; $f''(x) = \frac{8 - 24x^2}{\left(x^2+1\right)^3}$

7. If possible, sketch a graph of a function that has a local maximum and a local minimum and no points of inflection. If it is not possible, explain why not.

8. If possible, sketch a graph of a function that has two local maxima and no local minima. If it is not possible, explain why not.

9. Figures 17.21a and 17.21b show the graphs of the first and second derivatives of a function f that is continuous everywhere.

a. Find the critical values of f and of f'.

b. Use the x-values found in part a to divide the domain into open intervals, and create a table showing the signs of the derivatives over the domain of f.

c. Sketch a possible graph of $y = f(x)$.

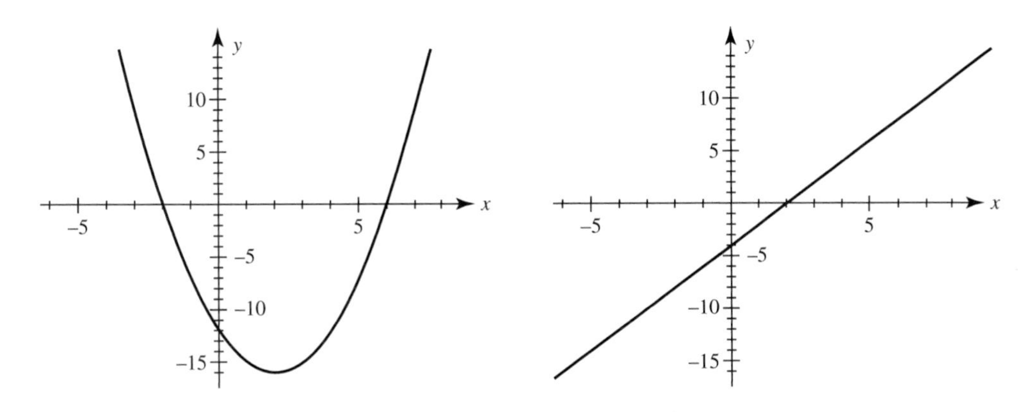

Figure 17.21a $y = f'(x)$ **Figure 17.21b** $y = f''(x)$

10. The function $f(x) = x^2 - 10{,}000 + \frac{1}{x}$ has derivatives $f'(x) = 2x - \frac{1}{x^2}$ and $f''(x) = 2 + \frac{2}{x^3}$. Explain why no graphing utility window can show all the key features of the graph of f. (First analyze the graph using the steps outlined in this chapter.)

CHAPTER 18

Companion to Antidifferentiation

18-A ANTIDIFFERENTIATION AS THE INVERSE OF DIFFERENTIATION

If $C(x) = 0.01x^2 - 5x + 1000$ is the function that represents the cost of producing x units of a product, then the function $C'(x) = 0.02x - 5$ (the derivative of the cost function) approximates the marginal cost, which is the cost to produce one more unit when x units are produced. In fact, most economists equate marginal cost with the derivative C'. Differentiation is an operation on a function; this operation yields another function that represents the instantaneous rate of change of the original function.

In practice, it may be easier to find the rate of change function than the original function. For example, statistical techniques can be used to find a linear function that approximates the marginal cost from data on the change in costs over a period of time. To find the cost function from the marginal cost requires the use of an operation that is the inverse of differentiation. This inverse operation is called *antidifferentiation*; a function found in this way is called an *antiderivative* of the original function. Thus, the cost function $C(x)$ is the antiderivative of the marginal cost function $C'(x)$.

In general, given a function $f(x)$, an antiderivative is another function $F(x)$ whose derivative is $f(x)$; that is, $F'(x) = f(x)$. The general antiderivative of $f(x)$ is represented by $\int f(x)\,dx$. For example, the general antiderivative of $2x$ could be represented by $\int 2x\,dx$. It is the most general function whose derivative is $2x$. This expression is commonly called an *indefinite integral*.

 The notation $\int f(x)\,dx$ for an antiderivative, or indefinite integral, always requires the differential "dx."

EXAMPLE 18.1 Find two additional antiderivatives of $f(x) = 2x + 3$, given that $F(x) = x^2 + 3x$ is one antiderivative. Graph all three antiderivatives. How do their graphs compare?

Solution Since the derivative of a constant is 0, there is a whole family of antiderivatives of f; each has a different constant term. Two other antiderivatives of $f(x)$ are $G(x) = x^2 + 3x + 10$ and $H(x) = x^2 + 3x - 32$. We can check that all three functions are antiderivatives of $f(x)$ by computing $F'(x)$, $G'(x)$, and $H'(x)$.

The graphs of F, G, and H are shown in Figures 18.1, 18.2, and 18.3. We see that all three graphs have exactly the same shape—each is a vertical translation of the graph of $F(x) = x^2 + 3x$.

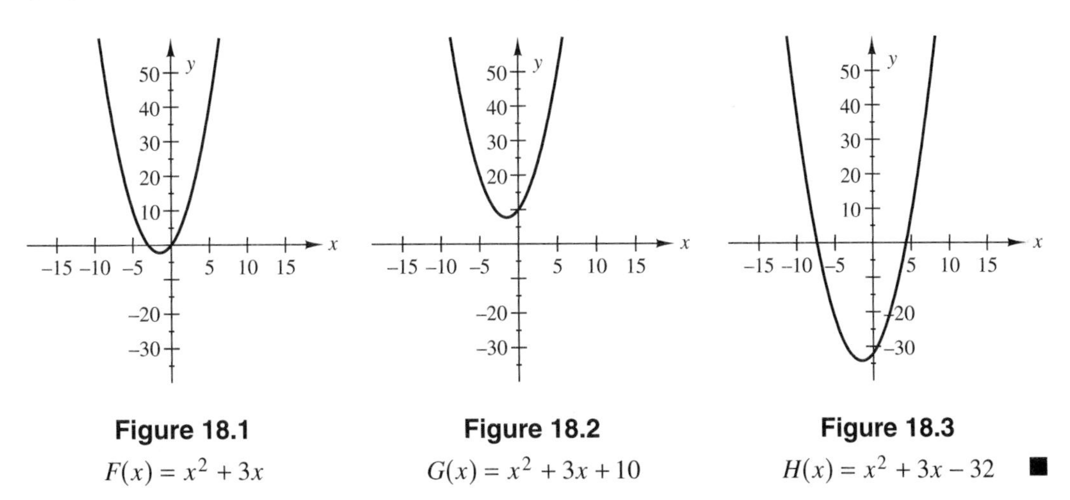

Figure 18.1
$F(x) = x^2 + 3x$

Figure 18.2
$G(x) = x^2 + 3x + 10$

Figure 18.3
$H(x) = x^2 + 3x - 32$ ■

The graph of every antiderivative of $f(x) = 2x + 3$ will have this same shape because every antiderivative of f will have the same first derivative (and thus the same second derivative), and the derivatives of a function determine the shape of its graph. Each of the functions F, G, and H represents a particular antiderivative of $f(x) = 2x + 3$. The most general antiderivative of $f(x)$ is $F(x) + C = x^2 + 3x + C$, where C represents an arbitrary constant (instead of a specific constant such as 0, 10, or –32).

Interpretations of the Antiderivative

We have seen that the derivative f' of a differentiable function f can have various interpretations, depending on the context of a problem or what the function f measures. At each point $(t, f(t))$ on the graph of f, the derivative $f'(t)$ is the slope of the tangent line to the graph at that point. If $f(t)$ measures the population of a bacteria colony after t hours have elapsed, then $f'(t)$ gives the growth rate of the colony at time t (in number of bacteria per hour). If $f(t)$ measures the distance (in meters) traveled by a falling object after t minutes, then $f'(t)$ measures the velocity (in meters per minute) of the object at time t. In all cases, the derivative f' measures the (instantaneous) rate of change of the function f.

In this section, we consider various interpretations of the antiderivative of a given function.

EXAMPLE 18.2 Suppose you drive on an interstate highway at a constant speed of 55 mph. Use this constant speed to obtain the distance traveled after t hours for $t = 0.5$, 1, 1.5, and 2. Find the function $s_1(t)$ that gives the distance traveled after t hours, and graph it.

Solution We multiply the rate (55 mph) by the length of the time interval (hours) to get the distance (miles).

hours traveled	Distance traveled
0.5	27.5 miles
1	55 miles
1.5	82.5 miles
2	110 miles

From the table, we see that the distance traveled after t hours is given by the function $s_1(t) = 55t$. This distance function is an antiderivative of the constant velocity function $v(t) = 55$. Figure 18.4 shows the graph of the velocity function $v(t) = 55$ and the associated distance function $s_1(t) = 55t$.

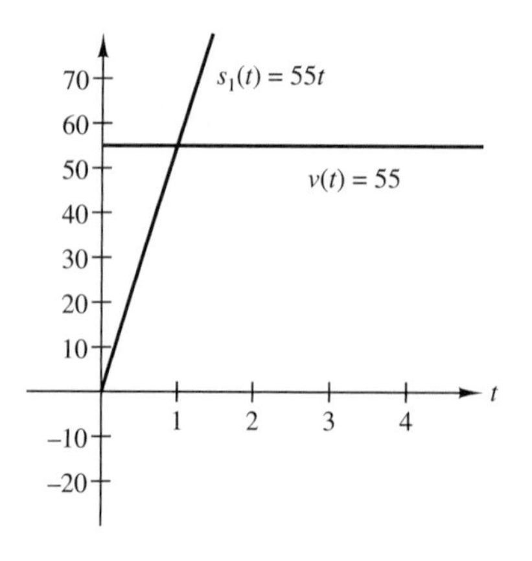

Figure 18.4 $s_1(t)$ is an antiderivative of $v(t)$ ■

EXAMPLE 18.3 Suppose you drive 10 miles before entering the interstate highway and then maintain a constant speed of 55 mph.

a. What is the function $s_2(t)$ that describes the total distance you travel by time t (in hours), assuming you enter the highway at time $t = 0$?

b. Is this function also an antiderivative of the velocity function, $v(t) = 55$?

c. Graph the functions $v(t)$ and $s_2(t)$. How do these graphs compare with those in Figure 18.4?

Solution

a. The distance you travel on the highway by time t is given by $55t$, since t is the time spent on the highway and 55 mph is the constant rate. Since you traveled 10 additional miles before entering the highway, the distance function is $s_2(t) = 55t + 10$.

b. The derivative of $s_2(t) = 55t + 10$ is $s'_2(t) = 55 = v(t)$, so $s_2(t)$ is an antiderivative of $v(t)$.

c. The graphs of $v(t) = 55$ and $s_2(t) = 55t + 10$ are shown in Figure 18.5. We can see that $v(t)$ is exactly the same as in Figure 18.4, and that the graphs of the two distance functions $s_1(t)$ and $s_2(t)$ are lines with the same slope and different y-intercepts. The graph of $s_2(t)$ is a vertical translation of the graph of $s_1(t)$.

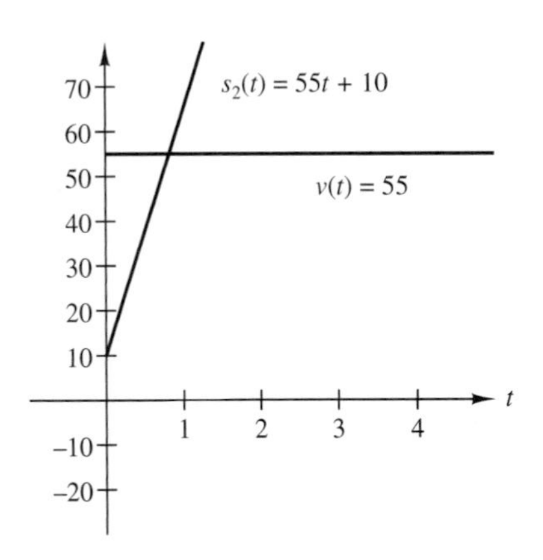

Figure 18.5 $s_2(t)$ is an antiderivative of $v(t)$ ■

In Examples 18.2 and 18.3, we see that the behavior of the function $v(t)$ tells us about the behavior of its antiderivatives. Since $v(t) = 55$, we know that the derivatives of $s_1(t)$ and $s_2(t)$ equal the constant 55, and so $s_1(t)$ and $s_2(t)$ must be linear functions whose graphs are lines with slope 55. In the next example, we match functions and their antiderivatives by looking at the behavior of their graphs.

EXAMPLE 18.4 Four graphs are shown in Figures 18.6 through 18.9.

a. Identify two pairs of graphs among these in which each function in a pair can be derived from the other by differentiation or antidifferentiation.

b. For each pair found in part a, indicate which function is the antiderivative of the other.

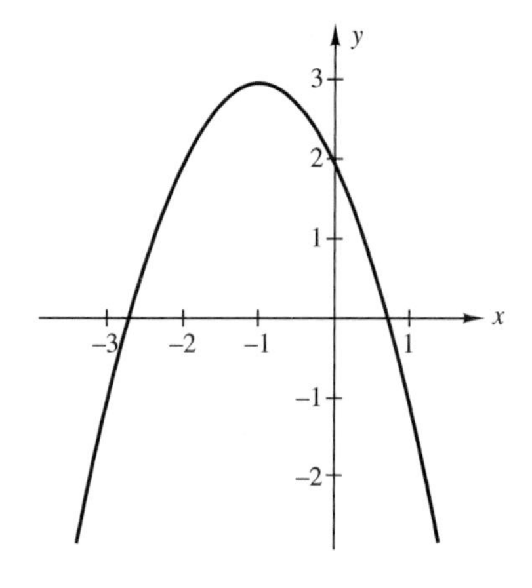

Figure 18.6 $y = f(x)$

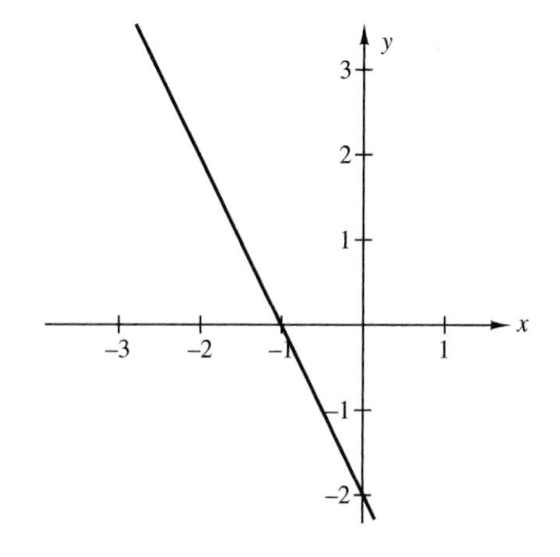

Figure 18.7 $y = k(x)$

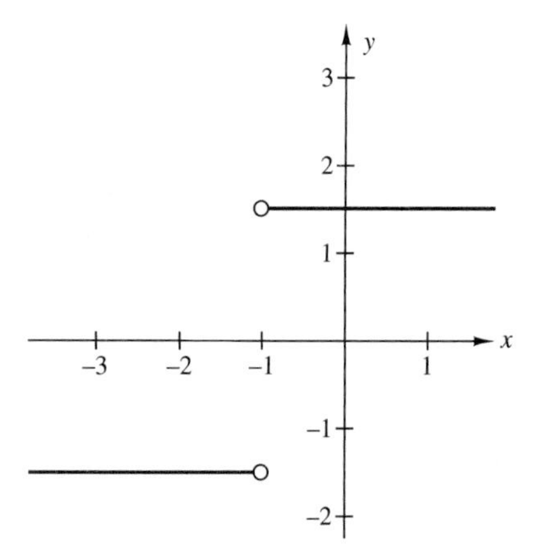

Figure 18.8 $y = g(x)$

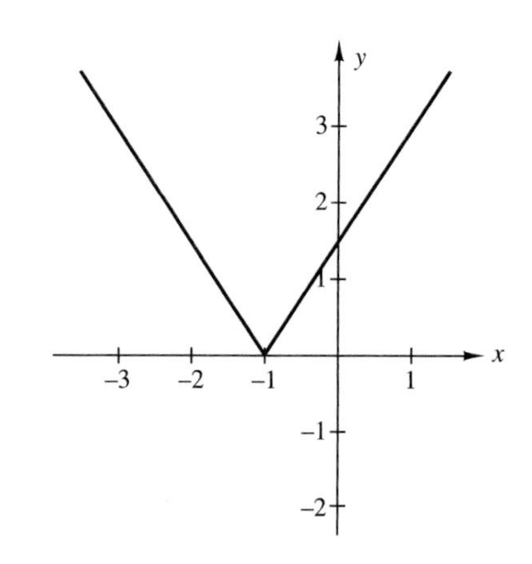

Figure 18.9 $y = h(x)$

Solution

a. The graph of $f(x)$ has a horizontal tangent line at the value $x = -1$; this tells us that $f'(-1) = 0$. Also, the graph of f is increasing in the interval $(-\infty, -1)$ and decreasing in the interval $(-1, \infty)$, so f' is positive for $x < -1$ and f' is negative for $x > -1$. Examining the other three graphs, we see that only the graph of $k(x)$ has all the properties of f'; thus $f'(x) = k(x)$. The remaining two graphs also must be related, so the two pairs of graphs that show a function and its antiderivative are $\{f(x), k(x)\}$ and $\{g(x), h(x)\}$.

b. We determined in part a that $f'(x) = k(x)$, which means that f is an antiderivative of k. If we examine the graphs of g and h, we see that both are piecewise linear and neither has a derivative at $x = -1$. The derivatives of g and h are just the slopes of the lines that make up their graphs. The derivative of g is 0 for all $x \neq -1$. The derivative of h is a constant (approximately -1.5) for all $x < -1$ and is a constant (approximately 1.5) for all $x > -1$. Thus $h' = g$ and so h is an antiderivative of g. ■

In Examples 18.2 and 18.3, we were given a function that was a constant instantaneous rate of change (namely, the distance traveled per unit of time, in miles per hour) and we used antidifferentiation to find a function that described the total distance traveled, in miles, after t hours. In general, when we know a function that may be interpreted as a rate of change with respect to a variable, we antidifferentiate in order to find a function of that variable that gives the total amount of change measured from some initial point.

EXAMPLE 18.5 For each of the following functions, describe in words what its antiderivative represents.

a. $f(t)$ represents the rate of change in temperature at time t of a metal rod (in degrees Celsius per minute).

b. $g(t)$ represents the velocity of an object at time t (in feet per second).

c. $h(t)$ represents the rate of change in employment at time t in an industrial city (in number of jobs per month).

d. $p(x)$ represents the marginal profit (in dollars per case) associated with selling x cases of bottled water.

Solution

a. $\int f(t)\,dt$ represents the temperature (in degrees Celsius) of the metal rod at time t, measured in minutes from an initial time.

b. $\int g(t)\,dt$ represents the displacement of the object at time t (in seconds), with displacement measured in feet relative to a starting point or origin.

c. $\int h(t)\,dt$ represents the employment level (number of jobs) in the city in month t, measured from some initial month.

d. $\int p(x)\,dx$ represents the profit in dollars associated with selling x cases of bottled water. ■

EXAMPLE 18.6 Records show that the number of births per year (in thousands) in Bloom County for the years 1960 through 1990 can be approximated by the function $b(t) = 0.09t^2 - t + 10$, where t is the number of years since 1960. An antiderivative of $b(t)$ is $B(t) = 0.03t^3 - 0.5t^2 + 10t$.

a. Verify that $B(t)$ is an antiderivative of $b(t)$.

b. What does the function $B(t)$ predict?

c. Find the total number of births between 1970 and 1980.

d. Find the function that represents the total population of Bloom County t years after 1960 if the population in 1960 was 30,000. Assume that the only change in population is due to births.

Solution

a. To verify that $B(t)$ is an antiderivative of $b(t)$, we differentiate $B(t)$ and find $B'(t) = 0.09t^2 - t + 10 = b(t)$.

b. The function $B(t)$ represents the total number of births in Bloom County for the t years since 1960.

c. $B(20)$ is the total number of births from 1960 to 1980 (a 20-year period), and $B(10)$ is the total number of births from 1960 to 1970. The difference $B(20) - B(10)$ is the number of births from 1970 to 1980. $B(20) - B(10) = [0.03(20)^3 - 0.5(20)^2 + 10(20)] - [0.03(10)^3 - 0.5(10)^2 + 10(10)] = 240 - 80 = 160$. Thus, the total number of births in Bloom County during this 10-year period was 160,000.

d. The total population $P(t)$ of Bloom County t years after 1960 equals the total population in 1960 plus the total number of births for the t years since 1960. So $P(t) = 30 + B(t) = 0.03t^3 - 0.5t^2 + 10t + 30$. Note that $P(t)$ is another antiderivative of $b(t)$, and $P(t)$ differs from $B(t)$ by the constant 30. ■

In part d of Example 18.5, we were told that the population in 1960 was 30,000, and we used that information to find a function that predicted the total population t years later. The initial population of 30,000 is called an *initial value* or *initial condition*, and this type of problem, where we are given a rate of change function and an initial condition, is called an *initial value problem.*

Exercises 18-A

1. The functions below represent the marginal revenue (in dollars per case) for three different brands of soup when x cases are sold. Find the revenue function for each brand if it is assumed that the revenue is zero when zero cases are sold.

a. $f(x) = 10$ **b**. $g(x) = 0.1x$ **c**. $h(x) = 10 - 0.1x$

2. The population of Metropolis is growing at the constant rate of 5000 persons per year. The current population is 100,000.

a. Find a function for the population of Metropolis t years from now.

b. Use the function found in part a to estimate the population of Metropolis 10 years from now.

c. How long will it take for the population to double?

3. Four graphs of functions are shown in Figures 18.10 through 18.13.

a. Identify two pairs of graphs among these in which each function in a pair can be derived from the other by using differentiation or antidifferentiation.

b. For each pair found in part a, indicate which function is the antiderivative of the other.

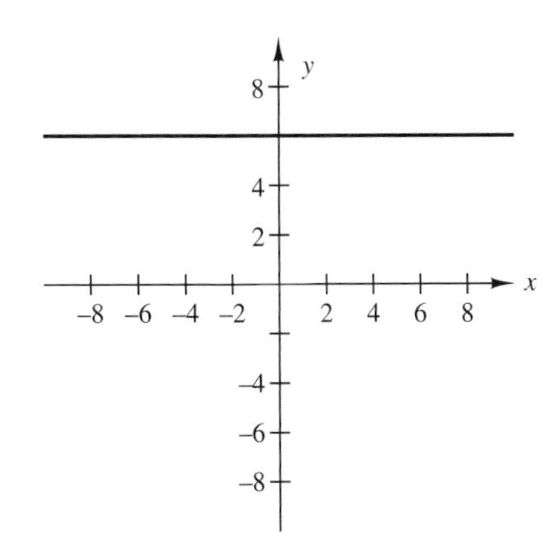

Figure 18.10 $y = f(x)$

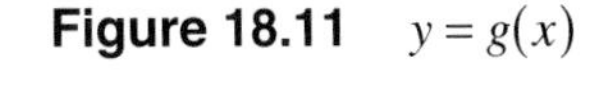

Figure 18.11 $y = g(x)$

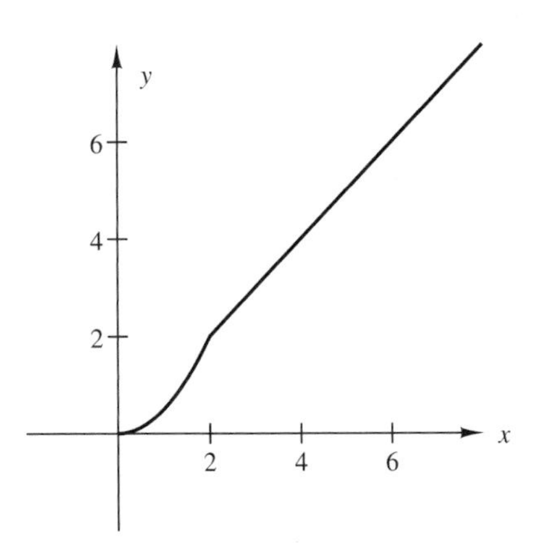

Figure 18.12 $y = h(x)$

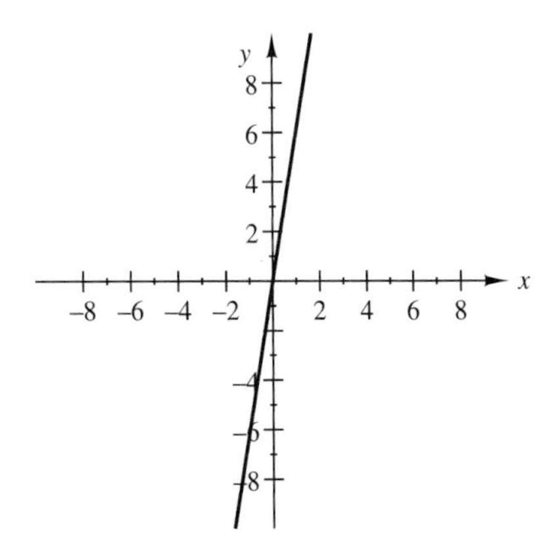

Figure 18.13 $y = k(x)$

4. The growth rate of a bacteria culture (in hundreds of bacteria per hour) for a 24-hour period is approximated by the function $g(t) = 0.5t^2 - 3t + 15$. Let $G(t) = \frac{1}{6}t^3 - \frac{3}{2}t^2 + 15t$.

a. Verify that $G(t)$ is an antiderivative of $g(t)$. What does the function $G(t)$ predict?

b. Find the difference $G(12) - G(2)$. What is the meaning of this difference?

c. Find a function that will predict the total number of bacteria in the culture after t hours if the initial population was 800.

d. Find the total number of bacteria predicted by the function in part c at the end of the 24-hour period.

5. The graph of a function $y = f(x)$ is shown in Figure 18.14 and the graph of one of the antiderivatives of f is shown in Figure 18.15.

a. Explain why the x-intercept of the graph of f has the same value as the x-coordinate of the minimum point on the graph of the antiderivative of f in Figure 18.15.

b. On the same set of axes in Figure 18.15, sketch the graphs of two other antiderivatives of f.

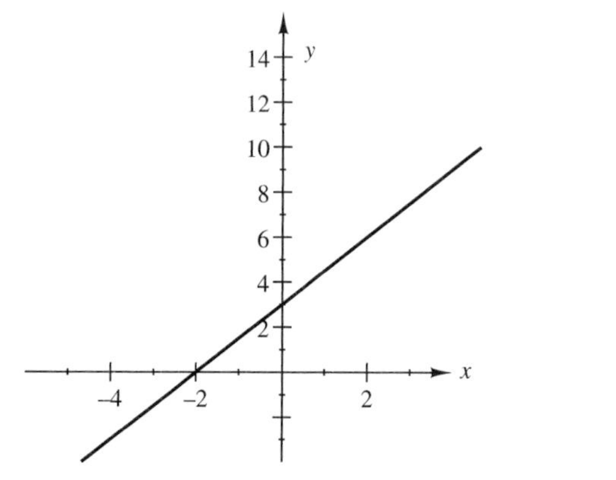

Figure 18.14 $y = f(x)$

Figure 18.15 An antiderivative of f

6. For each of the following functions, describe in words what its antiderivative represents.

a. $f(p)$ represents marginal demand (in razors per dollar) for electric razors sold at price p dollars each.

b. $g(t)$ represents the instantaneous rate of change in the number of bacteria per hour in a certain culture after t hours.

c. $a(t)$ represents the acceleration (in meters per sec^2) of an object at time t.

18-B RECOGNIZING ANTIDERIVATIVES

For some functions, finding an antiderivative involves remembering which function has its derivative equal to the function. For example, the antiderivative of $\cos x$ is $\sin x$, and the antiderivative of $\sec^2 x$ is $\tan x$; we can check this by differentiating $\sin x$ and $\tan x$.

Functions that involve sums, products, quotients, and roots of polynomials can sometimes be rewritten as a sum whose terms are a constant times a power of the variable—that is, terms of the form cx^n. When written in this form, antidifferentiation is straightforward. An antiderivative of cx^n is $c\left(\frac{1}{n+1}x^{n+1}\right) = \frac{cx^{n+1}}{n+1}$, for $n \neq -1$. We can verify this by differentiating $\frac{cx^{n+1}}{n+1}$; its derivative is cx^n.

EXAMPLE 18.7 Rewrite each of the following functions as a sum whose terms are a constant times a power of the variable:

a. $f(x) = \frac{x^2-7x+1}{2x}$ **b.** $g(x) = (x^2+3)(2x-1)$ **c.** $h(x) = \sqrt[3]{x^2} - 2$

d. $k(t) = \left(\frac{-3t^2}{\sqrt{t}}\right)^3$ **e.** $m(t) = \frac{(\sqrt{t}-2)(t+1)}{\sqrt[3]{t}}$

Solution

a. We can express $f(x)$ as a sum of three separate quotients with denominators $2x$; then we write each quotient in the form cx^n: $f(x) = \frac{x^2-7x+1}{2x} = \frac{x^2}{2x} - \frac{7x}{2x} + \frac{1}{2x} = \frac{1}{2}x - \frac{7}{2} + \frac{1}{2}x^{-1}$. Note that $\frac{7}{2} = \frac{7}{2}x^0$ is of the requested form.

b. We multiply the factors of $g(x)$: $g(x) = 2x^3 - x^2 + 6x - 3$.

c. We can write the first term with a rational exponent: $h(x) = \sqrt[3]{x^2} - 2 = x^{2/3} - 2$.

d. We first rewrite the quotient as a product and then use properties of exponents to simplify: $k(t) = \left(\frac{-3t^2}{\sqrt{t}}\right)^3 = \left(-3t^2t^{-1/2}\right)^3 = \left(-3t^{3/2}\right)^3 = -27t^{9/2}$.

e. We first rewrite $m(t)$ with rational exponents and then multiply the factors in the numerator. We write the quotient as a sum of four quotients and, finally, put each of these in the form ct^n.

$$m(t) = \frac{(\sqrt{t}-2)(t+1)}{\sqrt[3]{t}} = \frac{(t^{1/2}-2)(t+1)}{t^{1/3}} = \frac{t^{3/2}-2t+t^{1/2}-2}{t^{1/3}} = \frac{t^{3/2}}{t^{1/3}} - \frac{2t}{t^{1/3}} + \frac{t^{1/2}}{t^{1/3}} - \frac{2}{t^{1/3}}$$

$$= t^{3/2-1/3} - 2t^{1-1/3} + t^{1/2-1/3} - 2t^{-1/3} = t^{7/6} - 2t^{2/3} + t^{1/6} - 2t^{-1/3}$$ ■

Some functions cannot be rewritten as a sum whose terms are a constant times x to a power. For example, $\frac{1}{\sqrt{2x+1}}$ can be written as $(2x+1)^{-1/2}$ but cannot be further simplified. Similarly, $\frac{2x}{3x^2+1}$ can be written as $2x(3x^2+1)^{-1}$ but cannot be further simplified.

EXAMPLE 18.8 Determine which of the following functions can be rewritten as a sum whose terms are of the form cx^n. Rewrite the function in the desired form if possible. If it is not possible, explain why not.

a. $f(x) = \dfrac{1}{\sqrt[3]{x^2+5}}$ **b**. $g(x) = \dfrac{1}{\sqrt[5]{x^2}} + \dfrac{1}{6}$ **c**. $h(x) = \dfrac{5}{(x+3)^2}$ **d**. $k(x) = \left(\sqrt{x} + \dfrac{1}{x}\right)^2$

Solution

a. We can write $f(x)$ as $(x^2 + 5)^{-1/3}$, but we cannot write $f(x)$ as a sum of terms of the form cx^n since the cube root of a sum cannot be split.

b. We can write $g(x)$ as $\dfrac{1}{(x^2)^{1/5}} + \dfrac{1}{6} = x^{-2/5} + \dfrac{1}{6}$. This is the desired form.

c. We cannot write $h(x)$ as a sum of terms in the desired form. Although the denominator of $h(x)$ could be expanded, $h(x) = \dfrac{5}{(x+3)^2} = \dfrac{5}{x^2+6x+9}$, this quotient with a sum of terms in the denominator cannot be further simplified.

d. In order to write $k(x)$ as a sum of terms of the form cx^n, we first need to expand: $k(x) = \left(\sqrt{x} + \dfrac{1}{x}\right)^2 = x + \dfrac{2\sqrt{x}}{x} + \dfrac{1}{x^2}$. We simplify the middle term, $\dfrac{2\sqrt{x}}{x} = \dfrac{2}{\sqrt{x}}$, and then write each term in the desired form: $x + \dfrac{2}{\sqrt{x}} + \dfrac{1}{x^2} = x + 2x^{-1/2} + x^{-2}$. ■

There are several general rules for differentiating products, quotients, and powers of functions, but there are no such general rules for antidifferentiation. To check whether we have a correct antiderivative for a given function, we can differentiate the antiderivative. The following example illustrates this and also illustrates some common errors.

EXAMPLE 18.9 In parts a–e, determine if the function f is an antiderivative of the function g by differentiating $f(x)$ to see if $f'(x) = g(x)$.

a. $f(x) = x^{-2}$; $g(x) = -2x^{-1}$

b. $f(x) = \dfrac{2}{5}x^{5/2}$; $g(x) = x^{3/2}$

c. $f(x) = \pi^2 - 1$; $g(x) = 2\pi$

d. $f(x) = \dfrac{x^2}{2} \cdot \dfrac{2x^{3/2}}{3}$; $g(x) = x\sqrt{x}$

e. $f(x) = \dfrac{3x^2}{x^3+x}$; $g(x) = \dfrac{6x}{3x^2+1}$

Solution

a. We differentiate $f(x) = x^{-2}$ and find $f'(x) = -2x^{-3}$. Thus, $f(x) = x^{-2}$ is not an antiderivative of $g(x) = -2x^{-1}$. (An antiderivative of $g(x) = -2x^{-1}$ is $G(x) = -2 \ln x$.)

b. Since $f'(x) = \dfrac{2}{5} \cdot \dfrac{5}{2} x^{3/2} = x^{3/2}$, $f(x) = \dfrac{2}{5}x^{5/2}$ is an antiderivative of $g(x) = x^{3/2}$.

c. Since $f(x) = \pi^2 - 1$ is a constant, its derivative is 0. Therefore, $f(x) = \pi^2 - 1$ is not an antiderivative of $g(x) = 2\pi$. (An antiderivative of $g(x) = 2\pi$ is $G(x) = 2\pi x$.)

d. We first put $f(x)$ in simplest form and then differentiate: $f(x) = \frac{x^2}{2} \cdot \frac{2x^{3/2}}{3} = \frac{1}{3}x^2 \cdot x^{3/2} = \frac{1}{3}x^{7/2}$. So $f'(x) = \frac{7}{6}x^{5/2}$. Since $g(x) = x\sqrt{x} = x^{3/2}$, $f(x)$ is not an antiderivative of $g(x)$. (An antiderivative of $g(x) = x\sqrt{x}$ is $G(x) = \frac{2}{5}x^{5/2}$.)

e. We use the quotient rule to differentiate $f(x)$:

$$f'(x) = \frac{(x^3+x)(6x)-(3x^2)(3x^2+1)}{(x^3+x)^2} = \frac{6x^4+6x^2-9x^4-3x^2}{(x^3+x)^2} = \frac{-3x^4+3x^2}{(x^3+x)^2}.$$

Since $f'(x) \neq g(x)$, $f(x)$ is not an antiderivative of $g(x)$. (An antiderivative of $g(x) = \frac{6x}{3x^2+1}$ is $G(x) = \ln(3x^2 + 1.)$ ■

Although sums and differences of functions can be antidifferentiated term by term, there is no general rule for finding antiderivatives of functions that are products or quotients of other nonconstant functions.

Exercises 18-B

1. Rewrite each of the functions below as a sum whose terms are a constant times a power of the variable.

a. $f(x) = (2x^2 - 5)(3 - 4x)$

b. $g(x) = \frac{\sqrt{x} - x^3}{x}$

c. $h(z) = \frac{7}{z^2} - \frac{\sqrt{z}}{z} + 3$

d. $s(t) = \sqrt[3]{t^2} + \sqrt{t}$

e. $f(x) = \frac{3x^5 - 2\sqrt[3]{x} + 2}{x^4}$

f. $g(t) = (\sqrt{t} - 3)\left(\frac{1}{t} + \frac{2}{t^3}\right)$

2. For each of the functions below determine if it is possible to rewrite the function as a sum whose terms are of the form cx^n. If it is possible, rewrite it. If it is not possible, explain why not.

a. $f(x) = \sqrt{2x^3 + 8}$ **b.** $g(x) = \frac{1}{x^2 + x^4}$ **c.** $h(x) = \frac{1}{x^2} + \frac{1}{x^4}$

d. $k(x) = \sqrt{x^3 + x}$ **e.** $p(x) = \sqrt{x^3} + \sqrt{x}$

3. Consider the functions listed in parts a–f. For each of these functions, find an antiderivative among the functions listed in I–VIII. Confirm your choices by differentiating the antiderivatives.

a. $f(x) = 2x + 1$

b. $g(x) = \sqrt{2}$

c. $h(x) = 0$

d. $l(x) = \frac{1}{x^2}$

e. $m(x) = \frac{3}{2}\sqrt{x}$

f. $n(x) = \frac{x+1}{x}$

I. $t(x) = \sqrt{2}x + \sqrt{2}$

II. $s(x) = x\sqrt{x}$

III. $r(x) = \ln x + x$

IV. $y(x) = \frac{1}{x^3}$

V. $v(x) = 2 - \frac{1}{x}$

VI. $w(x) = 1000$

VII. $z(x) = x^2 + x + 4$

VIII. $F(x) = \frac{x^2 + x}{x^2}$

18-C SUBSTITUTION FOR ANTIDERIVATIVES

When we find the derivative of a composite function $F(g(x))$, we use the Chain Rule, which states that the derivative of $y = F(g(x))$ is $\frac{dy}{dx} = F'(g(x)) \cdot g'(x)$. For example, the derivative of $h(x) = (3x - 5)^5$ is $h'(x) = 5(3x - 5)^4 \cdot 3$.

We can restate the Chain Rule so it tells us about antiderivatives:

An antiderivative of a product of the form $F'(g(x) \cdot g'(x)$ is $F(g(x))$.

For example, the function $h(x) = 3 \sin^2 x \cos x = 3(\sin x)^2 \cos x$ is of the form $F'(g(x)) \cdot g'(x)$, where $F'(x) = 3x^2$ and $g(x) = \sin x$. The statement above tells us that an antiderivative of $h(x)$ is $F(\sin x)$, where $F(x) = x^3$. We conclude that the function $H(x) = (\sin x)^3$ is an antiderivative of $h(x) = 3 \sin^2 x \cos x$. We can check that this is true by differentiating to see that $H'(x) = h(x)$.

Recognizing functions as a product of the form $F'(g(x)) \cdot g'(x) = f(g(x)) \cdot g'(x)$ is always the first step in finding antiderivatives using this method. In the next examples, we "decompose" several functions to identify functions $f(x)$ and $g(x)$ so that we can recognize the original functions as products of the form $f(g(x)) \cdot g'(x)$. In the decomposition process, we often insert implied parentheses to make the composition of functions more visible.

EXAMPLE 18.10 Each of the functions below can be recognized as a product of the form $f(g(x)) \cdot g'(x)$. For each, identify $f(x)$ and $g(x)$.

a. $y = 2\sqrt{2x-1}$ **b.** $y = e^{-x^2}(-2x)$ **c.** $y = \frac{3}{3x - 4}$

d. $y = \tan^3 x \sec^2 x$ **e.** $y = 2x \sin x^2$ **f.** $y = \sin x \cos x$

Solution

a. The factor $\sqrt{2x-1}$ is a composite $f(g(x))$ with $g(x) = 2x - 1$ and $f(x) = \sqrt{x}$. Also, $g'(x) = 2$. Thus, $f(g(x)) \cdot g'(x) = (\sqrt{2x-1})(2) = 2\sqrt{2x-1}$, which is the original function.

b. The factor e^{-x^2} is a composite $f(g(x))$ with $g(x) = -x^2$ and $f(x) = e^x$. Since $g'(x) = -2x$, we have $f(g(x)) \cdot g'(x) = e^{-x^2}(-2x)$, the original function.

c. We first rewrite $\frac{3}{3x-4}$ as $3(3x - 4)^{-1}$. The factor $(3x - 4)^{-1}$ is a composite $f(g(x))$ with $g(x) = 3x - 4$ and $f(x) = x^{-1}$. Then $g'(x) = 3$ and $f(g(x)) \cdot g'(x) = (3x - 4)^{-1} \cdot 3$, which is the original function.

d. By definition, $\tan^3 x \sec^2 x = (\tan x)^3(\sec x)^2$. Both factors $(\tan x)^3$ and $(\sec x)^2$ are composites, so we have two possibilities for $g(x)$: $\tan x$ and $\sec x$. Since the derivative of $\tan x$ is $\sec^2 x$, $g(x) = \tan x$ is our best choice. Then $(\tan x)^3 = f(g(x))$, where $f(x) = x^3$. Thus, $f(g(x)) \cdot g'(x) = (\tan x)^3 \sec^2 x = \tan^3 x \sec^2 x$, the original function.

e. We first make explicit the implied parentheses: $2x \sin x^2 = 2x \sin(x^2)$. The factor $\sin(x^2)$ is a composite $f(g(x))$ with $g(x) = x^2$ and $f(x) = \sin x$. Since $g'(x) = 2x$, we have $f(g(x)) \cdot g'(x) = (\sin(x^2))(2x) = 2x \sin(x^2)$, the original function.

f. Both factors $\sin x$ and $\cos x$ may be interpreted as composite power functions; $\sin x = (\sin x)^1$ and $\cos x = (\cos x)^1$. If we let $f(g(x)) = (\sin x)^1$, then $g(x) = \sin x$ and $f(x) = x$. Since $g'(x) = \cos x$, $f(g(x)) \cdot g'(x) = \sin x \cos x$, the original function. ■

In parts d and f of Example 18.10, it was not immediately apparent which function to choose as f and which as g in order to recognize the given product as $f(g(x)) \cdot g'(x)$. When this happens, it is helpful to ask: Which factor in the product can serve as a derivative? Once we answer the question and choose that factor as $g'(x)$, we need to see if the remaining factors can be expressed as a composition $f(g(x))$. In part d, we noted that $\sec^2 x$ is the derivative of $\tan x$, so we chose $g'(x) = \sec^2 x$ and $g(x) = \tan x$; then the given product was of the desired form, $f(g(x)) = (g(x))^3 \cdot g'(x)$. In part f, either $\sin x$ or $\cos x$ can be a derivative; we chose $g'(x) = \cos x$ and $g(x) = \sin x$. With this choice, the given product was $(g(x))^1 \cdot g'(x)$. If instead we had chosen $g'(x) = \sin x$, then $g(x) = -\cos x$, and the given product is $f(g(x)) = -(g(x))^1 \cdot g'(x)$; for this choice, the function f is $f(x) = -x^1$.

When the general antiderivative of a function is represented as an indefinite integral $\int f(g(x)) \cdot g'(x)\, dx$, we can make a substitution (a change of variable) by naming the function $g(x)$ as a variable such as u. Since $u = g(x)$ implies that $du = g'(x)\, dx$, the product $f(g(x)) \cdot g'(x)\, dx$ takes the form $f(u)\, du$. The next two examples illustrate how such a change of variable gives a simpler form to the general antiderivative (indefinite integral).

EXAMPLE 18.11 Use the substitution technique to rewrite the general antiderivative for each function in Example 18.10 in the form $\int f(u)\,du$.

a. $\int 2\sqrt{2x-1}\,dx$ **b.** $\int e^{-x^2}(-2x)\,dx$ **c.** $\int \frac{3}{3x-4}\,dx$

d. $\int \tan^3 x \sec^2 x\,dx$ **e.** $\int 2x \sin x^2\,dx$ **f.** $\int \sin x \cos x\,dx$

Solution In each case, we choose $u = g(x)$, where $g(x)$ was identified in the solutions to Example 18.10.

a. We let $u = 2x - 1$. Then $du = 2\,dx$. Since $2\sqrt{2x-1} = (2x-1)^{1/2}(2)$, we can rewrite $\int 2\sqrt{2x-1}\,dx = \int (2x-1)^{1/2} \cdot 2\,dx$ as $\int u^{1/2}\,du$.

b. We let $u = -x^2$. Then $du = -2x\,dx$. We can rewrite $\int e^{-x^2}(-2x)\,dx$ as $\int e^u\,du$.

c. We let $u = 3x - 4$. Then $du = 3\,dx$. We can rewrite $\int \frac{3}{3x-4}\,dx = \int \frac{1}{3x-4} \cdot 3\,dx$ as $\int \frac{1}{u}\,du$.

d. We let $u = \tan x$. Then $du = \sec^2 x\,dx$. We can rewrite $\int \tan^3 x \sec^2 x\,dx = \int (\tan x)^3 \sec^2 x\,dx$ as $\int u^3\,du$.

e. We let $u = x^2$. Then $du = 2x\,dx$. We can rewrite $\int 2x \sin x^2\,dx = \int \left(\sin(x^2)\right) 2x\,dx$ as $\int \sin u\,du$.

f. We let $u = \sin x$. Then $du = \cos x\,dx$. We can rewrite $\int \sin x \cos x\,dx$ as $\int u\,du$.

■

In many instances the composite function to be antidifferentiated does not match the Chain Rule form exactly; that is, we are not given $\int f(g(x)) \cdot g'(x)\,dx$ in that exact form. In these cases, if the expression differs from the desired expression only by a constant factor, we can rewrite the expression as an equivalent expression that does match the Chain Rule. In the next two examples we illustrate how to use a change of variable (substitution) in such cases.

EXAMPLE 18.12 In the problems below, the change of variable $u = g(x)$ is given. For each function u, find $du = g'(x)\,dx$ and rewrite the indefinite integral in the form $\int f(u)\,du$.

a. $\int \sqrt{\cos x} \sin x\,dx$; $u = \cos x$

b. $\int \frac{x}{(x^2-3)^4} dx$; $\quad u = x^2 - 3$

c. $\int \sin(1-\pi x)$; $\quad u = 1 - \pi x$

Solution

a. We differentiate $u = \cos x$ to find $du = -\sin x\, dx$, so $-du = \sin x\, dx$. Substituting u for $\cos x$ and $-du$ for $\sin x\, dx$ in $\int \sqrt{\cos x} \sin x\, dx$, we get $\int \sqrt{u}(-du) = \int -u^{1/2}\, du$.

b. We first rewrite the integral: $\int \frac{x}{(x^2-3)^4} dx = \int (x^2-3)^{-4}\, x\, dx$. We differentiate $u = x^2 - 3$ and get $du = 2x\, dx$, which implies $\frac{1}{2} du = x\, dx$. We substitute u for $x^2 - 3$ and $\frac{1}{2} du$ for $x\, dx$ in the integral and obtain $\int u^{-4} \frac{1}{2} du = \frac{1}{2} \int u^{-4}\, du$.

c. The differential of $u = 1 - \pi x$ is $du = -\pi\, dx$, so $-\frac{1}{\pi} du = dx$. We substitute in the given integral to get $\int (\sin u)\left(-\frac{1}{\pi}\right) du = -\frac{1}{\pi} \int \sin u\, du$. ■

Note that in parts b and c of Example 18.12, we wrote the final form of the indefinite integral so that the constant term was in front of the integral sign. We can do this because of the rule of differentiation, which says that for any constant k, the derivative of $kF(x)$ equals $kF'(x)$.

EXAMPLE 18.13 In each of the following, identify a change of variable u, and rewrite the integral in the form of a constant times $\int f(u)\, du$ by using substitution.

a. $\int \sqrt{2x-1}\, dx$ $\quad$ b. $\int 3e^{-x^2} x\, dx$ $\quad$ c. $\int \sqrt{x^3 - 3x}\left(x^2 - 1\right) dx$

Solution

a. Since $\sqrt{2x-1} = (2x-1)^{1/2}$, we let $u = 2x - 1$. Then $du = 2\, dx$. Solving for dx, we get $dx = \frac{1}{2} du$. Substituting in the original integral, we get $\int \sqrt{2x-1}\, dx = \int \sqrt{u} \cdot \frac{1}{2} du = \frac{1}{2} \int u^{1/2}\, du$.

b. The function e^{-x^2} is a composite function $\exp(-x^2)$, so we let $u = -x^2$. Then $du = -2x\, dx$, so $x\, dx = -\frac{1}{2} du$. We substitute to get $\int 3e^{-x^2} x\, dx = \int 3e^u \left(-\frac{1}{2}\, du\right) = -\frac{3}{2} \int e^u\, du$.

c. The function $\sqrt{x^3 - 3x}$ is composite, so we let $u = x^3 - 3x$. Then $du = (3x^2 - 3)\, dx = 3(x^2 - 1)\, dx$. We solve for $(x^2 - 1)\, dx$ and get $(x^2 - 1)\, dx = \frac{1}{3}\, du$. Finally, we substitute and obtain $\int \sqrt{x^3 - 3x}(x^2 - 1)\, dx = \int \sqrt{u} \frac{1}{3} du = \frac{1}{3} \int \sqrt{u}\, du$.

■

Be careful when using substitution. Every expression in the original variable must be replaced by an equivalent expression in the new variable, and only constants may be factored outside the integral. After substitution, every term in the integral must be expressed in the new variable.

EXAMPLE 18.14 In a–c below, determine whether or not the indefinite integral can be rewritten as a constant times $\int f(u)\ du$ for some substitution function u and some function f.

a. $\int \sqrt{x^2 - 5x + 2}\, dx$ **b.** $\int 3x\sqrt{x^2 + 1}\, dx$ **c.** $\int e^{-x^2}\, dx$

Solution

a. If we let $u = x^2 - 5x + 2$, then $du = (2x - 5)\ dx$. Since the integral $\int \sqrt{x^2 - 5x + 2}\, dx$ does not contain a factor that is a multiple of $2x - 5$, we cannot use this substitution to put the integral in the form $k\int f(u)\, du$. In this case, there is no other reasonable choice for u.

b. We let $u = x^2 + 1$. Then $du = 2x\ dx$ and so $x\ dx = \frac{1}{2}\ du$. Since $\int 3x\sqrt{x^2 + 1}\, dx = \int \sqrt{x^2 + 1}(3x)\, dx$ differs from the form $\int u^{1/2}\ du = \int \sqrt{x^2 + 1}\ (2x)\ dx$ only in the constant factor, we can use substitution to rewrite $\int \sqrt{x^2 + 1}(3)(x\ dx) = \int \sqrt{u} \cdot 3 \cdot \frac{1}{2}\ du = \frac{3}{2}\int \sqrt{u}\, du$. ■

c. The given function is of the form e^u, where $u = -x^2$. If we differentiate u, we get $du = -2x\ dx$, so $-\frac{1}{2x}\ du = dx$. If we substitute for each term in the given integral, we then have an integral $\int e^u\left(-\frac{1}{2x}\right) du$, which cannot be integrated, since it contains functions in two variables, u and x. Because $-\frac{1}{2x}$ is not a constant, it cannot be factored outside the integral. So $\int e^{-x^2} dx$ cannot be rewritten in the form $k\int f(u)\ du$ with $u = -x^2$. There is no other reasonable choice for u. ■

Exercises 18-C

1. In a–f, each function can be recognized as a product $f(g(x)) \cdot g'(x)$. For each of the functions, identify $g(x)$ and $f(x)$.

a. $F(x) = \frac{1}{7}(7 - 4x)^6(-4)$ **b.** $F(x) = e^{0.5x}(0.5)$

c. $h(x) = -(4x-9)^{-2}(4)$ d. $h(x) = e^{x^2}(2x)$

e. $k(x) = 3x^2 \sec^2 x^3$ f. $s(x) = -\sin x \cos^4 x$

2. Use substitution to rewrite $\int f(g(x)) \cdot g'(x)\, dx$ for each of the functions in Exercise 1 in the form $\int f(u)\, du$, where u is the function $g(x)$ you found.

3. Rewrite each of the following indefinite integrals in the form $k\int f(u)\, du$ for the given u and an appropriate constant k.

a. $\int (4x-9)^5\, dx$; $u = 4x - 9$

b. $\int \frac{3}{(8-2x)^3}\, dx$; $u = 8 - 2x$

c. $\int x^2 \sqrt[3]{x^3 - 4}\, dx$; $u = x^3 - 4$

d. $\int \frac{\sin \sqrt{x}}{\sqrt{x}}\, dx$; $u = \sqrt{x}$

4. Rewrite each of the following indefinite integrals in the form $k\int f(u)\, du$, for an appropriate constant k. In each case, state what function the variable u represents.

a. $\int (3x-5)^3\, dx$ b. $\int \sqrt{5-4x}\, dx$ c. $\int e^{5t}\, dt$

d. $\int e^{1-0.5t}\, dt$ e. $\int \cot^3 x \csc^2 x\, dx$

5. For each of the indefinite integrals below, determine whether or not it can be rewritten as a constant times $\int f(u)\, du$ for some substitution function u and some function f.

a. $\int \frac{x}{\sqrt{3x^2+4}}\, dx$ b. $\int \frac{3}{(x^2+5)^3}\, dx$

c. $\int \sqrt[3]{4x^3+1}\, x^2\, dx$ d. $\int x e^{x^3}\, dx$

Chapter 18 Exercises

1. Four graphs of functions are given in Figures 18.16 through 18.19.

a. Identify two pairs of graphs among these in which each function in a pair can be derived from the other by differentiation or antidifferentiation.

b. For each pair found in part a, indicate which function is the antiderivative of the other.

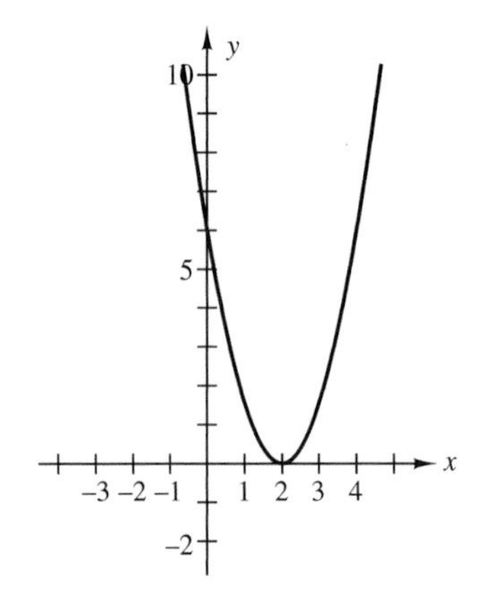

Figure 18.16 $y = f(x)$

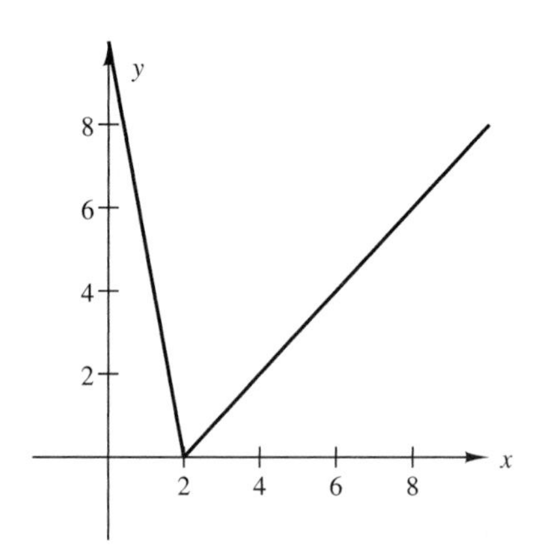

Figure 18.17 $y = g(x)$

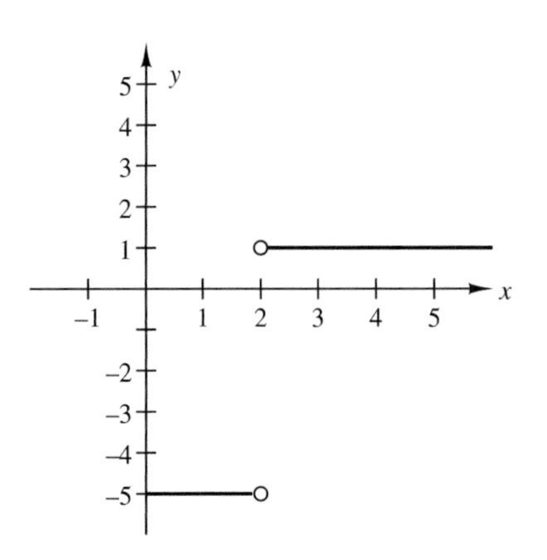

Figure 18.18 $y = h(x)$

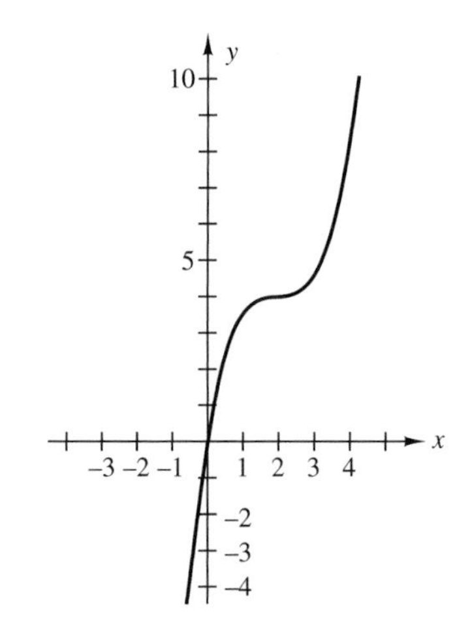

Figure 18.19 $y = k(x)$

2. The population of Metropolis is growing at the constant rate of 2000 people per year. The current population is 50,000.

a. Find a function for the population of Metropolis t years from now.

b. Use this function to estimate the population of Metropolis 10 years from now.

3. The rate at which a bacteria culture is growing at time t (in hundreds of bacteria per minute) is given by $f(t) = \frac{1}{2\sqrt{t+1}}$ for $t \geq 0$. Let $F(t) = \sqrt{t+1}$.

a. Verify that $F(t)$ is an antiderivative of $f(t)$.

b. What does the function $F(t)$ predict?

c. Find the function that predicts the number of bacteria at time t, if the initial population was 10,000.

4. Consider the functions listed in parts a–f. For each function, find an antiderivative among the functions listed in I–IX.

a. $f(x) = 4x - 2$

b. $g(x) = 2^3$

c. $h(x) = 0$

d. $l(x) = \frac{1}{\sqrt{x}}$

e. $m(x) = \frac{3}{2}\sqrt[3]{x}$

f. $n(x) = \frac{2x+4}{2x}$

I. $t(x) = \frac{9}{8}x^{4/3} - 2$

II. $s(x) = \frac{3}{2}x^2 - 2x$

III. $r(x) = x + 2 \ln x$

IV. $y(x) = 8x + 4$

V. $v(x) = \frac{2x}{\sqrt{x}}$

VI. $w(x) = \frac{1}{x^{3/2}}$

VII. $m(x) = \frac{x^2 + 4x}{x^2}$

VIII. $z(x) = 2x^2 - 2x - 4$

IX. $q(x) = \frac{2^4}{4}$

5. Rewrite each of the functions in a–d as a sum of terms that are a constant times a power of the variable.

a. $f(x) = (-3x^2 - 5)(3 - x)$

b. $g(x) = \frac{\sqrt{x} - 2x^3}{x^2}$

c. $h(z) = -\frac{4}{z^2} - \frac{\sqrt{z}}{z} + 4$

d. $s(t) = \sqrt[3]{t} + \sqrt[4]{t}$

6. Rewrite each of the functions in a–d as a sum whose terms are of the form cx^n, if possible. If it is not possible, explain why.

a. $f(x) = \frac{3x^5 - \sqrt[3]{x} + 3}{x^3}$

b. $g(x) = \left(\sqrt{x} + 3\right)\left(\frac{1}{x} - \frac{2}{x^2}\right)$

c. $h(x) = \sqrt[5]{3x^2 - x + 3}$

d. $k(x) = \frac{x}{x^2 + 1}$

7. Use substitution to rewrite each of the following indefinite integrals in the form $\int f(u)\,du$. Identify the function that u represents.

a. $\int \frac{1}{5}(7 - 2x)^5(-2)\,dx$

b. $\int e^{0.7x}(0.7)\,dx$

c. $\int -3(3x-8)^{-3}\,dx$

d. $\int 4xe^{2x^2}\,dx$

e. $\int -3t^2 \csc^2 t^3\,dt$

f. $\int \sin^3 t \cos t\,dt$

8. Use substitution to rewrite each of the following indefinite integrals in the form $k\int f(u)\,du$ for an appropriate constant k. Identify the function that u represents.

a. $\int (2x-5)^2\,dx$

b. $\int \sqrt{5-3x}\,dx$

c. $\int e^{2t}\,dt$

d. $\int e^{-0.4t}\,dt$

e. $\int \frac{\sec^2 x}{\tan^3 x}\,dx$

f. $\int x^3 \cos(3x^4+1)\,dx$

CHAPTER 19

Companion to Area and Riemann Sums

Integral calculus developed as a way to find the areas of regions with irregular shapes. The approach taken is similar to the method the ancient Greeks used to approximate the area of a circle. Figures 19.1 through 19.3 illustrate how the Greeks used inscribed regular polygons to approximate the area of a circle. To find the area of the polygons, they divided each polygon into congruent triangles and then used a formula to find the area of each triangle. The sum of the areas of the triangles is the area of the polygon. By increasing the number of sides of the inscribed polygons, they could improve the estimate of the area of the circle. We will use a similar approach to approximate the areas of regions with irregular shapes.

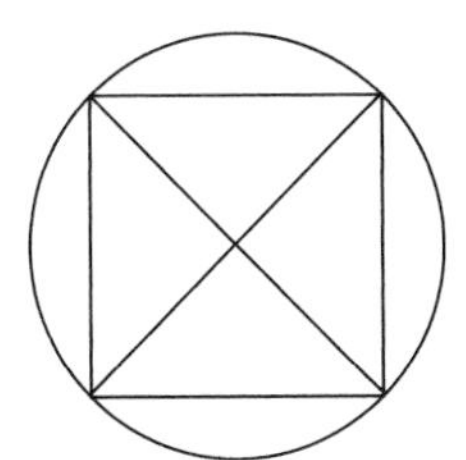

Figure 19.1

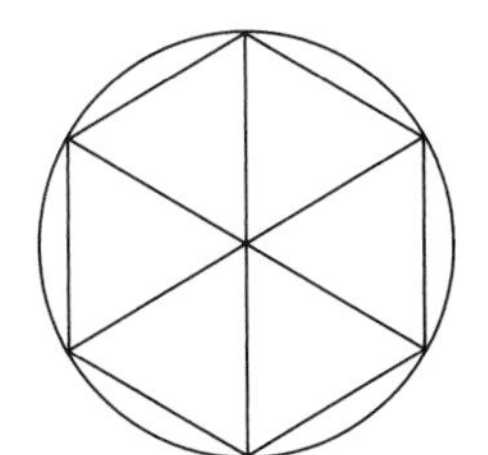

Figure 19.2

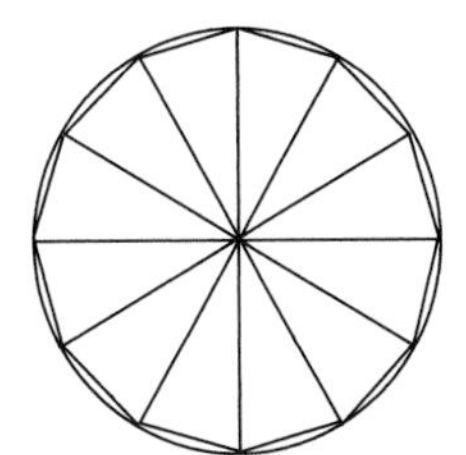

Figure 19.3

19-A EXACT AREAS AS SUMS OF BASIC GEOMETRIC SHAPES

Often the area of a region can be found by partitioning the region into basic shapes whose areas are easily found. The area of the whole region is then the sum of the areas of its parts. In each of the next two examples we find the area of a region by partitioning it into simple regions such as rectangles, triangles, circles, or trapezoids, whose areas we know how to find. (See Section 6-A.)

EXAMPLE 19.1 The diagram in Figure 19.4 shows the layout for the first floor of a new art museum. Find its total area.

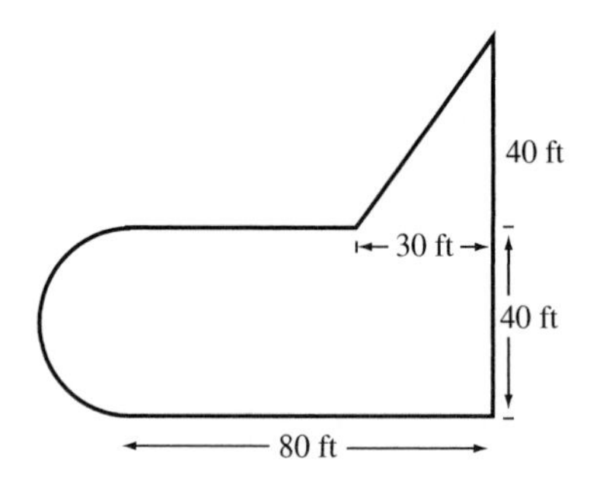

Figure 19.4

Solution We can break the figure into three regions—a semicircle, a rectangle, and a triangle—as shown in Figure 19.5. The total area is the sum of the areas of these regions. The area of the semicircle is $\frac{1}{2}\pi r^2 = \frac{1}{2}\pi(20)^2 = 200\pi$ ft^2. The area of the rectangle is $(40)(80) = 3200$ ft^2. The area of the triangle is $\frac{1}{2}(30)(40) = 600$ ft^2. Thus, the total area is $200\pi + 3200 + 600$, which is approximately 4428 ft^2.

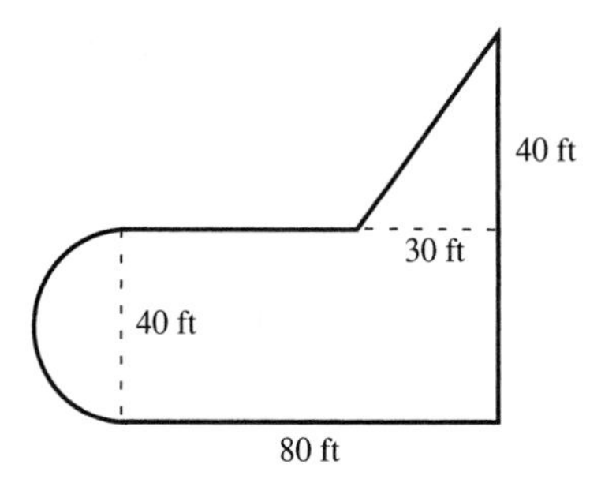

Figure 19.5 ■

EXAMPLE 19.2 A paper airplane has wings as shown in Figure 19.6. Find the area of the wings.

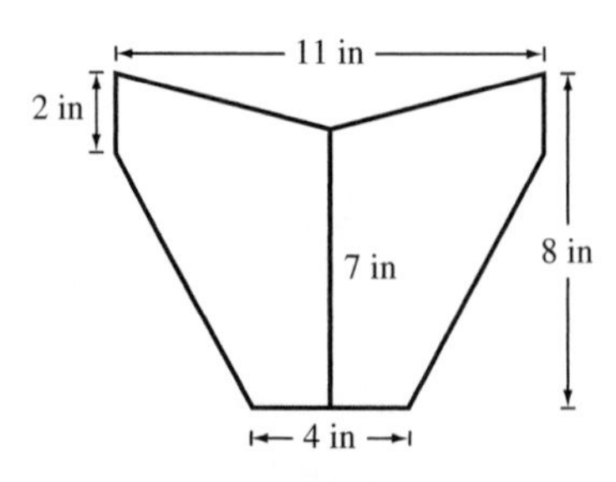

Figure 19.6

Solution We can partition the region that represents the two wings into three trapezoids:

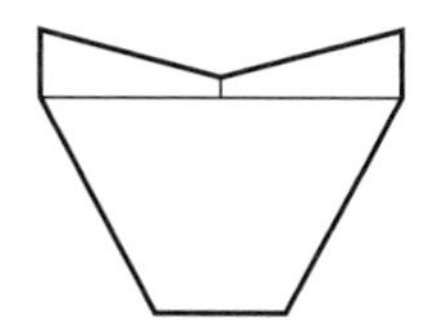

The larger trapezoid has horizontal bases, and the two smaller trapezoids of equal area have vertical bases. The bases of the larger trapezoid have lengths 4 in. and 11 in. Its height, the perpendicular distance between the two bases, is $8 - 2 = 6$ in. The area of the larger trapezoid is $\frac{1}{2}(b_1 + b_2)h = \frac{1}{2}(4 + 11)6 = 45$ in.^{2}. The bases of the smaller trapezoids have lengths 2 in. and $7 - 6 = 1$ in. and height $\frac{11}{2} = 5.5$ in. So the area of each smaller trapezoid is $\frac{1}{2}(2 + 1)(5.5)$ in.^{2}. The total area of the wings is $45 + 2\left(\frac{1}{2}(2+1)(5.5)\right) = 45 + 16.5 = 61.5$ in.^{2}. ■

> The bases of a trapezoid are always the parallel sides, and its height is the perpendicular distance between the two bases. A trapezoid may be oriented so that its bases are not horizontal; in most calculus applications, the bases appear in vertical position.

Area can represent a quantity other than a measure of a geometric shape. In fact, any quantity q that is the product of two positive numbers a and b ($q = ab$) can be represented by the area of a rectangle with sides of length a and b.

For example, distance = rate × time ($d = rt$). If the height of a rectangle represents the rate r and the width represents time t, then the area of the rectangle represents distance d. Similarly, if a population has a constant birthrate, then the number of births = birthrate × time ($B = rt$). Therefore, if the height of a rectangle represents the birthrate r and the width represents time t, the area of the rectangle represents the number of births B during that time period.

Example 19.3 illustrates another interpretation of area.

EXAMPLE 19.3 The graph in Figure 19.7 shows a company's average daily sales rate for each month (in thousands of dollars per day) for 2004. What is the total sales for the company for that year?

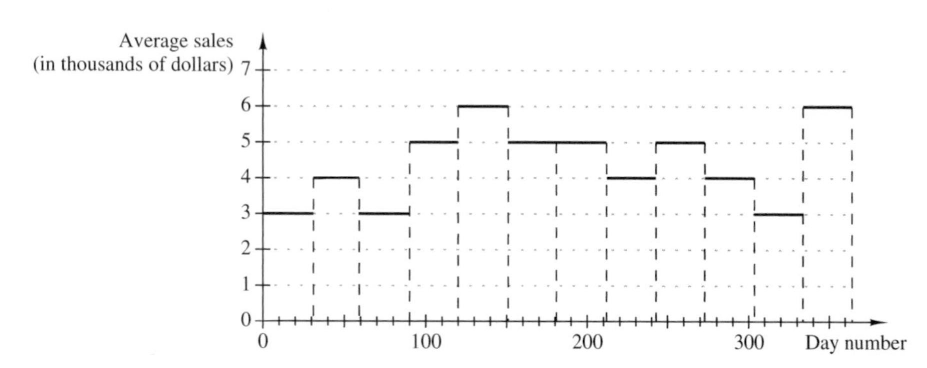

Figure 19.7 Average daily sales rate

Solution The graph in Figure 19.7 is a step function that defines 12 rectangles, one for each month of the year. The height of each rectangle is the average sales per day for a given month (in thousands of dollars), and the width of each rectangle represents the number of days in that month. Thus, for January the height of 3 represents average sales of \$3,000 per day. Since there are 31 days in January, the total sales in January were (\$3,000/day)(31 days) = \$93,000. Thus, the area of this rectangle (height times width) represents the total sales for the month of January. The total area under the graph (the sum of the areas of the 12 rectangles) represents the total sales for the year. The total sales is the sum (3)(31) + (4)(28) + (3)(31) + (5)(30) + (6)(31) + (5)(30) + (5)(31) + (4)(31) + (5)(30) + (4)(31) + (3)(30) + (6)(31) = 1613 thousands of dollars, or \$1,613,000 in total sales for the year. ■

Sigma Notation for Sums

In Example 19.3, if we let A_i represent the area of the ith rectangle (so A_i represents total sales in month i), then the total sales for the year is the sum of the 12 monthly sales totals: $A_1 + A_2 + \cdots + A_{12}$. This sum can be represented as $\sum_{i=1}^{12} A_i$. In this notation, called *sigma notation,* the Greek letter sigma ($\sum$) is an instruction and the letter i acts as a counter. The instruction is to sum the terms A_i starting with $i = 1$ and continue to sum the terms A_i, where i increases in steps of 1 until $i = 12$. If we let $S(t_i)$ represent the height of the ith rectangle (t_i is a day in month i) and Δt_i represent the number of days in month i (the width of each rectangle), then $S(t_i) \cdot \Delta t_i$ represents A_i, the total sales in month i. Thus, the sum $\sum_{i=1}^{12} S(t_i) \cdot \Delta t_i$ represents the total sales over the 12 months.

We can also represent more general sums using sigma notation. We let $f(i)$ represent a function of the counter i and let m and n be integers with $m \le n$. Then

$$\sum_{i=m}^{n} f(i) = f(m) + f(m+1) + f(m+2) + \cdots + f(n-1) + f(n).$$

Here we start the sum with $i = m$ and continue to sum the terms $f(i)$, where i increases in steps of 1 until $i = n$.

Examples 19.4 and 19.5 illustrate evaluation and representation of sums with sigma notation.

EXAMPLE 19.4 Expand and evaluate the following sums:

a. $\sum_{i=1}^{7} (i^2 + 1)\frac{1}{6}$ **b.** $\sum_{i=0}^{5} (2i + 3)\frac{2}{11}$

Solution

a. We start with the counter $i = 1$ and evaluate $f(1) = (1^2 + 1)\frac{1}{6} = 2\left(\frac{1}{6}\right)$, then $f(2) = (2^2 + 1)\frac{1}{6} = 5\left(\frac{1}{6}\right)$, and continue to evaluate $f(i)$, each time increasing i by 1 until $i = 7$. We add these values together to get

$$\sum_{i=1}^{7}(i^2+1)\frac{1}{6} = 2\left(\frac{1}{6}\right)+5\left(\frac{1}{6}\right)+10\left(\frac{1}{6}\right)+17\left(\frac{1}{6}\right)+26\left(\frac{1}{6}\right)+37\left(\frac{1}{6}\right)+50\left(\frac{1}{6}\right)$$

$$= \frac{2+5+10+17+26+37+50}{6} = \frac{147}{6} = 24.5.$$

b. We start with $i = 0$ and evaluate $f(0) = (2(0) + 3)\frac{2}{11} = 3\left(\frac{2}{11}\right)$; then we evaluate $f(1), f(2), f(3), f(4)$, and $f(5)$, and add them together:

$$\sum_{i=0}^{5}(2i+3)\frac{2}{11} = 3\left(\frac{2}{11}\right)+5\left(\frac{2}{11}\right)+7\left(\frac{2}{11}\right)+9\left(\frac{2}{11}\right)+11\left(\frac{2}{11}\right)+13\left(\frac{2}{11}\right)$$

$$= \frac{6+10+14+18+22+26}{11} = \frac{96}{11} \approx 8.73. \quad \blacksquare$$

To evaluate the sums in Example 19.4, we could have factored out the common factor in each term of the sum. For example, in part b, $\frac{2}{11}$ is a common factor of each term, so $\sum_{i=0}^{5}(2i+3)\frac{2}{11} = \frac{2}{11}\sum_{i=0}^{5}(2i+3) = \frac{2}{11}(3 + 5 + 7 + 9 + 11 + 13) = \frac{2}{11}(48) = \frac{96}{11}$.

EXAMPLE 19.5 Write each sum using sigma notation:

a. $(3)\left(\frac{1}{4}\right) + (3 \cdot 4)\left(\frac{1}{4}\right) + (3 \cdot 9)\left(\frac{1}{4}\right) + (3 \cdot 16)\left(\frac{1}{4}\right) + (3 \cdot 25)\left(\frac{1}{4}\right)$

b. $1 \cdot \frac{1}{3} + 2 \cdot \frac{1}{3} + 2^2 \cdot \frac{1}{3} + 2^3 \cdot \frac{1}{3} + 2^4 \cdot \frac{1}{3} + 2^5 \cdot \frac{1}{3}$

Solution

a. We look for a pattern in the terms and see that each term has a factor of 3 and a factor of $\frac{1}{4}$. All but the first term also have a factor that is a perfect square. Since we can write the first term as $(3 \cdot 1)\left(\frac{1}{4}\right)$, we see that every term has the same pattern of factors:

$$(3 \cdot 1)\left(\frac{1}{4}\right) + (3 \cdot 4)\left(\frac{1}{4}\right) + (3 \cdot 9)\left(\frac{1}{4}\right) + (3 \cdot 16)\left(\frac{1}{4}\right) + (3 \cdot 25)\left(\frac{1}{4}\right)$$

$$= (3 \cdot 1^2)\left(\frac{1}{4}\right) + (3 \cdot 2^2)\left(\frac{1}{4}\right) + (3 \cdot 3^2)\left(\frac{1}{4}\right) + (3 \cdot 4^2)\left(\frac{1}{4}\right) + (3 \cdot 5^2)\left(\frac{1}{4}\right)$$

$$= \sum_{i=1}^{5} 3i^2\left(\frac{1}{4}\right) = \sum_{i=1}^{5}\frac{3}{4}i^2 = \frac{3}{4}\sum_{i=1}^{5} i^2.$$

b. Every term in the sum $1 \cdot \frac{1}{3} + 2 \cdot \frac{1}{3} + 2^2 \cdot \frac{1}{3} + 2^3 \cdot \frac{1}{3} + 2^4 \cdot \frac{1}{3} + 2^5 \cdot \frac{1}{3}$, except the first term, has a factor of the form 2^i. If we write $1 = 2^0$, we see that every term in the sum is of the form $2^i \cdot \frac{1}{3}$. Thus the sum equals $2^0 \cdot \frac{1}{3} + 2^1 \cdot \frac{1}{3} + 2^2 \cdot \frac{1}{3} + 2^3 \cdot \frac{1}{3} + 2^4 \cdot \frac{1}{3} + 2^5 \cdot \frac{1}{3} = \sum_{i=0}^{5} 2^i\left(\frac{1}{3}\right) = \frac{1}{3}\sum_{i=0}^{5} 2^i$. ■

In addition to finding areas that are sums of rectangles, we can use geometric formulas to find areas of some regions of the plane that are bounded by lines and portions of circles.

EXAMPLE 19.6 Use geometric formulas to find the area of the region bounded by the curve $y = \sqrt{9 - x^2}$ and the line $y = -x + 3$.

Solution The curve $y = \sqrt{9 - x^2}$ is the upper half of the circle with equation $x^2 + y^2 = 9$; the line $y = -x + 3$ intersects the semicircle at the points (0, 3) and (3, 0). The region bounded by the semicircle and the line is shaded in Figure 19.8.

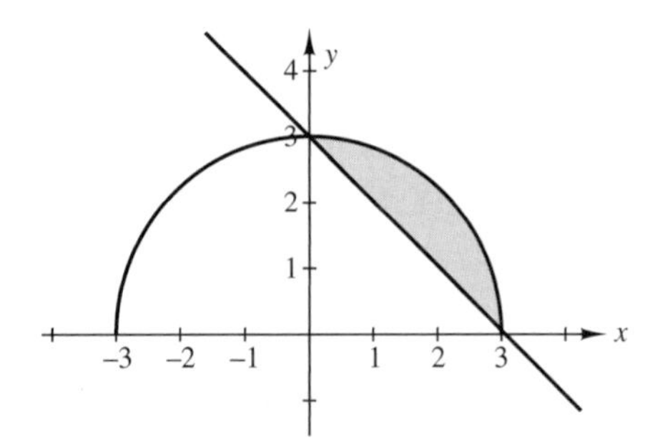

Figure 19.8

To evaluate the area we first find the area of the quarter-circle that lies in the first quadrant and then subtract the area of the triangle formed by the given line and the two coordinate axes. Since the area of a circle is given by πr^2 and $r = 3$, the area of the quarter-circle is $\frac{1}{4}\pi(3)^2 = \frac{9}{4}\pi$. The area of the triangle is $\frac{1}{2}bh = \frac{1}{2}(3)(3) = \frac{9}{2}$. We subtract these two results to get the area of the shaded region: $\frac{9}{4}\pi - \frac{9}{2} \approx 2.57$ square units. ■

To find the area between two curves, it is often necessary to find their points of intersection. This is usually done by solving simultaneously the pair of equations that define the curves. Some techniques for doing this are given in Appendix D.

Exercises 19-A

1. Find the area of each figure. All arcs are semicircles.

a.

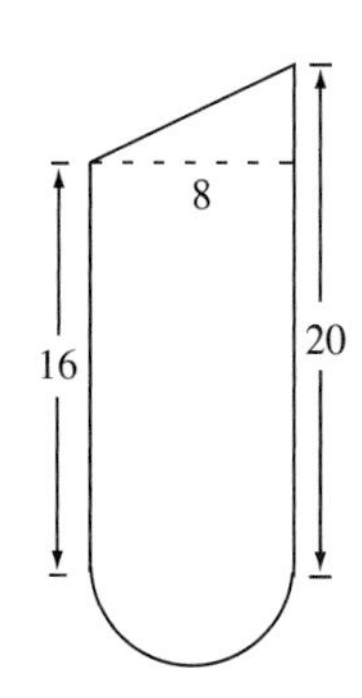

b.

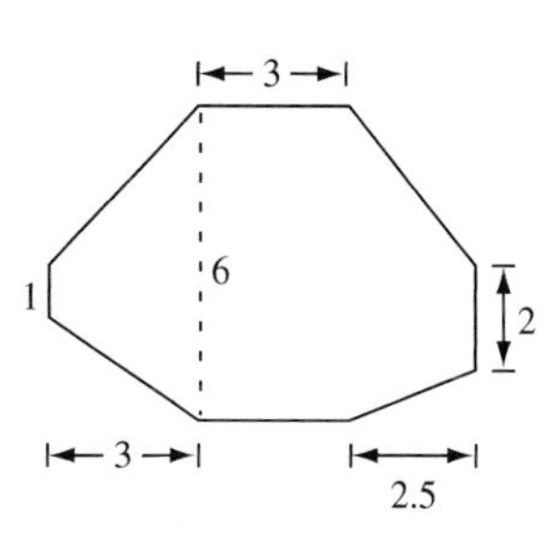

c.

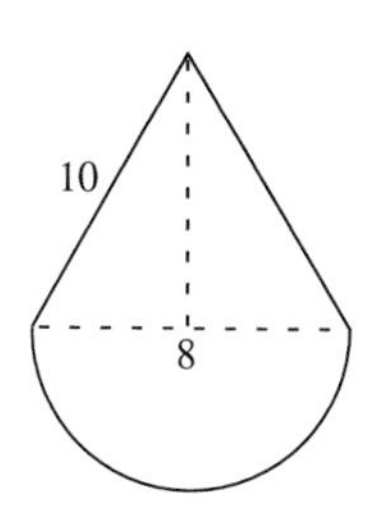

d.

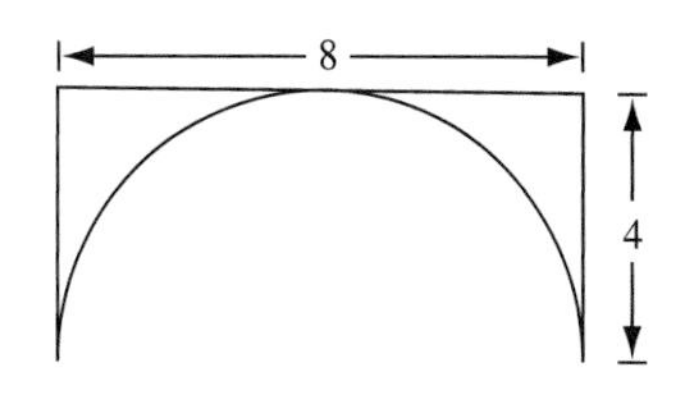

2. The following graph shows the daily average birthrate of gerbils in a pet shop for each month during 2004. Use the graph to find the total number of gerbils born in that pet shop in 2004.

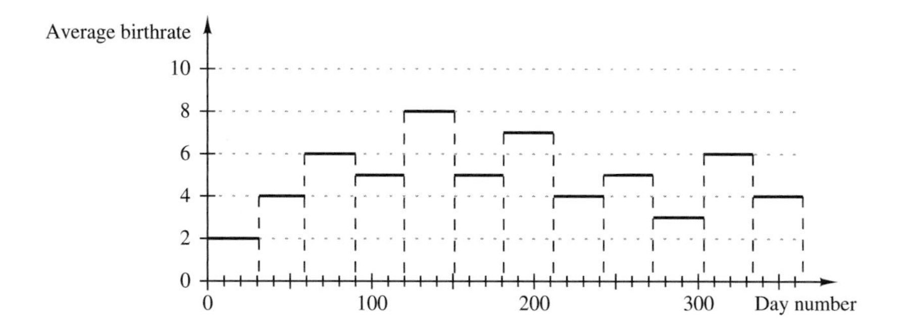

3. Find the area of the region bounded by the lines $y = x + 1$, $x = 1$, $x = 3$, and the x-axis. First sketch the region.

4. For each of the following lines, find the area of the region bounded by the curve $y = -\sqrt{16 - x^2}$ and the line.

a. the x-axis

b. $y = x - 4$

5. Expand and evaluate the following sums:

a. $\sum_{i=1}^{4}\left(\frac{1}{i}+3\right)\left(\frac{1}{2}\right)$ **b.** $\sum_{i=0}^{5}(3i-4)\left(\frac{2}{5}\right)$

6. Write each sum using sigma notation:

a. $(2)\left(\frac{3}{8}\right)+(3)\left(\frac{3}{8}\right)+(4)\left(\frac{3}{8}\right)+(5)\left(\frac{3}{8}\right)+(6)\left(\frac{3}{8}\right)+(7)\left(\frac{3}{8}\right)+(8)\left(\frac{3}{8}\right)$

b. $(2)\left(\frac{1}{4}\right)+(1)\left(\frac{1}{4}\right)+\left(\frac{2}{3}\right)\left(\frac{1}{4}\right)+\left(\frac{2}{4}\right)\left(\frac{1}{4}\right)+\left(\frac{2}{5}\right)\left(\frac{1}{4}\right)+\left(\frac{2}{6}\right)\left(\frac{1}{4}\right)$

19-B APPROXIMATIONS OF AREAS

When it is not possible to find an exact value for a desired area, we use approximation techniques to estimate the area. In this section we investigate some common approximation techniques from a geometric perspective.

EXAMPLE 19.7 Assume the area of each square in the grid in Figure 19.9 is 1 square unit.

a. Estimate the area of the region between the two curves by counting the number of squares entirely **inside** the region.

b. Estimate the area of the region between the two curves by counting all squares entirely inside the region plus those squares that are partially inside the region.

c. Find the average of the estimates of the area in parts a and b.

d. Estimate the area of the region between the two curves by estimating the fraction of the area of each square that is inside the region. Use one of the fractions 0, $\frac{1}{4}$, $\frac{1}{2}$, $\frac{3}{4}$, 1 for each estimate.

e. Compare the estimates in parts c and d.

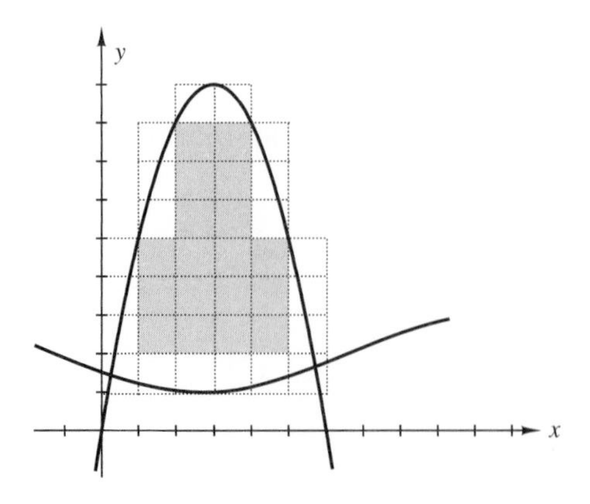

Figure 19.9

Solution

a. We count the shaded squares since these lie entirely inside the region. There are 18 shaded squares; thus, a lower estimate of the area is 18 square units.

b. An upper estimate of the area is 38 square units, since there are 38 squares entirely or partially inside the region.

c. The average of the lower and upper estimates is $\frac{18+38}{2}=\frac{56}{2}=28$ square units.

d. We count fractional estimates of squares partially inside the region, starting at the left of the region with the square where the two curves intersect and traveling clockwise around the figure; our fractional estimates are:

$$\frac{1}{4}, \frac{1}{2}, \frac{1}{2}, \frac{1}{4}, \frac{3}{4}, \frac{1}{2}, \frac{1}{4}, \frac{3}{4}, \frac{3}{4}, \frac{1}{4}, \frac{1}{2}, \frac{3}{4}, \frac{1}{4}, \frac{1}{2}, \frac{1}{2}, \frac{1}{4}, \frac{3}{4}, 1, 1, \frac{3}{4}.$$

The estimate of the area of the region is the sum of these fractional areas plus 18 (the area of the squares entirely inside the region), which is 29 square units.

e. The estimates found in parts c and d are relatively close to each other. (The estimate found in part c required much less work.) ■

Another method to approximate the area of a region displayed in a coordinate plane involves counting *lattice points*—that is, points that have integer coordinates. This method uses *Pick's Theorem*, which says that *the area of any polygon whose vertices lie on lattice points is given by the formula* $A=\frac{B}{2}+I-1$, where B is the number of lattice points on the boundary of the polygon and I is the number of lattice points in the interior of the polygon. By approximating the desired region with a lattice polygon, we can approximate the area inside the region by using Pick's formula. Example 19.8 illustrates the technique.

EXAMPLE 19.8 The points in Figure 19.10 have integer coordinates. Use Pick's Theorem to approximate the area inside the region between the two curves shown in Figure 19.10.

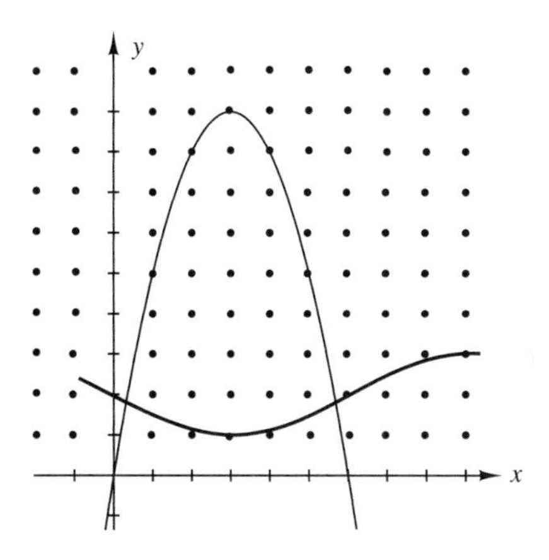

Figure 19.10

Solution Figure 19.11 shows a polygon that approximates the region and has its vertices on lattice points. The approximating polygon has 8 vertices on lattice points, so $B = 8$ in Pick's formula. We count 25 lattice points inside the polygon, so $I = 25$ in the formula. Therefore, the area of the polygon is $A = \frac{8}{2} + 25 - 1 = 28$. Since the area inside the polygon is approximately the same as the area inside the desired region, the area inside the desired region is approximately 28 square units.

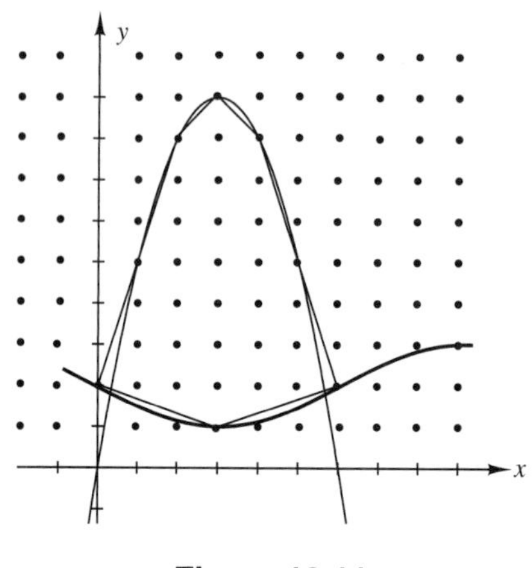

Figure 19.11 ■

We often use simple figures such as rectangles and trapezoids to approximate the area between the graph of a function and the horizontal axis. Figures 19.12, 19.13, and 19.14 show such approximations. In the first two cases the approximations appear to be quite good, since the shaded area outside the specified region is approximately equal to the unshaded area inside the region. In the third case, the approximation is larger than the exact area. The accuracy of the approximation in each case is due to the particular choice of rectangles or trapezoids.

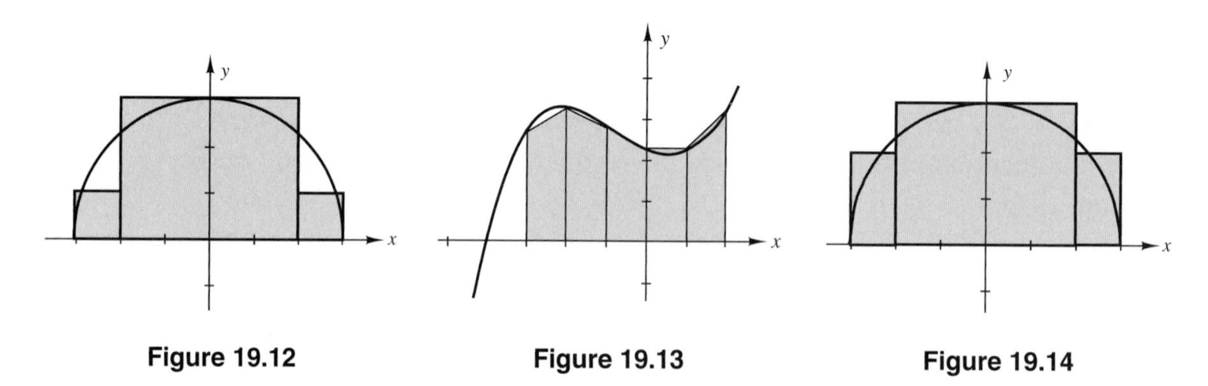

Figure 19.12 **Figure 19.13** **Figure 19.14**

In the next example, we approximate the area of a region bounded by a curve $y = f(x)$, the x-axis, and the lines $x = 1$ and $x = 5$ in two different ways: using rectangles and using trapezoids. We compare the errors introduced by each method.

EXAMPLE 19.9 Approximate the area between the curve $f(x) = x^2 - 4x + 5$ and the x-axis on the interval [1, 5] in the manner described in parts a and b. Find the error of each approximation, given that the exact area is $\frac{40}{3} = 13\frac{1}{3}$ square units.

a. Use four rectangles of equal width with the height of the rectangle determined by the value of f at the right endpoint of each subinterval.

b. Use four trapezoids of equal width with the bases of the trapezoids determined by the value of f at the endpoints of each subinterval.

c. Compare the approximations in parts a and b.

Solution Figures 19.15 and 19.16 picture the two approximations of the area of the region described.

a. Since the length of the interval [1, 5] is 4, the width of each rectangle is 1. We calculate their heights: $f(2) = 1, f(3) = 2, f(4) = 5, f(5) = 10$. (See Figure 19.15.) The sum of the areas of the four rectangles is $1(1) + 1(2) + 1(5) + 1(10) = 18$. This estimate differs from the exact value by $18 - \frac{40}{3} = \frac{54}{3} - \frac{40}{3} = \frac{14}{3} \approx 4.67$ square units.

b. The width h in the area formula $\frac{1}{2}(b_1 + b_2)h$ is 1 for each trapezoid. The length of base b_1 of each trapezoid is given by the value of f at the left endpoint of each subinterval. The length of base b_2 of each trapezoid is given by the value of f at the right endpoint of each subinterval. (See Figure 19.16.) The area of the first trapezoid is $\frac{1}{2}\left(f(1) + f(2)\right)h = \frac{1}{2}(2+1)1 = \frac{3}{2}$. We find the area of the other trapezoids in the same manner and sum to obtain an estimate for the area:

$$\tfrac{1}{2}(2+1)1 + \tfrac{1}{2}(1+2)1 + \tfrac{1}{2}(2+5)1 + \tfrac{1}{2}(5+10)1 = \tfrac{28}{2} = 14 \text{ square units}$$

This estimate differs from the exact value by $14 - \frac{40}{3} = \frac{42}{3} - \frac{40}{3} = \frac{2}{3} \approx 0.67$ square unit. Both estimates are larger than the exact area because the graph of f is concave up on [1, 5].

c. The approximation by trapezoids is better than the approximation by rectangles.

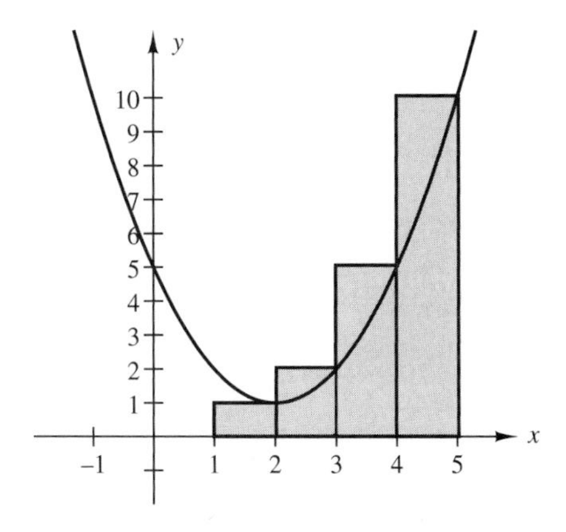

Figure 19.15

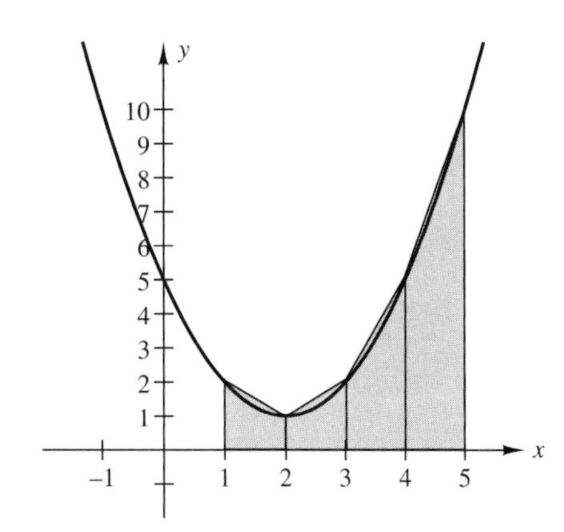

Figure 19.16 ■

In the calculation of the estimate using trapezoids in Example 19.9, each term has a factor of $\frac{1}{2}$ and also a factor of 1, the height of each trapezoid. In general, we can factor $\frac{1}{2}h = \frac{h}{2}$ out of each term. Also note that the values $f(2) = 1$, $f(3) = 2$, and $f(4) = 5$ were each used in two consecutive terms. We could have found this trapezoidal estimate by calculating this sum:

$$A = \frac{h}{2}\left(f(1) + 2f(2) + 2f(3) + 2f(4) + f(5)\right).$$

EXAMPLE 19.10

a. Approximate the area between the curve $y = x^2 - 4x + 5$ and the x-axis on the interval $[1, 5]$ as directed below.

(i) Use 8 rectangles of equal width with the height of each rectangle determined by the value of f at the right endpoint of each subinterval of $[1, 5]$.

(ii) Use 12 rectangles of equal width with the height of each rectangle determined by the value of f at the right endpoint of each subinterval of $[1, 5]$.

b. How does the approximation in Example 19.9 using 4 rectangles compare to the approximation obtained with 8 rectangles? With 12 rectangles?

Solution

a. (i) We divide the length of the interval by 8 to find that the width of each rectangle is $\frac{5-1}{8} = \frac{4}{8} = \frac{1}{2}$. The approximate area is the sum of the areas of the 8 rectangles: $A = \sum_{i=1}^{8} \frac{1}{2} \cdot f(x_i)$, where x_i is the right endpoint of the ith subinterval. Factoring out the $\frac{1}{2}$, we obtain $A = \frac{1}{2}\sum_{i=1}^{8} f(x_i)$. Thus, the area is:

$$A = \frac{1}{2}\left[f\left(\frac{3}{2}\right) + f(2) + f\left(\frac{5}{2}\right) + f(3) + f\left(\frac{7}{2}\right) + f(4) + f\left(\frac{9}{2}\right) + f(5)\right]$$

$$= \frac{1}{2}\left[\frac{5}{4} + 1 + \frac{5}{4} + 2 + \frac{13}{4} + 5 + \frac{29}{4} + 10\right] = \frac{1}{2}\left[\frac{124}{4}\right] = \frac{1}{2}[31] = 15.5 \text{ square units.}$$

(ii) The width of each rectangle is $\frac{5-1}{12} = \frac{4}{12} = \frac{1}{3}$. To find the approximate area we sum the areas of the 12 rectangles. Using the same procedure as in (i), we find that the approximate area is $14\frac{20}{27} \approx 14.74$ square units.

b. The error of the approximation with 4 rectangles is 4.67 square units; with 8 rectangles it is $\frac{31}{2} - \frac{40}{3} = \frac{13}{6} \approx 2.17$ square units. With 12 rectangles the error is $14\frac{20}{27} - 13\frac{1}{3} = 1\frac{11}{27} \approx 1.41$ square units. The more rectangles of equal width we use to approximate the area under the curve, the better the approximation. ■

Exercises 19-B

1. Assume that the area of each rectangle in Figure 19.17 is 1 square unit.

a. Estimate the area of the shaded region by counting the number of rectangles entirely inside the region.

b. Estimate the area of the shaded region by counting all rectangles entirely inside the region plus those rectangles that are partially inside the region.

c. Find the average of the estimates of the area in parts a and b.

d. Estimate the area of the shaded region by estimating the fraction of the area of each rectangle that is inside the region. Use one of the fractions 0, $\frac{1}{4}$, $\frac{1}{2}$, $\frac{3}{4}$, 1 for each estimate.

e. Compare the estimates in parts c and d.

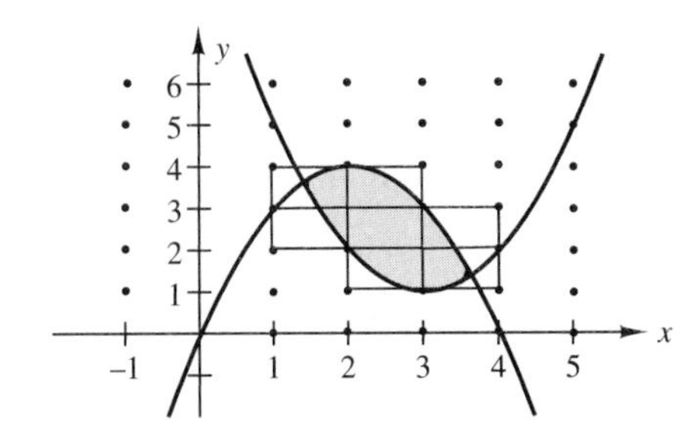

Figure 19.17

2. Assume that the area of each rectangle in Figure 19.18 is 1 square unit. Answer parts a–e in Exercise 1 for the shaded region in Figure 19.18.

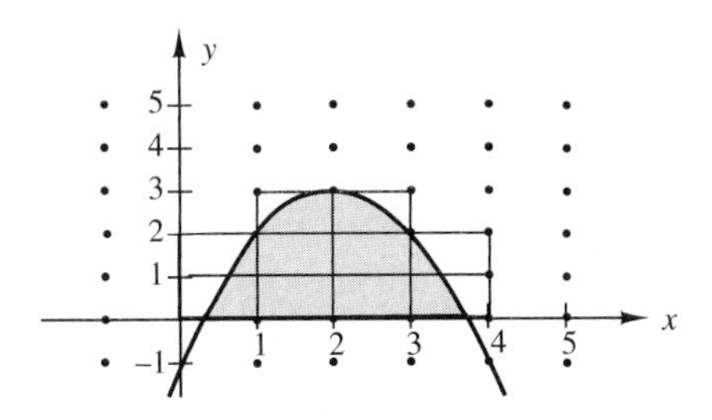

Figure 19.18

3. Approximate the shaded region in Exercise 1 with a polygon whose vertices are on lattice points. Then use Pick's formula to approximate the area of the region.

4. Approximate the shaded region in Exercise 2 with a polygon whose vertices are on lattice points. Then use Pick's formula to approximate the area of the region.

5. Approximate the area between the curve $f(x) = -x^2 + 4x + 5$ and the x-axis on the interval [2, 5] as described below. Find the error of each approximation. (The exact area is 18 square units.)

a. Use five rectangles of equal width with the height of each rectangle determined by the value of f at the right endpoint of each subinterval of [2, 5].

b. Use five trapezoids of equal width with the bases of the trapezoids determined by the values of f at the endpoints of each subinterval of [2, 5].

19-C RIEMANN SUMS AND THEIR INTERPRETATIONS

In Examples 19.9 and 19.10, we approximated the area of a region bounded by a curve and the x-axis by a sum of areas of rectangles. In general, if a curve is the graph of a function f and lies above the x-axis for $a < x < b$, then we can approximate the area between the curve and the x-axis over the interval from $x = a$ to $x = b$ by a sum of n rectangles, each like the one shown in Figure 19.19. The height of the ith rectangle is equal to $f(x_i)$, where x_i is a number in the ith subinterval, and the width of the ith subinterval is denoted by Δx_i. The approximating sum can be written $\sum_{i=1}^{n} f(x_i)\Delta x_i$.

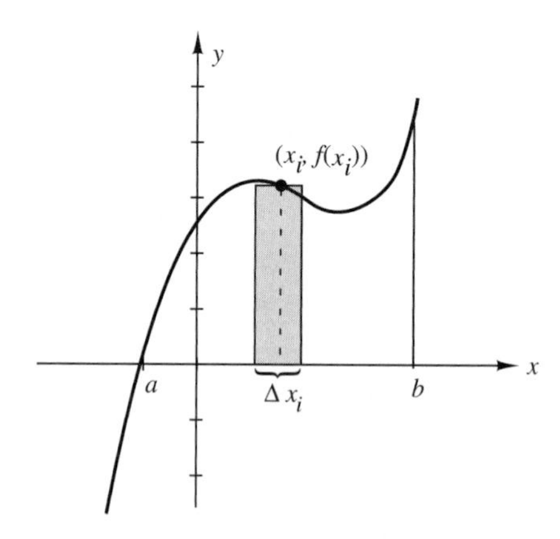

Figure 19.19

This kind of approximating sum is called a *Riemann sum*, after the German mathematician Georg Friedrich Bernhard Riemann. The value of the approximating sum depends on the following choices:

1. The number n of subintervals into which we subdivide the interval $[a, b]$—that is, the number of rectangles that are used.

2. How the points that determine the subintervals are spaced; these determine the widths of the rectangles—that is, the value of each Δx_i. It is often convenient to choose equally spaced points, but it is not necessary that they be equally spaced.

3. How the point x_i in each subinterval is chosen; these determine the height of each rectangle—that is, the value of $f(x_i)$.

We note that if the points that determine the subintervals are chosen to be equally spaced, then all Δx_i values are equal. In this case, we represent every Δx_i by Δx, with no i subscript. Then the Riemann sum can be written

$$\sum_{i=1}^{n} f(x_i)\Delta x_i = \sum_{i=1}^{n} f(x_i)\Delta x = f(x_1)\Delta x + f(x_2)\Delta x + \cdots + f(x_n)\Delta x = \Delta x \sum_{i=1}^{n} f(x_i).$$

The fact that Δx, the width of each rectangle, is the same allows us to factor Δx out of the sum.

Depending on what the variable x and the value $f(x)$ represent, the sum $\sum_{i=1}^{n} f(x_i)\Delta x_i$ can be interpreted in a variety of ways. It all depends on the interpretation of the product of the two quantities $f(x_i)$ and Δx_i.

Suppose a car travels for 6 hours and its average velocity is 30 mph during the first hour, 40 mph during the next 2 hours, and 50 mph during the last 3 hours. The total distance traveled is the sum of the distances traveled during the three time intervals: distance = (rate for first hour) $\times$ 1 + (rate for next 2 hours) $\times$ 2 + (rate for last 3 hours) $\times 3 = 30 \cdot 1 + 40 \cdot 2 + 50 \cdot 3 = 260$ miles. This total distance can be represented as the sum of the areas of the three rectangles shown in Figure 19.20.

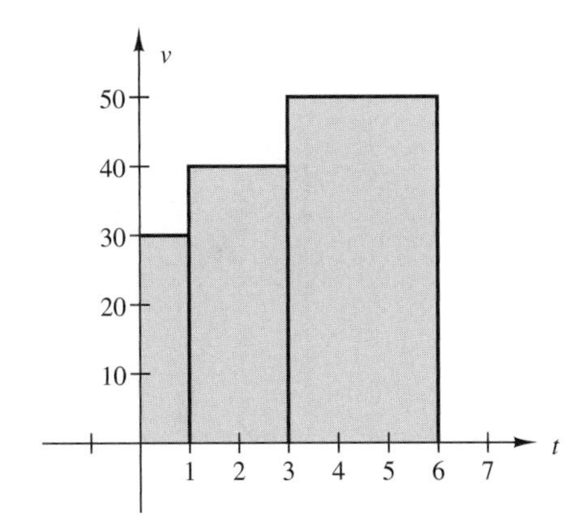

Figure 19.20

The velocity of a car usually varies continuously; suppose the velocity is given by a continuous function $v = f(t)$. How can we find the distance traveled by the car over the interval $0 \le t \le 6$? We can approximate the distances traveled in six 1-hour subintervals and add them. Although the actual velocity of the car varies throughout each 1-hour subinterval of [0, 6], we approximate the distance traveled by assuming that the velocity is constant in each subinterval, equal to the velocity at a time t in that 1-hour subinterval. For example, suppose we know the velocity of the car at the times shown in the following table:

Time t (h)	0.5	1.5	2.5	3.5	4.5	5.5
Velocity v (mph)	20	45	60	65	60	45

We use the velocity at time $t = 0.5$ as our estimate for the average velocity during the first hour, the velocity at time $t = 1.5$ for an estimate of the average velocity during the hour from $t = 1$ to $t = 2$, the velocity at time $t = 2.5$ for an estimate of the average velocity during the hour from $t = 2$ to $t = 3$, and so on. We use these estimates of the average velocity in the formula $d = rt$ for each 1-hour subinterval of time to compute an estimate for the distance traveled each hour. An estimate for the total distance is the sum of these hourly distance estimates:

$$20 \cdot 1 + 45 \cdot 1 + 60 \cdot 1 + 65 \cdot 1 + 60 \cdot 1 + 45 \cdot 1 = 295 \text{ miles.}$$

The estimate of the distance traveled each hour can be represented by the area of a rectangle with base length 1 and height v (the velocity) at the midpoint of the 1-hour subinterval. Then the total distance is estimated by the sum of the areas of these rectangles, which is the Riemann sum $\sum_{i=1}^{6} f(t_i)\ \Delta t_i$, where $\Delta t_i = 1$ for $i = 1, 2,$ 3, 4, 5, and 6 and $f(t_1) = 20$, $f(t_2) = 45$, $f(t_3) = 60$, $f(t_4) = 65$, $f(t_5) = 60$, and $f(t_6) = 45$. The graph of $v = f(t)$ with the six rectangles that approximate the distance is shown in Figure 19.21.

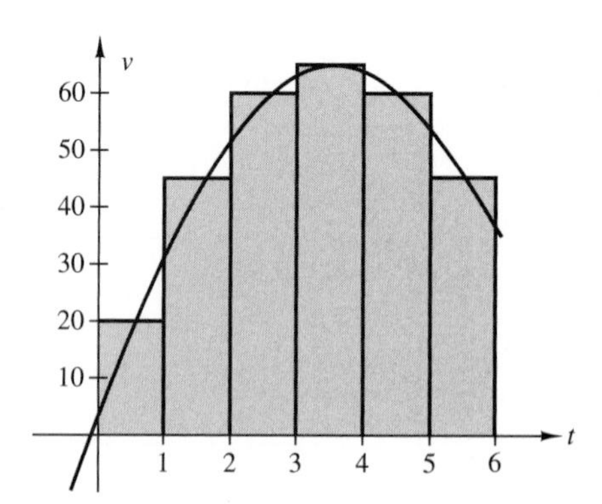

Figure 19.21 $v = f(t)$ and estimate of distance traveled

EXAMPLE 19.11 Suppose a car is traveling and $s(t) = 65t - \frac{5}{3}\left(t - \frac{7}{2}\right)^3$ gives the position (in miles) of the car from a starting point t hours after an initial time. The function $f(t) = 65 - 5\left(t - \frac{7}{2}\right)^2$ gives the car's velocity (in miles per hour). (Note that $s'(t) = f(t)$.)

a. Sketch the velocity curve $v = f(t)$ for $0 \le t \le 3$.

b. Estimate the distance traveled by the car from time $t = 0$ to time $t = 3$ hours using three 1-hour time intervals: from $t = 0$ to $t = 1$, from $t = 1$ to $t = 2$, and from $t = 2$ to $t = 3$. In each interval use the velocity at the right endpoint of the interval as an estimate for the average velocity throughout the interval.

c. Is the estimate in part b an overestimate or an underestimate of the exact distance traveled? Explain.

d. Find the exact distance traveled by the car during this 3-hour time period by finding the change in position of the car from time $t = 0$ to time $t = 3$.

Solution

a. The velocity curve is sketched in Figure 19.22.

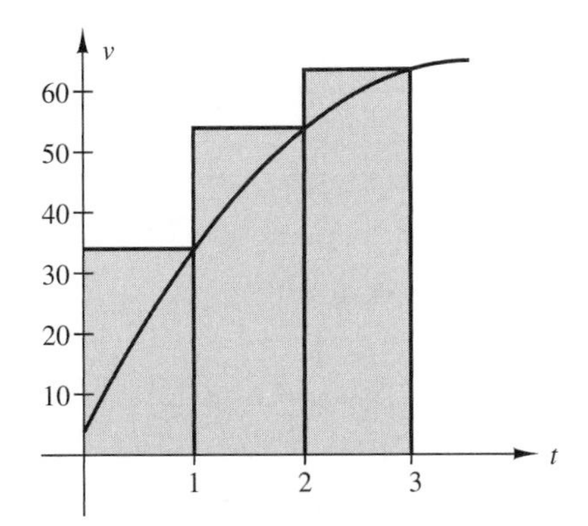

Figure 19.22

b. We need to compute the velocity at the right endpoints of the intervals—that is, at $t = 1$, $t = 2$, and $t = 3$.

At $t = 1$, $v = f(1) = 65 - 5\left(1 - \frac{7}{2}\right)^2 = 65 - 5\left(\frac{25}{4}\right) = \frac{135}{4} = 33.75$ mph.

At $t = 2$, $v = f(2) = 65 - 5\left(2 - \frac{7}{2}\right)^2 = 65 - \frac{45}{4} = \frac{215}{4} = 53.75$ mph.

At $t = 3$, $v = f(3) = 63.75$ mph.

An estimate for the total distance traveled is: $f(1) \cdot 1 + f(2) \cdot 1 + f(3) \cdot 1 = 33.75 + 53.75 + 63.75 = 151.25$ miles. This estimate is the sum of the areas of the three rectangles in Figure 19.22.

c. The estimate of 151.25 miles is an overestimate since each of the values $f(1)$, $f(2)$, and $f(3)$ that approximates the velocity over a subinterval of time is greater than the actual velocity for every instant of time in the interval, except the right endpoint.

d. The exact distance traveled by the car is the change in its position over the interval [0, 3]: $s(3) - s(0) = \left[65 \cdot 3 - \frac{5}{3}\left(3 - \frac{7}{2}\right)^3\right] - \left[65 \cdot 0 - \frac{5}{3}\left(0 - \frac{7}{2}\right)^3\right] = \left[195 + \frac{5}{24}\right] - \left[\frac{5}{3}\left(\frac{343}{8}\right)\right] = \frac{2970}{24} = 123.75$ miles. ■

Since all terms in a Riemann sum are products of two quantities (the change in a variable, Δx_i, and the value $f(x_i)$ of a function f for an x_i in that interval), a Riemann sum can be used to approximate various quantities that can be expressed as a product $f(x)\ \Delta x$.

For example, the product $N = r \times t$ states that the number N of bacteria produced in a time interval of length t equals the growth rate r times t. When the growth rate is not constant but is a function $r(t)$ of time, the number N can be approximated by the Riemann sum $N = \sum_{i=1}^{n} r(t_i)\Delta t_i$. Here, the interval of time (for which we want to approximate the number of bacteria produced) is split into n subintervals, and the length of the ith subinterval is denoted Δt_i. The number $r(t_i)$ is the growth rate at time t_i in the ith subinterval, and this constant approximates the growth rate for the whole i^{th} subinterval.

Another measurable quantity that is expressed as a product is the work expended to move an object: $W = F \times d$ (work equals force times distance moved). When force varies as a function $F(x)$ of position x, the amount of work can be approximated by a Riemann sum $W = \sum_{i=1}^{n} F(x_i)\ \Delta x_i$. Here, the distance the object is moved is split into n subintervals of lengths Δx_i, and the force $F(x_i)$ exerted at point x_i in the ith subinterval approximates the varying force throughout that interval.

We can also use Riemann sums to approximate volumes. Suppose we have a rod of length L with constant cross-sectional area A. (A standard circular rod with constant radius is an example; it is a thin cylinder.) Its volume V is the product $V = A \cdot L$. If the cross-sectional area $A(x)$ of a rod varies with the distance x from the end of the rod, the volume of such a rod can be approximated by a Riemann sum of the form $\sum_{i=1}^{n} A(x_i)\ \Delta x_i$.

In the next example we approximate the area of a cone-shaped rod.

EXAMPLE 19.12 A cone-shaped rod is 10 in. long and its cross-sectional area is $A(x) = x^2 + 1$ in.^{2} for $0 \le x \le 10$, where x is the distance from the left end of the rod. Approximate the volume of the rod using a Riemann sum in which the length is divided into six subintervals of equal length. Use the right endpoint of each subinterval to estimate the cross-sectional area for that subinterval.

Solution We subdivide the interval $0 \le x \le 10$ into six equal subintervals, each of length $\frac{10-0}{6} = \frac{5}{3}$. Thus the endpoints of the subintervals are 0, $\frac{5}{3}$, $\frac{10}{3}$, $\frac{15}{3}$, $= 5$, $\frac{20}{3}$, $\frac{25}{3}$, and $\frac{30}{3} = 10$. To approximate the total volume of the rod, we want to evaluate the Riemann sum $\sum_{i=1}^{6} A(x_i)\ \Delta x_i = \sum_{i=1}^{6} A(x_i)\frac{5}{3} = \frac{5}{3}\sum_{i=1}^{6} A(x_i)$. In this sum, the values $A(x_i)$ represent an estimate of the cross-sectional area for the ith subinterval and are determined by evaluating the area function $A(x) = x^2 + 1$ at the right endpoints of the subintervals. Therefore, the approximation to the volume of the rod is

$$V \approx \frac{5}{3}\left[A\left(\frac{5}{3}\right)+A\left(\frac{10}{3}\right)+A\left(\frac{15}{3}\right)+A\left(\frac{20}{3}\right)+A\left(\frac{25}{3}\right)+A\left(\frac{30}{3}\right)\right]$$

$$= \frac{5}{3}\left[\left(\frac{25}{9}+1\right)+\left(\frac{100}{9}+1\right)+\left(\frac{225}{9}+1\right)+\left(\frac{400}{9}+1\right)+\left(\frac{625}{9}+1\right)+\left(\frac{900}{9}+1\right)\right]$$

$$= \frac{5}{3}\left[\frac{2275}{9}+6\right] = \frac{5}{3}\left[\frac{2275}{9}+\frac{54}{9}\right] = \frac{5}{3}\left[\frac{2329}{9}\right] = \frac{11645}{27} \approx 431.30\,\text{in}^3. \quad \blacksquare$$

Suppose we know that the population growth rate p (in hundreds of people per year) in a town is constant for a period of t years. Then, during the same period of time, the total population growth in the town is $S = p \times t$ (in hundreds of people). If p is negative, then the population is decreasing and $S = p \times t$ is the population change, which is now negative. When the population rate of change $p(t)$ varies with time, we can approximate the change in population over an interval of time by a Riemann sum of the form $\sum_{i=1}^{n} p(t_i)\ \Delta t_i$. In the next example we estimate a net change in population when the population rate of change is at times negative.

EXAMPLE 19.13 The population growth rate (in number of people per year) at time t of a suburban area is given by $p(t) = -10t^3 + 190t^2 - 700t + 100$, where t represents years from 1960. (The graph of this function p is in Figure 19.23.) Use a Riemann sum to estimate the net change in population from 1960 to 1975 using five equal subintervals of time. To estimate the population growth rate for each subinterval:

a. Use the left endpoint of each subinterval.

b. Use the right endpoint of each subinterval.

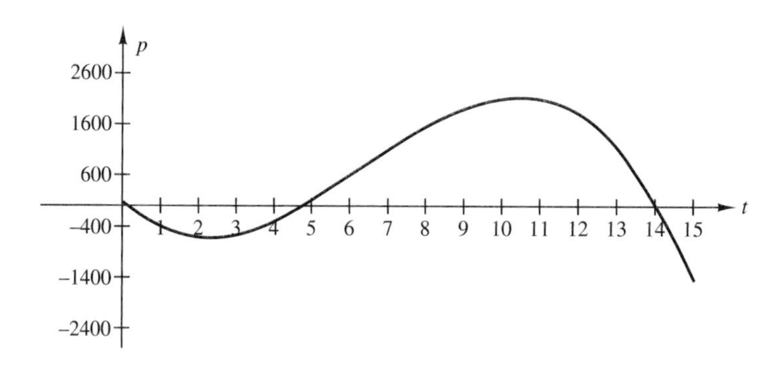

Figure 19.23

Solution

a. The interval from 1960 to 1975 corresponds to $0 \le t \le 15$. If we divide this interval into five equal subintervals, each has length $\frac{15-0}{5} = 3$; therefore, the left endpoints of the subintervals are 0, 3, 6, 9, and 12. The Riemann sum to estimate the population growth is $\sum_{i=1}^{5} p(t_i)\ \Delta t_i = \sum_{i=1}^{5} p(t_i)3 = 3\sum_{i=1}^{5} p(t_i) = 3[p(0) + p(3) + p(6) + p(9) + p(12)] = 3[100 - 560 + 580 + 1900 + 1780] = 3[3800] = 11{,}400$. In this estimate, the population grew by about 11,400 people from 1960 to 1975.

b. The right endpoints of the subintervals are 3, 6, 9, 12, and 15. So when we use the right endpoints the Riemann sum is $\sum_{i=1}^{5} p(t_i)\ \Delta t_i = \sum_{i=1}^{5} p(t_i)3 = 3\sum_{i=1}^{5} p(t_i) = 3[p(3) + p(6) + p(9) + p(12) + p(15)] = 3[-560 + 580 + 1900 + 1780 - 1400] = 6900$. In this estimate the population grew by 6900 people. ■

Note that the Riemann sums we used to approximate the population growth in Example 19.13 contain some negative terms. These are the terms $p(t_i)\ \Delta t_i$ where $p(t_i)$ is negative, which means that the population was actually decreasing at time t_i. For example, the second term of the Riemann sum in Example 19.13a is $p(3)3 = (-560)3 = -1680$. This is the estimate of the population change for the subinterval $3 \le t \le 6$ when we use left endpoints; that is, we estimate that from 1963 to 1966 the population decreased by 1680 people.

Earlier in this chapter we interpreted each positive term of a Riemann sum as the area of a rectangle. Likewise, we can interpret each negative term of a Riemann sum as the negative of the area of a rectangle. Figure 19.24 shows the rectangles that help us interpret the Riemann sum in Example 19.13a. The sum $3[p(0) + p(3) + p(6) + p(9) + p(12)]$ is the sum of the areas of the four rectangles that lie above the x-axis minus the area of the rectangle that lies below the x-axis. The Riemann sum in Example 19.13b, $3[p(3) + p(6) + p(9) + p(12) + p(15)]$, is the sum of the areas of the three rectangles in Figure 19.25 that lie above the x-axis minus the sum of the areas of the two rectangles that lie below the x-axis.

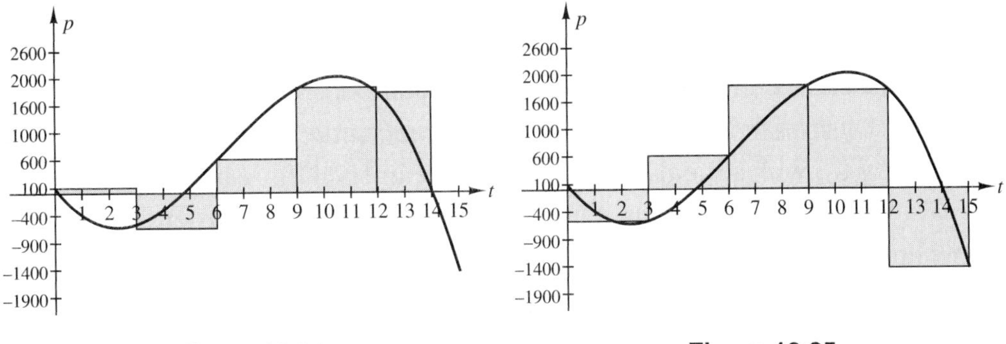

Figure 19.24

Figure 19.25

Exercises 19-C

1. Suppose $s(t) = t^2 + t$ gives the position from a starting point O (in feet) of a bicyclist at time t (in seconds). The function $f(t) = 2t + 1$ gives the bicyclist's velocity at time t (in feet per second).

a. Sketch the velocity function for $0 \leq t \leq 10$.

b. Estimate the distance traveled by the bicyclist from time $t = 0$ to time $t = 10$ by dividing the time interval [0, 10] into consecutive 2-sec subintervals. In each subinterval use the velocity at the left endpoint of the subinterval as an estimate for the average velocity throughout the subinterval.

c. Is the estimate you found in part b an overestimate or an underestimate of the exact distance traveled over the interval [0, 10]? Explain your answer.

d. Find the exact distance traveled by the bicyclist during the first 10 seconds by computing the change in position from $t = 0$ to $t = 10$ seconds.

2. A ball is thrown vertically into the air. The velocity of the ball (in ft/sec) at time t seconds after it is thrown is given by $v(t) = 128 - 32t$. Use a Riemann sum to approximate the displacement of the ball over the time interval $0 \leq t \leq 2$. Divide the interval [0, 2] into four subintervals of equal length, and use the right endpoint of each subinterval to estimate the velocity in the subinterval.

3. **a.** Repeat Exercise 2, but use the left endpoint of each subinterval to estimate the velocity in the subinterval.

b. Find the average of your two approximations of displacement (in part a and in Exercise 2).

c. The exact displacement of the ball for the interval $0 \le t \le 2$ is 192 ft. Which of the three approximations (Exercise 2 or this exercise part a or b) is the closest?

4. Suppose $g(t)$ denotes the temperature (in °F) in a room t hours after turning on a fan. The instantaneous rate of change in temperature (in °F/h) in the room at time t is $g'(t) = -2 - 4(t + 1)^{-1}$, $0 \le t \le 3$. Estimate the temperature change for the 3-hour interval $0 \le t \le 3$ with a Riemann sum. Divide the interval [0, 3] into four equal subintervals. Use the right endpoint of each subinterval to estimate the rate of change in temperature for the subinterval.

5. The rate at which a bacteria population is growing t hours after being placed in a refrigerator is $r(t) = -t^2 + 10t + 2475$ (in number of bacteria per hour). Use a Riemann sum to approximate the number of bacteria produced after 72 hours. Use four intervals of equal length and the right endpoint of each subinterval.

Chapter 19 Exercises

1. Find the exact area of each region outlined in Figures 19.26, 19.27, and 19.28. In Figure 19.28, the upper arcs are quarter-circles, and the lower arc is a semicircle.

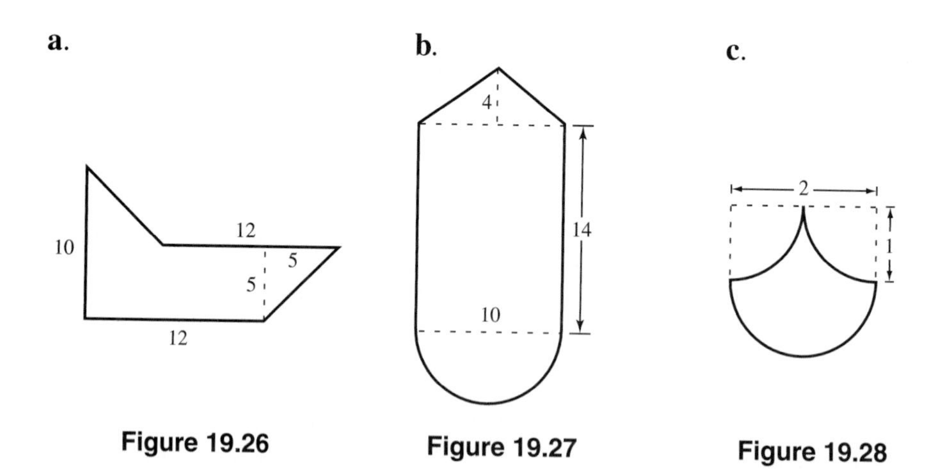

Figure 19.26 Figure 19.27 Figure 19.28

2. An athlete recovering from knee surgery starts daily walks on a treadmill. She walks for only 5 minutes on the first day and then increases her time on the treadmill by 5 minutes each day until she reaches 30 minutes per day. Her average speed each day is given in the following table.

Day	1	2	3	4	5	6	7
Speed (m/min)	0.05	0.06	0.06	0.08	0.09	0.09	0.1

a. Find the total distance the athlete walked on the treadmill during the seven days.

b. Graph the speed as a function of t, time (in minutes) spent on the treadmill from the beginning of the first day, over the interval $0 \le t \le 135$. (Note that the total time the athlete spent on the treadmill during the seven days is 135 minutes.) Explain how the answer obtained in part a can be interpreted as the sum of areas of rectangles. Draw the rectangles.

3. Use geometry to find the exact area of the region bounded by the lines $y = 2x - 1$, $x = 1$, $x = 3$, and the x-axis. First sketch the region.

4. Find the exact area of the region bounded by $y = \sqrt{25 - x^2}$, the line $y = x$, and the x-axis. Sketch the region and describe its shape.

5. Assume the area of each square in Figure 19.29 is 1 square unit.

a. Estimate the area of the shaded region by counting the number of squares entirely inside the region.

b. Estimate the area of the shaded region by counting all the squares that lie entirely inside the region plus those squares that are partially inside the region.

c. Find the average of the estimates of the area in parts a and b.

d. Estimate the area of the shaded region by estimating the fraction of the area of each square that is inside the region. Use one of the fractions $0, \frac{1}{4}, \frac{1}{2}, \frac{3}{4}, 1$ for each estimate.

e. Compare the estimates in parts c and d.

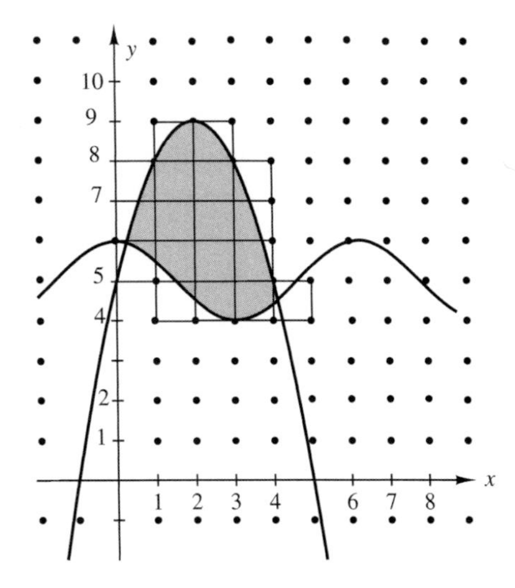

Figure 19.29

6. Approximate the shaded region in Figure 19.29 with a polygon that has its vertices at lattice points. Then use Pick's formula to approximate the area of the shaded region.

7. Sketch the curve $f(x) = -x^2 + 2x + 3$ on the interval [0, 3]. Approximate the area between the curve and the x-axis on the interval [0, 3] by dividing the interval into five subintervals of equal length and following the directions below. Find the error of each approximation. (The exact area is 9.)

a. Use five rectangles of equal width, each with the height of the rectangle determined by the value of f at the right endpoint of each subinterval.

b. Use five trapezoids of equal width with the bases of the trapezoids determined by the value of f at the endpoints of each subinterval.

8. The population growth rate at time t (in hundreds of people per year) of a certain region is given by $p(t) = 4t^2 - 15\sqrt{t}$, where t represents years from 1990. Use a Riemann sum to estimate the net change in population from 1990 to 2000 using six subintervals of equal length. Use the left endpoint of each subinterval to estimate the population growth rate for the subinterval.

9. Repeat Exercise 8 using 10 subintervals of equal length and the left endpoint of each subinterval. Compare the estimate with the actual change in population, which is 1017 people.

10. Let $r(t)$ be the rate of growth of a bacteria population (in number of bacteria per hour). Use sigma notation to express the Riemann sum that approximates the number of bacteria produced over the time interval $0 \le t \le 72$, by dividing the interval into 36 subintervals of equal length. Use the right endpoint of each subinterval.

11. In this chapter, we discussed several measurable quantities, each given by a formula that is the product of two other quantities (for example, $A = l \times w$, $N = r \times t$, $W = F \times d$). Find two other measurable quantities that are each calculated as the product of two quantities.

CHAPTER 20

Companion to the Definite Integral

20-A AREA UNDER A CURVE AS A DEFINITE INTEGRAL

Two different window designs are proposed for a room; both have the same width of 4 ft, and each has the same square base topped by an arc. The arc on the window in Figure 20.1a is a semicircle, with its highest point at 6 ft; the arc in Figure 20.1b is a parabola, with its highest point at 6.4 ft. The designers want to choose the design that allows in the most light. Our problem is to determine which window has the larger area (and will let in the most light).

Figure 20.1a

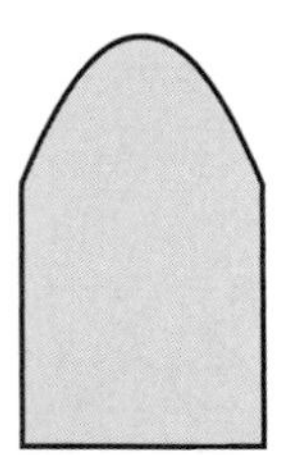

Figure 20.1b

If we introduce a coordinate system for each window so that the bottom is on the x-axis and the maximum height of the window is on the y-axis, then each of the areas is outlined by the graph of a function (the top arc), the x-axis, and the vertical lines $x = -2$ and $x = 2$. The semicircular arc has equation $f(x) = \sqrt{4 - x^2} + 4$, and the parabolic arc has equation $g(x) = -0.6x^2 + 6.4$. The area of the first window can be calculated exactly, since we can find the area of the 4×4 square base and the area of the semicircle of radius 2 that tops it and add them: $4^2 + \frac{\pi(2^2)}{2} = 16 + 2\pi \text{ ft}^2$. The area of the second window cannot be calculated so easily, since we do not have a formula for the area under a parabolic arc. In this section, we show how to extend the approximation techniques of Section 19-B so we can find exact areas of figures like the window topped by the parabolic arc.

The general situation is this: We are given a function $f(x)$ whose graph lies above the x-axis, and we want to find the exact area of the region bounded by the curve $y = f(x)$, the x-axis, and the vertical lines $x = a$ and $x = b$. In Section 19-B we saw how we can approximate the area of such a region using a Riemann sum obtained by partitioning the interval from $x = a$ to $x = b$ into n subintervals and forming n rectangles. Each rectangle has its base formed by one of the subintervals. The height of the ith rectangle is $f(x_i)$, where x_i is a number in the ith subinterval, and we denote the width of the ith rectangle as Δx_i. The Riemann sum that approximates the area of this region is

$$S = \sum_{i=1}^{n} f(x_i)\,\Delta x_i\,.$$

If we continue to decrease the width of the rectangles used to approximate such an area, the number of rectangles will increase without bound. Also, the smaller the width of the rectangles, the more accurate the estimate will be. The limit of the sums of the areas of the rectangles, as the largest width of any rectangle is decreased to 0, is the area under the graph of the function. In calculus this area is represented by the *definite integral*, $\int_a^b f(x)dx$. The integral symbol, $\int$, which looks somewhat like an elongated "S," is due to Gottfried Leibniz and can be thought of (as he did) as an infinite sum of infinitely thin rectangles. The function value $f(x)$ represents the height of each rectangle and dx represents the width of each rectangle. The two numbers a and b are called *limits of integration* and indicate the left and right endpoints of the interval over which the area is to be found. (See Figure 20.2a.)

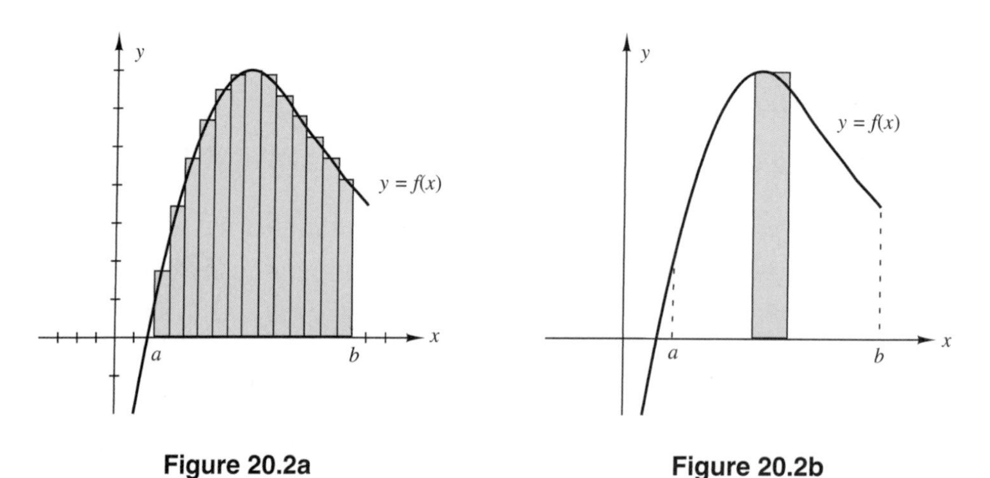

Figure 20.2a **Figure 20.2b**

When we find an area by expressing it as a definite integral, we visualize the area as a sum of areas of rectangles and draw one typical rectangle, as shown in Figure 20.2b.

There is a direct correlation between the sigma notation for the finite sum that approximates the area and the notation for the definite integral:

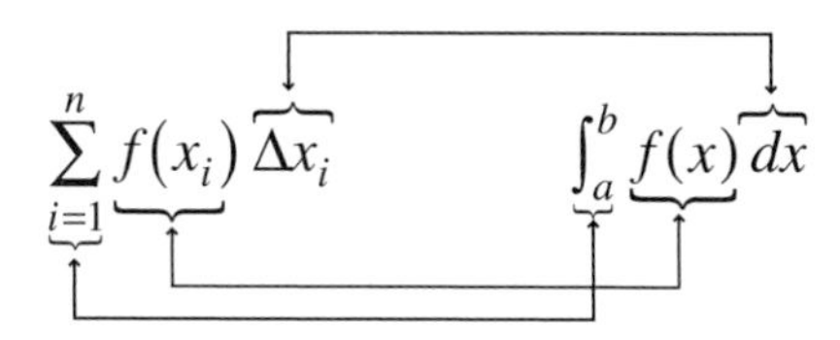

EXAMPLE 20.1 Let R be the region bounded by the curve $y = \sqrt{9-x^2}$ and the x-axis.

a. Give a Riemann sum of the form $\sum_{i=1}^{n} f(x_i)\ \Delta x_i$ that approximates the area of R.

b. Write a definite integral that gives the area of R.

Solution Figure 20.3 shows the semicircular region R for which we want to set up a Riemann sum and a definite integral.

a. One approximating rectangle, with height $\sqrt{9-x^2}$ and width Δx, is drawn in the region R. Since the graph of the function intersects the x-axis at $x = -3$ and $x = 3$, we construct n approximating rectangles throughout the shaded region from $x = -3$ to $x = 3$. Thus the Riemann sum that approximates the area of R is $\sum_{i=1}^{n} \sqrt{9-x_i^2}\,\Delta x_i$. If the widths Δx_i of the subintervals are all the same, then $\Delta x_i = \dfrac{b-a}{n}$

b. Since we construct approximating rectangles throughout the region from $x = -3$ to $x = 3$, the limits of integration for the definite integral are $a = -3$ and $b = 3$, and the desired integral is $\int_{-3}^{3} \sqrt{9-x^2}\,dx$. (We note that we can use the formula for the area of a circle to find this particular area.)

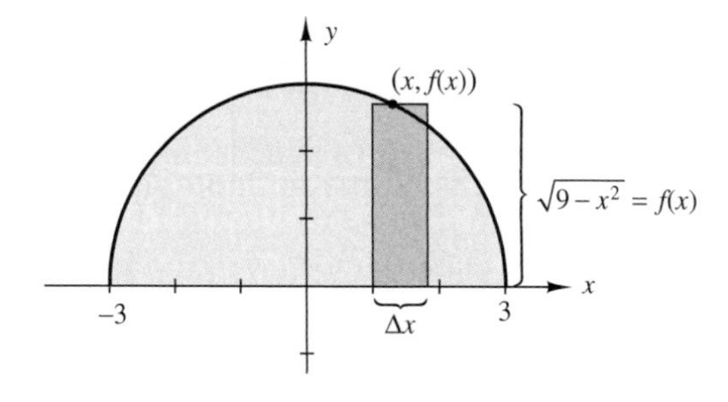

Figure 20.3 ■

EXAMPLE 20.2 Write a definite integral for the area of the shaded region under the curve $y = g(x)$ shown in Figure 20.4.

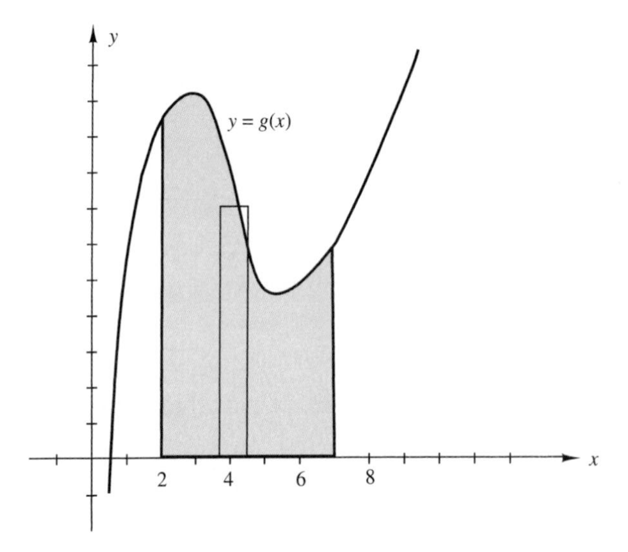

Figure 20.4

Solution We let $g(x)$ represent the height of each approximating rectangle and dx represent the width. The limits of integration are the left and right endpoints of the interval over which we wish to find the area, so $a = 2$ and $b = 7$. The area is given by the definite integral $\int_2^7 g(x)\,dx$. ■

EXAMPLE 20.3 Write a definite integral for the area of the region bounded above by the curve $y = 5x - 4 - x^2$, below by the x-axis, and on the right by the line $x = 3$.

Solution We first draw the region as shown in Figure 20.5. The curve $y = 5x - 4 - x^2$ is a parabola opening downward with x-intercepts $x = 1$ and $x = 4$. [These are the solutions of $0 = 5x - 4 - x^2 = -1(x^2 - 5x + 4) = -1(x - 4)(x - 1)$.]

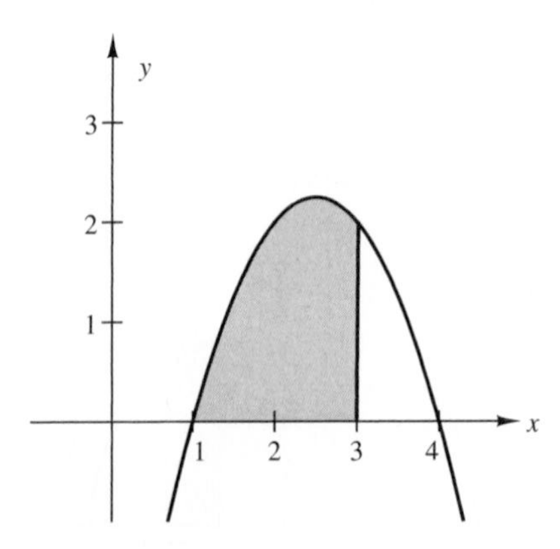

Figure 20.5 $y = 5x - 4 - x^2$

Since the line $x = 3$ is the right-hand boundary of the region, $b = 3$ is the upper limit of integration in the definite integral. From the picture we see that the lower limit of integration is $a = 1$. The definite integral is $\int_1^3 5x - 4 - x^2\,dx$. ■

Exercises 20-A

1. For each of the sums given below, write a definite integral that is approximated by that sum over the specified interval.

a. $\sum_{i=1}^{n} x_i^2 \, \Delta x_i$ on the interval [2, 5]

b. $\sum_{i=1}^{n} \sqrt{x_i - 3}\, \Delta x_i$ on the interval [6, 8]

2. Let R be the region bounded by the curve $y = \sin x$ and the x-axis on the interval $[0, \pi]$.

a. Give a Riemann sum of the form $\sum_{i=1}^{n} f(x_i)\, \Delta x_i$ that approximates the area of R.

b. Write a definite integral that gives the area of R.

3. Write a definite integral for the area of the region bounded by the curve $y = \sqrt{25 - x^2}$ and the x-axis.

4. Write a definite integral for the area of the shaded region under each curve in Figures 20.6 and 20.7.

a.

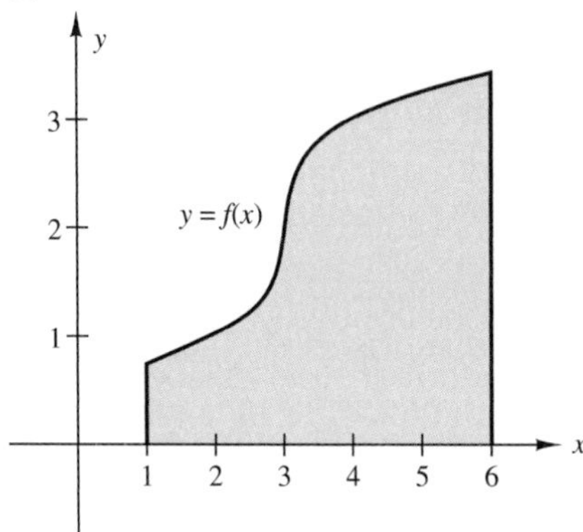

Figure 20.6

b.

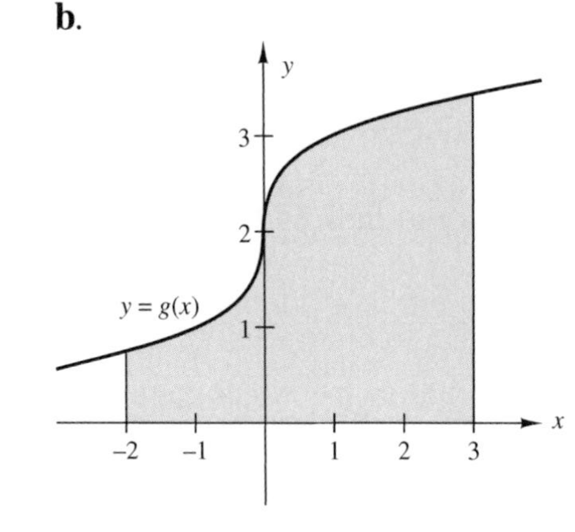

Figure 20.7

5. Write a definite integral for the area of the window in Figure 20.1b.

20-B OTHER INTERPRETATIONS OF THE DEFINITE INTEGRAL

The definite integral $\int_a^b f(x)\, dx$ is defined to be the limit of a sum of the form $\sum_{i=1}^{n} f(x_i)\, \Delta x_i$, where the limit is taken as the width Δx_i of the widest rectangle approaches 0. Each term in this sum is the product of two quantities, $f(x_i)$ and Δx_i. The interpretation of the limit of this sum, a definite integral, depends on the interpretation of the product of the two quantities $f(x_i)$ and Δx_i.

For example, when $f(t)$ represents the velocity at time t of a car traveling along a highway, $f(t_i)\ \Delta t_i$ approximates the distance the car travels during the time interval Δt_i. So the Riemann sum $\sum_{i=1}^{n} f(t_i)\ \Delta t_i$ approximates the distance traveled by the car over a time interval such as $0 \le t \le 2$. (See Section 19-C.) Then the limit of this sum is the definite integral $\int_0^2 f(t)\ dt$, the exact distance traveled in that time interval.

We saw in the previous section that for a continuous function $f(x)$ that is positive on an interval $[a, b]$, the definite integral $\int_a^b f(x)\ dx$ represents the area under the curve $y = f(x)$, above the x-axis, and between the vertical lines $x = a$ and $x = b$. In this case, the integral is the limit of sums of products $f(x_i) \cdot \Delta x_i$ that represent areas of rectangles of height $f(x_i)$ and width Δx_i. In the examples that follow we link the interpretation of $\int_a^b f(x)\ dx$ as an area to other interpretations of the definite integral.

EXAMPLE 20.4 Suppose the population growth rate at time t for high school students in an expanding school district is predicted to be $p(t) = 300 + 20t$ students per year over the next 4 years. We take the current year to be time $t = 0$, so $p(t) = 300 + 20t$, $0 \le t \le 4$.

a. Sketch the graph of $y = p(t)$ for $0 \le t \le 4$.

b. Use geometry to find $\int_0^4 p(t)\ dt$.

c. Interpret what the integral in part b represents in this context.

Solution

a. The graph of the linear function $p(t)$ is sketched in Figure 20.8.

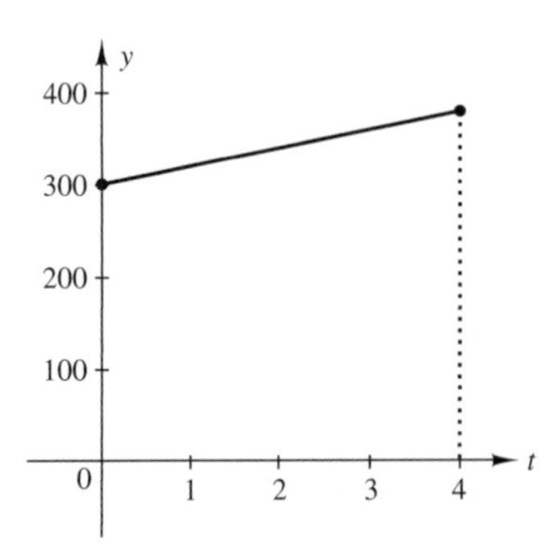

Figure 20.8 $y = p(t) = 300 + 20t$

b. Since $p(t) > 0$ for $0 \le t \le 4$, the definite integral is the area of the region bounded by $y = p(t)$, the t-axis, and the vertical lines $t = 0$ and $t = 4$. This region is a trapezoid with area $A = \frac{1}{2}(b_1 + b_2)h = \frac{1}{2}[p(0) + p(4)] \cdot 4 = 2[300 + 380] = 1360$.

c. The integral represents population growth over the time interval $0 \le t \le 4$, since the definite integral is the limit of sums of the products (population growth rate) × time. (Population growth rate is in number of students per year, and time is measured in years.) So the population of high school students in the district will grow by 1360 students in the next 4 years. ■

What does the definite integral represent geometrically when $f(x)$ is not always positive on the interval $[a, b]$?

Suppose the continuous function $f(x)$ satisfies $f(x) < 0$ for all x in $[a, b]$. Then $\int_a^b f(x)\,dx$ is still, by definition, the limit of sums of products $f(x) \cdot \Delta x$. But since f is negative, each of the terms $f(x) \cdot \Delta x$ is negative. Each term $f(x) \cdot \Delta x$ is the **negative** of the area of the rectangle with height $|f(x)|$ and base Δx. The definite integral then will be negative and represents the negative of the area of the region bounded by the curve $y = f(x)$, the x-axis, and the lines $x = a$ and $x = b$.

Figure 20.9 shows a function $f(x)$ that is negative on $[-4, 4]$. The integral $\int_{-4}^4 f(x)\,dx$ is the negative of the area of the triangular region bounded by the curve and the x-axis.

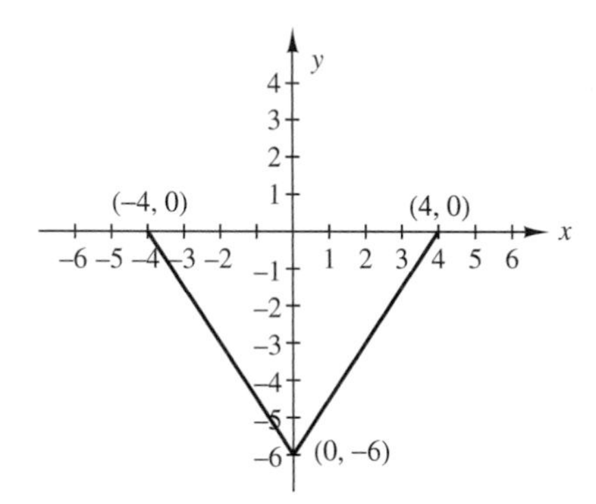

Figure 20.9 $y = f(x)$ on $[-4, 4]$

EXAMPLE 20.5 Suppose the function $f(x)$, whose graph is sketched in Figure 20.9, represents the rate of change of population (in hundreds) at time x in an inner city area over the 8-year period beginning 4 years ago. That is, we take the current year as time $x = 0$, so 4 years ago is $x = -4$.

a. Use geometry to find $\int_{-4}^4 f(x)\,dx$.

b. Interpret what the integral in part a represents in this context.

Solution

a. Since the graph of $y = f(x)$ lies entirely below the x-axis, the integral is the negative of the area of the region bounded by the x-axis and the curve. This region is a triangle with base $b = 8$ and height $h = 6$. The total area of the triangle is $\frac{1}{2} \cdot b \cdot h = \frac{1}{2} \cdot 8 \cdot 6 = 24$, so $\int_{-4}^{4} f(x)\, dx = -24$.

b. Since the population growth rate function is negative, this indicates the population is declining. The integral $\int_{-4}^{4} f(x)\, dx = -24$, which is negative, indicates the population declined by 2400 people over the 8-year period. ■

Now suppose that the continuous function f takes on both positive and negative values on the interval $[a, b]$. How do we interpret $\int_{a}^{b} f(x)\, dx$ in terms of the area of the region bounded by $y = f(x)$ and the x-axis? Suppose $f(x) < 0$ for x in $[a, c)$ and $f(x) > 0$ for x in $(c, b]$. Then $\int_{a}^{b} f(x)\, dx$ represents the difference of two areas: the area of the region bounded by the function $f(x)$ and the x-axis for the portion of the graph that lies above the x-axis, **minus** the area of the region bounded by the x-axis and the function $f(x)$ for the portion of the graph that lies below the x-axis. (See Figure 20.10.)

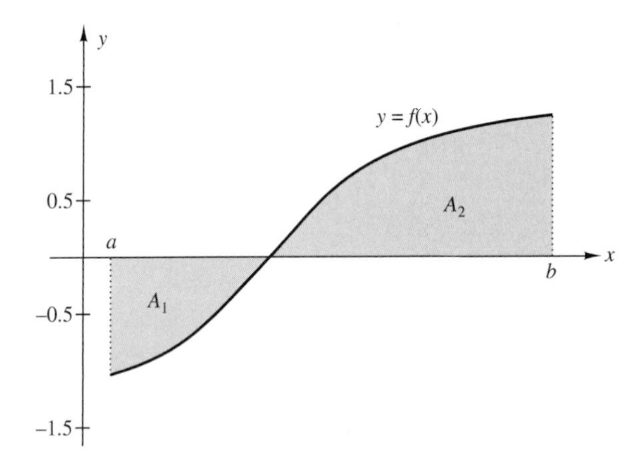

Figure 20.10 $\int_{a}^{b} f(x)\, dx = A_2 - A_1$

When $f(x)$ changes sign on an interval $[a, b]$, the definite integral $\int_{a}^{b} f(x)\, dx$ represents a *net change* in area or, in general, a net change in amount. Although area is always positive, net change in area or net change in amount can be positive, negative, or zero. The next three examples illustrate net change.

EXAMPLE 20.6 Use geometry to find $\int_0^5 4 - 3x\,dx$.

Solution We first sketch a graph of the function $f(x) = 4 - 3x$ on the interval [0, 5] to determine for what values of x the function lies above the x-axis and for what values of x it lies below the x-axis. The graph is sketched in Figure 20.11. Since the region below the x-axis is larger than the region above it, we expect the value of the integral to be negative.

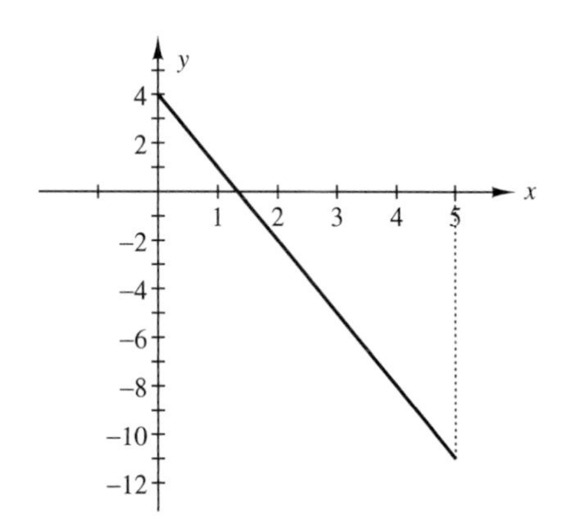

Figure 20.11 $f(x) = 4 - 3x$ for $0 \le x \le 5$

The integral is the area of the triangle that lies above the x-axis minus the area of the triangle that lies below the x-axis. The line $y = 4 - 3x$ has x-intercept $\frac{4}{3}$ so the triangle that lies above the x-axis has base length $\frac{4}{3}$ and height 4; the triangle that lies below the x-axis has base length $5 - \frac{4}{3} = \frac{11}{3}$ and height $|f(5)| = |-11| = 11$. Thus,

$$\int_0^5 4 - 3x\,dx = \left(\frac{1}{2}\right)\left(\frac{4}{3}\right)(4) - \left(\frac{1}{2}\right)\left(\frac{11}{3}\right)(11) = \frac{16}{6} - \frac{121}{6} = -\frac{105}{6} = -\frac{35}{2}. \qquad \blacksquare$$

EXAMPLE 20.7 A company makes novelty T-shirts. Suppose $P'(x) = 2 - 0.002x$ represents the marginal profit (in dollars per shirt) when production level is x shirts.

a. Find the net change in profit for the interval [0, 1000] by finding $\int_0^{1000} P'(x)\,dx$.

b. Find the net change in profit for the interval [0, 2000] by finding $\int_0^{2000} P'(x)\,dx$.

c. Suppose we know that $P(x) = -0.001x^2 + 2x - 500$ represents the profit function when production level is x shirts. Verify your results in parts a and b by using $P(x)$ to find the net change in profit on the interval for the intervals [0, 1000] and [0, 2000].

Solution Figure 20.12 shows the graph of the function $y = P'(x) = 2 - 0.002x$ on the interval [0, 2000].

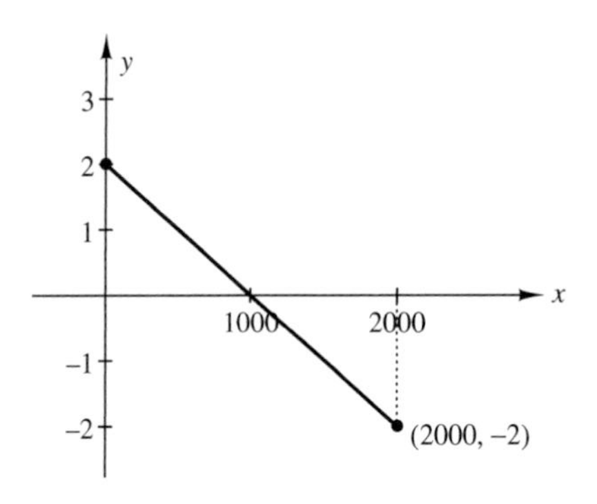

Figure 20.12 $y = P'(x) = 2 - 0.002x$ for $0 \le x \le 2000$

a. Since $P'(x) \ge 0$ for x in [0, 1000], $\int_0^{1000} P'(x)\,dx$ is the area of the triangle whose base is 1000 and whose height is 2. Thus, net change in profit = area = $\frac{1}{2} \cdot b \cdot h = \frac{1}{2}(1000)(2) = 1000$.

b. We see that $P'(x) > 0$ for x in (0, 1000) and $P'(x) < 0$ for x in (1000, 2000). Thus, the net change in profit over the interval [0, 2000] is the area of the region above the x-axis minus the area of the region below the x-axis, which is $\frac{1}{2}(1000)(2) - \frac{1}{2}(1000)(2) = 0$.

c. The net change in profit on the interval [0, 1000] is the difference between the profit when 1000 items are produced and when 0 items are produced: $P(1000) - P(0) = [-(0.001)(1000)^2 + 2(1000) - 500] - [-500] = -1000 + 2000 - 500 + 500 = 1000$. The net change in profit on the interval [0, 2000] is calculated similarly: $P(2000) - P(0) = [-(0.001)(2000)^2 + 2(2000) - 500] - [-500] = -500 + 500 = 0$. These results are the same as those found in parts a and b. ■

EXAMPLE 20.8 The rate of change of the temperature of a stone (in degrees Celsius) with respect to time t (in hours) is given by the function $T'(t) = 40 - 30t$, $0 \le t \le 2$.

a. Find the stone's net change in temperature over the time interval $0 \le t \le 1$.

b. Write the stone's net change in temperature over the time interval $0 \le t \le 1$ as a definite integral.

c. Find the stone's net change in temperature over the time interval $0 \le t \le 2$.

d. Write the stone's net change in temperature over the time interval $0 \le t \le 2$ as a definite integral.

Solution

a. The graph of $y = T'(t) = 40 - 30t$, $0 \le t \le 2$ is in Figure 20.13. The net change in temperature over the time interval $0 \le t \le 1$ can be interpreted as the area of the trapezoid formed by the line $y = 40 - 30t$, the lines $t = 0$ and $t = 1$, and the t-axis. This area is $\frac{1}{2}(b_1 + b_2)h = \frac{1}{2}(T'(0) + T'(1))(1 - 0) = \frac{1}{2}(40 + 10)(1) = 25$. So the net change in temperature is 25° for the first hour; that is, the stone is 25° warmer after 1 hour.

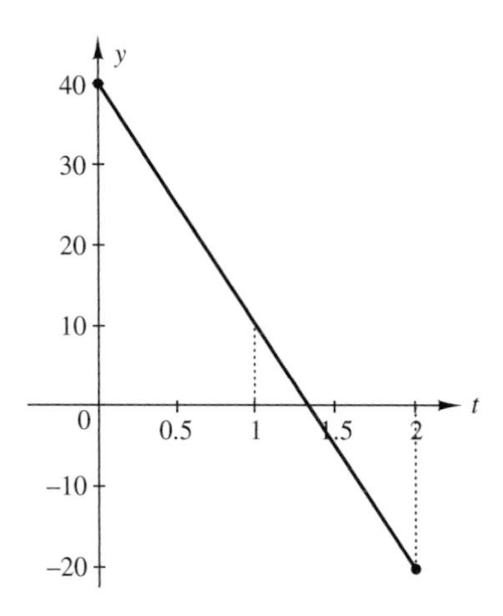

Figure 20.13 $y = T'(t) = 40 - 30t$ for $0 \le t \le 2$

b. The net change in temperature over the time interval $0 \le t \le 1$ can be represented as $\int_0^1 40 - 30t\, dt$.

c. We find that the graph of $y = T'(t) = 40 - 30t$ crosses the t-axis at $t = \frac{4}{3}$ by solving $T'(t) = 0$. Thus the net change in temperature over the time interval $0 \le t \le 2$ is the area of the triangular region above the t-axis bounded by the lines $y = 40 - 30t$ and $t = 0$, minus the area of the triangular region below the t-axis bounded by the lines $y = 40 - 30t$ and $t = 2$. We get $\left(\frac{1}{2}\right)\left(\frac{4}{3}\right)40 - \left(\frac{1}{2}\right)\left(\frac{2}{3}\right)20 = \frac{80}{3} - \frac{20}{3} = \frac{60}{3} = 20$. So the net change in the stone's temperature is 20° for the 2-hour period.

d. The net change in temperature over the time interval $0 \le t \le 2$ can be represented as $\int_0^2 40 - 30t\, dt$. ■

If a definite integral is set up specifically to find the area of a region in the plane, its value **must** be positive. In general, however, the value of a definite integral $\int_a^b f(x)\, dx$ can be positive, negative, or zero, depending on $f(x)$ and the interval $[a, b]$.

EXAMPLE 20.9 Consider the function $f(x) = 4 - (x + 1)^2$. For what value of $a > 0$ is $\int_0^a f(x)\, dx$ a maximum?

Solution We sketch the graph of $y = f(x) = 4 - (x + 1)^2$ for $x \geq 0$ to determine where the graph lies above the x-axis and where it lies below the x-axis. The graph is sketched in Figure 20.14. (Note that we can obtain the graph of $y = f(x)$ from the graph of $y = x^2$ using three transformations.)

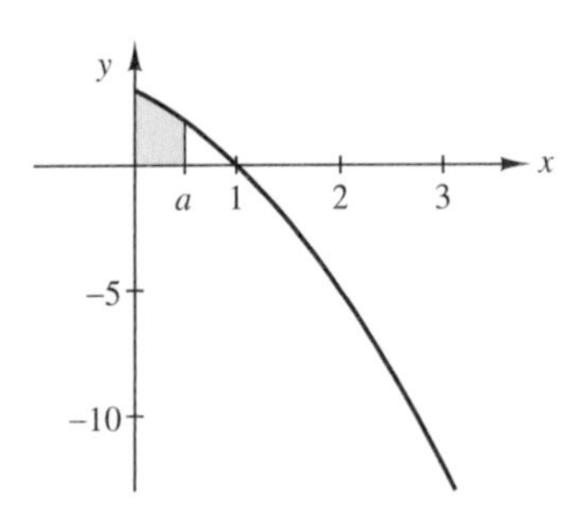

Figure 20.14 $f(x) = 4 - (x + 1)^2$

The graph lies above the x-axis for x between 0 and 1 and lies below the x-axis for $x > 1$. For $0 < a \leq 1$, the integral from 0 to a represents the area of the region (shaded in Figure 20.14) bounded by $f(x)$, the x-axis, and the lines $x = 0$ and $x = a$. As a increases from 0 to 1, the area under the curve above the x-axis increases. For $a > 1$, the integral from 0 to a represents the area of the region bounded by $f(x)$, the x-axis, and the lines $x = 0$ and $x = 1$, minus the area of the region bounded by $f(x)$, the x-axis, and the lines $x = 1$ and $x = a$. Thus, as a increases beyond 1, the value of the integral decreases; this means that the integral is a maximum when $a = 1$. ■

Exercises 20-B

1. Use geometry to find a value of $b > 0$ so that $\int_0^b 4 - 3x\,dx = 0$. Sketch the graph of $y = 4 - 3x$ for $x > 0$ to begin.

2. Sketch each function on the interval given by the limits of integration in the definite integral, and determine where on that interval the function lies above the x-axis and where it lies below. Use geometry to evaluate the integral.

a. $f(x) = 14;\ \int_{-2}^{3} 14\,dx$

b. $g(x) = -9;\ \int_1^7 -9\,dx$

c. $h(x) = 2x - 5;\ \int_0^4 2x - 5\,dx$

d. $k(x) = x - 4;\ \int_1^3 x - 4\,dx$

e. $m(x) = 7 - 3x;\ \int_{-1}^{4} 7 - 3x\,dx$

f. $p(x) = \sqrt{4 - x^2};\ \int_0^2 \sqrt{4 - x^2}\,dx$

3. Let $f(x) = 2$ and let $g(x) = x + 1$.

a. Sketch the graphs of $y = f(x)$ and $y = g(x)$ on separate axes.

b. Find a number $c > -1$ so that $\int_{-1}^{c} f(x)\,dx = \int_{-1}^{c} g(x)\,dx$.

c. If possible, find an interval $[a, b]$ so that $\int_{a}^{b} f(x)\,dx$ is negative. If it is not possible, explain why not.

d. If possible, find an interval $[a, b]$ so that $\int_{a}^{b} g(x)\,dx$ is negative. If it is not possible, explain why not.

e. If possible, find an interval $[a, b]$ so that $\int_{a}^{b} g(x)\,dx = 0$. If it is not possible, explain why not.

4. Consider the function $f(x)$ whose graph is given in Figure 20.15.

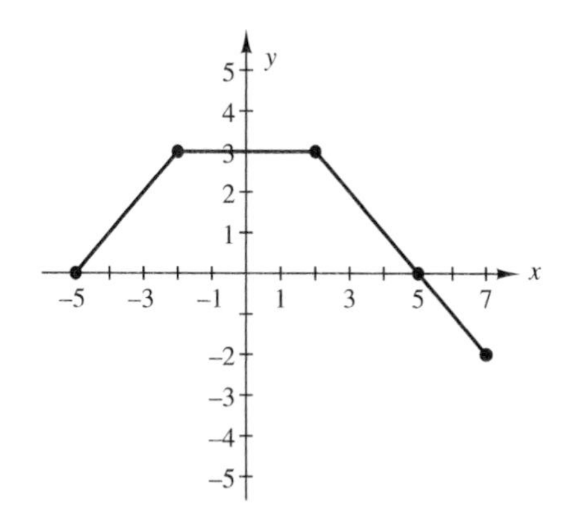

Figure 20.15

a. Find $\int_{-2}^{2} f(x)\,dx$.

b. Find $\int_{0}^{5} f(x)\,dx$.

c. Use your answer to part b to find $\int_{-5}^{5} f(x)\,dx$.

d. Determine which of the following integrals is larger, without evaluating the integrals: $\int_{0}^{5} f(x)\,dx$ or $\int_{0}^{7} f(x)\,dx$. Explain your answer.

5. A ball is thrown vertically into the air. The velocity of the ball (in feet per second) t seconds after it is thrown is given by $v(t) = 128 - 32t$.

a. Draw the graph of $y = v(t)$ for $0 \leq t \leq 6$.

b. Use geometry to find the displacement (the net change in vertical position, measured in feet from the ground) of the ball over the time interval $0 \leq t \leq 2$.

c. Set up a definite integral that represents the displacement of the ball over the time interval $0 \leq t \leq 2$.

d. Use geometry to find the displacement of the ball over the time interval $0 \le t \le 6$.

e. Set up a definite integral that represents the displacement of the ball over the time interval $0 \le t \le 6$.

6. Suppose $g(t)$ denotes the temperature (in degrees Fahrenheit) in a room t hours after turning on a fan. The instantaneous rate of change in temperature in the room is $g'(t) = -2 - 4(t+1)^{-1}, 0 \le t \le 3$.

a. Set up a definite integral that represents the temperature change for $0 \le t \le 3$.

b. Is the value of the integral in part a positive or negative? Explain why.

7. The instantaneous growth rate of the population of a metropolitan region at time t is given in hundreds of people per year by the function $p(t) = 25t^2 - 15\sqrt{t}$, where t represents years from 1980.

a. Set up a definite integral that represents the net change in population from 1980 to 1990.

b. Set up a definite integral that represents the net change in population from 1985 to 2000.

20-C THE FUNDAMENTAL THEOREM OF CALCULUS

The processes of differentiation and of finding a definite integral are linked by one of the most important theorems in calculus. The Fundamental Theorem of Calculus establishes the intimate connection between the definite integral and the antiderivative. This justifies the use of the integral notation for an antiderivative. The Fundamental Theorem of Calculus has two parts.

20-C.1 The Fundamental Theorem of Calculus, Part I

The first part of the Fundamental Theorem of Calculus shows how an integral can be used to provide an antiderivative for a function.

Part I of the Fundamental Theorem of Calculus: If $f(t)$ is continuous on the interval $[a, b]$, then the function F defined by

$$F(x) = \int_a^x f(t)\, dt,\ a \le x \le b$$

is continuous on $[a, b]$ and differentiable on (a, b) and $F'(x) = f(x)$.

In other words, Part I of the Fundamental Theorem of Calculus says that the function $F(x) = \int_a^x f(t)\,dt$ is an antiderivative of f on the interval (a, b).

Note that in the expression for $F(x)$, we are integrating a function of t, and x is a limit of integration. This means that for each value of x, we have a definite integral $\int_a^x f(t)\,dt$.

The expression $F(x)$ defined in Part I of the Fundamental Theorem of Calculus satisfies the definition of a function with domain $[a, b]$ because for each value of x in the interval $[a, b]$, there is a unique value of $F(x)$. If $f(t)$ lies entirely above the t-axis on $[a, b]$, then $F(x)$ can be interpreted as the area of the region bounded by the curve $y = f(t)$, the t-axis, and the lines $t = a$ and $t = x$. See Figure 20.16a for a picture of $f(t)$ and $F(x)$ when $f(t)$ is positive on $[a, b]$.

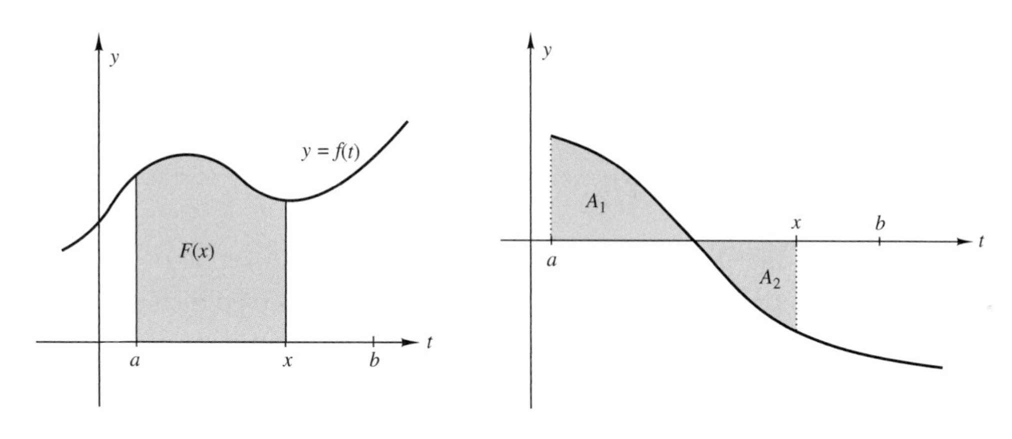

Figure 20.16a **Figure 20.16b** $F(x) = A_1 - A_2$

In general, $F(x)$ is the area of the region bounded by $f(t)$ and the t-axis for the portion of the graph that lies above the t-axis minus the area bounded by $f(t)$ and the t-axis for the portion that lies below the t-axis on the interval (a, x). See Figure 20.16b.

EXAMPLE 20.10 Let $F(x) = \int_1^x (t+3)\,dt$. Use geometry to find the value of $F(x)$ for the following x-values: $x = 2$, $x = 3.5$, $x = 11$.

Solution The function $f(t) = t + 3$ lies above the t-axis for $t \geq 1$, so $F(2)$ represents the area of the trapezoid formed by the line $f(t) = t + 3$, the t-axis, and the lines $t = 1$ and $t = 2$, as shown in Figure 20.17. Thus, $F(2) = \frac{1}{2}[f(1) + f(2)] \cdot (2 - 1) = \frac{1}{2}(4 + 5) \cdot 1 = \frac{9}{2}$.

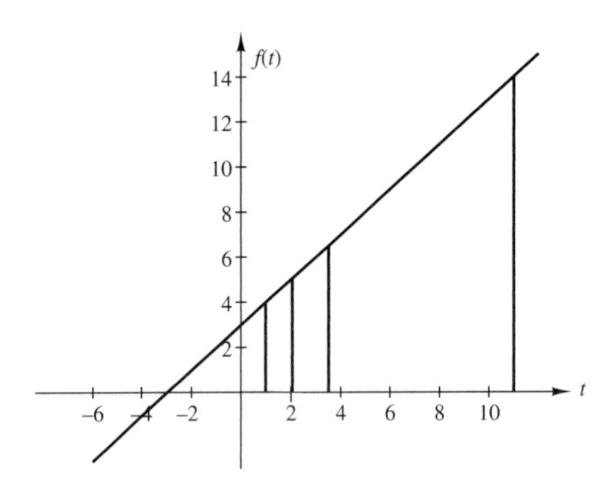

Figure 20.17 $f(t) = t + 3$

The value $F(3.5)$ is the area of the trapezoid formed by the line $f(t) = t + 3$, the t-axis, and the lines $t = 1$ and $t = 3.5$. Thus, $F(3.5) = \frac{1}{2}(4 + 6.5) \cdot 2.5 = 13.125$. Similarly, $F(11) = \frac{1}{2}(4 + 14) \cdot 10 = 90$. ■

EXAMPLE 20.11 Let $G(x) = \int_{-6}^{x} (t + 3)\, dt$. Refer to Figure 20.17 and use geometry to find the value of $G(x)$ for the following x-values: $x = -3$, $x = 0$, $x = 11$.

Solution Since the graph of $f(t) = t + 3$ lies below the t-axis on the interval $(-6, -3)$, $G(-3)$ is the negative of the area of the triangle formed by the line $f(t)$, the t-axis, and the line $t = -6$. This triangle has base $b = 3$ and height $h = 3$, so $G(-3) = -\frac{1}{2}bh = -\frac{1}{2}(3)(3) = -\frac{9}{2}$. The graph of f lies above the t-axis for t in the interval $(-3, 0)$, so $G(0)$ is the area of the triangle formed by the graph of f, the t-axis, and the line $t = 0$ minus the area of the triangle formed by the graph of f, the t-axis, and the line $t = -6$. (This is also the area of the triangle formed by the graph of f, the t-axis, and the line $t = 0$ plus $G(-3)$.) Each of these triangular regions has base $b = 3$ and height $h = 3$, so $G(0) = \frac{1}{2}(3)(3) - \frac{1}{2}(3)(3) = 0$. Since the graph of f lies above the t-axis for t in the interval $(-3, 11)$, $G(11)$ is the area of the triangle formed by the graph of f, the t-axis, and the line $t = 11$, plus $G(-3)$. Thus, $G(11) = \frac{1}{2}(14)(14) + \left(-\frac{9}{2}\right) = \frac{187}{2}$. ■

EXAMPLE 20.12 Let $H(x) = \int_{0}^{x} (t + 3)\, dt,\ x > 0$.

a. Use geometry and the formula for the area of a trapezoid to find a symbolic expression for $H(x)$.

b. Differentiate the formula found in part a.

Solution

a. We refer to the graph in Figure 20.17. We want to express the area of the trapezoid formed by the line $f(t) = t + 3$, the t-axis, and the lines $t = 0$ and $t = x$. This trapezoid has base $b_1 = 0 + 3 = 3$ and base $b_2 = x + 3$. The value of b_2 changes as x changes. The height h is x. Thus the area of the trapezoid is $\frac{1}{2}(b_1 + b_2) \cdot h = \frac{1}{2}(3 + x + 3) \cdot x = \frac{1}{2}x(x + 6)$. Therefore, $H(x) = \frac{1}{2}x(x + 6)$.

b. $H(x) = \frac{1}{2}x(x + 6) = \frac{1}{2}x^2 + 3x$, so $H'(x) = x + 3 = f(x)$. This verifies Part I of the Fundamental Theorem of Calculus for the function $f(t) = t + 3$ on the interval $[0, \infty)$. ■

20-C.2 The Fundamental Theorem of Calculus, Part II

Consider a car that is traveling along a straight highway. The area under the velocity curve, $v = f(t)$, $0 \le t \le 6$, is the exact distance traveled by the car during that time interval. It is obtained by taking sums of terms of the form *rate* × *time* over smaller and smaller time intervals. The definite integral $\int_0^6 f(t)\, dt$ is the limit, as the length of each time interval approaches 0, of the sum of all these distances. Thus, total distance d traveled by the car with velocity $v = f(t)$ from time $t = 0$ to time $t = 6$ is $d = \int_0^6 f(t)\, dt$.

Suppose the car is traveling along a straight highway and its *position* from its starting point can be measured at any time. If the car always travels in the same direction, then the total distance traveled by the car is also the *change in position* of the car from time $t = 0$ to time $t = 6$. Thus the total distance traveled by the car is

(position of the car at time $t = 6$) – (position of the car at time $t = 0$).

Let $s(t)$ represent the position of the car at time t. Then the velocity $v = f(t)$ of the car at time t is the rate of change of position with respect to time, so $v = f(t) = s'(t)$. Thus we have the following:

(distance traveled by the car from time $t = 0$ to time $t = 6$) =
change in position = $s(6) - s(0) = \int_0^6 f(t)\, dt = \int_0^6 s'(t)\, dt$.

In general, when we find the definite integral of a rate of change function over an interval, we are finding a change in amount over that interval. This result illustrates

Part II of the Fundamental Theorem of Calculus: If f is continuous on the interval $[a, b]$, then $\int_a^b f(x)\, dx = F(b) - F(a)$, where F is any antiderivative of f.

In other words, if we know an antiderivative for f on the interval (a, b) (that is, $F'(x) = f(x)$ on (a, b)), we can find the value of the definite integral $\int_a^b f(x)\,dx$ by evaluating the antiderivative at b and at a and taking the difference.

In the next example, we verify this statement for the function $f(x) = x + 3$ on the interval $[-1, 4]$.

EXAMPLE 20.13 Consider the definite integral $\int_{-1}^{4} (x+3)\,dx$. Verify Part II of the Fundamental Theorem of Calculus for this integral by evaluating the integral in two ways:

a. Use the Fundamental Theorem of Calculus to evaluate the integral.

b. Use geometry to find the area of the trapezoid whose area is expressed by the definite integral.

Solution

a. To use Part II of the Fundamental Theorem of Calculus, we need to find an antiderivative $F(x)$ of $f(x) = x + 3$. In Example 20.12b we found that $\frac{1}{2}x^2 + 3x$ is an antiderivative of $f(x)$. So we can let $F(x) = \frac{1}{2}x^2 + 3x$; then $\int_{-1}^{4} (x+3)\,dx =$ $F(4) - F(-1) = \left[\frac{1}{2}(4)^2 + 3(4)\right] - \left[\frac{1}{2}(-1)^2 + 3(-1)\right] = [20] - \left[-\frac{5}{2}\right] = 22.5.$

b. The trapezoid whose area we want is shaded in Figure 20.18. The area of the trapezoid formed by the line $y = x + 3$, the x-axis, and the lines $x = -1$ and $x = 4$ is $\frac{1}{2}(2+7)\cdot 5 = \frac{45}{2} = 22.5$. This agrees with our answer in part a.

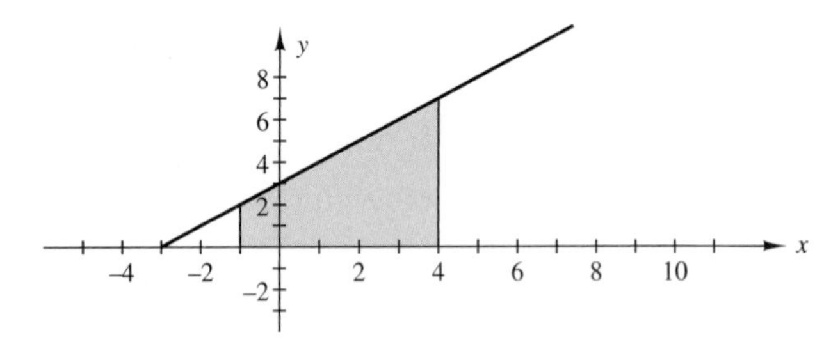

Figure 20.18 $f(x) = x + 3$ ■

Exercises 20-C

1. Let $f(t) = 3$ and let $F(x) = \int_{-2}^{x} f(t)\,dt = \int_{-2}^{x} 3\,dt$.

a. Sketch the graph of $f(t) = 3$ on the interval $[-2, 12]$.

b. Sketch the region bounded by $f(t) = 3$, the t-axis, and the lines $t = -2$ and $t = x$ for some x in the interval $[-1, 10]$.

c. Use a geometric formula to complete the table below.

x	0	1	2.5	4.7	5	12
$F(x)$						

d. Use geometry to find a symbolic formula for $F(x)$.

e. Differentiate the formula found in part d to show that $F'(x) = f(x)$.

f. What theorem is illustrated by part e?

2. Let $f(t) = 4t$ and let $F(x) = \int_0^x f(t)\,dt = \int_0^x 4t\,dt$.

a. Sketch the graph of $y = 4t$ on the interval $[0, 15]$.

b. Sketch the region bounded by $y = 4t$, the t-axis, and the lines $t = 0$ and $t = x$ for some x in the interval $[1, 12]$.

c. Use a geometric formula to complete the table below.

x	1	5.5	7	10	15
$F(x)$					

d. Use geometry to find a symbolic formula for $F(x)$.

e. Differentiate the formula found in part d to show that $F'(x) = f(x)$.

3. Let $f(t) = -2t$ and let $F(x) = \int_0^x f(t)\,dt = \int_0^x (-2t)\,dt$.

a. Sketch the graph of $y = -2t$ on the interval $[0, 20]$.

b. Sketch the region bounded by $y = -2t$, the t-axis, and the lines $t = 0$ and $t = x$ for some x in the interval $[1, 15]$.

c. Use a geometric formula to complete the table below.

x	1.5	3	6	7.8	20
$F(x)$					

d. Use geometry to find a symbolic formula for $F(x)$.

e. Differentiate the formula found in part d to show that $F'(x) = f(x)$.

4. Let $f(t) = 5t + 2$ and $F(x) = \int_1^x f(t)\ dt = \int_1^x 5t + 2\ dt$.

a. Sketch the graph of $y = 5t + 2$ on the interval [1, 15].

b. Sketch the region bounded by $y = 5t + 2$, the t-axis, and the lines $t = 1$ and $t = x$ for some x in the interval [2, 12].

c. Use a geometric formula to complete the table below.

x	1.5	5	8	10.5	15
$F(x)$					

d. Use geometry to find a symbolic formula for $F(x)$.

e. Differentiate the formula found in part d to show that $F'(x) = f(x)$.

5. Let $f(x) = 5$ and $F(x) = 5x$. Consider $\int_1^3 5\ dx$.

a. Verify that $F'(x) = f(x)$.

b. Use the Fundamental Theorem of Calculus, Part II to evaluate the integral.

c. What area does $\int_1^3 5\ dx$ represent? Sketch the region and use geometry to find its area.

6. Let $g(x) = -0.6x^2 + 6.4$; this is the function whose graph is the parabolic arc at the top of the window pictured in Figure 20.1b.

a. Let $G(x) = -0.2x^3 + 6.4x$. Verify that $G'(x) = g(x)$.

b. Use the Fundamental Theorem of Calculus to find the area of the window.

c. Which window, the one with the circular arc or the one with the parabolic arc, has the larger area?

7. For each function $f(x)$ and interval $[a, b]$ given below, graph the function on the interval and look at the graph to decide if $\int_a^b f(x)\ dx$ is positive, negative, or zero.

a. $f(x) = 2x^3 - 6x + 3$; $[a, b] = [-2, 2]$

b. $f(x) = 2 + x - x^4$; $[a, b] = [-2, 3]$

c. $f(x) = \tan x$; $[a, b] = [\frac{2\pi}{3}, \pi]$

d. $f(x) = \sin x$; $[a, b] = [-2\pi, 0]$

20-D CHANGE OF VARIABLE IN DEFINITE INTEGRALS

When we use the Fundamental Theorem of Calculus to evaluate a definite integral of the form $\int_a^b f(x)\,dx$, first we need to find an antiderivative of $f(x)$—that is, a function $F(x)$ satisfying $F'(x) = f(x)$. Then we evaluate $F(x)$ at $x = b$ and at $x = a$ and find the difference of these values. We often use substitution to find an antiderivative of $f(x)$. We then need to take extra care when we evaluate the antiderivative. We illustrate with an example.

EXAMPLE 20.14 Use the Fundamental Theorem of Calculus to evaluate the definite integral $\int_1^2 (2x+1)^3\,dx$.

Solution To use the Fundamental Theorem of Calculus, we first need to find an antiderivative of $f(x) = (2x + 1)^3$. We can use substitution (see Section 18-C) and let $u = 2x + 1$. Then $du = 2\,dx$, so $\frac{1}{2}du = dx$. We then substitute for $(2x + 1)^3$ and dx to obtain an indefinite integral in u. We integrate with respect to u, and finally we substitute $2x + 1$ for u in the answer to obtain an antiderivative as a function of x:

$$\int (2x+1)^3\,dx = \int u^3\left(\frac{1}{2}\right)du = \left(\frac{1}{2}\right)\frac{u^4}{4} + C = \frac{u^4}{8} + C = \frac{1}{8}(2x+1)^4 + C.$$

The function $\frac{1}{8}(2x + 1)^4 + C$ is the most general antiderivative of $(2x + 3)^3$. To evaluate the given definite integral, we can choose any antiderivative, so we choose $C = 0$ and use the antiderivative $F(x) = \frac{1}{8}(2x + 1)^4$. We evaluate $F(x)$ at $x = 2$ and at $x = 1$ and find the difference of these values: $F(2) = \frac{1}{8}(2 \cdot 2 + 1)^4 = \frac{1}{8}(5)^4 = \frac{625}{8}$ and $F(1) = \frac{1}{8}(2 \cdot 1 + 1)^4 = \frac{1}{8}(3)^4 = \frac{81}{8}$. Thus, $\int_1^2 (2x+1)^3\,dx = \frac{625}{8} - \frac{81}{8} = \frac{544}{8} = 68$. ■

We could evaluate $\int_1^2 (2x+1)^3\,dx$ another way. After we conclude that $\int (2x+1)^3\,dx = \int u^3\left(\frac{1}{2}\right) du$, we could work directly with the variable u and obtain the limits of integration of the original definite integral in terms of the variable u. Since $u = 2x + 1$, we evaluate u at the limits of integration $x = 1$ and $x = 2$. When $x = 1$, $u = 2(1) + 1 = 3$, and when $x = 2$, $u = 2(2) + 1 = 5$. We then change the definite integral in x to an equivalent definite integral in u:

$$\int_{x=1}^{x=2} (2x+1)^3\,dx = \int_{u=3}^{u=5} u^3\left(\frac{1}{2}\right) du = \left(\frac{1}{2}\right)\frac{u^4}{4}\Bigg|_{u=3}^{u=5}.$$

The vertical line symbol, with $u = 3$ and $u = 5$ following it, means "evaluated from $u = 3$ to $u = 5$." We now complete the problem by evaluating $\frac{1}{8}u^4$ for $u = 5$ and for $u = 3$, and then taking their difference $\frac{5^4}{8} - \frac{3^4}{8}$ to obtain $\frac{544}{8} = 68$, which agrees with our result using the first method.

When using substitution to evaluate a definite integral, we can use one of the following approaches.

1. Find an antiderivative in terms of the **original** variable and then evaluate this antiderivative at the original limits of integration; or

2. Find an antiderivative in terms of the new **substitution** variable, find the new limits of integration in terms of the new substitution variable, and evaluate the antiderivative at the new substitution variable limits of integration.

Before evaluating a definite integral in which substitution (change of variable) has been used, always make sure the limits of integration are consistent with the variable in the antiderivative.

Exercises 20-D

1. In a–d, use the given u substitution to rewrite the given integral as a definite integral of a function of u with appropriate limits of integration. Use the Fundamental Theorem of Calculus to evaluate the definite integral without converting back to the original variable x.

a. $u = 3x + 1$; $\int_0^3 \frac{1}{3x + 1}\,dx$

b. $u = x^2 + 2$; $\int_0^2 x\sqrt{x^2 + 2}\,dx$

c. $u = \ln x$; $\int_1^4 \frac{\ln x}{x}\,dx$

d. $u = 6x - 1$; $\int_{-1}^3 (6x - 1)^5\,dx$

2. For each part a–d in Exercise 1, use your indefinite integral in terms of u and the given u substitution to find an antiderivative in terms of x (the original variable). Then evaluate this antiderivative at the original limits of integration to find the value of the definite integral.

Chapter 20 Exercises

1. For each of the sums given below, write a definite integral that is approximated by that sum over the specified interval.

a. $\sum_{i=1}^{n} 2x_i \, \Delta x_i$ on the interval [2, 5]

b. $\sum_{i=1}^{n} \sqrt{3 - x_i} \, \Delta x_i$ on the interval [6, 8]

2. Let R be the region bounded by the curve $y = \cos x$ and the x-axis on the interval $\left[0, \frac{\pi}{2}\right]$.

a. Give a Riemann sum of the form $\sum_{i=1}^{n} f(x_i) \, \Delta x_i$ that approximates the area of R.

b. Write a definite integral that gives the area of R.

c. Sketch the graph of $y = \cos x$ on $[0, \pi]$, and find the value of $\int_0^{\pi} \cos x \, dx$.

3. Write a definite integral that represents the area of the region bounded by the curve $y = \sqrt{4 - x}$, the x-axis, and the y-axis.

4. Write a definite integral for the area of the shaded region shown in Figure 20.19.

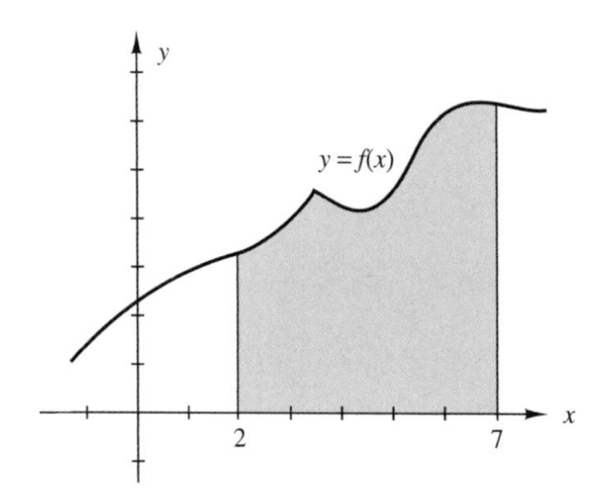

Figure 20.19

5. **a.** Sketch the graph of a continuous function $f(x)$ that is not constant on the interval [–2, 2] and that satisfies $\int_{-2}^{2} f(x) \, dx = 0$.

b. Explain why the integral of your function in part a is 0.

6. Consider the graph of $f(t)$ given in Figure 20.20.

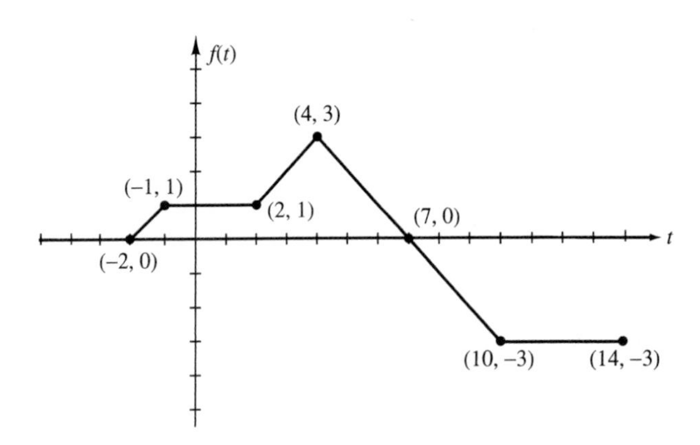

Figure 20.20

a. Find $\int_0^2 f(t)\,dt$. **b**. Find $\int_2^4 f(t)\,dt$.

c. Find $\int_4^{10} f(t)\,dt$. **d**. Find $\int_{-2}^{10} f(t)\,dt$.

7. Let $f(t)$ be the function sketched in Figure 20.20. Let $F(x) = \int_0^x f(t)\ dt$.

a. Find $F(2)$. **b**. Find $F(7)$. **c**. Find $F(14)$.

d. Is $F(x)$ increasing on the interval [0, 4]? Explain your answer.

e. Find the largest interval $[0, b]$ on which $F(x)$ is an increasing function.

f. For what value of x in the interval [0, 14] is $F(x)$ a maximum? Explain your answer.

8. Let $f(t)$ be the function sketched in Figure 20.20. Let $G(x) = \int_{-2}^x f(t)\ dt$.

a. Find $G(0)$. **b**. Find $G(2)$.

c. Find $G(7)$. **d**. Find $G(14)$.

e. Explain, in terms of areas of regions bounded by the graph in Figure 20.20, what the difference $G(14) - G(7)$ represents.

9. What is the relationship between $F(x)$ from Exercise 7 and $G(x)$ from Exercise 8? Explain your answer.

10. Because of changes in birthrates and population demographics, the population growth rate (in hundreds of students per year) at time t at an elementary school over the 10-year period from 1995 to 2005 was found to be given by

$$p(t) = \begin{cases} t+4 & 0 \le t \le 5 \\ 24-3t & 5 < t \le 10, \end{cases}$$

where t represents years from 1995.

a. Sketch the graph of $p(t)$ on the interval $0 \le t \le 10$.

b. Use geometry to find the net population growth over the period 1995 to 2000.

c. Use geometry to find the net population growth over the period 2000 to 2005.

d. Use geometry to find the net population growth over the period 2003 to 2005.

e. Interpret and compare your answers to parts b, c, and d above.

11. The production rate of a particular oil field (in thousands of barrels per year) is given by

$$p(t) = 80 - \frac{20}{3}t, \ 0 \le t \le 12.$$

a. Sketch the graph of $y = p(t)$, $0 \le t \le 12$.

b. Set up a definite integral that represents the total production during the first 6 years.

c. Use geometry to find the value of the definite integral in part b.

d. Set up a definite integral that represents the total production during the second 6-year period ($6 \le t \le 12$).

e. Use geometry to find the value of the definite integral in part d.

f. Interpret and compare your answers to parts c and e.

12. Let $f(x) = 3x + 6$ and $F(x) = \frac{3}{2}x^2 + 6x$. Consider $\int_{-2}^{4} 3x + 6\,dx$.

a. Verify that $F'(x) = f(x)$.

b. Use the Fundamental Theorem of Calculus, Part II, to evaluate the integral.

c. Use geometry to find the area represented by the definite integral.

Appendix

A ABOUT REAL NUMBERS

Numbers are everywhere in calculus. In fact, the word "calculus" comes from the same root word as "calculate." When we calculate, we use numbers and produce numbers. In calculus, functions transform numbers, variables represent numbers, points are described by numbers, and answers to problems are most often given as numbers. All of the numbers used in the first calculus course are *real numbers*. These numbers are used to measure quantity: distance, time, length, area, volume, velocity, pressure, cost, revenue, rate of inflation, temperature, growth rate, and so on.

The real numbers are most conveniently represented on an infinite number line (also called a coordinate line) in which each point on the line corresponds to exactly one number. To make this representation, we begin with an unmarked line, choose a point to represent 0 (zero), and choose a unit length. The number 0 acts as a dividing point on the number line, separating the positive and negative real numbers. The positive numbers are traditionally placed to the right of 0, and the negative numbers to the left of 0. In this representation, a positive real number r is represented by the point r units to the right of 0, and the negative real number $-r$ is represented by the point r units to the left of 0.

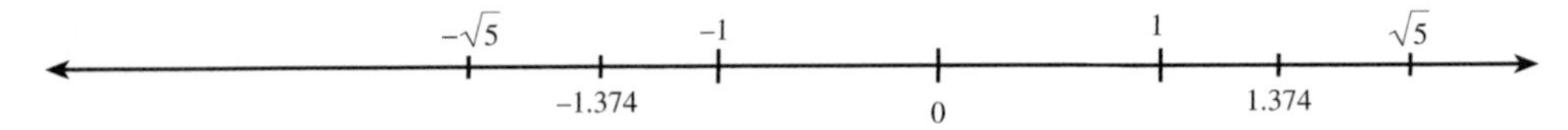

Negatives (Opposites) of Real Numbers

Every real number has an opposite, called its *negative*, and these two real numbers are at equal distances from 0 on the number line. Note that the negative of a positive real number is a negative number, and the negative of a negative real number is positive. For example, the negative of 1.374 is -1.374, and the negative of -1.374 is 1.374. The sum of any number and its negative is 0. For example, $1.374 + (-1.374) = -1.374 + 1.374 = 0$. For this reason, the negative of a number is often called its *additive inverse*. To obtain the negative of any number, we multiply it by -1. For example, the negative of $\sqrt{5}$ is $-1(\sqrt{5}) = -\sqrt{5}$ and the negative of $-\sqrt{5}$ is $-1(-\sqrt{5}) = \sqrt{5}$.

Signs of Real Numbers

The sign of a real number gives important information. For example, the sign can represent direction, can indicate debit or credit, or can indicate if a quantity is increasing or decreasing. A – sign in front of a number indicates that the number is negative. Positive real numbers are most often represented without any sign but can be represented with a + sign in front.

EXAMPLE A.1 In parts a–d, the meaning of a number is given. State the meaning of its negative.

a. 10°C represents 10 degrees above freezing on the Celsius scale.

b. 25 ft/sec represents the velocity of an object moving in an upward direction at a speed of 25 ft/sec.

c. –$32.50 represents a debit or loss.

d. 1.2% population growth rate means the population is increasing by 1.2% per year.

Solution

a. –10°C represents 10 degrees below freezing on the Celsius scale.

b. –25 ft/sec means the object is traveling downward at a speed of 25 ft/sec.

c. $32.50 represents a credit or profit.

d. –1.2% population growth rate means the population is decreasing by 1.2% per year. ■

When a variable, such as x, represents a quantity that can be positive, negative, or zero, the variable does not have a sign even though at times it might be negative. In this case, $-x$ cannot be assumed to be a negative number; it is the opposite of x. When x is positive, $-x$ is negative, and when x is negative, $-x$ is positive.

Do not assume that the symbol $-x$ always represents a negative number.

Reciprocals of Real Numbers

Every real number except 0 has a *reciprocal*—this is just 1 divided by the number. The reciprocal of a nonzero real number is also called its *multiplicative inverse*. This is because the product of any nonzero real number and its reciprocal equals 1. For example, the reciprocal of 2 is $\frac{1}{2}$, and the product $2\left(\frac{1}{2}\right) = 1$. The exponent -1 is used to designate the reciprocal of a number; that is, the notation x^{-1} means $\frac{1}{x}$. (In general, the notation x^{-n} means $\frac{1}{x^n}$.) For example, in exponential notation, the reciprocal of 2 is written as 2^{-1}, so $2^{-1} = \frac{1}{2}$ and $2(2^{-1}) = 2\left(\frac{1}{2}\right) = 1$. The reciprocal of 2^3 is written as 2^{-3}, and $2^{-3} = \frac{1}{2^3} = \frac{1}{8}$. The reciprocal of a nonzero fraction is obtained by interchanging the numerator and denominator. For example, the reciprocal of $\frac{4}{3}$ is $\frac{3}{4}$ (since their product equals 1). This can be written as $\left(\frac{4}{3}\right)^{-1} = \frac{3}{4}$. The quotient of two numbers $\frac{a}{b}$ can always be rewritten as ab^{-1}, a product of the numerator and the reciprocal of the denominator; this can be useful in recasting or simplifying expressions. For example, $\frac{24}{100}$ can be rewritten as 24×100^{-1} or 24×10^{-2}.

Categories of Real Numbers

There are several categories of real numbers, each distinguished by certain properties.

1. *Counting numbers*, or *natural numbers*, are those whole numbers used to answer a question such as: How many items are there? These numbers begin with 1 and continue in order:

1, 2, 3, 4, 5, 6, 7, 8, . . . (The ellipses . . . mean continue the pattern forever.)

2. *Integers* consist of counting numbers, their negatives, and 0:

. . . , −8, −7, −6, −5, −4, −3, −2, −1, 0, 1, 2, 3, 4, 5, 6, 7, 8, . . .

3. *Rational numbers* are those numbers that can be expressed as ratios of integers (i.e., they can be expressed as fractions $\frac{m}{n}$, with m and n integers and $n \neq 0$). Here are some examples:

$$\frac{1}{2},\quad \frac{6}{7},\quad \frac{19}{3},\quad -\frac{43}{7},\quad 3,\quad 17.5,\quad 0,\quad -\frac{33}{2},\quad 0.223,\quad 1.333\ldots.$$

In this list, the integers 3 and 0 fit the definition of rational number because we can rewrite 3 and 0 as $\frac{3}{1}$ and as $\frac{0}{1}$, respectively; they can be expressed as ratios of integers. Every integer is a rational number since it can be expressed as itself divided by 1. But why do the numbers 17.5 and 0.223 (which are numbers in terminating decimal form) and 1.333... (which is a number that has an infinitely repeating block of digits) satisfy the definition of rational number? It is because 17.5 and 0.223 can be rewritten as the fractions $\frac{175}{10}$ and $\frac{233}{1000}$, respectively, and the number 1.333... can be rewritten as $\frac{4}{3}$.

In general, a number written in decimal form is rational if its nonzero digits end after a finite number of decimal places (called a *terminating decimal*) or if it has a finite number of decimal digits followed by a block of digits that repeat forever (called a *repeating decimal*). Often, to indicate that a number in decimal form is a repeating decimal, the block that repeats has a bar over it. For example, the notation $1.2\overline{34}$ represents the rational number in which the block 34 repeats forever: 1.2343434343434... .

4. *Irrational numbers* are real numbers that are not rational. This means that it is impossible to represent such a number as a quotient of two integers. In decimal form, such numbers must have an infinite number of digits and cannot have a single block of digits that repeats forever. Certain irrational numbers turn up frequently in the definition and evaluation of functions in calculus. Examples of such numbers are $\sqrt{2}$ (the length of the diagonal of a square whose side has length 1), π (the constant that is the ratio of the circumference of a circle to its diameter), and e (the base of the natural exponential function). Every irrational number can be approximated by a rational number (in fact, as close an approximation as you specify). Calculators and computers display answers in decimal form with a finite number of digits, so these answers are often rational approximations to an exact answer. To show that a decimal number is an approximation rather than an exact answer, the symbol $\approx$ is used instead of =. For example, we can write $\sqrt{2} \approx 1.41412$, $\pi \approx 3.14159$, and $e \approx 2.71828$.

The real number line is a continuum. Every point on the line represents a unique number. We mark only a few values on a coordinate line to show the scale, but every real number, whether rational or irrational, corresponds to a point on the line.

Representation of Real Numbers

We can represent real numbers in many different ways, and it is important to be able to recognize them in these different forms and to be able to convert from one form to another. For example, the number that we call one-half can be represented as $\frac{1}{2}$ (rational, or fraction, form), 0.5 (terminating decimal form), 0.499999... (repeating decimal form), $\frac{1}{\sqrt{4}}$, $\frac{4}{8}$, 2^{-1}, and many other ways. To convert a rational number from fraction form to decimal form, we divide the denominator of the fraction into the numerator. (For example, we divide 2 into 1 to get 0.5; divide 3 into 4 to get 1.333... .) In the next example we show how to convert a rational number from decimal form to fractional form.

EXAMPLE A.2 Rewrite each of the following rational numbers as a fraction in lowest terms:

a. –23.789 **b.** –0.4999... **c.** $5.\overline{43}$

Solution

a. We multiply –23.789 by 1 in the special form $\frac{1000}{1000}$ to eliminate the decimal point:

$$-23.789 = -\frac{23,789}{1000}.$$

b. The number –0.4999... has a repeating block of the single digit 9. If we multiply –0.4999... by 10, the decimal point shifts one place to the right. We can subtract the original number from this product:

$$\begin{array}{rl} 10(-0.4999...) &= -4.9999... \\ -\ \ (-0.4999...) &= +0.4999... \\ \hline 9(-0.4999...) &= -4.5 \end{array}$$

To solve for –0.4999, we divide both sides of the last equation by 9: $-0.4999... = -\frac{4.5}{9}$. To obtain a quotient of integers, we multiply by $\frac{10}{10}$: $-\frac{4.5}{9} = -\frac{45}{90} = -\frac{1}{2}$.

c. The number $5.\overline{43}$ has repeating block 43, so we multiply by 100 to shift the decimal point two places to the start of the first repeat. Then from this product, we subtract the original number:

$$\begin{array}{rl} 100(5.434343...) &= 543.4343... \\ -\ \ 1(5.434343...) &= -\ 5.4343... \\ \hline 99(5.434343...) &= 538. \end{array}$$

Finally, we divide both sides of the last equation by 99:

$$5.434343... = \frac{538}{99}.$$

■

Parts b and c in Example A.2 show that the first step in the general procedure to change a rational number in the form of a repeating decimal into the quotient of two integers is to multiply the number by a suitable power of 10 so the decimal point is shifted to the right, to a position that separates the repeating block from its first repeat. For example, to convert the repeating decimal 2.31056565656... to a fraction, we first multiply by $10^5 = 100,000$ since 100,000(2.310565656...) = 231,056.5656... . Next we multiply the number by a suitable power of 10 so the decimal point is placed in a position at the beginning of the first repeating block. In our example of 2.310565656..., we multiply by 10^3 to obtain 2310.5656... . Then we subtract the number found in the second step from the number found in the first step, and solve for the number as a quotient of two integers. In our example of 2.310565656..., we get

$$\begin{array}{r} 100,000\,(2.3105656...) = 231,056.56... \\ -\ \ 1,000\,(2.3105656...) = -\ 2,310.56... \\ \hline 99,000\,(2.3105656...) = 228,746 \end{array}$$

So $2.3105656... = \frac{228,746}{99,000}$.

Calculations with numbers in fraction form can yield exact answers. Answers in decimal form often are rounded approximations (for example, 1.66667 for $\frac{5}{3}$). Repeated use of approximations in calculations may result in roundoff error. Fractions should be simplified before complicated calculations are carried out.

Scientific Notation

For very large and very small numbers, calculators will often display the number in *scientific notation,* a number between 1 and 10 times an integer power of 10. For example, the calculator display 1.0789467E–5 means 1.0789467×10^{-5}. (In the calculator display, the E may be uppercase or lowercase; it stands for exponent.) This number is the fraction $\frac{1.0789467}{100000}$, which has decimal form 0.000010789467. Scientific notation allows several significant digits of a number to be displayed in cases where truncation or rounding would drop these digits.

Special Properties of Zero

The number 0 is defined by its most important property: When 0 is added to any real number, the sum equals the number. For example, $0 + 2\pi = 2\pi + 0 = 2\pi$, $-3.025 + 0 = -3.025$, and $0 + 0 = 0$. In other words, adding 0 to a number does not change the number. No other real number has this property. For this reason, 0 is called the *additive identity* of the real numbers. Subtracting 0 from any real number also does not change the number.

On the other hand, multiplication by zero annihilates numbers: The product of 0 with any number is 0. For example, $-6.1 \cdot 0 = 0$, $0 \cdot \sqrt{3} = 0$, and $0 \cdot 0 = 0$. Also, when 0 is divided by any nonzero real number, the quotient equals 0. This is because division by a number can be rewritten as multiplication by the reciprocal of the number. For example, $\frac{0}{4} = 0\left(\frac{1}{4}\right) = 0$.

However, **division by zero is not possible!** We say that any quotient with 0 in the denominator is *undefined*. For example, the following expressions are undefined: $\frac{5}{0}, \frac{\sqrt{2}}{0}, \frac{0}{0}$. This means that these expressions cannot represent a real number. Expressions like these can occur when quotients of functions are evaluated for certain numbers—when this occurs, questions are asked about the behavior of the function near the numbers for which the quotient is not defined. This is a central consideration in the study of limits and derivatives of functions.

Exercises A

1. Rewrite each of the following rational numbers as a fraction in lowest terms:

a. $\frac{48}{60}$ **b**. -5.318 **c**. $2.2\overline{45}$ **d**. $\left(\frac{28}{35}\right)^{-1}$ **e**. 3.079999…

2. Write the following numbers in scientific notation:

a. 3,048,000,000,000 **b.** –0.00000200314

B ABOUT SYMBOLS

Symbols in mathematics allow us to be precise and concise. The first symbols we encounter in a mathematical context are number symbols, the digits 0, 1, 2, 3, 4, 5, 6, 7, 8, and 9. We then learn to concatenate, or string together, these digits to form larger numbers, incorporating commas for easier reading of large numbers and decimal points for noninteger numbers. Symbols for arithmetic operations, for dollars and cents, and for percent are all commonly used. Many of these symbols are so familiar to us that we don't think consciously about them as we use them. We associate a concept with them, however, and we know how to interpret them or calculate with them. As we learn new concepts and ideas in mathematics, it is critical that we also learn the symbols associated with these concepts and how to interpret and use the symbols.

The use of symbols requires us to be precise in our mathematical statements and instructions. The meaning of a symbol must be clear and widely understood for the symbol to be useful. In some cases, the origin and meaning of the symbol can be traced, but in others the origins are not clear. For example, the letter e, which represents the irrational constant approximately equal to 2.7183, is chosen for the first letter of the name of the mathematician Leonhard Euler. The symbols used to represent parallel and perpendicular lines, $\|$ and $\perp$, respectively, depict in pictures the ideas they represent.

Abbreviations of words or acronyms can also be thought of as symbols. For example, the abbreviation gcd for greatest common divisor can be thought of as a symbol for that concept. Some common abbreviations are lim for limit, min for minimum, max for maximum, sin for sine, cos for cosine, tan for tangent, cot for cotangent, sec for secant, csc for cosecant, log for logarithm, and exp for exponential. A glossary of some other mathematical symbols follows.

Glossary of Symbols

Numbers and Letters as Mathematical Symbols

0, 1, 2, 3, 4, 5, 6, 7, 8, 9
The ten digits that represent the smallest counting numbers and in strings or blocks form larger and smaller numbers.

t, x, y, z The letters at the end of the alphabet most often used to represent unknown numbers called variables.

a, b, c The letters at the beginning of the alphabet most often used to represent arbitrary, but fixed (or constant), numbers.

ε The Greek letter epsilon, often used to represent an arbitrary small positive number.

δ The Greek letter delta, often used to represent an arbitrary small positive number.

N The set of natural, or counting numbers.

R The set of real numbers.

Symbols for Special Numbers or Concepts

π The Greek letter pi represents the ratio of the length of the circumference of any circle to its diameter. This irrational constant has value approximately 3.14159.

e The irrational number e is the base for the natural logarithm. Its value is approximately 2.71828.

∞ The symbol for infinity represents the concept of increasing without bound.

Parentheses and Brackets

( ) Parentheses are generally used in pairs. In calculus, parentheses are frequently used in four distinct ways:

1. In mathematical expressions to remove ambiguity about what operations should be performed first. The operations inside the innermost parentheses are performed first. For example, $2(12 \div (5 - 2)) = 2(12 \div 3) = 2(4) = 8$.

2. To contain ordered pairs of real numbers that are coordinates of points in a Cartesian coordinate system. For example, $(2.4, -6.7)$ represents the point with x-coordinate 2.4 and y-coordinate -6.7.

3. To designate an interval on the real line when the endpoints of the interval are not included in the interval. For example, the set of real numbers $-2.5 < x < 4.8$ can also be written as the interval $(-2.5, 4.8)$. A single parenthesis is used when only one endpoint of an interval is not included. (See [ ] below.)

4. In function notation as described below under "Symbols for Functions."

{ } Braces (curly brackets) are used in pairs in two ways:

1. In set notation. For example, the set of integers between 1 and 6, inclusive, can be written $\{1, 2, 3, 4, 5, 6\}$. The set of real numbers between 1 and 6, inclusive, can be written $\{x \mid 1 \le x \le 6\}$. This notation is read as "the set of all real numbers x such that x is greater than or equal to 1 and less than or equal to 6."

2. In mathematical expressions to show what operations should be performed first, often in complicated expressions in which more than one set of parentheses needs to be used. For example, in the following expression, the use of braces and parentheses allows the reader to distinguish pairs of parentheses more easily:

$$\tfrac{1}{2}(8 + 6\{5 - 2(1 + 4)\}) = \tfrac{1}{2}(8 + 6\{5 - 10\}) = \tfrac{1}{2}(8 - 30) = \tfrac{1}{2}(-22) = -11.$$

[ ] Square brackets (often just called brackets) are generally used in pairs in two ways:

1. In interval notation to indicate that the endpoints of an interval are included in the interval. For example, the interval of real numbers $1.4 \le x \le 5.6$ can be written as $[1.4, 5.6]$. When one endpoint is included in the interval and the other is not, a parenthesis surrounds the missing point and a bracket surrounds the included point. For example, the interval of real numbers $-4.9 \le x < 7.1$ can be written as $[-4.9, 7.1)$.

2. To indicate precedence of operations. For example, the expression on page 518, $\frac{1}{2}(8+6\{5-2(1+4)\})$, could have been written with both brackets and braces, as well as parentheses, as follows: $\frac{1}{2}\{8+6[5-2(1+4)]\}$, and the computation is carried out as before. It is important to note that in this context, a bracket is always paired with another bracket, a brace is paired with another brace, and a parenthesis is paired with another parenthesis.

EXAMPLE B.1 Insert parentheses to make each statement true.

a. $10 - 5 + 1 = 4$

b. $20 \div 4 \cdot 5 + 3 = 40$

Solution

a. $10 - 5 + 1 = 5 + 1 = 6$. By enclosing $5 + 1$ in parentheses, we change the order of operations:

$$10-(5+1)=10-6=4.$$

b. $20 \div 4 \cdot 5 + 3 = 28$. By enclosing $5 + 3$ in parentheses, we obtain the desired answer:

$$20 \div 4 \cdot (5+3) = 20 \div 4 \cdot (8) = 5(8) = 40.$$

■

Symbols for Operations

$+$ The plus sign is an instruction to add two quantities.

$-$ The minus sign is an instruction to subtract the quantity following the sign from the quantity preceding the sign. It is also used to negate a number or expression.

$\pm$ This sign is a concise way of writing two distinct expressions. For example, the roots of the quadratic equation $ax^2 + bx + c = 0$ are usually written concisely as $x = \frac{-b \pm \sqrt{b^2-4ac}}{2a}$.

This means that the two roots are $x_1 = \frac{-b+\sqrt{b^2-4ac}}{2a}$ and $x_2 = \frac{-b-\sqrt{b^2-4ac}}{2a}$.

The symbol $\pm$ is also used to express the error in measurements or quantities. For example, if the length of a rope is given in inches as 18 ± 0.01, it means that the length of the rope is between 17.99 and 18.01 in. In other words, the rope measures 18 in. with an error of up to +0.01 or –0.01 in.

$\times$ or $\cdot$ The multiplication sign is an instruction to multiply two quantities.

$\div$ or / The division sign is an instruction to divide the quantity preceding the sign by the quantity following the sign. Writing a quotient with the slash is equivalent to writing it in stacked form. For example, $\pi/2$ is the same as $\frac{\pi}{2}$. Although the stacked form is easier to read, the slash is often used when entering functions into a calculator or computer. In this case, parentheses must be used for a multiterm numerator, a multiterm denominator, or a denominator with several factors. For example, consider these expressions in stacked form:

$$\frac{x^2+4}{x+3}; \quad \frac{x^2+4}{5x^2}.$$

We write these with a slash as follows:

$$(x^2+4)/(x+3); \qquad (x^2+4)/(5x^2).$$

^ This symbol represents exponentiation or raising to a power. For example, 3^5 means 3^5, or $3 \cdot 3 \cdot 3 \cdot 3 \cdot 3$. This sign is commonly used on calculators and computers to represent exponentiation.

$\sqrt{x}$ For any $x \geq 0$, this symbol represents the nonnegative square root of x; it is the nonnegative number a that satisfies $a^2 = x$. For example, $\sqrt{9} = 3$, since 3 is a nonnegative number and satisfies $3^2 = 9$. When there is a multiterm expression under the square root sign, as in $\sqrt{x^2+3}$, the expression under the sign is treated as if a set of parentheses were around it. To evaluate this expression for $x = 1$, we first calculate the quantity under the square root sign, $1^2 + 3 = 4$, and then compute $\sqrt{4} = 2$. We enter this expression into a graphing calculator or computer as √ (x ^ 2 + 3).

$\sqrt[n]{x}$ This symbol represents the nth root of x where n is an integer with $n > 2$; it is the number a that satisfies $a^n = x$. The symbol itself is called a radical. (Note that if n is even, x must be greater than or equal to 0, and in that case a is the nonnegative number that satisfies $a^n = x$.) For example, $\sqrt[3]{-125} = -5$; $\sqrt[4]{(-2)^8} = \sqrt[4]{256} = 4$; $\sqrt[6]{-2}$ is not defined.

Σ The capital Greek letter sigma is an instruction to add a specified collection of terms. See Section 19-A for a more complete discussion.

$n!$ This symbol is read as "n factorial." It means $1 \cdot 2 \cdot 3 \cdot \cdots \cdot (n-2) \cdot (n-1) \cdot n$, the product of all the positive integers up to and including the integer n. By definition, $0! = 1$.

$|x|$ This symbol represents the absolute value of an expression x. If $x \geq 0$, then $|x| = x$. If $x < 0$, then $|x| = -(x)$. In general, $|x|$ is greater than or equal to 0.

EXAMPLE B.2 Write each quotient using a slash, as if entering the expression in a graphing utility. Include only those parentheses required to make the new quotient equivalent to the original quotient.

a. $\frac{5x}{x+3}$

b. $\frac{2x-3}{4x}$

Solution

a. Since there is only one term in the numerator, we do not need parentheses around the numerator. Since the denominator includes the addition operation, parentheses are needed: $\frac{5x}{x+3} = 5x/(x+3)$.

b. Since the numerator is a difference and the denominator contains a product, both must be written inside parentheses: $\frac{2x-3}{4x} = (2x-3)/(4x)$. ■

EXAMPLE B.3 Evaluate the following, if possible. If not possible, explain why.

a. $\sqrt[3]{0.027}$ **b.** $\sqrt[4]{\frac{1}{16}}$ **c.** $\sqrt{-256}$

Solution

a. $\sqrt[3]{-0.027} = -0.3$, since $(-0.3)^3 = -0.027$.

b. $\sqrt[4]{\frac{1}{16}} = \frac{1}{2}$, since $\left(\frac{1}{2}\right)^4 = \frac{1}{16}$.

c. $\sqrt{-256}$ cannot be evaluated because there is no real number a for which $a^2 = -256$. ■

Symbols for Relations

$=$ Is equal to, or equals.

$\neq$ Is not equal to, or is unequal to.

$<$ Is less than.

$>$ Is greater than.

$\leq$ Is less than or equal to.

$\geq$ Is greater than or equal to.

$\approx$ Is approximately equal to, or approximately equals. For example, since π is an irrational number, its (exact) decimal form is nonterminating and nonrepeating. Thus we write $\pi \approx 3.1416$.

$\in$ Is an element of, or belongs to.

$\notin$ Is not an element of, or does not belong to.

$\cap$ Intersection. The intersection of two sets is the set of elements common to the two sets. For example, the intersection of two intervals is the set of numbers that lie in both intervals: $(2.3, 6.7) \cap [-3.5, 4.9] = (2.3, 4.9]$.

$\cup$ Union. The union of two sets is the set of elements that are in either set or both sets. For example, the union of two intervals is the set of numbers that lie in either interval: $(2.3, 6.7) \cup [-3.5, 4.9] = [-3.5, 6.7)$.

$\rightarrow$ Approaches, as used in $x \rightarrow a$; this means that x approaches the number a.

$\parallel$ Is parallel to.

$\perp$ Is perpendicular to.

Symbols for Functions

$f(x)$ A rule that assigns to each x in the domain of the function a unique value $f(x)$. A single letter, such as y, is often used to indicate the value $f(x)$. Function names (the letters that represent the function, such as f) are sometimes chosen to signify the quantity they represent. For example, a function that represents the height of a ball at time t might be represented as $h(t)$, or just h. See Chapter 2 for an extended discussion of functions.

$f^{-1}(x)$ The inverse function of the function f. This notation does not mean $\frac{1}{f(x)}$. See Chapter 14 for an extended discussion of inverse functions.

$f^n(x), n \neq 1$

Denotes $[f(x)]^n$, the nth power of $f(x)$. For example, $\sin^3 x$ means $(\sin x)^3$.

$\circ$ The symbol for composition of functions. For two functions f and g, the composition of f and g, denoted $f \circ g$, is defined by the rule $(f \circ g)(x) = f(g(x))$. See Section 4-C for a more detailed discussion of composition of functions.

$\sin(x)$ or $\sin x$

The sine function that represents the sine of the angle x, in radian measure. See Chapter 8 for abbreviations for all of the trigonometric functions.

$\exp(x)$ or e^x

The natural exponential function.

$\log(x)$ or $\log x$

The common logarithm, which is the logarithm to the base 10 of x.

$\ln(x)$ or $\ln x$

The natural logarithm, which is the logarithm to the base e of x.

$\text{abs}(x)$ The absolute value function. It is also denoted by $|x|$. If $x \geq 0$, then $\text{abs}(x) = x$; if $x < 0$, then $\text{abs}(x) = -(x)$.

Symbols for Words or Phrases Frequently Used in Mathematics

$\therefore$ Therefore, or thus.

$\exists$ There exists or there is at least one.

$\forall$ For all, or for every.

$\ldots$ The symbol of three dots (called ellipses) means "continuing in the same manner."

Symbols Used in Graphs

$\bullet$ The dot or filled circle emphasizes that the point lies on the graph.

$\circ$ The point represented by the open circle does not lie on the graph.

$\nearrow$ An arrow at the end of a line or curve indicates the line or curve extends forever in the direction of the arrow.

Calculus Symbols

Δx Represents the change in the variable x.

Δy Represents the change in the variable (or function) y.

dx Represents the differential of x.

dy Represents the differential of a function y.

$\lim_{x \to a} f(x)$

The limit of the function $f(x)$ as x approaches the number a ; a two-sided limit.

$\lim_{x \to a^-} f(x)$

The one-sided limit of the function $f(x)$ as x approaches the number a from the left.

$\lim_{x \to a^+} f(x)$

The one-sided limit of the function $f(x)$ as x approaches the number a from the right.

$\frac{d}{dx}$ or y' or D_x

The first derivative of a function with respect to x. The following all denote the first derivative of the function $y = f(x)$ with respect to x:

$$\frac{dy}{dx} = \frac{df}{dx} = y' = f'(x) = D_x y = D_x f = D_x f(x).$$

$\frac{d^2}{dx^2}$ or y'' or D^2_x

The second derivative of a function with respect to x. The following all denote the second derivative of the function $y = f(x)$ with respect to x:

$$\frac{d^2 y}{dx^2} = \frac{d^2 f}{dx^2} = y'' = f'' = D_x^2 y = D_x^2 f = D_x^2 f(x).$$

$\frac{d^n}{dx^n}$ or $y^{(n)}$ or D_x^n

The nth derivative of a function with respect to x.

$\int$ The indefinite integral, or antiderivative. This symbol always appears with a differential—for example, $\int f(x)\,dx$.

$\int_a^b$ The definite integral over the interval $[a, b]$. This symbol always appears with a differential—for example, $\int_a^b f(x)\,dx$.

Exercises B

1. Insert parentheses to make each statement true.

a. $2 + 3^2 - 3 = 22$

b. $20 - 2 \cdot 5 - 3 = 36$

2. Write each quotient using a slash. Insert only those parentheses required to make the new quotient equivalent to the original quotient.

a. $\frac{5x^3 y}{x - 12}$

b. $\frac{2x^2 - 3x}{4x^3}$

3. Evaluate the following, if possible. If not possible, explain why.

a. $\sqrt[5]{-32}$ **b.** $\sqrt[3]{0.064}$ **c.** $\sqrt[4]{\frac{16}{625}}$ **d.** $\sqrt{-81}$

C ABOUT ALGEBRAIC MANIPULATIONS

All calculations with numbers and with variables that represent numbers must follow basic rules. These rules allow us to simplify and combine numbers or algebraic expressions.

$a+b=b+a$	$ab=ba$	(commutative laws)
$a+(b+c)=(a+b)+c$	$a(bc)=(ab)c$	(associative laws)
$a(b+c)=ab+ac$		(distributive law)

For example,

$$3+2=2+3=5 \quad \text{and} \quad 3\cdot 2=2\cdot 3=6$$

$$5+\left(3+\sqrt{2}\right)=(5+3)+\sqrt{2}=8+\sqrt{2} \quad \text{and} \quad 5\left(3\sqrt{2}\right)=(5\cdot 3)\sqrt{2}=15\sqrt{2}$$

$$-3(6+\pi)=-3\cdot 6+(-3)\cdot\pi=-18-3\pi.$$

The associative law allows us to group terms in various ways in a sum or a product; as a result, parentheses are often omitted altogether. The sum $a+(b+c)$ can be written as $a+b+c$ and the product $a(bc)$ can be written as abc.

C-1 Use and Importance of Parentheses

Parentheses, braces, or brackets are used to indicate the order of operations on numbers and variables when more than one operation is to be performed. Calculations are done within innermost parentheses first. For example, $5+(3\cdot 4)$ and $(5+3)\cdot 4$ are two different numbers, since $5+(3\cdot 4)$ = $5+12=17$ and $(5+3)\cdot 4 = 8\cdot 4=32$.

To simplify notation and minimize the need for parentheses, there is a standard convention about the order in which arithmetic operations are to be performed.

For example, if we write $5+3\cdot 4$ without any parentheses, the product is to be computed first: $5+3\cdot 4 = 5+(3\cdot 4) = 17$.

The standard convention is that operations are performed in the following order:

1. Combine within parentheses, innermost first, and work outward.

2. Take powers and roots.

3. Perform multiplications and divisions (from left to right if there is more than one of these operations).

4. Perform additions and subtractions (from left to right if there is more than one of these operations).

EXAMPLE C.1 Evaluate the following:

a. $9-3-2$ **b.** $9-(3-2)$ **c.** $\left(2+5(3-(7-3))\right)\frac{2}{3}$ **d.** $-2+\sqrt{8+8}-4\cdot 3^2$

Solution Try these calculations with your calculator if your calculator has parentheses.

a. Subtract from left to right to obtain $9-3-2=6-2=4$.

b. Calculate within parentheses first: $9-(3-2)=9-1=8$.

c. Recall that parentheses always are used in pairs. This expression contains three pairs of nested parentheses. The innermost parentheses contain the difference $(7-3)$. We calculate this difference first; then work our way outward:

$$\left(2+5(3-(7-3))\right)\frac{2}{3}=(2+5(3-4))\frac{2}{3}=(2+5\cdot(-1))\frac{2}{3}=(2-5)\frac{2}{3}=(-3)\frac{2}{3}=-2.$$

d. We first evaluate roots and powers, then perform the multiplication, and finally add and subtract from left to right. The quantity under the square root is treated as if it were within parentheses. Thus,

$$-2+\sqrt{8+8}-4\cdot 3^2=-2+\sqrt{16}-4\cdot 3^2=-2+4-4\cdot 9=-2+4-36=2-36=-34. \quad \blacksquare$$

The same conventions apply to operations with algebraic expressions that involve variables (or unknowns). But very often we cannot simplify the expression within parentheses. For example, in $2(x+y)$, we cannot write $x+y$ as a single term. To eliminate parentheses in algebraic expressions, we use the basic laws of operations. For example, to write $2(x+y)$ without parentheses, we use the distributive law to write $2(x+y) = 2x+2y$. We use the same property when we write $-(x-y)=(-1)\cdot(x-y)=-x+y$. To eliminate parentheses in the expression $2(3x)^2$, we use the associative and commutative laws of multiplication to write

$$2(3x)^2=2\cdot(3x)\cdot(3x)=2\cdot 3\cdot x\cdot 3\cdot x=2\cdot 3\cdot 3\cdot x\cdot x=18x^2.$$

EXAMPLE C.2 Write each expression without parentheses.

a. $3x^3\left(x+3x^2-5x\right)$ **b**. $\left(-\frac{1}{2}\right)\left((x+y)+3x(x-1)\right)$

Solution

a. We start by combining like terms inside the parentheses: $3x^3\left(x+3x^2-5x\right) = 3x^3\left(-4x+3x^2\right)$. Then we multiply out, using the distributive law first and then the associative and commutative laws of multiplication and laws of exponents: $3x^3\left(-4x+3x^2\right)=3x^3\cdot(-4x)+3x^3\cdot 3x^2=-12x^4+9x^5$.

b. To eliminate the innermost parentheses (there are two pairs of them), we use the associative and distributive laws. Thus,

$$\left(-\frac{1}{2}\right)\left((x+y)+3x(x-1)\right)=\left(-\frac{1}{2}\right)\left(x+y+3x^2-3x\right)=\left(-\frac{1}{2}\right)\left(-2x+y+3x^2\right)=x-\frac{1}{2}y-\frac{3}{2}x^2. \quad \blacksquare$$

Terms with variables may be added and subtracted only if the variables are the same and their exponents are equal. For example, $3x^2+2x^2=5x^2$, but expressions such as $3x^2+2x^3$ and $3x^4-2y^4$ cannot be combined.

Parentheses and Substitution in Formulas

It is very important to always surround with parentheses any expression that contains more than one term when it is placed in a formula. For example, if we use the formula $A=b\cdot h$ to find the area of a rectangle of base x and height $x+2$, we need to write $A=x\cdot(x+2)$, which is $A=x^2+2x$. If we forget the parentheses, we get $x\cdot x+2=x^2+2$, which is not the same as x^2+2x. If, for example, $x=3.4$, the correct area is $A=3.4(3.4+2)=(3.4)^2+(3.4)\cdot 2=18.36$, while the incorrect formula gives $(3.4)^2+2=13.56$.

Exercises C-1

1. In each of the following, find the value of y when $x=-1$:

a. $y=(x+4)^2-2(x+4)$ **b**. $y=x+4^2-2(x+4)$

c. $y=(x+4)^2-2x+4$ **d**. $y=(x+4)^2(-2x+4)$

2. Use a graphing utility to graph the equations in parts a, b, and c of Exercise 1. Use the window $-8\le x\le 5, -5\le y\le 30$.

3. Multiply out and combine like terms to rewrite each equation in Exercise 1 without parentheses.

4. The volume of a cylinder of base radius r and height h is $V=\pi r^2 h$. Write V in terms of x if r and h are as follows:

a. $r=x-1,\ h=x+2$ **b**. $r=\frac{x}{2}+1,\ h=x^2+\frac{2}{x}$

C-2 Factoring Algebraic Expressions

We can use the distributive property to write $a\cdot b+a\cdot c=a\cdot(b+c)$. The expression on the left is a sum; the expression on the right is a product in factored form. To factor the expression on the left, we observe that a is a factor in every term on the left. The first step in factoring expressions is to look for a *common factor*: This is an *expression that appears as a multiplier in every single term of a sum.* The common factor can be a number, a letter (a variable), or any expression involving letters and numbers. For example, $\frac{x^2}{x+\sqrt{2}}$ is a common factor in the expression $\frac{3x^2}{x+\sqrt{2}}+\frac{x^3y}{x+\sqrt{2}}-\frac{x^2}{x+\sqrt{2}}$ since the first term is $3\cdot\left(\frac{x^2}{x+\sqrt{2}}\right)$, the second term is $xy\cdot\left(\frac{x^2}{x+\sqrt{2}}\right)$, and the third term is $(-1)\cdot\left(\frac{x^2}{x+\sqrt{2}}\right)$. Thus, $\frac{3x^2}{x+\sqrt{2}}+\frac{x^3y}{x+\sqrt{2}}-\frac{x^2}{x+\sqrt{2}}=\frac{x^2}{x+\sqrt{2}}\cdot(3+xy-1)=\frac{x^2}{x+\sqrt{2}}(2+xy)$.

EXAMPLE C.3 For each expression below, find all common factors that occur in every term and factor the expression as completely as possible.

a. $-2\sqrt{x}+x\sqrt{x}-x^2\sqrt{x}$ **b**. $5(x-3)^2(2x+1)^3-7x(x-3)^4(2x+1)$

c. $\frac{6x^2y}{x^{1/5}-2}+\frac{xy^3}{5\left(x^{1/5}-2\right)}$

Solution

a. The common factor is $\sqrt{x}$; thus $-2\sqrt{x}+x\sqrt{x}-x^2\sqrt{x}=\sqrt{x}\left(-2+x-x^2\right)$.

b. Each of the two terms in the given expression contains some powers of the factors $(x-3)$ and $(2x+1)$. We look for the largest exponents m, n for which $(x-3)^m$ and $(2x+1)^n$ are factors in every term. We find $m=2$ and $n=1$; thus $(x-3)^2(2x+1)$ is a common factor, and

$$5(x-3)^2(2x+1)^3-7x(x-3)^4(2x+1)=(x-3)^2(2x+1)\left(5(2x+1)^2-7x(x-3)^2\right).$$

c. We first note that xy is a factor in each term. Looking again, we see that each term also has $x^{1/5} - 2$ in the denominator. Thus, $\frac{xy}{x^{1/5}-2}$ is a common factor. The factored expression is $\frac{xy}{x^{1/5}-2}\left(6x+\frac{y^2}{5}\right)$. ■

See Section 3-B.1 for some other techniques of factorization.

Finding a Common Denominator

To add or subtract quotients, we need to find a common denominator. If all the terms in a given sum have different denominators with no factors in common, we take the product of the denominators as a common denominator. For example, to add $\frac{2x}{x^2+1} + \frac{5}{x} + \frac{x-2}{2x+1}$, we take $(x^2 + 1)x(2x + 1)$ as the common denominator. We then multiply each term in the sum by 1, written as an appropriate quotient so as to produce the common denominator. Thus,

$$\frac{2x}{x^2+1} + \frac{5}{x} + \frac{x-2}{2x+1} = \frac{2x}{x^2+1} \cdot \frac{x(2x+1)}{x(2x+1)} + \frac{5}{x} \cdot \frac{(x^2+1)(2x+1)}{(x^2+1)(2x+1)} + \frac{x-2}{2x+1} \cdot \frac{(x^2+1)x}{(x^2+1)x} =$$

$$\frac{2x \cdot x(2x+1)+5(x^2+1)(2x+1)+(x-2)(x^2+1)x}{(x^2+1)x(2x+1)}.$$

We use the distributive property (multiplying out the numerator) and then collect like terms in the numerator to write this as $\frac{x^4+12x^3+8x^2+8x+5}{(x^2+1)x(2x+1)}$.

If the denominators have some factors in common, we choose a common denominator without repeating those common factors. For example, to rewrite the difference $\frac{3}{x^2(x+1)} - \frac{2x+1}{5x(x+1)}$, we choose $5x^2(x + 1)$ as common denominator. Then

$$\frac{3}{x^2(x+1)} - \frac{2x+1}{5x(x+1)} = \frac{3}{x^2(x+1)} \cdot \frac{5}{5} - \frac{2x+1}{5x(x+1)} \cdot \frac{x}{x} = \frac{3\cdot 5-(2x+1)x}{5x^2(x+1)} = \frac{15-2x^2-x}{5x^2(x+1)}.$$

In general, to find a common denominator, we proceed as follows:

1. Factor each denominator as completely as possible.

2. Take as common denominator the product of all the distinct factors of all the denominators, each raised to the highest power that appears in any of the denominators.

EXAMPLE C.4 Find a common denominator and write as a single quotient:

$$\frac{3x-1}{2x-x^2} + \frac{2x+1}{(2-x)^2} - \frac{4}{x^3}.$$

Solution We factor the denominator of the first quotient as $x(2 - x)$; the other denominators are already completely factored. The distinct factors in the denominators are x and $(2 - x)$. The highest power to which the factor x occurs is 3, and the highest power to which the factor $(2 - x)$ occurs is 2. Thus the common denominator is $x^3(2-x)^2$. Then

$$\frac{3x-1}{2x-x^2} + \frac{2x+1}{(2-x)^2} - \frac{4}{x^3} = \frac{(3x-1)(2-x)x^2 + (2x+1)x^3 - 4(2-x)^2}{x^3(2-x)^2}$$

$$= \frac{(6x-3x^2-2+x)x^2+(2x+1)x^3-4(4-4x+x^2)}{x^3(2-x)^2} = \frac{-x^4+8x^3-6x^2+16x-16}{x^3(2-x)^2}. \blacksquare$$

Exercises C-2

1. For each of the following, find all common factors and factor the expression as completely as possible:

a. $3x^2\sqrt{x} - x^3\sqrt{x} + 5x\sqrt{x}$

b. $\dfrac{x^2y^3}{3(4x+3)} + \dfrac{2x^2y}{4x+3} - \dfrac{x^3y^2}{4x+3}$

c. $(2x+1)\left(x^5-1\right)-(2x+1)x^4 - x\sqrt{x}(2x+1)^2$

2. Find a common denominator and write as a single quotient:

$$\frac{4x+1}{x^2-3x}+\frac{x^2+1}{2x^3(x-3)}-\frac{x-5}{2x^3-6x^2}.$$

C-3 When Can I Cancel?

All cancellation in algebraic expressions boils down to one of two facts:

1. $a+(-a)=0$ for all a.

2. $\frac{a}{a}=1$ for all $a \neq 0$.

The a in these equations can be a number (such as 3 or $\sqrt{2}$ or π), or it can be a variable expression (such as xy or $3x^2$ or $(x-2)$). We cancel terms or factors when we need to simplify expressions and when we are solving equations.

Simplifying an Expression

In a sum, terms with opposite signs can be canceled using property 1 above. For example, $3x+4x^2-3x-1$ can be simplified to $4x^2-1$ since $3x - 3x = 3x + (-3x) = 0$.

A quotient can be simplified if there is a common factor in the numerator and denominator. If a is such a common factor, then the quotient can be written in the form $\frac{a \cdot b}{a \cdot c}$, and by property 2 it simplifies to $\frac{b}{c}$. For example, $\frac{2\pi r}{16r^3} = \frac{2r\cdot\pi}{2r\cdot 8r^2} = \frac{\pi}{8r^2}$.

EXAMPLE C.5 Simplify each expression as much as possible:

a. $4x^2 - 5xy - 4x^2 + 5xy^2$

b. $\dfrac{2x^3+3x^2}{x^2-3x}$

c. $\dfrac{3x^2+2}{x^2}$

Solution

a. The first and third terms are opposites, so $4x^2 - 5xy - 4x^2 + 5xy^2 = -5xy + 5xy^2$. This cannot be simplified further.

b. We factor the numerator and denominator and then simplify: $\dfrac{2x^3+3x^2}{x^2-3x} = \dfrac{x^2(2x+3)}{x(x-3)} = \dfrac{x(2x+3)}{x-3}$, for $x \neq 0$.

c. The expression $\frac{3x^2+2}{x^2}$ cannot be simplified because there is no factor common to all terms in the numerator and denominator. ■

Note that in part b of Example C.5, $\frac{2x^3+3x^2}{x^2-3x}$ is undefined when $x = 0$, but the simplified expression $\frac{x(2x+3)}{x-3}$ is defined for $x = 0$. To ensure that the two expressions are equal, we must note that $x \neq 0$ in the second expression. We can only cancel out a fraction equal to 1. If $x = 0$, $\frac{x}{x} = \frac{0}{0} \neq 1$.

Only common factors that appear in **every** term in the numerator and the denominator of a quotient can be canceled. To avoid making a mistake, always **factor numerator and denominator** first; then cancel if possible.

Solving Equations

To solve an equation we can:

1. Add (or subtract) the same expression on both sides.

2. Multiply (or divide) by the same **nonzero** factor on both sides.

EXAMPLE C.6 Solve each equation.

a. $2x^2 - 5x = 2 - 5x$ **b.** $x^3 + 2x = 3x^2 + 6$ **c.** $x\sqrt{x-3} = 2\sqrt{x-3}$

Solution

a. We first add $5x$ to both sides to eliminate the term $-5x$: $2x^2 - 5x + 5x = 2 - 5x + 5x$. This gives $2x^2 = 2$. Next we divide both sides of the equation by 2, $\frac{2x^2}{2} = \frac{2}{2}$, and obtain $x^2 = 1$. So the solutions are $x = \pm 1$.

b. We solve the equation by factoring each side to get $x(x^2 + 2) = 3(x^2 + 2)$. Since $x^2 + 2 \neq 0$ for all x, we divide both sides by $x^2 + 2$ and obtain $x = 3$.

c. It is tempting to divide both sides by $\sqrt{x-3}$ to obtain $x = 2$. However, $\sqrt{x-3}$ may equal 0, so we could be dividing by 0. Instead, we subtract $2\sqrt{x-3}$ from both sides to obtain $x\sqrt{x-3} - 2\sqrt{x-3} = 0$. Then we factor out $\sqrt{x-3}$ to get $\sqrt{x-3}(x-2) = 0$. Setting each factor equal to 0, we have $\sqrt{x-3} = 0$ and $x - 2 = 0$. The solutions to the equations are $x = 3$ and $x = 2$, respectively. ■

If we divide both sides of an equation by a factor that can be 0, we may lose solutions of the equation.

Exercises C-3

1. An expression on the left has been "simplified" to the expression on the right. Which of these simplifications are correct? If not correct, explain why.

a. $\frac{2x+3x+1}{3x+1} \stackrel{?}{=} 2x$ **b.** $5x+3x^2-2x-3x \stackrel{?}{=} 3x^2$ **c.** $\frac{1-x}{(3x-3)^2} \stackrel{?}{=} -\frac{1}{9(x-1)}$

d. $\frac{(4x+3)x^3}{x^3+x} \stackrel{?}{=} \frac{(4x+3)}{x}$ **e.** $\frac{x^4-2x+x}{x^2} \stackrel{?}{=} \frac{x^3-1}{x}$ **f.** $3x^2 - 3x + 4 \stackrel{?}{=} 4 + x$

2. Simplify the following expressions as much as possible:

a. $2x^2y^3 - 5x^3y^2 - 2x^2y^3$ **b.** $\dfrac{3x^4 + x^2}{3x^2 + 1}$ **c.** $\dfrac{5x^{1/3} - 7yx^{1/3}}{y + x^{1/3} - y}$

3. The equation on the right has been obtained from the equation on the left by simplification. Which of these are correct and which result in the loss of a solution?

a. $x^4 + 9x^2 = 2x^3 + 18x;\quad x^2 = 2x$ **b.** $x^2(x-1) = 3x(x-1);\quad x^2 = 3x$

c. $x^3 - 9x = x^3 + x;\quad -9x = x$ **d.** $\left(x^2+1\right)(x-1) = x\left(x^2+1\right);\quad x - 1 = x$

C-4 Working with Exponents

The definition of exponents and their rules of operation are reviewed in Sections 7-A and 13-A. Here are some basic tips to help you avoid some common mistakes.

Notation

An exponent applies only to the number or expression immediately to its left, unless parentheses are present. For example, in the expression $2x^3$ only x is raised to the third power, whereas in the expression $(2x)^3$ the expression $2x$ is raised to the third power: $(2x)^3 = 8x^3$.

Combining Exponents

To simplify expressions containing exponents we use the following rules:

1. We add exponents when we multiply two quantities with the same base:

$$a^m \cdot a^n = a^{m+n}.$$

For example, $2 \cdot 2^3 = 2^{1+3} = 2^4$, $a^3a^{-5} = a^{-2}$, and $b^3b^{\sqrt{2}} = b^{3+\sqrt{2}}$.

2. We subtract exponents when we divide two quantities with the same base:

$$\frac{a^m}{a^n} = a^{m-n}.$$

For example, $\dfrac{2^3}{2^4} = 2^{-1}$, $\dfrac{a^3}{a^{\sqrt{2}}} = a^{3-\sqrt{2}}$, $\dfrac{x^3}{x^{2/3}} = x^{3-(2/3)} = x^{7/3}$, and $\dfrac{y^3}{y^4} = y^{-1}$.

Note that rules 1 and 2 apply only when the quantities have the same base. The expression x^3y^2, for example, cannot be written with a single exponent because the bases x and y are different.

3. We multiply exponents when we raise an expression a^n to a power r:

$$\left(a^n\right)^r = a^{nr}, \text{ for } a \ge 0.$$

For example, $\left(2^3\right)^2 = 2^6$ and $\left(a^{1/2}\right)^4 = a^{4/2} = a^2$. When $a < 0$, this rule may not apply; for example, $((-2)^2)^{1/2} = 4^{1/2} = 2$.

> ☞ When a is negative and n is even, $\left(a^n\right)^{1/n} \neq a^1$. In fact, $\left(a^n\right)^{1/n} = |a|$, for all real numbers a.

Note that a^{n^r} is not the same as $\left(a^n\right)^r$. When parentheses are not present, by convention, it is understood that the exponent r affects only the number n. Thus, $a^{n^r} = a^{\left(n^r\right)}$. For example, $2^{3^2} = 2^{\left(3^2\right)} = 2^9 = 512$ whereas $\left(2^3\right)^2 = 2^6 = 64$.

In the following example we use rules 1 and 3 to factor an algebraic expression.

EXAMPLE C.7 Find a common factor for this expression and write it in factored form:

$$2x^{2/3} + 5yx^{1/3} - x^{5/3}.$$

Solution We see that $x^{1/3}$ is a factor in each term since $x^{2/3} = x^{1/3} \cdot x^{1/3}$ and $x^{5/3} = x^{1/3} \cdot x^{4/3}$. We factor the expression as follows: $2x^{2/3} + 5yx^{1/3} - x^{5/3} = x^{1/3}\left(2x^{1/3} + 5y - x^{4/3}\right)$. ■

Negative and Rational Exponents

A negative exponent indicates a reciprocal: By definition, $a^{-n} = \frac{1}{a^n}$. For example, $2^{-5} = \frac{1}{2^5} = \frac{1}{32}$, $\left(-\frac{1}{3}\right)^{-5} = \frac{1}{\left(-\frac{1}{3}\right)^5} = \frac{1}{-\frac{1}{243}} = -243$, and $3x^{-2} = 3 \cdot \frac{1}{x^2} = \frac{3}{x^2}$. Note that in the last equation, the exponent applies only to x.

A rational exponent that is not an integer indicates a root. More precisely, if m is an integer and n is an integer, $n > 1$, then $a^{m/n} = \left(\sqrt[n]{a}\right)^m$. When n is even, $a^{m/n}$ is undefined for $a < 0$.

For example, $4^{3/2} = \left(\sqrt{4}\right)^3 = 8$, $(-x)^{1/3} = \sqrt[3]{-x} = -\sqrt[3]{x}$, $4^{-1/2} = \frac{1}{4^{1/2}} = \frac{1}{\sqrt{4}} = \frac{1}{2}$, and $(-16)^{3/4}$ is undefined.

Sums Raised to a Power

Recall that $(a+b)^2 = a^2 + 2ab + b^2$, which is not the same as $a^2 + b^2$. So the exponent in $(a+b)^2$ cannot be distributed to each term. In general, a power applied to a sum or difference is not the same as the power applied to each term: $(a+b)^n$ is not equal to $a^n + b^n$, and $(a-b)^n$ is not equal to $a^n - b^n$. For example, $(9 + 16)^{1/2} = \sqrt{9+16} = \sqrt{25} = 5$, whereas $9^{1/2} + 16^{1/2} = \sqrt{9} + \sqrt{16} = 3 + 4 = 7$.

Products and Quotients Raised to a Power

If a product or quotient is raised to a power, then we can distribute the exponent; that is, $(ab)^n = a^n b^n$ and $\left(\frac{a}{b}\right)^n = \frac{a^n}{b^n}$. For example, $(4x)^{1/2} = 4^{1/2}x^{1/2} = 2x^{1/2}$, and $\left(\frac{8}{27}\right)^{2/3} = \frac{8^{2/3}}{27^{2/3}} = \frac{\left(\sqrt[3]{8}\right)^2}{\left(\sqrt[3]{27}\right)^2} = \frac{2^2}{3^2} = \frac{4}{9}$.

Expressions with Negative Exponents

To simplify an expression with terms that have negative exponents, we rewrite those terms as quotients and then find a common denominator to write the expression as a single quotient. For example, to solve the equation $30x^{-2} + 5x + x^{-1} = 0$, we first write $\frac{30}{x^2} + 5x + \frac{1}{x} = 0$ and then rewrite the equation as $\frac{30 + 5x^3 + x}{x^2} = 0$.

Exercises C-4

1. Identify all the expressions below that are equal to x^{2^3}.

a. x^6 **b.** $x^{(2^3)}$ **c.** $\left(x^2\right)^3$ **d.** x^3x^5 **e.** x^8 **f.** $x^2x^2x^2$

2. Identify all the expressions below that are equal to $5x^{-2}$.

a. $\frac{1}{5x^2}$ **b.** $\frac{1}{25x^2}$ **c.** $\frac{5}{x^2}$ **d.** $\frac{5}{\sqrt{x}}$ **e.** $5\left(\frac{1}{x^2}\right)$ **f.** $-5\sqrt{x}$

3. Simplify each expression and write the answer without radical signs or negative exponents.

a. $5x^2x^{-3}$ **b.** $\frac{a^3b^2a^5}{b^3\sqrt{a}}$ **c.** $\frac{(8x)^{-2/3}}{3x^{-1}}$ **d.** $\frac{(9+4)^{1/2}}{(9\cdot 4)^{1/2}}$ **e.** $\sqrt[3]{8x^{-4}}$ **f.** $\left(x^2y\right)^{-3}$

4. Find a common factor and write in factored form.

a. $3x^{3/7} + 5x^{1/7} - 4x^{15/7}$ **b.** $4x^{2/3} - \sqrt[3]{8x} + x^{4/3}$

D ABOUT SOLVING EQUATIONS

D-1 Completing the Square

In most applications, we *complete the square* to change a quadratic equation of the form $x^2 + bx = c$ into one that has a perfect square: $\left(x + \frac{b}{2}\right)^2 = d$. To complete the square in the equation $x^2 + 7x = -2$, we add the square of one-half the coefficient of x to both sides of the equation (i.e., add $\left(\frac{7}{2}\right)^2$ to both sides). This creates a perfect square on the left that can be written in the factored form $\left(x + \frac{7}{2}\right)^2$ and changes the original equation to the desired form. Here is the general process, illustrated with this example:

$$x^2 + bx = c \qquad\qquad x^2 + 7x = -2$$

$$x^2 + bx + \left(\frac{b}{2}\right)^2 = c + \left(\frac{b}{2}\right)^2 \qquad\qquad x^2 + 7x + \left(\frac{7}{2}\right)^2 = -2 + \left(\frac{7}{2}\right)^2$$

$$\left(x + \frac{b}{2}\right)^2 = c + \frac{b^2}{4} \qquad\qquad \left(x + \frac{7}{2}\right)^2 = -2 + \frac{49}{4} = \frac{41}{4}.$$

The process of completing the square is used to transform a quadratic equation in two variables into a standard form so it can be easily graphed (its graph, if there is more than one point, will be one of the conic sections—a circle, ellipse, parabola, or hyperbola).

EXAMPLE D.1 Complete the square in x to put the equation below in standard form for a conic. What is the graph of this equation?

$$x^2 + 1.6x + y^2 = -1$$

Solution We complete the square in x by adding $\left(\frac{1.6}{2}\right)^2 = (0.8)^2 = 0.64$ to both sides of the equation. Then we write the left side as a sum of two squared terms: $x^2 + 1.6x + 0.64 + y^2 = -1 + 0.64$. Since we created a perfect square, we can rewrite the equation as $(x + 0.8)^2 + y^2 = -0.36$. This equation has no graph, since the left side is always nonnegative and the right side is negative. ■

Derivation of the Quadratic Formula

The quadratic formula can be used to find all solutions to any quadratic equation of the form $ax^2 + bx + c = 0$, with $a \neq 0$. The formula can be derived by completing the square. We now show this and illustrate the steps with an example, side by side with the general derivation.

General Derivation		**Example** $(a = 3, b = 7, c = 1)$
$ax^2 + bx + c = 0, a \neq 0$		$3x^2 + 7x + 1 = 0$
$ax^2 + bx = -c$	Add $-c$ to both sides	$3x^2 + 7x = -1$
$x^2 + \frac{b}{a}x = \frac{-c}{a}$	Divide both sides by a	$x^2 + \frac{7}{3}x = \frac{-1}{3}$
$x^2 + \frac{b}{a}x + \frac{b^2}{4a^2} = \frac{-c}{a} + \frac{b^2}{4a^2}$	Complete the square	$x^2 + \frac{7}{3}x + \frac{49}{36} = \frac{-1}{3} + \frac{49}{36}$
$\left(x + \frac{b}{2a}\right)^2 = \frac{-c}{a} + \frac{b^2}{4a^2} = \frac{-4ac + b^2}{4a^2}$	Factor left; simplify right	$\left(x + \frac{7}{6}\right)^2 = \frac{37}{36}$
$x + \frac{b}{2a} = \pm\sqrt{\frac{-4ac + b^2}{4a^2}}$	Take square roots	$x + \frac{7}{6} = \pm\sqrt{\frac{37}{36}}$
$x = -\frac{b}{2a} \pm \frac{\sqrt{b^2 - 4ac}}{2a}$	Simplify the square root; solve for x	$x = -\frac{7}{6} \pm \frac{\sqrt{37}}{6}$
$x = \frac{-b \pm \sqrt{b^2 - 4ac}}{2a}$	Write as single fraction	$x = \frac{-7 \pm \sqrt{37}}{6}$

The solutions to the quadratic equation $ax^2 + bx + c = 0$, with $a \neq 0$, are given by the ***quadratic formula:***

$$x = \frac{-b \pm \sqrt{b^2 - 4ac}}{2a}.$$

This means that the two solutions are $x = \frac{-b + \sqrt{b^2 - 4ac}}{2a}$ and $x = \frac{-b - \sqrt{b^2 - 4ac}}{2a}$.

The expression $b^2 - 4ac$ under the radical in the quadratic formula is called the *discriminant.* It determines whether or not a quadratic equation has a real solution and if it does, how many.

If $b^2 - 4ac > 0$, there are two real solutions.
If $b^2 - 4ac = 0$, there is one real solution.
If $b^2 - 4ac < 0$, there are no real solutions.

Using the Quadratic Formula

Given any quadratic equation, the quadratic formula can be used to find its solutions.

EXAMPLE D.2 Use the quadratic formula to find all solutions to each quadratic equation.

a. $4x^2 - 3x = 4$ **b.** $2x(1 - x) = 3x + 7$

Solution

a. First, we subtract 4 from each side to write the equation in standard form $ax^2 + bx + c = 0$: $4x^2 - 3x - 4 = 0$. This equation has the same solutions as the original equation. We find them by substituting $a = 4$, $b = -3$, and $c = -4$ into the quadratic formula. The solutions are $x = \frac{-(-3) \pm \sqrt{(-3)^2 - 4(4)(-4)}}{2(4)} = \frac{3 \pm \sqrt{73}}{8}$. In decimal form, the solutions are $x \approx 1.443$ and $x \approx 0.693$.

b. First, we use the distributive property on the left side: $2x - 2x^2 = 3x + 7$. Then we add $-2x + 2x^2$ to both sides to put the equation in standard form: $0 = 2x^2 + x + 7$. Since the discriminant $b^2 - 4ac$ is $1 - 4 \cdot 2 \cdot 7 = -55$, the equation has no real solutions. ■

We can use solutions to a quadratic equation to factor the quadratic function into linear factors. For example, our solutions in part a of Example D.2 allow us to factor:

$$4x^2 - 3x - 4 = 4\left(x - \frac{3 + \sqrt{73}}{8}\right)\left(x - \frac{3 - \sqrt{73}}{8}\right)$$

Exercises D-1

1. Complete the square in x and y so that the quadratic equation below is in the form $(x - h)^2 + (y - k)^2 = \text{constant}$. (Hint: First divide the equation by 4.)

$$4x^2 + 2x + 4y^2 - 12y + 2 = 0$$

2. Use the quadratic formula to show that the equation $3x(x + 4) = 5x - 6$ has no real solutions.

3. Use the quadratic formula to factor the polynomial $2x^2 + 2x - 1$ into linear factors.

D-2 Solving Pairs of Equations

When the graphs of two equations intersect, their points of intersection lie on both graphs. This means that the coordinates of those points satisfy both equations. To find the coordinates, we solve the equations simultaneously. Pairs of equations that are linear or quadratic can always be solved exactly when there are solutions; some other pairs can also be solved exactly. We illustrate some solution techniques in the next example.

EXAMPLE D.3 Find the coordinates of the points of intersection of the graphs of each pair of equations.

a. $y - 1 = \frac{1}{2}x$; $3y - 5x = -2$

b. $y = 4(x-1)^2 - 1$; $y = -x^2 + 3$

c. $y = \sqrt{2x+5}$; $x = y - 2$

d. $y = \sin x$; $y = 1 - \sin x,\ 0 \le x \le \pi$

e. $y = 2\sin x$; $y = \cos x,\ 0 \le x < \frac{\pi}{2}$

f. $y = x^2$; $y = x - 1$

Solution

a. We can solve this pair by solving for y in the first equation and then substituting the result in the second equation: $y - 1 = \frac{1}{2}x$ gives $y = \frac{1}{2}x + 1$, so $3y - 5x = -2$ becomes $3\left(\frac{1}{2}x+1\right) - 5x = -2$. We expand the left side, combine like terms, and solve for x: $\frac{3}{2}x + 3 - 5x = -2$; $-\frac{7}{2}x + 3 = -2$; $-\frac{7}{2}x = -5$; $x = \frac{10}{7}$. To find the y-coordinate of the intersection point, we substitute $x = \frac{10}{7}$ into the equation $y = \frac{1}{2}x + 1$ and obtain $y = \frac{12}{7}$. So the point of intersection of the two graphs is $\left(\frac{10}{7}, \frac{12}{7}\right)$. It is a good idea to check that these coordinates satisfy the second equation $3y - 5x = -2$: $3\left(\frac{12}{7}\right) - 5\left(\frac{10}{7}\right) = \frac{36}{7} - \frac{50}{7} = -\frac{14}{7} = -2$, which is correct.

b. We set the y-variables of the equations equal to each other: $4(x-1)^2 - 1 = -x^2 + 3$. We expand the left side to obtain $4x^2 - 8x + 3 = -x^2 + 3$ and then combine like terms and put the equation in standard form: $5x^2 - 8x = 0$. This factors as $x(5x - 8) = 0$, so the solutions are $x = 0$ and $x = \frac{8}{5}$. We use the second equation $y = -x^2 + 3$ to find the corresponding y-coordinates: when $x = 0$, $y = 3$, and when $x = \frac{8}{5}$, $y = -\left(\frac{8}{5}\right)^2 + 3 = -\frac{64}{25} + \frac{75}{25} = \frac{11}{25}$. So the points of intersection are (0, 3) and $\left(\frac{8}{5}, \frac{11}{25}\right)$. Finally, these coordinates should be substituted into the first equation to check that they satisfy it (they do).

c. We use the second equation to substitute for x in the first equation: $y = \sqrt{2x+5} = \sqrt{2(y-2)+5}$. To solve for y, we square the left and right sides of this equation and simplify the right side: $y^2 = 2(y-2) + 5 = 2y + 1$. We subtract $2y + 1$ from the left and right sides of this equation to put it into standard form $y^2 - 2y - 1 = 0$. Then we use the quadratic formula to solve it: $y = \frac{-(-2)\pm\sqrt{(-2)^2 - 4(1)(-1)}}{2(1)} = \frac{2\pm\sqrt{8}}{2} = 1 \pm \sqrt{2}$. The value $y = 1 - \sqrt{2}$ is negative. This value cannot satisfy the first equation $y = \sqrt{2x+5}$ because a square root cannot be negative. Thus there is only one point of intersection of the graphs, and it has y-coordinate $1 + \sqrt{2}$. We use the second equation $x = y - 2$ to find the corresponding x-coordinate $x = -1 + \sqrt{2}$. Thus the graphs intersect at the point $(-1 + \sqrt{2}, 1 + \sqrt{2})$. Finally, we should check that these coordinates satisfy the first equation (they do). We note that the false intersection was introduced when we squared the equation.

d. We set the y-variables of the equations equal to each other: $\sin x = 1 - \sin x$. If we add $\sin x$ to both sides, we get $2 \sin x = 1$; dividing by 2 gives $\sin x = \frac{1}{2}$. The values of x in the interval $[0, \pi]$ that satisfy this equation are $x = \frac{\pi}{6}$ and $\frac{5\pi}{6}$. (See Chapter 8.) The points of intersection of the two graphs in this interval are $\left(\frac{\pi}{6}, \frac{1}{2}\right)$ and $\left(\frac{5\pi}{6}, \frac{1}{2}\right)$.

e. We set the y-variables of the equations equal to each other: $2 \sin x = \cos x$. To solve this, we use the definition of the tangent and inverse tangent functions (see Chapter 8). We divide both sides of the equation by $\cos x$ (this assumes that $\cos x \neq 0$, which is true when x is between 0 and $\frac{\pi}{2}$): $2\frac{\sin x}{\cos x} = 1$, which is the same as $2 \tan x = 1$, or $\tan x = \frac{1}{2}$. To solve this equation, we apply the inverse tangent function: $x = \tan^{-1}(\tan x) = \tan^{-1}\left(\frac{1}{2}\right) \approx 0.464$. This is the angle (in radian measure) between 0 and $\frac{\pi}{2}$ whose tangent equals $\frac{1}{2}$. To find the y-coordinate of this point of intersection, we use the second equation $y = \cos x$. We can draw a right triangle with acute angle x whose tangent equals $\frac{1}{2}$ and then use the Pythagorean theorem to find that the length of the hypotenuse is $\sqrt{5}$. From the picture, we can read that $\cos x = \frac{2}{\sqrt{5}}$. Thus the intersection point is $\left(\tan^{-1}\left(\frac{1}{2}\right), \frac{2}{\sqrt{5}}\right)$.

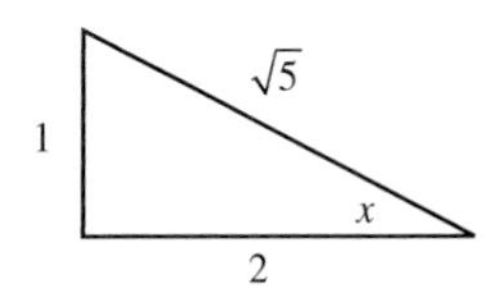

f. We set the y-variables of the equations equal to each other: $x^2 = x - 1$. We put this quadratic equation in standard form and then use the quadratic formula to solve. The equation $x^2 - x + 1 = 0$ has no real solutions since the discriminant $b^2 - 4ac = 1 - 4 \cdot 1 \cdot 1 = -3$. Thus the graphs have no points of intersection.

The graphs of the six pairs of functions are shown below.

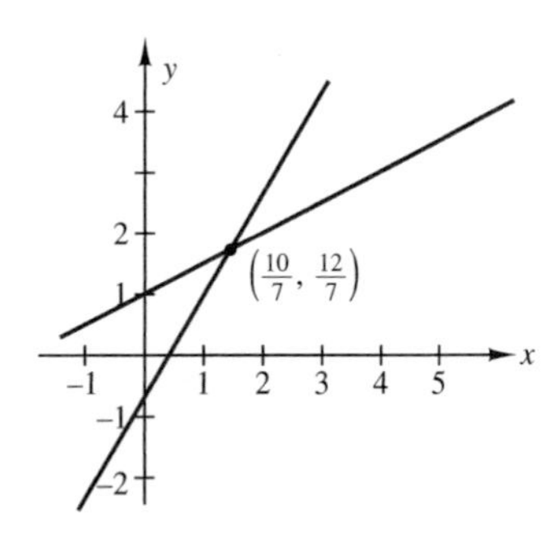

a. $y - 1 = \frac{1}{2}x$; $3y - 5x = -2$

(0, 3)

(8/5, 11/25)

b. $y = 4(x-1)^2 - 1$; $y = -x^2 + 3$

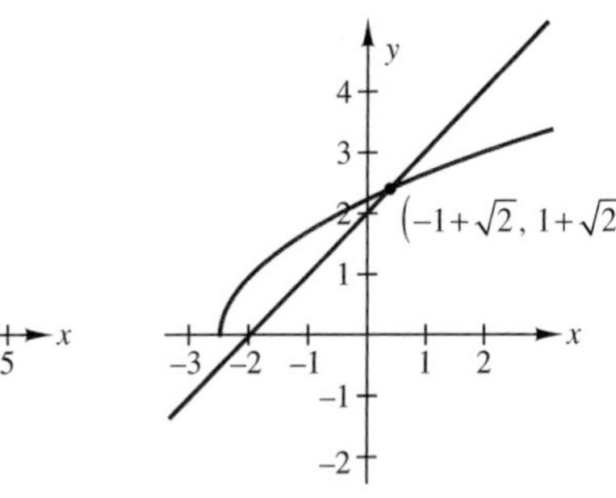

c. $y = \sqrt{2x+5}$; $x = y - 2$

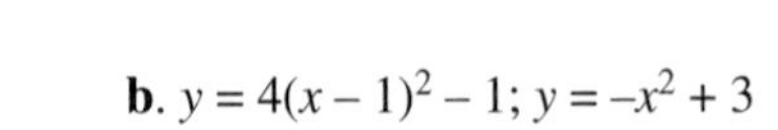

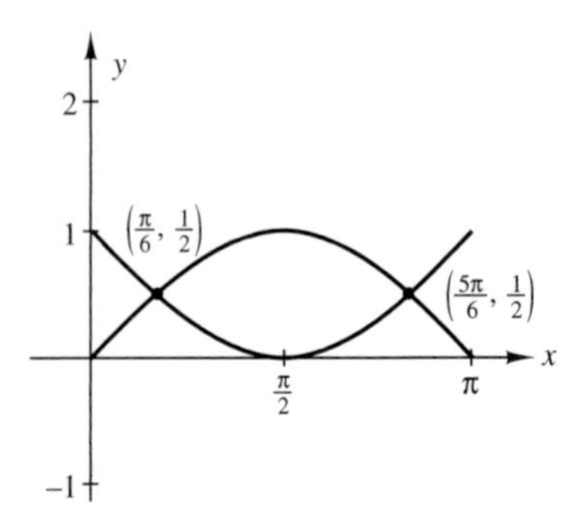

d. $y = \sin x$; $y = 1 - \sin x$, $0 \le x \le \pi$

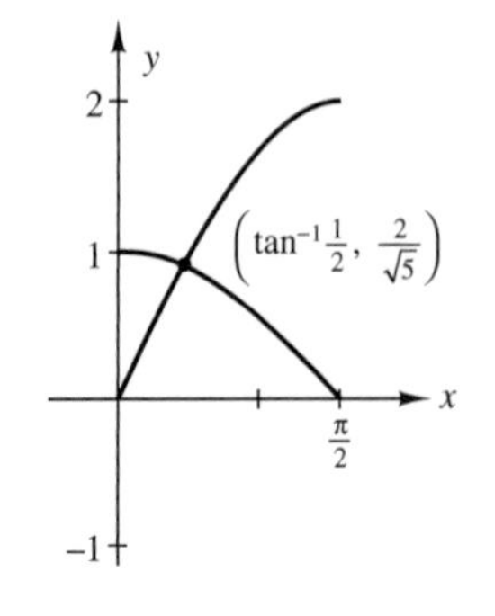

e. $y = 2 \sin x$; $y = \cos x$, $0 \le x \le \frac{\pi}{2}$

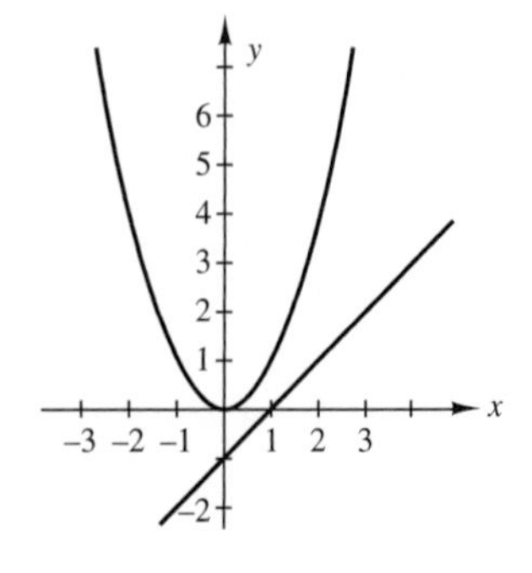

f. $y = x^2$; $y = x - 1$ ■

Exercises D-2

1. Find the coordinates of the points of intersection of the graphs of each pair of equations, and check your results. Graph each pair of equations on a single set of axes. If the graphs do not intersect, explain why.

a. $x - y = 6$ and $x + 4 = 2y$

b. $x = 3y + 4$ and $y = -(x - 3)^2 + 1$

c. $y = 2\sqrt{5 - 3x}$ and $y = -x + 5$

d. $y = 2 \sin x$ and $y = \tan x$, $0 \le x < \frac{\pi}{2}$

E ABOUT TRIANGLES

A more complete discussion of trigonometry, especially as used in calculus, is in Chapter 8.

Right Triangles

In a right triangle, the two sides adjacent to the right angle are called legs and the side opposite the right angle is called the hypotenuse. One of the most useful theorems is:

The Pythagorean Theorem. *The sum of the squares of the legs of a right triangle is equal to the square of the hypotenuse.*

In the right triangle shown, $x^2 + s^2 = t^2$.

The Pythagorean theorem is useful in many applications, especially "related rates" problems in calculus. (See Chapter 11.)

The Trigonometric Ratios. Trigonometry means "triangle measure." For more than 2000 years, trigonometry has been used to measure distances indirectly—it was used in ancient times to estimate the circumference of the earth, was used by ancient surveyors to carve tunnels through mountains accurately, and is used today in essentially the same way to drill subway tunnels under cities and to lay out plots in residential developments. The key to this technique of measuring is the use of trigonometric ratios.

For any acute angle θ, a right triangle with angle θ can be constructed:

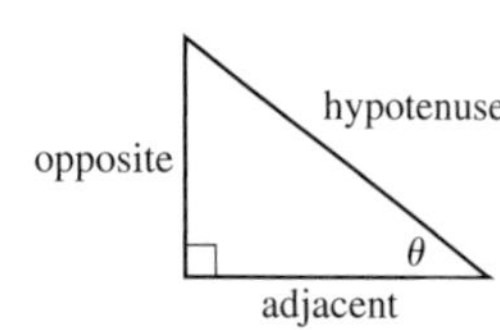

The values of the trigonometric functions sine, cosine, and tangent of θ are then defined as ratios of sides of that right triangle. (The size of the triangle does not affect the ratios since all right triangles with angle θ will be similar to each other.) These ratios are

$$\sin \theta = \frac{\text{leg opposite } \theta}{\text{hypotenuse}}, \ \cos \theta = \frac{\text{leg adjacent } \theta}{\text{hypotenuse}}, \ \tan \theta = \frac{\text{leg opposite } \theta}{\text{leg adjacent } \theta}.$$

The values of the other three trigonometric functions—secant, cosecant, and cotangent of θ—are then defined as reciprocals of the three ratios above (and are the remaining three ratios of pairs of sides of the triangle):

$$\sec\theta = \frac{1}{\cos\theta}, \quad \csc\theta = \frac{1}{\sin\theta}, \quad \cot\theta = \frac{1}{\tan\theta}.$$

The "co" in cosine, cotangent, and cosecant refers to the role of *complementary angles* in a right triangle. The two acute angles in a right triangle are complementary since their sum is 90°. If one of the acute angles is called θ and the other called θ', as in the right triangle pictured below, then the definitions of the trigonometric ratios lead to complementary relationships.

For example, $\sin\theta = \dfrac{\text{leg opposite }\theta}{\text{hypotenuse}} = \dfrac{\text{leg adjacent }\theta'}{\text{hypotenuse}} = \cos\theta'$.

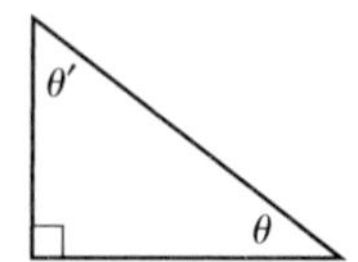

In a similar way, it can be shown that $\tan\theta = \cot\theta'$ and $\sec\theta = \csc\theta'$.

In calculus problems, an angle θ may be a variable, and as θ varies, so will the value of the six trigonometric ratios. If one of these ratios is called x, often it is necessary to express the other ratios in terms of x.

EXAMPLE E.1 For a given acute angle θ, we are told that $\sin\theta = x$. Express each of the trigonometric ratios for θ as a function of x.

Solution We begin by drawing a right triangle with angle θ and label the leg opposite θ with x. Then we label the hypotenuse 1 so that $\sin\theta = \frac{x}{1} = x$.

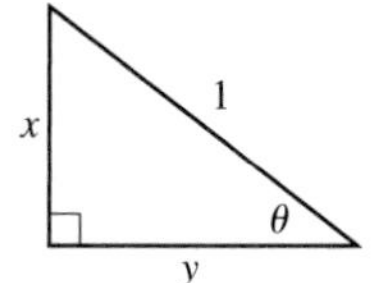

We name the other leg of the triangle y; then the Pythagorean theorem says that $x^2 + y^2 = 1$. We solve for y (length must be positive): $y = \sqrt{1-x^2}$. The values of the trigonometric ratios for θ are then

$$\sin\theta = x, \quad \cos\theta = \sqrt{1-x^2}, \quad \tan\theta = \frac{x}{\sqrt{1-x^2}},$$

$$\sec\theta = \frac{1}{\sqrt{1-x^2}}, \quad \csc\theta = \frac{1}{x}, \quad \cot\theta = \frac{\sqrt{1-x^2}}{x}.$$ ■

Arbitrary Triangles

An arbitrary triangle is one that does not have any special restriction on the lengths of its sides or the sizes of its angles. Relationships that are true for an arbitrary triangle are true for all triangles.

The Law of Cosines. This law is a generalization of the Pythagorean theorem and applies to any triangle. It allows us to find any of the angles of a given triangle if we know the lengths of its three sides; it also allows us to find the length of one side of a given triangle if we know the lengths of the two other sides and the angle between those two sides. The *Law of Cosines* says *the square of one side of a triangle equals the sum of the squares of the two other sides minus two times the product of the other sides times the cosine of the angle between these sides.*

Triangles are shown below, with sides x, s, and t and angle θ between x and s (θ is opposite side t).

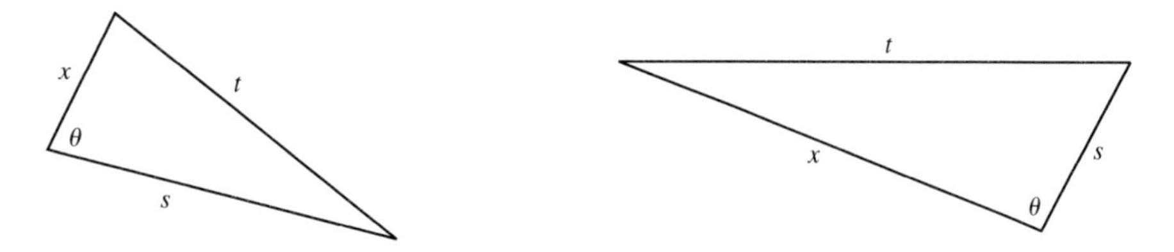

For these triangles, the Law of Cosines states $t^2 = x^2 + s^2 - 2xs \cos \theta$.

When the angle θ is a right angle, $\cos \theta = 0$; in this case, the equation for the Law of Cosines gives the Pythagorean theorem.

The Law of Sines. This law relates the sides of a triangle to the angles opposite them; it sets up a proportion. It says that *in any triangle, the three ratios formed by dividing the length of a side by the sine of the angle opposite that side are equal to one another.* To state the Law of Sines in equation form, the sides and vertices of the triangle are usually labeled as shown below, with uppercase letters denoting the vertices and corresponding lowercase letters denoting the opposite sides. By convention, we name an angle by its vertex.

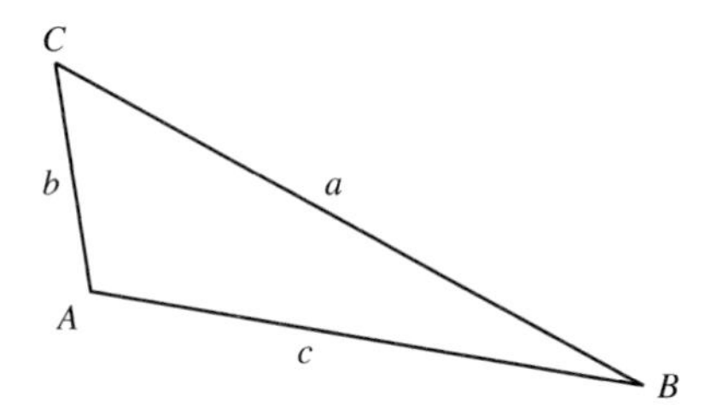

The Law of Sines: $\dfrac{a}{\sin A} = \dfrac{b}{\sin B} = \dfrac{c}{\sin C}$.

Exercises E

1. For a given acute angle θ, you are told that $\tan \theta = x$. Express each of the trigonometric ratios for θ as a function of x.

2. For a given acute angle θ, you are told that $\sec \theta = x$. Express each of the trigonometric ratios for θ as a function of x.

ANSWERS TO EXERCISES

Chapter 0 Exercises

1. See Chapter 6, Section 6-A, for how to place symbols on the diagrams of geometric figures.

a. (i) Rectangle; l = length, w = width
(ii) The area of a rectangle equals its length times its width.

b. (i) Parallelogram; b = length of base, h = height
(ii) The area of a parallelogram equals its base times its height.

c. (i) Circle; r = radius
(ii) The area of a circle equals π times the square of its radius.

d. (i) Rectangular box; l = length, w = width, h = height
(ii) The volume of a rectangular box equals the product of its length, width, and height.

e. (i) Circular cylinder; r = radius of base, h = height
(ii) The volume of a circular cylinder equals π times the product of its height and the square of its radius. This is also the product of its height and the area of its base.

f. (i) Circular cone; r = radius of base, h = height
(ii) The volume of a circular cone is one-third π times the product of its height and the square of its radius. This is also one-third the volume of a circular cylinder that has the same radius and height as the cone.

g. (i) Cube; s = length of each edge
(ii) The volume of a cube is the cube of the length of one edge.

h. (i) Right triangle; a and b are legs, c is the hypotenuse (side opposite the right angle).
(ii) The square of the hypotenuse of a right triangle equals the sum of the squares of its legs. This is the Pythagorean Theorem.

i. (i) Triangle; b = length of base, h = altitude
(ii) The area of any triangle is one-half its base times its altitude.

j. (i) The truck represents travel; d = distance, r = rate, t = time
(ii) Distance equals rate times time; distance is the product of the rate and time traveled.

k. (i) The thermometer represents temperature; F = degrees Fahrenheit, C = degrees Celsius (centigrade measure)
(ii) The Fahrenheit temperature equals $\frac{9}{5}$ the Celsius temperature plus 32.

l. (i) Circular sector (central wedge of a circle); θ = central angle, r = radius of circle
(ii) The area of a sector of a circle equals the square of the radius divided by 2 times the central angle (in radians).

m. (i) Sphere; r = radius
(ii) The volume of a sphere equals $\frac{4}{3}\pi$ times the cube of its radius.

2. **a.** $P = 2l + 2w$

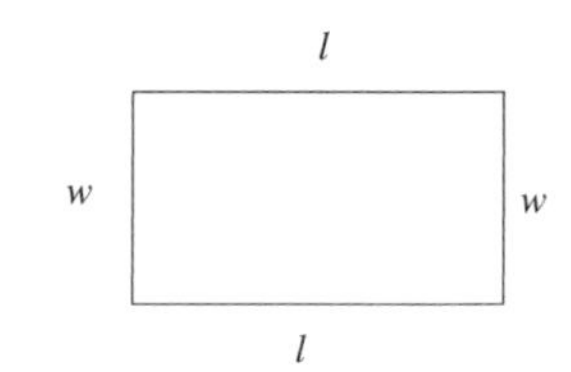

b. $A = 2lw + 2lh + 2wh$

c. $P = \frac{C}{2} + D$

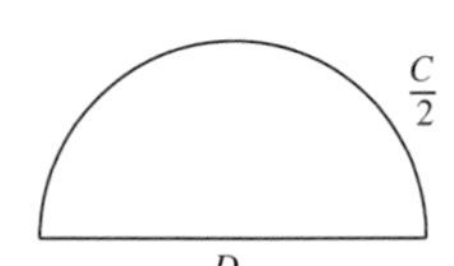

3.

C	F
0°	32°
100°	212°
20°	68°
37°	98.6°

4. There are many examples; here are two possibilities: $\sqrt[3]{\ }$ (cube root); | | (absolute value).

5. Some possibilities are $\leq$ (less than or equal to); $\geq$ (greater than or equal to); $\subset$ (is contained in); $\notin$ (does not belong to).

6. Here are a few: cm (centimeters); abs (absolute value); cos (cosine); ∞ (infinity); log (logarithm); ° (degrees).

Exercises 1-A

1. **a.**

B (–2.7, 3.5)
D (1.3, 2.5)
A (–1.2, –2.5)
C (1.5, –2.5)

b. *A*—III; *B*—II; *C*—IV; *D*—I
c. Both the x- and the y-coordinates are positive.
d. Both the x- and the y-coordinates are negative.

2. **a.** $(-1.2, 1)$; $(-1.2, 6.7)$; $(-1.2, -\pi)$; or any other pair $(-1.2, a)$
b. There are infinitely many such points; they lie on the vertical line passing through -1.2 on the x-axis. This line is parallel to the y-axis and perpendicular to the x-axis. See graph below.

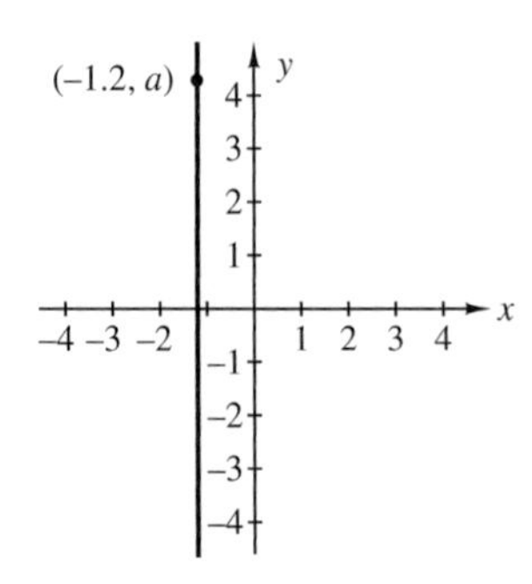

c. These are all the points on the horizontal line passing through 2.5 on the y-axis. This line is parallel to the x-axis and perpendicular to the y-axis. See graph below.

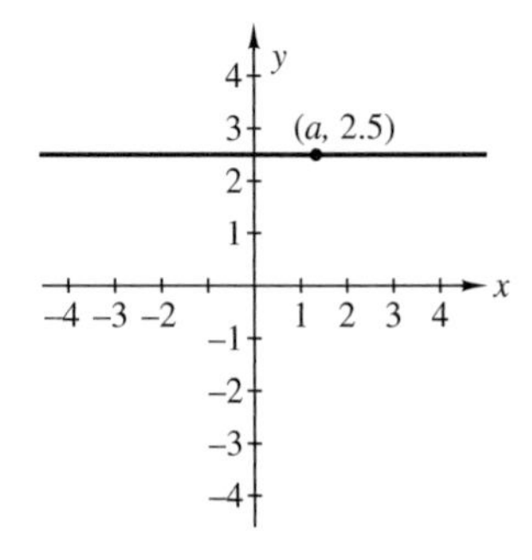

3. The fourth vertex is $(-1, 5)$. See graph below.

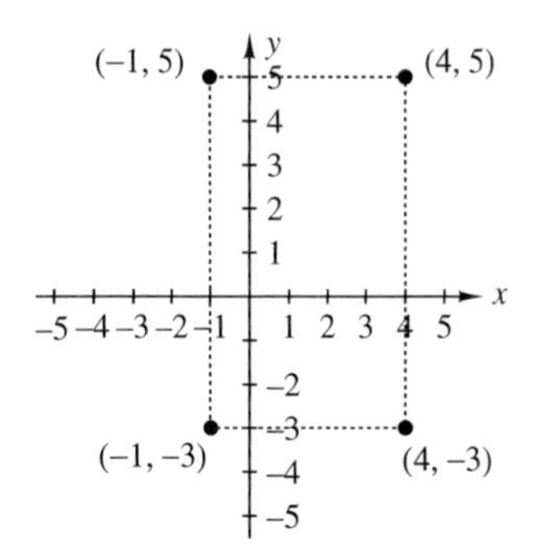

4. Since these coordinates are estimated values, your answers may be slightly different than those given.
a. The origin; $(0, 0)$
b. Quadrant II; $(-1.4, 2.8)$
c. Quadrant I ; $(2.3, 0.6)$
d. Quadrant IV; $(6, -0.3)$
e. Quadrant III; $(-0.15, -4.25)$
f. Quadrant II; $(-2.3, 8.3)$
g. Quadrant IV; $(0.2, -2.1)$

5. **a.** The Pentagon **b.** U.S. Capitol
c. Convention Center **d.** The Ellipse

6. **a.** 33 **b.** A4, B3, D2

Exercises 1-B

1. **a.** $x + y = 0$, or $y = -x$

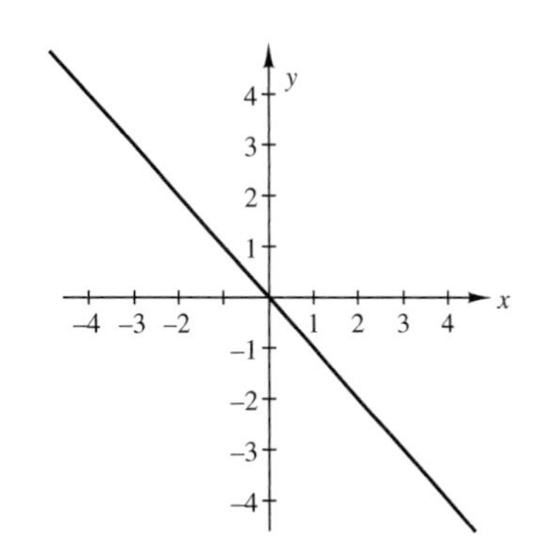

b. $x = 5$

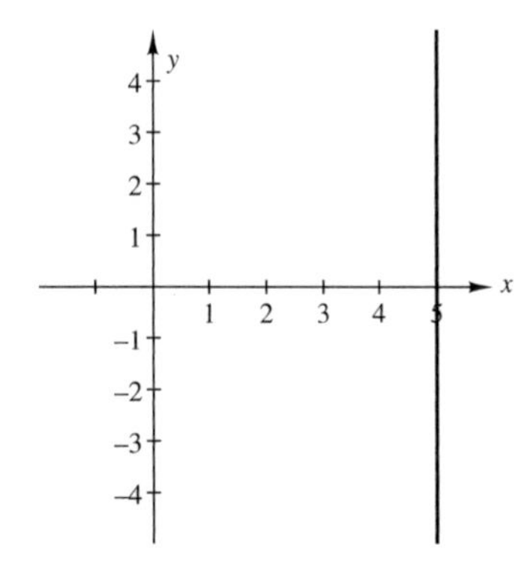

c. $y = -2.8$

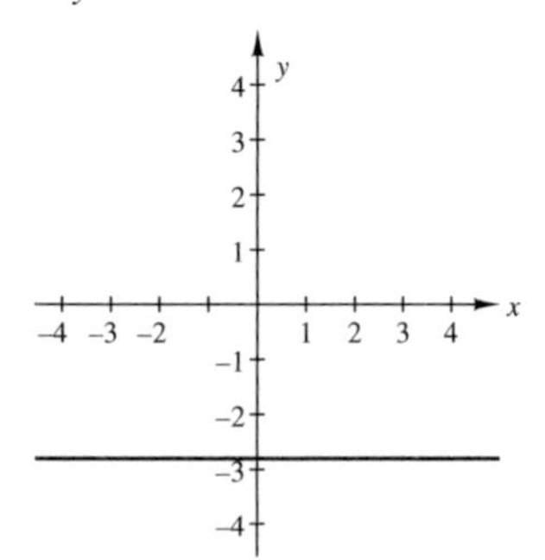

2. **a**. x-intercept: $x = -\frac{5.5}{2.1} = \frac{55}{21}$; y-intercept: $y = \frac{5.5}{3.7} = \frac{55}{37}$

   b. x-intercept: $x = -\frac{9}{2}$; y-intercepts: $y = \pm 3$
3. **a**. yes, no, yes **b**. no, yes, yes **c**. yes, no, yes
   d. no, yes, no, yes
4. We let x represent the number of years since 1990 to obtain the following graph:

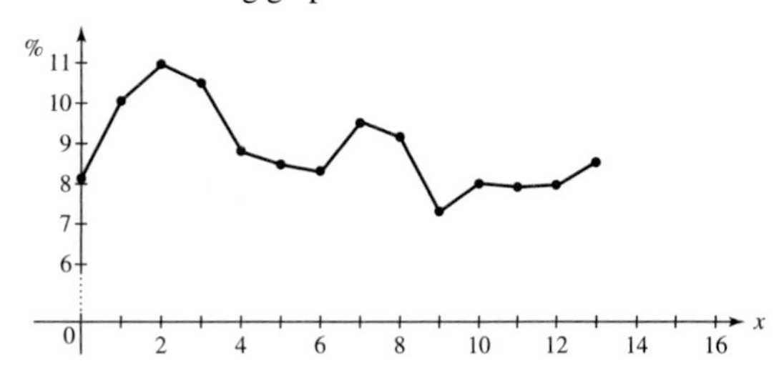

Exercises 1-C

1. **a**. no **b**. yes **c**. yes **d**. yes
2. **a**. $y = \frac{3}{2}x + \frac{5}{2}$

   slope: $\frac{3}{2}$; y-intercept: $\frac{5}{2}$

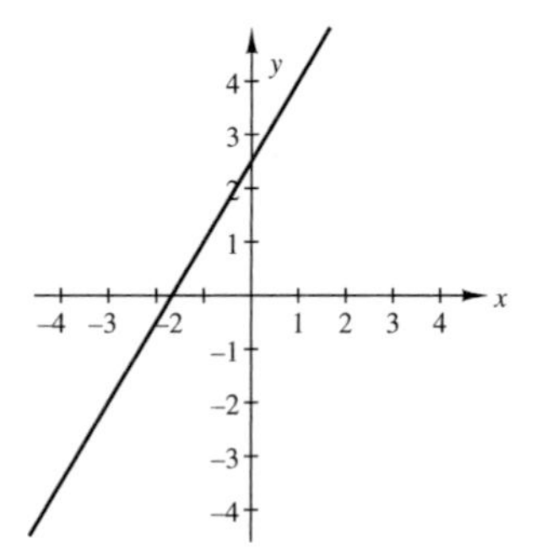

b. $y = -2.4x + 1.8$
slope: -2.4; y-intercept: 1.8

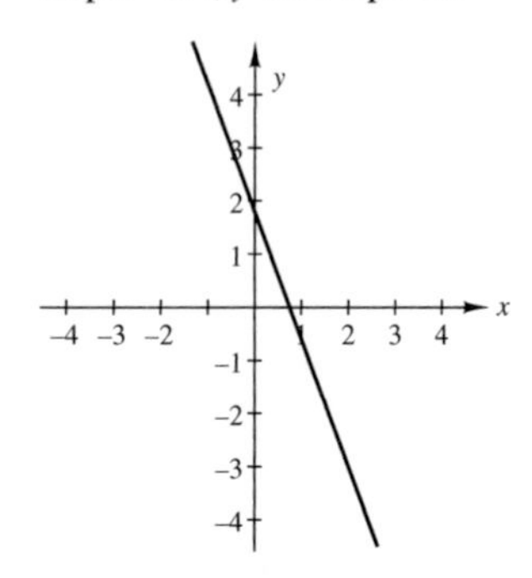

c. $y = 3$
slope: 0; y-intercept: 3

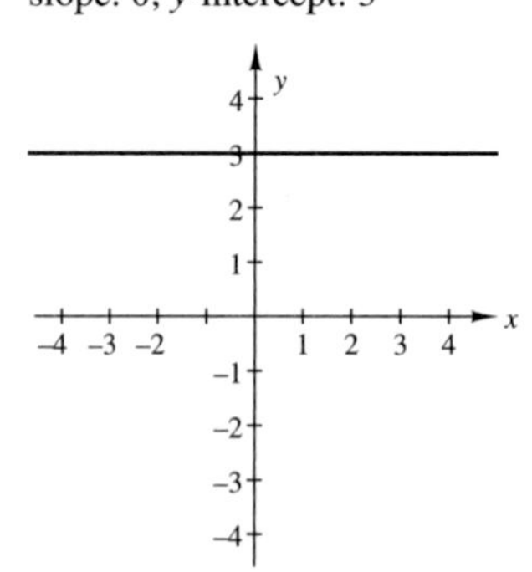

d. $x = -\frac{1}{2}$
slope: undefined; y-intercept: none

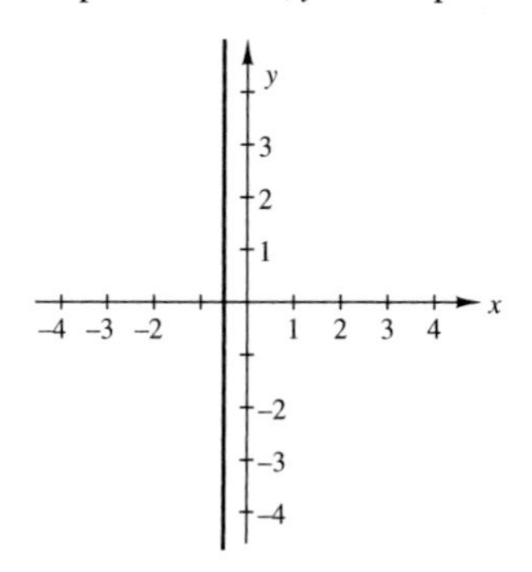

3. **a**. $m = -\frac{1}{7}$ **b**. $y = -\frac{1}{7}x + \frac{15}{7}$
   c. $\left(0, \frac{15}{7}\right)$ and $\left(3, \frac{12}{7}\right)$ are two possibilities.
   d.

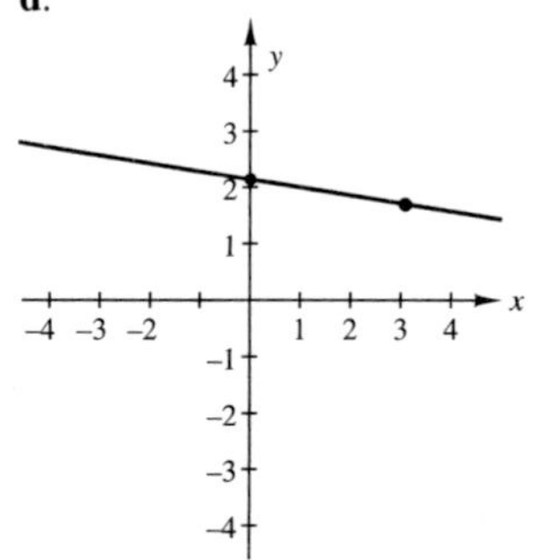

4. $L_1: m \approx \frac{2}{3}$; $L_2: m \approx \frac{1}{3}$; $L_3: m \approx -\frac{2}{3}$
5. **a**. (1.50, 1200), (2.50, 900)
   b. $m = -300$; For every $1 increase in price, the demand decreases by 300 cups.
   c. $d - 1200 = -300(p - 1.5)$; $d = -300p + 1650$

Exercises 1-D

1. **a**. $y = 2x + 1$; $y = 2x - \pi$ are two possibilities.
   b. $y = -\frac{1}{2}x$; $y = -\frac{1}{2}x - 1$ are two possibilities.
2. **a**. The line through A and C: $y = x$. The line through B and D: $y = -x$.
   b. yes **c**. $a = 1.25$
3. **a**. L_1: $y = 4$; L_2: $y = \frac{1}{2}x + 1$; L_3: $x = \frac{3}{2}$
   b. Intersection of L_1 and L_2: (6, 4); intersection of L_1 and L_3: $\left(\frac{3}{2}, 4\right)$

4. **a.** $\left(0, -\frac{7}{3}\right)$ is one point. Any point (a, b) that satisfies the equation is correct.
b. $\frac{5}{3}$ **c.** $-\frac{3}{5}$

Exercises 1-E

1. **a.** $|AB| = \sqrt{10}$; $|BC| = \sqrt{40} = 2\sqrt{10}$; $|AC| = \sqrt{90} = 3\sqrt{10}$
b. A, B, C lie on a line.
2. $\sqrt{36.04} \approx 6.003$
3. **a.**

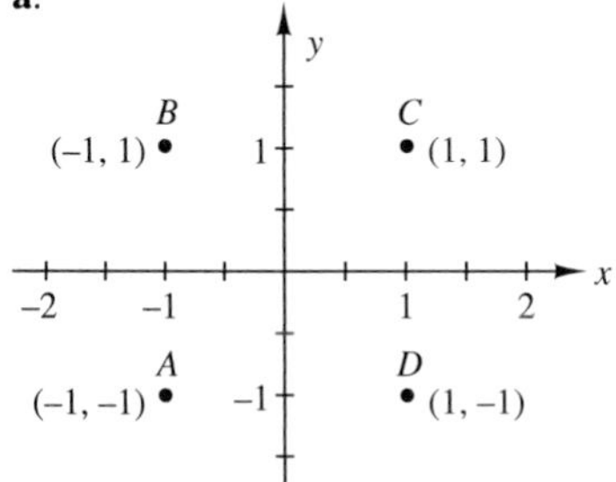

b. $|AB| = 2$; $|BC| = 2$; $|CD| = 2$; $|DA| = 2$; $|AC| = 2\sqrt{2}$; $|BD| = 2\sqrt{2}$
c. $ABCD$ is a square. It has all four sides equal and equal diagonals.

Exercises 1-F

1. **a.** $x^2 + y^2 = 5$ **b.** $(-1)^2 + 2^2 = 5$
c.

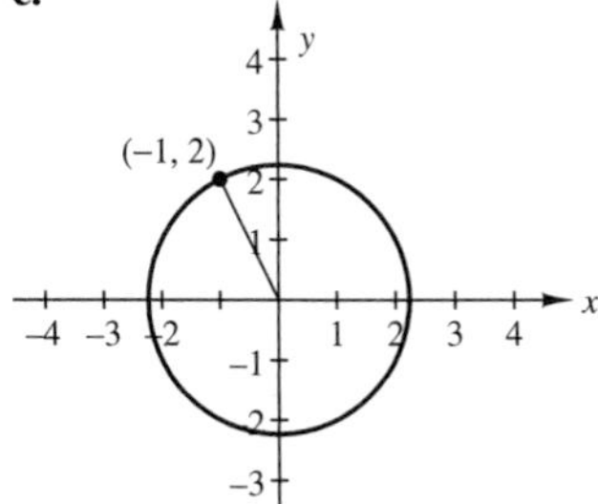

d. -2 **e.** $y = \frac{1}{2}x + \frac{5}{2}$
2. **a.** $5, 5, \sqrt{8}, \sqrt{41}$
b. P_1 falls on the circle; P_2 falls on the circle; P_3 falls inside; P_4 falls outside.
c. $|AP_1| + |P_1P_2| + |P_2B| = 2\sqrt{10} + \sqrt{2}$
3. **a.** $(x + 5)^2 + (y - 4)^2 = 1$

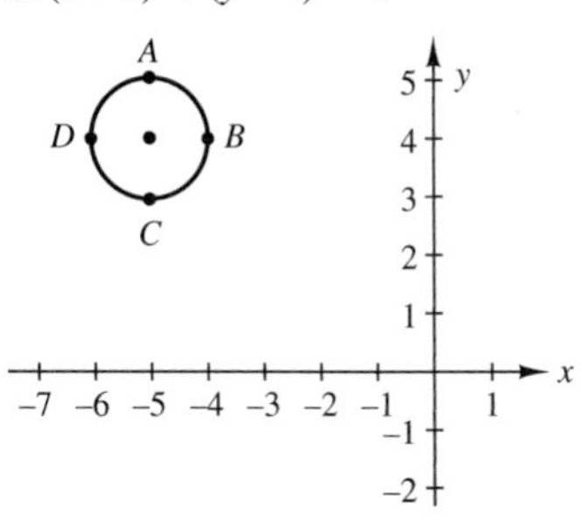

b. The tangent lines will have positive slope at the points on the circular arcs from $D(-6, 4)$ to $A(-5, 5)$ and from $C(-5, 3)$ to $B(-4, 4)$, not including the endpoints of the arcs.
c. The tangent lines will have negative slope at the points on the circular arcs from $C(-5, 3)$ to $D(-6, 4)$ and from $A(-5, 5)$ to $B(-4, 4)$, not including the endpoints of the arcs.
d. The tangent lines will be horizontal at $C(-5, 3)$ and $A(-5, 5)$; the tangent lines will be vertical at $D(-6, 4)$ and $B(-4, 4)$.
4. **a.** $(-1, 2)$ **b.** 3
c. $(x + 1)^2 + (y - 2)^2 = 9$
5. **a.** Center: $\left(-4, \frac{1}{2}\right)$; $r = \sqrt{3}$

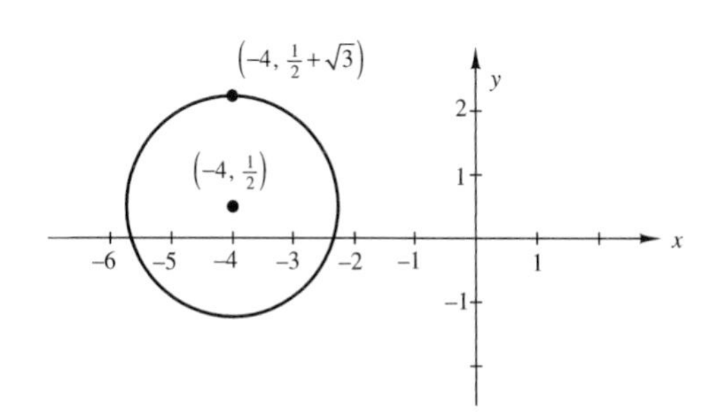

b. Center: $(0, 0)$; $r = \sqrt{\pi}$

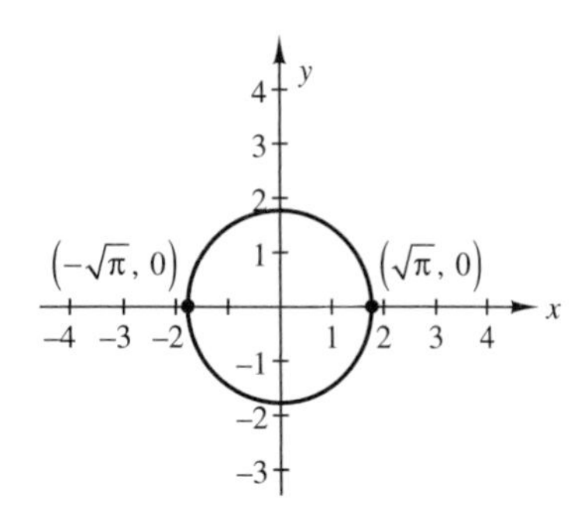

Chapter 1 Exercises

1. **a.** $y = 2$

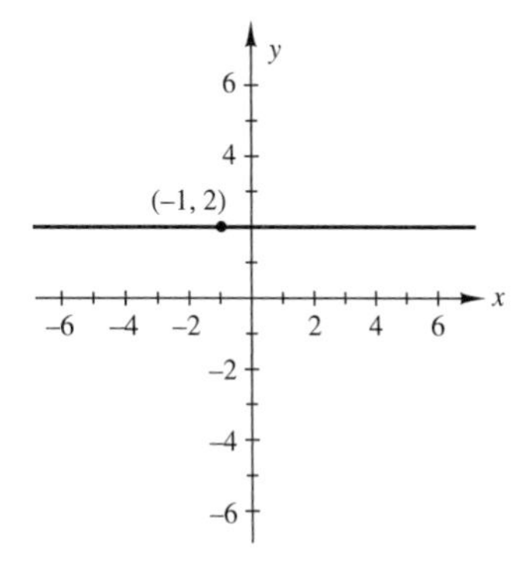

b. $x = -1$

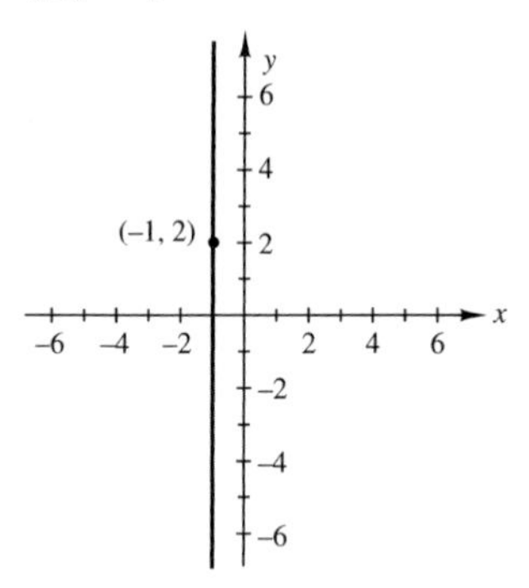

c. $y = 3$

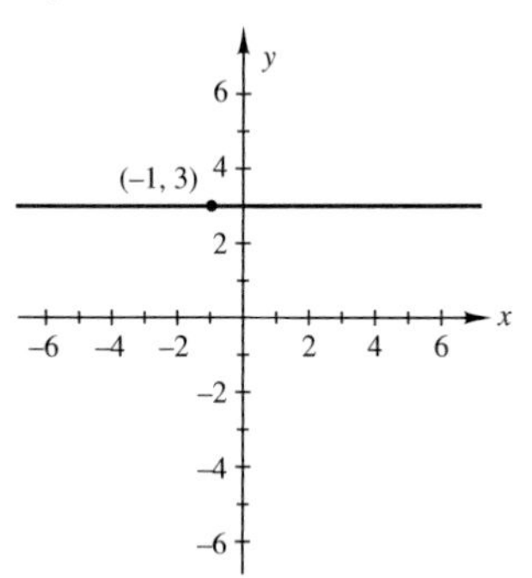

d. $y = -\frac{2}{5}x$

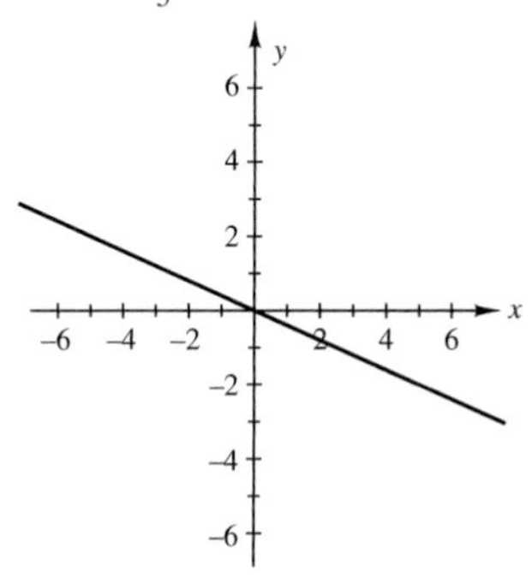

e. $y = 2.5x$

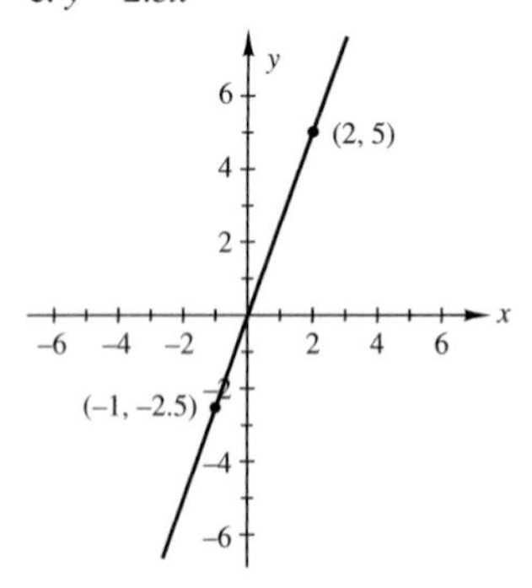

2. They are perpendicular.
3. L_1: $-2x + 2$; L_2: $y = -2.4$; L_3: $y = \frac{1}{2}x + 1$
4. **b.** The slopes of your lines will vary; the y-intercept of each line is -3.
   c. The three lines have the same y-intercept, $y = -3$.
5. **b.** The y-intercepts of your lines will vary; each line has slope 2.
   c. The three lines have the same slope, $m = 2$.
6. The point does not lie on the line.
7. **a.** Slope $= \frac{3}{4}$; y-intercept $= -\frac{9}{4}$; x-intercept $= 3$
   b. Slope $= 0$; y-intercept $= -3$; x-intercept—none
   c. Slope $= \frac{2}{3}$; y-intercept $= -\frac{1}{3}$; x-intercept $= \frac{1}{2}$
8. One possibility is shown below. Slope of line 1: 0; slope of line 2: undefined.

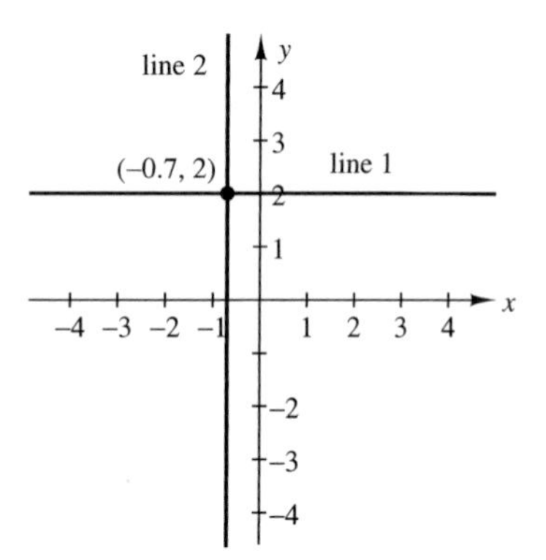

9. A line with slope $\frac{3}{2}$ is steeper than one with slope $\frac{2}{3}$.

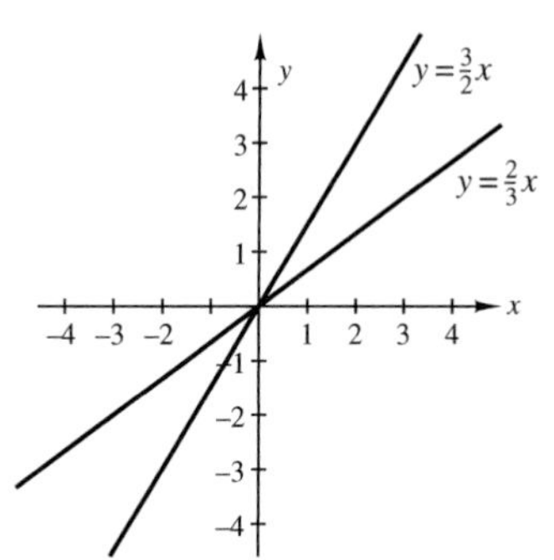

10. **a.** Three possibilities are shown below.

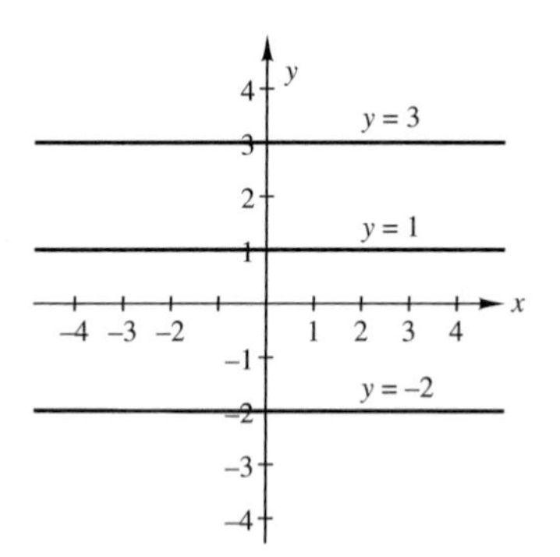

b. Undefined
c. $y = 1$; $y = 3$; $y = -2$. The line $x = 0$ (or any line $x = a$) is perpendicular to these.

11. **a.** $y = -\frac{1}{2}x + 7$; $y = \frac{1}{2}x + 2$
    b. no; $\left(-\frac{1}{2}\right)\left(\frac{1}{2}\right) \neq -1$.
12. **a.** 5; 5; 5; 5; $\sqrt{5}$; 10
    b. on; on; on; on; inside; outside
13. **a.** $(x + 1)^2 + (y + 3)^2 = 2$ **b.** $(x + 3)^2 + y^2 = 25$
    c. $(x - 4)^2 + (y + 1)^2 = 45$

14. **a**. 40 miles apart after 1 hour; 30 miles apart after 2 hours.
b. $D_A = 40t + 50$; $D_B = 50t$
c. A: y-intercept = 50, train A is intially 50 miles ahead of B; B: y-intercept = 0, train B is at the grain elevator at noon.
d.

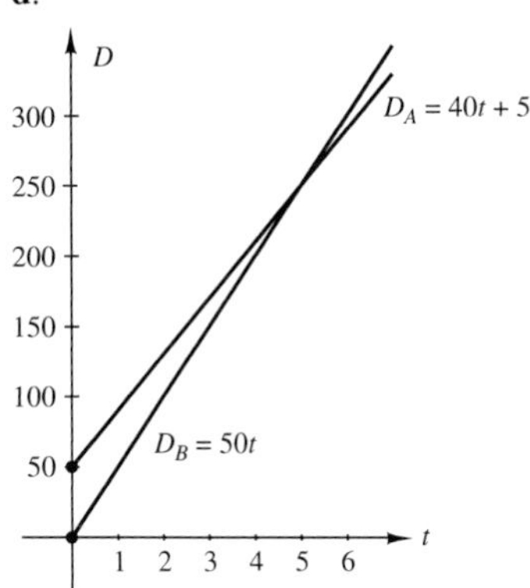

e. $m_A = 40$; $m_B = 50$. These slopes are the velocities of trains A and B, respectively.
f. Train B will catch up to train A 5 hours after they start.

15. **a**. $y = 3x$
b.

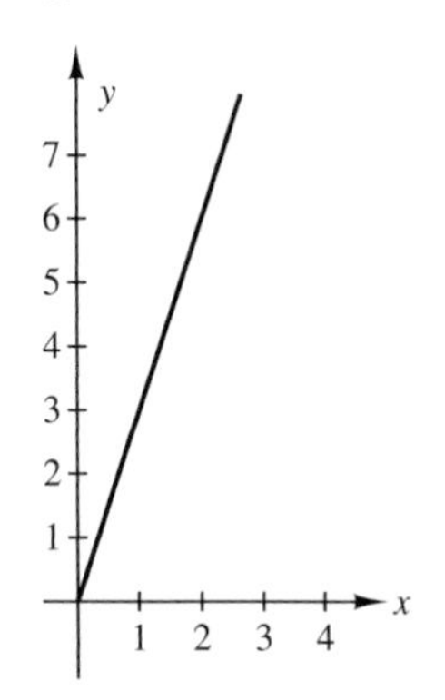

16. **a, b,** and **e.** The coordinates of P, Q, and R each satisfy the equation $y = -\frac{3}{2}x + 1$. (See following figure.)

The point $P\left(-1, \frac{5}{2}\right)$ is on the line, since $\frac{5}{2} = \left(-\frac{3}{2}\right)(-1) + 1$. Other verifications are similar.

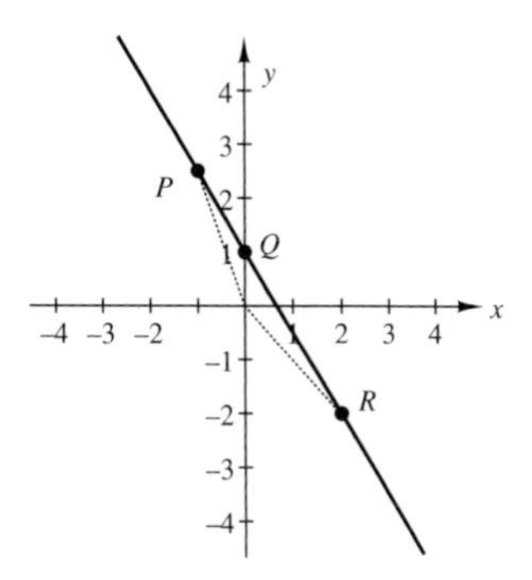

c. $|OP| = \sqrt{7.25}$; $|OQ| = 1$; $|OR| = 2\sqrt{2}$
d. $Q(0, 1)$ is closest.

17. **a.** $\sqrt{(x-0)^2 + \left(-\frac{3}{2}x + 1 - 0\right)^2} = \sqrt{\frac{13}{4}x^2 - 3x + 1}$

b. $\frac{\sqrt{13}}{6}$; $\frac{\sqrt{85}}{12}$; 1; $\frac{\sqrt{29}}{2}$; $\frac{\sqrt{5}}{2}$. The shortest of the distances is $\frac{\sqrt{13}}{6} \approx 0.6$.

c. One way is given in Exercise 18.

18. **a**. $y = \frac{2}{3}x$

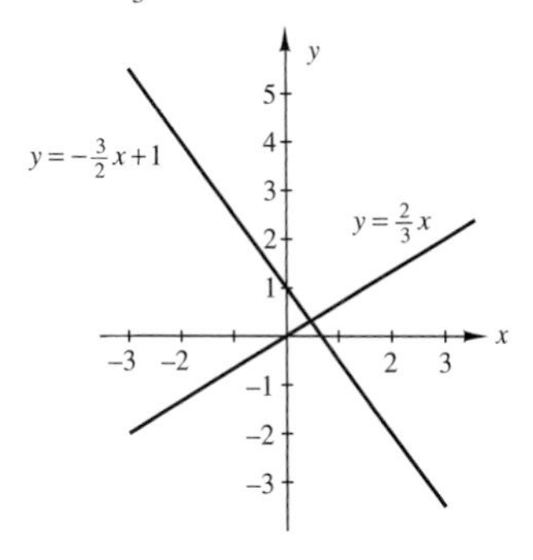

b. $\left(\frac{6}{13}, \frac{4}{13}\right)$ **c.** $\frac{2\sqrt{13}}{13}$

Exercises 2-A

1.

t	−3.4	−2.0	−1.2	−0.3	0.5	1.0	1.7
$g(t)$	36.445	15	9.32	20.27	−11.25	−3	5.14

2. **a**. $f(-0.1) = -\frac{15}{19} \approx -0.7895$; $f(0) = -\frac{1}{2}$; $f\left(\frac{1}{5}\right) = 0$; $f(2) = 2.25$; $f(x) = \frac{5x-1}{x+2}$
b. No. When $t = -2$ the denominator is 0, so the quotient is undefined.

3.

h	$f(3)$	$f(3+h)$	$f(3+h) - f(3)$	$\frac{f(3+h) - f(3)}{h}$
1	5	12	7	7
0.1	5	5.61	0.61	6.1
0.01	5	5.0601	0.0601	6.01
0.001	5	5.006001	0.006001	6.001
0.0001	5	5.00060001	0.00060001	6.0001

4. $G(-1) = -10$; $G(1.2) = 11.824$; $G(a) = 8a^3 - 2$; $G(t+1) = 8(t+1)^3 - 2$; $G(2+h) = 8(2+h)^3 - 2$

Exercises 2-B

1. **a.** $[-2, 3)$ **b.** $(-\infty, 2]$
   c. $(-\infty, -1)\cup[2, 4]$ **d.** $(1, \infty)$

2. **a.** (number line: –1, 3) $(-1, 3]$
   b. (number line: 4) $(-\infty, 4)$
   c.  (number line: 5.1) $[5.1, \infty)$
   d. (number line: 2, 4) $(-\infty, 2)\cup(2, 4)\cup(4, \infty)$

3. **a.** $\{x \mid x \neq 0\} = (-\infty, 0)\cup(0, \infty)$
   b. $\left\{t \middle| t \neq \frac{2}{5}\right\} = \left(-\infty, \frac{2}{5}\right)\cup\left(\frac{2}{5}, \infty\right)$
   c. $[0, \infty)$ **d.** $[2, \infty)$

4. **a.** Yes; $f(0) = -3$ is a real number.
   b. No; the equation $0 = \frac{3}{x-1}$ has no solution since a quotient with a nonzero numerator cannot equal 0.
   c. Yes; $g(-2) = (-2)^2 + 2 = 6$ is a real number.
   d. No; $g(x) = x^2 + 2 > 0$ for all real numbers x.

5. Domain = $[0, \infty)$; range = $(-\infty, 3]$

6. $A = \pi x^2$; domain = $(0, \infty)$; range = $(0, \infty)$

Exercises 2-C

1. **a.**

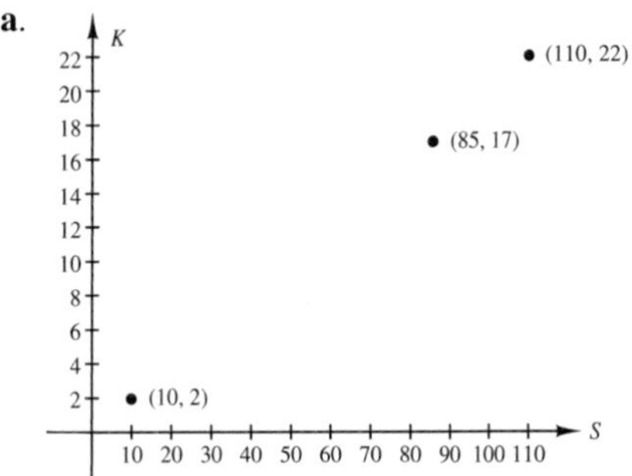

   b. Yes; the points are on a line because the three slopes $\frac{17-2}{85-10}$, $\frac{22-17}{110-85}$, and $\frac{22-2}{110-10}$ all equal $\frac{1}{5}$.
   c. Delilah keeps 20% of the selling price.
   d. $f(x) = \frac{20}{100}x = \frac{x}{5}$, or $f(x) = 0.2x$

2. **a.** A person should drink 12 oz plus an additional number of ounces equal to $\frac{8}{15}$ times the number of minutes he or she exercises.
   b.

m	$f(m) = L$
0	12
30	28
45	36
75	52

   c.

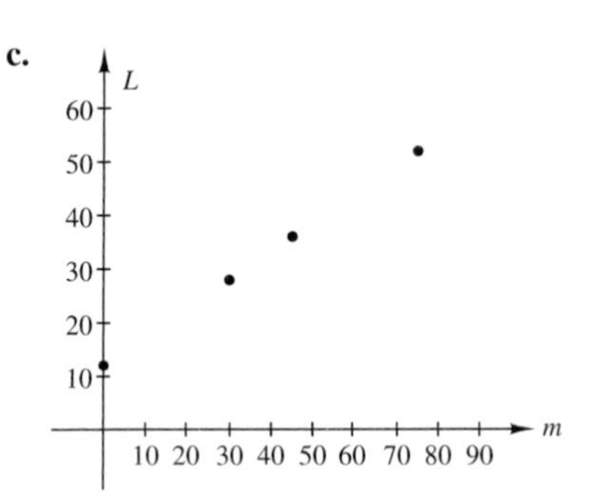

   d. The y-intercept gives the minimum amount that you should drink (no exercise). The slope $\frac{8}{15}$ says that for each 15 minutes of exercise, you should drink an additional 8 oz of liquid.

Exercises 2-D

1. **a.** $f(0) = 0$; $f(2.2) \approx 180$; $f(4) \approx 225$; these values represent the object's height at $t = 0$, $t = 2.2$, and $t = 4$ sec.
   b. $t \approx 0$ and $t \approx 7.4$
   c. $t = 3.5$; height ≈ 230 ft **d.** 1.5 and 6 sec
   e. No; there is no point on the graph with y-coordinate $y = 250$.
   f. Domain = $[0, 7.4]$; range = $[0, 230]$.

2. **a**, **d**, **e**, and **f** are all graphs of functions; the other graphs are not graphs of functions since they do not pass the vertical line test.

3. **a.** (i) 3 (ii) 0 (iii) 1 (iv) –3
   (v) –3 (vi) –1
   b. $[-1, 4]$ **c.** $[-1.7, 3]\cup\{-3\}$ **d.** –1.3
   e. – 0.4 and 1.3

4. Here is one possible graph.

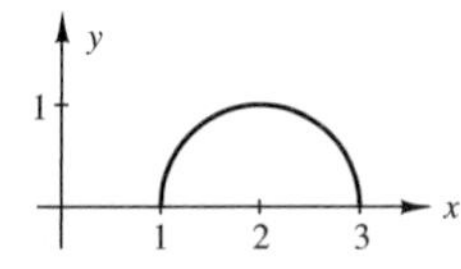

5. If a graph has two or more y-intercepts, then the vertical line $x = 0$ intersects the graph in more than one point; thus the graph fails the vertical line test.

Exercises 2-E

1. **a.** Linear

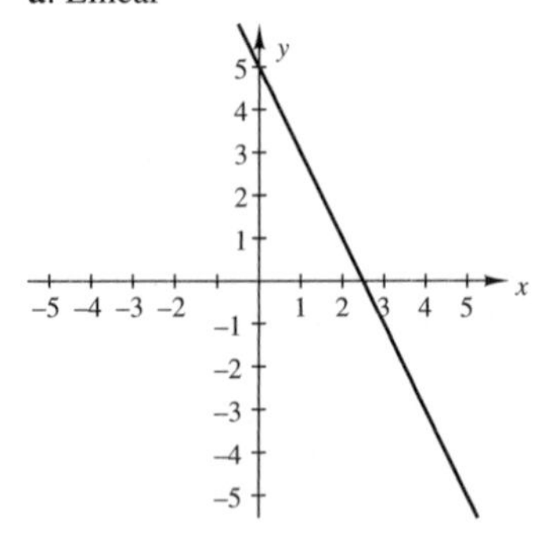

b. Half-parabola

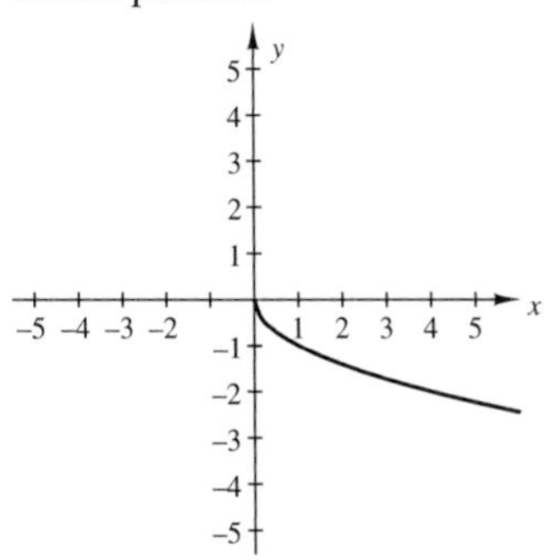

c. Parabola

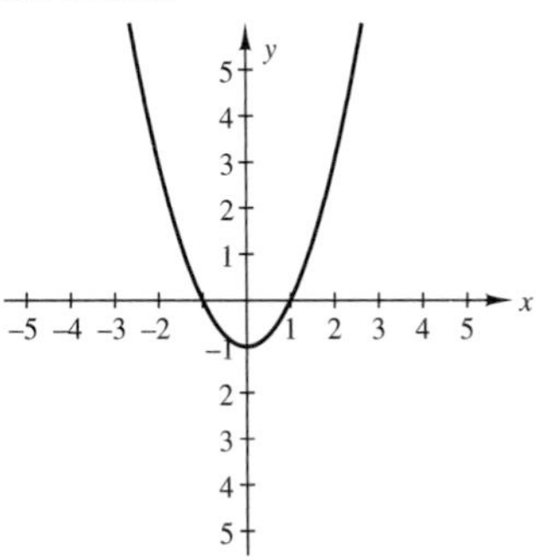

d. Linear

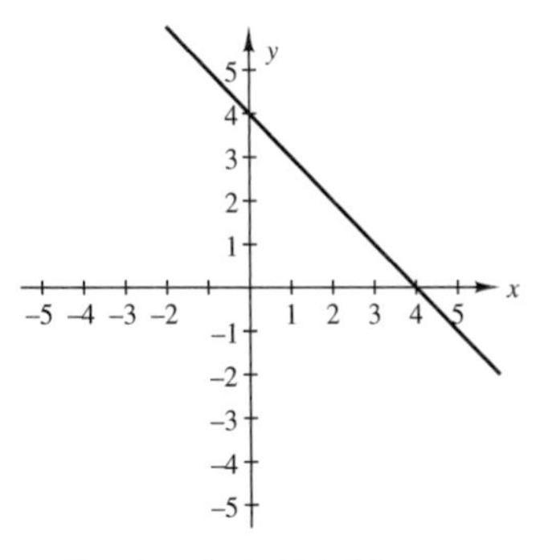

2. **a**. The domain is [24, 96).

b.

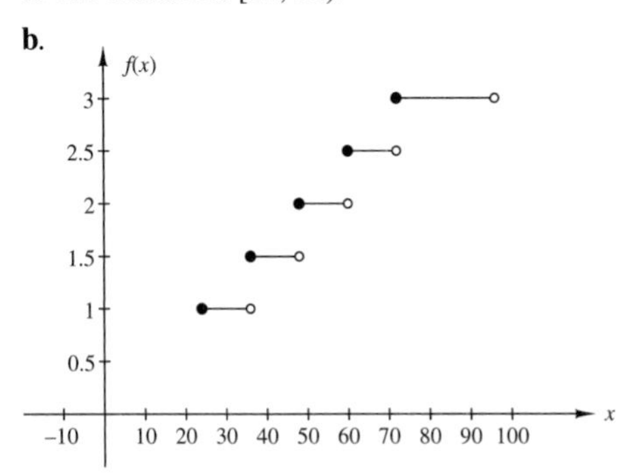

c.

$$f(x) = \begin{cases} 1 & \text{if } 24 \le x < 36 \\ 1.5 & \text{if } 36 \le x < 48 \\ 2 & \text{if } 48 \le x < 60 \\ 2.5 & \text{if } 60 \le x < 72 \\ 3 & \text{if } 72 \le x < 96 \end{cases}$$

d. 1.5 teaspoons

3. **a**. $y = \sqrt{4 - x^2} \ge 0$ and satisfies the equation $x^2 + y^2 = 4$, a circle with radius 2.

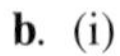

b. (i)

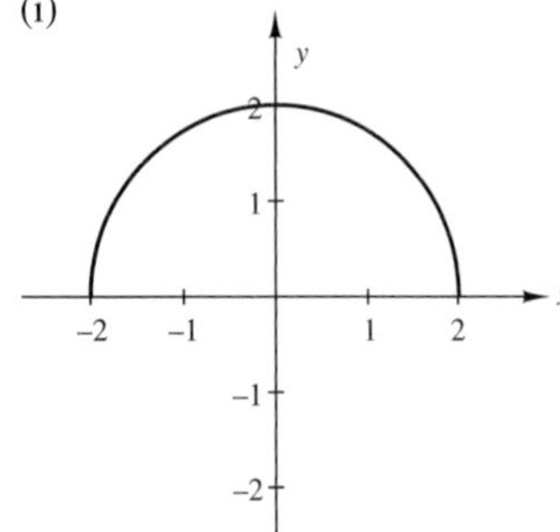

(ii)

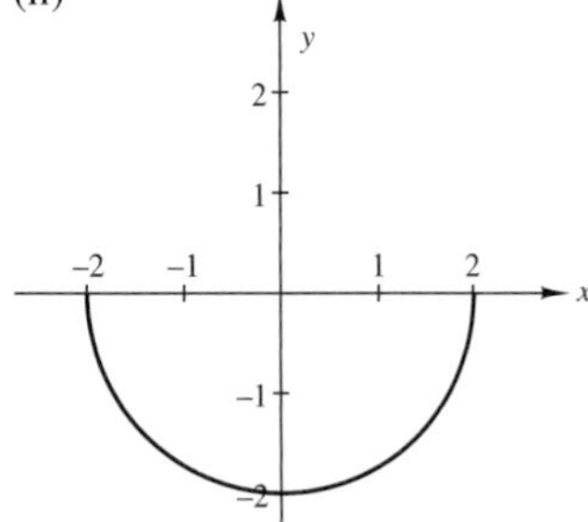

4. **a**.

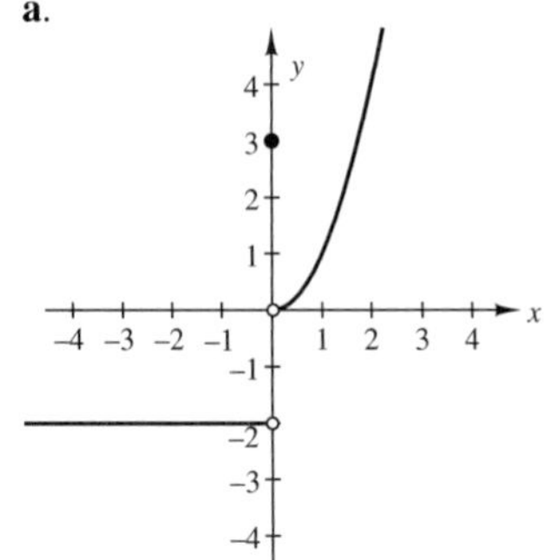

b. (i) -2 (ii) 3 (iii) $\frac{25}{4}$ (iv) 3

5. **a**. Domain = $\{x \mid x \ne -1\}$

b. $g(x) = x - 1$ for $x \ne -1$.

c.

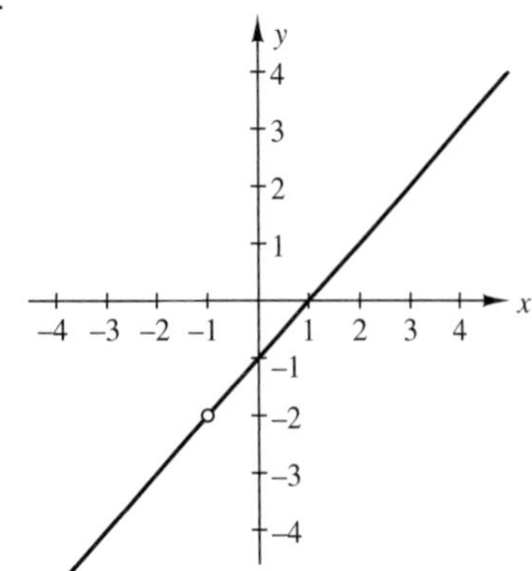

6. **a**. $F(x)=\dfrac{10}{x}$

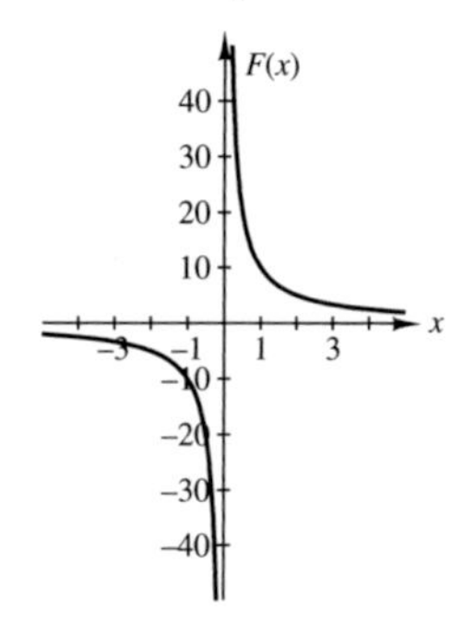

b. $y=-x^2+4$

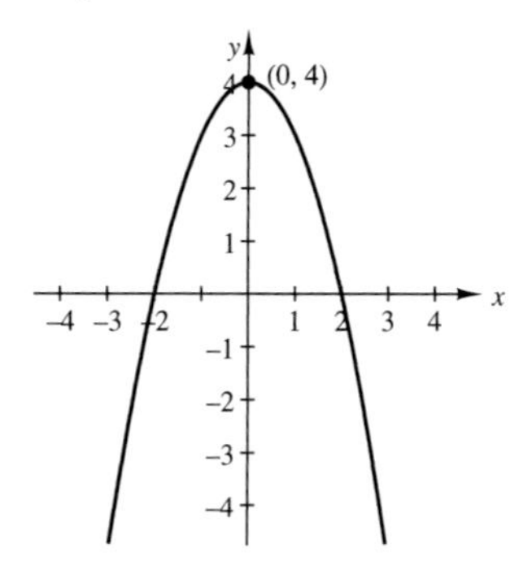

c. $y=\sqrt{3-x^2}$

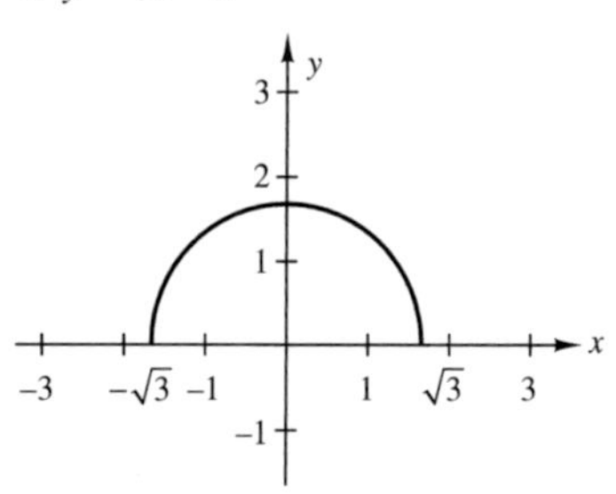

d. $G(t)=(t+2)^2-3$

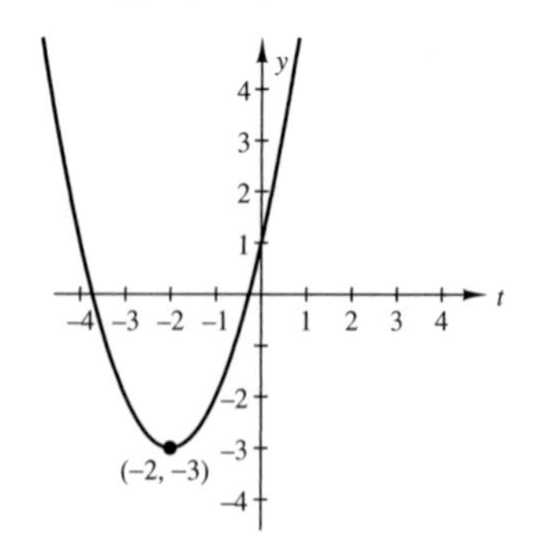

e. $h(x)=-2x+1$

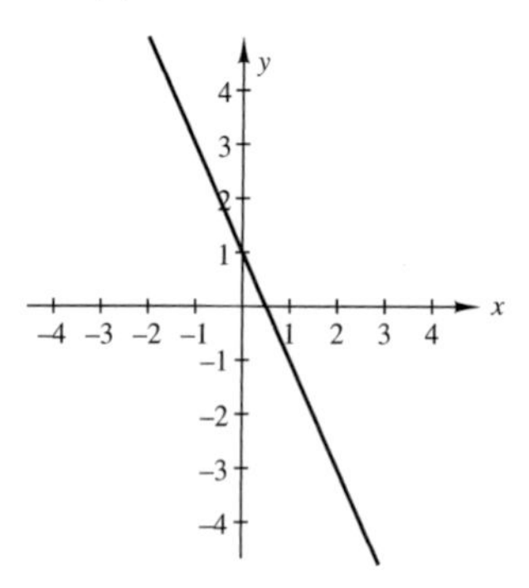

7. $h(2) = 3$; $h(-1.2) = 1.44$; $h(-\sqrt{3}) = 3$; $h(3.4) = 3.4$; $h\left(\frac{33}{28}\right) = \frac{61}{28}$; $h(-1)=0$

Exercises 2-F

1. **a**. $g(x)=|-x+3|$ **b**. $F(x)=-|x-3|$
c. $H(x)=\left|\frac{x}{3}+1\right|$ **d**. $f(x)=3-|x|$
e. $h(x)=|3x+1|$ **f**. $G(x)=3|x+1|$

2. **a**. Graph 3 **b**. Graph 5 **c**. Graph 4
d.

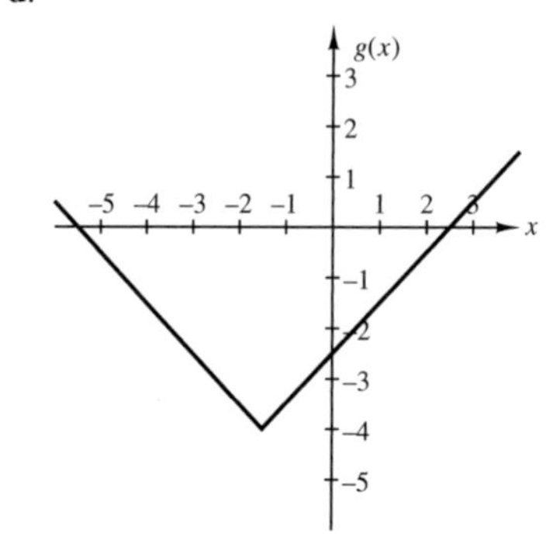

e.

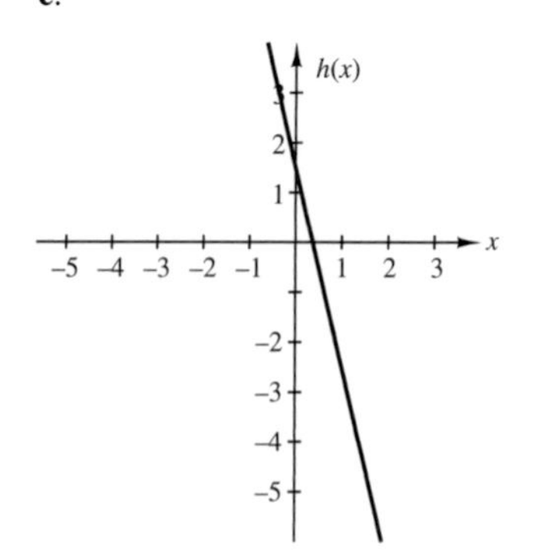

f.

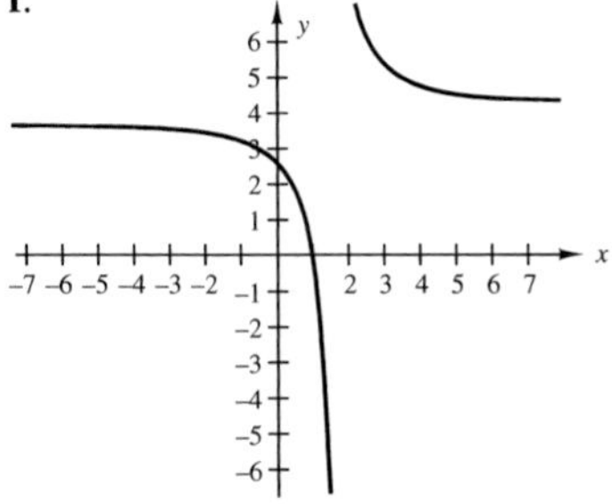

(Note that Graph 1 has correct asymptotes but the wrong y-intercept.)

g. Graph 6 **h**. Graph 2

3. **a.** Parabola

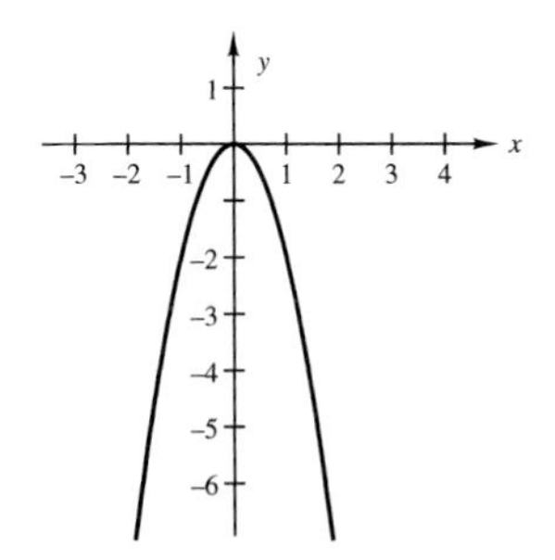

b. Half-parabola

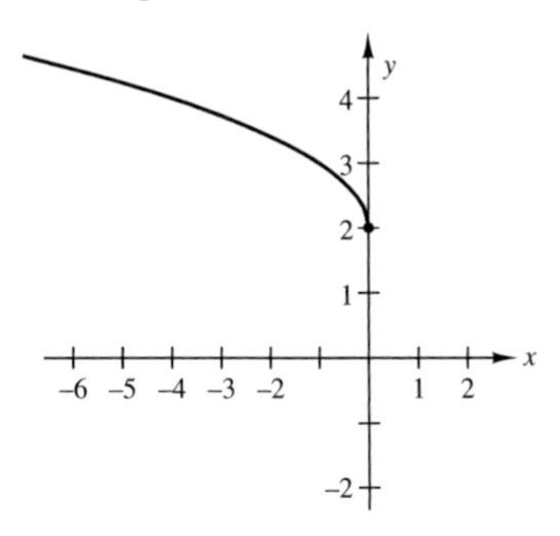

c. Line

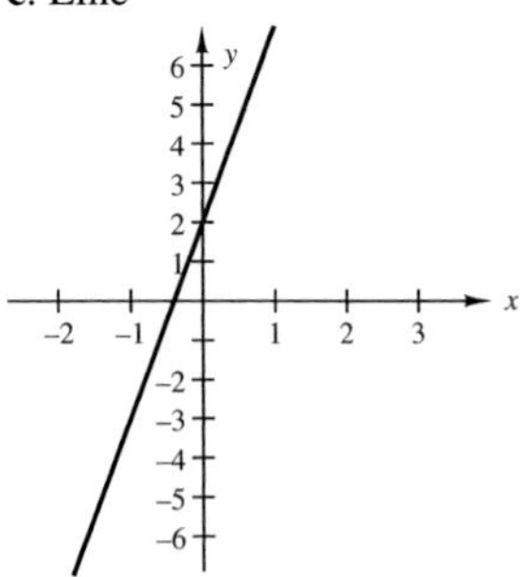

d. Half-parabola

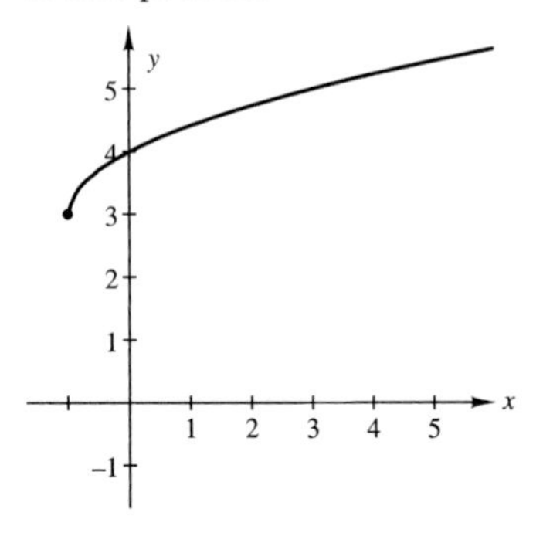

e. Semicircle

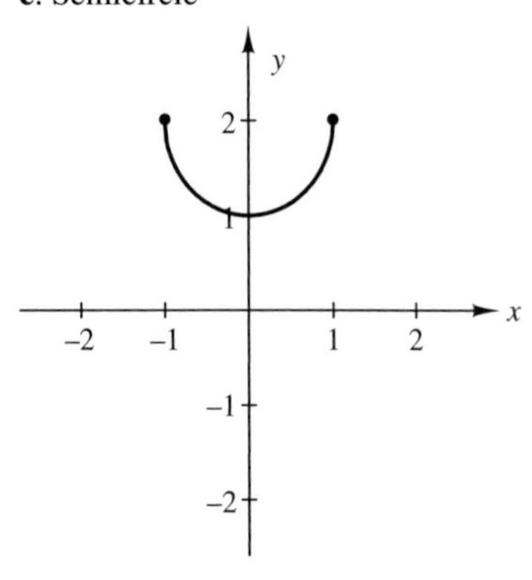

f. Reciprocal function (hyperbola)

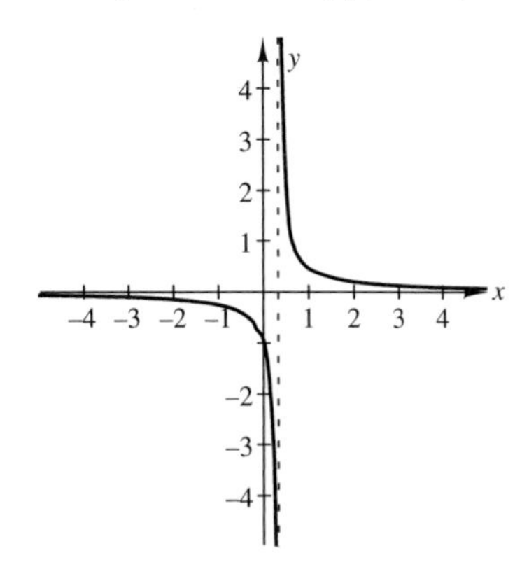

4. **a**. $g(x) = (x + 2)^3$

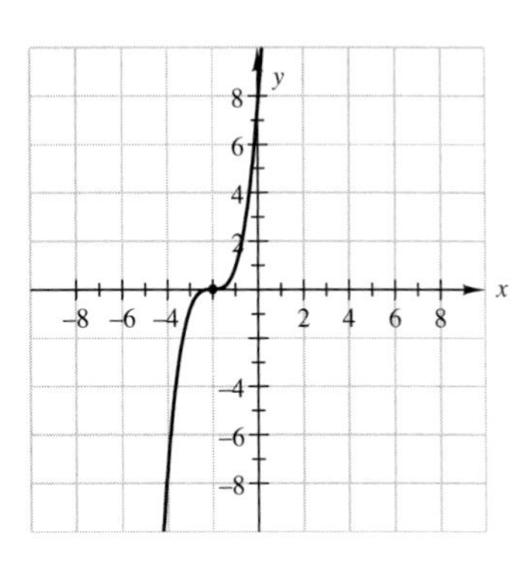

b. $h(x) = 2x^3 + 1$

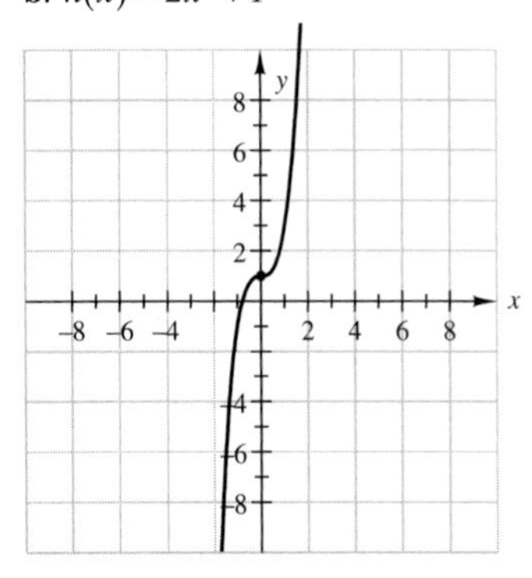

c. $F(x) = \frac{-x^3}{2}$

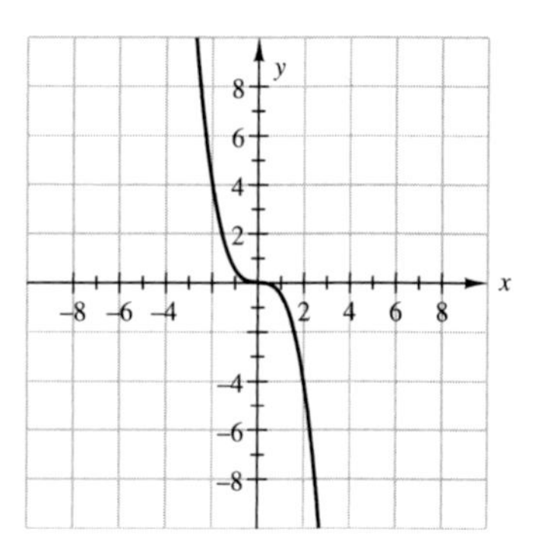

Chapter 2 Exercises

2. **a, b,** and **c** are linear functions; the functions g and h in **b** and **c** are the same.

   a. $f(x) = 2x + 1$

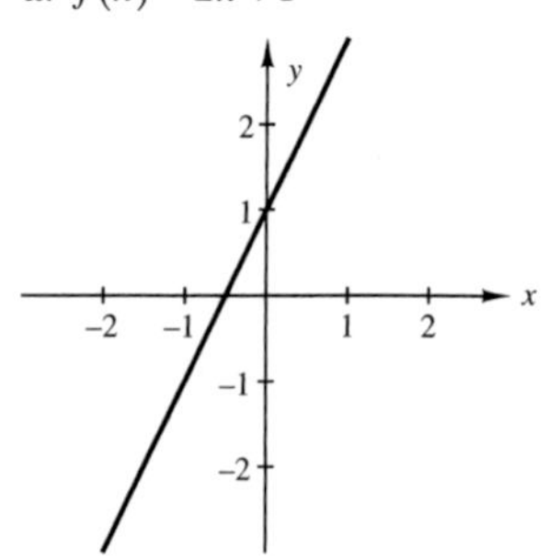

   b and **c**. $g(x) = -2 + 7x$; $h(t) = -2 + 7t$

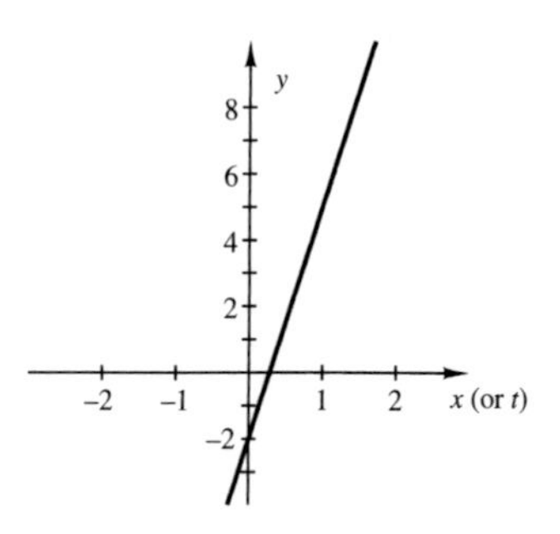

   d. $k(w) = 7 + w^2$

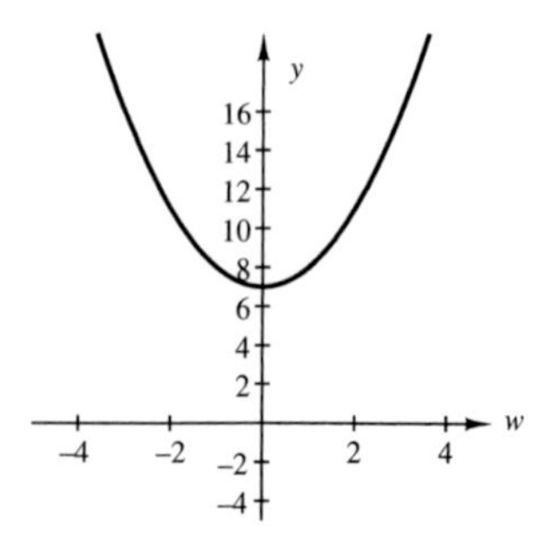

3. **a**. No vertical line intersects the graph at more than one point.

   b. 0; x-intercepts of the graph of the function

4. **a**. [0, 12] ; optimal selling price: $p = 5$

   b. (ii) $T = -0.2p^2 + 2.1p - 4$

5. **a**. −1.5, 1.5, 4.5

   b. [−2, −1.5) and (1.5, 4.5)

   c. (−1.5, 1.5) and (4.5, 5]

6. **a.** Any graph of a function that is not a line and that contains all the points is correct.

   b. Yes.

   (i) $f(x) = x + 1$

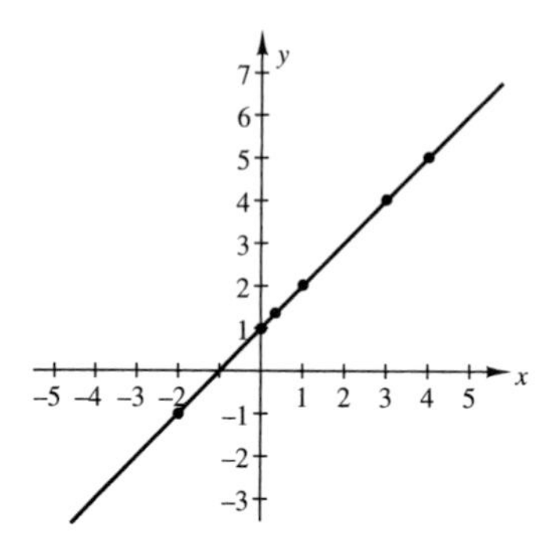

   (ii) $f(7) = 8$

7. **a**. $f(-1) = -2$; $f(-0.1) = -\frac{8}{49} \approx -0.16$; $f(1.2) = \frac{48}{31} \approx 1.55$; $f\left(\frac{3}{4}\right) = \frac{24}{23} \approx 1.04$; $f(0) = 0$; $f(-5)$ is undefined; $f(5) = 4$

   b. $(-\infty, -5) \cup (-5, \infty)$ **c**. Yes; $f\left(\frac{5}{3}\right) = 2$

   d. No; the equation $8 = \frac{8x}{5+x}$ has no solution.

8. **a**. 1 **b**. 1.8 **c**. 2 **d**. −4

   e. −2 and 2 **f**. None

9. **a**. Line; domain: $(-\infty, \infty)$

   range: $(-\infty, \infty)$

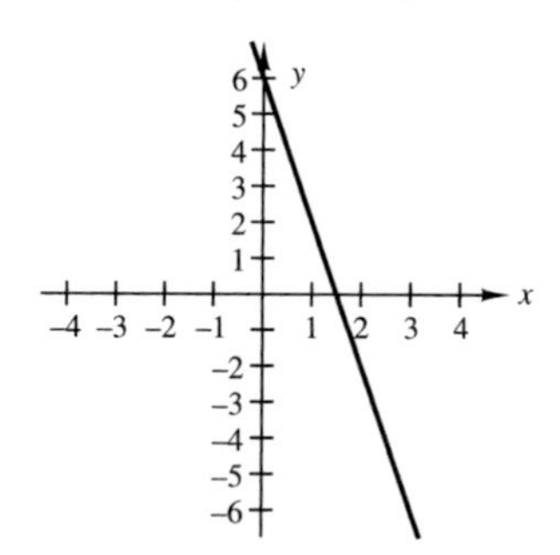

   b. Parabola; domain: $(-\infty, \infty)$

   range: $[2, \infty)$

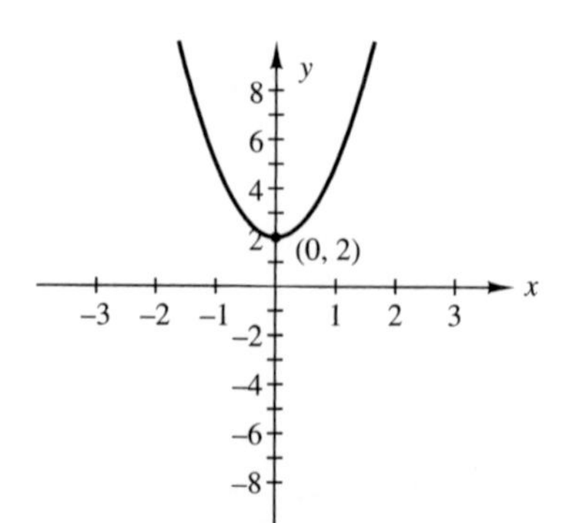

c. Parabola; domain: $(-\infty, \infty)$
range: $(-\infty, 4]$

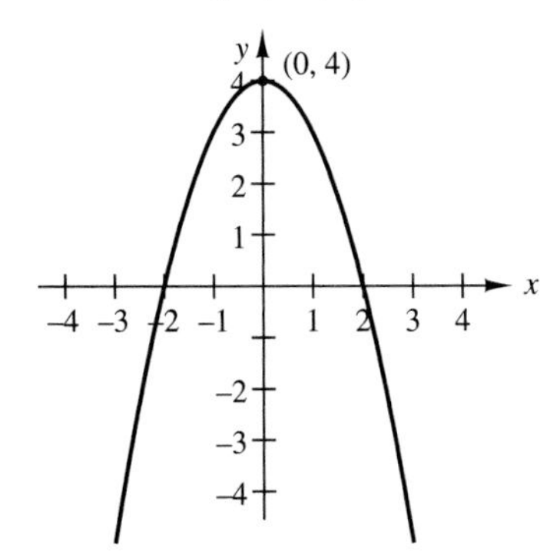

d. Absolute value;
domain: $(-\infty, \infty)$
range: $[-3, \infty)$

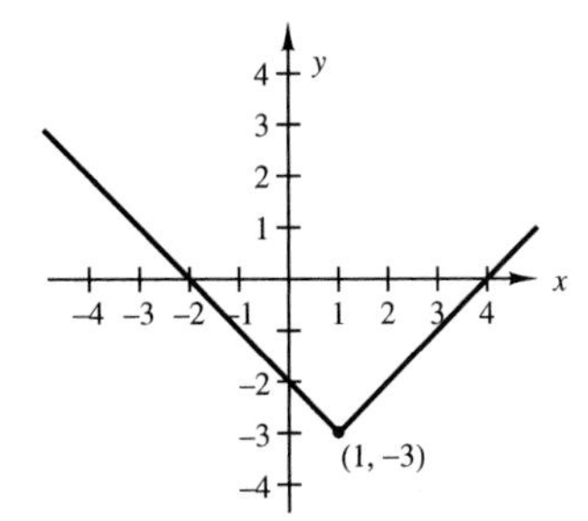

e. Half-parabola;
domain: $(-\infty, 1]$
range: $[0, \infty)$

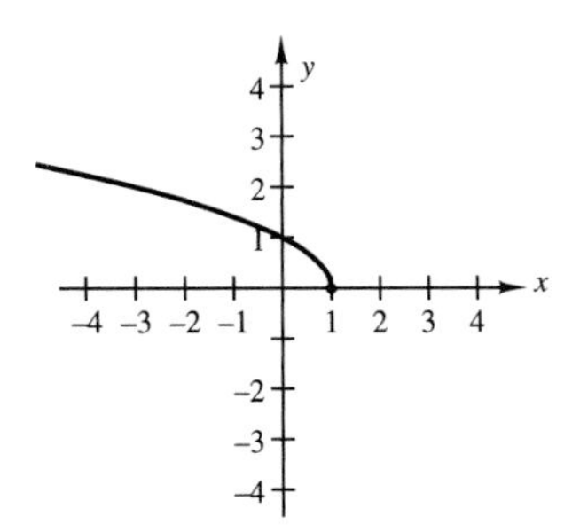

f. Half-parabola;
domain: $[1, \infty)$
range: $(-\infty, -3]$

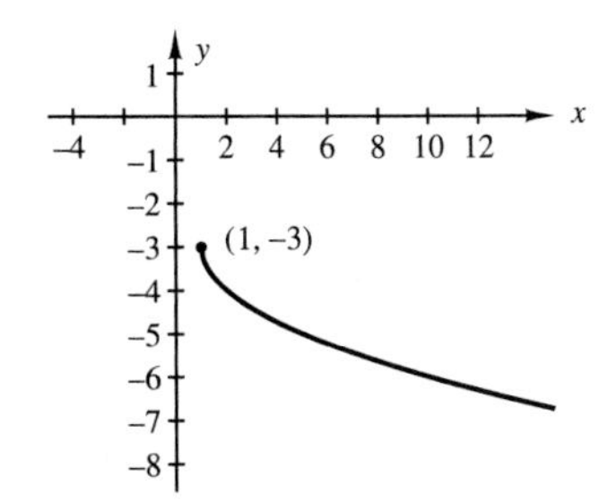

g. Semicircle; domain: $[-1, 1]$
range: $[2, 3]$

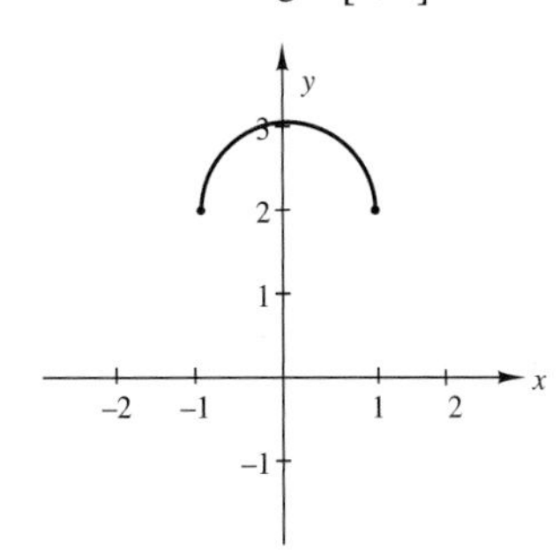

h. Reciprocal function; domain: $(-\infty, 0) \cup (0, \infty)$
range: $(-\infty, 2) \cup (2, \infty)$

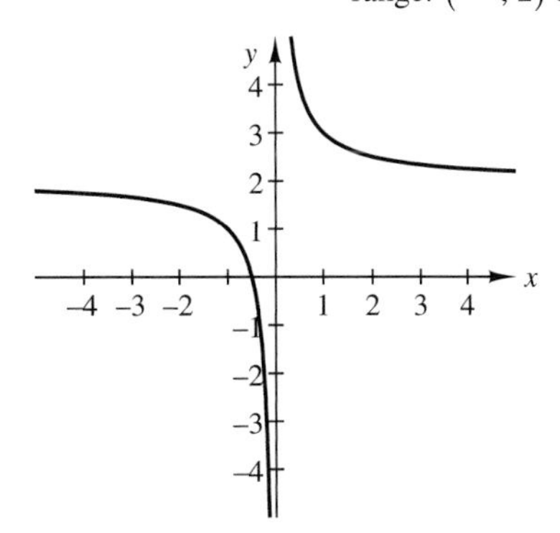

10. $f\left(\frac{1}{a}\right) = a^2 - a$; $f(-a) = \frac{1}{a^2} + \frac{1}{a}$; $-f(a) = -\frac{1}{a^2} + \frac{1}{a}$

11. **a.** $f(2000) \approx \$1.54$ is the average price for a gallon of regular gasoline at the start of 2000.
b. In general, the price of gasoline rose in the period 1986–2004, with price dips in 1997 and 2000–2001. From the graph, $1.75/gal is predicted for 2005.

12. **a.**

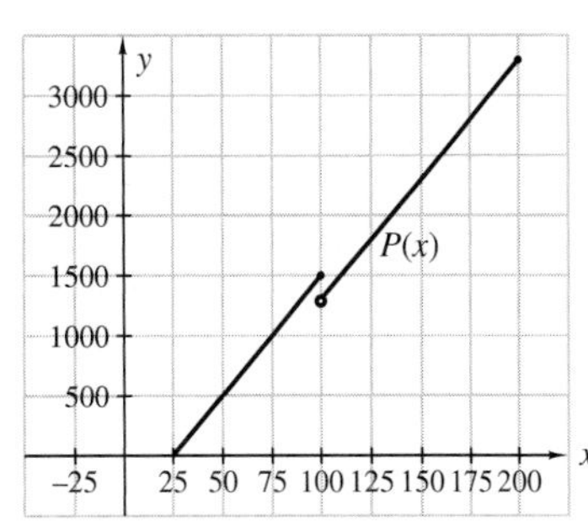

b. 600 **c**. 2400

13. $f(x) = 7.5 + 5x$; domain = $\{0, 1, 2, 3, 4, 5\}$

14. **a**. $F(0) = 1$; $F(-2) = 3$; $F\left(\frac{1}{2}\right) = \frac{3}{4}$; $F(1) = 0$; $F(2) = 0$;
$F\left(\frac{8}{9}\right) = \frac{17}{81}$; $F(\pi) = 0$ **b**. −2

15. $d = \sqrt{2x}$

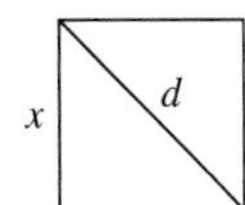

16.

h	$f(4)$	$f(4+h)$	$f(4+h)-f(4)$	$\frac{f(4+h)-f(4)}{h}$
1	23	35	12	12
0.1	23	24.11	1.11	11.1
0.01	23	23.1101	0.1101	11.01
0.001	23	23.011001	0.011001	11.001
0.0001	23	23.001100	0.001100	11.0001

17. **a**. Domain = [0, 200]; range = $[0, 80\sqrt{2}]$
b. When $h = 100$, $v = 80$; when $h = 120$, $v = 32\sqrt{5}$; when $h = 0$, $v = 80\sqrt{2}$.
c. $h = 200$ ft

Exercises 3-A

1. **a**. 0 **b**. 0 **c**. 0 **d**. −3
2. **a**. 4 **b**. 1.8 **c**. 6
   d. $(f \circ g)(5) = f(-4)$ is undefined. **e**. 1
3. **a**. $(f - g)(x) = x^2 + 3x - 11$; $(-\infty, \infty)$
   b. $\left(\frac{g}{h}\right)(x) = \frac{7-x^2}{\sqrt{x}}$; $(0, \infty)$
   c. $\left(\frac{f}{g+2}\right)(x) = \left(\frac{3x-4}{9-x^2}\right)$; $(-\infty, -3) \cup (-3, 3) \cup (3, \infty)$
   d. $(h \cdot f)(x) = \sqrt{x}(3x - 4)$; $[0, \infty)$
   e. $\left(\frac{h}{f}\right)(x) = \frac{\sqrt{x}}{3x-4}$; $\left[0, \frac{4}{3}\right) \cup \left(\frac{4}{3}, \infty\right)$
4. **a**. 14 **b**. −42 **c**. 2
   d. $(h \circ g)(4) = h(-9)$ is undefined.
   e. $(g \circ h)(-2)$ is undefined.
5. **a**. $17 - 3x^2$ **b**. $\sqrt{3x-4}$ **c**. $7 - x$
6. **a**. $C(x) = 19x + 380 - 0.05x^2$
   b. $\frac{C(x)}{x} = 19 + \frac{380}{x} - 0.05x$
   c. \$56.50; \$17.80; \$10.90 **d**. $P(x) = 31x - 0.15x^2 - 380$
7. **a**. $10{,}000\pi + 1000$ dollars ≈ \$32,415.90
   b. $A(t) = 4\pi t$ km² at t min **c**. 120π km² ≈ 377 km²
   d. $C(t) = 400\pi t + 1000$ dollars at t min
   e. $48{,}000\pi + 1000 \approx$ \$151,796.45
8. **a**. \$45,000
   b. $R(A) = 2A + 40$, with R and A in thousands of dollars
   c. \$35,500
   d. $C(A) = 18 + 1.4A$, with A and C in thousands of dollars
   e. \$31,000 **f**. $P(A) = 0.6A + 22$ **g**. \$82,000

Exercises 3-B

1. **a**. $f(x) = \frac{x+2}{x-3}$; $(-\infty, 3) \cup (3, \infty)$
   b. $g(x) = \frac{-2}{3(x+3)}$; $(-\infty, -3) \cup (-3, 0) \cup (0, \infty)$
   c. $h(x) = \frac{1}{4}$; $(-\infty, 2) \cup (2, \infty)$
   d. $v(x) = \begin{cases} -1 & \text{if } x < \frac{2}{3} \\ 1 & \text{if } x > \frac{2}{3} \end{cases}$; $\left(-\infty, \frac{2}{3}\right) \cup \left(\frac{2}{3}, \infty\right)$
   e. $w(x) = \frac{1}{x-3}$; $(-\infty, -3) \cup (-3, 3) \cup (3, \infty)$
2. **a**. $m(x) = \frac{\frac{1}{x+2} - \frac{1}{3}}{x-1}$, $x > -2$, $x \neq 1$
   b. $m(x) = \frac{-1}{3(x+2)}$, $x > -2$, $x \neq 1$
   c. The graph of $f(x) = \frac{1}{x+2}$ for $x > -2$ and secant lines through $\left(1, \frac{1}{3}\right)$ and $\left(x, \frac{1}{x+2}\right)$ for $x = -0.5$ and $x = 1.5$:

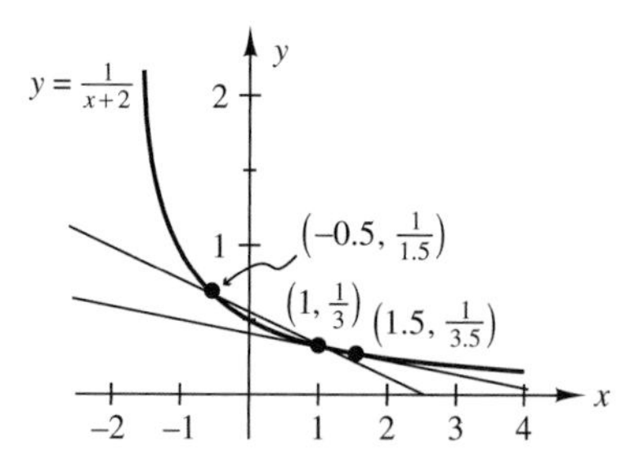

3. **a**. $n(x) = \frac{\frac{1}{x+2} - \frac{1}{4}}{x-2}$, $x > -2$, $x \neq 2$
   b. $n(x) = \frac{-1}{4(x+2)}$, $x > -2$, $x \neq 2$
4. **a**.

x	−1	0	3	5
$m(x)$	$-\frac{1}{3}$	$-\frac{1}{6}$	$-\frac{1}{15}$	$-\frac{1}{21}$
$n(x)$	$-\frac{1}{4}$	$-\frac{1}{8}$	$-\frac{1}{20}$	$-\frac{1}{28}$

5. **a**. $\frac{1}{\sqrt{x}+\sqrt{2}}$, $x \geq 2$, $x \neq 2$
   b. $\frac{-1}{25\sqrt{x}+5x}$, $x > 0$, $x \neq 25$
   c. $\frac{-1}{\sqrt{3-h}+\sqrt{3}}$, $h \leq 3$, $h \neq 0$
   d. $\frac{1}{\sqrt{x+1}+\sqrt{x}}$, $x \geq 0$

6. Two secant lines are shown; one with $x > 1$ and one with $x < 1$.

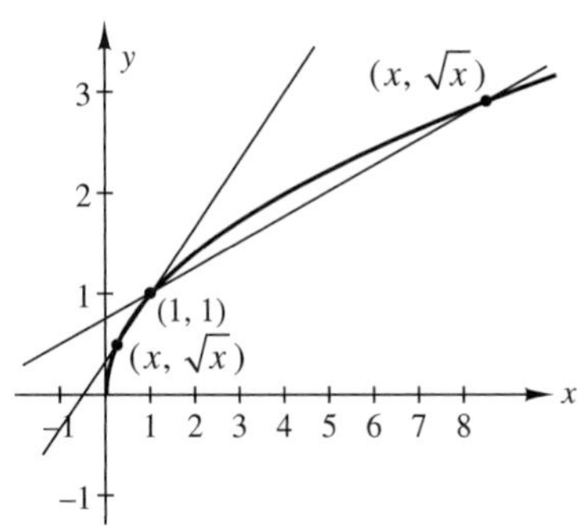

7. **a.** $-(3 + \sqrt{x}),\ x \ge 0,\ x \ne 9$

b. $\sqrt{5} + \sqrt{x},\ x \ge 0,\ x \ne 5$

c. $\sqrt{x} + 2,\ x \ge 0,\ x \ne 4$

8. **a.** $f(x) = x - 5,\ (-\infty, 2) \cup (2, \infty)$

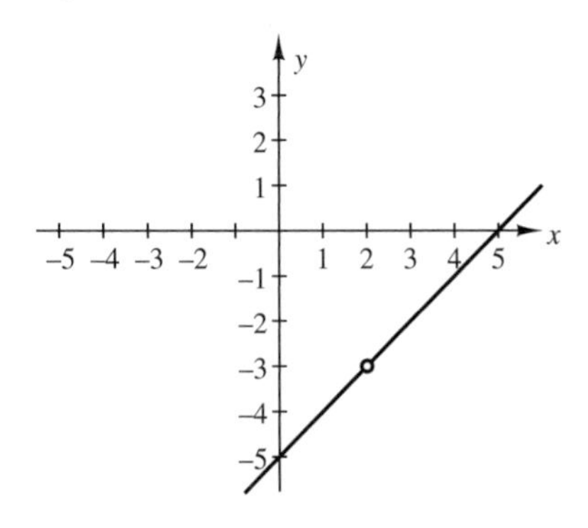

b. $g(x) = \begin{cases} -1 & \text{if } x < 2 \\ 1 & \text{if } x > 2; \end{cases}$ $(-\infty, 2) \cup (2, \infty)$

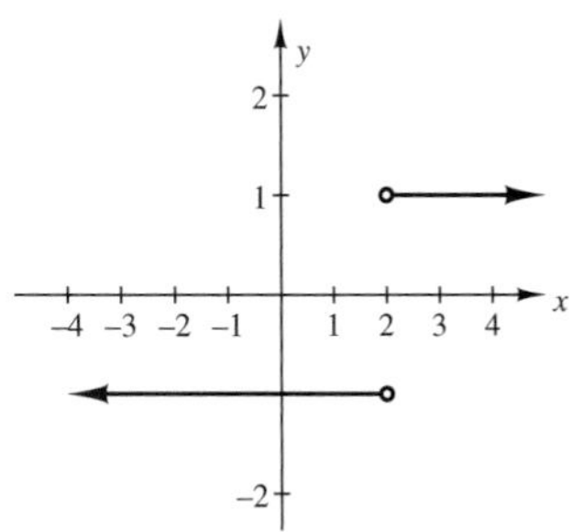

c. $h(x) = \dfrac{1}{x - 5};\ (-\infty, 2) \cup (2, 5) \cup (5, \infty)$

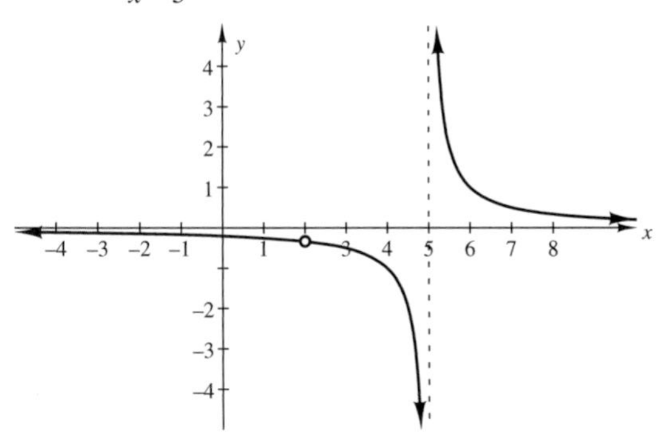

9. $Q(x) = (x - 3)(x^3 + 3x^2 + 9x + 27)$

Exercises 3-C.1

1. **a.** $\left(-\infty, -\frac{16}{3}\right)$ **b.** $\left(-\infty, \frac{1}{14}\right)$

c. $\left[-1.7, \frac{3.1}{3}\right] = \left[-\frac{17}{10}, \frac{31}{30}\right]$

d. $\left(\frac{2.3}{6}, \frac{1.1}{2}\right) = \left(\frac{23}{60}, \frac{11}{20}\right)$ **e.** $(-4, 2)$

2. **a.** $\left(-\infty, -\frac{10}{3}\right)$

b.

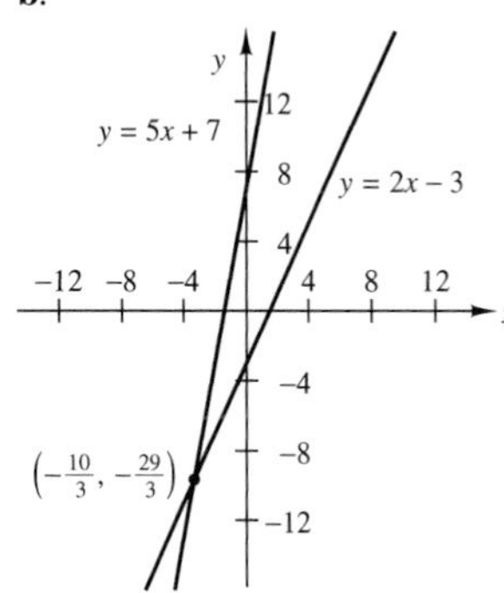

3. **a.** $\left(\frac{11}{9}, \infty\right)$

b.

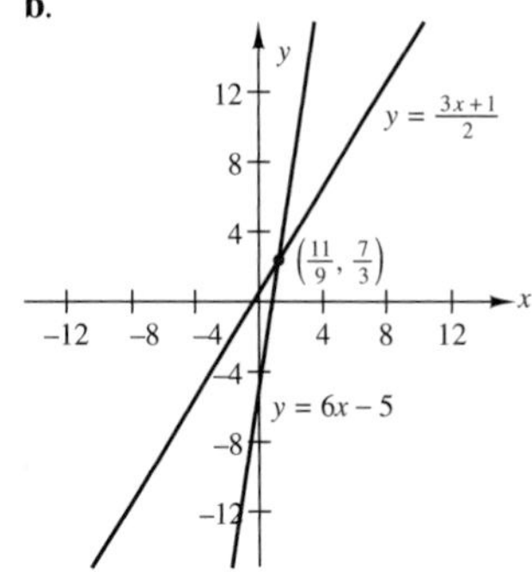

4. **a.** $\left(-2, \frac{5}{2}\right)$ **b.** $\left[\frac{1}{4}, \frac{3}{2}\right]$

5. No number can be simultaneously greater than 8 and less than –7.

Exercises 3-C.2

1. **a.** $x = -\frac{3}{2}$ or $x = \frac{5}{2}$ **b.** No solution **c.** $x = -0.25$

2. **a.** $(0.8, 1) \cup (1, 1.2)$

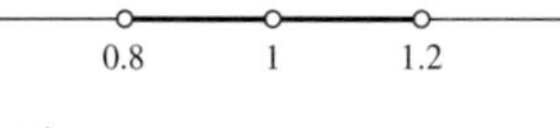

b. $\left(-\frac{7}{12}, -\frac{1}{3}\right) \cup \left(-\frac{1}{3}, -\frac{1}{12}\right)$

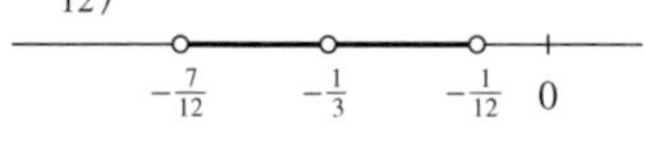

c. $[-1.4, -1.2]$

d. $(-\infty, \infty)$

e. $(-\infty, -3.1) \cup (-1.1, \infty)$

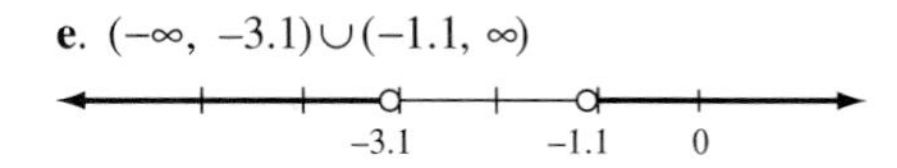

f. $\left(-\infty, -\frac{2}{3}\right] \cup [4, \infty)$

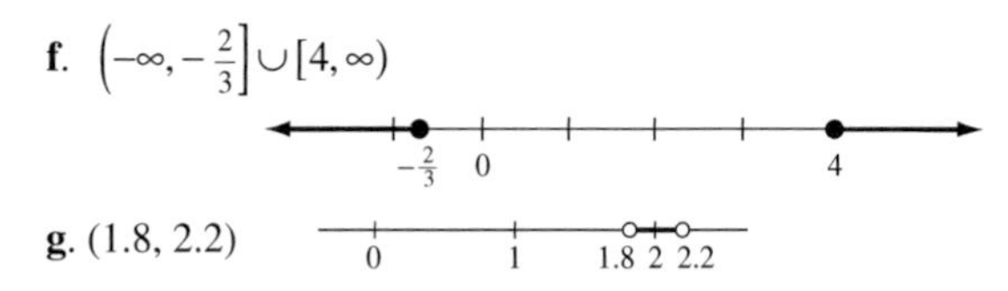

g. (1.8, 2.2)

h. No real numbers satisfy this inequality. The solution set is the empty set.

3. **a.** If x is within a distance of 0.3 from 1 and $x \neq 1$, then the value of $f(x)$ is within a distance of 0.1 from 2.

   b. If x is within a distance of $\frac{1}{5}$ from -3 and $x \neq -3$, then the distance from $g(x)$ to 0 is greater than 100.

4. **a.** The interval consists of all points that are within a distance of 0.4 from -2.

   b. $-2.4 < x < -1.6$ **c.** $(-2.4, -1.6)$ **d.** $|x + 2| < 0.4$

5. **a.**

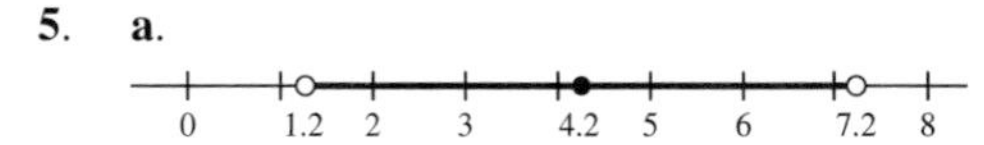

   b. (1.2, 7.2) **c.** $1.2 < x < 7.2$

6. **a.** $\varepsilon \approx 0.75$ **b.** $\varepsilon \approx 0.51$

7. **a.** $\delta \approx \frac{1}{3}$

   b. It is not possible, since $f(x) = 2$ for all x in the interval (2, 3) and for these values of x, $|f(x) - 2.5| = |2 - 2.5| = 0.5 > 0.25$.

8. **a.** $|x - 0.8| < 0.06$; (0.74, 0.86)

   b. $\left|x + \frac{14}{30}\right| \leq \frac{7}{30}$; $\left[-\frac{7}{10}, -\frac{7}{30}\right]$

   c. $|x - 21| < \frac{5}{2}$; $\left(\frac{37}{2}, \frac{47}{2}\right)$

9. **a.** If $|x + 3| < \frac{1}{2}$, then $|g(x) - 2| < \frac{3}{4}$.

   b. If $0 < |x - 4.5| < 3$, then $|h(x) + 5| < 1.7$.

Exercises 3-D

1. **a.** The statement is false, for example, $\sqrt{(-2)^2} = \sqrt{4} = 2 \neq -2$. To make the statement true, change the hypothesis to "If x is a nonnegative real number."

   b. The statement is true. **c.** The statement is true.

   d. The statement is false; for example, $-\sqrt{2}$ is real and $-(-\sqrt{2}) > 0$. To make the statement true, change the hypothesis to "If x is a positive real number."

2. **a.** If $a < b$, then $|a| < |b|$. If $|a| < |b|$, then $a < b$. The given biconditional statement is false.

   b. If $f(x) = x^2 - 4$ and $f(x) = 0$, then $x = 2$ or $x = -2$. If $f(x) = x^2 - 4$ and $x = 2$ or $x = -2$, then $f(x) = 0$. The given biconditional statement is true.

3. **a.** If two nonvertical lines have different slopes, the lines are not parallel.

   b. If two nonvertical lines are not parallel, the lines have different slopes.

4. **a.** Two nonvertical lines are parallel if and only if their slopes are equal. The biconditional statement is true, since both conditional statements are true.

   b. The contrapositive of Exercise 3a can be used to test whether two nonvertical lines intersect.

5. **a.** The hypothesis is not satisfied; the conclusion is satisfied.

   b. Both hypothesis and conclusion are satisfied.

   c. Both hypothesis and conclusion are not satisfied.

Chapter 3 Exercises

1. **i.** **a.** $(f + g)(x) = x^2 + \sqrt{2x - 3}$; domain $\left[\frac{3}{2}, \infty\right)$

   b. $(fg)(x) = x^2\sqrt{2x - 3}$; domain $\left[\frac{3}{2}, \infty\right)$

   c. $\left(\frac{g}{f}\right)(x) = \frac{\sqrt{2x-3}}{x^2}$; domain $\left[\frac{3}{2}, \infty\right)$

   d. $\frac{f(x)}{g(x)-1} = \frac{x^2}{\sqrt{2x-3}-1}$; domain $\left[\frac{3}{2}, 2\right) \cup (2, \infty)$

   ii. **a.** 5 **b.** 4 **c.** $\frac{1}{4}$ **d.** undefined

2. **i.** **a.** 1 **b.** $\sqrt{5}$

   c. $f(g(-4))$ is not defined. **d.** $\sqrt{29}$

   e. $2a - 3$ provided $a \geq \frac{3}{2}$; otherwise, $f(g(a))$ is not defined.

   f. $\sqrt{2a^2 + 4a - 1}$ provided $a \leq \frac{-2-\sqrt{6}}{2}$ or $a \geq \frac{-2+\sqrt{6}}{2}$; otherwise, $(g \circ f)(a + 1)$ is undefined.

   ii. The operation of composition is not commutative.

4. **a.** 0 **b.** undefined **c.** 3 **d.** $\frac{\sqrt{5}}{4}$

5. **a.** 6 **b.** undefined **c.** 3 **d.** 6

6. **a.** -6 **b.** undefined **c.** -3.5 **d.** 1 **e.** 5

   f. undefined **g.** -1 **h.** (1, 3] **i.** $[-5, 5]$

7. **a.** $H(x) = x^3 + x^2 + x + 1$; $x \neq 1$

   b. $G(x) = \frac{x^2 + 2x + 4}{8x^3}$; $x \neq 0, x \neq 2$

   c. $F(x) = \frac{-1}{3+\sqrt{x+4}}$; $x \geq -4, x \neq 5$

8. **a.** $h(x) = -x - 3$; $(-\infty, 1) \cup (1, \infty)$

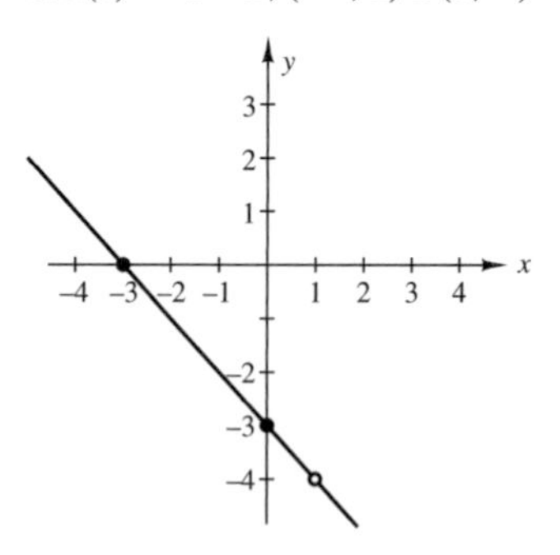

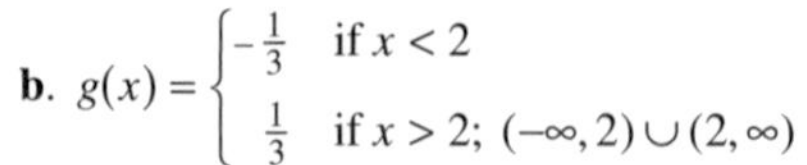

   b. $g(x) = \begin{cases} -\frac{1}{3} & \text{if } x < 2 \\ \frac{1}{3} & \text{if } x > 2; \end{cases}$ $(-\infty, 2) \cup (2, \infty)$

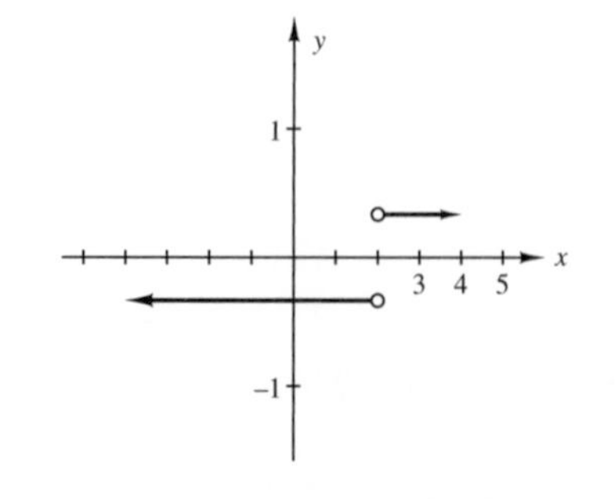

c. $f(x) = 3 + \sqrt{x+9}$; $[-9, 0) \cup (0, \infty)$

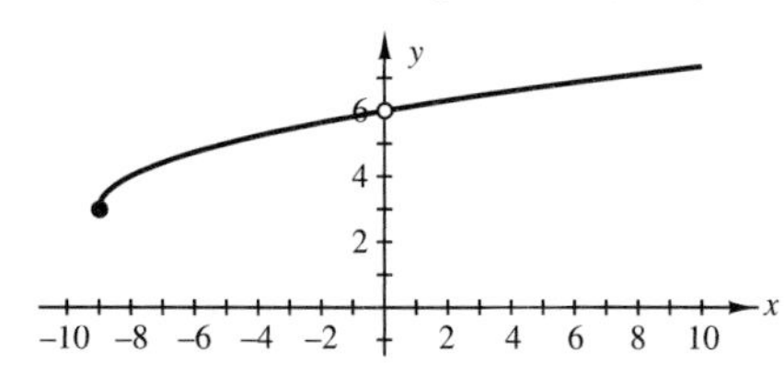

9. a. The interval consists of all points that are within a distance of 1 from −0.5.
   b. $-1.5 < x < 0.5$ c. $(-1.5, 0.5)$ d. $|x + 0.5| < 1$
10. a. The interval consists of all points different from 3 that are within a distance of 0.15 from 3.
   b. $(2.85, 3) \cup (3, 3.15)$ c. $0 < |x - 3| < 0.15$
11. a.
   b. $(0.7, 1.5) \cup (1.5, 2.3)$ 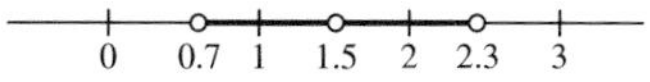
   c. $0.7 < x < 1.5$ or $1.5 < x < 2.3$
12. a. $\left(-4, -\frac{1}{2}\right) \cup \left(-\frac{1}{2}, 3\right)$
   b. $\left(-\frac{22}{3}, \frac{26}{3}\right)$ 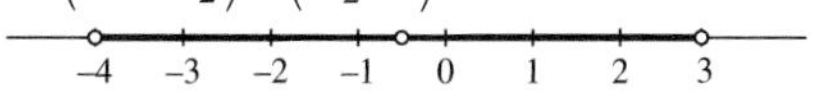
   c. No solution 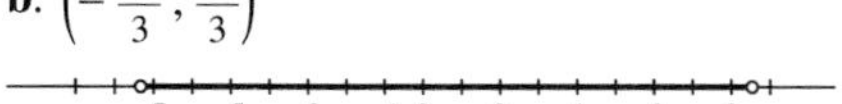

d. $\left(-\infty, \frac{7}{3}\right) \cup (3, \infty)$

e. $\left[\frac{3}{10}, \frac{59}{70}\right]$

f. No solution

g. $\left(-\frac{3}{2}, 3\right)$

13. a. If x is within a distance of 0.5 from −4, then the value of $f(x)$ is within a distance of 0.2 from 3.
   b. The value of $g(x)$ is within a distance of 0.15 from 7 whenever x is within a distance of 0.6 from 2.
   c. If x is within a distance of δ from 5, then the value of $f(x)$ is within a distance of 2 from −2.
   d. If x is within a distance of 0.3 from 1, then the value of $f(x)$ is within a distance of ε from −4.
14. a. If $0 < |x - 2| < \delta$, then $|f(x) - 5| < \varepsilon$.
   b. If $|f(x) - 5| \geq \varepsilon$, then $|x - 2| \geq \delta$ or $x - 2 = 0$.
15. a. $\varepsilon \approx 2.5$ b. $\delta \approx 0.25$

Exercises 4-A

1. Any graph of a function with no breaks, jumps, or holes will satisfy this question.
2. Any graph of a function with a break, jump, or hole will satisfy this question.
3. a. $f(x)$ is discontinuous at $x = 2$ because $f(x)$ is not defined for $x = 2$. Its graph has a vertical asymptote at $x = 2$.

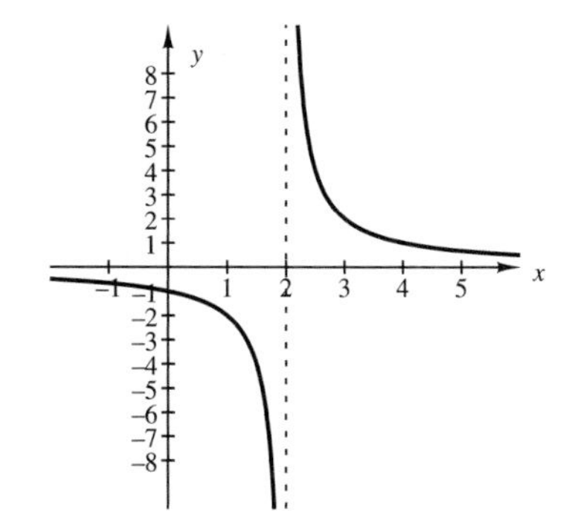

   b. $g(x)$ is discontinuous at $x = 2$ because $g(x)$ is not defined for $x = 2$. There is a hole in the graph at $x = 2$. See Graph 1.

   c. $h(x)$ is discontinuous at $x = 2$ because there is a jump in the graph at $x = 2$. See Graph 2.

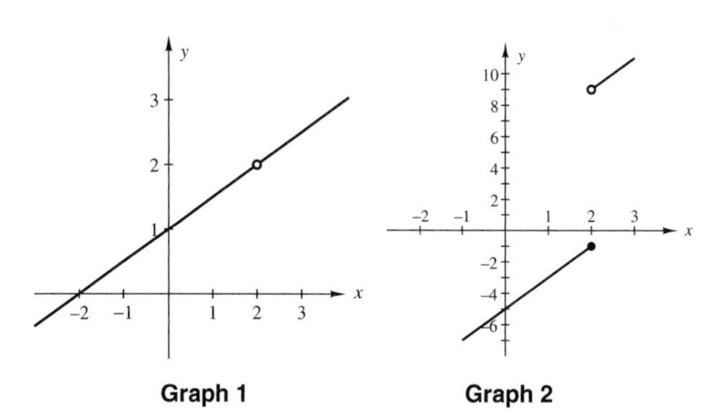

Graph 1 Graph 2

4. a. $f(x)$ is discontinuous at $x = 3$ and $x = -3$. There are vertical asymptotes at $x = 3$ and $x = -3$.

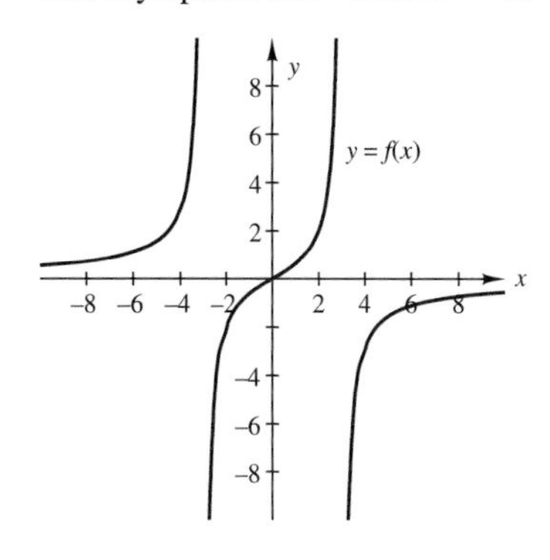

b. $g(x)$ is discontinuous at $x = 3$ and $x = -3$. There are vertical asymptotes at $x = 3$ and $x = -3$.

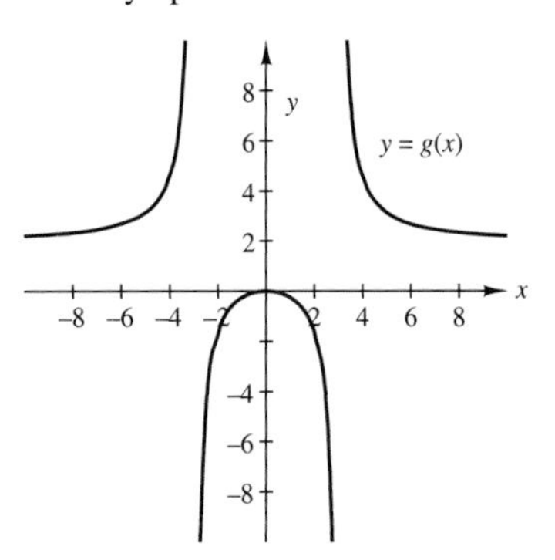

c. $h(x)$ is continuous for all x.

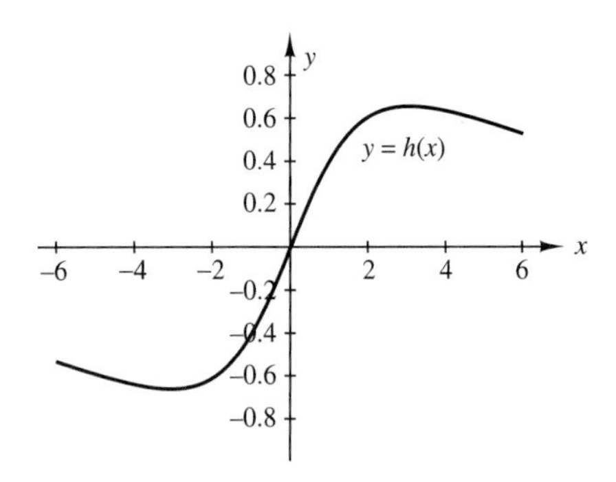

5. $f(x)$ is a polynomial because it is in the appropriate form $f(x) = a_n x^n + a_{n-1}x^{n-1} + \ldots + a_1 x + a_0$, with n a nonnegative integer and the a_i's real numbers; $n = 1$ in this case. $g(x)$ is not a polynomial because $\sqrt{x} = x^{1/2}$ has an exponent that is not an integer.
6. $f(x) = \frac{x^2}{3} = \frac{1}{3}x^2$ is a polynomial because it is in the appropriate form. But $g(x) = \frac{3}{x^2} = 3x^{-2}$ is not a polynomial, since the exponent of x is not a nonnegative integer.
7. **a**. The degree is 3, and the leading coefficient is $-\frac{1}{5}$.
   b. The degree is 0, and the leading coefficient is $3^2 - \sqrt{5}$.
   c. The degree is 4, and the leading coefficient is $-\sqrt{7}$.
8. **a**. 1 **b**. Slope of the line **c**. The y-intercept

Exercises 4-B.1

1. **a**. $f(x) = (8 - x)(8 + x)$; $x = 8$, $x = -8$
   b. $g(x) = x(2x + 3)$; $x = 0$, $x = \frac{-3}{2}$
   c. $h(x) = (x - 3)(x + 1)$; $x = 3$, $x = -1$
2. **a**. $x = 1$, $x = \frac{2}{5}$
   b. $x = \frac{1 \pm \sqrt{3}}{2}$, or $x \approx 1.37$, $x \approx -0.37$
   c. $x = 1$, $x = \frac{2}{3}$ **d**. No real solutions **e**. $x = 2.3$
4. Two linear factors of the quadratic polynomial are $x - 4$ and $x + 2$.
5. $\left(x + \frac{1}{4}\right)(x - 3)$ or $x^2 - \frac{11}{4}x - \frac{3}{4}$ or $a\left(x^2 - \frac{11}{4}x - \frac{3}{4}\right)$ for any $a \neq 0$.
6. The other solution is $2 - \sqrt{5}$.
7. **a**. $x = -2$ **b**. $x = 1$, $x = -1$
   c. No real zeros **d**. $x = \frac{1}{2}$
8. $f\left(\frac{3}{2}\right) = 0$, $f(x) = \left(x - \frac{3}{2}\right)(2x^2 + 4x + 6)$, and there are no other zeros of f.
9. $-2, \frac{3}{2}, 3$

Exercises 4-B.2

1. -2 and -1, or -1 and 0, or 0 and 1.
2. Three iterations on the interval $[-2, -1]$ show f has a zero in $[-1.625, -1.5]$. Any value in that interval will approximate the zero. Three iterations on the interval $[-1, 0]$ show f has a zero in $[-0.5, -0.375]$. Any value in that interval will approximate the zero. Three iterations on the interval $[0, 1]$ show f has a zero in $[0.875, 1]$. Any value in that interval will approximate the zero.
3. $f(x)$ is not continuous on $[-1, 1]$.
4. $f(x)$ is not continuous on $[-1, 1]$.
5. The function is not defined at $x = 0$.
6. -1.11 or 2.11
7. **a**. Three iterations on the interval $[2, 3]$ show f has a zero in the interval $[2.125, 2.25]$. Any value in that interval will approximate the zero.
   b. Three iterations on the interval $[-2, -1]$ show f has a zero in $[-1.75, -1.625]$. Any value in that interval will approximate the zero.
   c. Zoom and trace show the zeros are approximately -1.638 and 2.245.

Exercises 4-C

1. **a**. $\left(\frac{5}{3}, \infty\right)$ **b**. $\left[-\frac{1}{4}, \infty\right)$ **c**. $(-\infty, -5) \cup (5, \infty)$
   d. $(-\infty, -6] \cup [1, \infty)$ **e**. $\left[\frac{1 - \sqrt{61}}{6}, \frac{1 + \sqrt{61}}{6}\right]$
2. **a**. The domain is all real numbers.
   b. The domain is all real numbers except 0 and -3.
   c. The domain is all real numbers except 2 and -3.
   d. The domain is all real numbers except $\frac{5}{2}$ and -1.
3. **a**. $(-\infty, 7]$ **b**. $(-\infty, \infty)$
   c. $\left[-\frac{5}{2}, \frac{5}{2}\right]$ **d**. $(-\infty, \infty)$ **e**. $(-\infty, \infty)$

4. **a**. Yes; $(g \circ f)(1) = \frac{1}{3}$
   b. No; –3 is not in the domain of $f \circ g$ because –3 is not in the given domain of g.
   c. No; $(g \circ f)\left(\frac{1}{4}\right) = g\left(\frac{3}{2}\right)$ but $\frac{3}{2}$ is not in the given domain of g.
5. **a**. No; $(f \circ g)(0) = \sqrt{-\frac{1}{2}}$, which is not a real number.
   b. Yes; $(g \circ f)(0) = -\frac{1}{2}$.
   c. $f \circ g$ has domain $(2, \infty)$.
   d. $g \circ f$ has domain $[0,4) \cup (4, \infty)$.
6. **a**. Yes; $(g \circ f)(-4) = 9$.
   b. No; $(g \circ g)(2) = g(4)$ and 4 is not in the domain of g.
   c. Yes; $(f \circ g)(0) = 3$.
   d. $g \circ f$ has domain $[-5, 2]$.

Chapter 4 Exercises

1. $g(t)$, $h(z)$, and $j(y)$ are polynomials. $g(t)$ has degree 2 with leading coefficient $\frac{\sqrt{3}}{2}$, $h(z)$ has degree 0 with leading coefficient 7, and $j(y)$ has degree 3 with leading coefficient –8. $f(x)$ is not a polynomial because the domain of f does not contain $x = -\frac{4}{3}$; $i(x)$ is not a polynomial because it cannot be put in the form $i(x) = a_n x^n + a_{n-1}x^{n-1} + \cdots + a_1 x + a_0$ since the variable is an exponent.
2. **a**. $\frac{3}{2}$ **b**. $\frac{7 \pm \sqrt{89}}{2}$ **c**. 0 and $\frac{5}{2}$
   d. No solutions **e**. $\sqrt[3]{\frac{27}{0.125}} = 6$
   f. $C(x) = -5$ has no zeros.
4. The number of distinct real zeros is less than or equal to the degree of the polynomial.
5. **a**. The other real zeros are 3 and –3. $x^3 - 2x^2 - 9x + 18 = (x-2)(x-3)(x+3)$
   b. There are no other real zeros. $x^3 - 2x^2 + 4x - 8 = (x-2)(x^2+4)$
   c. The other real zeros are $\frac{3 \pm \sqrt{17}}{4}$.
   $2x^3 - 7x^2 + 5x + 2 = (x-2)\left(x - \frac{3+\sqrt{17}}{4}\right)\left(x - \frac{3-\sqrt{17}}{4}\right)$
6. Evaluate the polynomial at 3 or note that 3 is not a factor of the constant term –4.
7. By the Factor Theorem if $x - 2$ is a factor of the polynomial, then 2 must be a zero of the polynomial, but it is not.
8. **a**. There are two real zeros: $x \approx -2.36$, and $x \approx 1.5$.
   b. $x \approx 1.39$ is an estimate of the zero.
10. **a**. 39 **b**. 19 **c**. 441
    d. 48 **e**. 20.92 **f**. 54
11. **a**. $f(x) = 2x + 50{,}000$
    b. In 120 years, the population will exceed the limits of the city's resources.
12. **a**. The graphs of i, ii, and v have one zero. The graphs of iii and iv have no zeros.
    b. The discriminant is 0 for each quadratic in i, ii, and v. The discriminant of iii is –24, and the discriminant of iv is –4.
    c. When a quadratic function has a discriminant of 0, it has only one zero.
    d. When a quadratic function has a negative discriminant, it will have no real zeros.
    e. When the discriminant of a quadratic function is positive, it will have two distinct real zeros. This could be tested by graphing several such functions and looking at the zeros or finding the zeros of several such functions by using the quadratic formula.
13. The function in Figure 4.25 is discontinuous at $x = 2$. The function in Figure 4.26 is discontinuous at $x = -3$. The function in Figure 4.27 is discontinuous at $x = 1$. The function in Figure 4.28 is discontinuous at $x = 2.6$ (approximately).
14. **a**. $(-\infty, \infty)$ **b**. $(-\infty, \infty)$
    c. $\left[\frac{2}{3}, \infty\right)$ **d**. $(-\infty, -1] \cup \left[\frac{5}{2}, \infty\right)$
    e. $\left(-\infty, \frac{2-\sqrt{7}}{3}\right] \cup \left[\frac{2+\sqrt{7}}{3}, \infty\right)$ **f**. $(-\infty, \infty)$
15. **a**. $f(x)$ is discontinuous at $x = 3$ and $x = -3$.
    b. $g(x)$ is discontinuous at $x = 3$ and $x = -3$.
    c. $h(x)$ is discontinuous at $x = -4$.
    d. $s(t)$ is continuous everywhere.
    e. $v(t)$ is discontinuous at $t = 2$.
16. **a**. $f(x)$ has a vertical asymptote at $x = -3$ and a hole at $x = 3$.
    $g(x)$ has vertical asymptotes at $x = 3$ and $x = -3$.
    $h(x)$ has a hole at $x = -4$.
    $v(t)$ has a jump at $t = 2$.
    b. $f(x)$ has a removable discontinuity at $x = 3$, $h(x)$ has a removable discontinuity at $x = -4$.
17. **a**. No; since 1 is not in the domain of f, it is not in the domain of $g \circ f$.
    b. No; since $(f \circ g)(2) = f(-1)$ is not defined.
    c. Yes; $(f \circ g)(0) = 5$.
    d. $(-3, 0]$
    e. $(1, 3]$
18. **a**. Yes; $(f \circ g)(-3) = \sqrt{\frac{1}{2}}$.
    b. No; $(h \circ g)(-3) = h\left(\frac{1}{2}\right)$, which is not a real number.
    c. $[0, \infty)$
    d. $(-5, \infty)$
    e. $[4, \infty)$

Exercises 5-A

1. These are some possibilities:

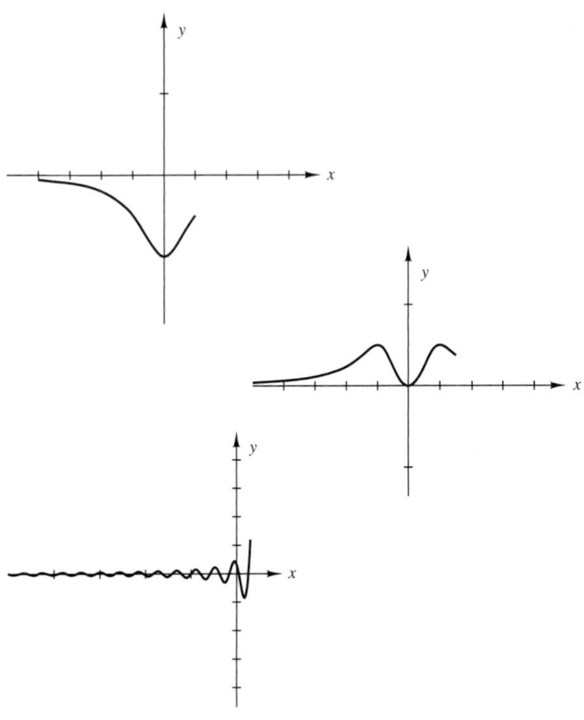

2. **a**. 1400
   b. The number of sick people increases toward 2000.
   c. 1800, 1900, 1995
   d. $y = 2000$; no more than this number caught the flu.
   e. 5 weeks
3. Yes. This is one possibility:

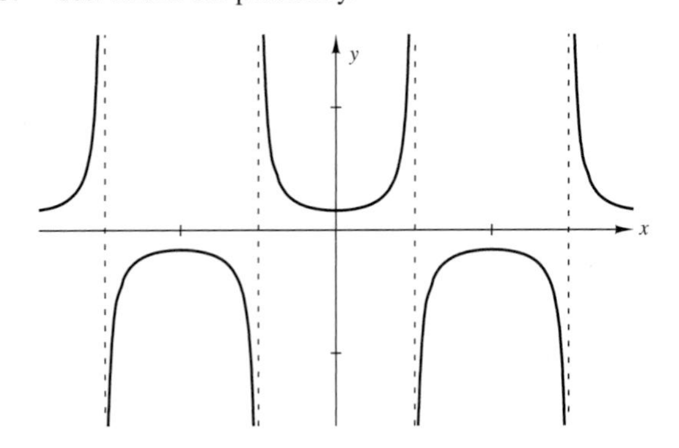

4. Yes. This is an example:

5. No. A horizontal asymptote occurs only when one (or both) of the following two statements is true: (i) As $x \to \infty$, $f(x)$ approaches a real number. (ii) As $x \to -\infty$, $f(x)$ approaches a real number.
6. **a**. 2 **b**. -1 **c**. 1 **d**. 1
   e. $-\infty$ **f**. ∞ **g**. ∞ **h**. $-\infty$
7. **a**. $y = g(x)$ has no vertical asymptotes; the vertical asymptotes of $y = h(x)$ are $x = -3$ and $x = 1$.
   b. The horizontal asymptotes of $y = g(x)$ are $y = 2$ and $y = -1$; $y = 1$ is the only horizontal asymptote of $y = h(x)$.

Exercises 5-B.1

1. **a**. $\dfrac{\frac{5}{x} - \frac{3}{x^2} + \frac{1}{x^3}}{1 - \frac{2}{x^3}}$ **b**. $f(x) \to 0$
   c. $f(x) \to 0$ **d**. $y = 0$
2. **a**. $g(x) = \dfrac{\frac{1}{x^2} - 3}{7 + \frac{1}{x^2}}$ **b**. $g(x) \to -\frac{3}{7}$
   c. $g(x) \to -\frac{3}{7}$ **d**. $y = -\frac{3}{7}$
3. **a**. $h(x) \to -\frac{2}{3} \approx -1.15$ **b**. $h(x) \to \frac{18}{28+\sqrt{3}} \approx 0.61$
4. **a**. $F(x) = \dfrac{\sqrt{1 - \frac{1}{x^3}}}{\frac{4}{x^2} - 1}$ **b**. $F(x) \to -1$ **c**. $y = -1$
5. **a**. $G(x) = \dfrac{\sqrt[3]{1 - \frac{2}{x^2} + \frac{1}{x^3}}}{7 - \frac{2}{x}}$ **b**. $G(x) \to \frac{1}{7}$
   c. $G(x) \to \frac{1}{7}$ **d**. $y = \frac{1}{7}$
6. **a**. $f(x) = -\dfrac{\sqrt{1 - \frac{3}{x^2}}}{\frac{2}{x} - 1}$ **b**. $f(x) \to 1$
7. As $x \to \infty$, $h(x) = \dfrac{3 + \frac{1}{x}}{\sqrt{1 + \frac{3}{x}} + \sqrt{1 - \frac{1}{x^2}}} \to \frac{3}{2}$.

Exercises 5-B.2

1. **a**. $f(x) \to -\frac{1}{2}$ **b**. $f(x) \to -\infty$ **c**. $f(x) \to \infty$
   d. No, the values of $f(x)$ do not increase or decrease without bound as $x \to 1$.
   e. Yes, because $f(x) \to -\infty$ as $x \to -1^-$ (or because $f(x) \to \infty$ as $x \to -1^+$).
2. **b**. (i) negative, (ii) negative, (iii) positive **c**. ∞
   d. (i) negative, (ii) positive, (iii) negative **e**. $-\infty$
3. **b**. (i) negative, (ii) positive, (iii) negative **c**. $-\infty$
   d. (i) negative, (ii) negative, (iii) positive **e**. ∞

4. **a**. 1 and 2.
b. $g(x) \to \infty$ as $x \to 1^-$; $g(x) \to -\infty$ as $x \to 1^+$; $g(x) \to -\infty$ as $x \to 2^-$; $g(x) \to \infty$ as $x \to 2^+$.
c. $x = 1, x = 2$

5. **a**. -3 and 3.
b. As $x \to -3$, $H(x) \to -\frac{2}{3}$; as $x \to 3^-$, $H(x) \to \infty$; as $x \to 3^+$, $H(x) \to -\infty$.
c. $x = 3$

6. **a**. $-\sqrt{5}$, 0, and $\sqrt{5}$.
b. $F(x) \to \frac{1}{10}$ as $x \to -\sqrt{5}$; $F(x) \to \infty$ as $x \to 0^-$; $F(x) \to -\infty$ as $x \to 0^+$; $F(x) \to -\infty$ as $x \to (\sqrt{5})^-$; $F(x) \to \infty$ as $x \to (\sqrt{5})^+$.
c. $x = 0$, $x = \sqrt{5}$.

7. **a**. 3.
b. $G(x) \to -\infty$ as $x \to 3^-$; $G(x) \to \infty$ as $x \to 3^+$
c. $x = 3$

Chapter 5 Exercises

1. **a**. (i) \$100,000 (ii) \$125,000 (iii) \$139,500
b. The profit increases toward \$140,000.
c. There is no value of x for which the profit is \$200,000; $P < 140{,}000$ for all x.
d. Not possible, since the profit is always less than 140,000.
e. $y = 140$

2. **a**. 6 **b**. 6 **c**. $-\infty$ **d**. 0
e. 2 **f**. $-\infty$ **g**. ∞

3. **a**. $y = f(x)$ has no horizontal asymptotes.
b. $y = 0$ **c**. $y = 2$

4. **a**. $y = f(x)$ has no vertical asymptotes.
b. $x = -3$ **c**. $x = -3$ and $x = 3$

5. **a**. $\frac{x^2-9}{x-3}$ **b**. $\frac{1}{x+3}$ **c**. $\frac{2x^2}{x^2-9}$

6. **a**. As $x \to \infty$, $G(x) \to 3$. **b**. As $x \to -\infty$, $G(x) \to 3$.
c. 0, -2
d. As $x \to 0^-$, $G(x) \to \infty$. As $x \to 0^+$, $G(x) \to -\infty$. As $x \to -2$, $G(x) \to 6$.
e. $x = 0$ **f**. $y = 3$

7. **a**. As $x \to \infty$, $F(x) \to 0$. **b**. As $x \to -\infty$, $F(x) \to 0$.
c. $-2, -3$
d. As $x \to -2$, $F(x) \to -1$; as $x \to -3^-$, $F(x) \to \infty$; as $x \to -3^+$, $F(x) \to -\infty$.
e. $x = -3$ **f**. $y = 0$

8. **a**. ∞ **b**. $-\infty$ **c**. ∞
d. $-\infty$ **e**. 0 **f**. 0

9. **a**. As $x \to \infty$, $f(x) \to -\frac{1}{2\sqrt{2}}$.
b. Yes, $y = -\frac{1}{2\sqrt{2}}$ is a horizontal asymptote.
c. The graph has no vertical asymptotes since $|f(x)|$ does not grow without bound near any number c.

10. **a**. $g(x) = \dfrac{3 - \frac{5}{x}}{-\sqrt{\frac{6}{x^2} + 1}}$ **b**. As $x \to -\infty$, $g(x) \to -3$.
c. $g(x) = \dfrac{3 - \frac{5}{x}}{\sqrt{\frac{6}{x^2} + 1}}$ **d**. As $x \to \infty$, $g(x) \to 3$.
e. $y = 3$ and $y = -3$ are horizontal asymptotes.
f. The graph has no vertical asymptotes since $|g(x)|$ does not grow without bound near any number c.

11. **a**. $h(x) = \dfrac{\sqrt[3]{\frac{4}{x^2} - 1}}{8 - \frac{5}{x}}$ **b**. As $x \to -\infty$, $h(x) \to -\frac{1}{8}$.
c. As $x \to \infty$, $h(x) \to -\frac{1}{8}$.
d. As $x \to \left(\frac{5}{8}\right)^-$, $h(x) \to -\infty$; as $x \to \left(\frac{5}{8}\right)^+$, $h(x) \to \infty$.
e. $y = -\frac{1}{8}$ **f**. $x = \frac{5}{8}$

Exercises 6-A

1. **a**. \$500 **b**. \$18.75 **c**. \$1,031.25
d. \$17.50 **e**. \$1,050 **f**. \$18.75
g. $R(x) = 20x$ for $25 \le x \le 50$
h. $p(x) = 20 - 0.25(x - 50)$ or $p = 32.5 - 0.25x$ for $x > 50$
i. $R(x) = (32.5 - 0.25x)x = 32.5x - 0.25x^2$ for $x > 50$
j. $R(x) = \begin{cases} 20x & 25 \le x \le 50 \\ 32.5x - 0.25x^2 & x > 50 \end{cases}$

2. **a**. \$450 **b**. \$495
c. \$900 **d**. \$850.85
e.

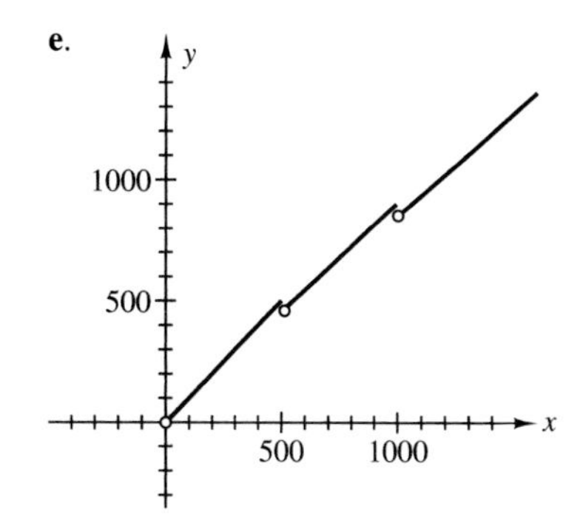

$$f(x) = \begin{cases} x & 0 < x \le 500 \\ 0.90x & 500 < x \le 1000 \\ 0.85x & 1000 < x \end{cases}$$

f. For dollar amounts exceeding \$450 up to \$500, it is less expensive to buy additional supplies to make the total exceed \$500. Beyond \$944.44 it is less expensive to buy additional supplies to make the total exceed \$1,000.

g. \$.85

3. **a**.

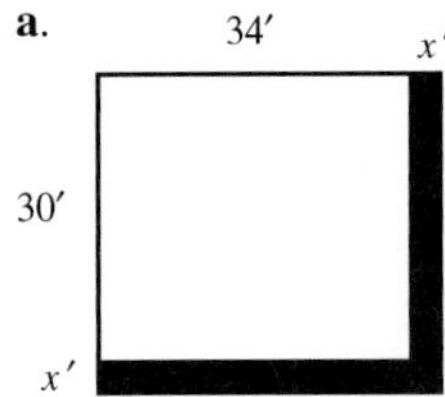

b. 65 sq ft **c**. \$1,300 **d**. $A(x) = 64x + x^2$
e. $C(x) = (64x + x^2)(20)$ **f**. \$2,640 **g**. 3 ft

Exercises 6-B

1. There are many possible answers such as 55 mph, 2 candy bars for \$1, or \$4.75 per hour.
3. **a**. Average rate of 30 miles per gallon **b**. \$20 per ticket
   c. Average rate of 40 miles per hour
4. **a**. −60 units per dollar **b**. −50 units per dollar
   c. −50 units per dollar
   d. A negative average rate of change in demand means that, on average, the demand (units purchased) decreases as the price increases.
5. **a**. \$21 per watch **b**. \$23 per watch **c**. \$19.10 per watch
6. **a**. 91° per minute **b**. 82.9° per minute
   c. 82.09° per minute **d**. 82.009° per minute
   e. 82.0009° per minute
   The average rate of change is approaching 82° per minute. The temperature is changing by 82° per minute at $t = 4$ minutes.
7. Estimates will vary. Values should be close to these.
   a. −\$1.67 per month **b**. \$1.00 per month
   c. \$.83 per month
8. **a**. 2 **b**. $2x + h$
   c. $\dfrac{2}{\sqrt{2(x+h)+3} + \sqrt{2x+3}}$ **d**. $\dfrac{-2}{(x+h-3)(x-3)}$
9. **a**. −2 **b**. $2(t_2 + t_1)$
   c. $\dfrac{-1}{\sqrt{5-t_2} + \sqrt{5-t_1}}$ **d**. $\dfrac{-3}{(t_2+1)(t_1+1)}$
10. **a**. 35°F
    b. The temperature during this time period was probably constant.
    c. No
    d. At 3:00 P.M. it was 42°F; At 7:00 P.M. it was 33°F.
    e. The temperature, on average, is decreasing by 3°F every hour.
11. **a**. 1 in. per month
    b. January–February: 2 in. per month
    February–March: 0 in. per month
    March–April: 3 in. per month
    April–May: 3 in. per month
    May–June: 1 in. per month
    c. $\frac{5}{3}$ in. per month
    d. The monthly rate of change for the period from January to February is close to the average rate of change for this period of time.

Exercises 6-C

1. **a**.

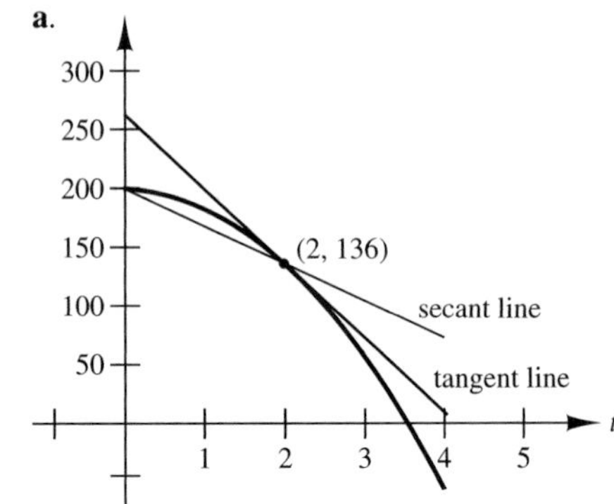

b. $s(0) = 200$ ft, $s(2) = 136$ ft
c. See the graph in part a.
d. Slope of secant line = −32; same as average velocity in Example 6.7.
e. See the graph in part a.
f. Slope of tangent line = −64. The velocity of the ball 2 sec after it is dropped is −64 ft/sec.

2. **a**. See the graph below. **b**. $C(10) = \$35$; $C(20) = \$70$
   c. See the graph below.

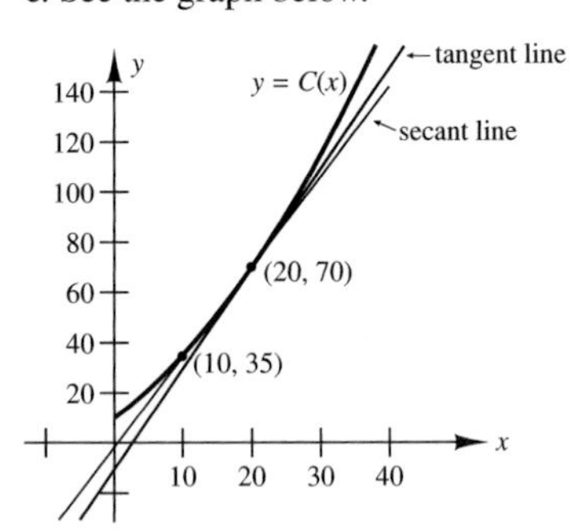

d. 3.5; the same as the average rate of change found in Example 6.8.
e. See the graph in part a.
f. Approximately 4. The marginal cost at a production of 20 bracelets is \$4.00 per bracelet.

3. **a**.

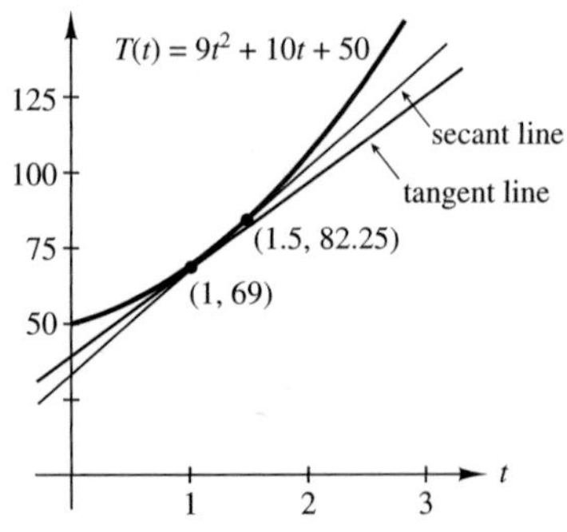

b. $T(1) = 69°$, $T(1.5) = 85.25°$
c. See the graph in part a.
d. 32.5; 32.5° is the average rate of change in temperature from time $t = 1$ min to time $t = 1.5$ min.
e. See the graph in part a.
f. Approximately 28. This number is the instantaneous rate of change in temperature at time $t = 1$ min.

Chapter 6 Exercises

1. **a**.

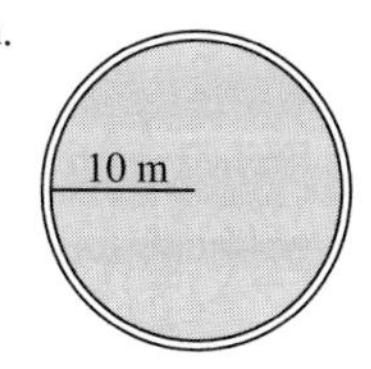

b. $144\pi\,m^2 \approx 452\,m^2$ **c**. $100\pi\,m^2 \approx 314\,m^2$
d. $44\pi\,m^2 \approx 138\,m^2$
e. $A = (10 + w)^2\pi - 100\pi = (20w + w^2)\pi\,m^2$
f. $V = 0.1(20w + w^2)\pi = (2w + 0.1w^2)\pi\,m^3$
g. $(2w + 0.1w^2)\pi$ tons
h. $C = 35(2w + 0.1w^2)\pi \approx 220w + 11w^2$ dollars
i. Approximately 3.8 m

2. **a**. \$590
b. $C = 10(30) + [30 - 0.10(x - 10)](x - 10) = -0.10x^2 + 32x - 10$ dollars
c. \$29.50 **d**. $\frac{-0.10x^2 + 32x - 10}{x} = -0.10x + 32 - \frac{10}{x}$
e. \$50 is saved by combining the orders.

3. **a**. \$400 per year **b**. \$640 per year
c. The average rate of change was the largest from 1994 to 1995; \$1,100 per year.
d. Approximately 4.4%
e. 1995; approximately 9.82%

4. **a**. An average of 185 births per month
b. An average of $\frac{9}{16}$ passes completed per attempt
c. An average of $\frac{3}{20}$ unemployed adults per surveyed adult

5. **a**. –\$22.4 billion per year **b**. –\$11.5 billion per year
c. \$65.85 billion per year **d**. –\$40.25 billion per year
e. Rising
f. 2001 or 2002; both had a rate of ≈ –\$240 billion per year.

6. **a**. \$103 per item; \$98.50 per item; \$98.05 per item
b. The marginal cost at $x = 10{,}000$ is approximately \$98 per item.

7. **a**. -3 **b**. $2x + h$
c. $\frac{1}{\sqrt{x+h-2} + \sqrt{x-2}}$ **d**. $\frac{-4}{(x+h+1)(x+1)}$

8. **a**. -3 **b**. $t_2 + t_1$
c. $\frac{1}{\sqrt{t_2 - 2} + \sqrt{t_1 - 2}}$ **d**. $\frac{-4}{(t_2+1)(t_1+1)}$

9. **a**. See graph below. **b**. 39.2 m
c. See graph below. **d**. 19.6
e. 19.6 m/sec **f**. See graph below.
g. Approximately 29.4. This number is the instantaneous velocity of the stone at time $t = 3$ seconds.

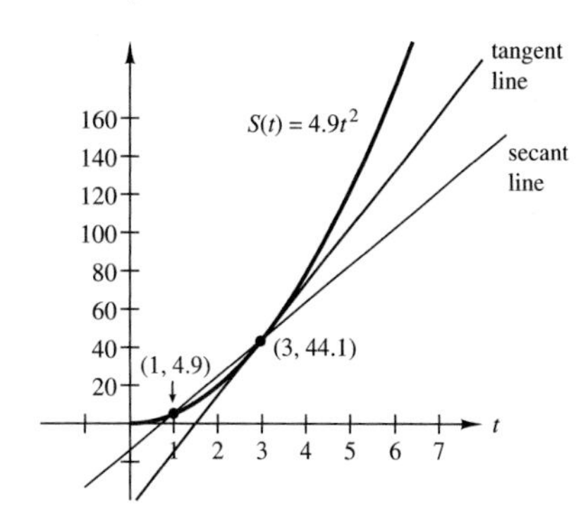

Exercises 7-A

1. **a**. 8 **b**. –8 **c**. $\frac{1}{8}$ **d**. $\frac{1}{64}$ **e**. a^6

2. **a**. 1 **b**. undefined **c**. undefined
d. 32 **e**. $\frac{1}{32}$
f. $\left(\sqrt{a}\right)^5$ if $a > 0$; undefined if $a \le 0$.

3. **a**. $\left(-\infty, \frac{5}{2}\right) \cup \left(\frac{5}{2}, \infty\right)$ **b**. $\left(-\frac{1}{4}, \infty\right)$
c. $[-3, 3]$

4. **a**. $f(x) = 2x^{-1/3} - \frac{1}{2}x^{-6} - \frac{1}{10}x^{-7} - 2x^{2/5}$
b. $g(x) = 3x^{5/2} - 2x^{1/2} + 6x^{-1/2}$
c. $h(x) = x^{7/3} - 2x^{4/3} + x^{1/3}$
d. $F(x) = x^2 + \frac{1}{2}x^{-3} - \frac{2}{3}x^{-1} - 12$
e. $G(x) = 4x^{-5/3} - 2x^{5/6} + 7x^{-13/6}$
f. $H(x) = \sqrt{5}x^{-4} + x^{-1/2}$
g. $W(x) = 4x^{-3/2} + 12 + 9x^{3/2}$
h. $S(x) = x^3 - 4x^{-1/2} + 4x^{-4}$

5. **a**. False **b**. False **c**. True **d**. True **e**. True
f. False. One value that shows equations a, b, and f are false is $x = 1$.

6. **a**. Not correct; $\sqrt{a+b} \ne \sqrt{a} + \sqrt{b}$. One value that shows equation a is false is $x = 1$.
b. Not correct; $\sqrt{a^2} \ne a$ when a is negative. One value that shows equation b is false is $x = -1$.
c. Correct

Exercises 7-B

1. **a.** $F = f \circ g$, where $f(x) = x^{1/2}$ and $g(x) = x + 5$.

   b. $G = \frac{f}{g}$, where $f(x) = 2x^4 - 1$ and $g(x) = x^{1/2} - 5$.

   c. $H = fg$, where $f(x) = \frac{1}{2}x^{-1/5} + x^3 - \frac{1}{3}x$ and $g(x) = 7x - \frac{4}{5}x^{-1}$.

   d. $B = f \circ g$, where $f(x) = x^{1/3}$ and $g(x) = x^5 - 2x^2 + x$.

2. **a.** $F = fg$, where $f(x) = \sqrt{x+5}$ and $g(x) = \sqrt[3]{x^2 - 1}$; $f = h \circ j$ and $g = k \circ m$, where $h(x) = x^{1/2}$, $j(x) = x + 5$, $k(x) = x^{1/3}$, and $m(x) = x^2 - 1$.

   b. $G = \frac{f}{g}$, where $f(x) = (3x+2)^7$ and $g(x) = 8 - x$; $f = h \circ j$, where $h(x) = x^7$ and $j(x) = 3x + 2$.

   c. $H = f \circ g$, where $f(x) = x^{1/2}$ and $g(x) = \frac{(x+8)^7}{(3x+2)^3}$; $g = \frac{h}{j}$, where $h(x) = (x+8)^7$ and $j(x) = (3x+2)^3$; $h = k \circ m$ and $j = p \circ q$, where $k(x) = x^7$, $m(x) = x + 8$, $p(x) = x^3$, and $q(x) = 3x + 2$.

   A better way: Write the function as $H(x) = \frac{(x+8)^{7/2}}{(3x+2)^{3/2}}$.

   Then $H = \frac{f}{g}$, where $f(x) = (x+8)^{7/2}$ and $g(x) = (3x+2)^{3/2}$; $f = h \circ k$ and $g = j \circ m$, where $h(x) = x^{7/2}$, $k(x) = x + 8$, $j(x) = x^{3/2}$, and $m(x) = 3x + 2$.

   d. $L = f - g$, where $f(x) = (3x - 8)^{4/3}$ and $g(x) = \frac{x}{x-1}$; $f = h \circ j$ and $g = \frac{k}{m}$, where $h(x) = x^{4/3}$, $j(x) = 3x - 8$, $k(x) = x$, and $m(x) = x - 1$.

   e. $N = \frac{f}{g}$, where $f(x) = 2 - x^{-1}$ and $g(x) = x + 1$.

   f. $p(x) = x^{3/2} + x^{-5/2}$

3. **a.** Product Rule, Chain Rule, Chain Rule

   b. Quotient Rule, Chain Rule

   c. Chain Rule, Quotient Rule, Chain Rule, Chain Rule; or write $H(x) = \frac{(x+8)^{7/2}}{(3x+2)^{3/2}}$, then the sequence of rules is Quotient Rule, Chain Rule, Chain Rule.

   d. The derivative of a difference is the difference of the derivatives. The Chain Rule would be used for the first term; the Quotient Rule for the second.

   e. Quotient Rule

   f. The derivative of a sum is the sum of the derivatives.

Exercises 7-C

1. **a.** $x = \frac{1 \pm \sqrt{5}}{2}$

   b. There are no horizontal tangent lines to this curve since $\frac{-5}{(2x-1)^2}$ cannot equal zero.

2. $x = \frac{6}{5}$

3. $x = -1$, $x = -\frac{1}{2}$, $x = -\frac{7}{10}$

4. **b.** $x = 0$

   c. $x = -1$

Chapter 7 Exercises

1. **a.** $\frac{1}{16\sqrt{2}}$ **b.** Undefined **c.** 8

   d. $\frac{1}{2}$ **e.** $\frac{1}{2\sqrt{2}}$

2. **a.** x^2 **b.** $4x^{2/3}$ **c.** $x^{2/3}$ **d.** $\frac{4}{x^4}$

3. **a.** True **b.** False **c.** False

   d. True **e.** False **f.** True; one value for which equations b, c, and e are false is $x = 4$.

4. **a.** $2x^{3/4} + x^{7/4}$ **b.** $\frac{2}{5}x^{-3} - \frac{1}{7}x^{2/3} + x^{3/2}$

   c. $x^{11/5} + \frac{2}{11}x^{26/5}$ **d.** $5x^{13/3} + \frac{1}{2}x^{-2/3} + \frac{1}{3}x^{7/3}$

   e. $\frac{\sqrt[3]{7}}{2} + \frac{\sqrt[3]{7}}{2}x^{-1} - x^{3/5} - x^{-2/5}$

   f. $3x^{-4} + 4\sqrt{3}x^{-3} + 4x^{-2}$

5. **a.** $F = \frac{f}{g}$, where $f(x) = x^2 + 2x$ and $g(x) = -5x + \sqrt{3}x^{1/2}$.

   b. $G = \frac{f}{g}$, where $f(x) = \sqrt[7]{2x^3 + 1}$ and $g(x) = \sqrt[5]{x - 3}$; $f = h \circ j$ and $g = k \circ m$, where $h(x) = x^{1/7}$, $j(x) = 2x^3 + 1$, $k(x) = x^{1/5}$, and $m(x) = x - 3$.

   c. $H = fg$, where $f(x) = x$ and $g(x) = \left(\sqrt{2} + \frac{1}{x^2}\right)^5$; $g = h \circ k$, where $h(x) = x^5$ and $k(x) = \sqrt{2} + x^{-2}$.

   d. $L = fg$, where $f(x) = \frac{1}{3}x^{-1} + 8x^{1/3}$ and $g(x) = \sqrt{x^5 - 2x}$; $g = h \circ k$, where $h(x) = x^{1/2}$ and $k(x) = x^5 - 2x$.

   e. $N = f + g$, where $f(x) = \sqrt[5]{\left(x + \frac{2}{x^5}\right)^2}$ and $g(x) = \frac{1}{9}x^{1/2}$; $f = h \circ j$, where $h(x) = x^{2/5}$ and $j(x) = x + 2x^{-5}$.

   f. $R = f - g$, where $f(x) = \frac{-7x + \frac{1}{2}}{x^3 - 4\sqrt{x}}$ and $g(x) = \frac{6}{5}x^{7/3}$; $f = \frac{h}{j}$, where $h(x) = -7x + \frac{1}{2}$ and $j(x) = x^3 - 4x^{1/2}$.

6. **b.** $x = 3$, $x = 1$, $x = -4$

7. **a.** $x = 0$, $x = 4$ **b.** $(1, 0)$, $(3, -22)$

   c. There are no points on the curve where the slope of the tangent line is -13.

8. $y' = \frac{3}{2}\frac{(x-1)}{\sqrt{x}}$; $y' = 0$ when $x = 1$.

Exercises 8-A

1.

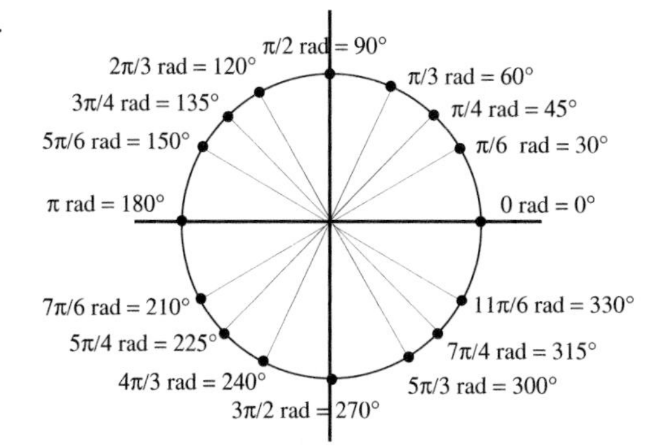

2.

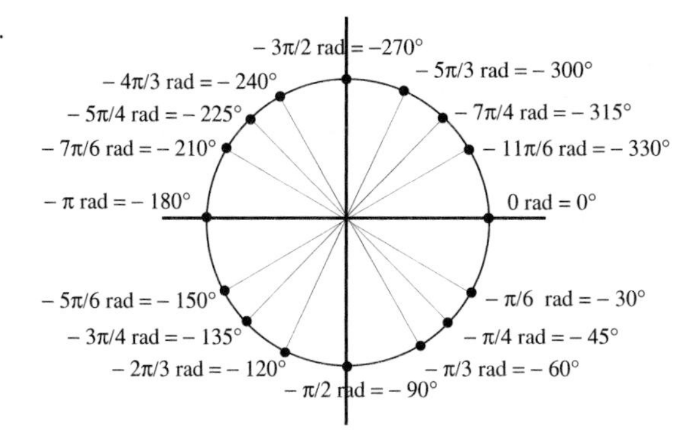

3. **a**. These are some possibilities:

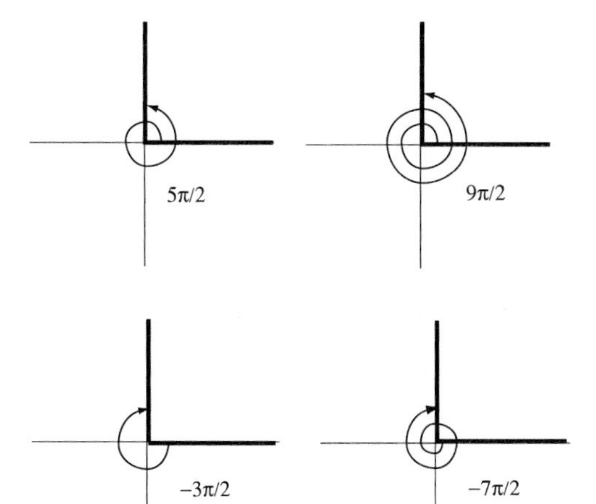

b. These are some possibilities:

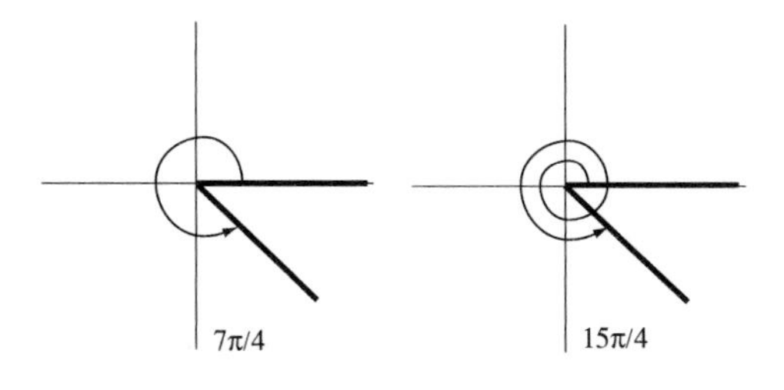

–9π/4

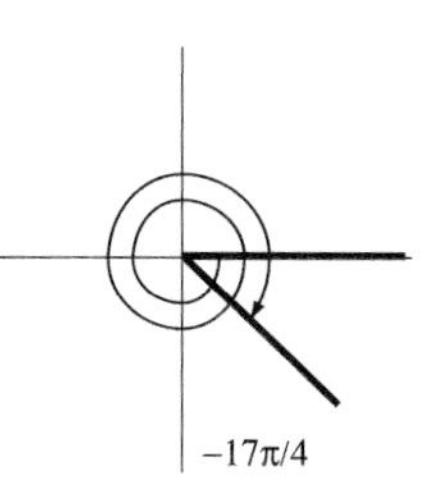

c. These are some possibilities:

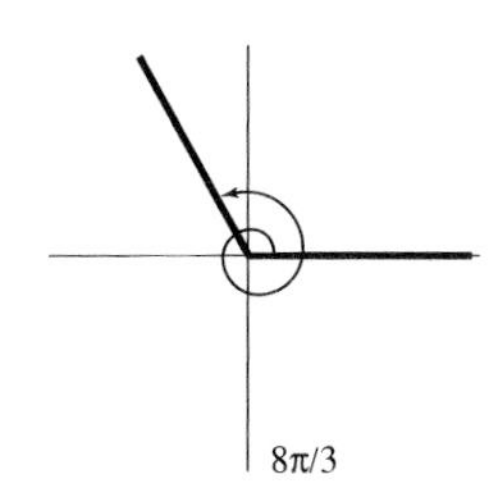

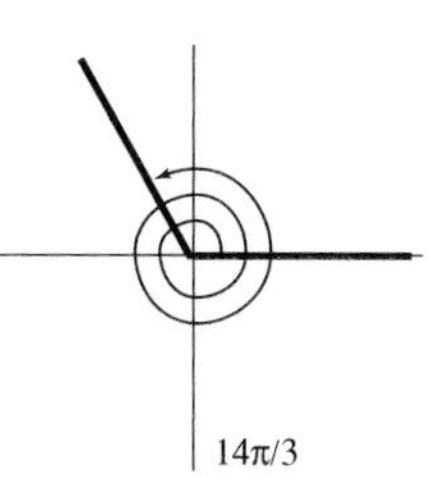

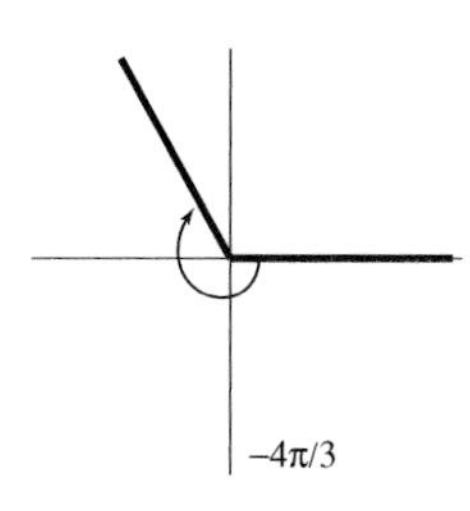

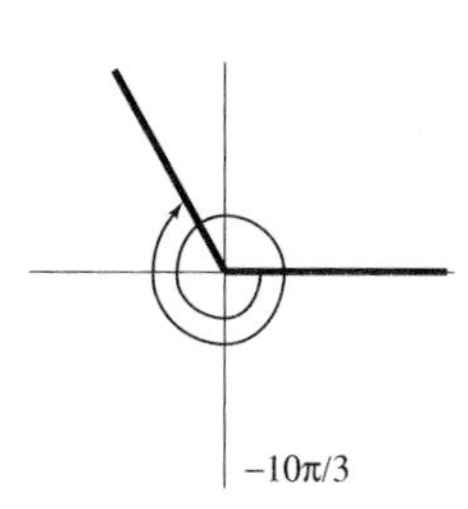

d. These are some possibilities:

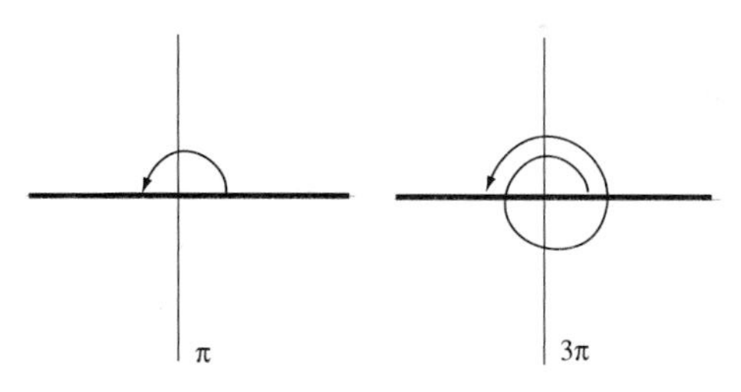

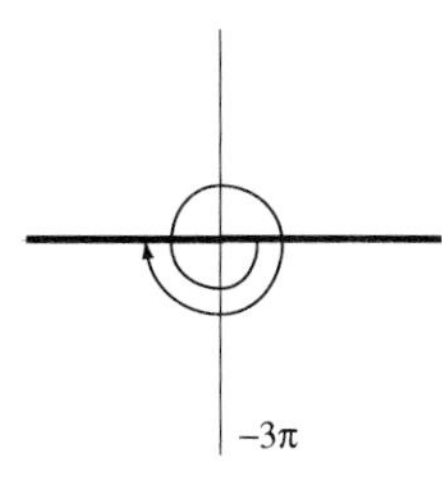

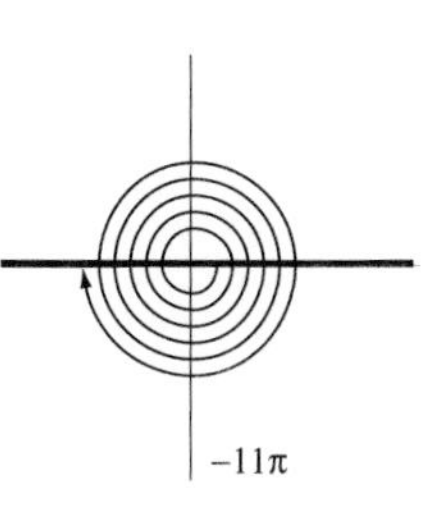

e. These are some possibilities:

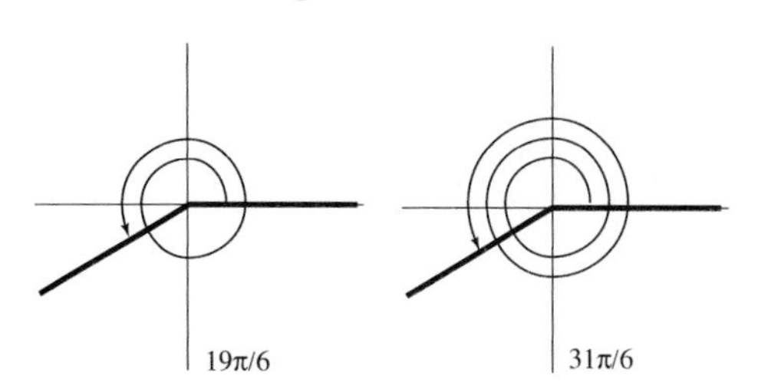

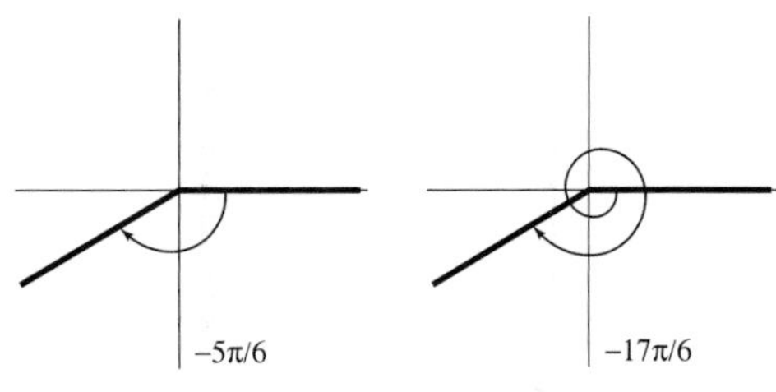

4. **a**. 1.5708 **b**. 3.1416 **c**. 4.7124
5. **a**. Second **b**. Third **c**. Third **d**. First
   e. First **f**. Third **g**. Second

Exercises 8-B

1.

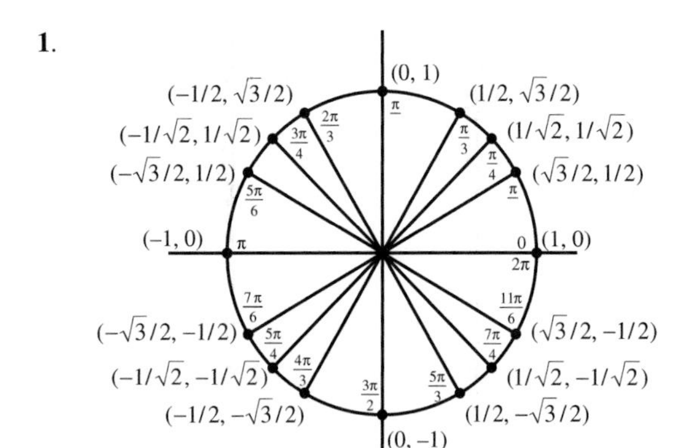

2. **a**. $\sin\theta = \frac{1}{2}$, $\cos\theta = -\frac{\sqrt{3}}{2}$
   b. $\theta_1 = -\frac{7\pi}{6}$, $\theta_2 = \frac{17\pi}{6}$ are some choices.
3. **a**. (i) $\sin\frac{4\pi}{3} = -\frac{\sqrt{3}}{2}$, $\cos\frac{4\pi}{3} = -\frac{1}{2}$ (ii) $\sin(-5\pi) = 0$, $\cos(-5\pi) = -1$ (iii) $\sin\frac{11\pi}{4} = \frac{1}{\sqrt{2}}$, $\cos\frac{11\pi}{4} = -\frac{1}{\sqrt{2}}$
   (iv) $\sin\left(-\frac{3\pi}{2}\right) = 1$, $\cos\left(-\frac{3\pi}{2}\right) = 0$
   (v) $\sin\left(-\frac{2\pi}{3}\right) = -\frac{\sqrt{3}}{2}$, $\cos\left(-\frac{2\pi}{3}\right) = -\frac{1}{2}$
   b. Some possible choices are: (i) $\theta_1 = -\frac{2\pi}{3}$, $\theta_2 = \frac{10\pi}{3}$
   (ii) $\theta_1 = -\pi$, $\theta_2 = \pi$ (iii) $\theta_1 = -\frac{5\pi}{4}$, $\theta_2 = \frac{3\pi}{4}$
   (iv) $\theta_1 = -\frac{7\pi}{2}$, $\theta_2 = \frac{\pi}{2}$ (v) $\theta_1 = -\frac{8\pi}{3}$, $\theta_2 = \frac{4\pi}{3}$
4. **a**. $\sin\theta \approx 0.8$, $\cos\theta \approx -0.5$
   b. $\sin\theta \approx -0.25$, $\cos\theta \approx -0.97$
5. **a**. All positive
   b. (i) Sine and cosecant are positive, the others negative. (ii) Tangent and cotangent are positive, the others negative. (iii) Cosine and secant are positive, the others negative.
6.

	θ	$\sin(\theta)$	$\cos(\theta)$	$\tan(\theta)$	$\cot(\theta)$	$\sec(\theta)$	$\csc(\theta)$
a.	-7π	0	-1	0	d.n.e.	-1	d.n.e.
b.	$\frac{10\pi}{3}$	$\frac{-\sqrt{3}}{2}$	$\frac{-1}{2}$	$\sqrt{3}$	$\frac{1}{\sqrt{3}}$	-2	$\frac{-2}{\sqrt{3}}$
c.	$\frac{11\pi}{4}$	$\frac{1}{\sqrt{2}}$	$\frac{-1}{\sqrt{2}}$	1	1	$-\sqrt{2}$	$\sqrt{2}$
d.	$\frac{-17\pi}{6}$	$\frac{-1}{2}$	$\frac{-\sqrt{3}}{2}$	$\frac{1}{\sqrt{3}}$	$\sqrt{3}$	$\frac{-2}{\sqrt{3}}$	-2

7.

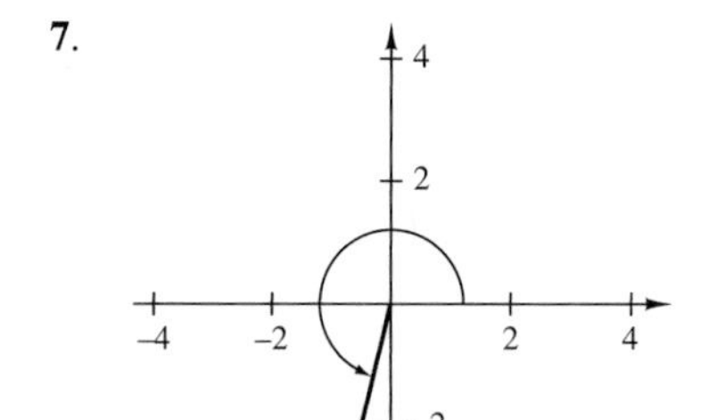

$\sin(\theta) = \frac{-4}{\sqrt{17}}$; $\cos(\theta) = \frac{-1}{\sqrt{17}}$; $\tan(\theta) = 4$; $\cot(\theta) = \frac{1}{4}$; $\sec(\theta) = -\sqrt{17}$; $\csc(\theta) = \frac{-\sqrt{17}}{4}$

Exercises 8-C

1. **a**. $\frac{\sqrt{8}}{3}$ and $-\frac{\sqrt{8}}{3}$
   b. These are two such angles (there are infinitely many):

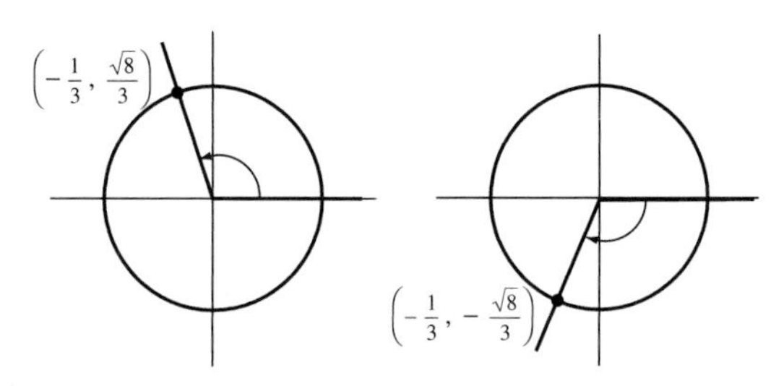

2. **a**. $\frac{\pi}{2}, \frac{\pi}{2} \pm \pi, \frac{\pi}{2} \pm 2\pi, \frac{\pi}{2} \pm 3\pi, \frac{\pi}{2} \pm 4\pi, \ldots$
   b. $0, \pm 2\pi, \pm 4\pi, \pm 6\pi, \pm 8\pi, \ldots$
   c. $\pm\pi, \pm 3\pi, \pm 5\pi, \pm 7\pi, \pm 9\pi, \ldots$
   d. $\frac{\pi}{2}, \frac{\pi}{2} \pm 2\pi, \frac{\pi}{2} \pm 4\pi, \frac{\pi}{2} \pm 6\pi, \ldots$
   e. $\frac{3\pi}{2}, \frac{3\pi}{2} \pm 2\pi, \frac{3\pi}{2} \pm 4\pi, \frac{3\pi}{2} \pm 6\pi, \frac{3\pi}{2} \pm 8\pi, \ldots$
   f. $\frac{\pi}{4}, \frac{\pi}{4} \pm \pi, \frac{\pi}{4} \pm 2\pi, \frac{\pi}{4} \pm 3\pi, \ldots$
3. **a**. $\frac{5\pi}{6}, \frac{5\pi}{6} \pm 2\pi, \frac{5\pi}{6} \pm 4\pi, \frac{5\pi}{6} \pm 6\pi, \ldots$ and $\frac{7\pi}{6}, \frac{7\pi}{6} \pm 2\pi, \frac{7\pi}{6} \pm 4\pi, \frac{7\pi}{6} \pm 6\pi, \ldots$
   b. No solution **c**. $\frac{\pi}{2}, \frac{\pi}{2} \pm 2\pi, \frac{\pi}{2} \pm 4\pi, \ldots$
   d. $\frac{\pi}{4}, \frac{\pi}{4} \pm \pi, \frac{\pi}{4} \pm 2\pi, \frac{\pi}{4} \pm 3\pi, \ldots$
   e. $\frac{\pi}{4}, \frac{\pi}{4} \pm \pi, \frac{\pi}{4} \pm 2\pi, \frac{\pi}{4} \pm 3\pi, \ldots$ and $\frac{3\pi}{4}, \frac{3\pi}{4} \pm \pi, \frac{3\pi}{4} \pm 2\pi, \frac{3\pi}{4} \pm 3\pi, \ldots$
   f. No solution
4. **a**. $\frac{\sqrt{5}}{3}$ **b**. $-\frac{2}{3}$ **c**. $\frac{\sqrt{5}}{3}$ **d**. $\frac{2}{\sqrt{5}}$ **e**. $-\frac{2}{\sqrt{5}}$ **f**. $-\frac{2}{3}$ **g**. $\frac{\sqrt{5}}{3}$
5. Use the identity $\sec\theta = \frac{1}{\cos\theta}$.
6. **a**. $-\frac{\sqrt{105}+3}{16}$ **b**. $\frac{3-\sqrt{105}}{16}$ **c**. $\frac{\sqrt{15}+3}{4}$ **d**. $\frac{\sqrt{15}-3}{4}$
7. **a**. $\frac{\sqrt{7}-3\sqrt{15}}{16}$ **b**. $\frac{\sqrt{7}+3\sqrt{15}}{16}$ **c**. $\frac{-1+\sqrt{7}}{4}$ **d**. $\frac{-1-\sqrt{7}}{4}$
8. **b**. $\frac{\sqrt{3}-1}{2\sqrt{2}} = \frac{\sqrt{6}-\sqrt{2}}{4}$

9. $\tan(2\theta) = \frac{2\tan\theta}{1-\tan^2\theta}$

10. a. $\cos^3\theta = \cos\theta \cdot \cos^2\theta = \cos\theta \cdot (1-\sin^2\theta)$ b. $\cos^4\theta = \frac{3}{8} + \frac{1}{2}\cos(2\theta) + \frac{1}{8}\cos(4\theta)$

Exercises 8-D

1. y-intercept: 0; x-intercepts: $0, \pm\pi, \pm 2\pi, \pm 3\pi, \pm 4\pi, \ldots$ (all integer multiples of π)

2. a. Increasing: $\ldots, \left(-\frac{5\pi}{2}, -\frac{3\pi}{2}\right), \left(-\frac{\pi}{2}, \frac{\pi}{2}\right), \left(\frac{3\pi}{2}, \frac{5\pi}{2}\right), \left(\frac{7\pi}{2}, \frac{9\pi}{2}\right), \ldots$
   Decreasing: $\ldots, \left(-\frac{7\pi}{2}, -\frac{5\pi}{2}\right), \left(-\frac{3\pi}{2}, -\frac{\pi}{2}\right), \left(\frac{\pi}{2}, \frac{3\pi}{2}\right), \left(\frac{5\pi}{2}, \frac{7\pi}{2}\right), \ldots$
   b. Increasing: $\ldots, (-3\pi, -2\pi), (-\pi, 0), (\pi, 2\pi), (3\pi, 4\pi), \ldots$
   Decreasing: $\ldots, (-4\pi, -3\pi), (-2\pi, -\pi), (0, \pi), (2\pi, 3\pi), \ldots$

3. a. $x \approx \frac{\pi}{12} \approx 0.26$; $\frac{\pi}{12} \approx 0.26$ radians, 15° b. $x \approx 0.25$ radians $\approx 14.3°$
   c. For part a, $\sin(0.26) \approx 0.25708$; $\sin(15°) \approx 0.25882$
   For part b, $\sin(0.25) \approx 0.24740$; $\sin(14.3°) \approx 0.24700$
   d. For part a, the differences $\sin(x) - 0.25$ are 0.00708 and 0.00882.
   For part b, the differences $\sin(x) - 0.25$ are –0.00260 and –0.00300.

4. a. $\{x \mid x \neq 0, \pm\pi, \pm 2\pi, \pm 3\pi, \pm 4\pi, \ldots\}$
   b. $\pm\frac{\pi}{2}, \pm\frac{3\pi}{2}, \pm\frac{5\pi}{2}, \ldots$
   c. $x = 0, \pm\pi, \pm 2\pi, \pm 3\pi, \pm 4\pi, \ldots$
   d.

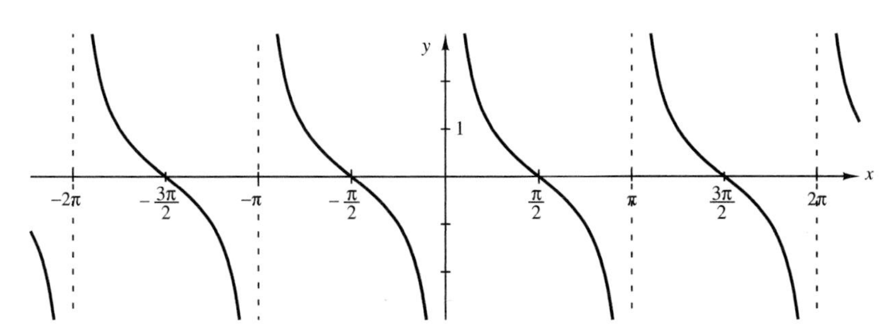

   e. $(-\infty, \infty)$
   f. π

5. a. $\{x \mid x \neq 0, \pm\pi, \pm 2\pi, \pm 3\pi, \pm 4\pi, \ldots\}$
   b. None
   c. $x = 0, \pm\pi, \pm 2\pi, \pm 3\pi, \pm 4\pi, \ldots$
   d.

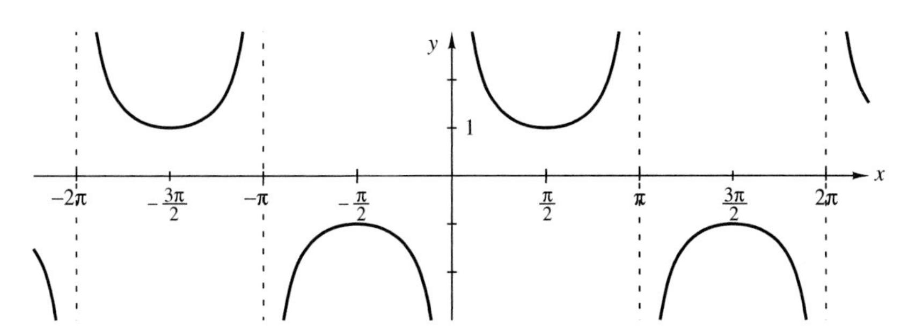

   e. $(-\infty, -1] \cup [1, \infty)$
   f. 2π

Exercises 8-E

1. **a**. Amplitude = 3, period = π

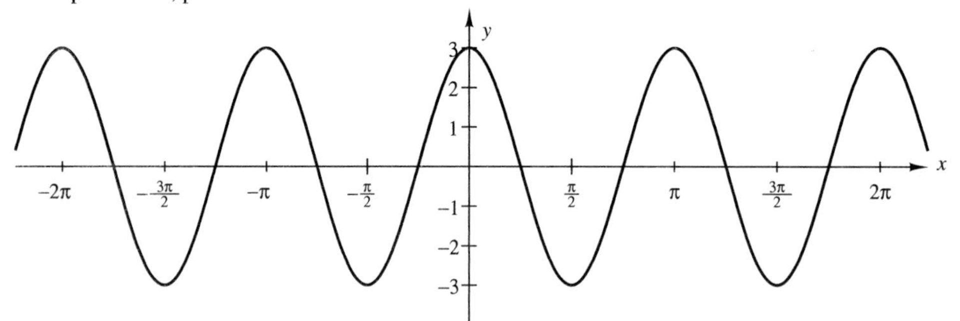

b. Amplitude = 1, period = 4π

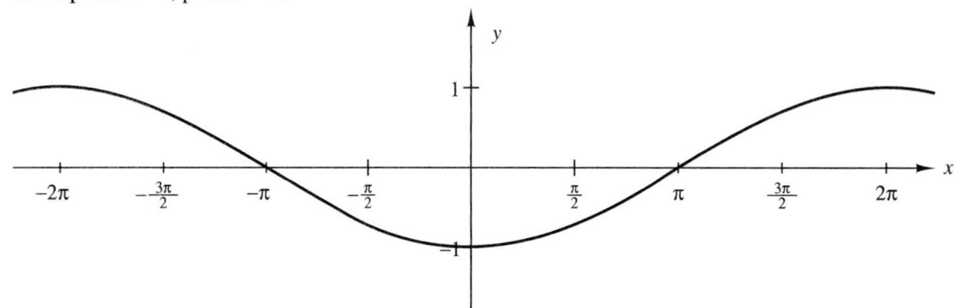

c. Amplitude = $\frac{3}{2}$, period = $\frac{\pi}{3}$

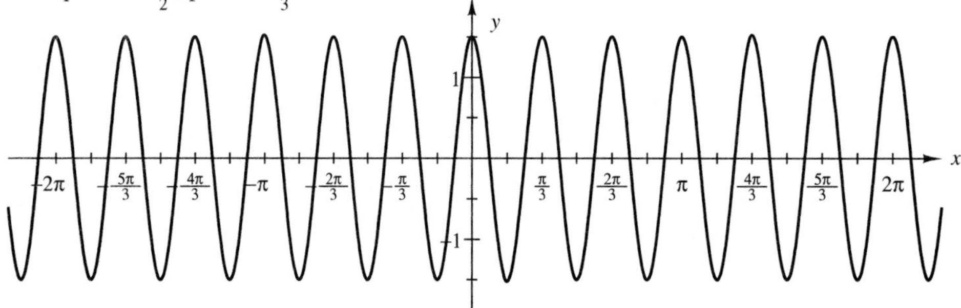

d. Amplitude = 8 , period = 6π

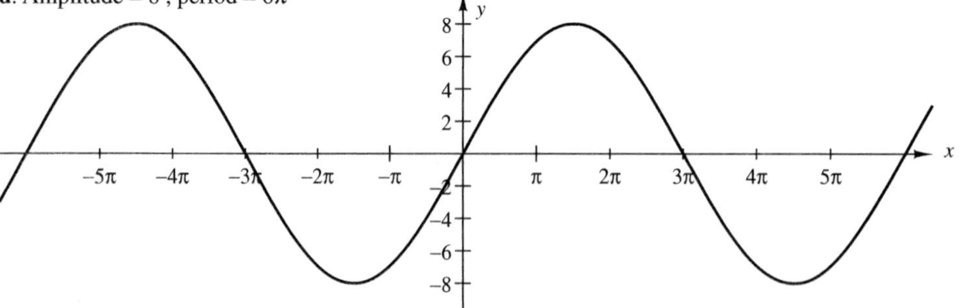

2. **a**. 0 **b**. $\frac{1}{2}$ **c**. 1 **d**. 0 **e**. $\sqrt{\frac{3}{2}}$

3. **a**. 5100 **b**. 5100 **c**. 3900
d. Periodic with period 12
e. Highest on June 21 of each year (when $t = 3, 15, 27, 39$); lowest on December 21 of each year (when $t = 9, 21, 33, 45$)
f.

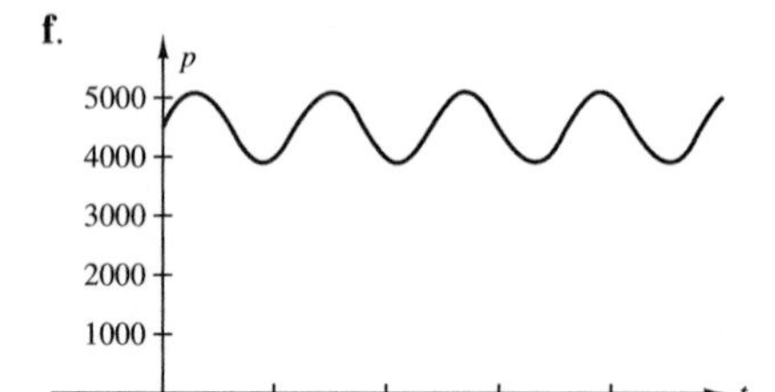

4. **a.** $\sec(\tan(5x))$; -1.02924 **b.** $(\sin(-\cos(x)))^3$; -0.4551
c. $\cos(\sin(x^3))$; 0.99224 **d.** $\frac{\tan(3x^2)}{6}$; 0.68862

5. **a.** $F(x) = (\tan(x^3))\left(\cos\left(\frac{x}{2}\right)\right)$; the Product Rule and the Chain Rule
b. $G(x) = \sin(\tan(x) - 1)$; the Chain Rule
c. $H(x) = \cos\left(\frac{x^2+7x}{3}\right)$; the Chain Rule
d. $K(x) = 2x + \cos(x^5)$; the rule that says the derivative of a sum is the sum of the derivatives, the Chain Rule, and the Power Rule
e. $R(x) = \cos(x^3) - \tan(x+1)$; the rule that says the derivative of a difference is the difference of the derivatives and the Chain Rule
f. $T(x)=(\sin(x+2))^4$; the Chain Rule

6. **a.** $f(x) = \sin(\cos x)$ **b.** $(\sin 2)(\cos x)$ and $\sin(2\cos x)$
c. $(\sin 2)(\cos x)$

7. **a.** $\left\{x \mid x \neq \frac{\pi}{3} \pm \frac{\pi}{2}, \frac{\pi}{3} \pm \frac{3\pi}{2}, \frac{\pi}{3} \pm \frac{5\pi}{2}, \ldots\right\}$
b. $(-\infty, 3) \cup (3, \infty)$

Chapter 8 Exercises

1. **a.** $\frac{7\pi}{12}$ radians; $105°$
b. $\sin\gamma \approx 0.9$, $\cos\gamma \approx -0.25$

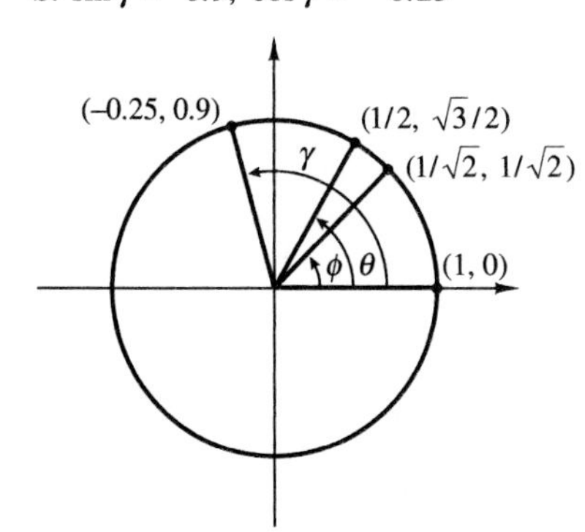

c. $\sin\gamma = \frac{\sqrt{3}+1}{2\sqrt{2}} = \frac{\sqrt{6}+\sqrt{2}}{4}$; $\cos\gamma = \frac{1-\sqrt{3}}{2\sqrt{2}} = \frac{\sqrt{2}-\sqrt{6}}{4}$

2. **a.** (i) $-\frac{\sqrt{24}}{5}$ (ii) 5 (iii) $-\sqrt{24}$
(iv) $-\frac{1}{\sqrt{24}}$ (v) $-\frac{5}{\sqrt{24}}$
b.

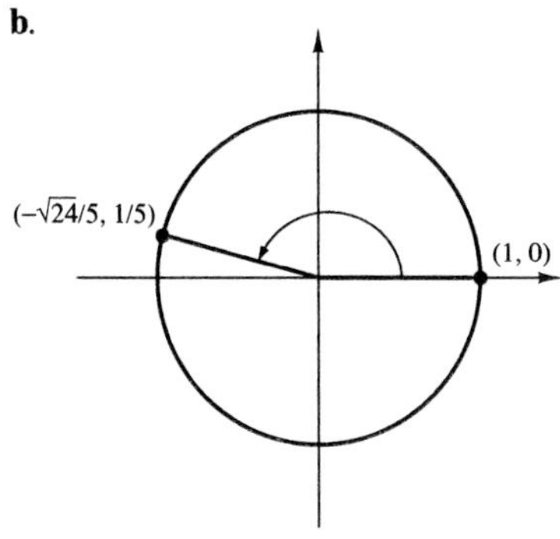

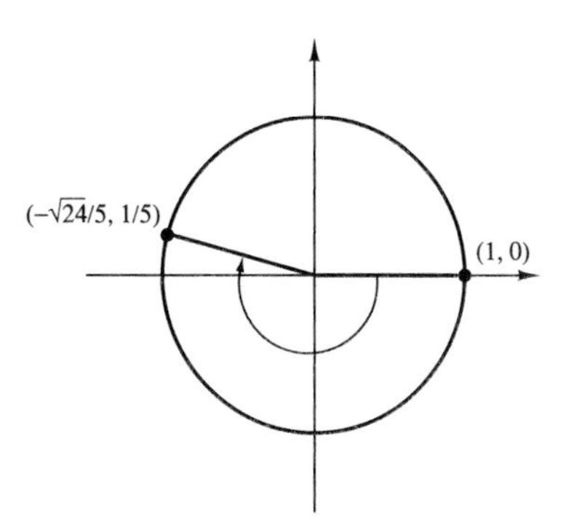

3. **a.** $-\frac{2}{\sqrt{2}} = -\sqrt{2}$ **b.** $-\sqrt{3}$ **c.** 1 **d.** -1 **e.** 1

4. **a.** The graph is symmetric with respect to the origin; that is, if it is turned $180°$ about the origin, it looks exactly the same.
b. Translating the graph π units to the left produces the same result as reflecting it about the x-axis.
c. Translating the graph π units to the right produces the same result as reflecting it about the x-axis.

5. **a.** The graph is symmetric with respect to the y-axis; that is, if it is reflected about the y-axis, it looks exactly the same.
b. Translating the graph π units to the left produces the same result as reflecting it about the x-axis.
c. Translating the graph π units to the right produces the same result as reflecting it about the x-axis.

6. **a.** $\left\{x \mid x \neq \frac{1}{3} \pm \frac{\pi}{6}, \frac{1}{3} \pm \frac{3\pi}{6}, \frac{1}{3} \pm \frac{5\pi}{6}, \frac{1}{3} \pm \frac{7\pi}{6}, \ldots\right\}$
b. $\left\{x \mid x \neq \pm\frac{\pi}{4}, \pm\frac{3\pi}{4}, \pm\frac{5\pi}{4}, \pm\frac{7\pi}{4}, \ldots\right\}$
c. $(-\infty, \infty)$

7. **a.** $\ldots, \left(-\frac{3\pi}{2}, -\frac{\pi}{2}\right), \left(-\frac{\pi}{2}, \frac{\pi}{2}\right), \left(\frac{\pi}{2}, \frac{3\pi}{2}\right), \left(\frac{3\pi}{2}, \frac{5\pi}{2}\right), \ldots$
b. The function is never decreasing.
c. (i) False (ii) True (iii) True (iv) True (v) False

8. **a.** $-\frac{1}{\sqrt{2}}$ **b.** $-\frac{1}{2}$ **c.** 1 **d.** 2 **e.** -2 **f.** 1
g. Undefined **h.** Undefined **i.** 1 **j.** 3 **k.** 1

9. **a.** False **b.** False **c.** True **d.** True **e.** False
f. True **g.** False **h.** True **i.** True **j.** True

10. (i) and (v)

11. **a.** $F(x) = \cos(x^5)$; the Chain Rule and Power Rule
b. $G(x) = (\cos x)^5$; the Chain Rule
c. $H(x) = \cos(\tan(3x)) - 1$; the Chain Rule (twice)
d. $K(x) = \frac{\cos(7x)}{\sec(5x-2)}$; the Quotient Rule and the Chain Rule (twice)

12. $A = \frac{1}{3}$, $k = 4$

13. $A = -7$, $k = \frac{1}{3}$

14. **a**. sin $3x$ in black; csc $3x$ shaded

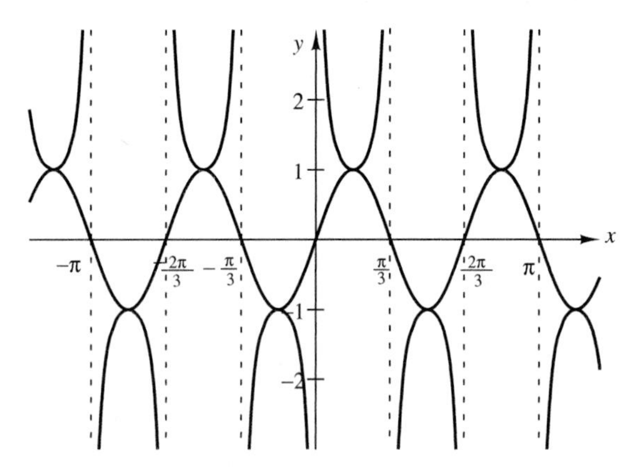

b. Domain = $(-\infty,\infty)$; range = $[-1, 1]$

c. Domain = $\left\{x \mid x \neq 0, \pm\frac{\pi}{3}, \pm\frac{2\pi}{3}, \pm\frac{3\pi}{3}, \pm\frac{4\pi}{3}, \ldots\right\}$;

range = $(-\infty, -1] \cup [1, \infty)$

15. **a**. 6 **b**. 8 **c**. 2 **d**. 5 **e**. 1 **f**. 7 **g**. 3 **h**. 4

16. **a**. By Identity 18, $\sin(\pi - x) =$
$\sin(\pi)\cos(x) - \cos(\pi)\sin(x) = 0 - (-1)\sin(x) = \sin(x)$.

d. By Identity 17, $\cos(\pi + x) =$
$\cos(\pi)\cos(x) - \sin(\pi)\sin(x) = (-1)\cos(x) - 0 = -\cos(x)$.

17. **a**. $\tan(\theta + \pi) = \tan(\theta)$

b. $\tan(\theta - \pi) = \tan(\theta)$

c. $\tan\left(\theta + \frac{\pi}{2}\right) = -\cot(\theta)$

d. $\tan\left(\theta - \frac{\pi}{2}\right) = -\cot(\theta)$

e. $\tan^2\theta = \dfrac{1-\cos(2\theta)}{1+\cos(2\theta)}$

18. $2\csc(2t) = 2\cdot\dfrac{1}{\sin(2t)} = 2\cdot\dfrac{1}{2\sin(t)\cos(t)} = \dfrac{1}{\sin(t)}\cdot\dfrac{1}{\cos(t)} =$

$\csc(t)\cdot\sec(t)$

Exercises 9-A

1. a, b, d, e, f, g, h

2. **a**. $y = \pm\sqrt{x^2 - 1}$ domain: $(-\infty, -1] \cup [1, \infty)$

b. $y = -3 \pm \sqrt{16-(x-2)^2}$ domain: $[-2, 6]$

c. $y = -\dfrac{4}{x}$ domain: $(-\infty, 0) \cup (0, \infty)$

d. $y = \frac{x}{2} + \frac{1}{4}$ domain: $(-\infty, \infty)$

e. $y = \dfrac{-2 \pm \sqrt{-2x+10}}{2}$ domain: $(-\infty, 5]$

f. $y = 3x - 2$ domain: $[0, \infty)$
$y = -3x - 2$ domain: $[0, \infty)$

g. $y = \dfrac{-2 \pm 3\sqrt{4-(x-3)^2}}{2}$ domain: $[1, 5]$

4. Figure 9.13 is not the graph of a function. Here are two possible implicit functions with the same domain as the original graph.

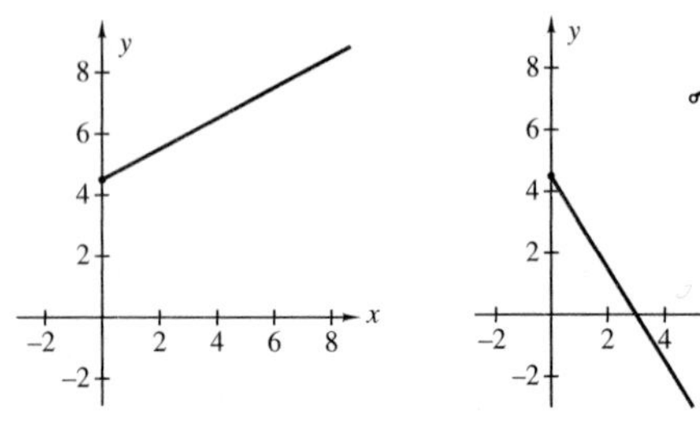

Figure 9.14 is the graph of a function.
Figure 9.15 is not the graph of a function. Here are two possible implicit functions with the same domain as the original graph.

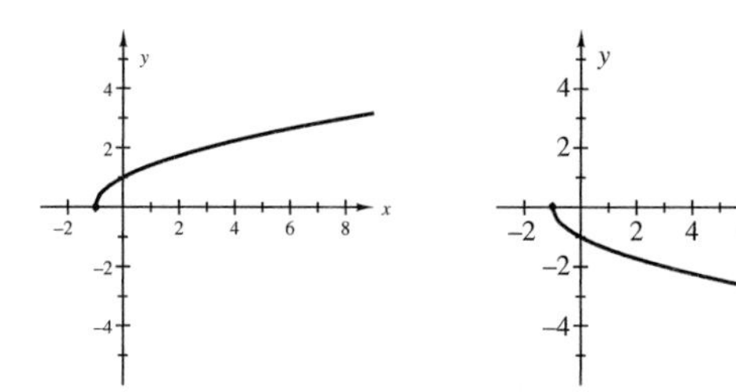

Figure 9.16 is not the graph of a function. Here are two possible implicit functions with the same domain as the original graph.

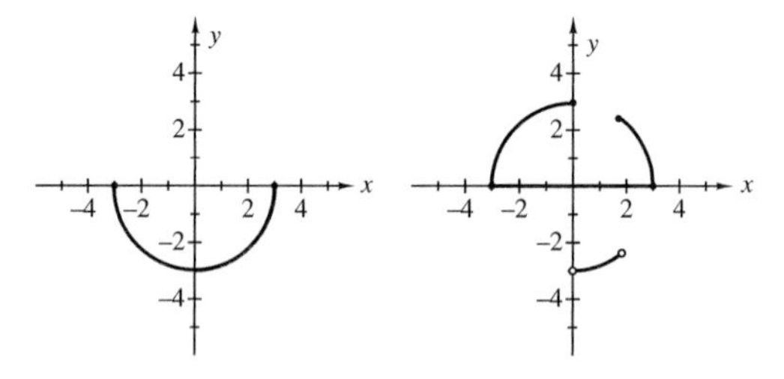

Exercises 9-B

1. **a**. $\dfrac{dy}{dx} = \dfrac{x}{2y}$ **b**. $y' = \dfrac{3}{4}$ **c**. $\dfrac{dy}{dx} = \dfrac{2x-3}{y-3}$

d. $y' = \dfrac{y^2 - \frac{1}{4}}{1-2xy}$ **e**. $\dfrac{dy}{dx} = \dfrac{y^2+1}{\frac{x}{y} + xy}$ **f**. $y' = \dfrac{2xy-1}{2y - \frac{x}{y}}$

2. **a**. $\frac{1}{4}$ **b**. $\frac{3}{4}$ **c**. 1 **d**. $-\frac{5}{4}$ **e**. 2 **f**. $\frac{6}{7}$

3. **a**. $y - 2 = \frac{1}{4}(x-1)$ **b**. $y - 2 = \frac{3}{4}(x-1)$

c. $y - 2 = x - 1$ **d**. $y - 2 = -\frac{5}{4}(x-1)$

e. $y - 2 = 2(x-1)$ **f**. $y - 2 = \frac{6}{7}(x-1)$

Chapter 9 Exercises

1. Equations a, c, d, e, and f are equations in which y is defined implicitly as a function of x.

2. a. $y = -\sqrt{9-(x+4)^2}$ domain: $[-7, -1]$, range: $[-3, 0]$

   $y = \sqrt{9-(x+4)^2}$ domain: $[-7, -1]$, range: $[0, 3]$

   b. $y = \frac{3}{x^3-2}$ domain: $\left(-\infty, \sqrt[3]{2}\right) \cup \left(\sqrt[3]{2}, \infty\right)$, range: $(-\infty, 0) \cup (0, \infty)$

   c. $y = \frac{7}{x-3} - 4$ domain: $(-\infty, 3) \cup (3, \infty)$, range: $(-\infty, -4) \cup (-4, \infty)$

   d. $y = 2\sqrt{4-(x+1)^2}$ domain: $[-3, 1]$, range: $[0, 4]$

   $y = -2\sqrt{4-(x+1)^2}$ domain: $[-3, 1]$, range: $[-4, 0]$

   e. $y = 1 - \sqrt{4-x}$ domain: $(-\infty, 4]$, range: $(-\infty, 1]$

   $y = 1 + \sqrt{4-x}$ domain: $(-\infty, 4]$, range: $[1, \infty)$

4. a. $\frac{dy}{dx} = \frac{2x+3}{y}$ b. $\frac{dy}{dx} = -\frac{3}{x}$ c. $y' = \frac{3x}{1-x}$

   d. $\frac{dy}{dx} = \frac{y^3 - y^4}{(2y-1)(xy^2+2)}$ e. $y' = \frac{y^4 - 2y^3}{2xy^3 + x(1-y^2)}$

5. a. -7 b. $-\frac{3}{2}$ c. -6 d. $\frac{1}{6}$ e. $-\frac{3}{4}$

6. a. $y + 1 = -7(x-2)$ b. $y + 1 = -\frac{3}{2}(x-2)$

   c. $y + 1 = -6(x-2)$ d. $y + 1 = \frac{1}{6}(x-2)$

   e. $y + 1 = -\frac{3}{4}(x-2)$

7. For each of the given graphs, here are two possible functions that are implicitly defined by the equation whose graph is shown and have the same domain as the graph.

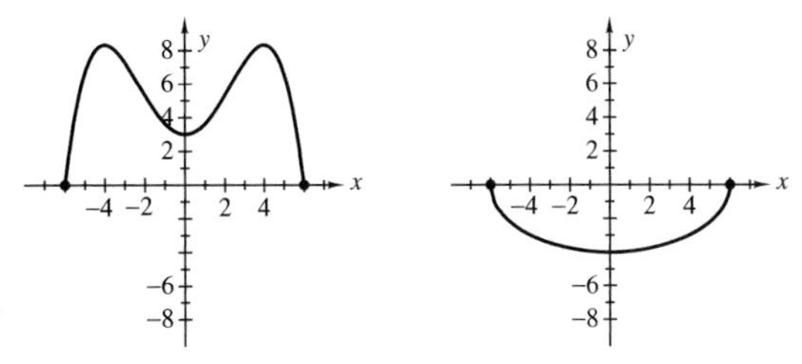

Figure 9.18

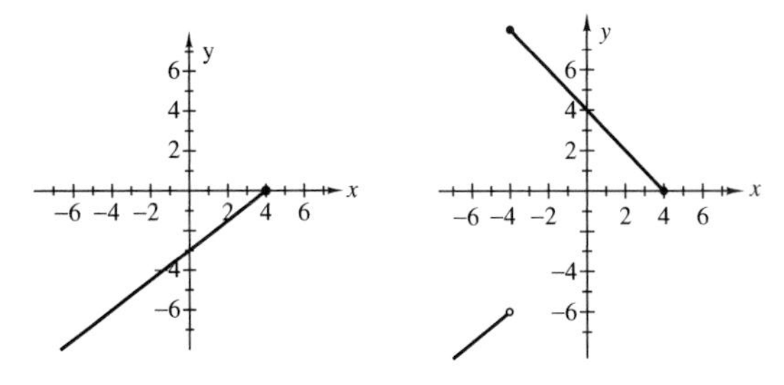

Figure 9.19

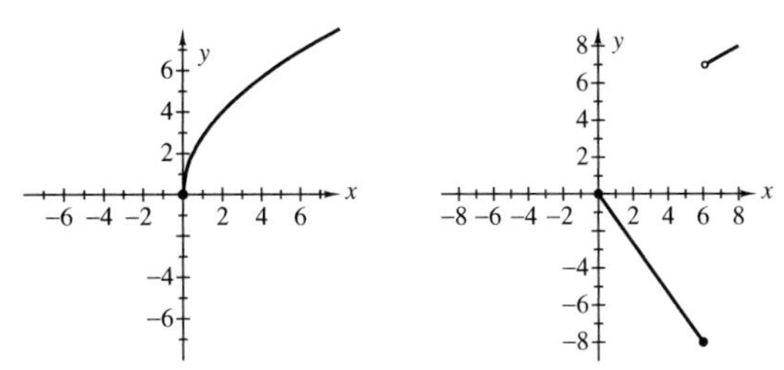

Figure 9.20

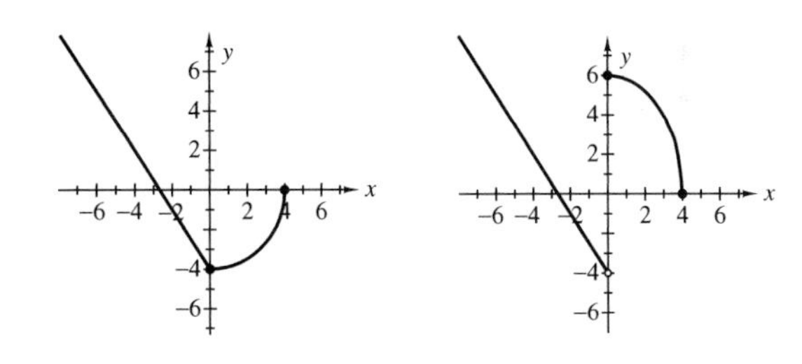

Figure 9.21

Exercises 10-A

1. a. $f^{(8)}(x) = 8! = 40{,}320$ b. $g^{(8)}(x) = 0$

   c. $h^{(8)}(x) = (100)(9!)x = 36{,}288{,}000x$

   d. $k^{(8)}(x) = 10 \cdot 9 \cdot 8 \cdot 7 \cdot 6 \cdot 5 \cdot 4 \cdot 3 \cdot (3x-3)^2 \cdot 3^8$
   $= 1{,}814{,}400(3x-3)^2 \cdot 3^8$

2. a. $F'(x) = 10\sqrt{2}\,x^9 - 86$

   $F''(x) = 9 \cdot 10\sqrt{2}\,x^8 = 90\sqrt{2}\,x^8$

   $F^{(3)}(x) = 8 \cdot 9 \cdot 10\sqrt{2}\,x^7 = 720\sqrt{2}\,x^7$

   $F^{(4)}(x) = 7 \cdot 8 \cdot 9 \cdot 10\sqrt{2}\,x^6 = 5040\sqrt{2}\,x^6$

   $F^{(5)}(x) = 6 \cdot 7 \cdot 8 \cdot 9 \cdot 10\sqrt{2}\,x^5 = 30{,}240\sqrt{2}\,x^5$

   b. $H'(x) = -2x^{-2}$

   $H''(x) = 2 \cdot 2x^{-3} = 4x^{-3}$

   $H^{(3)}(x) = -3 \cdot 2 \cdot 2x^{-4} = -12x^{-4}$

   $H^{(4)}(x) = 4 \cdot 3 \cdot 2 \cdot 2x^{-5} = 48x^{-5}$

   $H^{(5)}(x) = -5 \cdot 4 \cdot 3 \cdot 2 \cdot 2x^{-6} = -240x^{-6}$

   c. $K'(x) = 8x^7 + 7x^6$

   $K''(x) = 7 \cdot 8x^6 + 6 \cdot 7x^5 = 56x^6 + 42x^5$

   $K^{(3)}(x) = 6 \cdot 7 \cdot 8x^5 + 5 \cdot 6 \cdot 7x^4 = 336x^5 + 210x^4$

   $K^{(4)}(x) = 5 \cdot 6 \cdot 7 \cdot 8x^4 + 4 \cdot 5 \cdot 6 \cdot 7x^3 = 1680x^4 + 840x^3$

   $K^{(5)}(x) = 4 \cdot 5 \cdot 6 \cdot 7 \cdot 8x^3 + 3 \cdot 4 \cdot 5 \cdot 6 \cdot 7x^2$
   $= 6720x^3 + 2520x^2$

3. a. $F^{(6)}(x) = 5 \cdot 6 \cdot 7 \cdot 8 \cdot 9 \cdot 10\sqrt{2}\,x^4 = 151{,}200\sqrt{2}\,x^4$

   $F^{(7)}(x) = 4 \cdot 5 \cdot 6 \cdot 7 \cdot 8 \cdot 9 \cdot 10\sqrt{2}\,x^3 = 604{,}800\sqrt{2}\,x^3$

   $F^{(8)}(x) = 3 \cdot 4 \cdot 5 \cdot 6 \cdot 7 \cdot 8 \cdot 9 \cdot 10\sqrt{2}\,x^2$
   $= 1{,}814{,}400\sqrt{2}\,x^2$

$F^{(9)}(x) = 10!\sqrt{2}\ x = 3{,}628{,}800\sqrt{2}\ x$
$F^{(10)}(x) = 10!\sqrt{2} = 3{,}628{,}800\sqrt{2}$
$F^{(n)}(x) = 0$ for all $n \geq 11$

b. $H^{(n)}(x) \neq 0$ for all n

c. $K^{(6)}(x) = 3 \cdot 4 \cdot 5 \cdot 6 \cdot 7 \cdot 8x^2 + 7!x$
$= 20{,}160x^2 + 5040x$
$K^{(7)}(x) = 8!x + 7! = 40{,}320x + 5040$
$K^{(8)}(x) = 8! = 40{,}320$
$K^{(n)}(x) = 0$ for all $n \geq 9$

4. Substituting $n = 1$ in the formula gives $g^{(1)}(x) = (-1)^1 1!\,x^{-(1+1)} = -1!\,x^{-2}$, which equals $g'(x)$ found in Example 10.3. Other verifications are similar.
5. $H^n(x) = (-1)^n\,2 \cdot n!\,x^{-n-1}$
6. **a.** $G'(x) = \cos x$
$G''(x) = -\sin x$
$G^{(3)}(x) = -\cos x$
$G^{(4)}(x) = \sin x$
$G^{(5)}(x) = \cos x$

b. Look at the remainder when n is divided by 4.

$$G^{(n)}(x) = \begin{cases} \sin x & \text{if } n = 4k \\ \cos x & \text{if } n = 4k+1 \\ -\sin x & \text{if } n = 4k+2 \\ -\cos x & \text{if } n = 4k+3 \end{cases}$$

c. $G^{(20)}(x) = \sin x$, since $20 = 4 \cdot 5$; $G^{(102)}(x) = -\sin x$, since $102 = 4 \cdot 25 + 2$.

7. $f^{(6)}(x) = -945(2x)^{-11/2} = \dfrac{-945}{\sqrt{(2x)^{11}}}$

$f^{(6)}(2) = \dfrac{-945}{2^{11}} = \dfrac{-945}{2048}$; $f^{(6)}(0)$ does not exist (division by 0).

Exercises 10-B

1. **a.** f is increasing on [0, 6]; f is decreasing on [6, 8].
b. f', the rate of new infections, is increasing on (0, 5) and (6, 8); it is decreasing on (5, 6).
c. f'', the growth rate, is positive on (0, 5) and (6, 8); it is negative on (5, 6).
2. **a.** 2 **b.** 1 **c.** 2 **d.** 1 **e.** 3
3. **a.** f has constant slope 3.
b.

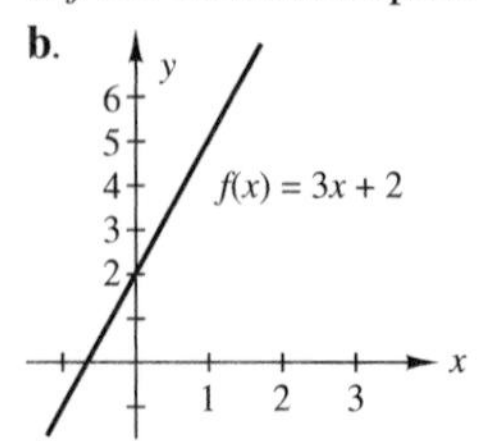

c. g is concave up on (2, 10) and concave down on (0, 2).
d.

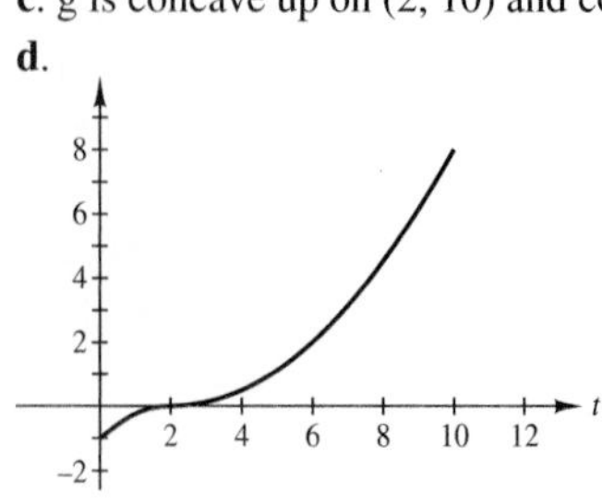

4. **a.** The velocity is 0 at time $t = \frac{15}{8} = 1.875$ sec; at this time the ball is at its highest point, 56.25 ft in the air.
b. -32 ft/sec^2
c. The ball will hit the ground at time $t = \frac{15}{4} = 3.75$ sec.
d. -32 ft/sec^2
e. The answers to parts b and d are the same because the acceleration due to gravity is constant.
5. **a.** The concentration is about 0.0077%.
b. 0.00148%/hour. The concentration is increasing 2 hours after the drug is taken.
c.

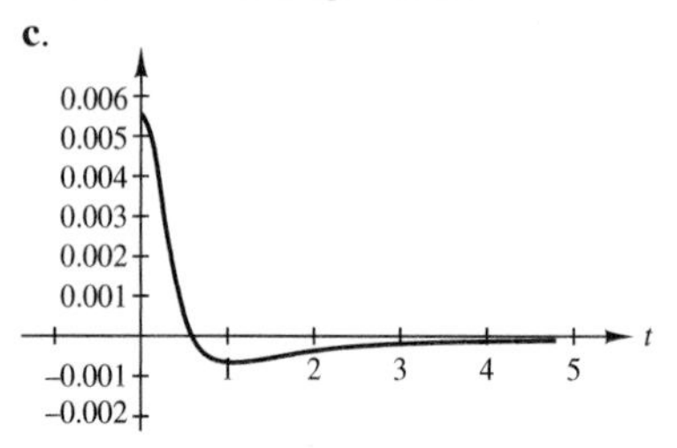

d. The rate of change of the concentration is decreasing 2 hours after the drug is taken.

Chapter 10 Exercises

1. **a.** $F^{(7)}(x) = 8!x - 7! \cdot 2 = 40{,}320x - 10{,}080$
b. $G^{(7)}(x) = 0$
c. $H^{(7)}(x) = -7!(x+2)^{-8} = -5040(x+2)^{-8}$
d. $K^{(7)}(x) = 1575(1+2x)^{-9/2}$
e. $R^{(7)}(x) = -3^7 \cdot \cos(3x) = -2187\cos(3x)$
2. **c.** $H^{(n)}(x) = (-1)^n n!(x+2)^{-n-1}$
e.

$$R^{(n)}(x) = \begin{cases} 3^n \sin(3x) & \text{if } n = 4k \\ 3^n \cos(3x) & \text{if } n = 4k+1 \\ -3^n \sin(3x) & \text{if } n = 4k+2 \\ -3^n \cos(3x) & \text{if } n = 4k+3 \end{cases}$$

3. $\frac{9}{128}$
4. **a.** $f^{(5)}(x) = (-1)^5\,2^5 \cdot 2 \cdot 3 \cdot 4 \cdot 5\,(2x)^{-6}$
b. $f^{(n)}(x) = (-1)^n\,2^n\,n!\,(2x)^{-n-1}$
5. **a.** The unemployment is the highest on day t_2.
b. The unemployment is increasing on $[0, t_2]$.
c. The unemployment is decreasing on $[t_2, t_3]$.
d. The rate of change of unemployment is increasing on $[0, t_1]$.
e. The rate of change of unemployment is decreasing on $[t_1, t_3]$.
f. **b.** $f'(t)$ is positive on $(0, t_2)$.
c. $f'(t)$ is negative on (t_2, t_3).
d. $f''(t)$ is positive on $(0, t_1)$.
e. $f''(t)$ is negative on (t_1, t_3).
6. **a.** The rate of change of the population is increasing on $[0, t_1)$ and decreasing for $t > t_1$.
b. The rate of change of the population is increasing on $[0, t_1)$ and for $t > t_1$. The rate of change of the population is never decreasing.
c. The rate of change of the population is decreasing for $t > 0$.

d. The rate of change of the population is neither increasing nor decreasing; that is, it is constant.

7. **a**. The profit is increasing when four lots are produced each month, since $P'(4) > 0$.
b. The rate of change of profit per lot is decreasing when four lots are produced each month. When four lots are produced each month, the profit per lot is increasing, but the rate at which it is increasing is becoming slower.

8. Graph 2 is the graph of f.
Graph 1 is the graph of g.

Exercises 11-A

1. **a.**

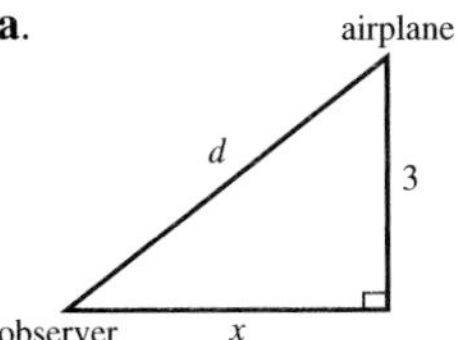

b. d = the distance (in miles) from the observer to the airplane at time t; x = the distance (in miles) along the ground from the observer to the point on the ground directly below the airplane at time t.
c. $d^2 = x^2 + 3^2$

2. **a**.

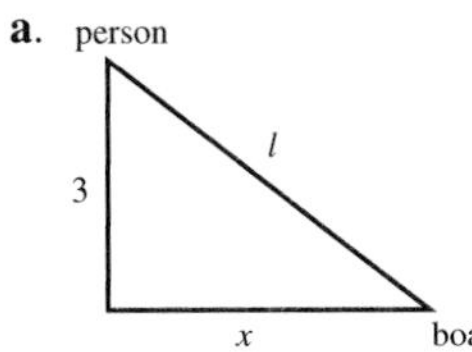

b. l = length of the rope (in feet) from the person's hands to the boat at time t; x = the distance (in feet) from the boat to the point at water level directly below the person's hands at time t.
c. $l^2 = 3^2 + x^2$

3. **a**. r = radius of the oil in the tank at time t; h = height (in meters) of the oil in the tank at time t; V = volume (in cubic meters) of the oil in the tank at time t.

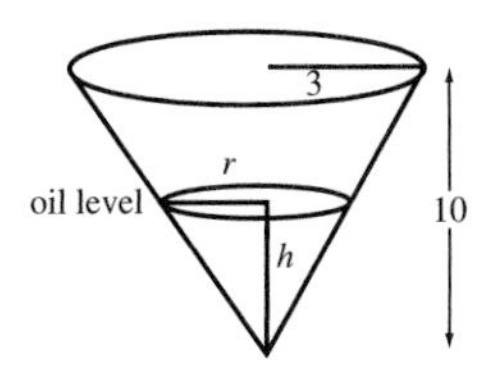

b. $\frac{r}{3} = \frac{h}{10}$, so $V = \frac{10}{9}\pi r^3$ or $V = 0.03\pi h^3$

4. **a**.

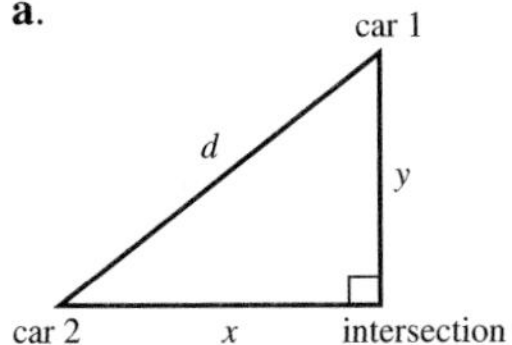

b. d = the distance between the two cars at time t; x = the distance from car 2 to the intersection at time t; y = the distance from car 1 to the intersection at time t.
c. $d^2 = x^2 + y^2$

5. F = gravitational force of the earth on a rocket traveling in space at time t; m = the mass of the rocket; d = the distance from the rocket to the center of the earth at time t.
$F = K \cdot \frac{m}{d^2}$, K = proportionality constant.

6. n = the number of students who decorate the gym for an event; h = the number of hours they spent on that job.
$h = \frac{k}{n}$, k a constant.

Exercises 11-B

1. **a**. h = the height of the balloon at time t. $\frac{dh}{dt} = 10$ ft/sec.

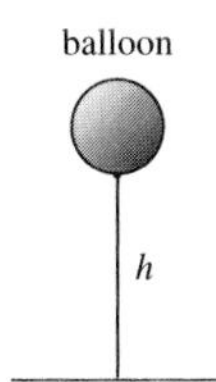

b. x = the distance of the police car from the entrance to the highway at time t. $\frac{dx}{dt} = 75$ mph when $t = \frac{1}{60}$ hour.

x

entrance to highway

car

c. V = the volume of the salt pile. $\frac{dV}{dt} = 45$ ft^3/min. The radius r of the pile and the height h of the pile are also changing over time.

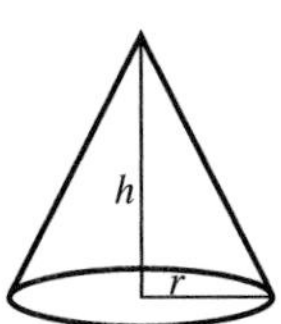

2. **a**. $\frac{dx}{dt} = -6$ **b**. $\frac{dx}{dt} = \frac{-16}{5\sqrt[3]{11^2}}$ **c**. $\frac{dy}{dt} = -\frac{52}{9}$

3. **a**. $\frac{dA}{dt} = 12\pi$ cm^2/sec **b**. $\frac{dA}{dt} = 42\pi$ cm^2/sec **c**. no
d. $\frac{dA}{dt}$ depends on r and $\frac{dr}{dt}$; thus as r changes, so does $\frac{dA}{dt}$.

Chapter 11 Exercises

1. a.

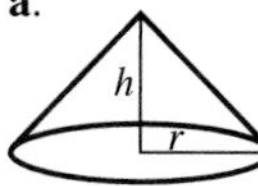

b. V = volume of the pile of salt at time t; h = height of the pile of salt at time t; r = radius of the pile of salt at time t

c. $V = \frac{1}{3}\pi r^2 h$

d. $V = \frac{1}{3}\pi r^3$ or $V = \frac{1}{3}\pi h^3$

2. a. See Figure 11.3; $\frac{dx}{dt} = 4$ ft/sec.

b. $\frac{ds}{dt}$ = the rate of change of the length of the man's shadow

c. $\frac{ds}{dt} = 2$ ft/sec

3. a. See Figure 11.4; $\frac{dV}{dt} = 3$ m^3/min. b. $\frac{dh}{dt}$

c. $\frac{dh}{dt} = \frac{12}{25\pi}$ m/min

4. a. See answer 1 in Exercises 11-A; $\frac{dx}{dt} = 350$ mph.

b. $\frac{dd}{dt} = 175\sqrt{3}$ mph

5. a. See answer 2 in Exercises 11-A; $\frac{dl}{dt} = -30$ ft/min.

b. $\frac{dx}{dt} = -3\sqrt{109}$ ft/min. The boat is approaching the dock at a rate of $3\sqrt{109}$ ft/min.

6. a. V = volume of helium in the balloon; $\frac{dV}{dt} = 6$ m^3/min.

b. $\frac{dr}{dt} = \frac{3}{50\pi}$ m/min

7. a. See answer 4 in Exercises 11-A;

$\frac{dy}{dt} = 30$ mph, $\frac{dx}{dt} = 20$ mph.

b. 10 miles c. $\frac{dd}{dt} = 10\sqrt{13}$ mph

8. a. $\frac{dC}{dt} = \$1,400$ per month b. $\frac{dR}{dt} = \$3,500$ per month

c. $\frac{dP}{dt} = \$2,100$ per month

9. a. B is directly proportional to w.

b. $\frac{dB}{dt} = -\frac{703.08}{68^2} \approx -0.15$ units per week

c. The man's Body Mass Index is decreasing.

10. $d = \frac{k}{p}$, where k is a constant, and the price p and the demand d are variables.

Exercises 12-A

1. a. $y = 1$ b. 1 c. 1

d. $f(-0.05) \approx 1.00125$. The error of the tangent approximation is ≈ 0.00125.

e. $f(0.1) \approx 1.00499$. The error of the tangent approximation is ≈ 0.00499.

2. a. $y = x$ b. -0.05 c. 0.1

d. $g(-0.05) \approx -0.04998$. The error of the tangent approximation is ≈ 0.00002.

e. $g(0.1) \approx 0.09983$. The error of the tangent approximation is ≈ 0.00017.

3. a.

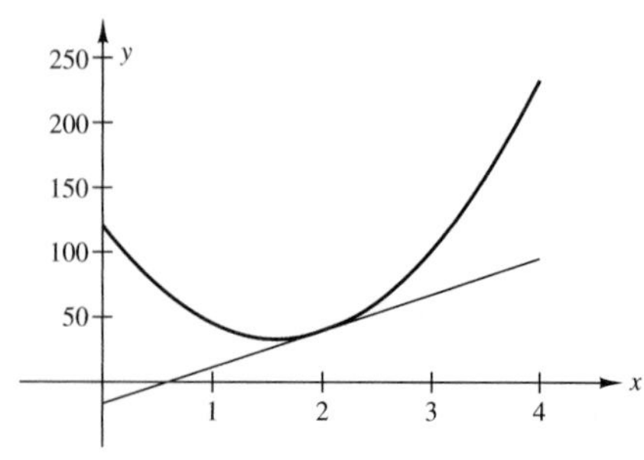

Graph 1

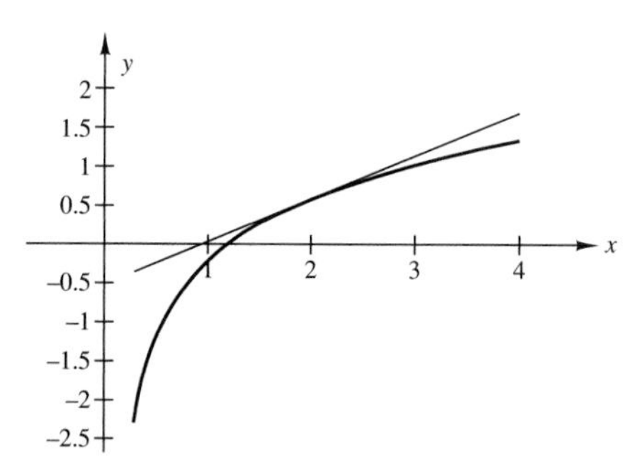

Graph 2

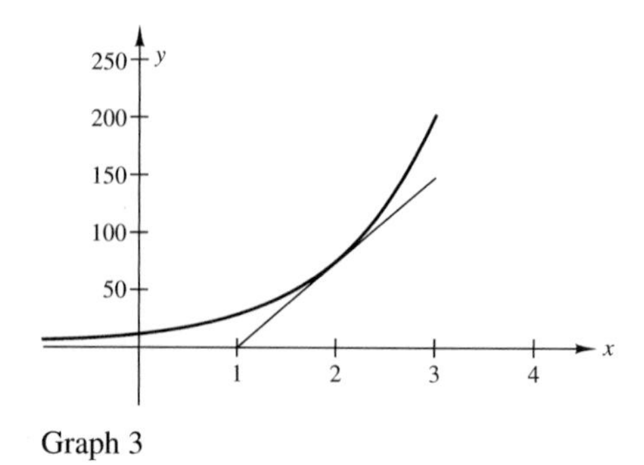

Graph 3

b. The tangent line approximation is best for Graph 2.

Exercises 12-B

1. a.

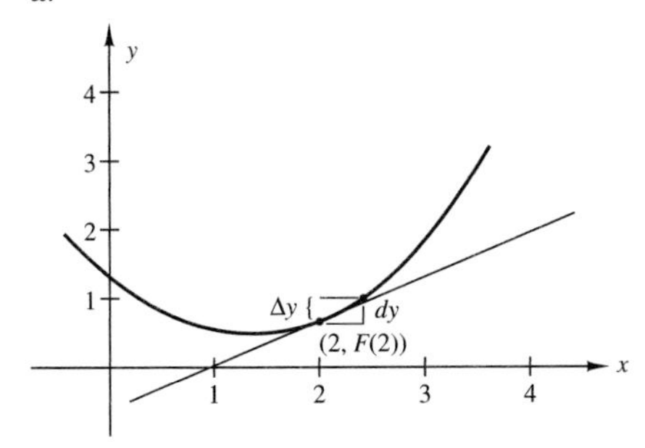

b.

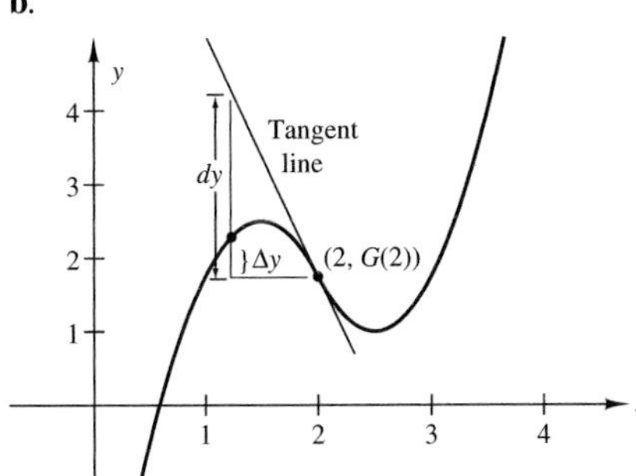

2. a. From the graphs in the answer to Exercise 1, you can see that the differential dy is a good approximation to Δy for the function $F(x)$ and Δx given in part 1a (the lengths of the two differences are very close in measure), but dy is not a good approximation to Δy for the function $G(x)$ and Δx given in part 1b (dy is more than three times as much as Δy).
   b. For Figure 12.10, $\Delta x \approx 0.7$; for Figure 2.11, $\Delta x \approx 0.3$.
3. a. $dy = -12$; $\Delta y = -12.25$; $dy - \Delta y = 0.25$
   b. $dy = -1.2$; $\Delta y = -1.2025$; $dy - \Delta y = 0.0025$
   c. $dy = -0.12$; $\Delta y = -0.120025$; $dy - \Delta y = 0.000025$
4. a. 7976.024 (to three decimal places)
   b. 8024.024 (to three decimal places)
   c. Let $f(x) = x^3$ and use dy to approximate $\Delta y = f(x + \Delta x) - f(x)$ for $x = 20$ and $\Delta x = 0.02$ (and for $\Delta x = -0.02$). Then approximate $f(x + \Delta x)$ by $dy + f(x)$.

Chapter 12 Exercises

1. a. 0 b. −\$50,000 c. \$50,000
   d. $y = 100\,(x - 100) + 15{,}000$ (y in thousands of dollars)
   e. \$15,300,000
2. a. $dy = 39.2\pi \approx 123.15$ in.^{3}; this is the increase in volume.
   b. $\Delta y = \frac{4}{3}\pi(7.2)^3 - \frac{4}{3}\pi(7)^3 \approx 126.70$ in.^{3}
   c. $dy - \Delta y \approx -3.55$ in.^{3}
3. a. (i)

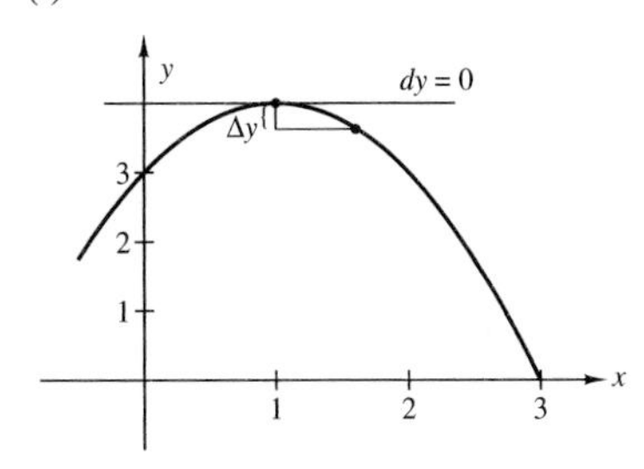

b. (i)

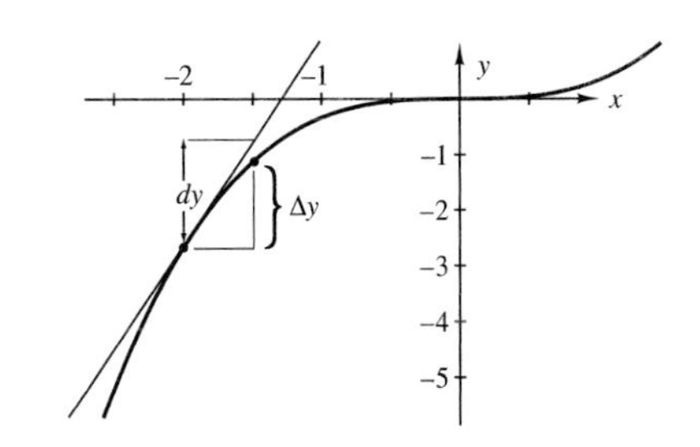

c. (i)

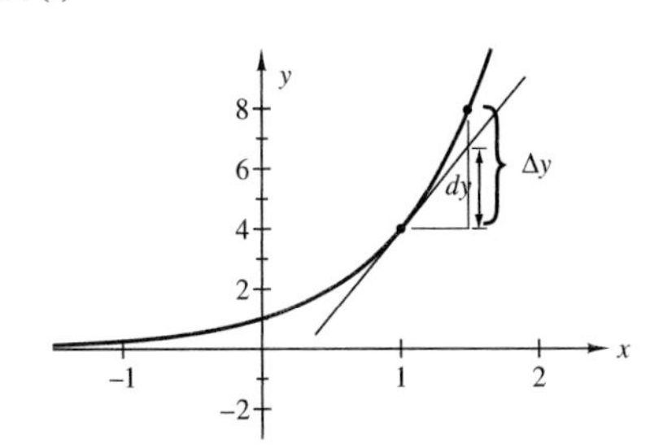

d. (i)

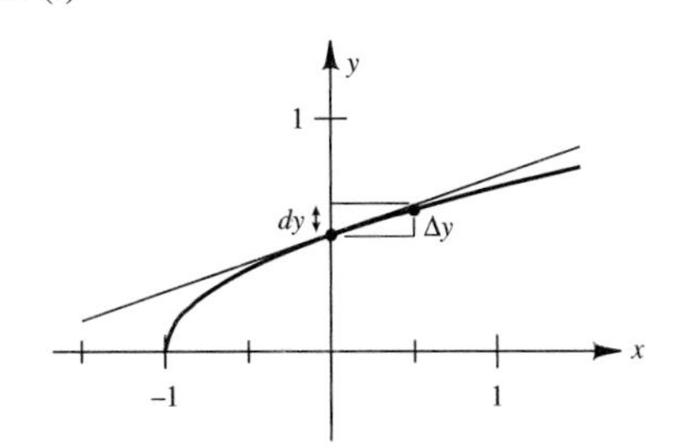

4. a. $\Delta x \approx 0.5$ b. $\Delta x \approx 0.5$ c. $\Delta x \approx 0.3$
5. a. In all cases, the tangent line gives a good approximation since the maximum error on the given interval $[-0.5, 0.5]$ is ≈ 0.04.
   b. The tangent line to the function in (iii) gives the best approximation.
   c. (i) Difference ≈ 0.02526
   (ii) Difference ≈ 0.02195
   (iii) Difference ≈ 0.01832

Exercises 13-A

1. **a.** Graph 3 **b.** Graph 1 **c.** Graph 4 **d.** Graph 2

2.

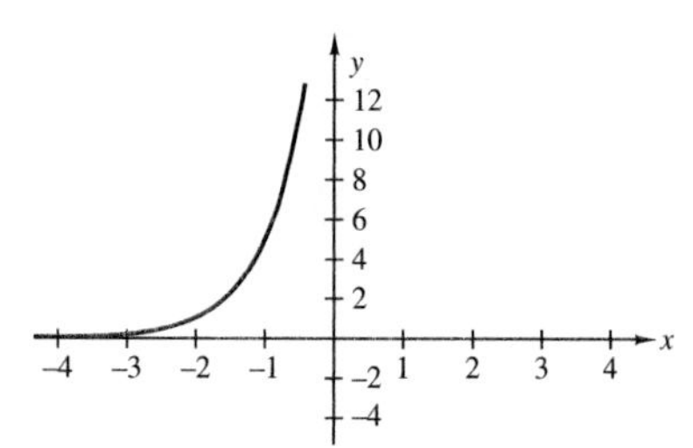

3. $y = b^x - 3$

4. **a.**

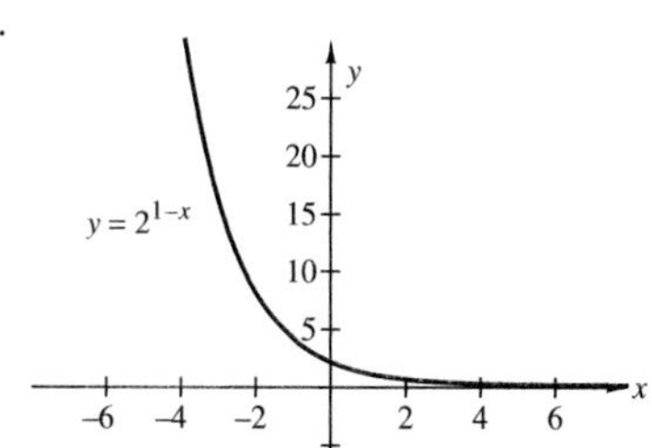

b.

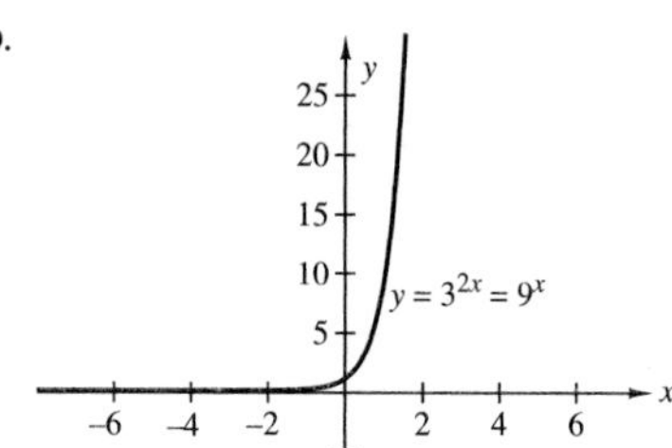

c.

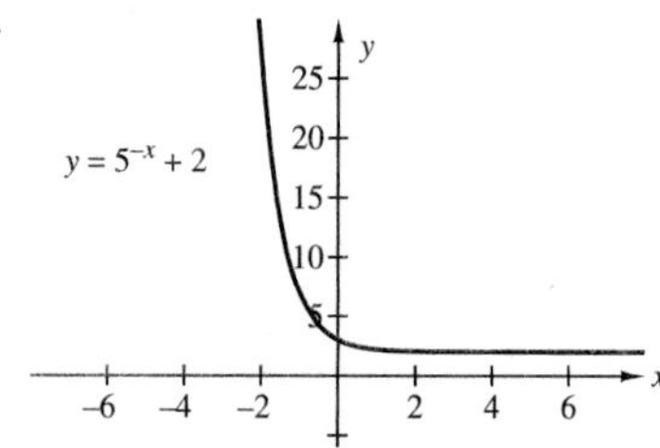

d.

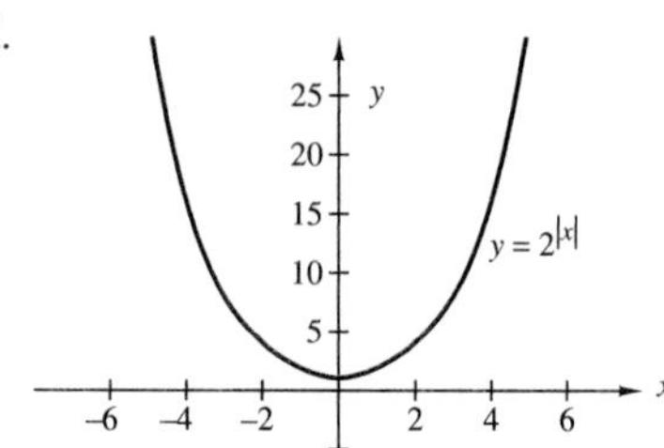

e.

y

25

20

15

10

5

$y = (2)2^x = 2^{x+1}$

x

-6 -4 -2 2 4 6

5. **a.** 2^9 **b.** $3^{-7/2}$ **c.** $\left(\frac{2}{3}\right)^{10/3}$

6. **a.** $y = 1.10^t$ is a mathematical model for the value of \$1 invested at 10% compounded annually.

b. \$2.08 at 5% growth rate; \$4.18 at 10% growth rate

7. **a.** $y = .90^t$ gives the real value of \$1 after t years at an annual inflation rate of 10%.

b. \$.21 at 10% inflation rate; \$.40 at 6% inflation rate

8. **a.** 2^{t+t^2} **b.** $2^{3t-2-3t^2}$ **c.** $2^{3t/4+2}$

9. **a.** False; $3^5 + 4^5 = 243 + 1024 = 1267$, whereas $7^5 = 16{,}807$.

b. False; $2^{-1/2} = \frac{1}{\sqrt{2}}$, whereas $-\frac{1}{2^2} = -\frac{1}{4}$.

c. True; $\frac{1}{a} = a^{-1}$ (rule 2).

d. False; $2^3 + 2^4 = 8 + 16 = 24$, whereas $2^7 = 128$.

e. False; $\frac{1}{2} + \frac{1}{3} = \frac{5}{6}$, whereas $\frac{1}{2+3} = \frac{1}{5}$.

f. False; $2^{-1} = \frac{1}{2} \neq \sqrt{2}$.

10. **a.** Not equal; for example, $7^{-1} = \frac{1}{7}$, whereas $-\frac{1}{7^1} = -\frac{1}{7}$.

b. Equal; $a^x a^y = a^{x+y}$ (rule 3).

c. Not equal; for example, $\frac{1}{3^1 - 4^1} = -1$, whereas $3^{-1} - 4^{-1} = \frac{1}{3} - \frac{1}{4} = \frac{1}{12}$.

d. Equal; $\frac{1}{a^x} = a^{-x} = \left(\frac{1}{a}\right)^x$ (rule 2).

e. Equal; $\frac{a^x}{b^x} = \left(\frac{a}{b}\right)^x$ (rule 7) and $a^{-x} = \left(\frac{1}{a}\right)^x$ (rule 2).

f. Not equal; for example, $(5^1)^2 = 5^2 = 25$, whereas $5^{1^2} = 5^1 = 5$.

g. Not equal; $2^1 \cdot 2^{1^2} = 2 \cdot 2 = 4$, whereas $2^{1^3} = 2^1 = 2$.

11. \$1,079.46

12. 4 million; 64 million

13. After 5 years, it is worth $\$1.00(0.96)^5 \approx \$.82$.
After 10 years, it is worth $\$1.00(0.96)^{10} \approx \$.66$.

Exercises 13-B

1. **a.** 2000 mg **b.** 1637.5 mg **c.** 735.8 mg

2. **a.** \$1,109.82 **b.** \$1,112.67 **c.** \$1,112.77

3. **a.** Graph 4 **b.** Graph 2 **c.** Graph 3 **d.** Graph 1

4. **a.**

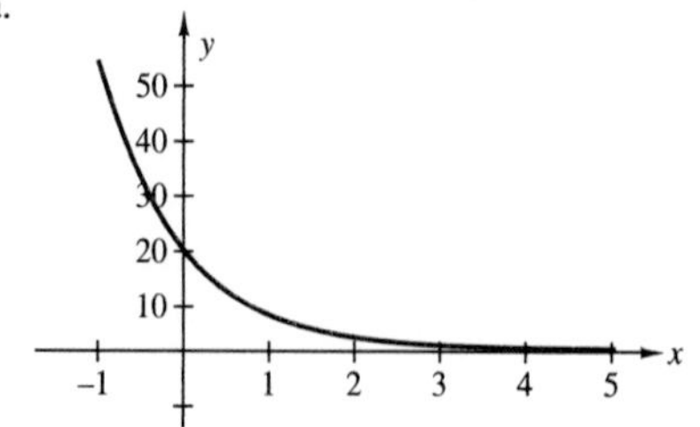

b. $y = b^x + 3$

5. **a.** $e^{3t/2}$ **b.** e^{-t+1} **c.** e^{2et-3t}

6. **a.** $e^{\cos t}$ **b.** $\cos(e^t)$ **c.** e^{t^2+1} **d.** $e^{2t} + 1$

Chapter 13 Exercises

1. **a.** (i) $x^{3/4}$ (ii) $x^{11/6}$ (iii) $x^{5/6}$
   b. Only $\sqrt[3]{x}\ \sqrt[4]{x^6}$ is defined for $x < 0$. When $x < 0$, the function equals $-(-x)^{11/6}$.
2. **a.** e^{-3t^2} **b.** $3^{\frac{2t^2+t}{3}}$ **c.** 2^{2t+16}
3. **a.** $6e^t - 3$ **b.** $3e^{3e^t}$ **c.** $3e^{1/t}$
   d. $\frac{1}{3e^t} = \frac{1}{3}e^{-t}$ **e.** $3e^{2t-3}$
   All functions have domains $(-\infty, \infty)$ except part c, which has domain $(-\infty, 0) \cup (0, \infty)$.
4. **a.** ≈ 143 **b.** ≈ 150 **c.** $N(t)$ approaches 150.
5. **a.** $y = (0.8)^x$

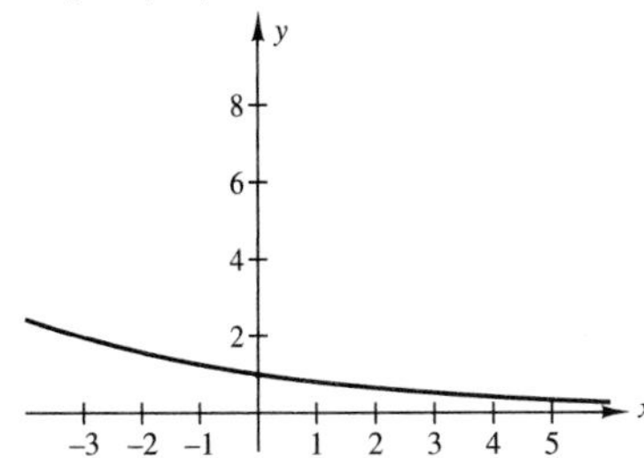

b. $y = -3^x$

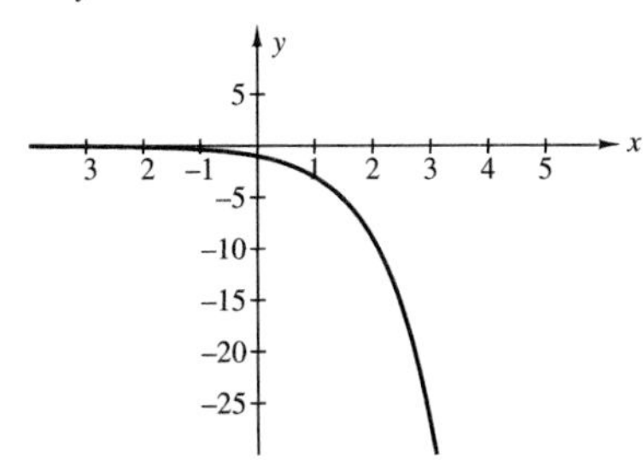

c. $y = 2^{x+1}$

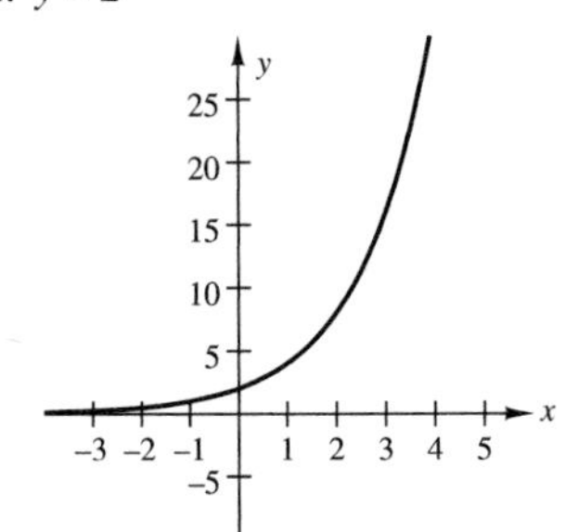

d. $y = 3 \cdot 5^x$

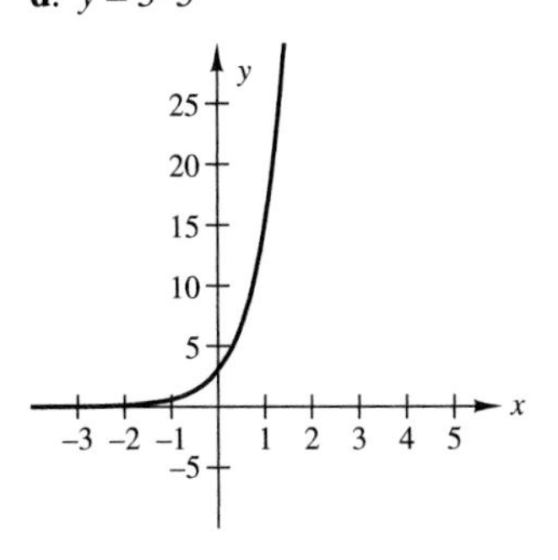

e. $y = 2 - 4^x$

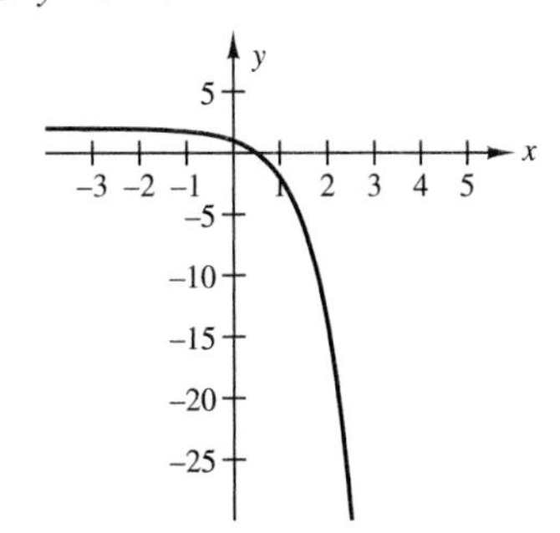

f. $y = \frac{2}{e^x} = 2e^{-x}$

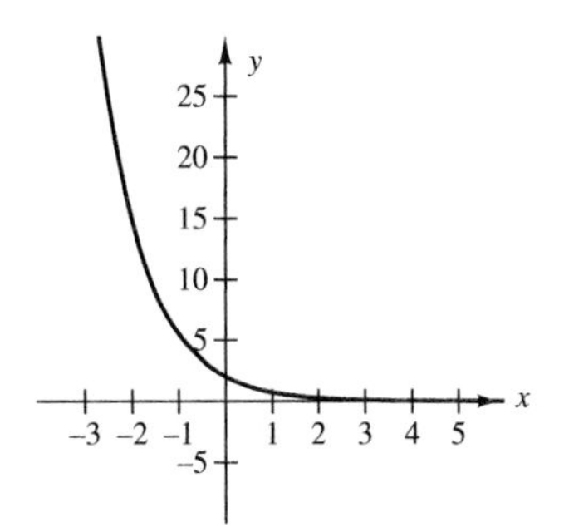

g. $y = 4 + e^x$

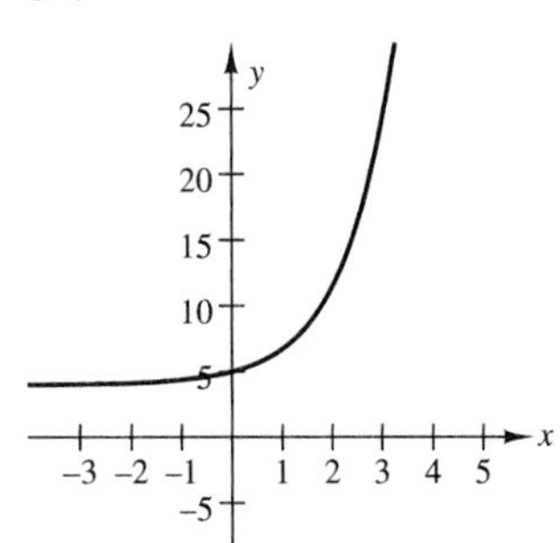

6. **a.** 4 **b.** 1 **c.** 2 **d.** 5 **e.** 6
   f. 2 **g.** 3 **h.** 1 **i.** 4
7. **a.** \$1,716.54 **b.** \$1,718.84 **c.** \$1,720.00
8. **a.** Not correct; $2^5 - 2^3 = 24$; $2^2 = 4$
   b. Correct; $\frac{a^x}{a^y} = a^{x-y}$ (rule 4)
   c. Not correct for $x < 0$; $\sqrt{x^2} = |x|$
   d. Not correct; $3^{-2} = \frac{1}{3^2} = \frac{1}{9}$; $\frac{1}{3^{1/2}} = \frac{1}{\sqrt{3}}$
   e. Correct; $2^{-3}3^{-2} = \frac{1}{2^3} \cdot \frac{1}{3^2}$ (rule 2).
   f. Not correct; $x^{-1} + y^{-1} = \frac{1}{x} + \frac{1}{y}$; $\frac{1}{x} + \frac{1}{y} \neq \frac{1}{x+y}$

Exercises 14-A

1. **a**. Table is not a function.
   b. Table is a function that is not one-to-one.
2. **a**. Not a function.
   b. A function that is one-to-one.
3. **a**. One person would be assigned two different Social Security numbers.
   b. Two different people would be assigned the same Social Security number.
4. **a**. f is one-to-one. **b**. g is one-to-one.
   c. h is one-to-one. **d**. F is not one-to-one.
   e. G is one-to-one. **f**. H is not one-to-one.

Exercises 14-B

1. **a**. f is one-to-one. **b**. f is not one-to-one.

x	$f^{-1}(x)$
−6	2
−13	0
$\frac{1}{3}$	−1
$\frac{1}{8}$	−24
$\frac{1}{16}$	−44

2. For the function given in part 1a,
   a. $\{2, 0, -1, -24, -44\}$ **b**. $\{-6, -13, \frac{1}{3}, \frac{1}{8}, \frac{1}{16}\}$
   c. $\{-6, -13, \frac{1}{3}, \frac{1}{8}, \frac{1}{16}\}$ **d**. $\{2, 0, -1, -24, -44\}$
3. **a**.

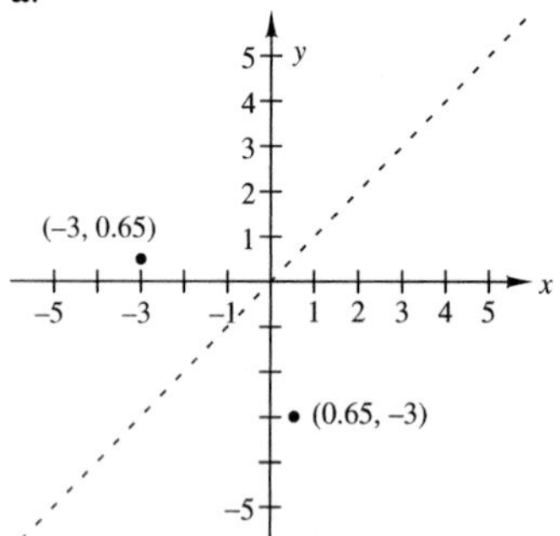

b.

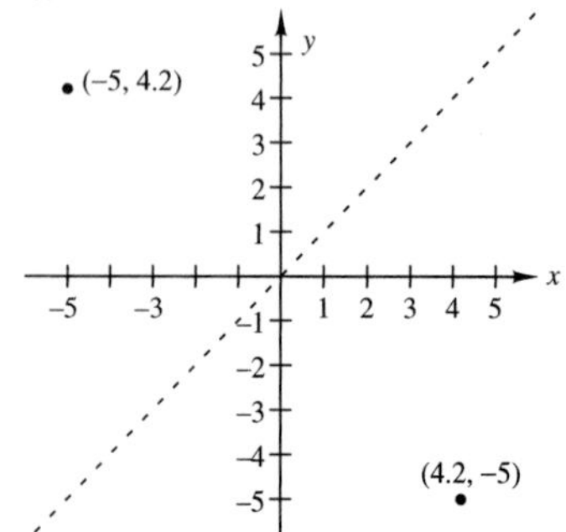

c.

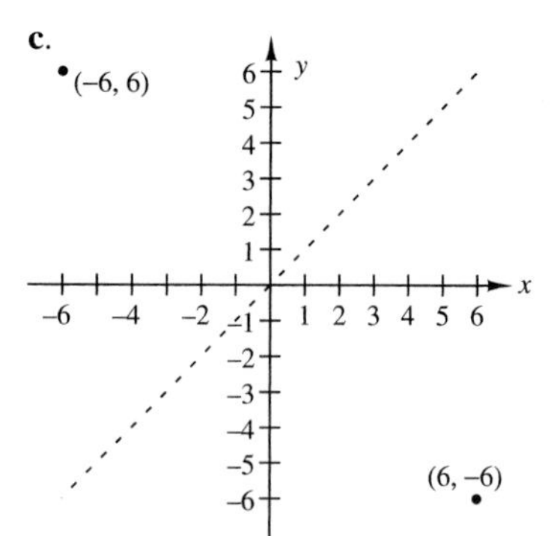

d.

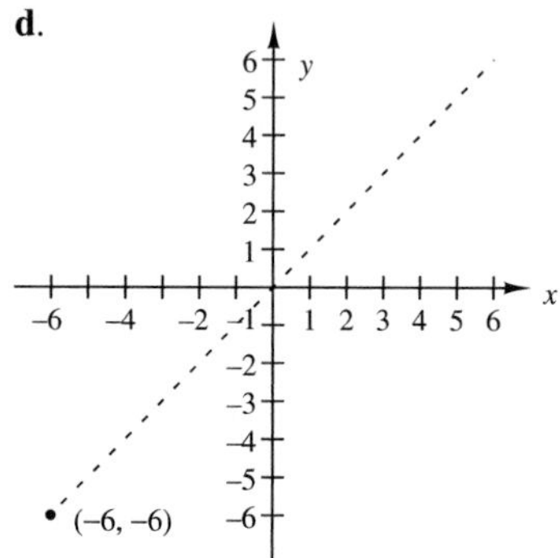

e.

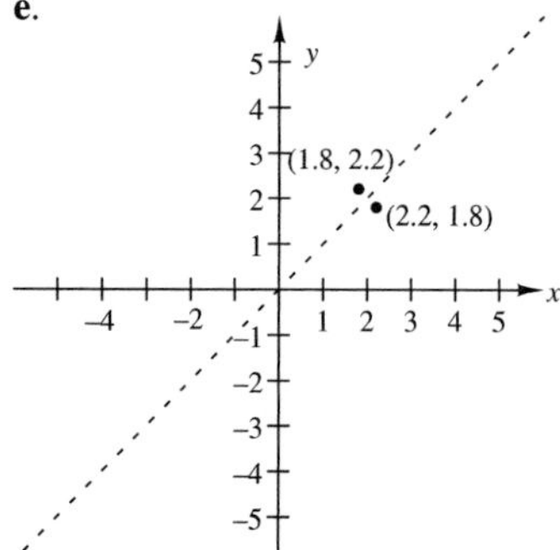

4. **a**. Not one-to-one since it fails the horizontal line test.
   b.

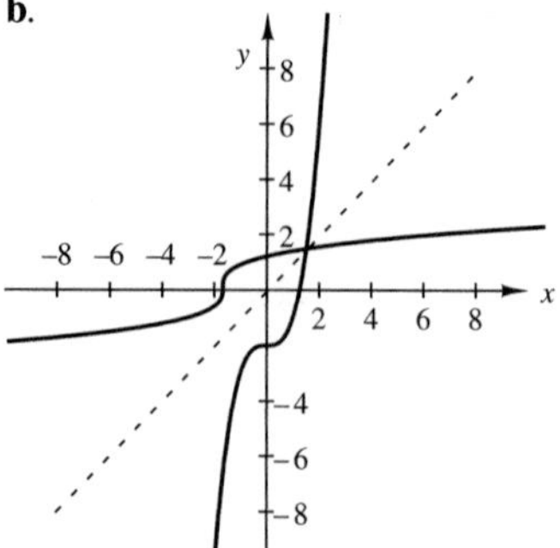

5. **a**. $f(x)$ and $g(x)$ are not inverses of each other.
   b. $f(x)$ and $g(x)$ are inverses of each other.
   c. $f(x)$ and $g(x)$ are inverses of each other.
6. **a**. (i) and (iii) are pairs of inverse functions because their graphs are reflections of each other across the line $y = x$.
   b. Confirm by showing that the composite of the functions in each pair is the identity function.

Exercises 14-C

1. **a**. $f^{-1}(x)=\frac{x-9}{2}$ **b**. g is not one-to-one.
   c. $h^{-1}(x)=\left(\frac{x-9}{2}\right)^{1/3}$ **d**. F is not one-to-one.
   e. $G^{-1}(x)=\left(\frac{x-9}{2}\right)^{2}$ **f**. $H^{-1}(x)=\left(\frac{x-9}{2}\right)^{3}$
3. **a**. and **d**.

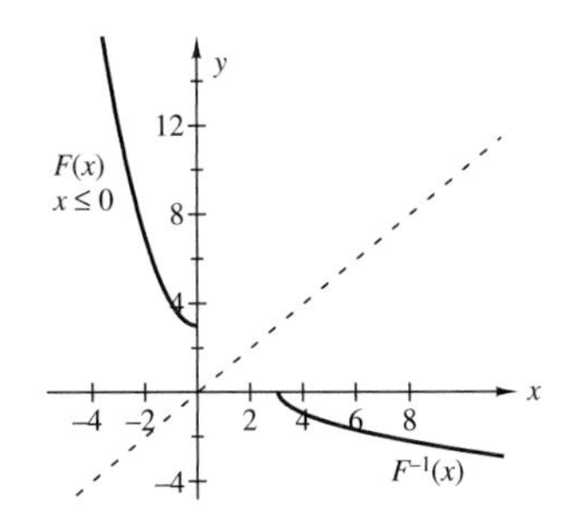

   c. $F^{-1}(x)=-\sqrt{x-3}$
4. **a**.

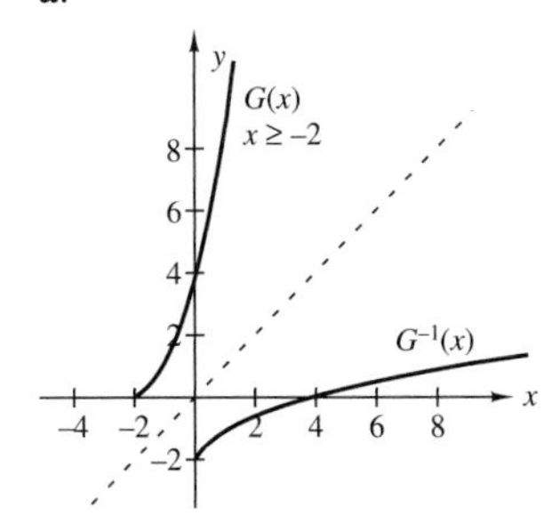

   c. $G^{-1}(x)=\sqrt{x}-2$. See graph in part a.
5. **a**. $-\frac{\pi}{3}$ **b**. $\frac{5\pi}{6}$ **c**. $\frac{\pi}{4}$ **d**. $-\frac{\pi}{3}$
   e. Not possible; $\cos x \le 1$ **f**. $-\frac{\pi}{6}$
6. **a**. $\frac{\sqrt{2}}{2}$ **b**. Not defined **c**. $\frac{\sqrt{2}}{2}$
   d. -2.4 **e**. $\frac{-\pi}{2}$ **f**. $\frac{\pi}{4}$
7. Argue from this picture:

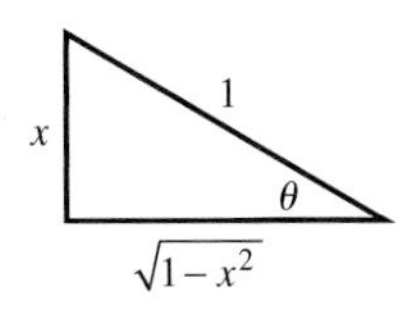

8. **a**. $\cos(\sin^{-1}(0.2)) \approx 0.9798$; $\sin\left(\cos^{-1}\left(-\frac{1}{3}\right)\right) \approx 0.9428$
   b. The results are the same.

Chapter 14 Exercises

1. $f(x)=x^2+3$; $g(x)=\frac{1}{2}x+3$
2. **a**. Not one-to-one **b**. One-to-one

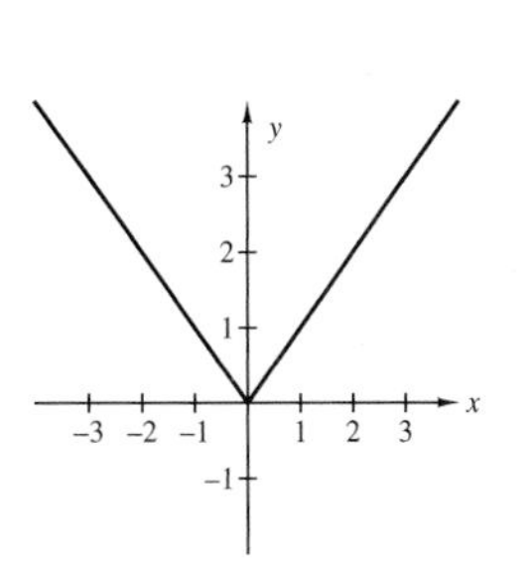

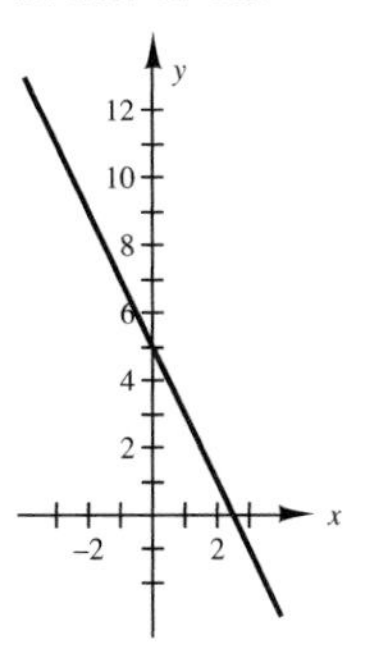

   c. One-to-one **d**. One-to-one

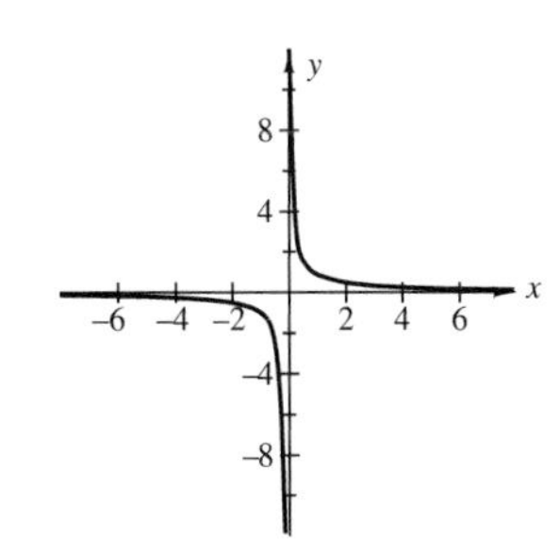

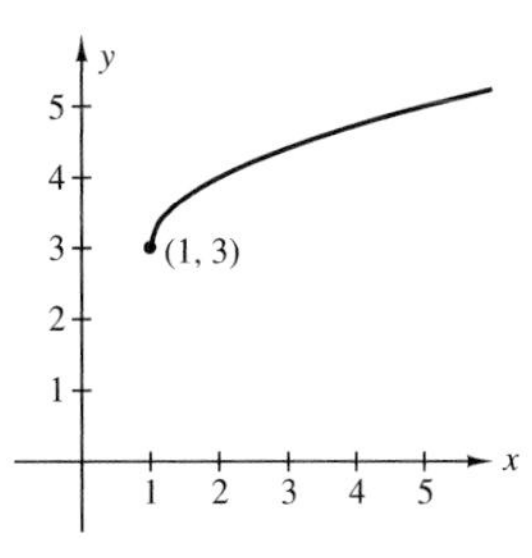

3. **b**. $g^{-1}(x)=\frac{5-x}{2}$ **c**. $h^{-1}(x)=\frac{1}{x}$

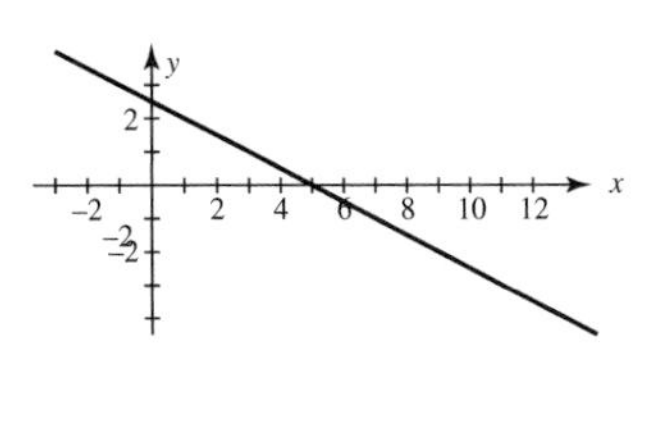

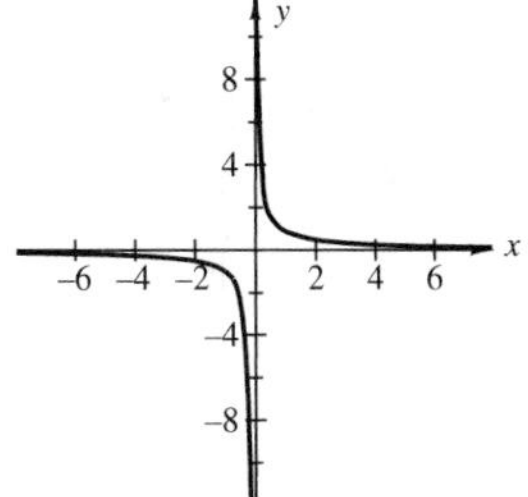

   d. $F^{-1}(x)=(x-3)^2+1$, restricted to domain $[3,\infty)$

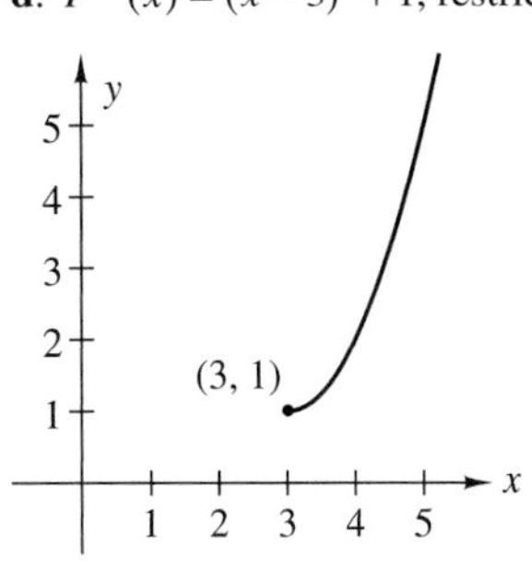

4. **a**. f and g are inverses of each other.
   b. f and g are not inverses of each other.
   c. f and g are inverses of each other.
   d. f and g are not inverses of each other.
5. $C(x)=F^{-1}(x)=\frac{5}{9}(x-32)$
6. **a**. $f(x)=\frac{x}{144}$; x in square inches, $f(x)$ in square feet
   b. $f^{-1}(x)=144x$; f^{-1} converts area in square feet to area in square inches.

7. **a.** −7 **b.** $\frac{1}{2}$ **c.** 344 **d.** $\sqrt[3]{6}$ **e.** 7 **f.** −2
8. **b.** Yes; f satisfies definition (2) for a one-to-one function.
9. **a.** One-to-one function

   b. Function not one-to-one

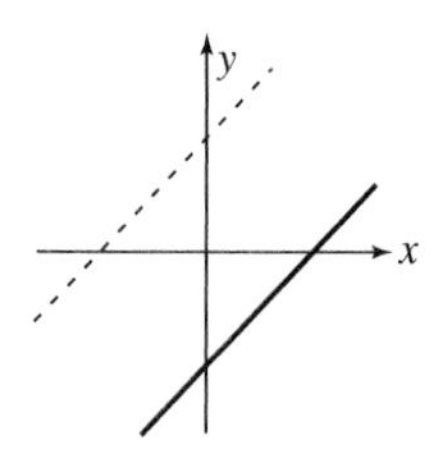

c. One-to-one function **d.** Not a function

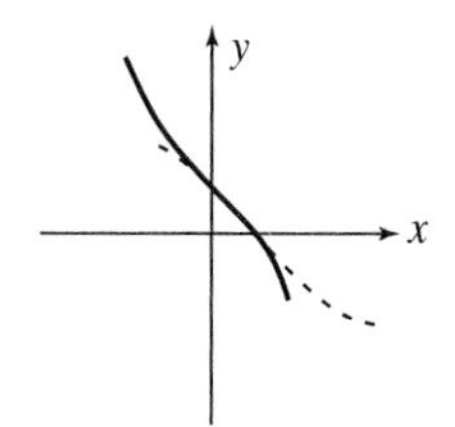

e. One-to-one function

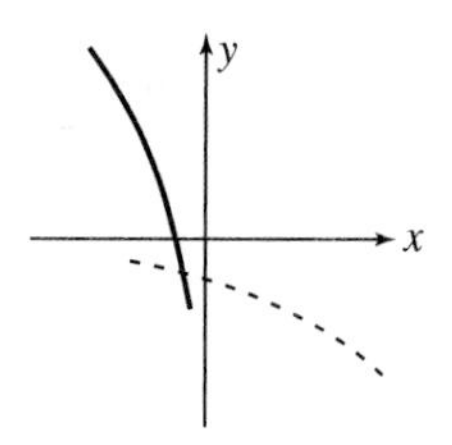

10. **a.** $\frac{\pi}{7}$ **b.** $\frac{\pi}{4}$ **c.** $-\frac{\pi}{4}$

    d. $\frac{1}{2}$ **e.** 0.47 **f.** Not possible
11. **a.** Argue from this picture:

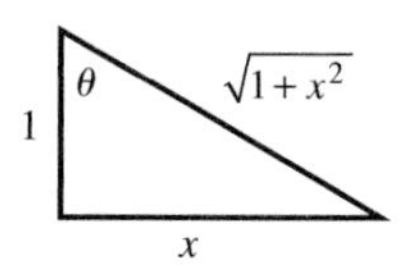

b. $\cos(\tan^{-1} 3) = \frac{1}{\sqrt{10}} = \frac{10}{\sqrt{10}} \approx 0.3162$; $\sin(\tan^{-1}(-2)) = \frac{-2}{\sqrt{5}} = \frac{-2\sqrt{5}}{5} \approx -0.8944$

c. The results are the same.

Exercises 15-A

1. Statements will vary slightly; we give a statement for b.

   b. The logarithm to the base a of x to the rth power is r times the logarithm to the base a of x.
2. **a.** 3 **b.** 2 **c.** −2 **d.** $\frac{1}{5}$
3. **a.** $x + 1$, for $x > -1$ **b.** xe^x **c.** $3x(x + 1)$

   d. x^4, for $x > 0$ **e.** $(x-2)^3$, for $x > 2$

   f. $\tan x$, for x in intervals of the form $(0 + 2k\pi, \pi/2 + 2k\pi)$, k an integer.
4. All expressions in parts a–d are defined for $x > 0$.

   a. Incorrect; for $x = 1$, $\ln(1) + 1 = 0 + 1 = 1$, while $\ln(1 + 1) = \ln 2 \approx 0.69$.

   b. Incorrect; for $x = 2$, $(\ln 2)\, 2^2 = 4 \ln 2 \approx 2.77$, while $\ln(2^3) = \ln 8 = 3 \ln 2 \approx 2.08$.

   c. Correct

   d. Incorrect; for $x = 2$, $\log_2 2 + \log_5 2 \approx 1.43$, while $\log 2 \approx 0.3$.

   e. Incorrect; for $x = 10$, $(\log 10)^3 = 1^3 = 1$, while $3 \log 10 = 3$.

   f. Incorrect; for $x = 5$, $\frac{\log_5 5}{\log_5 25} = \frac{1}{2}$, while $\log_5\left(\frac{5}{25}\right) = \log_5\left(\frac{1}{5}\right) = -1$.
5. **a.** $\ln(x+1) + \ln(x-1) - \ln(x+3)$

   b. $\log_2 4 + \frac{1}{2}\log_2(x+1) = 2 + \frac{1}{2}\log_2(x+1)$

   c. $2 - 4\ln x$

   d. $3\log_5 5(x^2 + 1) = 3 + 3\log_5 (x^2 + 1)$

   e. $2\log_3 x + \log_3(x + 1)$

   f. $x - 1 + \ln(x - 1)$
6. **a.** $\ln\left(\frac{x^2(x-1)^5}{x+2}\right)$ **b.** Not possible

   c. $\log_9\left(\frac{\sqrt{t}}{2}\right)$ **d.** Not possible

   e. $\log_2\left(\frac{x^4}{x+1}\right)$ **f.** $\ln(\cot x)$
7. **a.** $F(x) = \ln(e^{(2\cos(x))}) = 2\cos(x)$.

   b. $K(x) = \log_7((\tan(x))^4) = 4\log_7(\tan(x))$. Use the Chain Rule to differentiate.

   c. $H(x) = \ln((\sin(4x))^{1/2}) = \frac{1}{2}\ln(\sin(4x))$. Use the Chain Rule (twice) to differentiate.

   d. $G(x) = (\ln(x^4))(e^{-x}) = (4\ln(x))(e^{-x})$. Differentiate G with the Product Rule, and use the Chain Rule to differentiate the factor e^{-x}.

   e. $N(x) = \frac{4^{(5x-2)}}{\ln(x^5)} = \frac{4^{(5x-2)}}{5\ln x}$. Differentiate N with the Quotient Rule, and use the Chain Rule to differentiate the numerator.

Exercises 15-B

1. **a.** $(-\infty, 0) \cup (0, \infty)$ **b.** $(0, \infty)$

   c. $(-\infty, 0)$ **d.** $\left(\frac{2}{5}, \infty\right)$

   e. $(-\infty, -6) \cup (3, \infty)$
2. **a.** −2, 0, undefined, 1, undefined, $-\frac{1}{2}$

b.

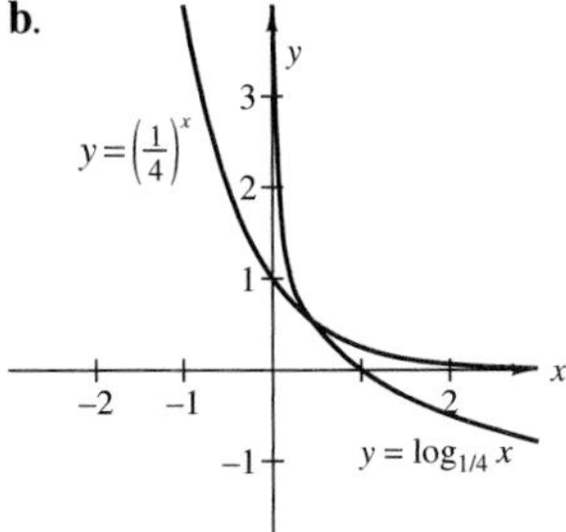

domain of f^{-1}: $(0, \infty)$
range of f^{-1}: $(-\infty, \infty)$

c.

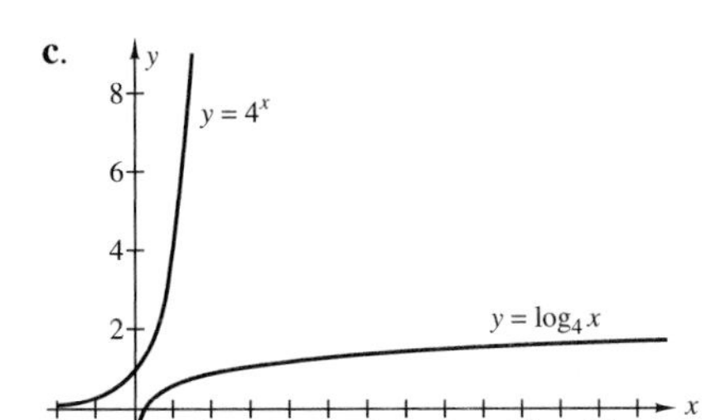

domain of g^{-1}: $(0, \infty)$
range of g^{-1}: $(-\infty, \infty)$

3. **a.**

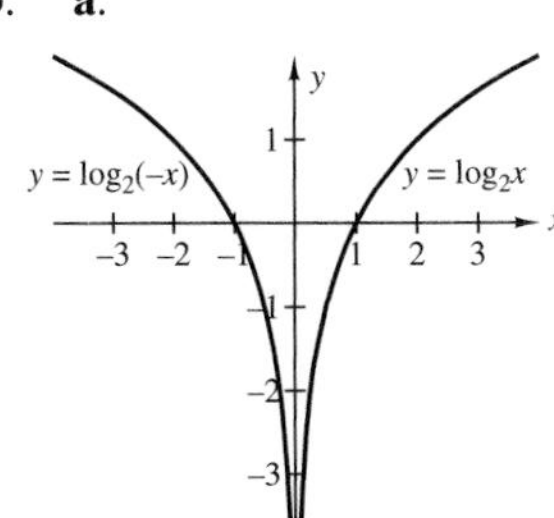

b.

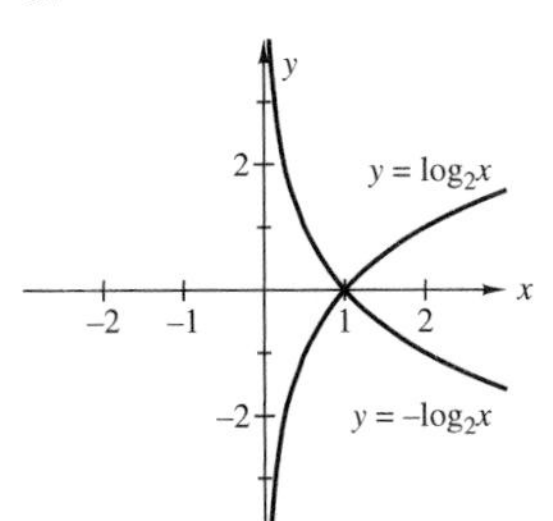

4. **a.** $y = 2^{\log_2 x} = x, x > 0$ **b.** $y = \log(x + 3)$

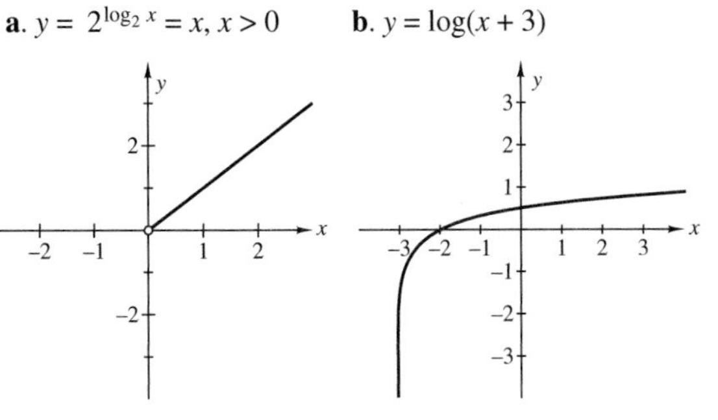

c. $y = \log x + 3$ **d.** $y = 2 - \ln x$

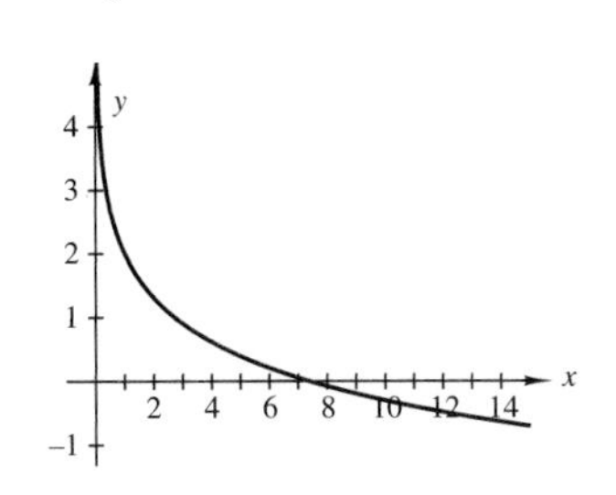

e. $y = \log_5 5^x = x$

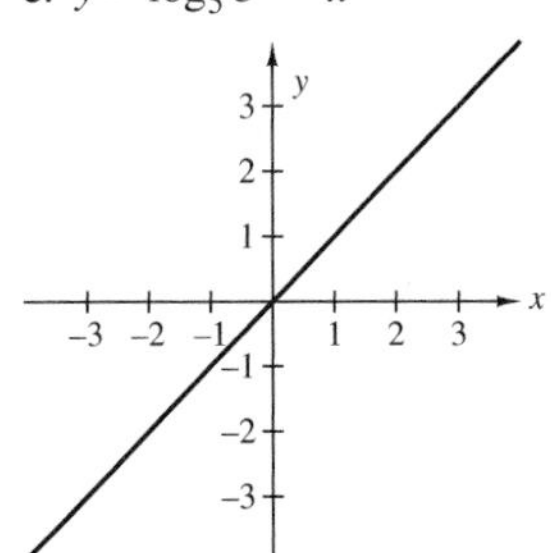

5. **a.**

x	$f(x) = \sqrt{x}$	$g(x) = \ln x$
0.5	0.70711	−0.69315
1	1	0
5	2.23607	1.60944
10	3.16228	2.30259
25	5	3.21888
50	7.07107	3.91202
100	10	4.60517
1,000	31.62278	6.90776
10,000	100	9.21034

b.

c. Values of $f(x) = \sqrt{x}$ grow much faster than those of $g(x) = \ln x$.

6. **a.**

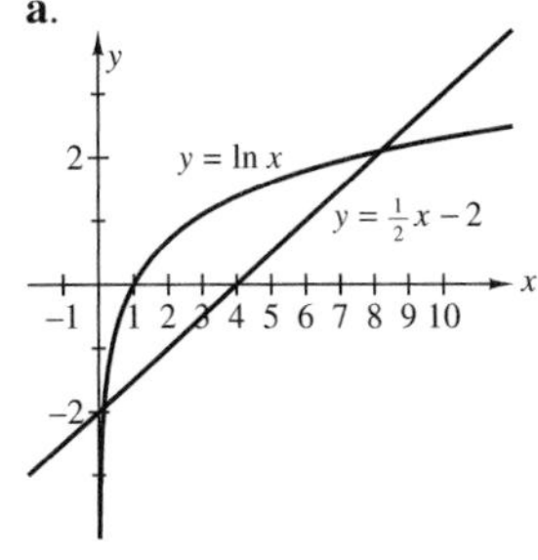

b. 2 **c.** 2

Exercises 15-C

1. **a.** $x = 35$ **b.** $x = \pm\frac{1}{\sqrt{3}}$ **c.** $t = 1, t = -4$

d. $t = 1$ **e.** $x = \frac{1}{\sqrt{3}}$ **f.** $z = \frac{1}{3}\ln 2$

g. $x = -\frac{1}{2}\ln 5$ **h.** $x = 13$

i. $x = \frac{3\ln 3 + \ln 10}{\ln 3 - 4\ln 10} \approx -0.69016$ or $x = \frac{3\log 3 + 1}{\log 3 - 4}$

j. $x = 4\ln 2 \approx 2.77$ or $x = \frac{4}{\log_2 e}$

2. **a.** 0.03 g/liter **b.** 0.55 miles or more
3. $\approx$ 22 years
4. **a.** $f(t) = 10\, e^{-0.000121t}$ **b.** Approximately 2378 years old

Chapter 15 Exercises

1. **a.** 3 **b.** 2 **c.** –3 **d.** x **e.** –3 **f.** 4

2. **a.** $y = \ln(1-x) - \ln(1+x)$
b. $y = \log_2(2x+1) + \log_2(2x+1)$
c. $y = 2\ln(x+2) - 3\ln(x+1)$
d. $y = \ln(x+4) + \ln(x-1)$
e. $y = \frac{1}{2}\log x + \log(2x+1)$

3. **a.** $\log_8 x^2(x-1)^3$ **b.** $\ln\frac{2\sqrt{x+2}}{(x+1)^3}$
c. $\log\frac{2x+3}{(x+3)\cdot 25x^2}$

4. **a.** $y = -\ln x$ **b.** $y = \ln(-x)$

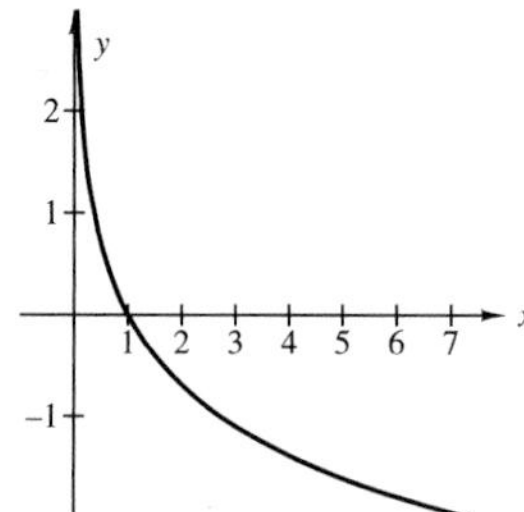

Domain: $(0, \infty)$; asymptote: $x = 0$

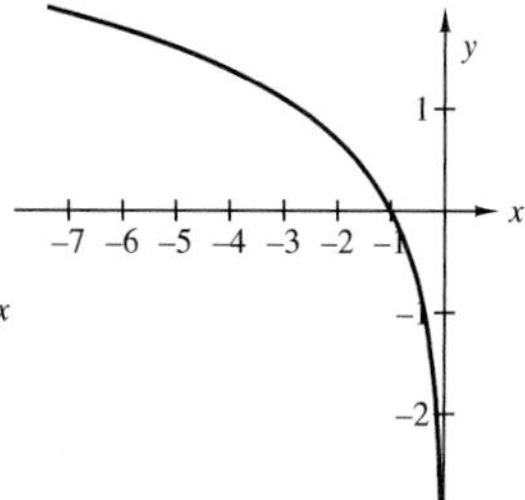

Domain: $(-\infty, 0)$; asymptote: $x = 0$

c. $y = \log_2(x+3)$

d. $y = \log_3(1-x)$

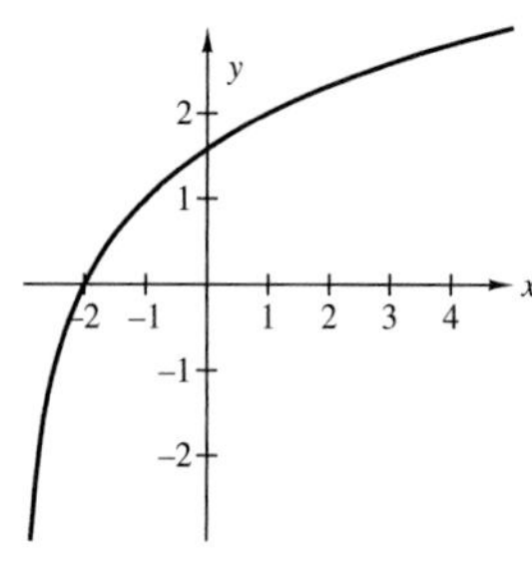

Domain: $(-3, \infty)$; asymptote: $x = -3$

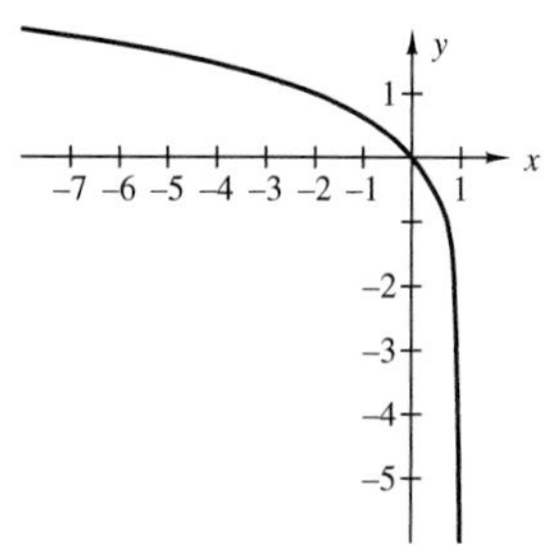

Domain: $(-\infty, 1)$; asymptote: $x = 1$

e. $y = \log x - 2$

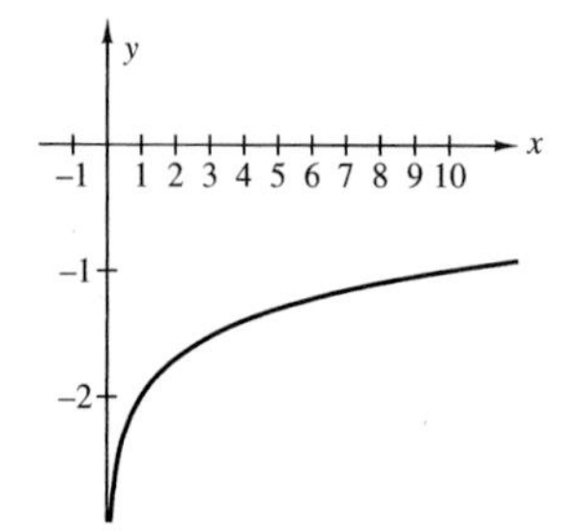

Domain: $(0, \infty)$; asymptote: $x = 0$

5. **a.** $(-\infty, 0)$ **b.** $(-\sqrt{2}, \sqrt{2})$
c. $(-5, 3)$ **d.** $(-1, 2) \cup (2, \infty)$

7. **a.** $t = \frac{1}{3}\ln 100$ **b.** $x = \frac{1}{7}$ **c.** $x = \frac{3\ln 2 + \ln 7}{2\ln 7 - \ln 2}$
d. $t = 1$ **e.** $z = \frac{e^3+2}{5}$ **f.** $x = 4$

8. **a.** Approximately 6640
b. $t = 7.68$, or approximately 3:41 P.M.

9. 6.9%

10. **a.** \$168.75 **b.** 1.39 years

11. **a.**

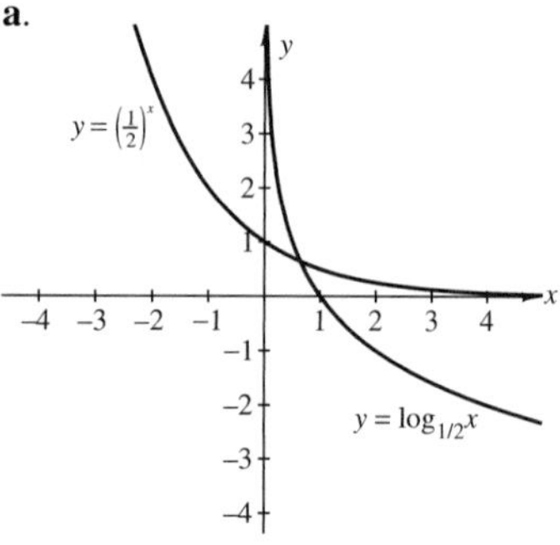

b.

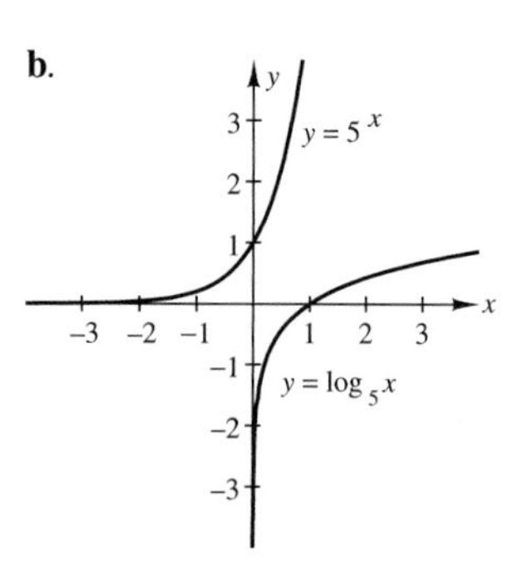

12. **a.** $F(x) = e^{(\sin(x) + \ln(x))} = xe^{\sin(x)}$; differentiate with the Product Rule and use the Chain Rule on the factor $e^{\sin(x)}$.

b. $G(x) = \ln\left(\frac{x^6}{e^x}\right) = 6\ln(x) - x$; differentiate with the Difference Rule.

c. $K(x) = \log_3(\cot(x))^2 = 2\log_3(\cot(x))$; differentiate with the Chain Rule.

d. $M(x) = e^{(2\ln(x))} = x^2$

e. $N(x) = \frac{3^{(-x-2)}}{\ln(x^4)} = \frac{3^{(-x-2)}}{4\ln(x)}$; differentiate with the Quotient Rule and use the Chain Rule on the numerator.

Another way: $N(x) = \frac{3^{(-x-2)}}{4\ln(x)} = \frac{1}{(4\ln(x))(3^{(x+2)})} =$ $[(4\ln(x))(3^{(x+2)})]^{-1} = [(4\ln(x))(9\cdot 3^x)]^{-1}$; differentiate with the Chain Rule and then the Product Rule.

13. **a.** $y = e^{-x} + 1$

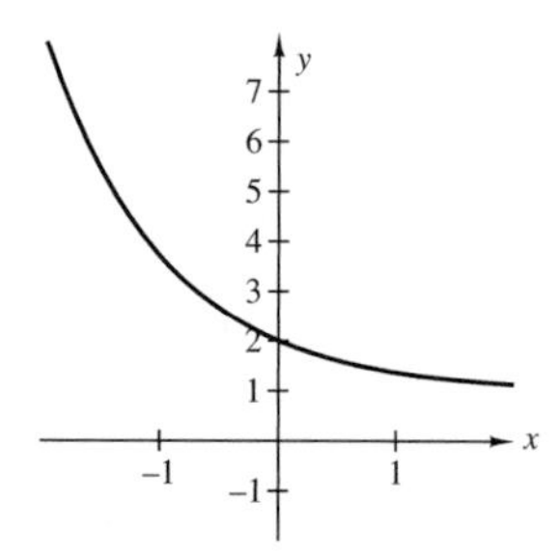

b. The graph passes the horizontal line test.

c. $f^{-1}(x) = -\ln(x-1)$

d.

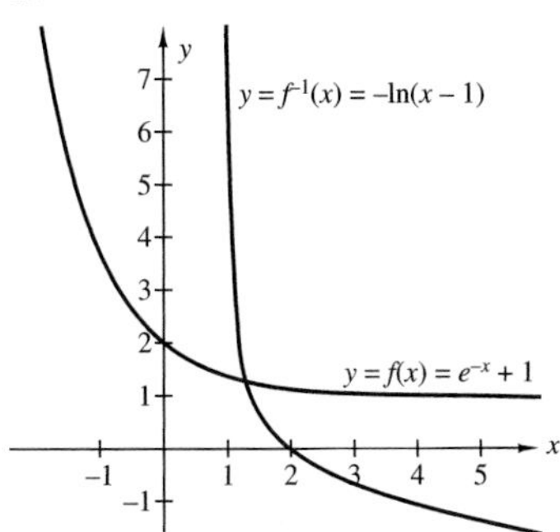

14. **a.**

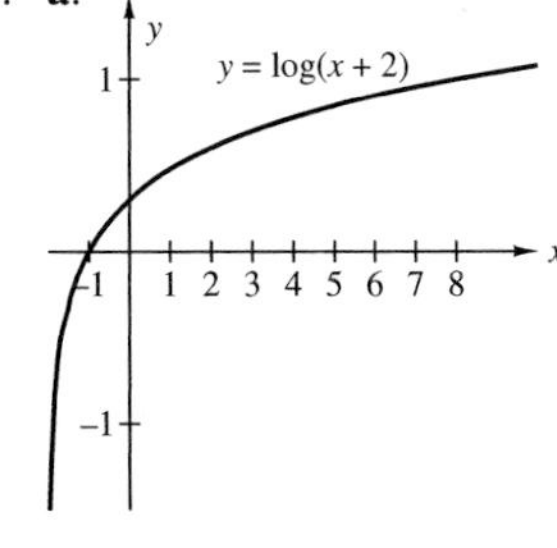

b. The graph passes the horizontal line test.

c. $g^{-1}(x) = 10^x - 2$

d.

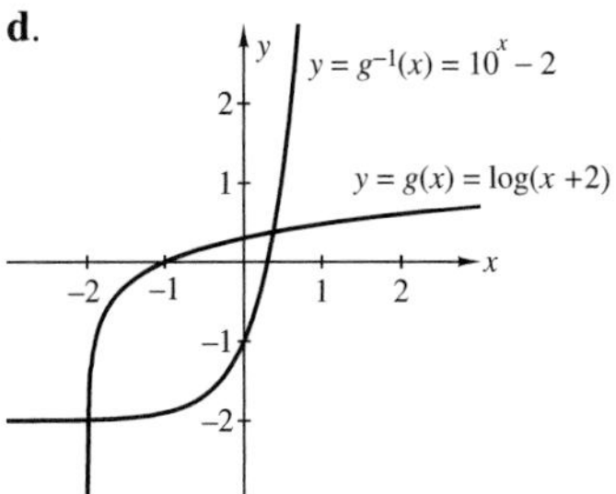

15. **a.** 5 **b.** 3 **c.** 4 **d.** 6 **e.** 1 **f.** 2

16. **a.** $f(t) = 8e^{-0.000121t}$

b. approximately 11,457 years old

Exercises 16-A

1. **a.** No minimum or maximum
   b. No minimum; maximum = 3 = $f(0)$
   c. Minimum = 2 = $f(-2)$; no maximum
   d. Minimum = 2 = $f(-2) = f(2)$; maximum = 3 = $f(0)$
2. **a.** No maximum or minimum
   b. Minimum ≈ -1 at $x \approx -0.8$; maximum ≈ 5 at $x \approx -3.25$
   c. Minimum ≈ -1 at $x \approx -0.8$; maximum ≈ 5 at $x \approx -3.25$
   d. Minimum ≈ -1 at $x \approx -0.8$; maximum ≈ 7.75 at $x \approx 0.5$
3. **a.** 5 at $x \approx -3.25$ and 7.75 at $x \approx 0.5$
   b. -1 at $x \approx -0.8$ and 3.5 at $x \approx 1.5$
4. **a.** No absolute maximum or absolute minimum
   b. Absolute minimum = $G(-0.8) \approx 1.5$; absolute maximum = $G(0) = 2$
   c. No absolute minimum; absolute maximum = 2 = $G(0)$
   d. No absolute minimum or absolute maximum
5. Relative maximum at $x = 0$
6. **a.** No extreme values **b.** No extreme values

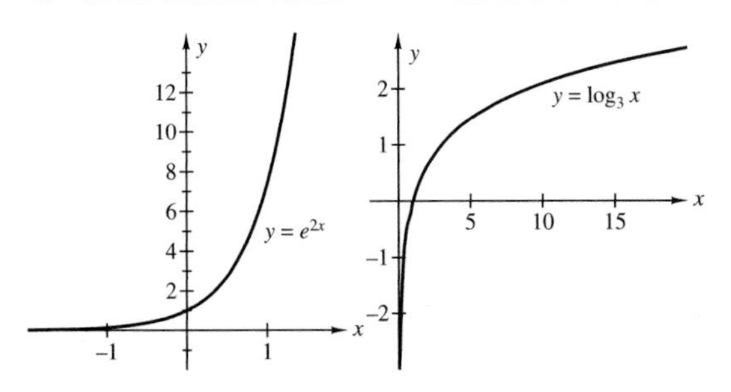

c. Absolute minimum of $y = 2$ at $x = 1$

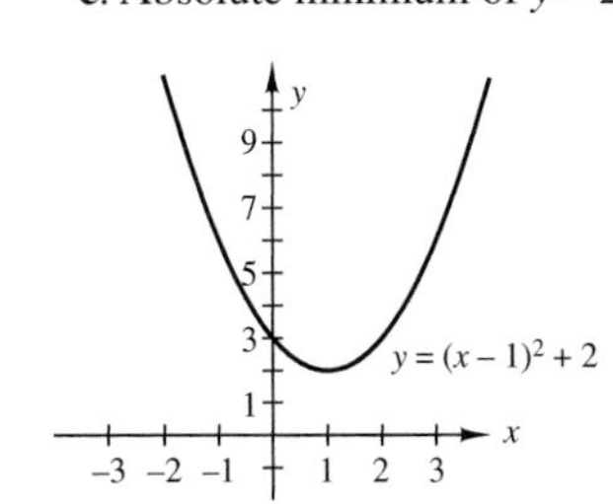

7. **a.** $-\frac{1}{3}, 1$ **b.** None **c.** 0, 2 **d.** $-\sqrt{3}, 0, \sqrt{3}$
   e. $-3, -2$ **f.** 0, 3

Exercises 16-B

1. $f(x) = 7.2x + \frac{432}{x}$ (x is the length of the side along the neighbor's lot), $0 < x < 40$
2. $P(x) = 14,000 + 416x - 3x^2$ (where x dollars is the price increase), $0 \le x < \frac{500}{3}$; or $P(p) = -13,587 + 710p - 3p^2$ (where p is the selling price), $49 \le p < \frac{647}{3}$
3. $C(x) = 110x + 390\sqrt{250,000 + (750 - x)^2}$ (x is the length of underground pipe), $0 \le x \le 750$
4. $C(r) = q\left(\frac{24}{r} + 4\pi r^2\right)$ (r is the radius of the can and q is the cost per square inch of constructing the cardboard side) ; $0 < r$
5. $V(x) = (11 - 2x)(8.5 - 2x)x$ (x is the side length of each cut out square); $0 < x < 4.25$

6. 1. $y = \frac{432}{x} + 7.2x$; minimum $\approx 7.7 \times 5.8$ yd

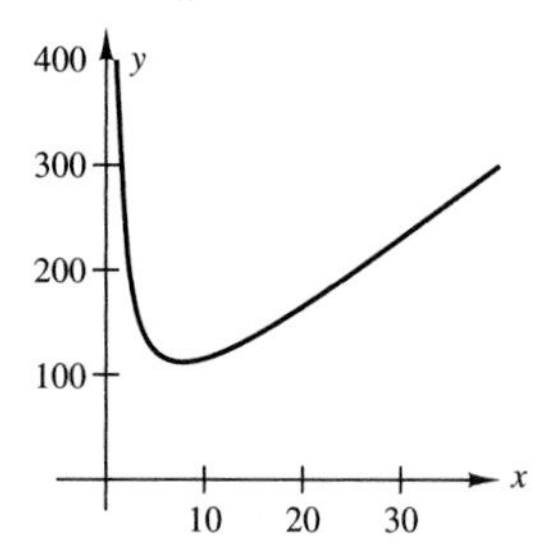

2. $y = 14{,}000 + 416x - 3x^2$; maximum selling price $\approx$ $69 + 49 = \$118.00$

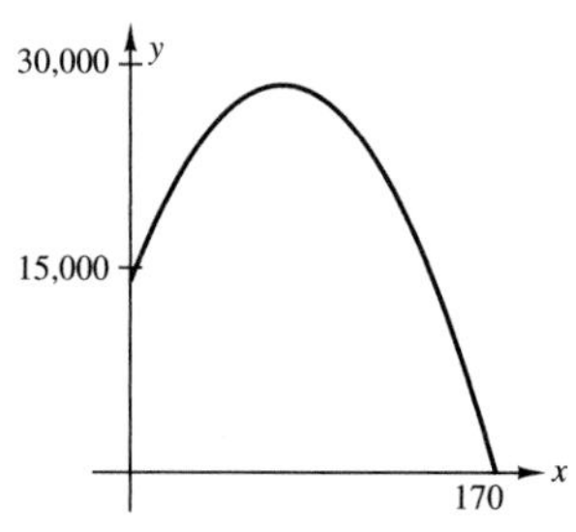

3. $y = 110x + 390\sqrt{250{,}000 + (750 - x)^2}$; minimum when underground pipe is $\approx$ 600 yd

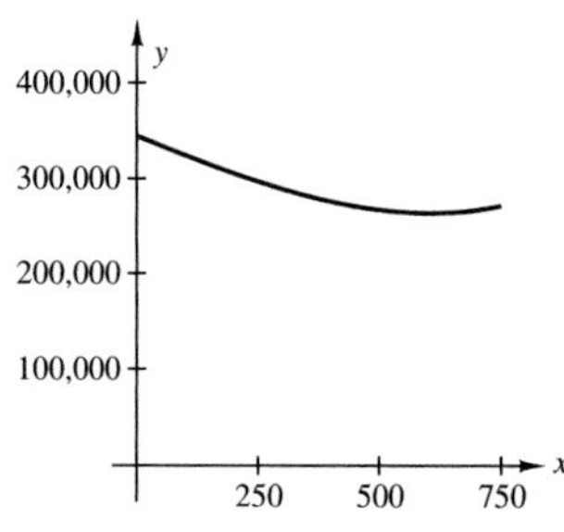

4. $y = \frac{24}{r} + 4\pi r^2$; minimum when radius $r \approx 1$ in. and height ≈ 3.8 in.

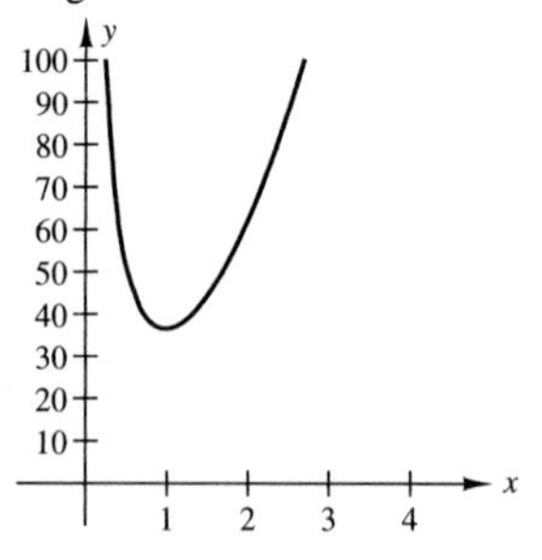

5. $y = (11 - 2x)(8.5 - 2x)x$; maximum when $x \approx 1.6$ in.

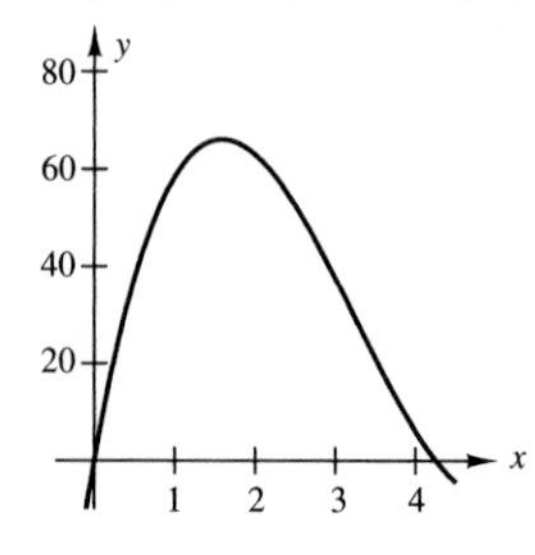

7. **a.** $S(h) = 2\pi\sqrt{\frac{21.64}{\pi h}} \cdot h + \frac{43.28}{h} = 2\sqrt{21.64\pi h} + \frac{43.28}{h}$

Chapter 16 Exercises

1. **a.** $0, \pm\frac{\pi}{2}, \pm\pi, \pm\frac{3\pi}{2}, \ldots$ **b.** 0 **c.** $\frac{1}{e}$ **d.** $0, 4, \frac{24}{7}$

2. **a.** (i) The population will be largest approximately 8.5 years from now.

(ii) The population will be growing most rapidly approximately 3.5 years from now.

b. $\frac{20 + \sqrt{940}}{6} \approx 8.44$

3. **a.** $t = \frac{17}{6}$

b. At about 9:50 A.M. (approximately $2\frac{5}{6}$ hr after 7 A.M.)

c. At 7 A.M.

d.

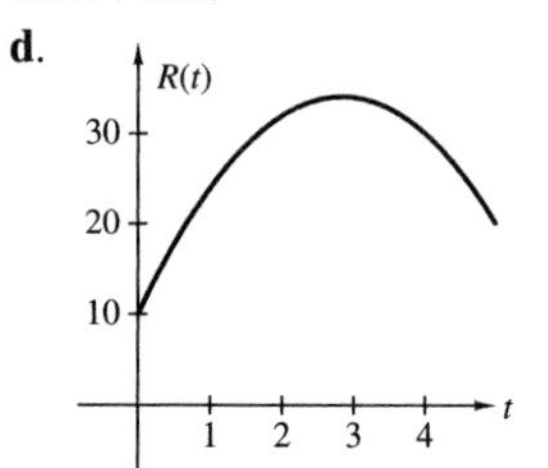

4. **a.** Absolute minimum $\approx -3 \approx g(-3.2)$; no absolute maximum

b. No absolute minimum; absolute maximum $\approx 3 \approx g(-0.8)$

c. Absolute minimum $= 0 = g(-2)$; absolute maximum $\approx 3 \approx g(-0.8)$

d. Absolute minimum $= g\left(\frac{1}{2}\right) \approx -5$; absolute maximum $\approx 3 \approx g(-0.8)$

5. Relative minimum ≈ -3, at $x \approx -3.2$; relative maximum ≈ 3, at $x \approx -0.8$

6. **a.** March through July **b.** March; 350,000 new jobs

c. A minimum of 80,000 new jobs in February and in July

7. **a.** 1903, 1907, 1910, 1914 **b.** 1918; yes

8. **a.** Here is one example: **b.** Here is one example:

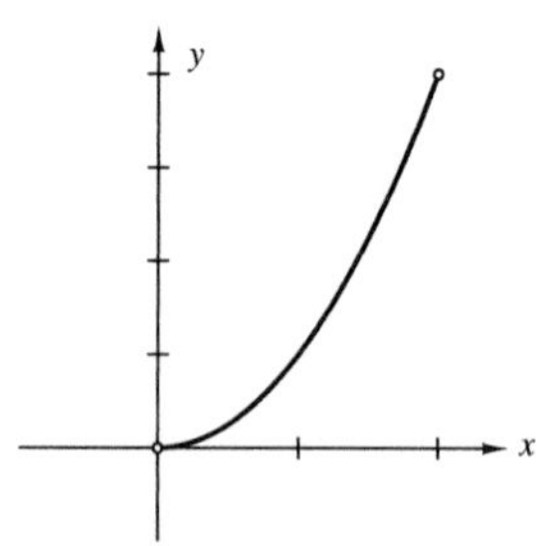

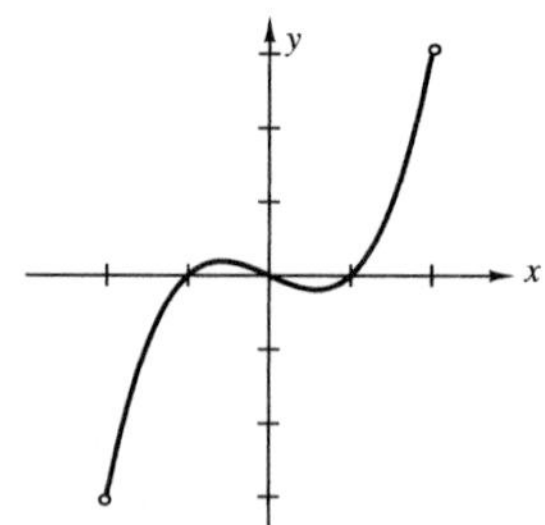

c. Here is one example:

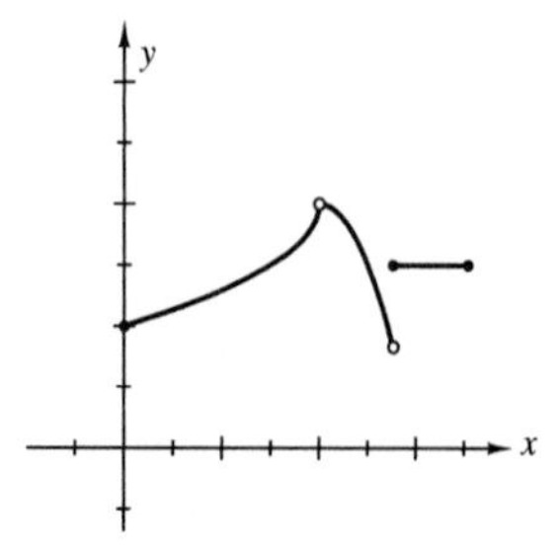

9. $f(x) = x^2 + \frac{20}{x}, x > 0$

10. $P(x) = 4400 + 148x - 40x^2, 0 \le x \le 12$

Exercises 17-A

1.

	Critical values	Derivative positive	Derivative negative
a.	$-\frac{1}{3}$, 1	$\left(-\infty, -\frac{1}{3}\right) \cup (1, \infty)$	$\left(-\frac{1}{3}, 1\right)$
b.	None	$(-\infty, \infty)$	Nowhere
c.	2, 0	$(-\infty, 0) \cup (2, \infty)$	$(0, 2)$
d.	0	$(0, \infty)$	$(-\infty, 0)$
e.	0, 3	$(-\infty, 0) \cup (0, 3)$	$(3, \infty)$

2.

	Second derivative positive	Second derivative negative
a.	$\left(\frac{1}{3}, \infty\right)$	$\left(-\infty, \frac{1}{3}\right)$
b.	$(0, \infty)$	$(-\infty, 0)$
c.	$(-1, 0) \cup (0, \infty)$	$(-\infty, -1)$
d.	$\left(-\frac{1}{\sqrt{3}}, \frac{1}{\sqrt{3}}\right)$	$\left(-\infty, -\frac{1}{\sqrt{3}}\right) \cup \left(\frac{1}{\sqrt{3}}, \infty\right)$
e.	$\left(0, 3-\sqrt{3}\right) \cup \left(3+\sqrt{3}, \infty\right)$	$(-\infty, 0) \cup \left(3-\sqrt{3}, 3+\sqrt{3}\right)$

3. **a.** $(-\infty, \infty)$ **b.** $f'(x) = \frac{9x+5}{(x+1)^{1/5}}$ **c.** $x = -\frac{5}{9}$
   d. $x = -1$ **e.** $(-\infty, -1) \cup \left(-\frac{5}{9}, \infty\right)$ **f.** $\left(-1, -\frac{5}{9}\right)$

4. **a.** $(-\infty, 4) \cup (4, \infty)$ **b.** $f'(x) = \frac{x^2-8x-20}{(x-4)^2}$
   c. $x = 10$; $x = -2$
   d. None, since $x = 4$ is not in domain.
   e. $(-\infty, -2) \cup (10, \infty)$ **f.** $(-2, 4) \cup (4, 10)$

5. **a.** $(-\infty, \infty)$ **c.** $x = -2$ **d.** No values of x
   e. $(-\infty, -2) \cup (-2, \infty)$ **f.** No values of x

Exercises 17-B

1. **a.** $x = -1$ **b.** $x = 1$ **c.** $(-1, 1)$ **d.** $(-4, -1) \cup (1, 6)$
   e. $(-4, 1) \cup (1, 6)$ **f.** Nowhere

2. **a.** $f(x)$ **b.** $g(x)$

   c. $h(x)$ **d.** $k(x)$

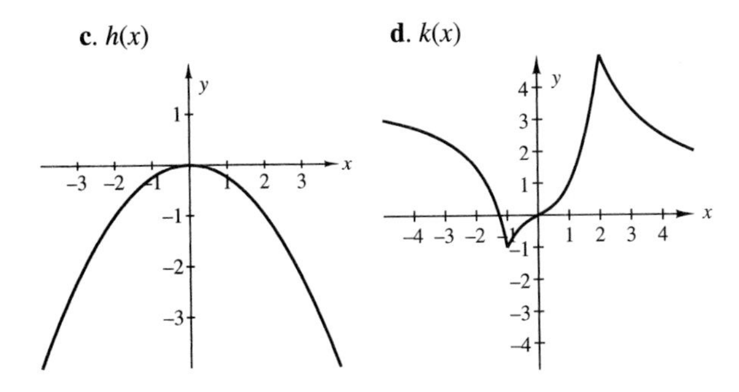

3. **a.**

	f'	f''	Summary
$-4 < x < -1$	+	+	f is increasing and concave up
$x = -1$	+		(inflection point)
$-1 < x < 0$	+	−	f is increasing and concave down
$x = 0$	0	−	(relative maximum)
$0 < x < 1$	−	−	f is decreasing and concave down
$x = 1$	−		(inflection point)
$1 < x < 4$	−	+	f is decreasing and concave up

b.

	f'	f''	Summary
$-4 < x < -2$	+	+	f is increasing and concave up
$x = -2$	d.n.e.	d.n.e.	
$-2 < x < 0$	+	−	f is increasing and concave down
$x = 0$	0		(relative maximum)
$0 < x < 2$	−	−	f is decreasing and concave down
$x = 2$	d.n.e.	d.n.e.	
$2 < x < 4$	−	+	f is decreasing and concave up

4. Many graphs are possible. Answers to all parts except d depend on your graph.
   d. Yes, here is one graph:

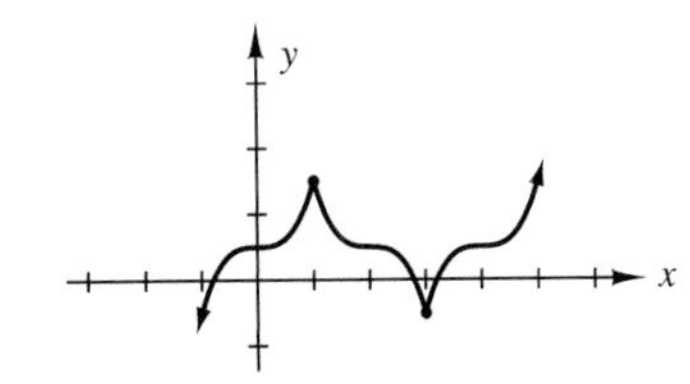

5. **a.** $x \approx -0.6$, $x \approx 0.6$
   b. $x \approx -1.1$, $x \approx -0.15$
   c.

	f'	f''
$x < -1.1$	+	+
$x = -1.1$	+	0
$-1.1 < x < -0.6$	+	−
$x = -0.6$	0	−
$-0.6 < x < -0.15$	−	−
$x = -0.15$	−	0
$-0.15 < x < 0.6$	−	+
$x = 0.6$	0	+
$x > 0.6$	+	+

   d.

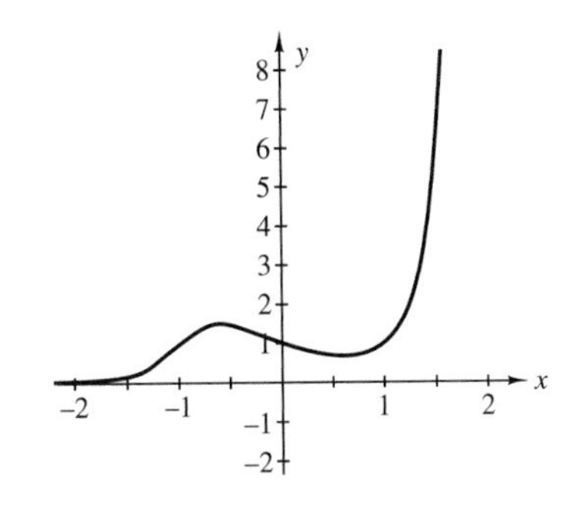

Exercises 17-C

1. Many graphs are possible. Here is one example:

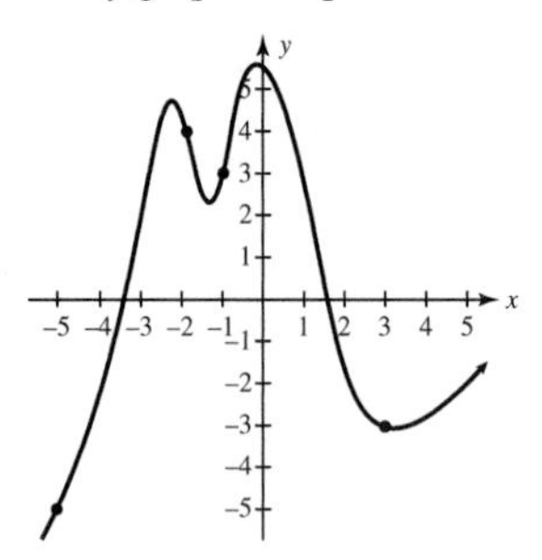

2.

	f	f'	f''	Summary
$x<-2$		+	−	f increasing and concave down
$x=-2$	0	0	0	(inflection point)
$x>-2$		+	+	f increasing and concave up

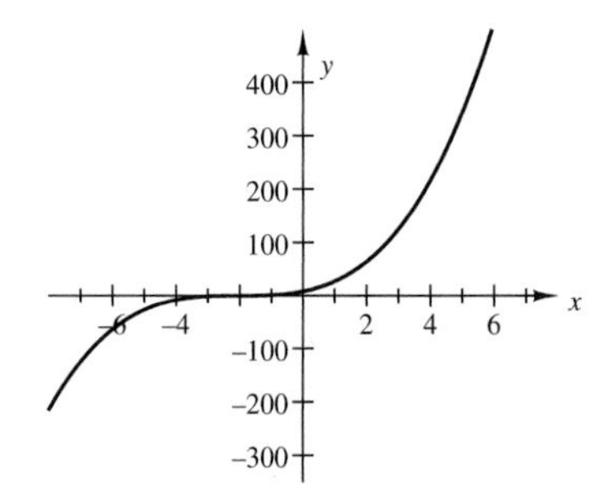

3.

	f	f'	f''	Summary
$x<0$		−	+	f decreasing and concave up
$x=0$	0	0	+	(relative minimum)
$0<x<\frac{1}{2}-\frac{\sqrt{3}}{6}$		+	+	f increasing and concave up
$x=\frac{1}{2}-\frac{\sqrt{3}}{6}$		+	0	(inflection point)
$\frac{1}{2}-\frac{\sqrt{3}}{6}<x<\frac{1}{2}$		+	−	f increasing and concave down
$x=\frac{1}{2}$	$\frac{1}{16}$	0	−	(relative maximum)
$\frac{1}{2}<x<\frac{1}{2}+\frac{\sqrt{3}}{6}$		−	−	f decreasing and concave down
$x=\frac{1}{2}+\frac{\sqrt{3}}{6}$		−	0	(inflection point)
$\frac{1}{2}+\frac{\sqrt{3}}{6}<x<1$		−	+	f decreasing and concave up
$x=1$	0	0	+	(relative minimum)
$x>1$		+	+	f increasing and concave up

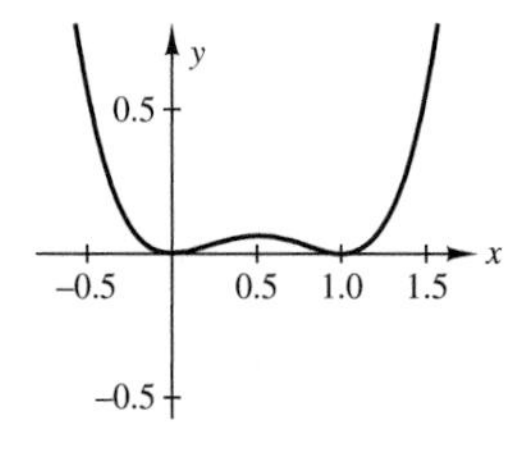

4.

	f	f'	f''	Summary
$x<-2$		+	−	f increasing and concave down
$x=-2$	−1	0	−	(relative maximum)
$-2<x<4$		−	−	f decreasing and concave down
$x=4$	d.n.e.	d.n.e.	d.n.e.	(vertical asymptote)
$4<x<10$		−	+	f decreasing and concave up
$x=10$	23	0	+	(relative minimum)
$x>10$		+	+	f increasing and concave up

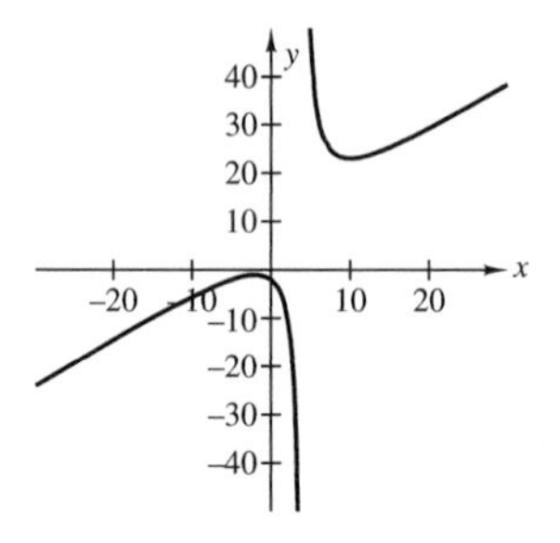

5. Here is one possible graph:

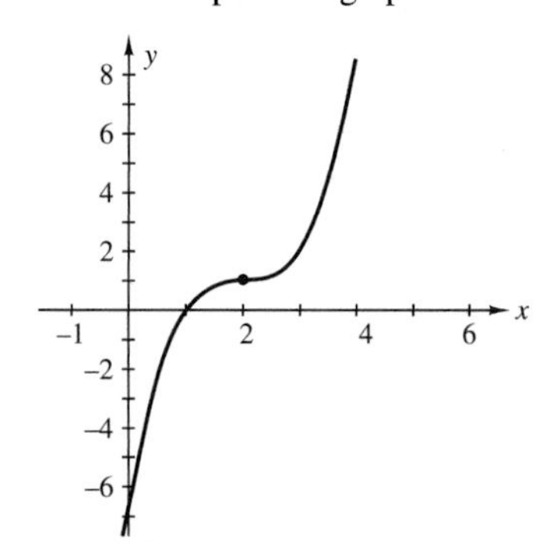

6. Here is one possible graph:

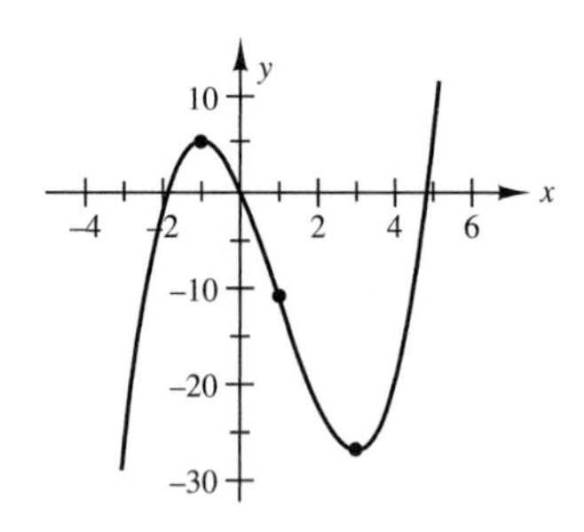

Chapter 17 Exercises

1. a.

	f	f'	f''
$x<2$		−	+
$x=2$			
$x>2$		−	−

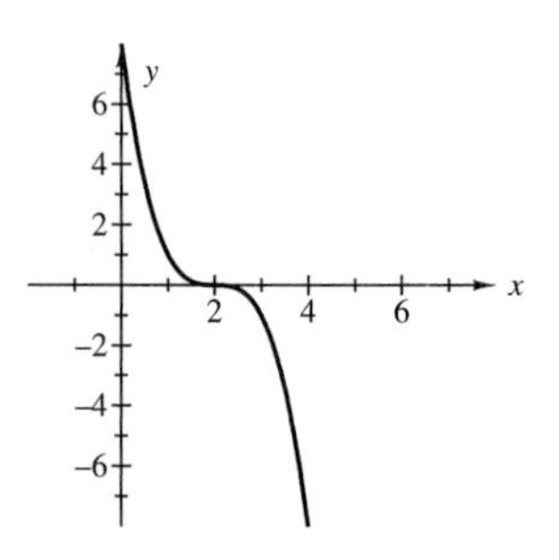

b.

	f	f'	f''
all x	$-$	$-$	$-$

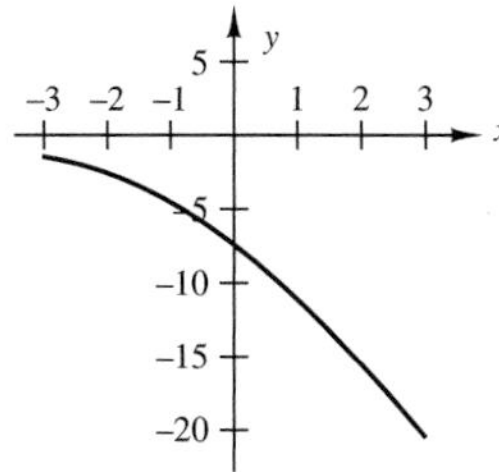

c.

	f	f'	f''
$x<2$			$-$
$x=2$	-1	d.n.e.	d.n.e.
$x>2$		$+$	

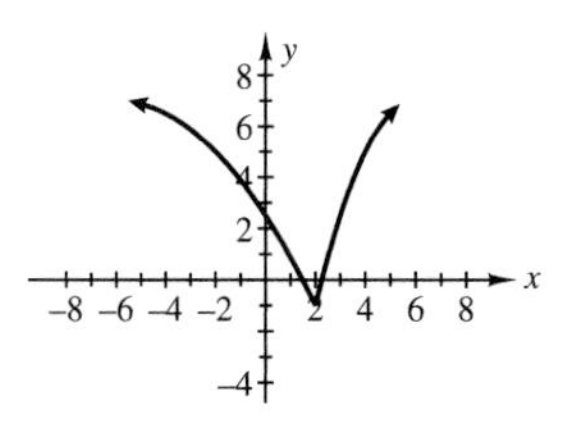

2. **a.** (i) None (ii) None
(iii) $(-1, 0) \cup (1, \infty)$ (iv) $(-\infty, -1) \cup (0, 1)$
b. (i) $(1.25, 2.5)$ (ii) $(-\infty, 0) \cup (2.5, 3.25)$
(iii) $(0.5, 1.25)$ (iv) $(0, 0.5) \cup (3.25, \infty)$
c. (i) None (ii) None
(iii) None (iv) None
d. (i) None (ii) $(-\infty, 1)$
(iii) $(2, \infty)$ (iv) $(1, 2)$
e. (i) $(1, \infty)$ (ii) $(-\infty, -1)$
(iii) $(0, 1)$ (iv) $(-1, 0)$
f. (i) None (ii) $(-\infty, -2) \cup (0, 2)$
(iii) None (iv) $(-2, 0) \cup (2, \infty)$

3. **a.**

	f	f'	f''
$x<-1$		$-$	$-$
$x=-1$		d.n.e.	d.n.e.
$-1<x<0$		$-$	$+$
$x=0$		0	0
$0<x<1$		$-$	$-$
$x=1$		d.n.e.	d.n.e.
$x>1$		$-$	$+$

b.

	f	f'	f''
$x<0$		$+$	$-$
$x=0$		0	$-$
$0<x<0.5$		$-$	$-$
$x=0.5$		$-$	0
$0.5<x<1.25$		$-$	$+$
$x=1.25$		0	$+$
$1.25<x<2.5$		$+$	$+$
$x=2.5$		$+$	0
$2.5<x<3.25$		$+$	$-$
$x=3.25$		0	$-$
$x>3.25$		$-$	$-$

c.

	f	f'	f''
$(-\infty, \infty)$		$-$	0

d.

	f	f'	f''
$x<1$		$+$	$-$
$x=1$		0	$-$
$1<x<2$		$-$	$-$
$x=2$		$-$	0
$x>2$		$-$	$+$

e.

	f	f'	f''
$x<-1$		$+$	$-$
$x=-1$		0	$-$
$-1<x<0$		$-$	$-$
$x=0$		d.n.e.	d.n.e.
$0<x<1$		$-$	$+$
$x=1$		0	$+$
$x>1$		$+$	$+$

f.

	f	f'	f''
$x<-2$		$+$	$-$
$x=-2$		0	$-$
$-2<x<0$		$-$	$-$
$x=0$		d.n.e.	d.n.e.
$0<x<2$		$+$	$-$
$x=2$		0	$-$
$x>2$		$-$	$-$

4.

	x-intercepts	y-intercepts
a.	0	0
b.	0, 2, 4	0
c.	0.5	1
d.	0	0
e.	None	None
f.	0, 4	0

5.

	Vertical asymptotes	Horizontal asymptotes
a.	$x=-1, x=1$	$y=-1, y=1$
b.	None	None
c.	None	None
d.	None	$y=0$
e.	$x=0$	None
f.	None	None

6. **a.**

	f	f'	f''	Summary
$x<-\frac{1}{3}$		$+$	$-$	f increasing and concave down
$x=-\frac{1}{3}$		0	$-$	(relative maximum)
$-\frac{1}{3}<x<\frac{1}{3}$		$-$	$-$	f decreasing and concave down
$x=\frac{1}{3}$		$-$	0	(inflection point)
$\frac{1}{3}<x<1$		$-$	$+$	f decreasing and concave up
$x=1$		0	$+$	(relative minimum)
$x>1$		$+$	$+$	f increasing and concave up

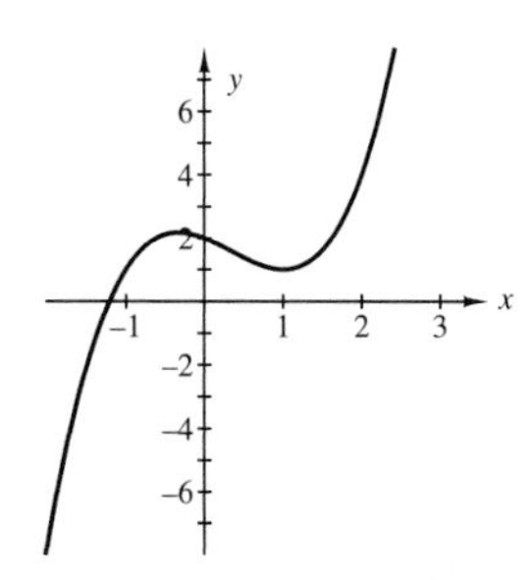

b.

	f	f'	f''	Summary
$x<0$		+	–	f increasing and concave down
$x=0$		+	0	(inflection point)
$x>0$		+	+	f increasing and concave up

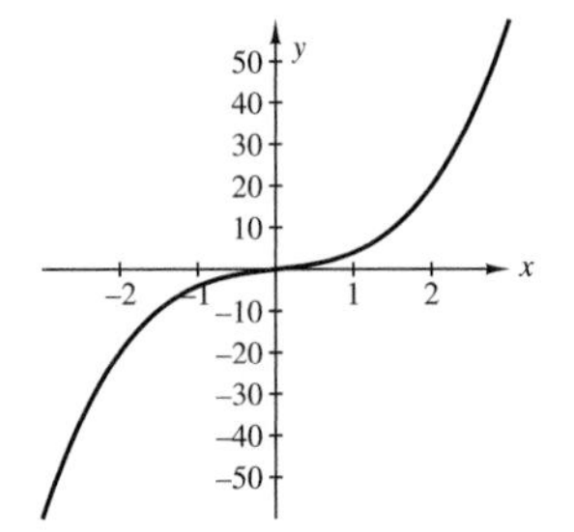

c.

	f	f'	f''	Summary
$x<-1$		+	–	f increasing and concave down
$x=-1$		+	0	(inflection point)
$-1<x<0$		+	+	f increasing and concave up
$x=0$	0	D.N.E.	D.N.E.	(relative maximum)
$0<x<2$		–	+	f decreasing and concave up
$x=2$		0	+	(relative minimum)
$x>2$		+	+	f increasing and concave up

A graphing utility does not show the change in concavity at (–1, –6)

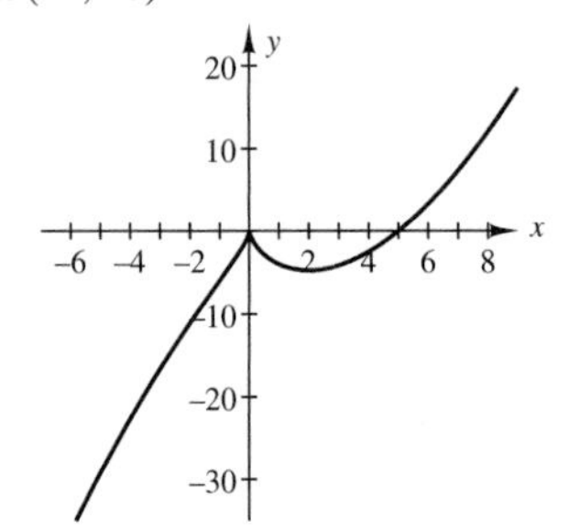

d.

	f	f'	f''	Summary
$x<-\frac{1}{\sqrt{3}}$		–	–	f decreasing and concave down
$x=-\frac{1}{\sqrt{3}}$		–	0	(inflection point)
$-\frac{1}{\sqrt{3}}<x<0$		–	+	f decreasing and concave up
$x=0$		0	+	(relative minimum)
$0<x<\frac{1}{\sqrt{3}}$		+	+	f increasing and concave up
$x=\frac{1}{\sqrt{3}}$		+	0	(inflection point)
$x>\frac{1}{\sqrt{3}}$		+	–	f increasing and concave down

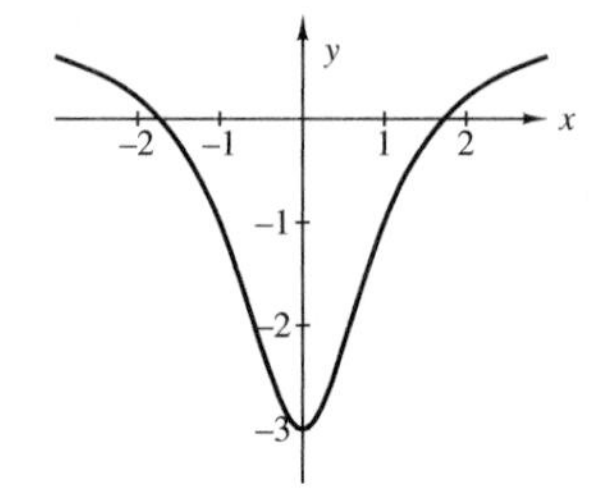

7. It is possible.
8. It is possible.
9. **a.** Critical values of f: $x=-2$, $x=6$; critical value of f': $x=2$

b.

	f'	f''
$x<-2$	+	–
$x=-2$	0	–
$-2<x<2$	–	–
$x=2$	–	0
$2<x<6$	–	+
$x=6$	0	+
$x>6$	+	+

c.

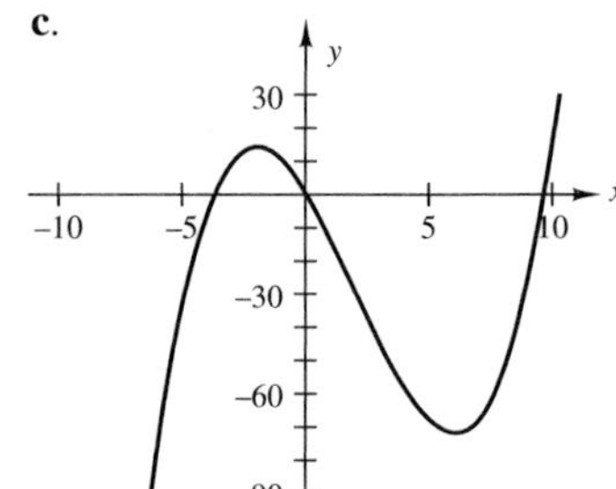

Exercises 18-A

1. **a.** $R(x)=10x$ **b.** $R(x)=0.05x^2$
   c. $R(x)=10x-0.05x^2$
2. **a.** $P(t)=100{,}000+5000t$ **b.** $P(10)=150{,}000$
   c. 20 years
3. **a.** Figures 18.10 and 18.13; Figures 18.11 and 18.12
   b. $k(x)$ is the antiderivative of $f(x)$; $h(x)$ is the antiderivative of $g(x)$.
4. **a.** $G'(t)=\frac{1}{2}t^2-3t+15$. $G(t)$ predicts the amount (in hundreds) by which the bacteria population has grown after t hours.
   b. $G(12)-G(2)=226\frac{2}{3}$. The bacteria population increased by 22,667 between the second hour ($t=2$) and the twelfth hour ($t=12$).
   c. $P(t)=G(t)+8=\frac{t^3}{6}-\frac{3t^2}{2}+15t+8$ (in hundreds)
   d. 180,800
5. **b.** Graphs of all other antiderivatives have the same shape as the graph shown in Figure 18.15 and are vertical translations of that graph. Two such graphs are shown here.

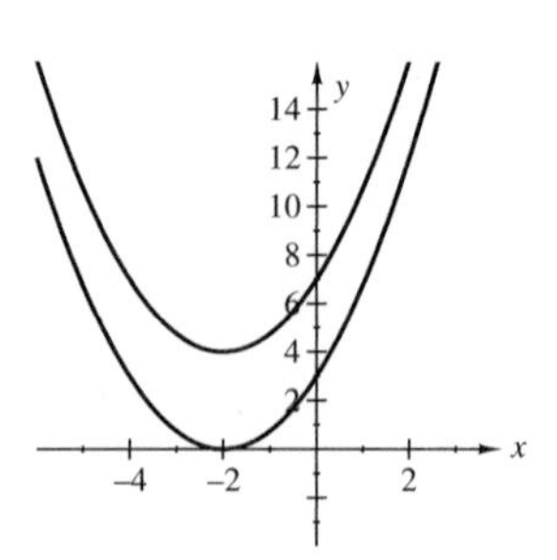

6. **a.** Its antiderivative represents the demand for the electric razors at price p.
   b. Its antiderivative represents the number of bacteria in the culture after t hours or the amount by which the number of bacteria in the culture has increased or decreased after t hours.
   c. Its antiderivative represents the velocity of the object at time t, in meters per second.

Exercises 18-B

1. a. $f(x) = -8x^3 + 6x^2 + 20x - 15$ b. $g(x) = x^{-1/2} - x^2$
   c. $h(z) = 7z^{-2} - z^{-1/2} + 3$ d. $s(t) = t^{2/3} + t^{1/2}$
   e. $f(x) = 3x - 2x^{-11/3} + 2x^{-4}$
   f. $g(t) = t^{-1/2} + 2t^{-5/2} - 3t^{-1} - 6t^{-3}$
2. a. Not possible; $(2x^3 + 8)^{1/2}$ cannot be simplified.
   b. Not possible; $(x^2 + x^4)^{-1}$ cannot be simplified.
   c. $h(x) = x^{-2} + x^{-4}$
   d. Not possible; $(x^3 + x)^{1/2}$ cannot be simplified.
   e. $p(x) = x^{3/2} + x^{1/2}$
3. a. VII b. I c. VI d. V e. II f. III

Exercises 18-C

1. a. $g(x) = 7 - 4x$ $f(x) = \frac{1}{7}x^6$
   b. $g(x) = 0.5x$ $f(x) = e^x$
   c. $g(x) = 4x - 9$ $f(x) = -x^{-2}$
   d. $g(x) = x^2$ $f(x) = e^x$
   e. $g(x) = x^3$ $f(x) = \sec^2 x$
   f. $g(x) = \cos x$ $f(x) = x^4$
2. a. $\int \frac{1}{7}u^6 du,\ u = 7 - 4x$ b. $\int e^u du,\ u = 0.5x$
   c. $\int -u^{-2} du,\ u = 4x - 9$ d. $\int e^u du,\ u = x^2$
   e. $\int \sec^2 u\, du,\ u = x^3$ f. $\int u^4 du,\ u = \cos x$
3. a. $\frac{1}{4}\int u^5\, du$ b. $-\frac{3}{2}\int u^{-3}\, du$
   c. $\frac{1}{3}\int u^{1/3}\, du = \frac{1}{3}\int \sqrt[3]{u}\, du$ d. $2\int \sin u\, du$
4. a. $u = 3x - 5;\ \frac{1}{3}\int u^3\, du$ b. $u = 5 - 4x;\ -\frac{1}{4}\int \sqrt{u}\, du$
   c. $u = 5t;\ \frac{1}{5}\int e^u\, du$ d. $u = 1 - 0.5t;\ -2\int e^u\, du$
   e. $u = \cot t;\ -\int u^3\, du$
5. a. $\frac{1}{6}\int u^{-1/2}\, du;\ u = 3x^2 + 4$ b. Not possible
   c. $\frac{1}{12}\int u^{1/3}\, du;\ u = 4x^3 + 1$ d. Not possible

Chapter 18 Exercises

1. a. $f(x)$ and $k(x)$; $g(x)$ and $h(x)$
   b. $k(x)$ is the antiderivative of $f(x)$; $g(x)$ is the antiderivative of $h(x)$.
2. a. $p(t) = 2000t + 50{,}000$ b. $p(10) = 70{,}000$
3. a. $F'(t) = \frac{1}{2}(t+1)^{-1/2}$
   b. $F(t)$ predicts the amount by which the bacteria population has increased after t minutes.
   c. $f(t) = \sqrt{t+1} + 9999$
4. a. VIII b. IV c. IX d. V e. I f. III
5. a. $f(x) = 3x^3 - 9x^2 + 5x - 15$ b. $g(x) = x^{-3/2} - 2x$
   c. $h(z) = -4z^{-2} - z^{-1/2} + 4$ d. $s(t) = t^{1/3} + t^{1/4}$
6. a. $f(x) = 3x^2 - x^{-8/3} + 3x^{-3}$
   b. $g(x) = x^{-1/2} - 2x^{-3/2} + 3x^{-1} - 6x^{-2}$
   c. Not possible; $(3x^2 - x + 3)^{1/5}$ cannot be simplified.
   d. Not possible; $x(x^2 + 1)^{-1}$ cannot be simplified.
7. a. $u = 7 - 2x;\ \int \frac{1}{5}u^5\, du$ b. $u = 0.7x;\ \int e^u\, du$
   c. $u = 3x - 8;\ \int -u^{-3}\, du$ d. $u = 2x^2;\ \int e^u\, du$
   e. $u = t^3;\ \int -\csc^2 u\, du$ f. $u = \sin t;\ \int u^3\, du$
8. a. $u = 2x - 5;\ \frac{1}{2}\int u^2\, du$ b. $u = 5 - 3x;\ -\frac{1}{3}\int \sqrt{u}\, du$
   c. $u = 2t;\ \frac{1}{2}\int e^u\, du$ d. $u = -0.4t;\ -2.5\int e^u\, du$
   e. $u = \tan x;\ \int \frac{1}{u^3} du$ f. $u = 3x^4 + 1;\ \frac{1}{12}\int \cos u\, du$

Exercises 19-A

1. a. $8\pi + 144 \approx 169.1$ square units
   b. 38.5 square units
   c. $8\pi + 4\sqrt{84} \approx 61.8$ square units
   d. $32 - 8\pi \approx 6.87$ square units
2. 1800 3. 6 square units
4. a. $8\pi \approx 25.1$ square units b. $4\pi - 8 \approx 4.57$ square units
5. a. $\frac{169}{24}$ b. $\frac{42}{5}$ 6. a. $\sum_{i=2}^{8} \frac{3}{8} i$ b. $\sum_{i=1}^{6} \frac{1}{4}\left(\frac{2}{i}\right)$

Exercises 19-B

All answers in square units.

1. a. 1 b. 7 c. 4 d. ≈ 4; your answer may vary.
2. a. 4 b. 10 c. 7 d. ≈ 7
3. 4 4. 7; your answer may vary slightly.
5. a. 15.12; error = 2.88 b. 17.82; error = 0.18

Exercises 19-C

1. a.

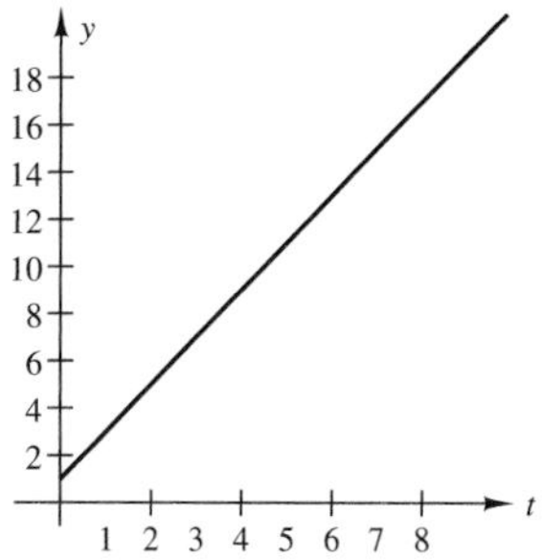

   b. 90 ft c. An underestimate d. 110 ft
2. 176 ft
3. a. 208 ft b. 192 ft
4. ≈ −10.59°
5. 35,640

Chapter 19 Exercises

1. **a**. 85 sq. u. **b**. $160 + \frac{25\pi}{2} \approx 199.3$ sq. u. **c**. 2 sq. u.
2. **a**. 11.3 miles

   b.

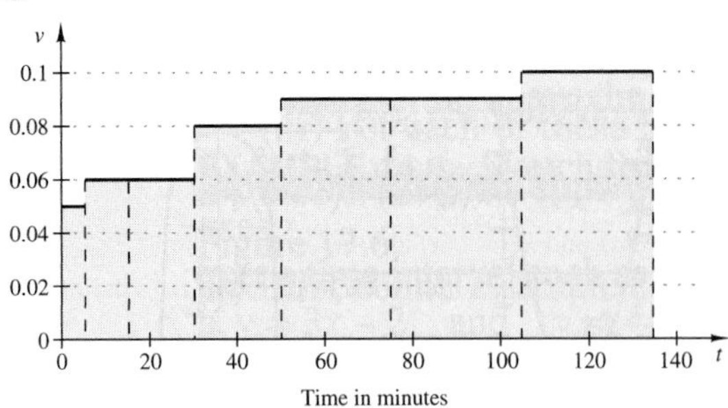

   The sum of the areas of the shaded rectangles is 11.3 square units.

3. 6 square units
4. $\frac{25\pi}{8}$ square units

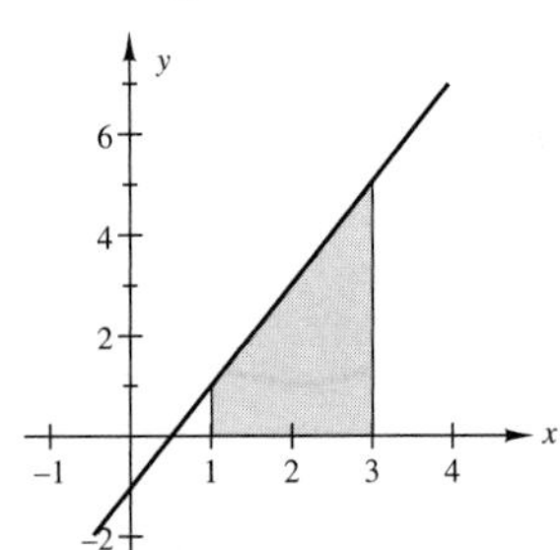

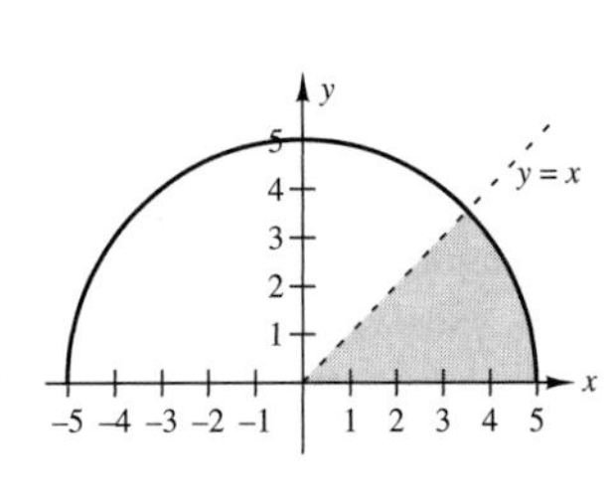

5. **a**. 5 sq. u. **b**. 18 sq. u. **c**. 11.5 sq. u. **d**. 11 sq. u.; your answer may vary.
6. 11.5 sq. u.; your answer may vary.
7. **a**. 7.92 sq. u.; error $\approx$ 1.08 sq. u.

   b. 8.82 sq. u.; error $\approx$ 0.18 sq. u.

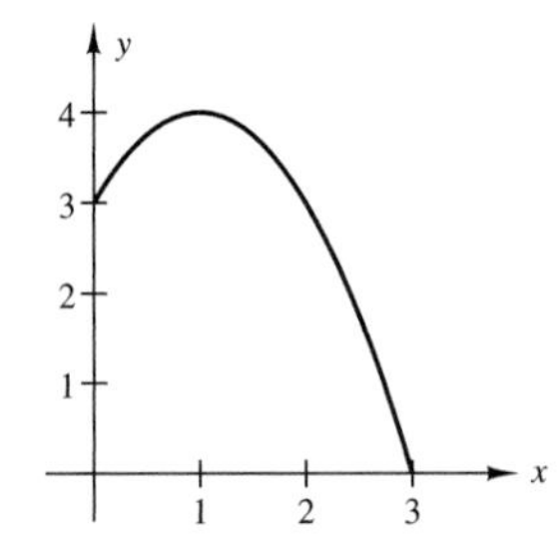

8. 748
9. 850; error $\approx$ 167
10. $\sum_{i=1}^{36} r(2i)\cdot 2$

Exercises 20-A

1. **a**. $\int_2^5 x^2 dx$ **b**. $\int_6^8 \sqrt{x-3}\, dx$
2. **a**. $\sum_{i=1}^{n} \sin(x_i)\Delta x_i$ on the interval $[0, \pi]$ **b**. $\int_0^{\pi} \sin x\, dx$
3. $\int_{-5}^{5} \sqrt{25 - x^2}\, dx$
4. **a**. $\int_1^6 f(x)dx$ **b**. $\int_{-2}^{3} g(x)dx$
5. $\int_{-2}^{2} (-0.6x^2 + 6.4)dx$

Exercises 20-B

1. $b = \frac{8}{3}$
2. **a**. 70 **b**. −54

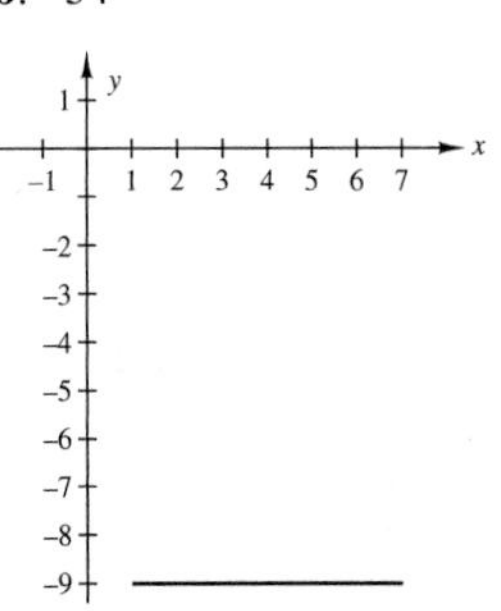

c. −4

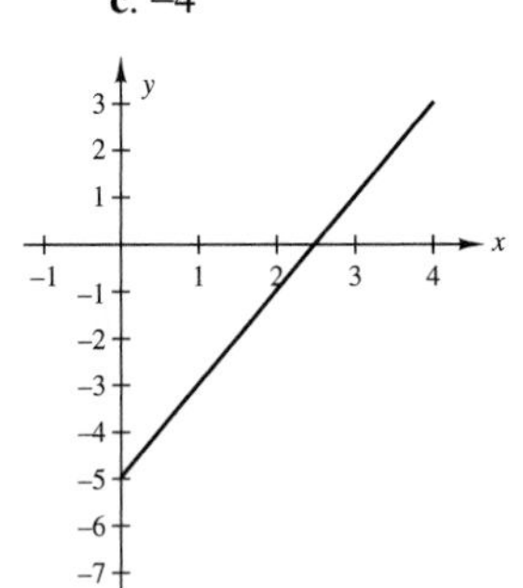

d. −4

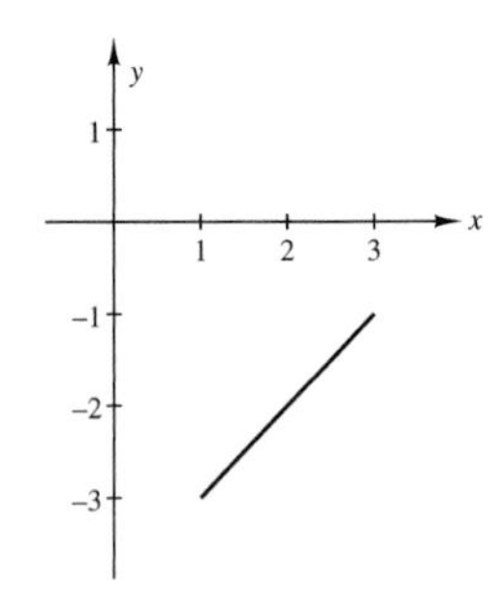

e. $\frac{25}{2}$

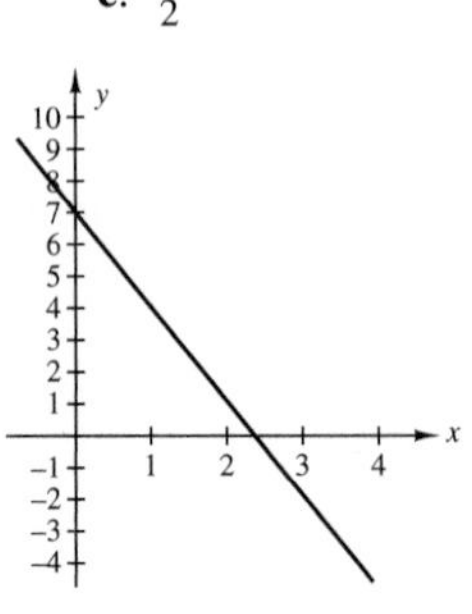

f. π

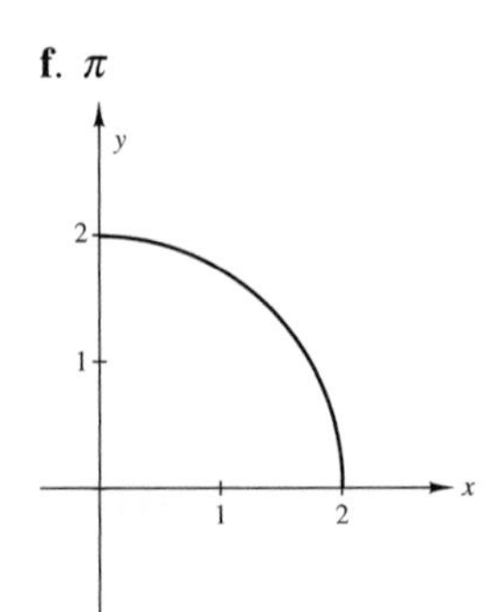

3. **a.**

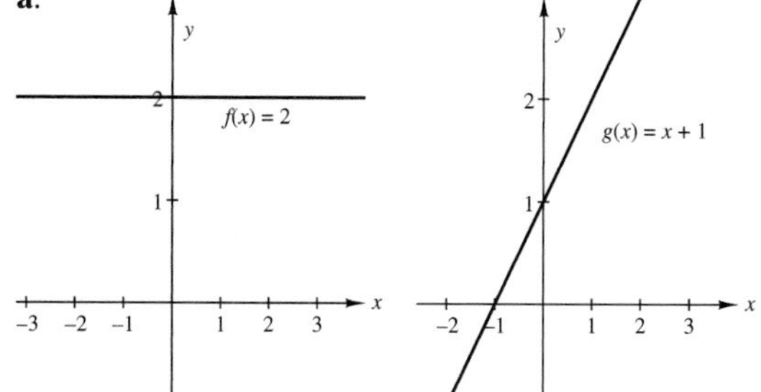

b. $c = 3$ **c.** Not possible

d. $[-2, -1]$ is one possible interval.

e. $[-2, 0]$ is one possible interval.

4. **a.** 12 **b.** $\frac{21}{2}$ **c.** 21 **d.** $\int_0^5 f(x)\,dx$ is larger.

5. **a.**

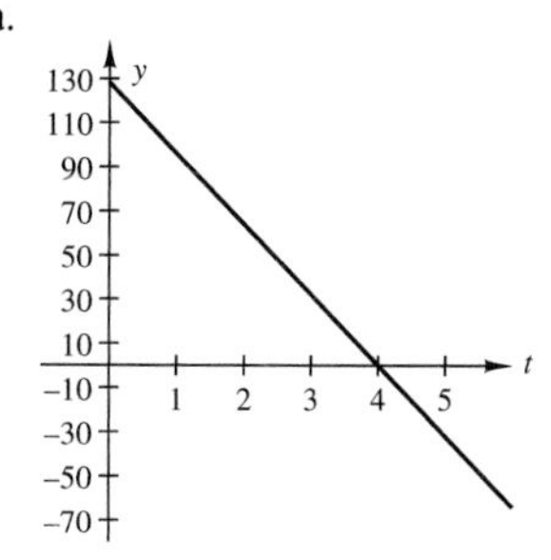

b. 192 ft **c.** $\int_0^2 128 - 32t\,dt$ **d.** 192 ft **e.** $\int_0^6 128 - 32t\,dt$

6. **a.** $\int_0^3 -2 - 4(t+1)^{-1}\,dt$

b. Negative, because $g'(t) < 0$ for $0 \le t \le 3$

7. **a.** $\int_0^{10} 25t^2 - 15\sqrt{t}\,dt$ **b.** $\int_5^{20} 25t^2 - 15\sqrt{t}\,dt$

Exercises 20-C

1. **a** and **b**.

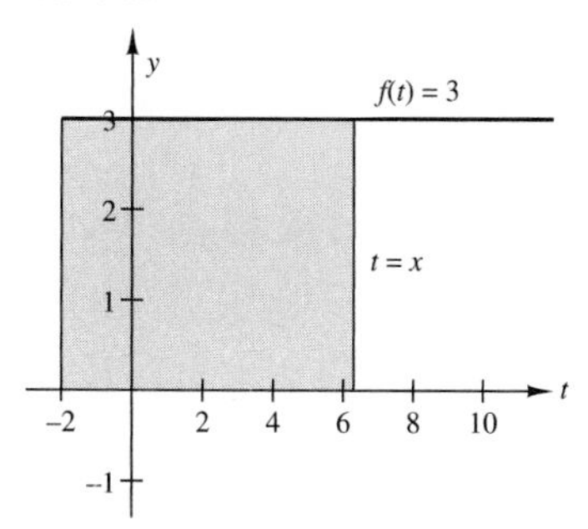

c.

x	0	1	2.5	4.7	5	12
$F(x)$	6	9	13.5	20.1	21	42

d. $F(x) = 3(x + 2)$ **e.** $F'(x) = 3$

f. The Fundamental Theorem of Calculus, Part I

2. **a** and **b**.

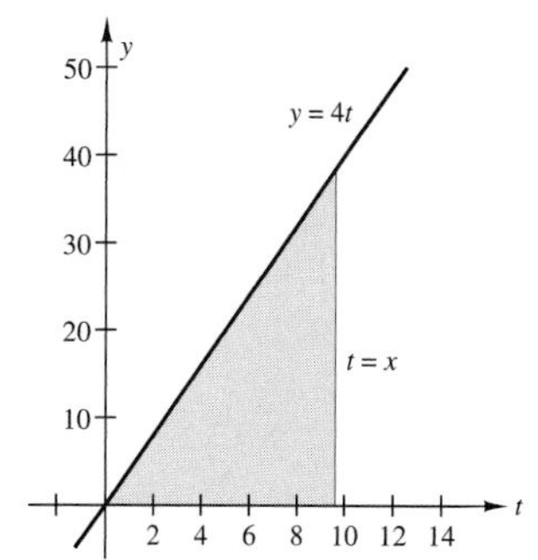

c.

x	1	5.5	7	10	15
$F(x)$	2	60.5	98	200	450

d. $F(x) = \frac{1}{2} \cdot x \cdot 4x = 2x^2$ **e.** $F'(x) = 4x$

3. **a** and **b**.

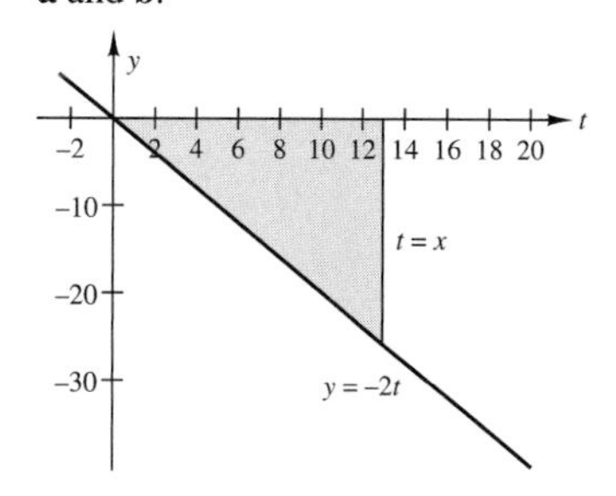

c.

x	1.5	3	6	7.8	20
$F(x)$	−2.25	−9	−36	−60.84	−400

d. $F(x) = -\frac{1}{2} \cdot x \cdot 2x = -x^2$ **e.** $F'(x) = -2x$

4. **a** and **b**.

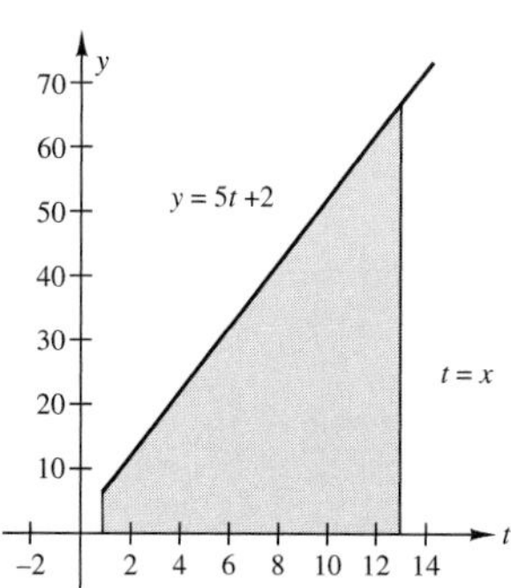

c.

x	1.5	5	8	10.5	15
$F(x)$	4.125	68	171.5	292.125	588

d. $F(x) = \frac{1}{2}(7 + 5x + 2)(x - 1) = \frac{5}{2}x^2 + 2x - \frac{9}{2}$

e. $F'(x) = 5x + 2$

5. **b.** 10

c.

Area = $2 \times 5 = 10$

6. **b**. Area = 22.4 square ft

   c. The window with the parabolic arc has the larger area.

7. **a**. The integral is positive.

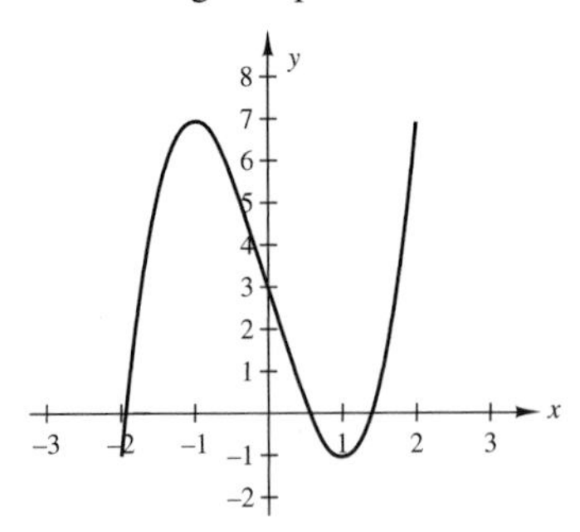

   b. The integral is negative.

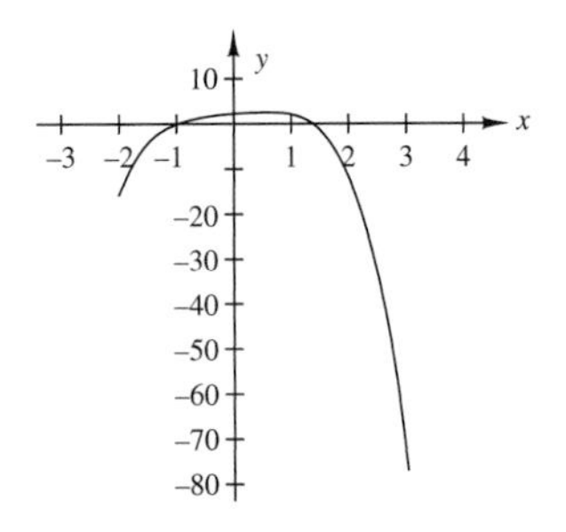

   c. The integral is negative.

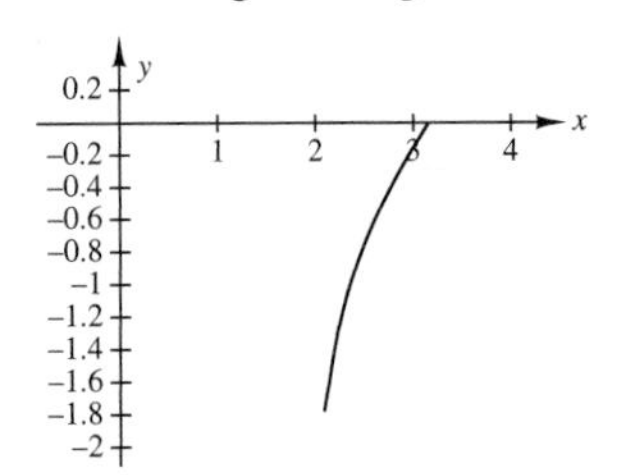

   d. The integral is 0.

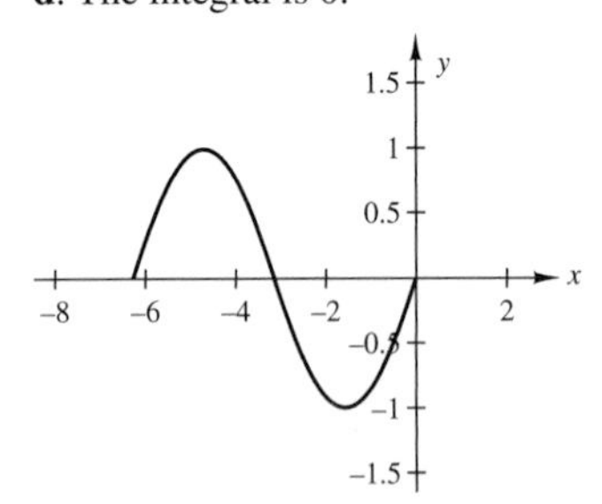

Exercises 20-D

1. **a**. $\frac{1}{3}\int_1^{10} \frac{1}{u}\, du = \frac{\ln 10}{3}$

   b. $\frac{1}{2}\int_2^{6} u^{1/2}\, du = \frac{6^{3/2} - 2^{3/2}}{3} \approx 3.96$

   c. $\int_0^{\ln 4} u\, du = \frac{1}{2}(\ln 4)^2 = 2(\ln 2)^2$

   d. $\frac{1}{6}\int_{-7}^{17} u^5\, du = 667{,}220$

Chapter 20 Exercises

1. **a**. $\int_2^5 2x\, dx$ **b**. $\int_6^8 \sqrt{3-x}\, dx$

2. **a**. $\sum_{i=1}^{n} \cos(x_i)\Delta x_i$ **b**. $\int_0^{\pi/2} \cos x\, dx$ **c**. 0

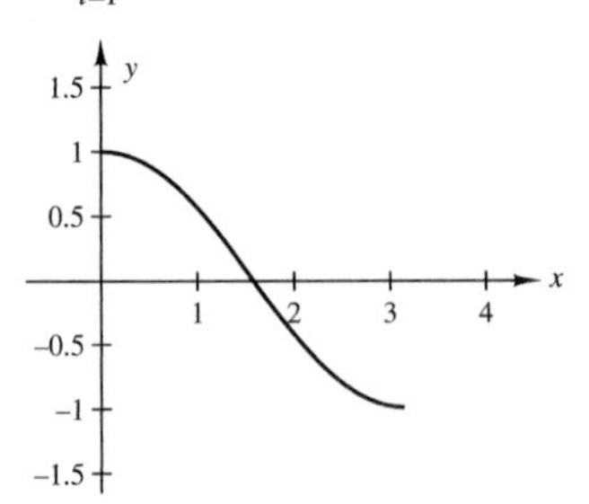

3. $\int_0^4 \sqrt{4-x}\, dx$

4. $\int_2^7 f(x)\, dx$

5. Here is one possible function:

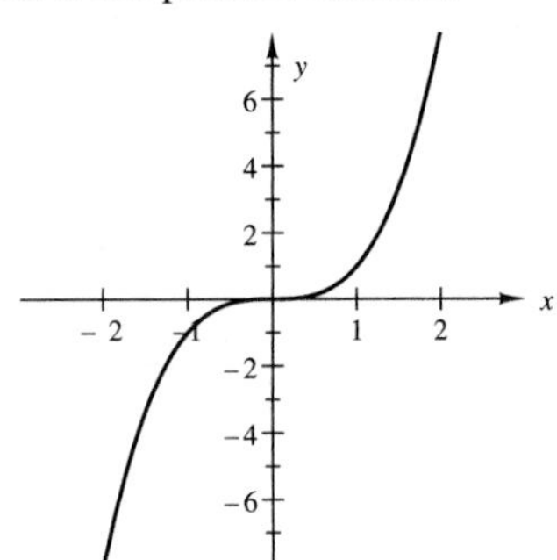

6. **a**. 2 **b**. 4 **c**. 0 **d**. $\frac{15}{2}$

7. **a**. 2 **b**. $\frac{21}{2}$ **c**. –6 **d**. Yes **e**. [0, 7] **f**. $x = 7$

8. **a**. $\frac{3}{2}$ **b**. $\frac{7}{2}$ **c**. 12 **d**. $-\frac{9}{2}$

   e. $G(14) - G(7)$ is the negative of the area of the region bounded by the graph of $y = f(t)$, the t-axis, the line joining the points (7, 0) and (10, –3), the line joining the points (10, –3) and (14, –3), and the line $t = 14$.

9. $F(x) + \frac{3}{2} = G(x)$

10. **a**.

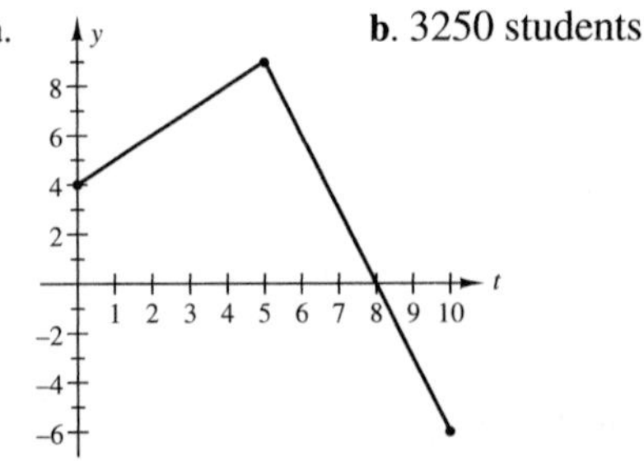

   b. 3250 students

   c. 750 students **d**. decreased by 600 students

   e. Since the growth rate is increasing from 1995 to 2000 and decreasing from 2000 to 2005, the net population growth is larger for the 5-year period 1995 to 2000. The growth rate is negative over the period 2003 to 2005, so the net population growth is also negative over that period.

11. **a.**

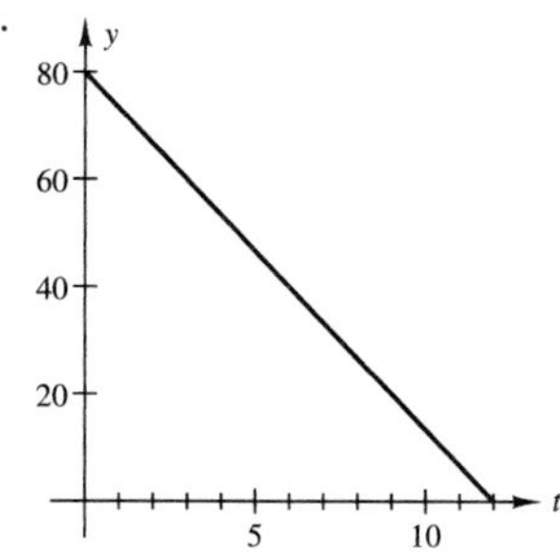

b. $\int_0^6 80 - \frac{20}{3}t\,dt$ **c.** 360,000 barrels

d. $\int_6^{12} 80 - \frac{20}{3}t\,dt$ **e.** 120,000 barrels

f. Since the production rate is decreasing, the production during the first 6-year period is greater than the production during the second 6-year period.

12. **b.** 54 **c.** 54

Appendix Exercises

Exercises A

1. **a.** $\frac{4}{5}$ **b.** $-\frac{2659}{500}$ **c.** $\frac{247}{110}$ **d.** $\frac{5}{4}$ **e.** $\frac{77}{25}$

2. **a.** 3.048×10^{12} **b.** 2.00314×10^{-6}

Exercises B

1. **a.** $(2 + 3)^2 - 3 = 22$ **b.** $(20 - 2)\cdot(5 - 3) = 36$

2. **a.** $5x^3y/(x - 12)$ **b.** $(2x^2 - 3x)/(4x^3)$

3. **a.** -2 **b.** 0.4 **c.** $\frac{2}{5}$ **d.** Not possible

Exercises C-1

1. **a.** 3 **b.** 9 **c.** 15 **d.** 54

2. **a.**

b.

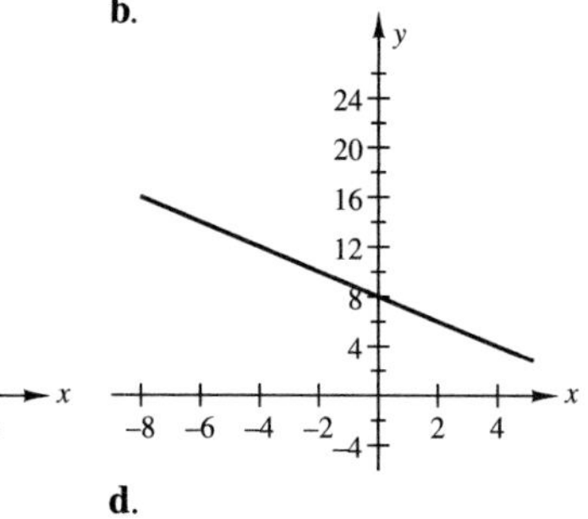

c.

d.

3. **a.** $x^2 + 6x + 8$ **b.** $-x + 8$ **c.** $x^2 + 6x + 20$
d. $-2x^3 - 12x^2 + 64$

4. **a.** $\pi(x - 1)^2(x + 2) = \pi(x^3 - 3x + 2)$

b. $\pi\left(\frac{x}{2}+1\right)^2\left(x^2+\frac{2}{x}\right) = \pi\left(\frac{x^4}{4}+x^3+x^2+\frac{x}{2}+2+\frac{2}{x}\right)$

Exercises C-1

1. **a.** $x\sqrt{x}(3x - x^2 + 5)$ **b.** $\frac{x^2y}{4x+3}\left(\frac{y^2}{3} + 2 - xy\right)$

c. $(2x+1)\left[x^5 - 1 - x^4 - x\sqrt{x}(2x+1)\right]$

2. $\frac{2x^2(4x+1)+x^2+1-x(x-5)}{2x^3(x-3)} = \frac{8x^3+2x^2+5x+1}{2x^3(x-3)}$

Exercises C-3

1. **a.** Incorrect; $(3x + 1)$ is not a factor of all terms in the numerator, so it cannot be canceled.
b. Correct **c.** Correct
d. Incorrect; the correct cancellation gives $\frac{(4x+3)x^2}{x^2+1}$, and this cannot be simplified further.
e. Correct **f.** Incorrect; cannot be simplified

2. **a.** $-5x^3y^2$ **b.** x^2
c. $5 - 7y$ when $x \neq 0$; not defined for $x = 0$

3. **a.** Correct **b.** Solution $x = 1$ is lost.
c. Correct **d.** Correct

Exercises C-4

1. b, d, and e **2.** c and e

3. **a.** $\frac{5}{x}$ **b.** $\frac{a^{15/2}}{b}$ **c.** $\frac{x^{1/3}}{(3)8^{2/3}} = \frac{x^{1/3}}{12}$
d. $\frac{13^{1/2}}{6}$ **e.** $\frac{2}{x^{4/3}}$ **f.** $\frac{1}{(x^2y)^3} = \frac{1}{x^6y^3}$

4. **a.** $x^{1/7}(3x^{2/7} + 5 - 4x^2)$ **b.** $x^{1/3}(4x^{1/3} - 2 + x)$

Exercises D-1

1. $\left(x + \frac{1}{4}\right)^2 + \left(y - \frac{3}{2}\right)^2 = \frac{29}{16}$

2. The discriminant is $49 - 72 = -23$, so there are no real solutions.

3. $2x^2 + 2x - 1 = 2\left(x - \frac{-1-\sqrt{3}}{2}\right)\left(x - \frac{-1+\sqrt{3}}{2}\right)$

Exercises D-2

1. **a.** (16, 10) **b.** (4, 0) and $\left(\frac{5}{3}, -\frac{7}{9}\right)$
c. Do not intersect **d.** (0, 0) and $\left(\frac{\pi}{3}, \sqrt{3}\right)$

Exercises E

1. $\sin\theta = \frac{x}{\sqrt{x^2+1}}$, $\cos\theta = \frac{1}{\sqrt{x^2+1}}$, $\tan\theta = x$,
$\cot\theta = \frac{1}{x}$, $\sec\theta = \sqrt{x^2+1}$, $\csc\theta = \frac{\sqrt{x^2+1}}{x}$

2. $\sin\theta = \frac{\sqrt{x^2-1}}{x}$, $\cos\theta = \frac{1}{x}$, $\tan\theta = \sqrt{x^2-1}$,
$\cot\theta = \frac{1}{\sqrt{x^2-1}}$, $\sec\theta = x$, $\csc\theta = \frac{x}{\sqrt{x^2-1}}$

Index